Lecture Notes in Computer Science 7669

Commenced Publication in 1973
Founding and Former Series Editors:
Gerhard Goos, Juris Hartmanis, and Jan van Leeuwen

Runhe Huang   Ali A. Ghorbani
Gabriella Pasi   Takahira Yamaguchi
Neil Y. Yen   Beihong Jin (Eds.)

Active Media Technology

8th International Conference, AMT 2012
Macau, China, December 4-7, 2012
Proceedings

 Springer

Volume Editors

Runhe Huang
Hosei University, Tokyo 184-8584, Japan
E-mail: rhuang@hosei.ac.jp

Ali A. Ghorbani
University of New Brunswick, Fredericton, NB, Canada E3B 5A3
E-mail: ghorbani@unb.ca

Gabriella Pasi
University of Milano Bicocca, 20126 Milano, Italy
E-mail: pasi@disco.unimib.it

Takahira Yamaguchi
Keio University, Yokohama 223-8522, Japan
E-mail: yamaguti@ae.keio.ac.jp

Neil Y. Yen
University of Aizu, Fukushima 965-8580, Japan
E-mail: neil219@gmail.com

Beihong Jin
Chinese Academy of Sciences, Beijing 100190, China
E-mail: beihong@iscas.ac.cn

ISSN 0302-9743   e-ISSN 1611-3349
ISBN 978-3-642-35235-5   e-ISBN 978-3-642-35236-2
DOI 10.1007/978-3-642-35236-2
Springer Heidelberg Dordrecht London New York

Library of Congress Control Number: 2012952396

CR Subject Classification (1998): H.4, I.2, H.3, H.2.8, H.5, C.2, J.1, I.2.11, K.4

LNCS Sublibrary: SL 3 – Information Systems and Application, incl. Internet/Web
and HCI

Typesetting: Camera-ready by author, data conversion by Scientific Publishing Services, Chennai, India

Printed on acid-free paper

Springer is part of Springer Science+Business Media (www.springer.com)

Preface

This volume contains the papers selected for presentation at the 2012 International Conference on Active Media Technology (AMT 2012), held as part of the 2012 World Intelligence Congress, a special event of the Turing Centennial Celebration, jointly with other international conferences (BI12, WI12, IAT12, and ISMIS12) at Fisherman's Wharf, Macau, China, during December 4–7, 2012. Organized by the Web Intelligence Consortium (WIC), by the IEEE Computational Intelligence Society Task Force on Brain Informatics (IEEE TF-BI), and by the University of Macau, this conference constitutes the eighth of the AMT series since the initial conference at Hong Kong Baptist University in 2001 (followed by AMT 2004 in Chongqing, China, AMT 2005 in Kagawa, Japan, AMT 2006 in Brisbane, Australia, AMT 2009 in Beijing, China, AMT 2010 in Toronto, Canada, and AMT 2011 in Lanzhou, China).

In this great digital era, we are witnessing many rapid scientific and technological developments in human-centered and seamless computing environments, interfaces, devices, and systems with applications ranging from business and communication to entertainment and learning. These developments are collectively best characterized as *active media technology (AMT)*, a new area of intelligent information technology and computer science that emphasizes the proactive and seamless roles of interfaces and systems as well as new media in all aspects of digital life. An AMT-based system offers services to enable the rapid design, implementation, and support of customized solutions.

The AMT conference series aims to explore and present state-of-the-art research in a wide spectrum of realms related to AMT. After a rigorous evaluation process, 43 papers were accepted for the AMT conference this year. These papers are from 16 countries comprising China, Japan, Germany, The Netherlands, Australia, USA, Iran, Algeria, Italy, Sweden, UK, Korea, Romania, France, Vietnam, and Ireland. The topics of these papers centered on the main themes of AMT 2012, encompassing: awareness services via active media technology; agent-based software engineering and multi-agent systems; data mining, ontology mining and web reasoning; social applications of active media; human-centered computing, personalization, and adaptation; information retrieval; machine learning and human-centric robotics; network; semantic computing for active media systems; smart digital art, media and their applications; pervasive/ubiquitous devices and intelligent systems; Web-based social networks; and e-learning, e-commerce and Web services. The core issues of AMT have been investigated using various criteria that were carefully identified from both theoretical and practical perspectives to ensure the successful incorporation of technologies.

Apart from the 43 high-quality papers accepted by the main conference of AMT 2012, another 22 papers were selected for oral presentations in the workshop and special sessions. The 2012 Workshop on Meta-synthesis and Complex Systems

(MCS 2012) aims at providing opportunities to facilitate interdisciplinary studies for those researchers who are interested in systems sciences, complex problem solving and advanced modeling, knowledge-oriented technology and integration, decision sciences and advanced support technologies, etc. The Special Session on Social Knowledge Discovery and Management deals with the process of automatic extraction of interesting and useful knowledge from very large data in the social world, which consists of a range of strategies and practices to identify, create, represent, distribute, and enable the adoption of novel insights and experiences for human-centered support. The Special Session on Human–Computer Interaction and Knowledge Discovery from Big Data goes further in supporting end users to interactively identify, analyze, and understand information from huge masses of high-dimensional data, especially those weakly-structured and non-standardized ones.

Here we would like to express our gratitude to all members of the Conference Committee for their precious and unfailing support. AMT 2012 had a very exciting program with a number of features, ranging from keynote talks, technical sessions, workshops, and social programs. A program of this kind would not have been possible without the generous dedication of the Program Committee members and the external reviewers who reviewed the papers submitted to AMT 2012, and of our keynote speakers, Edward Feigenbaum of the Stanford University and Sankar Pal of Indian Statistical Institute. Special thanks are also due to the Local Organizing Chairs, Ryan Leang Hou U and Feng Wan, the Workshop Chairs, Xijin Tang and Xiaoji Zhou, the Special Session Chairs, Andreas Holzinger and Andreas Auinger, Juzhen Dong, and Yuwei Yang, for their generous efforts. We are extremely grateful for their strong support and dedication. In addition, we also thank all the authors who contributed their research results to this volume. We also would like to thank the sponsors of this conference.

Finally, we extend our highest appreciation to Springer *Lecture Notes in Computer Science* (LNCS/LNAI) for their generous support. We thank Alfred Hofmann and Anna Kramer of Springer for their help in coordinating the publication of this special volume in an emerging and interdisciplinary research field.

September 2012

Runhe Huang
Ali A. Ghorbani
Takahira Yamaguchi
Gabriella Pasi
Neil Y. Yen
Beihong Jin

Conference Organization

Conference General Chairs

Takahira Yamaguchi   Keio University, Japan
Gabriella Pasi   University of Milano - Bicocca, Italy

Program Chairs

Runhe Huang   Hosei University, Japan
Ali A. Ghorbani   University of New Brunswick, Canada

Workshop/Special Session Organizing Chairs

Beihong Jin   Chinese Academy of Sciences, China
Neil Y. Yen   University of Aizu, Japan

Local Organizing Chairs

Ryan Leang Hou U   University of Macau, Macau SAR
Feng Wan   University of Macau, Macau SAR

Publicity Chairs

Qin (Christine) Lv   University of Colorado Boulder, USA
Lars Schwabe   University of Rostock, Germany
Daniel Tao   QUT, Australia
Jian Yang   Beijing University of Technology, China
Shinichi Motomura   Maebashi Institute of Technology, Japan

WIC Co-Chairs/Directors

Ning Zhong   Maebashi Institute of Technology, Japan
Jiming Liu   Hong Kong Baptist University, Hong Kong

IEEE CIS-TFBI Chair

Ning Zhong   Maebashi Institute of Technology, Japan

WIC Advisory Board

Edward A. Feigenbaum	Stanford University, USA
Setsuo Ohsuga	University of Tokyo, Japan
Benjamin Wah	Chinese University of Hong Kong, Hong Kong
Philip Yu	Univ. of Illinois, Chicago, USA
L.A. Zadeh	Univ. of California, Berkeley, USA

WIC Technical Committee

Jeffrey Bradshaw	UWF/Institute for Human and Machine Cognition, USA
Nick Cercone	York University, Canada
Dieter Fensel	University of Innsbruck, Austria
Georg Gottlob	Oxford University, UK
Lakhmi Jain	University of South Australia, Australia
Jianhua Ma	Hosei University, Japan
Jianchang Mao	Yahoo! Inc., USA
Pierre Morizet-Mahoudeaux	Compiegne University of Technology, France
Hiroshi Motoda	Osaka University, Japan
Toyoaki Nishida	Kyoto University, Japan
Andrzej Skowron	Warsaw University, Poland
Jinglong Wu	Okayama University, Japan
Xindong Wu	University of Vermont, USA
Yiyu Yao	University of Regina, Canada

Program Committee

Mina Akaishi	Hosei University, Japan
Aijun An	York University, Canada
Bill Andreopoulos	Technische Universität Dresden, Germany
Gloria Bordogna	National Research Council of Italy, Italy
Antonio Chella	Università degli Studi di Palermo, Italy
Yiqiang Chen	CAS, China
Mao Chengsheng	Lanzhou University, China
Chin-Wan Chung	KAIST, Korea
Zhiming Ding	Chinese Academy of Sciences, China
Adrian Giurca	BTU, Cottbus, Germany
William Grosky	University of Michigan, USA
Bin Guo	NWPU, China
Masahito Hirakawa	Shimane University, Japan
Wolfgang Huerst	Utrecht University, The Netherlands
Hiroshi Ishikawa	Kagawa University, Japan
Hanmin Jung	KISTI, Korea
Yeong Su Lee	Munich University, Germany
Jing Li	USTC, China

Table of Contents

Ubiquitous Intelligent Devices and Systems

Active Media Based Information Retrieval and Processing

Semantic Computing for ActiveMedia, Social Networks, and AMT-Based Systems

International Workshop on Meta-synthesis and Complex Systems

Special Session on Social Knowledge Discovery and Management

Special Session on Human-Computer Interaction and Knowledge Discovery from Big Data

Movie Genre Classification Using SVM
with Audio and Video Features

Yin-Fu Huang and Shih-Hao Wang

Department of Computer Science and Information Engineering
National Yunlin University of Science and Technology
{huangyf,g9817731}@yuntech.edu.tw

Abstract. In this paper, we propose a movie genre classification system using a meta-heuristic optimization algorithm called Self-Adaptive Harmony Search (i.e., SAHS) to select local features for corresponding movie genres. Then, each one-against-one Support Vector Machine (i.e., SVM) classifier is fed with the corresponding local feature set and the majority voting method is used to determine the prediction of each movie. Totally, we extract 277 features from each movie trailer, including visual and audio features. However, no more than 25 features are used to discriminate each pair of movie genres. The experimental results show that the overall accuracy reaches 91.9%, and this demonstrates more precise features can be selected for each pair of genres to get better classification results.

Keywords: Movie genre classification, feature selection, harmony search algorithm, multimedia data mining.

1 Introduction

In recent years, films have become a large portion of the entertainment industry. Every year, there are about 4500 films released around the world and these films approximate 9000 hours of video length [20]. Till now, movie genre classification is still done by man power and has no standardization. In terms of data mining, we would like to explore the hidden differences between movie genres by analyzing the information and/or features in videos.

In order to gather up the enough knowledge for classification, a generated feature set commonly contains the abundant information with some probably redundant features. For solving this problem, dimensionality reduction techniques are frequently employed and they can be classified into two approaches. One kind is to transform the matrix of a feature set from a high-dimensional space to a lower-dimensional space through the linear combination of matrix using the techniques such as principal component analysis (PCA) [9], non-negative multi-linear principal component analysis (NMPCA) [17], non-negative tensor factorization (NTF) [16], et al. Another kind is called the feature selection which finds an optimum subset from the original feature set using the search algorithms such as genetic algorithm (GA) [10], ant colony

R. Huang et al. (Eds.): AMT 2012, LNCS 7669, pp. 1–10, 2012.
© Springer-Verlag Berlin Heidelberg 2012

optimum (ACO) [3], harmony search (HS) [4], et al. Both these two approaches can effectively reduce the dimensions of a feature set. In our work, we extensively collect useful features and aim to evaluate which features are more relevant. Here, a feature selection approach called the self-adaptive harmony search (SAHS) [7] algorithm is adopted to obtain a better feature subset. In general, feature selection approaches are to select a global feature set for all movie genres. However, in order to achieve a global optimization, we adopt a local selection strategy based on each pair of movie genres since it can derive a more relevant local feature set than a global selection strategy.

For the prediction, an SVM classifier is adopted since, in general, it presents better performances than other classifiers [6, 8, 18] while a kernel function and parameters are appropriately chosen. In this paper, we match each local feature set with an SVM classifier and use the majority voting method to determine the prediction of each movie. The experimental results verify that more precise features can be selected for each pair of genres to get better classification results.

The remainder of the paper is organized as follows. In Section 2, we propose the system architecture and describe the visual and audio features used to discriminate movie genres. In Section 3, the SAHS algorithm and correlation measuring method are proposed to derive an optimum feature subset from the original feature set. In Section 4, the experimental results based on different local features for each pair of genres are presented. Finally, we make conclusions in Section 5.

2   System Overviews

In this section, we proposed an automatic movie classification system. As we know, features extracted from movies play an important role in movie classification. If relevant features are selected, movies would be easily distinguished based on them. Here, a meta-heuristic optimization algorithm called Self-Adaptive Harmony Search (i.e., SAHS) algorithm [7] is employed to select features for corresponding movie genres. The feature selection mechanism is applied on each one-against-one classifier (or SVM) to enable ambiguous genres to be classified more precisely. As illustrated in Fig. 1, the system architecture consists of two parts: training phase and test phase.

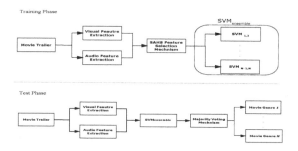

Fig. 1. System architecture

In the training phase, retrieving the visual and audio features from the training set is the first step. Visual features can be classified into temporal and spatial features. Temporal features are the ones relevant to video timing, which are the results of shot boundary detection, such as shot number, average shot number, total frame number, and average shot length. On the other hand, spatial features are more well-known than temporal features; they are MPEG-7 feature descriptors [2, 13, 14] all extracted from key-frames directly. For audio features, spectrum, compactness, Zero Crossing Rate (ZCR), Root Mean Square (RMS), Linear Prediction Coefficients (LPC), and MFCC are extracted in the system. For these collected features, the feature selection mechanism is used to pick up relevant features for each pair of genres, and then produces their optimum feature set (i.e., a local feature set). Thus, for N-genre videos to be classified, $C(_2^N)$ local feature sets would be generated in the mechanism. Next, each one-against-one SVM is trained by the corresponding local feature set. Finally, these trained SVMs are combined together to form an SVM ensemble model used in the test phase.

For the test phase, we also retrieve the visual and audio features from the test set first. Then, they are fed into the SVM ensemble model. Finally, through the majority voting on $C(_2^N)$ local predictions, the genre of a test sample could be determined.

2.1   Visual Features

In general, visual signals can be measured in two different ways: temporal and spatial features. To extract these visual features as shown in Fig. 2, a video should be preprocessed using shot boundary detection [12] to redefine the video as a series of video shots from which temporal features can be got. Then, key-frames are selected from each video shot, and spatial features can be extracted from these key-frames.

Fig. 2. Visual feature extraction

Temporal Features. In general, the rhythms of each film categories are totally different from others. For example, an action movie with fighting scenes, gun wars, car crashes, and explosion scenes always has a faster tempo. Thus, the temporal features as illustrated in Table 1 are viewed as the related ones in movie classification work.

Table 1. Temporal features

No.	Feature description	Dim.	Overall statistics	Total number
1	Shot number	1	1	1
2	Average shot number	1	1	1
3	Total frame numbers	1	1	1
4	Average shot length	1	1	1

Spatial Features. The spatial features defined by *MPEG-7 feature descriptors* [14] are widely used in existing movie classification work. MPEG-7 is formally known as the multimedia content description interface in the ISO/IEC 15938 standard developed by Moving Picture Experts Group (i.e., MPEG). It addresses multimedia contents with various modalities including image, video, audio, speech, graphics, and their combinations. The ultimate goal and objective of MPEG-7 is to provide interoperability among systems and applications used in generation management, distribution, and consumption of audio-visual content descriptions. As illustrated in Table 2, we only use motion vector, average color histogram, and average lighting key defined in Motion Activity and Color Layout descriptors as spatial features.

Table 2. Spatial features

No.	Feature description	Dim.	Overall statistics	Total number
5~10	Motion vector	1	6	6
11~74	Average color histogram	1	64	64
75	Average lighting key	1	1	1

2.2   Audio Features

In general, audio signals can be measured in two different ways: time domain and fre-quency domain. For the audio features in the time domain, the samples of audio signals are directly processed along time so that we can observe some characteristics about the amplitude such as intensity and rhythm. For the frequency domain, each amplitude sample is transformed from the time domain to a corresponding frequency band in a spectrum through the discrete Fourier transform (i.e., DFT). More detailed characteris-tics about acoustics such as timbre are presented by estimating the spectrum. All audio features are extracted using the well-known software "jAudio" [15].

Intensity. In acoustics, intensity as illustrated in Table 3 is also called loudness, vo-lume, or energy. It is the most obvious feature and means the loud volume heard by human perception. It is commonly measured by the amplitude within a frame in the time domain. In general, the unit of intensity is represented in decibel (i.e., dB). The Root Mean Square (i.e., RMS) amplitude is a measure of the power of a signal, which can be expressed as follows:

$$P(n) = \sqrt{\frac{1}{N} \sum_{i=0}^{N-1} S_n^2(i)}$$

(1)

where $S_n(i)$ is the ith sample of the nth audio frame and N is the total sample numbers in the frame. Besides, the fraction of low energy windows is to measure how many windows (or frames) are quiet relative to the others in the audio signal. First, the mean of the RMS amplitude of the last 100 windows is calculated, and the fraction of these 100 windows with RMS amplitude below the mean can be found.

Table 3. Intensity features

No.	Feature description	Dim.	Overall statistics	Total number
76~83	RMS amplitude	1	8	8
84~91	Fraction of low energy windows	1	8	8

Timbre. Timbre is an audio feature used to distinguish the sounds which have the same intensity and pitch. Different timbre is presented in different structures of amplitude spectrum on each or all frequency bands. By estimating the spectrum of audio signals, we can derive some timbre characteristics. In our work, we consider total 178 timbre features as illustrated in Table 4. Spectral centroid estimates the centroid frequency of spectrum; spectral flux estimates the distance of spectrum between adjacent frames; spectral variability estimates the variation degree of the neighboring peaks of spectrum; zero crossing counts the number of the signals crossing the zero line; compactness is closely related to harmonic spectral smoothness but is estimated through amplitudes; MFCC offers a description of the spectral shape based on Mel-frequency; finally linear prediction coefficients are calculated using autocorrelation and Levinson-Durbin recursion [11].

Table 4. Timbre features

No.	Feature description	Dim.	Overall statistics	Total number
92~99	Spectral centroid	1	8	8
100~107	Spectral flux	1	8	8
108~115	Spectral variability	1	8	8
116~123	Zero crossing	1	8	8
124~131	Compactness	1	8	8
132~209	MFCC	13	6	78
210~269	LPC	10	6	60

Rhythm. Rhythm may be generally defined as a "movement marked by the regulated succession of strong and weak elements, or of opposite or different conditions" [23]. In our viewpoint, rhythm is a time-related characteristic in sounds, and consists of beats and tempos. A beat is related to notes and commonly produced by striking or hitting. It is usually measured by the peaks of amplitude. As illustrated in Table 5, the strongest beat is the strongest bin in the beat histogram.

Table 5. Rhythm features

No.	Feature description	Dim.	Overall statistics	Total number
270~277	Strongest beat	1	8	8

3   Feature Selection Using SAHS

The purpose of feature selection is to select the most relevant features to facilitate the classification, using subset selection algorithms. These selection algorithms could be categorized into three different approaches: wrappers, filters, and embedded. The wrapper approach utilizes the performance of a target classifier to evaluate feature subsets; however it is computationally heavy since a target classifier must be iteratively trained on each feature subset. The filter approach evaluates the performance of feature subsets through an independent filter model before a target classifier is applied. The embedded approach must utilize the specific classification algorithm containing the evaluation body inside. Here, we use the filter approach to select the most relevant features since its computational loading is acceptable and the selected classification algorithm could be general. In our work, the feature selection model consists of two parts: Self-Adaptive Harmony Search (i.e., SAHS) algorithm [7] and relative correlations, as illustrated in Fig. 3. Once the original feature set is given, the SAHS algorithm starts to iteratively search the better solution that would be evaluated later by the relative correlations. Finally, the best solution will be output as the final feature subset.

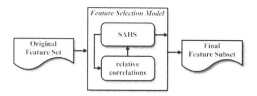

Fig. 3. Feature selection model

3.1   Relative Correlations

When each new harmony symbolizing a selected feature subset is generated from the SAHS algorithm, the relative correlations are used to evaluate the performance of the selected feature subset. The correlations of the selected feature subset are conducted in two phases; one is the intra-correlation evaluating the mutual correlation between features within the subset, and another is the inter-correlation comparing each feature inside the subset with the corresponding class. If a subset has better performance, it must possess the property of lower intra-correlation and higher inter-correlation. The lower intra-correlation means the features within the subset are relevant, whereas the higher inter-correlation means each feature within the subset is discriminative for the corresponding class.

Here, the well-known measuring formula called mutual information [19] is adopted, which evaluates the degree of the mutual dependence between two variables. The definition for discrete random variables is shown as follows.

$$I(X;Y) = \sum_{x \in X} \sum_{y \in Y} p(x,y) \, log \left(\frac{p(x,y)}{p_1(x)p_2(y)} \right) \qquad (2)$$

where $p(x,y)$ is the joint probability density of X and Y, and $p_1(x)$ and $p_2(y)$ are the marginal probability distribution functions of X and Y, respectively. The value of $I(X;Y)$ is nonnegative; if $I(X;Y)$ is zero, it means X and Y are independent; otherwise, a high value indicates a strong dependence between X and Y.

The intra-correlation within the feature subset is shown as follows.

$$RI(S) = \frac{1}{C(|S|,2)} \sum_{i=1}^{|S|} \sum_{j=i+1}^{|S|} I(x_i; x_j) \qquad (3)$$

where $C(|S|,2)$ is the number of 2-combination on the cardinality of feature subset S. The overall correlation within subset S is divided by $C(|S|,2)$ to present the average correlation between features within subset S. The inter-correlation between the feature subset and the corresponding class is shown as follows.

$$RT(S,y) = \frac{1}{|S|} \sum_{i=1}^{|S|} I(x_i; y) \qquad (4)$$

where $|S|$ is the cardinality of feature subset S, and y is the output class. The overall correlation is divided by the cardinality to derive the average correlation between features and the corresponding class.

Finally, the relative overall correlation combining both the intra-correlation and inter-correlation [5] is shown as follows.

$$RC(S,y) = \frac{k \times RT(S,y)}{\sqrt{k + k \times (k-1) \times RI(S)}} \qquad (5)$$

where k is the cardinality of feature subset S. It not only indicates a higher RC value has a better selected feature subset (i.e., it has lower intra-correlation and higher inter-correlation), but also balances the effect of average correlation on different cardinalities of candidate feature subsets. As a result, the RC is chosen as the objective function in the feature selection model.

4   Experimental Results

In our work, we collect 223 movie trailers from the Apple Movie Trailers website [21], including 35 for action, 14 for animation, 46 for comedy, 28 for documentary, 67 for drama, 7 for musical, and 26 for thriller. The genres of these 223 movies are defined by the Internet Movie Database (IMDb) [22]. In the SAHS algorithm, the parameters HMS and HMCR are set with 50 and 0.99 to present the optimum performance. Next, we adopt the well-known LIBSVM developed by Chang [1] as an SVM classifier. The kernel function used here is RBF (i.e., Radial Basis Function) since it is more accurate and effective than the other kernel ones. The parameters γ and C are determined   by   the   optimum   performance   of   6X6   combinations   between

$[2^{-4},...,2^1]$ and $[2^{-2},...,2^3]$. Moreover, the feature vector is normalized as the range $[-1, 1]$. For the classification results, a confusion matrix is employed to present the entire statistic on the correct and false predictions after the cross validation. Precision, recall, and accuracy are estimated as the performance evaluation and showed as follows.

$$Pr\,ecision = \frac{N_C}{N_C + N_F}$$

$$Re\,call = \frac{N_C}{N_C + N_M}$$

$$Accuracy = \frac{total_C}{total_M}$$

(6)

4.1 Classification Results Using the Feature Selection

The classification results using the local feature selection are presented here. Each feature set selected based on the local selection strategy is for each pair of all seven genres. In the experiments, for seven genres, 21 one-against-one local feature sets are generated for classification. The number of features in each local feature set is much less than the original 277 features, as illustrated in Table 6. We find that Color Histogram, Motion Vector, MFCC, and LPC are four critical feature types appearing in most pairs. For the classification results illustrated in Table 7, except the recall in

Table 6. Number of features in each local feature set

Genre	action	animation	comedy	documentary	drama	musical	thriller
action	*	13	11	21	10	5	3
animation	-	*	15	12	19	2	11
comedy	-	-	*	11	14	3	15
documentary	-	-	-	*	2	3	10
drama	-	-	-	-	*	2	11
musical	-	-	-	-	-	*	5
thriller	-	-	-	-	-	-	*

Table 7. Confusion matrix by the local selection strategy

Predicted / Actual	action	animation	comedy	documentary	drama	musical	thrill	*Recall (%)*
action	35	0	0	0	0	0	0	100
animation	0	13	0	0	1	0	0	92.9
comedy	0	0	45	0	1	0	0	97.8
documentary	0	0	1	23	4	0	0	82.1
drama	0	0	0	0	66	0	1	98.5
musical	0	0	0	0	1	6	0	85.7
thriller	1	0	0	0	8	0	17	65.4
Precision (%)	97.2	100	97.8	100	81.5	100	94.4	

thrill, we get good precision and recall in each genre. We find that it is usual to identi-fy thrill movies as drama movies, even when we use both visual and audio features in genre classification. However, the overall accuracy is around 91.9%, and this demon-strates that, without considering the other genres, more precise features can be selected for each pair of genres to get better classification results.

4.2   Comparisons among All Methods

Finally, the qualitative comparisons between our methods and the other start-of-art methods are illustrated in Table 8. Although we extract two feature types (i.e., visual and audio) including 277 features from movie trailers, more movie genres could be classified more precisely using the local feature selection. In fact, no more than 25 features as shown in Table 6 are used to discriminate each pair of movie genres.

Table 8. Comparisons among all methods

	Z. Rasheed et al. [18]	H.Y. Huang et al. [6]	S.K. Jain et al. [8]	Ours
No. of genres	4	3	5	7
Feature types	Visual	Visual	Visual-Audio	Visual-Audio
Feature dim.	4	4	21	277
Feature selection	N	N	N	Y
Classifier	Mean Shift Classifi-cation	2-Layer Neural Network	Neural Network	SVMs
Accuracy	83%	80.2%	87.5%	91.9%

5   Conclusions

In this paper, a movie genre classification system is proposed. Totally, 277 visual and audio features are extracted from movie trailers. By employing the SAHS (self-adaptive harmony search) algorithm on these 277 features, the feature selection model can effectively find the optimum feature subsets for corresponding movie genres. From the results of the SAHS algorithm, we find audio features are more relevant than visual features in discriminating movie genres. Finally, the experimental results show that the overall accuracy reaches 91.9%, and this demonstrates more precise features can be selected for each pair of genres to get better classification results. In the future, we hope high-level features or semantics can be explored to classify movie genres more precisely.

Acknowledgments. This work was supported by National Science Council of R.O.C. under Grant NSC100-2218-E-224-011-MY3.

References

1. Chang, C.-C., Lin, C.-J.: LIBSVM: a library for support vector machines, software available at http://www.csie.ntu.edu.tw/~cjlin/libsvm
2. Chang, S.-F., Sikora, T., Puri, A.: Overview of the MPEG-7 standard. IEEE Transactions on Circuits and Systems for Video Technology 11(6), 688–695 (2001)
3. Deriche, M.: Feature selection using ant colony optimization. In: Proc. the 6th International Multi-Conference on Systems, Signals and Devices, pp. 1–4 (2009)
4. Diao, R., Shen, Q.: Two new approaches to feature selection with harmony search. In: Proc. the IEEE International Conference on Fuzzy Systems, pp. 1–7 (2010)
5. Hall, M.A., Smith, L.A.: Practical feature subset selection for machine learning. In: Proc. the 21st Australian Computer Science Conference, pp. 181–191 (1998)
6. Huang, H.Y., Shih, W.S., Hsu, W.H.: Movie classification using visual effect features. In: Proc. the IEEE Workshop on Signal Processing Systems, pp. 295–300 (2007)
7. Huang, Y.-F., Wang, C.-M.: Self-adaptive harmony search algorithm for optimization. Expert Systems with Applications 37(4), 2826–2837 (2010)
8. Jain, S.K., Jadon, R.S.: Movies genres classifier using neural network. In: Proc. the 24th International Symposium on Computer and Information Sciences, pp. 575–580 (2009)
9. Jin, X., Bie, R.: Random forest and PCA for self-organizing maps based automatic music genre discrimination. In: Proc. the International Conference on Data Mining, pp. 414–417 (2006)
10. Lanzi, P.L.: Fast feature selection with genetic algorithms: a filter approach. In: Proc. the International Conference on Evolutionary Computation, pp. 537–540 (1997)
11. Levinson, N.: The Wiener RMS error criterion in filter design and prediction. J. Math. Phys. 25, 261–278 (1947)
12. Lienhart, R.: Comparison of automatic shot boundary detection algorithms. In: Proc. Storage and Retrieval for Image and Video Databases VII. SPIE, vol. 3656, pp. 1–12 (1998)
13. Martinez, J.M., Koenen, R., Pereira, F.: MPEG-7: the generic multimedia content description standard, part 1. IEEE Multimedia 9(2), 78–87 (2002)
14. Martinez, J.M.: MPEG-7 overview (version 10), ISO/IEC JTC1/SC29/WG11 N6828 (2004)
15. McEnnis, D., McKay, C., Fujinaga, I., Depalle, P.: jAudio: a feature extraction library. In: Proc. the 6th International Conference on Music Information Retrieval, pp. 600–603 (2005)
16. Panagakis, Y., Kotropoulos, C.: Music genre classification via topology preserving non-negative tensor factorization and sparse representations. In: Proc. the 35th IEEE International Conference on Acoustics, Speech, and Signal Processing, pp. 249–252 (2010)
17. Panagakis, Y., Kotropoulos, C., Arce, G.R.: Non-negative multilinear principal component analysis of auditory temporal modulations for music genre classification. IEEE Transactions on Audio, Speech, and Language Processing 18(3), 576–588 (2010)
18. Rasheed, Z., Sheikn, Y., Shah, M.: On the use of computable features for film classification. IEEE Transactions on Circuits and System for Video Technology 15(1), 52–64 (2005)
19. Silviu, G.: Information Theory with Applications. McGraw-Hill (1977)
20. Wactlar, H.D.: The challenges of continuous capture, contemporaneous analysis, and customized summarization of video content. In: Proc. the Workshop on Defining a Motion Imagery Research and Development Program, pp. 1–9 (2001)
21. Apple iTunes Movie Trailers Website at http://trailers.apple.com/trailers/
22. Internet Movie Database (IMDb) at http://www.imdb.com/
23. The Compact Edition of the Oxford English Dictionary II, p. 2537. Oxford University Press (1971)

Automatic Player Behavior Analyses from Baseball Broadcast Videos

Yin-Fu Huang and Zong-Xian Yang

Department of Computer Science and Information Engineering
National Yunlin University of Science and Technology
{huangyf,g9817723}@yuntech.edu.tw

Abstract. In this paper, we present a baseball player behavior analysis system by combining pitch types and swing events. We use eight kinds of semantic scenes detected from baseball videos in our previous work. For the pitch types, we use the characteristic of the ball in a pitch scene to identify the ball trajectory, and then 39 features are extracted to feed into a trained SVM for classifying pitch types. For the swing events, we use moving objects in the batter region to determine whether a swing occurs. Then, the event following the swing is detected using an HMM, based on the after-swing scene sequence. Next, the experimental results show that both pitch type recognition and swing event detection have accuracy rates 91.5% and 91.1%. Finally, we analyze and summarize player behavior by combining pitch types and swing events.

Keywords: Baseball broadcast video, player behavior analysis, pitch type recognition, event detection, SVM, HMM.

1 Introduction

In the past decade, the increasing amount of multimedia technologies has grown rapidly. With the increasing amount of multimedia information, many researchers focus on the highlight extraction, the event detection, the indexing, the retrieval and so forth on sport videos, to get the information that users want. To analyze video contents, different technologies such as pattern recognition, artificial intelligence, and computer vision are applied to videos. With the processing of sport videos, it would be convenient and effective for users to get more interesting information in a huge amount of videos.

For the processing of a baseball video, the characteristics and visual features in a baseball game were used to extract a ball trajectory in [2, 4]. Chen et al. [3] detected a strike zone utilizing shaping and visualization in a broadcast baseball video. The superimposed caption in a baseball video and rule-based decision approach [10, 12], and even the web-casting [6] were used to detect events. Furthermore, HMM could be also used to detect events, based on scene sequences [7, 8, 9]. In [11], ball trajectories were analyzed and even classified into pitch types using a random forest tree. However, player behavior has not analyzed yet.

R. Huang et al. (Eds.): AMT 2012, LNCS 7669, pp. 11–21, 2012.

In this paper, we analyze baseball player behavior by combining pitch types and swing events. First, the semantic scenes detected in the previous work are extracted. Among the detected scenes, we extract ball trajectories from pitch scenes [2]. Then, we extend and improve the method in [11] to extract the relevant features in a trajectory, and classify trajectories into pitch types. To detect swing events, a batter region should be found first and the event following the swing is detected using HMM, based on the after-swing scene sequence. Finally, we analyze and summarize player behavior by combining pitch types and swing events.

The remainder of the paper is organized as follows. In Section 2, the system overview of our baseball player behavior analysis is introduced first. In Section 3, we describe the extraction of a baseball trajectory and the relevant features, and then classify trajectories into pitch types using SVM. In Section 4, we present the swing detection and after-swing event detection using HMM. The experimental results are presented in Section 5. Finally, we make conclusions in Section 6.

2   System Architecture

In the previous work [5], we have segmented a baseball video into scenes and classified them into "pitch", "infield-hitting", "outfield-hitting", "field", "player", "running", "close-up", and "others".

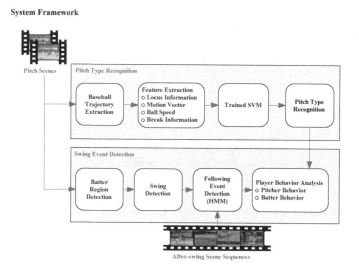

Fig. 1. System architecture

In this paper, we propose a baseball player behavior analysis system as shown in Fig. 1. The system consists of two parts: 1) pitch type recognition and 2) swing event detection. First, the pitch scenes detected in the previous work are loaded. Then, we use the characteristics of the ball in a pitch scene to extraction the ball track [2], and extract the related features such as "locus information", "motion vector", "ball speed", and "break information". Next, these features are fed into a trained SVM to classify pitch types. For the swing event detection, a batter region should be detected first. Then, we use moving pixels in the batter region to determine whether a swing occurs. Next, the event following the swing is detected using an HMM, based on the after-swing scene sequence. Finally, we analyze and summarize player behavior by combining pitch types and swing events.

3   Pitch Type Recognition

In this section, we derive a baseball trajectory using the algorithm proposed by Chen et al. [2]. Then, we extract the features in a trajectory, some of which are extended from those used by Takahashi et al. [11]. Finally, a trained SVM is used to classify trajectories into pitch types.

3.1   Baseball Trajectory Extraction

According to the characteristics and visual features in a baseball game, the framework to derive a baseball trajectory can be depicted in Fig. 2. First, the moving objects of each frame are located in pitch scenes. The ball candidates of each frame are detected in terms of position, color, size, and compactness filters. The motion of a ball may be a parabolic curve, due to the gravitational influence in a trajectory. Therefore, we utilize this physical characteristic to track a ball on the distribution of X-axis and Y-axis. Finally, a baseball trajectory is identified.

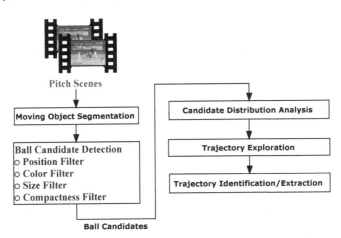

Fig. 2. Process of baseball trajectory extraction

3.2   Feature Extraction

The shape of a ball trajectory varies with its pitch type. Not only the locus information is included, but also the motion vectors, ball speed, and break information in the trajectory are selected as features in the system. Totally, 39 features are extracted from the ball trajectory as follows.

- Locus slop and locus curvature (i.e., first-order differential coefficient and second-order differential coefficient): Entire, first, second, third, and last parts
- Motion vectors (for horizontal and vertical): Last 12 frames in the ball trajectory
- Speed ratio of the first part to the second part
- Ball speed
- Break length
- The frame number with the break point
- The distance between the release point and the break point

We divide the ball trajectory into four parts, and the slope and curvature of each part are calculated, respectively. Since the closer a ball is to a home plate and the more movement a ball varies, the motion vectors in the last 12 frames are collected. Besides, the speed ratio is obtained from the ball speeds calculated from the motion vectors for the first and second part of the trajectory. The ball speed can be derived using the trajectory length expressed with total frame number. For the distance between the pitcher mound and home plate 18.44 m, and a video with 30 frames/sec, the ball speed can be calculated as follows:

$$Ball\ speed(km/h) = \frac{0.01844\ (km)}{(Trajectory\ length/30/3600)(h)} \tag{1}$$

Finally, the break length is the greatest distance between the trajectory and the straight line from the release point and the front of a home plate.

3.3 Support Vector Machines

The number of pitch types to be recognized depends on a pitcher, and the trajectory, movement, speed, and position of a pitching ball could vary with different pitchers. Hence, in our system, a classifier was created for each pitcher. Although there are various types of pitches in professional baseball, we could classify pitch types into nine types by using support vector machines (i.e., libSVM toolkit [1]). In the training phase, the radial basis function (i.e., RBF) was selected for the kernel function and five-fold cross validation was conducted on the training data.

4   Swing Event Detection

For the swing event detection, a batter region is detected first. Then, the moving objects in the batter region are used to determine whether a swing occurs. Next, the event following the swing is detected using an HMM, based on the after-swing scene sequence.

4.1   Batter Region Detection

Since a pitch scene is always taken by a fixed camera, we can consider two batter regions are also fixed. To detect a batter is right-handed or left-handed, the intensity differences between frames within each batter region are used. In other words, the batter region occupied by a batter would have larger intensity differences between frames than another batter region.

4.2   Swing Detection

Next, the moving objects in the detected batter region are used to determine whether a swing occurs. In general, the more moving objects the batter region has, the more likely a swing occurs. When a ball is close to a home plate (i.e., the last three frames in the trajectory), if the batter swings a bat, the number of the moving objects in the batter region would be more than a threshold Ts.

4.3   Following Event Detection

Here, five kinds of after-swing events are identified using an HMM method. As shown in Fig. 3, five trained HMMs representing different events are used to detect after-swing events. We feed each of these HMMs with after-swing scene sequences, and each HMM would produce the probability of the corresponding event. Finally, the detected event of the scene sequences would be what the HMM with the maximal probability represents.

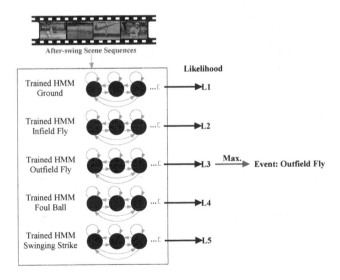

Fig. 3. HMMs used to detect after-swing events

5   Experimental Results

In the experiments, the training/test videos were recorded from the broadcast TV programs of "Chinese Professional Baseball League". These games include different teams, pitchers, baseball fields, and all the videos are compressed in the MPEG-1 format with frame size 352*240 and frame rate 29.97fps. Two performance measures of precision and recall were applied to analyze the accuracy of the pitch type recognition and swing event detection.

$$\text{Precision} = \frac{N_C}{N_C + N_F} \tag{2}$$

$$\text{Recall} = \frac{N_C}{N_C + N_M} \tag{3}$$

$$\text{Accuracy} = \frac{total_c}{total} \tag{4}$$

where NC is the number of pitch types (or swing events) that were retrieved and correctly detected, NF is the number of pitch types (or swing events) that were retrieved but falsely detected, and NM is the number of missed pitch types (or swing events), totalc is the number of all pitch types (or swing events) correctly predicted, and total is the number of all pitch types (or swing events).

5.1   Pitch Type Recognition

The recognition results of seven pitchers are presented in Table 1. Their dominant arms include left and right, each pitcher has four or five pitch types, and the number of pitches for each pitcher is more than 300. The total accuracy for all the pitch type recognition is 91.5%. Besides, we also show the detailed recognition results for pitcher A in Table 2. For pitcher A, two-seam fastball accounts for over 60% of all pitched. Two-seam fastball is better recognized than fastball or breaking ball so that it has high precision and recall. Changeup ball has low recall caused by limited training data (accounting for only 2% of all pitched).

Table 1. Precision and recall for each pitcher

Pitcher	Dominant arm	Number of pitch types	Number of pitches	Accuracy
A	right	4	469	93.8%
B	left	4	335	91.9%
C	right	5	389	91.8%
D	right	5	342	88.0%
E	right	4	327	90.2%
F	left	5	312	91.3%
G	left	4	343	92.7%
Total	right/left	9	2517	91.5%

Besides, we also compare the recognition results between the Takahashi et al. method [11] and our method, as shown in Table 3. In this paper, we extend the feature extraction model used by Takahashi et al., and use a different classification method, thereby achieving significant accuracy.

Table 2. Results for pitcher A

	Two-seam Fastball	Slider	Circle-changeup	Changeup	Recall
Two-seam Fastball	280	0	4	0	98.6%
Slider	11	90	1	0	88.2%
Circle-changeup	5	5	63	0	86.3%
Changeup	3	0	0	7	70.0%
Precision	93.6%	94.7%	92.6%	100.0%	

Table 3. Comparisons between two methods

	Number of pitch types	Number of pitchers	Number of pitches	Number of features	Methods	Accuracy
Takahashi et al.	9	5	1538	36	Random forest	88.7%
Ours	9	7	2517	39	SVM	91.5%

5.2   Swing Event Detection

Here, the 4-state HMM is used to detect five after-swing events. The datasets used to train and test the HMM are shown in Table 4. In the experiments, five kinds of after-swing events could be detected, including "Ground", "Infield Fly", "Outfield Fly", "Foul Ball", and "Swinging Strike".

Table 4. Events in the training set and test set

	Training set	Test set
Ground	70	35
Infield Fly	11	6
Outfield Fly	59	29
Foul Ball	83	41
Swinging Strike	48	24
Total events	271	135

We present the swing event detection results as shown in Table 5. The total accuracy for all the swing event detection is 91.1%. From the results, we found that an outfield fly event has high accuracy owing to its salient scene sequence. However, an infield fly event has low accuracy since its scene sequence is very similar to a ground event and only 17 events exist in the dataset.

Table 5. Precision and recall of event detection

	Ground	Infield Fly	Outfield Fly	Foul Ball	Swinging Strike	Recall
Ground	31	4	0	0	0	88.6%
Infield Fly	3	3	0	0	0	50.0%
Outfield Fly	0	0	29	0	0	100.0%
Foul Ball	0	0	0	37	4	90.2%
Swinging Strike	0	1	0	0	23	95.8%
Precision	91.2%	37.5%	100.0%	100.0%	85.2%	

5.3   Behavior Analysis

In this section, we summarize and analyze player behavior by combining the pitch type recognition and swing event detection. Table 6 presents the pitch result distribution for pitcher A. We observed that two-seam fastball causes major foul ball events (40.2%) and ground events (28%); slider/circle-changeup causes major swinging strike events (35.8%/39.3%). Changeup even causes complete swinging strike events (100%) although the number of changeup is only 10 as indicated in Table 2.

Table 6. Pitch result distribution for pitcher A

	Ground	Infield Fly	Outfield Fly	Foul Ball	Swinging Strike	Swing rate
Two-seam Fastball	28%	3.4%	20.5%	40.2%	7.7%	39.3%
Slider	22.6%	3.8%	15.1%	22.6%	35.8%	47.3%
Circle-changeup	10.7%	14.3%	10.7%	25.0%	39.3%	45.9%
Changeup	0.0%	0.0%	0.0%	0.0%	100.0%	37.5%

In general, the majority of pitch types in a baseball game is fastball. Table 7 and Table 8 present the summarize results of two kinds of fastball. In the summarization, we observed that two-seam fastball causes more ground events than four-seam fastball. But, four-seam fastball causes more outfield fly events than two-seam fastball, especially for pitcher D (30.8%) and F (32.0%).

Table 7. Two-seam fastball summarization

Pitcher	Ground	Infield Fly	Outfield Fly	Foul Ball	Swinging Strike	Swing rate
A	28%	3.4%	20.5%	40.2%	7.7%	39.3%
C	27.1%	2.3%	13.5%	45.1%	12.0%	46.5%

For batter part, Table 9 presents the swinging result distribution for batter 1. We observed that four-seam fastball and knuckle curve cause more outfield fly events (45.0% and 50.0%), and it indicates batter 1 can grasp these pitch types easily. On the contrary, slider causes more swinging strike (50.0%), and it indicates batter 1 cannot grasp this pitch type. Besides, according to the swing rate, we found that batter 1 prefers two-seam fastball (54.5%), knuckle curve (66.7%), and split-finger (60.0%) to the other pitch types.

Table 8. Four-seam fastball summarization

Pitcher	Ground	Infield Fly	Outfield Fly	Foul Ball	Swinging Strike	Swing rate
B	22.3%	3.2%	24.5%	42.6%	7.4%	42.0%
D	12.1%	1.1%	30.8%	40.7%	15.4%	67.9%
E	23.2%	2.4%	24.4%	42.7%	7.3%	40.4%
F	21.3%	5.3%	32.0%	40.0%	14.7%	41.4%
G	22.8%	4.3%	15.2%	44.6%	13.0%	37.5%

Table 9. Swinging result distribution for batter 1

	Ground	Infield Fly	Outfield Fly	Foul Ball	Swinging Strike	Swing rate
Four-seam Fastball	10.0%	0.0%	45.0%	40%	5.0%	42.6%
Two-seam Fastball	16.7%	8.33%	25.0%	33.3%	16.7%	54.5%
Slider	12.5%	0.0%	0.0%	37.5%	50.0%	47.1%
Curve	0.0%	0.0%	0.0%	0.0%	0.0%	0.0%
Changeup	0.0%	0.0%	33.3%	50.0%	16.7%	35.3%
Circle-changeup	0.0%	0.0%	50.0%	0.0%	50.0%	40.0%
Knuckle Curve	0.0%	0.0%	50.0%	50.0%	0.0%	66.7%
Fork	0.0%	0.0%	0.0%	0.0%	0.0%	0.0%
Split-finger	33.3%	0.0%	0.0%	33.3%	33.3%	60.0%

Table 10 presents the swinging result distribution for slider. In the results, batter 1 has more swinging strike event (50.0%); batter 2 and batter 3 have more ground event (38.5% and 37.5%). Besides, batter 4 grasps slider easily whereas batter 1 cannot. In summary, the swing characteristic of each batter can be obtained, which a coach can use to arrange in a baseball game.

Table 10. Swinging result distribution for slider

Batter	Ground	Infield Fly	Outfield Fly	Foul Ball	Swinging Strike	Swing rate
Batter 1	12.5%	0.0%	0.0%	37.5%	50.0%	47.1%
Batter 2	38.5%	0.0%	15.4%	46.2%	0.0%	54.2%
Batter 3	37.5%	0.0%	12.5%	12.5%	37.5%	57.1%
Batter 4	0.0%	0.0%	33.3%	66.7%	0.0%	37.5%
Batter 5	22.2%	0.0%	22.2%	33.3%	22.2%	47.4%
Batter 6	0.0%	0.0%	33.3%	33.3%	33.3%	42.9%

6   Conclusions

In this paper, we propose a baseball player behavior analysis system using both pitch types and swing events. For the pitch types, we extract the features from ball trajectories including locus information, motion vectors, ball speed, and break information, and then classify them into nine pitch types. For the swing events, we use the moving objects in the batter region to determine whether a swing occurs, and then a swing event is detected using an HMM, based on the after-swing scene sequence. The experimental results show that both pitch type recognition and swing event detection have high accuracy (i.e., 91.5% and 91.1%). Finally, the pitch types and swing events are combined to analyze player behavior, and this information will be useful for coach to arrange in a baseball game.

Acknowledgments. This work was supported by National Science Council of R.O.C. under Grant NSC100-2218-E-224-011-MY3.

References

1. Chang, C.-C., Lin, C.-J.: LIBSVM: a library for support vector machines (2001), software available at http://www.csie.ntu.edu.tw/~cjlin/libsvm
2. Chen, H.-T., Chen, H.-S., Hsiao, M.-H., Tsai, W.-J., Lee, S.-Y.: A trajectory-based ball tracking framework with visual enrichment for broadcast baseball videos. Journal of Information Science and Engineering 24(1), 143–157 (2008)
3. Chen, H.-T., Tsai, W.-J., Lee, S.-Y.: Stance-based strike zone shaping and visualization in broadcast baseball video: providing reference for pitch location positioning. In: Proc. IEEE International Conference on Multimedia and Expo, pp. 302–305 (2009)
4. Chu, W.-T., Wang, C.-W., Wu, J.-L.: Extraction of baseball trajectory and physics-based validation for single-view baseball video sequences. In: Proc. IEEE International Conference on Multimedia and Expo, pp. 1813–1816 (2006)
5. Huang, Y.-F., Tung, L.-H.: Semantic scene detection system for baseball videos based on the MPEG-7 specification. In: Proc. ACM Symposium on Applied Computing, pp. 1–9 (2008)
6. Huang, Y.-F., Chen, L.-W.: An event-based video retrieval system by combining broadcasting baseball video and web-casting text. In: Proc. ACM Symposium on Applied Computing, pp. 846–852 (2011)
7. Huang, Y.-F., Huang, J.-J.: Semantic event detection in baseball videos based on a multi-output hidden Markov model. In: Proc. ACM Symposium on Applied Computing, pp. 929–936 (2011)
8. Lien, C.-C., Chiang, C.-L., Lee, C.-H.: Scene-based event detection for baseball videos. Journal of Visual Communication and Image Representation 18(1), 1–14 (2007)
9. Mochizuki, T., Tadenuma, M., Yagi, N.: Baseball video indexing using patternization of scenes and hidden Markov model. In: Proc. IEEE International Conference on Image Processing, pp. 1212–1215 (2005)
10. Su, Y.-M., Hsieh, C.-H.: Semantic events detection and classification for baseball videos. In: Proc. IEEE International Symposium on Industrial Electronics, pp. 1332–1336 (2009)

11. Takahashi, M., Fujii, M., Yagi, N.: Automatic pitch type recognition from baseball broadcast videos. In: Proc. the 10th IEEE International Symposium on Multimedia, pp. 15–22 (2008)
12. Zhang, D., Chang, S.-F.: Event detection in baseball video using superimposed caption recognition. In: Proc. the 10th ACM International Conference on Multimedia, pp. 315–318 (2002)

Hot Topic Detection in News Blogs from the Perspective of W2T

Erzhong Zhou[1], Ning Zhong[1,2], Yuefeng Li[3], and Jia-jin Huang[1]

[1] International WIC Institute, Beijing University of Technology
Beijing 100124, P.R. China
{zez2008,hjj}@emails.bjut.edu.cn
[2] Department of Life Science and Informatics, Maebashi Institute of Technology
460-1 Kamisadori-Cho, Maebashi 371-0816, Japan
zhong@maebashi-it.ac.jp
[3] Faculty of Science and Technology, Queensland University of Technology
Brisbane QLD 4001, Australia
y2.li@qut.edu.au

Abstract. News blog hot topics are important for the information recommendation service and marketing. However, information overload and
personalized management make the information arrangement more difficult. Moreover, what influences the formation and development of blog
hot topics is seldom paid attention to. In order to correctly detect news
blog hot topics, the paper first analyzes the development of topics in a
new perspective based on W2T (Wisdom Web of Things) methodology.
Namely, the characteristics of blog users, context of topic propagation
and information granularity are unified to analyze the related problems.
Some factors such as the user behavior pattern, network opinion and
opinion leader are subsequently identified to be important for the development of topics. Then the topic model based on the view of event
reports is constructed. At last, hot topics are identified by the duration, topic novelty, degree of topic growth and degree of user attention.
The experimental results show that the proposed method is feasible and
effective.

1 Introduction

A blog is an online diary which provides information sharing and opinion interaction service. A topic is defined as a seminal event or activity, along with all
directly related events and activities [3]. A blog hot topic is a topic which users
widely discuss and consecutively pay attention to in blogspace. Nowadays, most
scholars pay more attention to overcoming the technique bottleneck because information overload and personalized management increase difficulties in topic
detection.

Hokama et al. [6] adopted an extended agglomerative hierarchical clustering
technique regarding the timestamp to detect topics, and evaluated the topic hotness by counting the number of articles related to each topic. He et al. [5] used

R. Huang et al. (Eds.): AMT 2012, LNCS 7669, pp. 22–31, 2012.

incremental term frequency inverse document frequency model and incremental clustering algorithm to detect new events, and considered the frequency and consecutive time of news reports to evaluate the topic hotness. Chen et al. [3] first extracted hot words based on the distribution over time and life cycle, then identified key sentences and grouped the key sentences into clusters that represent hot topics. As for evaluating blog topics, the methods mentioned above have many shortcomings. First, most of evaluation strategies focus on not users but media. Second, the hotness evaluation measures are too simple, and many factors which influence topics are not carefully analyzed. For example, a temporal social network is formed during the topic propagation and opinion interaction. Roles of blog users in topic propagation need analyzing because different kinds of users have different impacts on blog topics. Hence, what determines the formation and development trend of a blog topic is still an unsolved problem.

W2T (Wisdom Web of Things) proposed by Zhong et al. is an extension of the Wisdom Web in the Internet of Things age [13]. Besides, the W2T methodology gives a perspective on how to unify humans' models, information networks and granularity for analyzing the intersectional problems between the social world and cyber world [14]. In order to identify the important factors for detecting hot topics in news blogs, the paper is based on the W2T methodology to analyze the formation and development of a blog topic. Namely, the context of topic propagation is first decomposed into different kinds of complex networks related to users. Then the influence of different kinds of users and network opinions on blog topics is measured in related information networks corresponding to the special information granularity. The paper concludes that the user motivation and behavior pattern determine the burst and temporal features of a news blog topic. Moreover, the user behavior pattern, network opinion and opinion leader play a vital role in different phases of a blog topic.

The rest of the paper is organized as follows. Section 2 analyzes some problems related to the news blog topics. Section 3 describes the proposed method. Section 4 tests the feasibility and effectiveness of the proposed method by some experiments. Section 5 gives conclusions and future work.

2   Problem Analysis

In order to construct an effective topic model to represent news blog topics and reasonably evaluate the topics, the paper has to cope with the two problems ignored by the traditional methods. First, the characteristics of news blog topics need to be analyzed on the basis of the motivations and features of blog users. Second, the context of topic propagation needs to be decomposed into the related complex networks, and the key factors that determine the development trends of news blog topics are identified in different complex networks.

2.1   Characteristics of a News Blog

The characteristics of information are influenced by the related medium. For example, there is no limitation on the writing style of a blog post, and users often publish a topic from different views or levels. A blog topic model is consequently dynamic and hierarchical. Blogs can be categorized by the types of topics, and users visit different kinds of blogs with different motivations [8]. Hence, the development of a blog topic is related to the characteristics of the related blog. A news blog is a kind of temporal topic blog. A news blog topic is derived from the news event in the daily life, and often shows the burst and temporal features. Users take part in the topic interaction for the sake of knowing the truth of a news event or expressing their own personal feeling. The relationships among members in a topic group are weak because the group is mainly maintained by user interests [11]. The topic group dissolves when users lose interests in the related topic.

2.2   Development of a Blog Topic

A blog topic can experience the birth, growth, maturity and fade [12]. At the beginning, one blogger sponsors an issue, and other bloggers are attracted to join. When the topic enters into the growth phase, users actively express opinions or spread the topic. While the topic group rapidly grows up, some ordinary users can become opinion leaders that are influential in giving birth to the public opinion. When the topic grows up into the maturity, the public opinion emerges. At last, the influence of opinion leaders gradually becomes weak and then the topic fades away.

According to the description mentioned above, the user, topic and opinion are interrelated with each other in blogspace. Hence, the blog community, topic network and opinion network are extracted from the blogspace to analyze the context of topic propagation. The blog community is composed of users with the similar interest. The topic network is composed of topics with different granularities and the evolutionary relationship between topics. The opinion network is composed of opinions and interaction relationships among users.

As shown in Figure 1, the structure of the topic network can present compositions of each topic and the evolution of different topics. Besides, the boundary of a blog community is identified by blog topics. As a result, the information granularity not only determines the structure of the topic network but also identifies the range of the blog community. According to the social network analysis theory, the structure of an online community is related to user interaction patterns and relationships among users. The obvious phenomena are the most of users are active in the early phase. The phenomena can be explained as follows. Users lack the authoritative information and are very curious about the event in the early phase. Hence, the effect of sheep flock occurs in the community. When a public opinion comes into being, the phenomenon of spiral of silence emerges. Namely, users often keep silent when their opinions are in the minority. Moreover, the novelty of topics gets weaker and weaker with the lapse of time. The structural changes of a blog community consequently reflect the development trends of topics. As far as the user role is concerned, some users become opinion leaders during the topic propagation and opinion interaction. Opinion leaders are influential during the evolution of network opinions. An active user plays the role of a disseminator. However, the authority of such an active user is often weak in comparison with the opinion leader. Ordinary users often lurk after the public opinion emerges. Hence, the position of a blog user in the social network indicates the user role [10]. An opinion is a product of topic propagation. The formation and structural change of an opinion network all result from the interaction between users. The opinion expression is also the motivation of topic interaction, and the opinion network contributes to understanding the user behavior and evolution of network opinions. On the other hand, users also publish opinions on a topic from different views, and the opinion distribution in the opinion network can reflect the state of the topic. Besides, the structure and sentimental polarity distribution of the opinion network contribute to recognizing the opinion leaders [2]. Hence, the user behavior pattern, network opinion and opinion leader all impact on the development of a blog topic. Moreover, opinion leaders determine the development trend of the blog topic to a great extent.

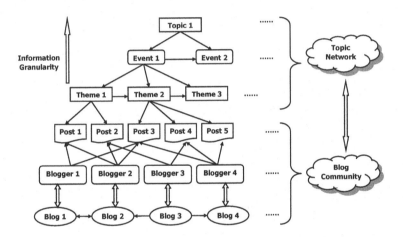

Fig. 1. The structure of a topic network in blogspace

3   Proposed Method

The information granularity needs considering for the construction of a topic model so as to analyze the structural characteristics and evolution of the topic. Hence, this section presents a new topic detection approach based on the different views of event reports and evolutionary relationship between events. As for the topic hotness, the changes of user interests can influence the vitality of the topic, and user interests need correctly evaluating. In order to detect the current and forthcoming hot topics, the growth state of a topic is measured, and opinion leaders are identified. Finally, hot topics are identified by the duration, topic novelty, degree of topic growth and degree of user attention.

3.1   Hot Topic Detection Algorithm

The hot topic detection algorithm comprises the topic detection and hotness evaluation, and is presented in Algorithm 1.

Algorithm 1. Hot Topic Detection

Input: S is a post set, n is the number of time units, d is the threshold of hotness.
Output: T_{set} is a hot topic set.
1. **for** each post i in S **do**
2.   Extract keywords from i, and construct the theme model of i;
3. **end for**
4. **for** time unit $j = 0$ to n **do**
5.   Construct event models, and add the models into event list EL_j within j;
6.   **if** j is equal to 1 **then**
7.   Construct topic models within j manually;
8.   **else**
9.   Construct a new topic model, or update the previous topic model according to the related event model in EL_j;
10.   **end if**
11.   Analyze the structure of the blog community constructed by interaction relationships among users;
12.   Detect network opinions on each topic according to the topic models;
13.   Identify opinion leaders with respect to each topic;
14.   Evaluate all topics within j by the duration, topic novelty, degree of topic growth and degree of user attention;
15.   Select the topic whose hotness is more than d, and add new hot topics into T_{set};
16. **end for**
17. Return T_{set}.

3.2   Topic Detection Based on the View of Event Reports

Considering the information granularity in a topic network, a topic can be considered to be a cluster of related events, and an event can be considered to be a cluster of related themes. As for the literal expression of a theme, the theme can be expressed by a group of related keywords. Hence, the topic can be divided into three layers. The models in three layers are as follows:

- A theme model is defined as $Theme = \{Keyword_1,\ Keyword_2,\ \ldots,\ Keyword_s\}$, where $Keyword_i$ denotes the ith keyword;
- An event model is defined as $Event = \{Theme_1,\ Theme_2,\ \ldots,\ Theme_m\}$, where $Theme_i$ denotes the ith theme;
- A topic model is defined as $Topic = \{Event_1,\ Event_2,\ \ldots,\ Event_n\}$, where $Event_i$ denotes the ith event.

The event model is the key to constructing the topic model. The single pass clustering algorithm is widely used in event detection and tracking [1], which can be explained as follows. A report is merged into an event if the content similarity between two objects is above a threshold. Otherwise, the report is considered to be the first report of a new event. Topic detection based on the view of event reports first adopts single pass clustering algorithm to extract events which occur in a given time interval. Then evolutionary relationship between events is identified by the content similarity between events and event distribution in different topics. At last, the topic model is created or updated by detecting the new event and tracking the previous event.

A keyword is a basic element for a theme model. The keyword is identified according to the weight of the word. The weight of a word is evaluated by the following equation in the Term Frequency Inverse Document Frequency (TFIDF) method [9]:

$$Weight(t_k, r) = TF(t_k, r) * \log(\frac{N}{N_k + 0.5}) \tag{1}$$

where r is a blog post, t_k is a word, $Weight(t_k, r)$ is the weight of t_k in r, $TF(t_k, r)$ is the frequency of t_k in the text of r, N is the number of blog posts, and N_k is the number of blog posts where t_k appears.

A theme model is the basis of an event model. When two events are compared, each theme in one event model is compared with that in the other event model. Then the average of similarities between themes of events is regarded as a proof of the similarity evaluation of two events. However, theme models abstracted from different posts can own different event attributions. Sometimes, one theme model can be a subset of the other theme model in terms of event attributes. Hence, the intersection between theme models is essential to compare the similarity between theme models. On the other hand, the size of a theme model has a great impact on the intersection between theme models. When the size of a theme model is small, the intersection between theme models is also small even if one theme model is the subset of the other one. As for that phenomenon, the minimum of two theme models is also taken into consideration by using the following equation, in order to correctly compare the similarity between theme models:

$$sim(d_i, d_j) = \alpha * \frac{|d_i \cap d_j|}{|d_i \cup d_j|} + \beta * \frac{|d_i \cap d_j|}{min(|d_i|, |d_j|)} \tag{2}$$

where d_j denotes the keyword set of the jth theme, $Sim(d_i, d_j)$ is the similarity between the ith theme and the jth theme, $d_i \cap d_j$ is the intersection between d_i and d_j, $d_i \cup d_j$ is the union of d_i and d_j, $|d_i|$ is the size of set d_i, α and β are coefficients, and $min(y, z)$ is the minimum between y and z.

In general, one event may evolve into other events. Moreover, main attributes of events often change. However, time is the key to the identification of the evolutionary relationship between events. If events occur very closely and the content similarity between events is high, the events are likely to be correlated with each other. The evolutionary relationship between two events is identified by the following norm. If the most of similar events which occur before the target event belong to the same topic, the target event evolves from the related events.

3.3 Topic Hotness Evaluation Based on User Interests

The growth state indicates the development trend of the topic. Replies, repliers, opinion leaders and network opinions all have great changes during the development of the blog topic. Hence, the growth degree of a topic is measured by using the following equation:

$$Growth(x) = (\mu_1 * \sum_{i=1}^{n} f(m_i) + \mu_2 * \sum_{i=1}^{n} f(l_i) + \mu_3 * \sum_{i=1}^{n} f(c_i) + \mu_4 * \sum_{i=1}^{n} f(s_i)) * \frac{1}{n} \quad (3)$$

$$f(p_i) = \begin{cases} 1, & i = 1 \\ norm(\frac{p_i}{p_{i-1}+0.1}), & i > 1 \end{cases} \quad (4)$$

$$norm(y) = \begin{cases} 1, & y \geq 1 \\ y, & otherwise \end{cases} \quad (5)$$

where $Growth(x)$ is the growth degree of topic x, m_i is the number of repliers of topic x within the ith time unit, l_i is the number of opinion leaders of topic x within the ith time unit, c_i is the number of replies of topic x within the ith time unit, s_i is the number of opinions on topic x within the ith time unit, n is the total number of time units, μ_1 is the growth coefficient of a user, μ_2 is the growth coefficient of an opinion leader, μ_3 is the growth coefficient of a reply, and μ_4 is the growth coefficient of an opinion. When opinions are counted, repetitive opinions of which the same user publishes are ignored.

The degree of user participation, position in the social network and influence on neighbors' opinions are assessed to recognize the opinion leader. Besides, the user role during the topic propagation and evolution of network opinions need analyzing. The influence of a user is evaluated by the following equation:

$$Opind_n(w, x) = \sum_{i=1}^{n} \frac{nop_i(w, x)}{Tnop_i(x)} * (\varphi * \frac{nos_i(w, x)}{Tnos_i(x)} + \psi * cen_i(w) + \delta * \frac{bp_i(w, x)}{Tp_i(x)}) \quad (6)$$

where $Opind_n(w, x)$ is the influence of user w on topic x within the nth time unit, i denotes the ith time unit, $nop_i(w, x)$ is the number of opinions that user w publishes on topic x within the ith time unit, $Tnop_i(x)$ is the total number of opinions that all users publish on topic x within the ith time unit, $cen_i(w)$ is the central degree of user w in the social network within the ith time unit, $nos_i(w, x)$ is the number of users who have the same opinion as user w within the ith time unit, $Tnos_i(x)$ is the total number of users who publish opinions on topic x within the ith time unit, $bp_i(w, x)$ is the number of posts that user w publishes with respect to topic x and are recommended by the website within the ith time unit, $Tp_i(x)$ is the total number of posts that all users publish with respect to topic x within the ith time unit, φ is the emotion coefficient, ψ is the position coefficient, and δ is the quotation coefficient.

The topic hotness reflects user interests on a topic. If a topic is very new, the user is more likely to be interested in the topic. If a topic spreads a long time, the topic can have a great probability to attract users. The representations of user interests can also be reflected from the quotation number and reply number of posts. Hence, the topic hotness is evaluated by using the following equation:

$$Hotness(x) = \frac{cu}{n} * Growth(x) * (\lambda * sc(x) + \xi * qu(x)) * \Delta t(x)^{-k} \quad (7)$$

where $Hotness(x)$ denotes the hotness of topic x, n is the total number of time units, cu is the number of consecutive time units which topic x occurs in, $Growth(x)$ is the growth degree of topic x, $sc(x)$ is the total number of replies of topic x, $qu(x)$ is the quotation number of posts that belong to topic x, $\Delta t(x)$ is the time difference between the publication date of topic x and the current time, k is the decay coefficient, λ is the comment coefficient, and ξ is the quotation coefficient of a post.

4   Experiments and Discussion

Experiments are performed in order to validate the feasibility and effectiveness of the proposed method. The test sample set includes 1520 posts and 202290 related replies published in the China Sina blog website. The publication date of the test sample is between November 9, 2011 and January 18, 2012. The training sample set includes 12 blog hot topics listed by the China Sina blog website in 2011 and 17910 plain texts for the keyword extraction from the China sogou laboratory. The Chinese word segmentation software ICTCLAS is applied.

As for the user behavior, anonymous users frequently participate in the opinion interaction and are likely to publish negative opinions [7]. Hence, the characteristics of anonymous users need analyzing for knowing the structure of a blog community. According to the related study, most of bloggers are inclined to have a pseudonymous username [7]. As a result, the characteristics of anonymous users can be inferred by observing the behavior patterns of the pseudonymous users. The number of replies and average length of replies for different kinds of pseudonymous users categorized by the total number of replies are counted. As shown in Figure 2, the pseudonymous user is more likely to continue replying when the length of the user reply is long. A reply threshold is consequently set to estimate the range of anonymous users. Experiment 1 evaluates the topic hotness on the basis of the proposed method but does not take the growth degree of a topic into account. Experiment 2 adopts the proposed method to evaluate the topic hotness. Experiment 3 applies the hot topic detection method based on the agglomerative hierarchical clustering algorithm [4], and topics are evaluated by the total number of posts and replies. As shown in Figure 3, the performance of the proposed method is good. The agglomerative hierarchical clustering algorithm is based on the vector space model, and most of blog posts are not normalized so that the accuracy of the topic detection method based on agglomerative hierarchical clustering algorithm is low. However, the proposed method focuses on keywords of a post. Hence, blog posts that are not normalized do not cause a great impact on the proposed method. Moreover, the precision of hot topic detection is improved by measuring the growth state of a blog topic.

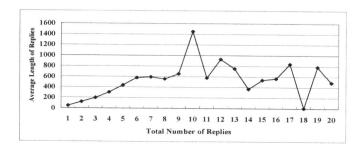

Fig. 2. Reply characteristics of pseudonymous users

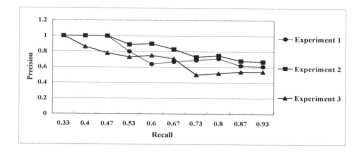

Fig. 3. Performance comparison of three experiments

5   Conclusions and Future Work

The problem solving strategy of W2T lays emphasis on human nature. The mechanism of information propagation in news blogs causes that the user motivation and behavior pattern determine the burst and temporal features of topics. Hence, users cannot be separated from topics in blog topic detection. Experimental results prove the proposed method is feasible and effective. The user behavior pattern, network opinion and opinion leader are identified to be important for the development trend of a blog topic. However, there are still some shortcomings. Experiments do not consider synonymous words. Although the virtual network breaks some barriers of the physical world, the human nature formed in the physical world still has a great impact on the virtual network. The behavior model of a blog user needs the further research. The shortage of word processing is going to be improved in the future. We will pay more attention to the latest researches on the model of the social contact in blogspace.

Acknowledgments. The study was supported by Beijing Natural Science Foundation (4102007) and National Natural Science Foundation of China (60905027).

References

1. Allan, J., Papka, R., Lavrenko, V.: On-line New Event Detection and Tracking. In: Proceedings of the Twenty-first Annual International ACM SIGIR Conference, pp. 37–45 (1998)
2. Bodendorf, F., Kaiser, C.: Detecting Opinion Leaders and Trends in Online Social Networks. In: Proceedings of the Fourth International Conference on Digital Society, pp. 124–129 (2010)
3. Chen, K.Y., Luesukprasert, L., Chou, S.C.T.: Hot Topic Extraction Based on Timeline Analysis and Multidimensional Sentence Modeling. IEEE Transactions on Knowledge and Data Engineering 19(8), 1016–1025 (2007)
4. Dai, X.Y., Chen, Q.C., Wang, X.L., Xu, J.: Online Topic Detection and Tracking of Financial News Based on Hierarchical Clustering. In: Proceedings of the Ninth International Conference on Machine Learning and Cybernetics, vol. 6, pp. 3341–3346 (2010)
5. He, T., Qu, G., Li, S., Tu, X., Zhang, Y., Ren, H.: Semi-automatic Hot Event Detection. In: Li, X., Zaïane, O.R., Li, Z. (eds.) ADMA 2006. LNCS (LNAI), vol. 4093, pp. 1008–1016. Springer, Heidelberg (2006)
6. Hokama, T., Kitagawa, H.: Detecting Hot Topics about a Person from Blogspace. In: Proceedings of the Sixteenth European-Japanese Conference on Information Modeling and Knowledge Bases, pp. 290–294 (2006)
7. Kilner, P.G., Hoadley, C.M.: Anonymity Options and Professional Participation in an Online Community of Practice. In: Proceedings of the 2005 Conference on Computer Support for Collaborative Learning, pp. 272–280 (2005)
8. Qiu, H.M.: The Social Network Analysis of Blogosphere. Harbin Institute of Technology, Harbin (2007)
9. Salton, G., Buckley, C.: Term-weighting Approaches in Automatic Text Retrieval. Information Processing & Management 24(5), 513–523 (1988)
10. Song, X.D., Chi, Y., Hino, K., Tseng, B.: Identifying Opinion Leaders in the Blogosphere. In: Proceedings of the Sixteenth ACM Conference on Information and Knowledge Management, pp. 971–974 (2007)
11. Sun, W.J., Qiu, H.M.: A Social Network Analysis on Blogospheres. In: Proceedings of 2008 International Conference on Management Science and Engineering, pp. 1769–1773 (2008)
12. Zhang, Y.: A Study on the Phenomenon of Public-opinion-spreading Through Bulletin Board System. Jilin University, Changchun (2011)
13. Zhong, N., Ma, J.H., Huang, R.H., Liu, J.M., Yao, Y.Y., Zhang, Y.X., Chen, J.H.: Research Challenges and Perspectives on Wisdom Web of Things (W2T). Journal of Supercomputing (2010)
14. Zhong, N., Bradshaw, J.M., Liu, J.M., Taylor, J.G.: Brain Informatics. IEEE Intelligent Systems 26(5), 16–21 (2011)

A Clustering Ensemble Based on a Modified Normalized Mutual Information Metric

Hamid Parvin, Behzad Maleki, and Sajad Parvin

Islamic Azad University, Nourabad Mamasani Branch, Mamasani Nourabad
Iranhamidparvin@mamasaniiau.ac.ir,
b.maleki@ut.ac.ir, s.parvin@iust.ac.ir

Abstract. It has been proved that ensemble learning is a solid approach to reach more accurate, stable, robust, and novel results in all data mining tasks such as clustering, classification, regression and etc. Clustering ensemble as a sub-field of ensemble learning is a general approach to improve the performance of clustering task. In this paper by defining a new criterion for clusters validation named Modified Normalized Mutual Information (MNMI), a clustering ensemble framework is proposed. In the framework first a large number of clusters are prepared and then some of them are selected for the final ensemble. The clusters which satisfy a threshold of the proposed metric are selected to participate in final clustering ensemble. For combining the chosen clusters, a co-association based consensus function is applied. Since the Evidence Accumulation Clustering (EAC) method can't derive the co-association matrix from a subset of clusters, Extended Evidence Accumulation Clustering (EEAC), is applied for constructing the co-association matrix from the subset of clusters. Employing this new cluster validation criterion, the obtained ensemble is evaluated on some well-known and standard datasets. The empirical studies show promising results for the ensemble obtained using the proposed criterion comparing with the ensemble obtained using the standard clusters validation criterion.

Keywords: Clustering Ensemble, Stability Measure, Cluster Evaluation.

1   Introduction

Nowadays, usage of recognition systems has found many applications in almost all fields [15]-[28]. Many researches are done to improve their performance. Most of these algorithms have provided good performance for specific problem, but they have not enough robustness for other problems. Because of the difficulty that these algorithms are faced to, recent researches are directed to the combinational methods. Ensemble learning has been proved to be a solid way to reach more accurate and stable results in data mining. Classifier ensemble as a sub-field of ensemble learning is a general method to improve the performance of classification. At first glance, usage of ensemble learning in clustering sounds similar to the widely prevalent use of combining multiple classifiers to solve difficult classification problems, using techniques such as bagging and boosting.

R. Huang et al. (Eds.): AMT 2012, LNCS 7669, pp. 32–42, 2012.
© Springer-Verlag Berlin Heidelberg 2012

Data clustering or unsupervised learning is an important and very difficult problem. The objective of clustering is to partition a set of unlabeled objects into homogeneous groups or clusters [3], [4] and [10]. There are many applications that use clustering techniques to discover latent structures of data, such as data mining [11], information retrieval [2], image segmentation [9], linkage learning [15], and machine learning. In real-world problems, clusters can appear with different shapes, sizes, degrees of data sparseness, and degrees of separation. Clustering techniques require the definition of a similarity measure between patterns. Some of the most typical clustering algorithms include: (a) Hierarchical clustering algorithms build clusters based on distance connectivity, (b) Centroid clustering algorithms such as k-means algorithm represents each cluster by a single mean vector, and (c) Density clustering algorithms such as DBSCAN define clusters as connected dense regions in the data space.

DBSCAN (stands for Density-Based Spatial Clustering of Applications with Noise) is a data clustering algorithm proposed by Ester et al. [32]. It is named density-based clustering because it searches for some partitions beginning at the estimated density distribution of corresponding nodes. DBSCAN is one of the most common clustering algorithms. Hierarchical clustering is another approach in clustering algorithms that seeks to build a hierarchy of partitions. It uses a number of the merge (or split) operators to reach the goal. The operators are employed in a greedy manner. The results of hierarchical clustering are usually presented in a dendrogram [33].

Since there is no prior knowledge about cluster shapes, choosing a specific clustering method is not easy [29]. Studies in the last few years have tended to combinational methods. Cluster ensemble methods attempt to find better and more robust clustering solutions by fusing information from several primary data partitions [8].

Fern and Lin [8] have offered a clustering ensemble framework that selects a few of the base partitionings to make a thinner but better ensemble than using all primary the base partitionings. The ensemble selection approach is designed based on quality and diversity, the only two factors that have been proven to effect cluster ensemble performance. Their method tries to select a subset of the base partitionings which simultaneously has both the highest quality and the most diversity. The Sum of Normalized Mutual Information, SNMI [5], [6] and [30], is used to measure the quality of each individual partition with respect to other partitions. Also, the Normalized Mutual Information, NMI, is employed to measure the diversity among partitions. Although the ensemble size in this method is relatively small, this method achieves significant performance improvement over full ensembles. Law et al. proposed a multi-objective data clustering method based on the selection of individual clusters produced by several clustering algorithms through an optimization procedure [13]. This technique chooses the best set of objective functions for different parts of the feature space from the results of base clustering algorithms. Fred and Jain [7] have offered a new clustering ensemble method which learns the pairwise similarities between points in order to facilitate a proper partition of the data without the a priori knowledge of the number and the shape of the clusters. This method which is based on cluster stability evaluates the primary clustering results instead of final clustering.

Alizadeh et al. discuss the drawbacks of the common approaches and then have proposed a new asymmetric criterion to assess the association between a cluster and a partition which is called Alizadeh-Parvin-Minaei criterion, APM. The APM criterion compensates the drawbacks of the common method. Also, a clustering ensemble method is proposed which is based on aggregating a subset of primary clusters. This method uses the Average APM as fitness measure to select a number of clusters. The clusters which satisfy a predefined threshold of the mentioned measure are selected to participate in the clustering ensemble. To combine the chosen clusters, a co-association based consensus function is employed [12], [31].

To evaluate a cluster, the NMI method has many weaknesses that are described in [31]. Alizadeh et al. propose another version of NMI named max method. They also show that the max method also has some drawbacks, so they propose another metric named APMM, which is first of their author names [12].

This paper proposes a new measure to evaluate a cluster in that it is desired to evaluate the average similarity of the cluster with other clusters by eliminating its complement. We employ this criterion to select the more robust clusters in the final ensemble. To aggregate the final partitionings into consensus partitioning, a number of well-known methods are employed to make a decisive conclusion.

Rest of this paper is organized as follows. In section 2, we explain the proposed method. Section 3 demonstrates results of our proposed method against traditional comparatively. Finally, we conclude in section 4.

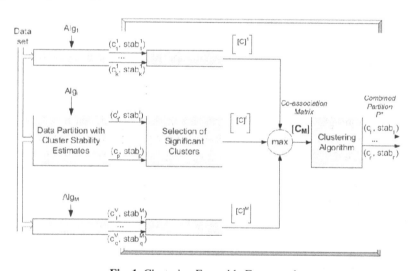

Fig. 1. Clustering Ensemble Framework

2   Proposed Method

In this section, first our proposed clustering ensemble method is briefly outlined, and then its phases are described in detail. The main idea of our proposed clustering ensemble framework is similar to Max and APMM [20] to utilize a subset of the best performing primary clusters in the ensemble, rather than using all of clusters. Only the clusters which satisfy a stability criterion are better to participate in the consensus function.

The proposed framework is depicted in Fig 1. It has four steps. In the first step B partitionings are extracted out of dataset. The partitioning i is denoted by $partitioning_i$. The $partitioning_i$ is obtained by a k-means algorithm with a new initialization of the seed points. Note that the $partitioning_i$ is to extract $k(i)$ clusters out of dataset. Then each partitioning is broken in some distinct partitions (or clusters). It means $partitioning_i$ converted to $k(i)$ clusters denoted by $c_1^i, c_2^i, ...$ and $c_{k(i)}^i$ respectively. After obtaining a pool of clusters, in the second step, a stability value is computed as a tag for each of them. The stability value of the cluster c_j^i is denoted by $stab_j^i$.

The manner of computing stability for each cluster is described in the sections 2.2 in more detail. A subset of stable clusters having a good diversity is selected by a thresholding scheme in the third step. This step is explained in detail in section 2.3. In the next step, the selected clusters are used to construct the consensus partitioning. This is done in two subparts: (a) to extract a co-association matrix from them (section 2.4) along with (b) a linkage clustering. Since the original EAC method [8] cannot truly identify the pairwise similarities between dataitems when there is only a subset of clusters, we use a method explained in [1] to construct the co-association matrix from the base selected clusters. This method is called EEAC: Extended Evidence Accumulation Clustering method. Finally, we use a hierarchical clustering algorithm, like single-link method, to extract the final clusters out of this matrix. For more generality, some heuristic consensus functions are also used as aggregators of selected clusters [30]. These heuristic consensus functions that are based on hypergraph partitioning and have first introduced by Strehl and Ghosh, are HyperGraph Partitioning Algorithm (HGPA), Meta-Clustering Algorithm (MCLA) and Cluster-based Similarity Partitioning Algorithm (CSPA) [30].

In the first step B partitionings are extracted out of dataset by B independent runnings of the k-means algorithm. The $partitioning_i$ is obtained by the i-th running of the k-means algorithm with a new initialization of the seed points. To produce the diverse cluster as much as possible the k-means algorithms are run, aiming at extracting different number of clusters out of dataset. It means that the $partitioning_i$ extracts $k(i)$ clusters out of dataset. As it is mentioned the proposed method tries to select a subset of well-performing clusters (or equivalently partitions) instead of a subset of clusterings (or equivalently partitionings). So each partitioning is broken in some distinct partitions clusters (or equivalently partitions).

Second step is stability computation. Since the goodness of a cluster C_i is determined by all of the data points, the goodness function $g_j = (C_i, D)$ depends on both the cluster C_i and the entire dataset D, instead of C_i alone. The stability as a measure of cluster goodness is used in [1], [13] and [20]. A stable cluster is the one that has a high likelihood of recurrence across multiple applications of a clustering algorithm. Stable clusters are usually preferable, since they are robust with respect to minor changes in the dataset [14].

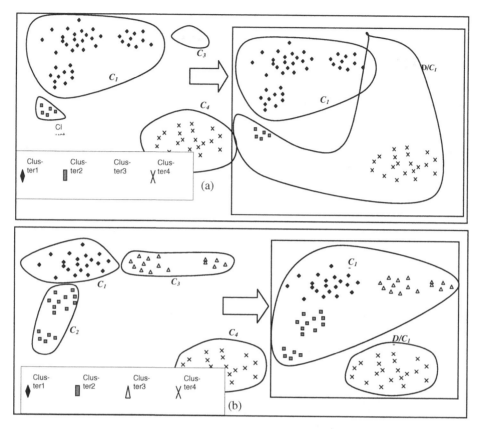

Fig. 2. Computing the stability of Cluster 1 of the partition in Fig. 2 (a) considering the partition in the Fig. 2 (b) of the reference set using NMI method

Now assume that the stability of cluster C_i is to be computed. In this method first a set of partitionings over dataset is provided which is called the reference set. One can consider the partitionings obtained in the first step as reference set for decreasing the runtime. In this notation D is dataset and $P_w(D)$ is a partitioning over D. Now, the problem is: "How many times is the cluster C_i repeated in the reference partitions?" Assume that the NMI between the cluster C_i and a reference partition $P_w(D)$ is denoted by $NMI(C_i, P_w(D))$. While the most of previous works only compare *a partition with another partition* [18], however, the stability used in [14] evaluates the similarity between *a cluster and a partition* by transforming the cluster C_i to a partition and after that by employing the common partition-to-partition *NMI*. To illustrate this method let $P_1 = P^a = \{C_i, D/C_i\}$ be a partition with two clusters, where D/C_i denotes the set of data points in D that are not in C_i. Then we may assume a second partition $P_2 = P^b = \{C_w^*, D/C_w^*\}$, where C_w^* denotes the union of all "positive" clusters in $P_w(D)$ and others are in D/C_w^*. A cluster C_r in $P_w(D)$ is positive cluster for C_i if more than half of its data points also belongs to C_i.

Now, define $NMI(C_i, P_w(D))$ by $NMI(P^a, P^b)$ which is calculated as [9]:

$$NMI(P^a, P^b) = \frac{-2\sum_{i=1}^{ka}\sum_{j=1}^{kb} n_{ij}^{ab} log\left(\frac{n_{ij}^{ab}.n}{n_i^a n_j^b}\right)}{\sum_{i=1}^{ka} n_i^a log\left(\frac{n_i^a}{n}\right) + \sum_{i=1}^{kb} n_i^b log\left(\frac{n_i^b}{n}\right)} \tag{1}$$

where n is the total number of samples and n_{ij}^{ab} denotes the number of shared patterns between clusters $C_i^a \in P^a$ and $C_j^b \in P^b$; n_i^a is the number of patterns in the cluster i of partition a; also n_j^b are the number of patterns in the cluster j of partition b. This computation is done between the cluster C_i and all partitions available in the reference set. This method is named NMI method. Fig. 2 illustrates the NMI method.

After producing P_1, if we assume a second partition $P_2 = P^b = \{C_w^*\} \cup Cs_w^*$, where C_w^* denotes the same clusters in $P_w(D)$ defined by APM [1] and for each of other data we consider a cluster. The set of these clusters is denoted by Cs_w^*. Fig. 3 shows the method explained above which is named Edited APM, EAPM.

NMI_h in the paper shows the stability of cluster C_i with respect to the hth partition in reference set. The total stability of cluster C_i is defined as:

$$Stab(C_i) = \frac{\sum_{j=1}^{B} NMI_j}{B} \tag{2}$$

This procedure is applied for each cluster available in the pool clusters obtained in the first step. It means this procedure must be iterated q times, where q is computed as equation 3.

$$q = \sum_{i=1}^{B} k(i) \tag{3}$$

Third step is simply done be a thresholding. It means that the clusters with higher stability values are selected for next step and other are omitted.

In forth step, the selected clusters are used to produce final clusters in a co-association based model. In the step it is to construct the co-association matrix and then to apply a hierarchical clustering. To construct the co-association matrix from the selected clusters EEAC is employed. In the EAC method the m primary partitions from dataset are accumulated in a $n \times n$ co-association matrix. Each entry in this matrix is computed from equation 4.

$$C_{ij} = \frac{n_{ij}}{m_{ij}} \tag{4}$$

where m_{ij} counts the number of clusters shared by objects with indices i and j in the pool of all clusters obtained in the first step. It is worthy to note that the maximum possible value of m_{ij} computed as equation 3. Also n_{ij} is the number of partitions where this pair of objects is simultaneously present in the selected clusters. Note that the value of n_{ij} is at most as many as the number of selected clusters which is less than the value of m_{ij}.

3   Experimental Study

After producing the consensus partition, the most important question is "how good a partition is?". The evaluation of a partition is very important as it is mentioned. Here the NMI between the consensus partition and real labels of the dataset is considered

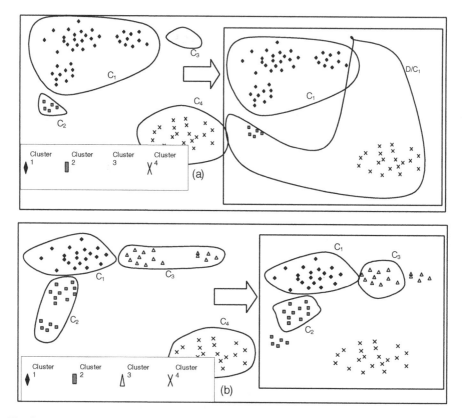

Fig. 3. Computing the stability of Cluster 1 of the partition in Fig. 3 (a) considering the partition in the Fig. 3 (b) of the reference set using EAPM method

as an evaluation metric of the consensus partition. Also accuracy between the consensus partition and real labels of the dataset is considered as another metric.

The proposed method is examined over 9 different standard datasets and one artificial dataset. It is tried for datasets to be diverse in their number of true classes, features and samples. A large variety in used datasets can more validate the obtained results. More information is available in [14].

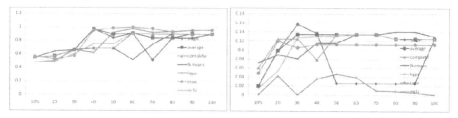

Fig. 4. The horizontal axis stands for the rate of stable clusters that are selected. The vertical axis stands for the NMI values between labels of Iris (left) and Ionosphere (right) datasets and the consensus partitions obtained by different consensus functions over the selected clusters.

To be more general and fair, all experiments are averaged over 10 independent runs. In all experimentations there are 120 independent partitions obtained by 120 independent runs of k-means clustering algorithm with different initialized seed points and different k parameter, ranging from k to 2*k. After selecting a subset of clusters, to extract the final partition from them, the real number of clusters is served by the consensus functions.

As it is known in fuzzy k-means clustering algorithm, each data point belongs to all clusters with different membership values. To extract the final partition from output of fuzzy k-means algorithm as consensus function, each data point is assigned to the most membership value.

As it is inferred from the Fig. 4, the best ratio of selection of the stable clusters is 60% and the best option for consensus function is CSPA for Iris dataset.

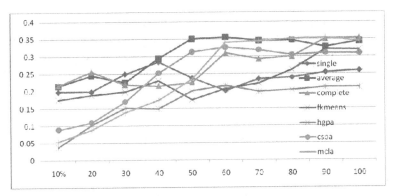

Fig. 5. The horizontal axis stands for the rate of stable clusters that are selected. The vertical axis stands for the averaged NMI values for all ten datasets.

Fig. 4 also makes it clear that the best ratio of selection of the stable clusters is 30% and the best option for consensus function is Single-Linkage for Ionosphere dataset.

To see whether the use of a subset of the most stable clusters can affect the quality of the final cluster or not, consider Fig. 5. To make a general decisive conclusion, the results for all ten datasets are averaged and the final results are illustrated in the Fig. 5. The Averaged-Linkage consensus function over 50% of the most stable clusters generally reaches the maximum for all dataset.

Table 1 shows the performance of the proposed method comparing with most common base and ensemble methods.

Table 1. Experimental results

Metric Evalua-tion	Dataset										
	Breast Cancer	Iris	Bupa	SAHeart	Ionos-phere	Glass	Halfrings	Galaxy	Yeast	Wine	Norm. Wine
NMI	95.73	76.13	54.33	63.36	70.60	**47.76**	74.48	**31.27**	42.93	69.38	85.17
MAX	96.49	84.87	**57.42**	**63.87**	57.75	44.35	74.55	29.85	51.27	70.00	94.44
APM	95.46	**90.00**	55.07	63.85	70.66	45.79	54.00	30.65	**53.10**	**70.23**	96.63
EAPM	**96.93**	88.67	54.78	63.20	**71.23**	43.93	**88.00**	30.65	50.47	**70.23**	**97.19**

4   Conclusion and Future Works

In this paper a new clustering ensemble method that is based on a subset of total prima-ry spurious clusters is offered. Since the worth of the base partitions is not identical and also existence of a subset of them may yet result to a better performance, here an ap-proach to choose a subset of more effective partitions is offered. A common metric based on that this subset is derived is normalized mutual information. Recently some drawbacks of NMI criterion are discussed and some alternative criterions, such as APM and Max, are proposed. In the paper while mentioning some drawbacks unhandled by APM and Max, a new metric that is named EAPM is proposed to solve the new raised drawbacks. The empirical studies over several datasets robustly show that the quality of the proposed method is usually better than other ones. The experiments confirm that the EAPM criterion does slightly better than NMI criterion generally; however it signifi-cantly outperforms the NMI criterion in the case of synthetic datasets. Because of the symmetry which is concealed in NMI criterion and also in NMI based stability, it yields to lower performance whenever symmetry is also appeared in the dataset. The experi-ments also show that the EAPM criterion does better than Max and APM criterions.

References

1. Ayad, H., Kamel, M.S.: Cumulative Voting Consensus Method for Partitions with a Vari-able Number of Clusters. IEEE Trans. on Pattern Analysis and Machine Intelligence 30(1), 160–173 (2008)
2. Bhatia, S.K., Deogun, J.S.: Conceptual Clustering in Information Retrieval. IEEE Trans. Systems, Man, and Cybernetics 28(3), 427–536 (1998)
3. Dudoit, S., Fridlyand, J.: Bagging to improve the accuracy of a clustering procedure. Bio-informatics 19(9), 1090–1099 (2003)
4. Faceli, K., Carvalho, A.C.P.L.F., Souto, M.C.P.: Multi-objective Clustering Ensemble. In: Proceedings of the Sixth International Conference on Hybrid Intelligent Systems, HIS 2006 (2006)
5. Fred, A., Jain, A.K.: Data Clustering Using Evidence Accumulation. In: Proc. of the 16th Intl. Conf. on Pattern Recognition, ICPR 2002, Quebec City, pp. 276–280 (2002)
6. Fred, A., Jain, A.K.: Combining Multiple Clusterings Using Evidence Accumulation. IEEE Trans. on Pattern Analysis and Machine Intelligence 27(6), 835–850 (2005)
7. Fred, A., Jain, A.K.: Learning Pairwise Similarity for Data Clustering. In: Proc. of the 18th Int. Conf. on Pattern Recognition, ICPR 2006 (2006)
8. Fred, A., Lourenco, A.: Cluster Ensemble Methods: from Single Clusterings to Combined Solutions. SCI, vol. 126, pp. 3–30 (2008)
9. Frigui, H., Krishnapuram, R.: A Robust Competitive Clustering Algorithm with Applica-tions in Computer Vision. IEEE Trans. Pattern Analysis and Machine Intelligence 21(5), 450–466 (1999)
10. Jain, A.K., Murty, M.N., Flynn, P.: Data clustering: A review. ACM Computing Sur-veys 31(3), 264–323 (1999)
11. Judd, D., Mckinley, P., Jain, A.K.: Large-Scale Parallel Data Clustering. IEEE Trans. Pat-tern Analysis and Machine Intelligence 19(2), 153–158 (1997)

12. Alizadeh, H., Minaei-Bidgoli, B., Parvin, H.: A New Asymmetric Criterion for Cluster Validation. In: San Martin, C., Kim, S.-W. (eds.) CIARP 2011. LNCS, vol. 7042, pp. 320–330. Springer, Heidelberg (2011)
13. Law, M.H.C., Topchy, A.P., Jain, A.K.: Multiobjective data clustering. In: Proc. of IEEE Conference on Computer Vision and Pattern Recognition, Washington, D.C., vol. 2, pp. 424–430 (2004)
14. Newman, C.B.D.J., Hettich, S., Merz, C.: UCI repository of machine learning databases (1998), http://www.ics.uci.edu/~mlearn/MLSummary.html
15. Parvin, H., Minaei-Bidgoli, B., Alinejad, H.: Linkage Learning Based on Differences in Local Optimums of Building Blocks with One Optima. International Journal of the Physical Sciences, IJPS, 3419–3425 (2011)
16. Daryabari, M., Minaei-Bidgoli, B., Parvin, H.: Localizing Program Logical Errors Using Extraction of Knowledge from Invariants. In: Pardalos, P.M., Rebennack, S. (eds.) SEA 2011. LNCS, vol. 6630, pp. 124–135. Springer, Heidelberg (2011)
17. Minaei-Bidgoli, B., Parvin, H., Alinejad-Rokny, H., Alizadeh, H., Punch, W.F.: Effects of resampling method and adaptation on clustering ensemble efficacy, Online (2011)
18. Fouladgar, H., Minaei-Bidgoli, B., Parvin, H.: On Possibility of Conditional Invariant Detection. In: König, A., Dengel, A., Hinkelmann, K., Kise, K., Howlett, R.J., Jain, L.C. (eds.) KES 2011, Part II. LNCS, vol. 6882, pp. 214–224. Springer, Heidelberg (2011)
19. Parvin, H., Minaei-Bidgoli, B.: Linkage Learning Based on Local Optima. In: Jędrzejowicz, P., Nguyen, N.T., Hoang, K. (eds.) ICCCI 2011, Part I. LNCS, vol. 6922, pp. 163–172. Springer, Heidelberg (2011)
20. Parvin, H., Helmi, H., Minaei-Bidgoli, B., Alinejad-Rokny, H., Shirgahi, H.: Linkage Learning Based on Differences in Local Optimums of Building Blocks with One Optima. International Journal of the Physical Sciences 6(14), 3419–3425 (2011)
21. Qodmanan, H.R., Nasiri, M., Minaei-Bidgoli, B.: Multi objective association rule mining with genetic algorithm without specifying minimum support and minimum confidence. Expert Systems with Applications 38(1), 288–298 (2011)
22. Parvin, H., Minaei-Bidgoli, B., Alizadeh, H.: A New Clustering Algorithm with the Convergence Proof. In: König, A., Dengel, A., Hinkelmann, K., Kise, K., Howlett, R.J., Jain, L.C. (eds.) KES 2011, Part I. LNCS, vol. 6881, pp. 21–31. Springer, Heidelberg (2011)
23. Parvin, H., Minaei, B., Alizadeh, H., Beigi, A.: A Novel Classifier Ensemble Method Based on Class Weightening in Huge Dataset. In: Liu, D., Zhang, H., Polycarpou, M., Alippi, C., He, H. (eds.) ISNN 2011, Part II. LNCS, vol. 6676, pp. 144–150. Springer, Heidelberg (2011)
24. Parvin, H., Minaei-Bidgoli, B., Alizadeh, H.: Detection of Cancer Patients Using an Innovative Method for Learning at Imbalanced Datasets. In: Yao, J., Ramanna, S., Wang, G., Suraj, Z. (eds.) RSKT 2011. LNCS, vol. 6954, pp. 376–381. Springer, Heidelberg (2011)
25. Parvin, H., Minaei-Bidgoli, B., Ghaffarian, H.: An Innovative Feature Selection Using Fuzzy Entropy. In: Liu, D., Zhang, H., Polycarpou, M., Alippi, C., He, H. (eds.) ISNN 2011, Part III. LNCS, vol. 6677, pp. 576–585. Springer, Heidelberg (2011)
26. Parvin, H., Minaei, B., Parvin, S.: A Metric to Evaluate a Cluster by Eliminating Effect of Complement Cluster. In: Bach, J., Edelkamp, S. (eds.) KI 2011. LNCS, vol. 7006, pp. 246–254. Springer, Heidelberg (2011)
27. Parvin, H., Minaei-Bidgoli, B., Ghatei, S., Alinejad-Rokny, H.: An Innovative Combination of Particle Swarm Optimization, Learning Automaton and Great Deluge Algorithms for Dynamic Environments. International Journal of the Physical Sciences 6(22), 5121–5127 (2011)

28. Parvin, H., Minaei, B., Karshenas, H., Beigi, A.: A New N-gram Feature Extraction-Selection Method for Malicious Code. In: Dobnikar, A., Lotrič, U., Šter, B. (eds.) ICANNGA 2011, Part II. LNCS, vol. 6594, pp. 98–107. Springer, Heidelberg (2011)
29. Roth, V., Lange, T., Braun, M., Buhmann, J.: A Resampling Approach to Cluster Validation. In: Intl. Conf. on Computational Statistics, COMPSTAT (2002)
30. Strehl, A., Ghosh, J.: Cluster ensembles - a knowledge reuse framework for combining multiple partitions. Journal of Machine Learning Research 3, 583–617 (2002)
31. Alizadeh, H., Minaei, B., Parvin, H.: A New Criterion for Clusters Validation. In: Iliadis, L., Maglogiannis, I., Papadopoulos, H. (eds.) EANN/AIAI 2011, Part II. IFIP AICT, vol. 364, pp. 110–115. Springer, Heidelberg (2011)
32. Ester, M., Kriegel, H.P., Sander, J., Xu, X.: A density-based algorithm for discovering clusters in large spatial databases with noise. In: International Conference on Knowledge Discovery and Data Mining, pp. 226–231. AAAI Press (1996)
33. Sibson, R.: SLINK: an optimally efficient algorithm for the single-link cluster method. The Computer Journal 16(1), 30–34 (1973)

Influence of Erroneous Pairwise Constraints in Semi-supervised Clustering

Tetsuya Yoshida

Graduate School of Information Science and Technology,
Hokkaido University
N-14 W-9, Sapporo 060-0814, Japan
yoshida@meme.hokudai.ac.jp

Abstract. Side information such as pairwise constraints is useful to improve the clustering performance in general. However, constraints are not always error free in general. When erroneous constraints are specified as side information, treating them as hard constraints could have the disadvantage since strengthening incorrect or erroneous constraints can lead to performance degradation. In this paper we conduct extensive experiments to investigate the influence of erroneous pairwise constraints over various document datasets. Several state-of-the-art semi-supervised clustering methods with graph representation were evaluated with respect to the type of constraints as well as the number of constraints. Experimental results confirmed that treating pairwise constraints as hard constraints is vulnerable to erroneous ones. However, the results also revealed that the influence of erroneous constraints depends on how the constraints are exploited inside a learning algorithm.

1 Introduction

In order to effectively utilize a small amount of supervised information such as labeled instances, semi-supervised learning has been attracting much interest these days [2]. In semi-supervised learning, unlabeled data can also be utilized to improve the performance of learning systems significantly as well [2]. On the other hand, although labeled data is not required in clustering, sometimes constraints over the data assignment into clusters might be available as domain knowledge about the data to be clustered. when such information is available, it is desirable to utilize the available constraints as semi-supervised information and to improve the performance of clustering [1].

Various research efforts have been conducted on semi-supervised clustering [1,6,11]. In most approaches, pairwise constraints among instances are considered as additional or side information. For example, in exploratory data analysis, a user is allowed to iteratively provide feedback based on the constructed clusters, and the feedback can be incorporated in the form of constraints. From the constructed clusters, a user can specify a pair of instances which should be in the same cluster, or a pair of instances which should not be in the same cluster.

R. Huang et al. (Eds.): AMT 2012, LNCS 7669, pp. 43–52, 2012.
© Springer-Verlag Berlin Heidelberg 2012

Side information such as pairwise constraints is useful to improve the clustering performance in general. However, constraints are not always error free in general. When erroneous constraints are specified as side information, treating them as hard constraints could have the disadvantage since strengthening incorrect or erroneous constraints can lead to performance degradation. In this paper we conduct extensive experiments to investigate the influence of erroneous pairwise constraints over various document datasets. Several state-of-the-art semi-supervised clustering methods with graph representation [6,5,11,16,15] were evaluated with respect to the type of constraints as well as the number of constraints.

2   Semi-supervised Clustering with Pairwise Constraints

2.1   Pairwise Constraints

We use a bold italic lower case letter to denote a vector, and a bold normal upper case letter to denote a matrix. For instance, $\boldsymbol{h}$ represents a vector, and $\mathbf{W}$ represents a matrix.

For a given dataset X, when supervised information on clustering is specified as a set of constraints, *semi-supervised clustering* (or, constrained clustering) is conducted by finding nonempty mutually disjoint subsets of X, whose union is X, under the specified constraints. The subsets correspond to a partition of X, and each subset is called a cluster.

There can be various forms of constraints. Based on previous work [13,14,11], we consider the following two kinds of constraints, which are defined as must-link constraints and cannot-link constraints in [13].

Definition 1 (Pairwise Constraints). *For a given dataset X and a partition T (a set of clusters), must-link constraints C_{ML} and cannot-link constraints C_{CL} are sets of pairs of instances such that:*

$$(x, y) \in C_{ML} \Rightarrow \exists t \in T, (x \in t \land y \in t) \tag{1}$$

$$(x, y) \in C_{CL} \Rightarrow \exists s, t \in T, s \neq t, (x \in s \land y \in t) \tag{2}$$

2.2   Graph Representation for Pairwise Relations

When a pairwise relation among instances are specified, by connecting each pair of instances with an edge, the entire data can be represented as a graph.

Formally, when a weight is assigned to each edge of a graph, it is called an edge-weighted graph. An edge-weighted graph $G(V, E, W)$ is defined as a graph with the weight on each edge in E among the vertices V. When the number of vertices is n, the weights over the edges can be represented as an n by n matrix $\mathbf{W}$. In matrix $\mathbf{W}$, w_{ij} stands for the weight on the edge for the pair of vertices (i, j). All the weights are assumed to be non-negative and real values in $[0, 1]$. For any pair of vertices (i, j) which is not in E, we set its weight to zero.

3   Semi-supervised Clustering with Graph Representation

3.1   Spectral Learning

As described in Section 2.2, when similarities among instances can be specified, the entire instances can be represented as a weighted graph. Among various graph based machine learning methods, spectral clustering has been widely utilized due to its high clustering performance [12]. Based on the (weighted) adjacency matrix of the graph, the framework of spectral learning [10], which encompasses spectral clustering, constructs a representation of the instances based on the eigenvectors of the adjacency matrix. Under this framework, minimization of the following objective function is conducted to find out a matrix $\mathbf{H}$:

$$J_1 = \mathrm{tr}(\mathbf{H}^t \mathbf{L} \mathbf{H}) \tag{3}$$
$$such \ that \ \ \mathbf{H}^t \mathbf{D} \mathbf{H} = \mathbf{I}$$

$$\mathbf{D} = diag(d_1, \ldots, d_n) \ \ (d_i = \sum_{j=1}^{n} w_{ij}) \tag{4}$$

$$\mathbf{L} = \mathbf{D} - \mathbf{W} \tag{5}$$

where tr in eq.(3) represents the trace of a matrix, and $\mathbf{I}$ represents the unit matrix. $diag()$ in eq.(4) represents a diagonal matrix with the specified diagonal elements. The matrix $\mathbf{L}$ in eq.(5) is called a graph Laplacian [3]. Column vectors $\boldsymbol{h}_1, \ldots, \boldsymbol{h}_l$ of $\mathbf{H}$ are the eigenvectors with the smallest positive eigenvalues. The real valued matrix $\mathbf{H}$ corresponds to the projected representation of the given dataset onto the subspace spanned by the eigenvectors. In many approaches [12,10], instead of $\mathbf{L}$, the normalized Laplacian $\mathbf{L}_{sym} = \mathbf{D}^{-\frac{1}{2}} \mathbf{L} \mathbf{D}^{-\frac{1}{2}}$ is utilized for balancing the size of clusters.

3.2   Semi-Supervised Clustering Methods with Graph Representation

Graph-Based Semi-supervised Clustering (GBSSC): We have proposed a graph-based approach for semi-supervised clustering [16,15]. A set of vertices connected by must-links in a graph G are contracted into a vertex based on graph contraction in graph theory [7], and weights are re-defined over the contracted graph G'. Next, weights on the contracted graph are further modified based on cannot-links. Finally, a projected representation of the given data is constructed with respect to G' as in spectral learning in Section 3.1, and clustering is conducted over the projected representation.

When constructing the projected representation, the following objective function is optimized based on cannot-links [15]:

$$J_2 = tr(\mathbf{H}^t \mathbf{L} \mathbf{H}) + \lambda \ tr(\mathbf{H}^t \mathbf{S} \mathbf{H}) \tag{6}$$
$$such \ that \ \ \mathbf{H}^t \mathbf{D} \mathbf{H} = \mathbf{I}$$

where $\lambda \geq 0$ is a regularization parameter.

The matrix $\mathbf{S}$ is defined based on cannot-links as follows [15]:

$$s_{ik} = \begin{cases} 1 & if \ (i,k) \in C_{CL} \\ \frac{(1-w'_{ik})}{m_{ik}} \left(\displaystyle\sum_{(v_i,v_j) \in C_{CL}} w'_{jk} + \displaystyle\sum_{(v_k,v_j) \in C_{CL}} w'_{ji} \right) & otherwise \end{cases} \tag{7}$$

where m_{ik} represents the number of constraints which are considered in two summations in the second line in eq.(7), and w'_{ik} is the weight in the contracted graph G'. When m_{ik} is zero, s_{ik} is set to zero. This method is called Graph-based Semi-Supervised Clustering (GBSSC).

Subspace Trick: For the two class case, an approach for pairwise constraints under the framework of spectral clustering was proposed in [6,5] was named as a "subspace trick". The basic idea of subspace trick is to utilize a linear mapping which is constructed based on the pairwise constraints, and to map the representation constructed by spectral learning in Section 3.1 into the subspace where the pairwise constraints are enforced.

For the two class case. both must-links and cannot-links can be dealt with in subspace trick. However, for the general multi-class setting, although must-links can still be treated, it is not possible to deal with cannot-link constraints [6,5].

PCP: Mapping of the given instances into another space is also pursued in PCP[11]. Using the specified pairwise constraints, it conducts metric learning based on semi-definite programming (SDP) and learns the kernel matrix $\mathbf{K}$ which minimizes the following objective function:

$$J_3 = \mathbf{K} \bullet \mathbf{L}_{sym} \tag{8}$$

$$such \ that \begin{cases} \mathbf{K}_{ii} = 1 \ \forall i \\ \mathbf{K}_{ij} = 1 \ if \ (i,k) \in C_{ML} \\ \mathbf{K}_{ij} = 0 \ if \ (i,k) \in C_{CL} \end{cases}$$

where $\bullet$ represents the inner product of matrices [9]D After learning the kernel matrix $\mathbf{K}$ with SDP, kernel k-means clustering [8] is conducted over $\mathbf{K}$.

3.3   Erroneous Pairwise Constraints

Both types of constraints in Definition 1 can be treated as hard or soft constraints. When constraints are treated as hard constraints, the cluster assignment needs to satisfy the constraints. On the other hand, the cluster assignment does not necessarily need to satisfy the constraints if they are treated as soft constraints. Better performance would be obtained if the constraints are treated as hard constraints, instead of soft constraints. However, constraints are not always error free in general. When erroneous constraints are specified, treating them as hard constraints could have the disadvantage since strengthening incorrect or erroneous constraints can lead to performance degradation.

A must-link requires that two instances should be assigned to the same cluster. On the other hand, a cannot-link corresponds to the negation of cluster assignment, in the sense that their assigned clusters are not the same. Thus,

Table 1. 20 Newsgroup dataset

dataset	included groups
Multi5	comp.graphics, rec.motorcycles,rec.sport.baseball, sci.space talk.politics.mideast
Multi10	alt.atheism, comp.sys.mac.hardware,misc.forsale, rec.autos,rec.sport.hockey, sci.crypt,sci.med, sci.electronics,sci.space,talk.politics.guns
Multi15	alt.atheism, comp.graphics, comp.sys.mac.hardware, misc.forsale, rec.autos, rec.motorcycles, rec.sport.baseball, rec.sport.hockey, sci.crypt, sci.electronics, sci.med, sci.space, talk.politics.guns, talk.politics.mideast, talk.politics.misc

Table 2. TREC datasets

dataset	# attr.	#classes
hitech	126372	6
sports	126372	7
la12	31372	6

Table 3. TREC datasets (unbalanced)

dataset	# attr.	#classes	#instances
tr11	6429	9	414
tr12	5804	8	313
tr23	5832	6	204

there remains more freedom for the assignment of instances on which a cannot-link is specified. For instance, for a pair of instances $(u, v) \in C_{CL}$, the instance v can be assigned to any cluster except for the cluster to which the instance u is assigned. Thus, the assignment cannot be determined solely based on the cannot-link; rather, it is necessary to consider the relation with other instances.

In order to deal with the general multi-class problem within the framework of spectral clustering, GBSSC [16,15] treats must-links as hard constraints and cannot-links as soft constraints. Subspace [6,5] also treats must-links as hard constraints, but cannot-link constraints cannot be dealt with for the general multi-class problem. On the other hand, when learning a kernel matrix using SDP in PCP [11], pairwise constraints are utilized as hard constraints. However, for determining the cluster assignment of instances, the constraints are not treated as hard constraints when clustering is conducted over the learned matrix.

4   Performance Evaluations

4.1   Experimental Settings

Datasets: Based on the previous work, we evaluated semi-supervised clustering methods on 20 Newsgroup data (20NG)[1] and TREC datasets[2]. For 20NG, we created three sets of groups, as shown in Table 1. Following the similar procedure in [16], 50 documents were sampled from each group in order to create one dataset, and 10 datasets were created for each set of groups. For each dataset, we conducted stemming using porter stemmer[3] and MontyTagger[4], removed stop words, and selected 2,000 words with large mutual information [4].

---

[1] http://people.csail.mit.edu/~jrennie/20Newsgroups/. 20news-18828 was utilized.

[2] http://glaros.dtc.umn.edu/gkhome/cluto/cluto/download

[3] http://www.tartarus.org/~martin/PorterStemmer

[4] http://web.media.mit.edu/~hugo/montytagger

For the TREC datasets, we utilized 6 datasets in Table 2 and Table 3[5]. For the datasets in Table 2, we followed the same procedure in 20NG and created 10 samples for each dataset. On the other hand, since the datasets in Table 3 are very unbalanced with respect to the number of instances per class, we did not conduct sampling for these datasets. The number of instances in these datasets are shown in the rightmost column in Table 3.

Evaluation Measures: For each dataset, the cluster assignment was evaluated with respect to Normalized Mutual Information (NMI). Let T, $\hat{T}$ stand for the random variables over the true and assigned clusters. NMI is defined as $NMI = \frac{I(\hat{T};T)}{(H(\hat{T})+H(T))/2}(\in [0,1])$ where $H(T)$ is Shannon Entropy. NMI corresponds to the accuracy of assignment. The larger NMI is, the better.

Compared Methods: We compared the semi-supervised clustering methods in Section 3.2, namely: GBSSC, subspace, and PCP. Since all the compared methods are partitioning based clustering methods, we assume that the number of clusters k is specified.

Parameters: The parameters under the pairwise constraints in Definition 1 are: 1) the number of constraints, and 2) the pairs of instances for constraints. As for 2), pairs of instances were randomly sampled from each dataset to generate the constraints. Thus, the main parameter is 1), the number of constraints, for C_{ML} and C_{CL}. We set $|C_{ML}| = |C_{CL}|$[6], and varied the number of constraints.

Cosine similarity was utilized to define the pairwise similarities in each dataset. The dimension l of the subspace was set to the number of clusters k. In addition, following the procedure in [11], m-nearest neighbor graph was constructed for PCP with $m = 10$. As in [15], the parameter λ in eq.(6) was set to as $\lambda = \lambda_0 \cdot {}_kC_2$ (with $\lambda_0 = 0.02$).

Synthetic Erroneous Constraints: We evaluated the influence of erroneous constraints by inserting incorrect must-links or incorrect cannot-links randomly[7]. An erroneous (incorrect) must-link was generated by randomly selecting a pair of instances with different labels and treating the pair as a must-link. An erroneous cannot-link was generated by randomly selecting a pair of instances with the same label and treating the pair as a cannot-link. The percentage of erroneous constraints were varied from 0% to 30% in the following experiments.

Evaluation Procedure: For each number of constraints, the pairwise constraints C_{ML} and C_{CL} were generated randomly based on the ground-truth labels in the datasets, and clustering was conducted with the generated constraints. Clustering with the same number of constraints was repeated 10 times with different initial configuration in clustering. In addition, the above process was also repeated 10 times for each number of constraints. Thus, for each dataset and the number of constraints, 100 runs were conducted. Furthermore, this process was repeated over 10 samples for the datasets in Table 1 and Table 2, Thus,

---

[5] These datasets are already preprocessed and provided as count data.
[6] $|\cdot|$ represents the cardinality of a set.
[7] With erroneous must-links, cannot-links were set as error free, and vice versa.

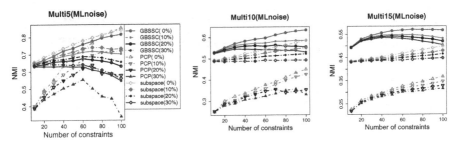

Fig. 1. Erroneous must-links (20NG)

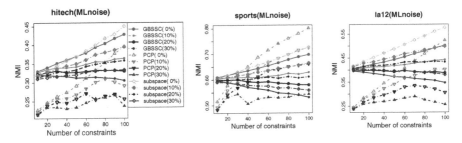

Fig. 2. Erroneous must-links (TREC in Table 2)

the average of 1,000 runs is reported for the datasets in Table 1 and Table 2, and the average of 100 runs for Table 3.

4.2 Results of Erroneous Must-Links

The results with erroneous must-links are shown in Fig. 1, Fig. 2, and Fig. 3. The horizontal axis in these figures corresponds to the number of constraints, and the vertical one corresponds to the clustering accuracy (*NMI*). In the legend in the figures, solid lines with circles correspond to GBSSC, dashed lines with triangles correspond to PCP, and dotted lines with diamonds correspond to subspace trick. Also, 30% in the legend corresponds to the situation where 30% are erroneous must-links (e.g., GBSSC(30%)).

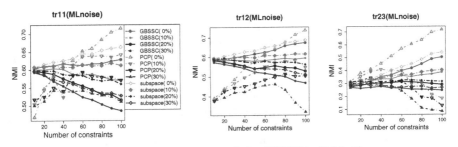

Fig. 3. Erroneous must-links (TREC in Table 3)

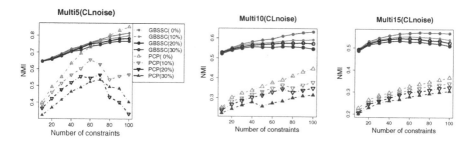

Fig. 4. Erroneous cannot-links (20NG)

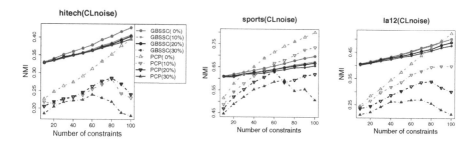

Fig. 5. Erroneous cannot-links (TREC in Table 2)

The results in these figures show that, as expected, the performance of all the methods degraded as the percentage of erroneous must-links increased. However, the influence of the erroneous must-links differed depending on how the constraints are utilized inside the learning algorithms. In terms of the performance degradation, GBSSC and subspace trick, both of which treat must-links as hard constraints, showed similar tendency with respect to the ratio of erroneous must-links. However, the performance of PCP, which treats must-links as soft constraints for the cluster assignment, got even much worse with erroneous must-links (cf., see PCP(30%)). This rather surprising result would be due to a bad encoding of the (erroneous) constraints in the learned kernel matrix in PCP.

The above results confirmed that treating must-links as hard constraints is vulnerable to erroneous ones. However, the results also revealed that the influence of erroneous constraints depended on how the constraints are exploited inside a learning algorithm: learning a kernel matrix with SDP (as in PCP) can exploit the available side information (pairwise constraints) with the increased running time, but it can be a double-edged sword when erroneous constraints are included, even if they are treated as soft constraints.

4.3   Results of Erroneous Cannot-Links

The results with erroneous cannot-links are shown in Fig. 4, Fig. 5, and Fig. 6. As in Section 4.2, 30% in the legend corresponds to the situation where 30%

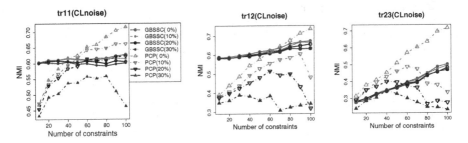

Fig. 6. Erroneous cannot-links (TREC in Table 3)

are erroneous cannot-links. Subspace is not shown in these figures for erroneous cannot-links, since it does not handle cannot-links for the general multi-class problem.

The results in these figures show that GBSSC is robust with respect to erroneous cannot-link constraints, since its performance was not so much affected by them. This would be because cannot-links are treated as soft constraints in GBSSC. However, the performance of PCP got much worse when the noise ratio increased. Note that both GBSSC and PCP treat cannot-links as soft constraints for the cluster assignment. As described in the results in Section 4.2, the difference in the influence of erroneous cannot-links would be due to a bad encoding of the (erroneous) constraints in the learned kernel matrix in PCP.

On the other hand, when no erroneous constraints are inserted (i.e., for the lines with 0%), PCP could exploit the constraints more effectively due to the utilization of SDP (but with the increased running time), and its performance improved as the number of constraints increased, especially in Fig. 5 and Fig. 6.

4.4   Discussions

Experimental results confirmed that treating pairwise constraints as hard constraints is vulnerable to erroneous ones in semi-supervised clustering. Since both GBSSC [16,15] and subspace [6,5] treat must-links as hard constraints under the framework of spectral clustering, their performance degraded as the ratio of erroneous constraints increased. In terms of the clustering accuracy (NMI), GBSSC outperformed subspace in 20NG datasets, especially with many labels (Multi10 and Multi15). On the other hand, GBSSC could not outperform in TREC datasets, especially for the very unbalanced clusters in Table 3. Further investigation of the influence of unbalanced clusters is left for future work.

However, the results also revealed that the influence of erroneous constraints depends on how the constraints are exploited inside a learning algorithm. Although PCP treats pairwise constraints as soft constraints for the cluster assignment, since the constraints are utilized as hard ones when learning the kernel matrix in eq.(8), its performance got even much worse with erroneous must-links compared with GBSSC and subspace.

5   Concluding Remarks

This paper reported the performance evaluation of erroneous pairwise constraints in semi-supervised clustering over various document datasets. Performance of several state-of-the-art semi-supervised clustering methods were analyzed with respect to the type of constraints as well as the number of constraints. Experimental results confirmed that treating pairwise constraints as hard constraints is vulnerable to erroneous ones. However, the results also revealed that the influence of erroneous constraints depends on how the constraints are exploited inside a learning algorithm. Exploiting the available constraints more aggressively (e.g., with SDP) can lead to much worse performance degradation even if they are treated as soft constraints.

Acknowledgments. This work is partially supported by the grant-in-aid for scientific research (No. 24300049) funded by MEXT, Japan, and the Murata Science Foundation.

References

1. Basu, S., Davidson, I., Wagstaff, K. (eds.): Constrained Clustering: Advances in Algorithms, Theory, and Applications. Chapman & Hall/CRC Press (2008)
2. Chapelle, O., Schölkopf, B., Zien, A. (eds.): Semi-Supervised Learning. MIT Press (2006)
3. Chung, F.: Spectral Graph Theory. American Mathematical Society (1997)
4. Cover, T., Thomas, J.: Elements of Information Theory. Wiley (2006)
5. De Bie, T., Cristianini, N.: Fast sdp relaxations of graph cut clustering, transduction, and other combinatorial problem. Journal of Machine Learning Research 7, 1409–1436 (2006)
6. De Bie, T., Suykens, J., De Moor, B.: Learning from General Label Constraints. In: Fred, A., Caelli, T.M., Duin, R.P.W., Campilho, A.C., de Ridder, D. (eds.) SSPR&SPR 2004. LNCS, vol. 3138, pp. 671–679. Springer, Heidelberg (2004)
7. Diestel, R.: Graph Theory. Springer (2006)
8. Girolami, M.: Mercer kernel-based clustering in feature space. IEEE Transactions on Neural Networks 13(3), 780–784 (2002)
9. Harville, D.A.: Matrix Algebra From a Statistican's Perspective. Springer (2008)
10. Kamvar, S.D., Klein, D., Manning, C.D.: Spectral learning. In: Proc. of IJCAI 2003, pp. 561–566 (2003)
11. Li, Z., Liu, J., Tang, X.: Pairwise constraint propagation by semidefinite programming for semi-supervised classification. In: Proc. ICML 2008, pp. 576–583 (2008)
12. Ng, A.Y., Jordan, M.I., Weiss, Y.: On Spectral Clustering: Analysis and an algorithm. In: NIPS 14, pp. 849–856 (2001)
13. Wagstaff, K., Cardie, C., Rogers, S., Schroedl, S.: Constrained k-means clustering with background knowledge. In: Proc. ICML 2001, pp. 577–584 (2001)
14. Xing, E.P., Ng, A.Y., Jordan, M.I., Russell, S.: Distance metric learning, with application to clustering with side-information. In: Proc. NIPS 15, pp. 505–512 (2003)
15. Yoshida, T.: Pairwise Constraint Propagation for Graph-Based Semi-supervised Clustering. In: Kryszkiewicz, M., Rybinski, H., Skowron, A., Raś, Z.W. (eds.) IS-MIS 2011. LNCS (LNAI), vol. 6804, pp. 358–364. Springer, Heidelberg (2011)
16. Yoshida, T., Okatani, K.: A Graph-Based Projection Approach for Semi-supervised Clustering. In: Kang, B.-H., Richards, D. (eds.) PKAW 2010. LNCS (LNAI), vol. 6232, pp. 1–13. Springer, Heidelberg (2010)

User Correlation Discovery and Dynamical Profiling Based on Social Streams

Xiaokang Zhou and Qun Jin

Graduate School of Human Sciences, Waseda University,
2-579-15 Mikajima, Tokorozawa-shi, Saitama, Japan
{xkzhou@ruri.,jin@}waseda.jp

Abstract. In this study, we try to discover the potential and dynamical user correlations using those reorganized social streams in accordance with users' current interests and needs, in order to assist the information seeking process. We develop a mechanism to build a Dynamical Socialized User Networking (DSUN) model, and define a set of measures (such as interest degree, and popularity degree) and concepts (such as complementary tie, weak tie, and strong tie), which can discover and represent users' current profiling and dynamical correlations. The corresponding algorithms are developed respectively. Based on these, we finally discuss an application scenario of the DSUN model with experiment results.

Keywords: Social Stream, Stream Metaphor, User Profiling, Information Seeking, SNS.

1   Introduction

Recently, SNS (Social Network Service), such as facebook and twitter, is becoming increasingly popular. These so-called social media have become an important part of our life, which makes it possible for us to share our feelings, experiences and knowledge with each other timely and promptly. It may be less effective to only provide users with the sought information for satisfying their broad and ambiguous requirements [1]. As more and more of the populations are being engaged into this vortex of the social networking revolution, information search should also be socialized [2]. Based on these, it is essential to extend our social circle to obtain more timely and accurate information in this stream world. Therefore, in this paper, we try to utilize these social streams to discover the potential user correlations in order to assist the information seeking process.

In our previous study [3], we have introduced and defined a set of metaphors to represent a variety of social stream data with a hierarchical structure. We have developed two algorithms to assist users' information seeking which can best fit users' current interests and needs [4]. Moreover, we proposed a Dynamical Socialized User Networking Model (DSUN) which represents users' profiling and relationship to enhance social learning [5].

R. Huang et al. (Eds.): AMT 2012, LNCS 7669, pp. 53–62, 2012.
© Springer-Verlag Berlin Heidelberg 2012

In this study, to facilitate the information seeking process, we further delve into the mining and discovery of dynamical and potential user correlations in this stream environment. In details, we introduce the structure of the DSUN model and develop a mechanism to build the model in accordance with the reorganized social streams which implicate users' current interests and needs. Based on these, we introduce a set of measures (such as interest degree, and popularity degree) and develop a series of concepts (such as complementary tie, weak tie, and strong tie) to discover and represent users' potential profiling and dynamical correlation.

The rest of this paper is organized as follows. We give a brief overview on the related issues and works in Section 2. In Section 3, we give the definition and structure of the DSUN model, followed by a mechanism to build this model in accordance to the organized social streams. In Section 4, we introduce a set of measures and develop a series of ties to describe the DSUN model in details. Two algorithms are conceived and developed to create the ties respectively. We show an application scenario to demonstrate our proposed method with experiment results in Section 5. Finally, we conclude this study and give promising perspectives on future work in Section 6.

2   Related Work

There are many applications of social streams in SNS (such as twitter). Signorini Alessio et al. show that twitter can be used as a measure of public interest or concern about health-related events by examining the twitter stream to track rapidly-evolving public sentiment and activity [6]. Analysis of twitter communications in [7] showed the experimental evidence that twitter can be used as an educational tool to help engage students and to mobilize faculty into a more active and participatory role. A study addressed in [8] examined the impact of posting social, scholarly, or a combination of social and scholarly information to twitter on the perceived credibility of the instructor, which may have implications for both teaching and learning.

For the information recommendation aspects, the web mining [9] approaches have been extensively used for information recommendation. As an additional area, semantic web mining [10] was proposed. A new document representation model was presented in [11], which is based on implicit users' feedback to achieve better results in organizing web documents, such as clustering and labeling. Identifying relevant websites from user activities [12] is another attempt to organize web pages. LinkSelector [13] is a web mining approach focusing on structure and usage. By this approach, hyperlinks-structural relationships were extracted from existing web sites and theirs access logs.

3   User Networking and Profiling Based on Social Streams

3.1   Metaphors for Streams

We introduce a set of metaphors to represent these kinds of data streams with a hierarchical structure as follows [3].

Drop: Drop is a minimum unit of data streams, such as a message posted to the microblog (e.g., twitter) by a user, or a status change in SNS (e.g., facebook).

Stream: Stream is a collection of drops in timeline, which contains the messages, activities and actions of a user.

River: River is a confluence of streams from different users which are formed by following or subscribing his/her followers/friends. It could be extended to followers' followers.

Ocean: Ocean is a combination of all the streams.

The following definitions are used to seek information that satisfies users' current interests and needs. [4]

Heuristic Stone: It represents one of a specific user's current interests which may be changed dynamically.

Associative Ripple: It is a meaningfully associated collection of the drops related to some topics of a specific user's interests, which are formed by the heuristic stone in the river.

3.2   Dynamical Socialized User Networking Model

The Dynamically Socialized User Networking (DSUN) model is constructed based on the analysis of social streams, in order to discover and represent users' potential profiling and dynamical correlation. It can be defined and expressed as:

$$G_{DSUN}(V, E, W) \tag{1}$$

where $V = \{v_1, v_2, v_3, ..., v_n\}$: a non-empty set of vertexes in the network, each of which indicates a user in the system. $E = \{e_{ij}: <v_i, v_j> \mid if\ correlation\ exists\ between\ v_i$ $and\ v_j\}$: a collection of edges that connect the vertex in V, which represent all the relationships among every vertex in the network. $W = \{w_{ij} \mid if\ \exists\ e_{ij} \in E\}$: the weight w_{ij}, appending on the corresponding edge, is developed to identify the strength of specific relation between two users. This value is also employed to dynamically construct the model.

The structure of vertex and the edge are defined as follows.

Vertex v_i *(id, heuristic stone)*: Each vertex is used to store the corresponding user's information, such as the user's id, user's interests or needs represented as the heuristic stones.

Edge $e_{ij} <v_i, w_{ij}, h, v_j>$: Vertex v_i is the head of this edge, which indicates the user who is the beneficiary that receive the information in the system; v_j is the tail of this edge, which indicates the user who is the benefactor that provides the information in the system; w_{ij} is the relationship weight between these two users, which can be calculated in accordance with the organized social streams; h indicates the heuristic stone that this relationship is based on.

Following these definitions, we go further to discuss how to construct the DSUN model and how this model can be constructed dynamically. The Formula (2) is employed to calculate the weight wij appending on each edge and expressed as:

$$w_{ij} = \sum_{i=1}^{n}(Tf_n * (ImR_{ij} + ExR_{ij})) \tag{2}$$

To construct the DSUN model, in this formula, two major factors are taken account into the quantification of the weight: the time factor and the relationship factor. That is, *Tf* denotes a time fading factor, and the relationship between two users is further categorized into two major types: *ImR$_{ij}$* denotes the *Implicit Relationship*, while *ExR$_{ij}$* denotes the *Explicit Relationship*. These relationships can be defined in details as follows.

Implicit Relationship: The implicit relationship means those relations that cannot be viewed directly from the interface. That is, it indicates the relations based on the so-called social streams. The information hidden in these stream data can be extracted and organized to analyze the potential user correlations.

Explicit Relationship: The explicit relationship means those relations that can be viewed directly from the interface. That is, it indicates the relations based on interactions generated among users in an SNS environment. For instance, two users follow each other; the reply from one user to another; one user mentions other users in his/her posted contents. These interactions around a specific topic during a specific period can be used to calculate the dynamical user correlations.

The quantifications of these two relationships will be further discussed respectively. According to our previous study, the design of *associative ripple* is used to organize the raw social stream data into meaningful contents in accordance with *heuristic stone* which represents a specific user's current intention. In details, the whole timeline has been divided into several time slices, each of which will contain an *associative ripple*. An *associative ripple* consists of a series of circles, ranging from the center to outside, in order to represent the relation degree between the centers and drops. That is, the drops distributing in different circles mean they are clustered to the center in some degrees. For each *drop* to a specific ripple, the closer to the center, the more relative information in accordance with the owner of this ripple it may contain. Following the discussion given above, the users whom those related drops belong to are further involved into the analysis of the potential user correlations. In other words, for a specific user and one of his/her ripples, the amount of drops belonging to other users, which converge to the center of this ripple in different scales, are taken account into the calculation of correlation between this user and others. Thus, the *Implicit Relationship* can be quantified as:

$$ImR_{ij} = \frac{\sum Cw * AR_n(Cir, \ v_j)}{\sum AR_n(v_i) * Cw} \tag{3}$$

where $AR_n(v_i)$ denotes the total amount of drops that are clustered in the user v_i's ripple, $AR_n(Cir, \ v_j)$ denotes the amount of drops that are extracted from the user v_j in a specific circle *Cir*. The variable *Cw*, which shall be assigned to the drops appending on the circle, is the weight depending on the different *Cir*. That is, its value, ranging from 1 to 0, descends from the center to outside, in order to indicate the relevance to the ripple center from high to low.

As to the *explicit relationship*, three mentioned major types of it shall be taken account into. Thus, the formula can be simplified as:

$$ExR_{ij} = T_1 + T_2 + T_3 \tag{4}$$

In details, T_1 represents the reply relationship, and can be expressed as:

$$T_1 = \frac{\sum R_e(v_i, \ v_j)}{\sum R_e(v_i)} \tag{5}$$

The value of T_1, ranging from 0 to 1, indicates the interested degree from the user v_i to the user v_j based on the reply times. That is, $R_e(v_i, \ v_j)$ indicates the times that the user v_i replies the user v_j during a specific period.

Similarly, T2 represent the mention relationship, and can be expressed as:

$$T_2 = \frac{\sum M_e(v_i, \ v_j)}{\sum M_e(v_i)} \tag{6}$$

The value of T_2, ranging from 0 to 1, indicates the interested degree from the user v_i to the user v_j based on the mention times. That is, $M_e(v_i, \ v_j)$ indicates the times that the user v_i mentions the user v_j in his/her posted messages during a specific period.

Finally, T_3 represents the following and followed relationship, and can be expressed as:

$$T_3 = \begin{cases} 1, & \text{if these two users have followed each other} \\ 0, & \text{otherwise} \end{cases} \tag{7}$$

As mentioned above, in the *associative ripple* generation process, the whole timeline has been divided into a series of time slices evenly. However, after the clustering in each ripple, the drops distribute in a time slice never following the timeline, but based on the relevance. But, the generated ripples according to a specific heuristic stone in a sequence still follow the timeline, which can be further considered for the time fading factor. Thus, the coefficient Tf can be expressed as:

$$Tf(v_i)_k = 1 + \frac{1-k}{n} \tag{8}$$

where, n is the amount of time slices that has been divided from the whole timeline, and k, which is an integer and ranges from 1 to n, indicates the number of each time slice. The granularity of a time slice depends on the value of n. that is, take one month (30 days) as the timescale for example, if n is 5, each time slice shall be nearly one week (6 days), while if n is 30, each time slice shall be one day. Accordingly, the time fading velocity shall be 1/5 and 1/30 respectively. The value of Tf, descending from 1 to 0 evenly according to the variable n, indicates the importance of time influence from the recent to the past.

Based on these discussed above, finally, the value of w_{ij} between vertex v_i and v_j with a specific threshold will determine whether the relation should exist between these two users. And the DSUN model will be constructed timely and dynamically based on users' current interests and individual needs, which are collected by the heuristic stone and extended by the associative ripples. That is, in different periods, users will have different interests or needs which may change dynamically. The requirements of the social relationships among all the users based on these may also change quickly. Since we have caught and involved those reorganized social streams related to users' current interests and ever-changing needs into the network building, using the mechanism we discussed above, the DSUN model, which will change dynamically in different periods, can represent those dynamical relationships and further provide some potential information (we will discuss it in the following section).

4   User Correlation Discovery

4.1   Measures for the DSUN Model

To describe our DSUN model in details, firstly, we introduce a set of measures and concepts for the model as follows.

Interest Degree: Interest degree is the out-degree of each vertex. That is, the interest degree of each vertex is the number of the arcs that take this vertex as the tail, which can be recorded as $ID(v)$. The interest degree of a specific vertex indicates the number of other users from whom this user may get helpful information related to his/her current interests or needs. The higher interest degree is, the more extensive interests this user may have.

Popularity Degree: Popularity degree is the in-degree of each vertex. That is, the popularity degree of each vertex is the number of the arcs that take this vertex as the head, which can be recorded as $PD(v)$. The popularity degree of a specific vertex indicates the number of other users who may get information related to his/her current interests or needs from this user. The higher popularity degree is, the more contribution this user may have.

Tie: Tie indicates a relationship among a series of vertexes, which may have several types. There may be a variety of ties in a DSUM model. One tie in the model may contain different amount of vertexes, from one vertex to all vertexes. For instance, in special situations, a tie that contains only one vertex is a weakest tie, which means this isolated user may have no similarity with other users; on the other hand, a tie that contains all the vertexes in the model is a super strong tie, which means all users communicate well in this system presented by the model.

4.2   User Correlation Discovery Based on DSUN Model

We further define three important types of ties as follows, which are used to describe users' correlations.

Complementary Tie: For each $<u, v>$, if $< u, v >\in E$, there always exists $< v, u >$ $\in E$, then $<u, v>$ is a complementary tie, which can be recorded as $<u, v, u>$.

Weak Tie: Weak tie is a connected component in a network, which is a maximum connected sub-graph. The complementary tie in the DSUN model indicates a cycle between two vertexes, which means these two users are interested in each other. In other words, they may be complementary each other by sharing information. One DSUN model may have several weak ties, which divide the whole network into several parts. In this situation, we call it weakly connected. If a DSUN model has only one tie, it is a strong connected network. The algorithm for creating the weak tie is based on the Depth-First Search algorithm, which is described in Figure 1.

Strong Tie: Strong tie is a loop in the DSUN model. The strong tie in a DSUN model shows a path in which the first and the last vertex is the same one. That means the users indicated in this loop may have some characters or profiling in common so that they can discuss and share together more effectively. Moreover, their motivations can also be promoted to be more active by stimulating each other. The strong ties are created from weak ties. The algorithm is also based on the Depth-First Search algorithm, which is described in Fig. 1.

```
input: the DSUN G
output: the component set{G₁, G₂, G₃, ..., Gₙ}
Begin
Γ=Φ;                                      //the initial component set is empty
Stack s;
AM[ ][ ];                                 //the adjacency matrix
vertex[V];                                // initialization of vertex set
Init( );
s.push (v1);                              //start from the first vertex
vertex.delete(v1);
while(!vertex.empty)  {
  while(s.is_not_empty) {                 //depth-first search for a component
    index = findNeighbor(s.peek( )-1);    // seeking the neighbor vertex
    if (index != -1 && !s.search(index)){ //if finding a unvisited neighbor
      vistedEdge(s.peek( )-1, index);     //visit the edge
      s.push(index+1);                    //record the vertex
      vertex.delete(index+1);
    }
    if(index != -1 && s.search(index)) {  //if the neighbor has been visited
      vistedEdge(s.peek( ), index);
      for (int j = 0; j <s.getSize( ); j++)
      temp.add(s.pick(j));
    }
    if(-1==index)   {                     //if there is no neighbor
      for (int j = 0; j <s.getSize( ); j++)
      temp.add(s.pick(j));
      s.pop( );
    }
  }
  Γ=Γ∪{temp};                             //create the component set
}
clean( );                                 //free the space
output Γ;                                 //out put the results
End
```

Fig. 1. Algorithm for creating weak ties

5   Scenario and Experiment Analysis

We assume a class with 47 students, where the professor requires these students to use microblog (we are using StatusNet[1] as a test bed for a lecture called *Introduction to*

---

[1] http://nislab.human.waseda.ac.jp/statusnet/

Information Systems) to publish messages, to raise questions, and discuss with the professor and/or other students. This will show if they understand the lecture and help the professor to monitor and regulate the progress of learning activity. We try to use our proposed method to benefit both students and professor for the facilitation of information seeking and discovery in this learning process.

Each drop means a message from one student comes together to form a stream. In this case, the professor follows 47 students' streams. Take a 90 minutes class for example, this class can be divided into several sessions, at any time in each session, we can generate several associative ripples by students using his/her own heuristic stones. A DSUN model can also be built dynamically based on these associative ripples. In this experiment, we only consider the timeline factor and the interaction factor to create a DSUN model based on the messages posted by students in one class, which is shown as follows.

```
input: the DSUN G & the component set{G₁, G₂, G₃, ..., Gₙ}
output: the strong component set{SG₁, SG₂, SG₃, ..., SGₙ}
Begin
Γ=Φ;                                      //the initial strong component set is empty
Stack s;
AM[ ][ ];                                 //the adjacency matrix
graph[ ];                                 // initialization of component set
Init( );
for (g = graph[0]; i<gragh.gphnum; ++i) { //start from the first component
s.push(v);
while(s.is_not_empty)    {                 //depth-first search for a component
index = findNeighbor(s.peek( )-1);        // seeking the neighbor vertex
if (index != -1) {                         //if finding a neighbor
vistedEdge(s.peek( )-1, index);           //visit the edge
if (!s.search(index+1))                    // if the neighbor is unvisited
s.push(index+1);                          //record the vertex
else {
  s.push(index+1);
  for (int j = 0; j <s.getsearchSize( ); j++)
    temp.add(s.pick(j));
  }
}
if(-1==index)                              //if there is no neighbor
s.pop( );
}
Γ=Γ∪{temp};                               //create the strong component set
}
clean( );                                 //free the space
output Γ;                                  //out put the results
End
```

Fig. 2. Algorithm for creating strong ties

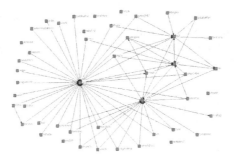

Fig. 3. A DSUN model built from a class

The DSUN model shown in Figure 3 contains totally one professor, two teaching assistants and 47 students, which demonstrate their dynamical correlations in accordance with the reorganized stream data that consist of messages posted in this class. Based on what have discussed in Section 4, we can further get a series of empirical analysis results. Firstly, potential user profiling can be discovered according to the measures. For example, from the popularity degree calculated in each vertex, users *ta1* and *utti58* is the top two users, which means they have the most contributions in this class. Thus all students may go to see these two users' posted messages to find sort of valuable information.

Learning groups can also be divided based on the ties timely and dynamically. For instance, we can create a group {*ta1, naga, tf, ta2*} according to the strong tie, so these four users can discuss and share information better in this group for they may have some same interests or needs at this moment. It can also benefit the professor in his/her teaching process. That is, the professor can decide whether he/she should change his/her teaching style, or what he/she should talk about in next session timely, which are expected to make more students into one group or divide students into more groups, in order to make the discussion more effective and enhance the quality of teaching.

6   Conclusion

In this paper, we have proposed a method to mine and discover the dynamical and potential user correlations in accordance with their personal and current interests and needs based on social streams.

Firstly, we have introduced a set of stream metaphors, definitions, and the structure of our DSUN model. We have further developed a mechanism to build the model with an algorithm using those reorganized social streams. After that, we introduced a set of measures and developed a variety of ties to present users' current profiling and correlation in accordance with users' current interests and needs. Two algorithms were developed to create the weak tie and strong tie respectively. Based on these, we have finally showed an application scenario to demonstrate our proposed model and method, which are expected to benefit user correlation discovery and further facilitate information seeking process.

As for future work, an improved prototype system with the developed algorithms will be implemented. Moreover, we will try to improve the correlation mining results. For instance, we will try to optimize a user relationship tree which can help to recommend the specific user with more related users and information.

Acknowledgements. The work has been partly supported by 2011 and 2012 Waseda University Grants for Special Research Project No. 2011B-259 and No. 2012B-215, and 2010–2012 Waseda University Advanced Research Center for Human Sciences Project "Distributed Collaborative Knowledge Information Sharing Systems for Growing Campus."

References

1. Carpineto, C., Osiński, S., Romano, G., Weiss, D.: A Survey of Web Clustering Engines. ACM Computing Surveys (CSUR) 41(3) (2009)
2. Chi, E.H.: Information Seeking Can Be Social. Computer 42(3), 42–46 (2009)
3. Chen, H., Zhou, X.K., Man, H.F., Wu, Y., Ahmed, A.U., Jin, Q.: A Framework of Organic Streams: Integrating Dynamically Diversified Contents into Ubiquitous Personal Study. In: 2nd International Symposium on Multidisciplinary Emerging Networks and Systems. Xi'an (2010)
4. Zhou, X.K., Chen, H., Jin, Q., Yong, J.M.: Generating Associative Ripples of Relevant Information from a Variety of Data Streams by Throwing a Heuristic Stone. In: ACM ICUIMC 2011 (5th International Conference on Ubiquitous Information Management and Communication), Seoul, Korea (2011)
5. Zhou, X., Jin, Q.: Dynamical User Networking and Profiling Based on Activity Streams for Enhanced Social Learning. In: Leung, H., Popescu, E., Cao, Y., Lau, R.W.H., Nejdl, W. (eds.) ICWL 2011. LNCS, vol. 7048, pp. 219–225. Springer, Heidelberg (2011)
6. Signorini, A., Segre, A.M., Polgreen, P.M.: The Use of Twitter to Track Levels of Disease Activity and Public Concern in the US during the Influenza A H1N1 Pandemic. PLOS ONE 6(5) (2011)
7. Junco, R., Heiberger, G., Loken, E.: The effect of Twitter on college student engagement and grades. Journal of Computer Assisted Learning 27(2), 119–132 (2011)
8. Johnson, K.A.: The effect of Twitter posts on students' perceptions of instructor credibility. Learning Media and Technology 36(1), 21–38 (2011)
9. Srivastava, J., Cooley, R., Deshpande, M., Tan, P.-N.: Web Usage Mining: Discovery and Applications of Usage Patterns from Web Data. ACM SIGKDD 1(2), 12–23 (2000)
10. Stumme, G., Hotho, A., Berendt, B.: Semantic Web Mining State of the Art and Future Directions. Elsevier Web Semantics: Science, Services and Agents on the World Wide Web 4(2), 124–143 (2006)
11. Poblete, B., Baeza-Yates, R.: Query-Sets: Using Implicit Feedback and Query Patterns to Organize Web Documents. In: Proc. WWW 2008, Beijing, pp. 41–48 (2008)
12. Bilenko, M., White, R.W.: Mining the Search Trails of Surfing Crowds: Identifying Relevant Websites From User Activity. In: Proc. WWW 2008, Beijing, pp. 51–60 (2008)
13. Fang, X., Liu Sheng, O.R.: LinkSelector: A Web Mining Approach to Hyperlink Selection for Web Portals. ACM Transactions on Internet Technology 4(2), 209–237 (2004)

Extraction of Human Social Behavior from Mobile Phone Sensing

Minshu Li, Haipeng Wang, Bin Guo, and Zhiwen Yu

School of Computer Science, Northwestern Polytechnical University, Xi'an, China
liminshu1@163.com, haipeng.wang@gmail.com,
{guob,zhiwenyu}@nwpu.edu.cn

Abstract. With lots of sensors built in, mobile phones become a pervasive platform for seamlessly sensing of human behaviors. In this paper, we investigate how to use location data and communication records collected from mobile phones to obtain human social interaction features and activity patterns. Social Interaction features refer to the temporal and spatial interactive information, and activity patterns include movement patterns. Meanwhile, the similarities and differences of human behaviors at different ages, as well as distinct occupations are analyzed. The results indicate that different population has a diversity of social interaction and activity patterns, and human social behaviors are highly associated with age and occupation. Furthermore, we make a correlation analysis about social temporal interaction, social spatial interaction and social activity, which lead us to conclude that the three elements are interrelated among young people but not middle-ages. Our work could be a cornerstone for research of personalized psychological health assistance based on mobile phone data.

Keywords: Social behavior, mobile phone data, pervasive computing, health computing.

1   Introduction

Following Aristotle, "Man is by nature a social animal; an individual who is unsocial naturally and not accidentally is either beneath out notice or more than human" [1]. Every day people interact with others and it has become an indispensable part of life. **Social behavior** is complex and vital to human health, both mentally and physically. E.g. lacking of social participation increases the risk of Alzheimer's (AD) disease-like dementia [2]. Social communication contributes to maintaining the cognitive performance and mood [3]. In this way, the change in user's **social behavior** could indicate the change in user's lifestyle, physical health and psychological health. Furthermore, from a user's **social behavior**, we can find a lot of useful information about himself, such as his character and so on. Social intelligence can be extracted from the "digital footprints" left by people, while this technology will convert the understandings of our lives and bring about innovative services in areas like human health [4].

With the development of ICT (information communication technology) techniques, mobile phone has become portable and been used on daily basis. Since the underlying

R. Huang et al. (Eds.): AMT 2012, LNCS 7669, pp. 63–72, 2012.

sensing technologies are commonplace and readily available at present, mobile phones can be used for sensing enormous human activity features unobtrusively and precisely, For example, via call log and Bluetooth proximity data we can obtain inter-personal interaction information, and GPS and WLAN data are widely applied in navigation and positioning. These 'digital footprints' are capable of reflecting user's real-time **social behavior** factually without disturbing [5]. And this technology elimi-nates the dependence on self-reporting, which is prevalent in traditional social science. By using mobile phone records, we mining user's **social behavior** in order to help people aware of their own behavior more, as well as provide useful information for doctors or health institutes. In addition, what we have done may also lay the groundwork for physical and psychological health analysis.

Our work is based on the Nokia open project "Mobile Data Challenge [6]". It con-sists of mobile phone sensing data collected from 38 subjects during one year. The research topics and main results are as follows: (1) extracting two kinds of the user's semantic information to mine user's **social behavior**: *social interaction features* from call log and Bluetooth proximity data, along with *activity patterns* from GPS and WLAN data.; (2) analyze human's **social behavior** at different ages and distinct oc-cupations using Nokia's data, the result indicate that different population has obvious diversities in social interaction and *activity patterns*, human's **social behavior** is as-sociated with ages and occupations; (3) make a correlation analysis among temporal interaction, spatial interaction and *activity patterns* at different ages and distinct occu-pations, results demonstrate that they are related to each other for youngers while not for middle-ages.

This article consists of six sections. Section 2 introduces the related work. Data processing and feature extraction are presented in Section 3. In Section 4 we make a comparison of different population on *social interaction features* in order to observe the particularity in different populations. Some understandings of **social behavior** are presented in Section 5. The last section summarizes the whole article.

2   Related Work

Mobile sensor data are now being used to understand a broad spectrum of sensing and modeling questions. Researches focus on the influence upon user's physical world as well as psychological world. For example, in Houston [7], small groups of users share their step counts and progress toward daily step count goals with each other via their mobile phones. Madan et al. [5] found that weight changes of users are correlated with exposure to peers who gained weight in the same period and also understood the link between specific behaviors and changes in political opinions using mobile phone data. They demonstrated that phone-based features can be used to predict changes in health, such as common cold, stress, and automatically identify symptomatic days.

Currently many researchers employ mobile phones on social interaction study, like E-Shadow [8], a distributed mobile phone-based local social networking, provides mobile phone software that facilitates abundant social interactions. The software maps proximate users' local profiles to their human owners and enables user communication and content sharing as human appear to be literally wired for social interaction. E-SmallTalker [9] leverages Bluetooth Service Discovery Protocol to exchange user-defined contents without establishing a connection. Social Contact Manager [10] has demonstrated how to merge the complementary features of online and opportunistic social networks, to automatically collect rich information (e.g., profile, face-to-face meeting contexts) about their contacts. Just-for-Us [11] uses GPS to help people locate their friends in physical proximity.

All these work use merely mobile phone sensors to promote users' social interactions, without relating to mining users' own **social behavior** and user's social activities. This work, however aims at exploring user's movement patterns to be able to provide better service.

3   Data Choosing and Processing

Social behavior includes social interaction and *activity patterns*. *Social interaction features* depict user's interactions and contact activities with other people. It's mainly having two forms: temporal interaction and spatial interaction. *Activity patterns* refer to the space rang of user's daily activities. Temporal interaction can be reflected by the call log data, spatial interaction can be reflected by Bluetooth data, while the GPS&WLAN data embody *activity patterns*--GPS data records user's outdoor activity track while WLAN data records indoor activity track, together they form a complete activity track.

We try to use those mobile phone data (call log, Bluetooth, GPS&WLAN) to mine user's **social behavior** and shed light upon the similarities and differences of users at different ages and distinct occupations. Due to the activities of daily life diversified in different time periods, and in consideration of accuracy and completeness as well as rationality, we analyze the semantic information separately according to different time periods. In the selection of features, we consider total interaction, weekday versus weekend interaction, and daytime versus night interaction (here, 'night' represents interactions between 5pm and 8am every day). Therefore, time periods are divided into four parts-Workdays Daytime (WD), Workday Night (WN), Holiday Daytime (HD), and Holiday Night (HN)-based on which we mine and analyze user's semantic information. Since the number of days is different for workday and weekend, we make an average daily statistics. The procedure of data processing and **social behavior** feature extractions are presented as follows.

3.1   Social Interaction Features

Temporal Interaction
To obtain user's temporal interaction, we mainly use the call log data for analysis. In the call log data, main parameters, including direction of the call/short message, contact time and type of the call, are completely recorded to demonstrate temporal-domain interactions. Specifically, direction of the call/short message refers to whether the call is incoming, or outgoing; contact time means when the user make a call (e.g.,

in the evening, at weekends) [12]; type of call means voice call or short message. The statistic from call log data indicates the frequency of user's interaction. The unsuccessful interactive data is ignored since it is meaningless for our evaluation.

Spatial Interaction

As Bluetooth proximity sensing allows us to quantify time spent in face-to-face proximity for individuals, as opposed to relying on binary responses to represent of social ties, which makes it perfect for spatial interaction analysis. Take an example, if a user spends a lot of time at home or in office, the number of Bluetooth devices he/she meets usually remains constant, and so is the records number. On the contrary, if he/she has a lot of outdoor activities, the number of Bluetooth devices may increase while records number rarely changes compared to former situation. Bluetooth proximity data records information of other Bluetooth device user come across every moment. Some researchers have demonstrated that one Bluetooth device is able to scan all Bluetooth devices around it within 10.2 seconds [13]. Therefore, user's interaction with others can be obtained. If we take a statistical number of Bluetooth device and records, we can get user's initiative of interaction.

3.2   Activity Patterns

In order to get user's activity range, we apply a cluster analysis to the combined GPS & WLAN data. Specially, we only care about user's specific activity locations but not the traffic data between them. Research shows that the limit velocity of human being is 10m/s. on that assumption, the points whose velocities are under 10m/s are chosen for cluster analysis. Since the number of user's specific activity locations is unknown, we use a Mean-Shift cluster [14]. Every record of the data includes an active point consist of latitude and longitude. Applying the Mean-Shift cluster to those active points, we are able to find user's regular movement patterns. Moreover, the radius of cluster can be adjusted to the local regional partition situation. We transform latitude and longitude angle into mileage, so that it becomes easier to confirm the radius of cluster using GPS data [15]. In our experiment, the radius is set to 0.00045 corresponding to 50m.

4   Analysis Results of Mobile Phone Data

In this section, we use the data processing method mentioned above to analyze the similarities and diversities of users at different ages and distinct occupations from the following aspects: temporal interaction, spatial interaction and activity. In order to avoid interference of various factors, parameters out of our consideration are kept constant. Here, we divided users into two groups: (1) people with distinct occupation but in the same age (full-time students in age group 22 to 33 and full-time workers in age group 22 to 33); (2) people with the same occupation but at different age (full-time workers in age group 22 to 33 and full-time works in age group 33 to 44).

4.1   Social Interaction Features

Social Interaction features refer to the temporal and spacial interactive information. Here we would like to make a combination analysis of these two forms of social interaction. And in the following section, we will make a detail analysis separately. Fig.1 (a) describes temporal interaction whileFig.2 (b) describes spatial interaction. We find that no matter it's on temporal or on spatial interactive, users of different ages and occupations have more social interaction at night than in the daytime and users of different ages and occupations have more social interaction in holiday than during workday. In general, the social interaction of students is less than that of workers, but their spatial interaction on workday daytime is more than that of workers, which may be caused by their surroundings. Furthermore, middle-ages interacts less than younger and their interaction on workday is more than that of on holiday, which is just the reverse compared to younger.

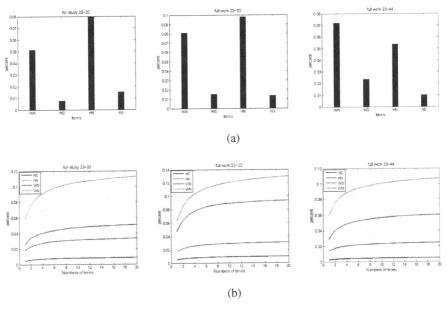

(a)

(b)

Fig. 1. Social interaction features combine analysis. The left chart is full-time student in age group 22 to 33, the middle is full-time workers in age group 22-33, and the right is full-time workers in age group 33-44. Each color stands for one time period. WN means Workday Night, WD means Workday Daytime, HN means Holiday Night, HD means Holiday Daytime.

Temporal Interaction

Call log data reflects user's temporal interaction with others. Due to users' various habits, it shows an obvious diversity on the number of call. Therefore, we firstly normalize the data by calculating the ratio of different parts to total calls, based on what we observed the diversity of users' interaction with others. Part(a) of Fig.2 and Fig.3 generally describe the ratio of incoming calls, outgoing calls and short messages to the total number in different time periods(daytime, night, workday and holiday), while part(b) make a detail description of interaction features of different users.

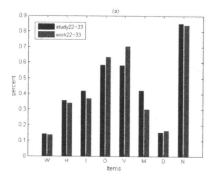
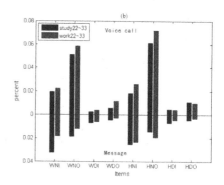

Fig. 2. Call log statistic between different occupations. Vertical axis indicates the ratio and horizontal refers to different items. In detail, W means Workday, H means Holiday, I means Incoming, O means Outgoing, V means Voice call, M means message, D means Daytime, N means Night. Blue stands for students in age group 22 to 33, red stands for workers in age group 22 to 33. Part (b) refers to voice call and the bottom part refers to short message.

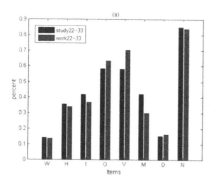
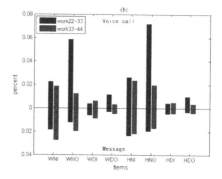

Fig. 3. Call log statistic between different ages

Fig.2 and Fig.3 shows that those three types of users' communication at night are far more than workday that in the daytime and the communication in holiday are more than that in. Namely, user rarely communicates in the daytime or during workday, and outgoing calls are more than incoming calls. Compared to users of different occupations, the communication mode of full-time students and full-time workers are distinguishing. The short message on holiday night of workers significantly increases with respect to workday night, while for students, there's no apparent change. However, for both workers and students, phone calls on holiday nights are more than on workday nights. The phone calls at holiday daytime of students increase significantly compared with workday daytime, while for workers there's no evident change. Among users in different age groups, the comparison of full-time worker in age group 22 to 33 with age group 33 to 44 demonstrates that the former's outgoing calls are more than incoming calls, while for the latter the situation is just in the opposite. However, the age group 22 to 33 likes making phone calls better than sending short messages. The value of workday daytime is almost the same as that of in holiday daytime, but both of them are quite small. However, the value of holiday night is more than workday night.

Spatial Interaction

Bluetooth proximity data explores special interactive features as well as outdoor activities. In particular, if user usually stays at some fixed places like home or office, etc. , the probability of discovering other Bluetooth devices is tiny, in other words, the number of scanned other Bluetooth devices remains small. However, if user has a lot of outdoors activities, the probability increases and the number become large.

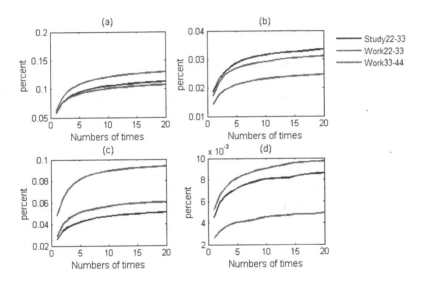

Fig. 4. Bluetooth statistics. (a) Refers to workday night, (b) refers to workday daytime, (c) refers to holiday night, and (d) refers to holiday daytime. Vertical axis means cumulative percent to total number. Horizontal axis means numbers of scanned Bluetooth devices. And blue stands for full-time students in age group 22 to 33, green stands for full-time workers in age group 22 to 33, red stands for full-time workers in age group 33 to 44.

In Fig.4,subplot (a),(b),(c) and (d) respectively describes the Bluetooth proximate data of three kinds of users during different time periods((a)workday night, (b)workday daytime, (c)holiday night and (d)holiday daytime). In addition, we make an average daily statistics. Before going further, we would like to make an explanation. The horizontal axis refers to specific times of scanned Bluetooth device during some periods. For example, horizontal coordinate 5 in subplot (a) means the statistic of Bluetooth devices have been scanned by our Bluetooth for five times and so on. And the vertical axis stands for the ratio to total scanned Bluetooth device number. The probability is overlaid to facilitate the observation so that we could put it in this way: the area surrounded by the curve and horizontal axis reflects the number of scanned Bluetooth devices of specific users in different age groups. Fig4 shows that the number of scanned Bluetooth devices at night is significantly larger than that by daytime. That means, user has more outdoor activities at night. And the number of workday is slightly more than holiday. The spatial interaction of users in age group 33 to 44 is obviously less than users in age group 22 to 33 in any time period. Meanwhile, workers interact more than students except in workday daytime. Maybe it is because students can meet many classmates in the classroom. The most significant difference in the number of scanned Bluetooth devices can be found at holiday night.

4.2   Activity Patterns

User's movement patterns, which we call user aliveness, can be excavated via GPS & WLAN data. More users' activity locations, higher user's aliveness is, vice versa. On the other hand, user's movement patterns indicate the spatial interaction. It means more spatial interaction less narrow activity range.

As shown in Fig.5, the number of user's activity places at night is more than that in the daytime, namely, daytime activity range is wider than that of night. And it is the same with workday to holiday. Meanwhile, the number of daytime activity places is almost not exceeding 5, which indicates user's daytime activity place is relatively fixed. In regard to the number of activity places, there is no obvious difference between users in different age groups. And it is also true for users with different occupations.

We make a statistic about places that users of different occupations usually go to at different periods of time and project the place into the physical location by Google earth. As shown in table1 difference exists on types of activity places with respect to users in different ages and distinct occupations; it varies with not only occupation but also with time period.

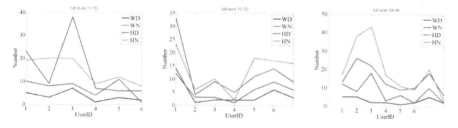

Fig. 5. Movement patterns of users. The left chart is full-time student in age group 22 to 33, the middle is full-time workers in age group 22-33, and the right is full-time workers in age group 33-44. Each color stands for one time period. WN means Workday Night, WD means Workday Daytime, HN means Holiday Night, HD means Holiday Daytime.

Table 1. Statistics of user's activity locations

Population	Time	Place	Time	Place
Full work 33-44	WN	Uptown, scenic spot, etc.	HN	Beach, park, etc.
	WD	Station, industrial park, etc.	HD	Gym, Mall, etc.
Full work 22-22	WN	Mall, chapel, etc.	HN	leisure square, dining-hall, etc.
	WD	Company, Library, etc.	HD	Mall, Station, etc.
Full study	WN	School, industrial park, etc.	HN	School, industrial park, etc.
	WD	School, Station, etc.	HD	School, Chapel, outskirts, etc.

5   Understanding of Social Behavior

5.1   Diversity of Social Behaviors

In previous section, we make a comparison among different types of users on the aspect of social interaction and social activity. Both similarity and diversity exist, which indicates that we can obtain user's **social behavior** by mobile phone data, and we find that whether in social interaction or in social activity, different population shows higher aliveness at night than in the daytime. During workday, people prefer face-to-face interaction. While on holidays they tend to make phone calls. Population in age group 22 to 33 interacts more than the age group 33 to 44. As well as workers interacts more than students.

5.2   Correlation among Social Behaviors

At the same time, combining the three aspects, we make a correlation analysis about them by making a statistic for every user on call log data, Bluetooth data and GPS data. Since the value of different data diversifies greatly, we project all data onto the same scale space. As shown in Fig.6, the aliveness of three aspects of students' tend to be consistent as compared to workers', so do younger's as compared to middle-ages'. In other words, for users of specific type their aliveness in temporal interaction, spatial interaction and activity are related to each other. If in one aspect the aliveness is high, it is the same in the other aspects and vice versa. The age group 33 to 44 is not following this rule. It's maybe because middle-aged workers have more steady jobs and life habits.

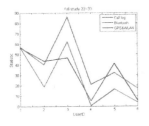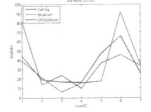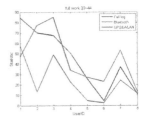

Fig. 6. Analysis

6   Conclusion

In this paper, we have investigated how to use mobile phone data, including call log, Bluetooth, GPS and WLAN, to mine the *social interaction features* and the *activity patterns* of user's **social behavior**. Moreover, we analyzed the similarity, the diversity among social temporal interaction, social spatial interaction and activity as well as their relevance. The results show that different populations have significant difference in *social interaction features* and *activity patterns*. At the same time, we find that youngers show a high consistency between social interaction and social activity, while it is not obvious among middle-aged workers. Analysis results could provide reference for stress detection and psychological health assistance among different population as well as personalized service.

References

1. Vinciarelli, A.: Capturing order in social interactions. IEEE Signal Processing Magazine (2009)
2. Barclay, L., et al.: Loneliness increases risk for AD-Like dementia. Archives General Psychiatry 64, 234–240
3. Waterworth, J.A., et al.: Ageing in a networked society – social inclusion and mental stimulation. In: ACM PETRA (2009)
4. Guo, B., Zhang, D., Yu, Z., Calabrese, F.: Extracting Social and Community Intelligence from Digital Footprints. Journal of Ambient Intelligence and Humanized Computing (2012)
5. Madan, A., Cebrian, M., Moturu, S., Farrahi, K., Pentland, A.: Sensing the health state of a community. IEEE Pervasive Computing (forthcoming)
6. Nokia Mobile Data Challenge, http://research.nokia.com/page/12000
7. Consolvo, S., Everitt, K., Smith, I., Landay, J.A.: Design Requirements for Technologies that Encourage Physical Activity. In: Proc. of CHI (2006)
8. Teng, J., Zhang, B., Li, X., Bai, X., Xuan, D.: E-Shadow: Lubricating Social Interaction using Mobile Phones. In: ICDCS (2011)
9. Yang, Z., Zhang, B., Dai, J., Champion, A., Xuan, D.: E-SmallTalker: A Distributed Mobile System for Social Networking in Physical Proximity. In: Proc. of ICDCS (2010)
10. Guo, B., Zhang, D., Yang, D.: "Read" More from Business Cards: Toward a Smart Social Contact Management System. In: WI (2011)
11. Kjeldskov, J., Paay, J.: Just-for-Us: A Context-Aware Mobile Information System Facilitating Sociality. In: Proc. of MobileHCI (2005)
12. Guo, B., Zhang, D., Yu, Z., Zhou, X.: Hybrid SN: Interlinking Opportunistic and Online Communities to Augment Information Dissemination. In: UIC (2012)
13. Chen, Y., Li, Q., Liu, J., et al.: Sensing Surrounding Contexts using Dynamic Bluetooth Information. Journal of Software, 137–146 (2011)
14. Comaniciu, D., Meer, P.: Mean shift: a robust approach toward feature sapce analysis. IEEE Transaction on Pattern Analysis and Machine Intelligence 24(5), 603–619 (2002)
15. Chen, S., Li, J.: Data analysis and processing of GPS module. GNSS Word of China, 87–91 (June 2011)

Continuity of Defuzzification on L^2 Space for Optimization of Fuzzy Control[*]

Takashi Mitsuishi[1], Takanori Terashima[2], Nami Shimada[1],
Toshimichi Homma[3], Kiyoshi Sawada[1], and Yasunari Shidama[4]

[1] University of Marketing and Distribution Sciences, Kobe, Japan
[2] Center for Green Computing, Nagoya Institute of Technology, Nagoya, Japan
[3] Osaka University of Economics, Osaka, Japan
[4] Shinshu University, Nagano, Japan
takashi_mitsuishi@red.umds.ac.jp

Abstract. The purpose of this study is to consider the fuzzy optimal control based on the functional analysis. We used a mathematical approach to compute optimal solutions. The feedback of fuzzy control is evaluated through approximate reasoning using the center of sums defuzzification method or the height method on IF-THEN fuzzy rules. The framework consists of two propositions: To guarantee the convergence of optimal solution, a set of fuzzy membership functions (admissible fuzzy controller) which are selected out of continuous function space is compact metrizable. And assuming approximate reasoning to be a functional on the set of membership functions, its continuity is proved. Then, we show the existence of a fuzzy controller which minimizes (maximizes) the integral performance function of the nonlinear feedback fuzzy system.

1 Introduction

We have examined the automatic and computational determination of fuzzy membership functions, which give optimal controls in fuzzy control system. We also have been studying to find algorithms that compute optimal solutions. In this study, we discussed optimization of fuzzy control which is different from conventional method such as classical control and modern control known for a linear matrix inequality approach [4]. We consider fuzzy optimal control problems as problems of finding the minimum (maximum) value of the performance function with feedback law constructed by approximate reasoning with the way of functional analysis [5,6]. These methods use the center of gravity defuzzification method which is the most prevalent. However many defuzzification methods are proposed in recent years. The center of sums method among the many methods is focused in this study. This method is similar to the weighted average method (Nakamori model) and the height method.

This study is organized in the following way: A nonlinear feedback control using approximate reasoning is presented. Then, the approach to the control

---

[*] This work was supported by JSPS KAKENHI Grant Numbers 24700235, 23730395.

R. Huang et al. (Eds.): AMT 2012, LNCS 7669, pp. 73–81, 2012.

based on functional analysis is discussed, and some propositions are obtained. Finally, the mathematical properties are applied to optimal control problem.

2   Fuzzy Control

2.1   Feedback System

In this study we assume that the feedback part in this system is calculated by approximate reasoning through the center of sums defuzzification method. Using an idea and framework mentioned in the following section 3 and 4, the existence of optimal control based on fuzzy rules is designed.

$\mathbb{R}^n$ denotes the n-dimensional Euclidean space with the usual norm $\|\cdot\|$. Let $f(v_1, v_2) : \mathbb{R}^n \times \mathbb{R} \to \mathbb{R}^n$ be a (nonlinear) vector valued function which is Lipschitz continuous. In addition, assume that there exists a constant $M_f > 0$ such that

$$\|f(v_1, v_2)\| \leq M_f (\|v_1\| + |v_2| + 1)$$

for all $(v_1, v_2) \in \mathbb{R}^n \times \mathbb{R}$. Consider a system given by the following state equation:

$$\dot{x}(t) = f(x(t), u(t)). \tag{1}$$

where $x(t)$ is the state and the control input $u(t)$ of the system is given by the state feedback

$$u(t) = \rho(x(t)).$$

For a sufficiently large $r > 0$, $B_r = \{x \in \mathbb{R}^n : \|x\| \leq r\}$ denotes a bounded set containing all possible initial states x_0 of the system. Let T be a sufficiently large final time. Then, we have

Proposition 1. *Let $\rho : \mathbb{R}^n \to \mathbb{R}$ be a Lipschitz continuous function and $x_0 \in B_r$. Then, the state equation*

$$\dot{x}(t) = f(x(t), \rho(x(t)))$$

has a unique solution $x(t, x_0, \rho)$ on $[0, T]$ with the initial condition $x(0) = x_0$ such that the mapping

$$(t, x_0) \in [0, T] \times B_r \mapsto x(t, x_0, \rho)$$

is continuous.

For any $r_2 > 0$, denote by Φ the set of Lipschitz continuous functions $\rho : \mathbb{R}^n \to \mathbb{R}$ satisfying

$$\sup_{v \in \mathbb{R}^n} |\rho(v)| \leq r_2.$$

Then, the following a) *and* b) *hold.*
a) *For any $t \in [0, T], x_0 \in B_r$ and $\rho \in \Phi$,*

$$\|x(t, x_0, \rho)\| \leq r_1,$$

where
$$r_1 = e^{M_f T} r + (e^{M_f T} - 1)(r_2 + 1). \qquad (2)$$

b) *Let $\rho_1, \rho_2 \in \Phi$. Then, for any $t \in [0, T]$ and $x_0 \in B_r$,*

$$\|x(t, x_0, \rho_1) - x(t, x_0, \rho_2)\|$$
$$\leq \frac{e^{\Delta_f (1 + \Delta_{\rho_1}) t} - 1}{1 + \Delta_{\rho_1}} \sup_{u \in [-r_1, r_1]^n} |\rho_1(u) - \rho_2(u)|, \qquad (3)$$

where Δ_f and Δ_{ρ_1} are the Lipschitz constants of f and ρ_1 [6].

2.2 Approximate Reasoning Using Center of Sums Method or Height Method

In this section we briefly explain the approximate reasoning using the center of sums method or the height method [13] which decides feedback output in the previous nonlinear system for the convenience of the reader.

Assume the feedback law ρ consists of the following m IF-THEN type fuzzy control rules.

> RULE 1: *IF x_1 is A_{11} and ... and x_n is A_{1n}*
> *THEN y is B_1*
>
> ...
>
> RULE i: *IF x_1 is A_{i1} and ... and x_n is A_{in}*   (4)
> *THEN y is B_i*
>
> ...
>
> RULE m: *IF x_1 is A_{m1} and ... and x_n is A_{mn}*
> *THEN y is B_m*

Here, m is the number of fuzzy production rules, and n is the number of premise variables $x_1, x_2, \ldots, x_n$. Let $\mu_{A_{ij}}$ ($i = 1, 2, \ldots, m$; $j = 1, 2, \ldots, n$) be membership functions defined on a certain closed interval of the fuzzy set A_{ij}. Let μ_{B_i} ($i = 1, 2, \ldots, m$) be membership functions defined on a certain closed interval of the fuzzy set B_i. For simplicity, we write "IF" and "THEN" parts in the rules by the following notation:

$$\mathcal{A}_i = (\mu_{A_{i1}}, \ldots, \mu_{A_{in}}) \quad (i = 1, \ldots, m),$$

$$\mathcal{A} = (\mathcal{A}_1, \ldots, \mathcal{A}_m), \quad \mathcal{B} = (\mu_{B_1}, \ldots, \mu_{B_m}).$$

Then, the IF-THEN type fuzzy control rules (4) is called a fuzzy controller, and is denoted by $(\mathcal{A}, \mathcal{B})$. In the rules, the tuple of premise variable $x = (x_1, x_2, \ldots, x_n)$ is called input information given to the fuzzy controller $(\mathcal{A}, \mathcal{B})$, and y is called a control variable.

Mamdani method [1] is widely used in fuzzy controls because of its simplicity and comprehensibility. Nakamori fuzzy model (the weighted average method)

[3] does not infer each rule, and defines the inference result of all rules as a weighted average of agreement degree. The following center of sums method is similar to the weighted average method except that the weight is the area instead of the agreement degree of premise part. This method is faster than many defuzzification methods and not restricted to symmetrical membership functions. The height method is a special case that all forms of membership functions are the same.

In this study, when an input information $x = (x_1, x_2, \ldots, x_n) \in \mathbb{R}^n$ is given to the fuzzy controller $(\mathcal{A}, \mathcal{B})$ in other words, the IF-THEN rules (4), then one can obtain the amount of operation from the controller through the following procedures:

Procedure 1: The degree of each rule i is calculated by

$$\alpha_{\mathcal{A}_i}(x) = \bigotimes_{j=1}^{n} \mu_{A_{ij}}(x_j) \quad (i = 1, 2, \ldots, m).$$

Here, $\otimes$ means scaling down calculation "product $\prod$" or clipping calculation "minimum $\bigwedge$" [1,2].

Procedure 2: The inference result of i-th rule is calculated by

$$\beta_{A_i B_i}(x, y) = \alpha_{\mathcal{A}_i}(x) \otimes \mu_{B_i}(y) \quad (i = 1, 2, \ldots, m).$$

Procedure 3: Defuzzification stage. Center of sums method:

$$CoS_{AB}(x) = \frac{\sum_{i=1}^{m} \bar{y}_i \int \beta_{A_i B_i}(x, y) dy}{\sum_{i=1}^{m} \int \beta_{A_i B_i}(x, y) dy}.$$

Height method:

$$HM_{AB}(x) = \frac{\sum_{i=1}^{m} \bar{y}_i \alpha_{\mathcal{A}_i}(x)}{\sum_{i=1}^{m} \alpha_{\mathcal{A}_i}(x)}.$$

Here, $\bar{y}_i$ is the defuzzified value of individual fuzzy set in consequent part of i-th rule given by

$$\bar{y}_i = \frac{\int y\mu_{B_i}(y) dy}{\int \mu_{B_i}(y) dy} \quad (i = 1, 2, \ldots, m).$$

This is the centroid of the membership function. If all the consequent part fuzzy sets have the same form like a triangle or a trapezoid, we can consider the center of sums defuzzification method to be the height method.

3   Admissible Fuzzy Controller

In this section, we introduce two sets of fuzzy membership functions and study their topological properties. Then we can show that a set of admissible fuzzy controllers is compact and metrizable with respect to an appropriate topology on fuzzy membership functions.

In the following fix $r > 0$, a sufficiently large $r_2 > 0$ and a final time T of the control (1). Put r_1 be the positive constant determined by (2). We also fix $\Delta_{ij} > 0$ ($i = 1, 2, \ldots, m$; $j = 1, 2, \ldots, n$). Let $C[-r_1, r_1]$ be the Banach space of all continuous real functions on $[-r_1, r_1]$. Denote by $L^2[-r_2, r_2]$ the Hilbert space of all square integrable, Legesgue measurable real function on $[-r_2, r_2]$. We consider the following two sets of fuzzy membership functions.

$$F_{\Delta_{ij}} = \{\mu \in C[-r_1, r_1] : 0 \leq \mu(x) \leq 1 \text{ for } \forall x \in [-r_1, r_1],$$
$$|\mu(x) - \mu(x')| \leq \Delta_{ij}|x - x'| \text{ for } \forall x, x' \in [-r_1, r_1]\}$$

and

$$G = \{\mu \in L^2[-r_2, r_2] : 0 \leq \mu(y) \leq 1 \text{ a.e. } y \in [-r_2, r_2]\}.$$

The set $F_{\Delta_{ij}}$, which is more restrictive than G, contains asymmetric triangular, trapezoidal and bell-shaped fuzzy membership functions with gradients less than positive value Δ_{ij}. Consequently, if $\Delta_{ij} > 0$ is taken large enough, $F_{\Delta_{ij}}$ contains almost all fuzzy membership functions which are used in practical applications. We assume that the fuzzy membership function $\mu_{A_{ij}}$ in "IF" parts of the rules (4) belongs to the set $F_{\Delta_{ij}}$. On the other hand, we also assume that the fuzzy membership function μ_{B_i} in "THEN" parts of the rules (4) belongs to the set G. In the following, we endow the space $F_{\Delta_{ij}}$ with the norm topology on $C[-r_1, r_1]$, and endow the space G with the weak topology on $L^2[-r_2, r_2]$. Then for all $i = 1, 2, \ldots, m$ and $j = 1, 2, \ldots, n$, $F_{\Delta_{ij}}$ is a compact subset of $C[-r_1, r_1]$, and G is compact metrizable for the weak topology on $L^2[-r_2, r_2]$ [5]. Put

$$\mathcal{L} = \prod_{i=1}^{m} \left\{ \prod_{j=1}^{n} F_{\Delta_{ij}} \right\} \times G^m,$$

where G^m denotes the m times Cartesian product of G. Then, every element $(\mathcal{A}, \mathcal{B})$ of $\mathcal{L}$ is a fuzzy controller given by the IF-THEN type fuzzy control rules (4). By the Tychonoff theorem we can have readily:

Proposition 2. $\mathcal{L}$ is compact and metrizable with respect to the product topology on $C[-r_1, r_1]^{mn} \times L^2[-r_2, r_2]^m$.

To avoid making the denominator of the fractional expressions in the procedure 3 in the previous section equal to 0, for any $\delta > 0$, $\sigma > 0$, we consider the set:

$$\mathcal{L}_{\delta\sigma} = \left\{ (\mathcal{A}, \mathcal{B}) \in \mathcal{L}; \ \forall x \in [-r_1, r_1]^n, \sum_{i=1}^{m} \alpha_{\mathcal{A}_i}(x) \geq \sigma, \right.$$

$$\left. \forall i = 1, 2, \ldots, n, \int_{-r_2}^{r_2} \mu_{B_i}(y)dy \geq \delta \right\}, \tag{5}$$

which is a slight modification of $\mathcal{L}$. If δ and σ are taken small enough, it is possible to consider $\mathcal{L} = \mathcal{L}_{\delta\sigma}$ for practical applications. We say that an element $(\mathcal{A}, \mathcal{B})$ of $\mathcal{L}_{\delta\sigma}$ is an admissible fuzzy controller. Then, we have the following:

Proposition 3. *The set $\mathcal{L}_{\delta\sigma}$ of all admissible fuzzy controllers is compact and metrizable with respect to the product topology on $C[-r_1, r_1]^{mn} \times L^2[-r_2, r_2]^m$.*

Proof. We first note that a sequence $\{(\mathcal{A}^k, \mathcal{B}^k)\} \subset \mathcal{L}$ converges to $(\mathcal{A}, \mathcal{B})$ in $\mathcal{L}$ for the product topology if and only if, for each $(i = 1, 2, \ldots, m)$,

$$\|\alpha_{\mathcal{A}_i^k} - \alpha_{\mathcal{A}_i}\|_\infty = \sup_{x \in [-r_1, r_1]^n} |\alpha_{\mathcal{A}_i^k}(x) - \alpha_{\mathcal{A}_i}(x)| \to 0$$

and

$$\|\mu_{B_i^k} - \mu_{B_i}\|_\infty = \sup_{y \in [-r_2, r_2]} |\mu_{B_i^k}(y) - \mu_{B_i}(y)| \to 0$$

Assume that a sequence $\{(\mathcal{A}^k, \mathcal{B}^k)\}$ in $\mathcal{L}_{\delta\sigma}$ converges to $(\mathcal{A}, \mathcal{B}) \in \mathcal{L}$. Fix $x \in [-r_1, r_1]^n$. Then, it is easy to show that

$$\forall i = 1, 2, \ldots, m, \quad \int_{-r_2}^{r_2} \mu_{B_i}(y)dy = \lim_{k \to \infty} \int_{-r_2}^{r_2} \mu_{B_i^k}(y)dy \geq \delta$$

and

$$\sum_{i=1}^m \int_{-r_2}^{r_2} \beta_{\mathcal{A}_i B_i}(x, y)dy = \lim_{k \to 0} \sum_{i=1}^m \int_{-r_2}^{r_2} \beta_{\mathcal{A}_i^k B_i^k}(x, y)dy \geq \sigma.$$

And this implies $(\mathcal{A}, \mathcal{B}) \in \mathcal{L}_{\delta\sigma}$. Therefore, $\mathcal{L}_{\delta\sigma}$ is a closed subset of $\mathcal{L}$, and hence it is compact metrizable.

4   Continuity as Functional

In this section, the continuity of approximate reasoning as functional on the set of membership functions $\mathcal{L}_{\delta\sigma}$ is shown.

4.1   Center of Sums Method

Assume that a sequence $(\mathcal{A}^k, \mathcal{B}^k) \subset \mathcal{L}_{\delta\sigma}$ converges to $(\mathcal{A}, \mathcal{B})$ for the product topology. It is already shown that the calculations in the procedure 1 and 2 are continuous even if minimum operation or product operation [9]. That is, for each $i = 1, 2, \ldots, m, j = 1, 2, \ldots, n$,

$$\mu_{A_{ij}}^k \to \mu_{A_{ij}} \text{ and } \mu_{B_i}^k \to \mu_{B_i} \text{ for } k \to \infty$$

implies that

$$\|\alpha_{\mathcal{A}_i^k} - \alpha_{\mathcal{A}_i}\|_\infty = \sup_{x \in [-r_1, r_1]^n} |\alpha_{\mathcal{A}_i^k}(x) - \alpha_{\mathcal{A}_i}(x)| \to 0$$

and

$$\|\beta_{\mathcal{A}_i^k B_i^k} - \beta_{\mathcal{A}_i B_i}\| \to 0.$$

Moreover, the defuzzified value $\bar{y}_i$ by the center of gravity method (centroid method) is continuous on $\mathcal{L}_{\delta\sigma}$ [5]. It follows from routine calculation that

$$|CoS_{\mathcal{A}^k\mathcal{B}^k}(x) - CoS_{\mathcal{AB}}(x)| \leq \frac{4r_2^2}{m\delta^2}\left\{\sum_{i=1}^{m}|\bar{y}_i^k - \bar{y}_i|\right.$$

$$\left.\sum_{i=1}^{m}\left|\int_{-r_2}^{r_2}\beta_{\mathcal{A}_i^k\mathcal{B}_i^k}(x,y)dy - \int_{-r_2}^{r_2}\beta_{\mathcal{A}_i\mathcal{B}_i}(x,y)dy\right|\right\}$$

by noting that $\sum_{i=1}^{m}\int_{-r_2}^{r_2}\beta_{\mathcal{A}_i^k\mathcal{B}_i^k}(x,y)dy \geq m\delta$. Hence, $(\mathcal{A}^k, \mathcal{B}^k) \to (\mathcal{A}, \mathcal{B})$ implies that

$$\sup_{x\in[-r_1,r_1]^n}|CoS_{\mathcal{A}^k\mathcal{B}^k}(x) - CoS_{\mathcal{AB}}(x)| \to 0$$

for $k \to \infty$. Therefore the continuity of the center of sums defuzzification method on $\mathcal{L}_{\delta\sigma}$ is obtained.

4.2   Height Method

In the same way as the area method, the following inequality is obtained:

$$|HM_{\mathcal{A}^k\mathcal{B}^k}(x) - HM_{\mathcal{AB}}(x)|$$

$$= \left|\frac{\sum_{i=1}^{m}\bar{y}_i^k\alpha_{\mathcal{A}_i^k}(x)}{\sum_{i=1}^{m}\alpha_{\mathcal{A}_i^k}(x)} - \frac{\sum_{i=1}^{m}\bar{y}_i\alpha_{\mathcal{A}_i}(x)}{\sum_{i=1}^{m}\alpha_{\mathcal{A}_i}(x)}\right|$$

$$\leq \frac{m}{\sigma^2}\sum_{i=1}^{m}|\bar{y}_i^k - \bar{y}_i| + \frac{2mr_2}{\sigma^2}\sum_{i=1}^{m}\|\alpha_{\mathcal{A}_i^k} - \alpha_{\mathcal{A}_i}\|_{\infty}.$$

It is easy to lead that the functional HM is continuous on $\mathcal{L}_{\delta\sigma}$ from above inequality.

5   Application to Optimization

In this section, using an idea and framework mentioned in the previous section, the existence of optimal control based on fuzzy rules in the admissible fuzzy controller will be established. The performance index of this control system (1) for the feedback law ρ in the previous section is evaluated with the following integral performance function:

$$J = \int_{B_r}\int_0^T w(x(t,\zeta,\rho),\rho(x(t,\zeta,\rho)))dtd\zeta, \tag{6}$$

where $w : \mathbb{R}^n \times \mathbb{R} \to \mathbb{R}$ is a positive continuous function.

The following theorem guarantee the existence of an admissible fuzzy controller $(\mathcal{A}, \mathcal{B})$ which minimizes the previous function (6). In the following, approximate reasoning $CoS_{\mathcal{AB}}$ and $HM_{\mathcal{AB}}$ are written as $\rho_{\mathcal{AB}}$ because the difference between them does not influence the theorem.

Theorem 1. *The mapping*

$$(\mathcal{A}, \mathcal{B}) \mapsto \int_{B_r} \int_0^T w(x(t, \zeta, \rho_{AB}), \rho_{AB}(x(t, \zeta, \rho_{AB}))) dt d\zeta$$

has a minimum value on the compact metric space $\mathcal{L}_{\delta\sigma}$ defined by (5).

Proof. It is sufficient to prove the continuity of performance function J on $\mathcal{L}_{\delta\sigma}$. Assume that $(\mathcal{A}^k, \mathcal{B}^k) \to (\mathcal{A}, \mathcal{B})$ in $\mathcal{L}_{\delta\sigma}$ and fix $(t, \zeta) \in [0, T] \times B_r$. Then it follows from the section 4 that

$$\lim_{k \to \infty} \sup_{x \in [-r_1, r_1]^n} |\rho_{A^k B^k}(x) - \rho_{AB}(x)| = 0. \tag{7}$$

Hence, by (3) of proposition 1, we have

$$\lim_{k \to \infty} \|x(t, \zeta, \rho_{A^k B^k}) - x(t, \zeta, \rho_{AB})\| = 0. \tag{8}$$

Further, it follows from (7), (8) and (2) of proposition 1 that

$$\lim_{k \to \infty} \rho_{A^k B^k}(x(t, \zeta, \rho_{A^k B^k})) = \rho_{AB}(x(t, \zeta, \rho_{AB})). \tag{9}$$

Noting that $w : \mathbb{R}^n \times \mathbb{R} \to \mathbb{R}$ is positive and continuous, it follows from (8), (9) and the Lebesgue's dominated convergence theorem [12] that the mapping is continuous on the compact metric space $\mathcal{L}_{\delta\sigma}$. Thus it has a minimum (maximum) value on $\mathcal{L}_{\delta\sigma}$, and the proof is complete.

6   Conclusion

This study presents the analysis of fuzzy approximate reasoning using the center of sums defuzzification method and the height method. Since the feedback calculation constructed by fuzzy inference is also Lipschitz continuous for premise variable in IF-THEN rules (state variable), the state equation has a unique solution. The details are omitted on account of limited space. Continuity on the set of membership functions guarantees the existence of a rule set (IF-THEN rules), in other words a tuple of membership functions, which minimize the integral cost function with compactness of the set of membership functions. The compact membership function set $\mathcal{L}_{\delta\sigma}$ in this paper includes the class of triangular, trapezoidal, and bell-shaped functions used in an application generally. The compactness of the family of sets of membership functions will be useful tool of successive approximation of the solution.

A human operator's control knowledge to decision making system has various complexities. Although one of them is fuzziness of its linguistic expression, fuzzy logic control can communicate it to the system. On the other, these properties will build the framework of optimization for the decision making system from the observation of real phenomena. In knowledge management known for data mining and classification field, experts judged and extracted the outliers previously. Its automation with a human linguistic knowledge is not so easy, and includes the problems of ambiguousness (fuzziness). Therefore the inference system in this study is useful in the capacity of a subsystem to find the outliers from large-scale data.

References

1. Mamdani, E.H.: Application of fuzzy algorithms for control of simple dynamic plant. Proc. IEE 121(12), 1585–1588 (1974)
2. Mizumoto, M.: Improvement of fuzzy control (IV) - Case by product-sum-gravity method. In: Proc. 6th Fuzzy System Symposium, pp. 9–13 (1990)
3. Nakamori, Y., Ryoke, M.: Identification of fuzzy prediction models through hyperellipsoidal clustering. IEEE Transactions on Systems, Man and Cybernetics SMC-24(7), 1153–1173 (1994)
4. Tanaka, K., Sugeno, M.: Stability Analysis of Fuzzy Systems and Construction Procedure for Lyapunov Functions. Transactions of the Japan Society of Mechanical Engineers (C) 58(550), 1766–1772 (1992)
5. Mitsuishi, T., Wasaki, K., Kawabe, J., Kawamoto, N.P., Shidama, Y.: Fuzzy optimal control in L^2 space. In: Proc. 7th IFAC Symposium Artificial Intelligence in Real-Time Control, pp. 173–177 (1998)
6. Mitsuishi, T., Kawabe, J., Wasaki, K., Shidama, Y.: Optimization of Fuzzy Feedback Control Determined by Product-Sum-Gravity Method. Journal of Nonlinear and Convex Analysis 1(2), 201–211 (2000)
7. Mitsuishi, T., Shidama, Y.: Continuity of Fuzzy Approximate Reasoning and Its Application to Optimization. In: Orgun, M.A., Thornton, J. (eds.) AI 2007. LNCS (LNAI), vol. 4830, pp. 529–538. Springer, Heidelberg (2007)
8. Mitsuishi, T., Shidama, Y.: gCompactness of Family of Fuzzy Sets in L^2 Space with Application to Optimal Control. IEICE Transactions on Fundamentals of Electronics, Communications and Computer Sciences E92-A(4), 952–957 (2009)
9. Mitsuishi, T.: Continuity of Approximate Reasoning Using Center of Sums Defuzzification Method. In: Proc. of IEEE 35th International Convention of Information Communication Technology, Electronics and Microelectronics MIPRO 2012, pp. 1172–1175 (2012)
10. Miller, R.K., Michel, A.N.: Ordinary Differential Equations. Academic Press, New York (1982)
11. Riesz, F., Sz.-Nagy, B.: Functional Analysis. Dover Publications, New York (1990)
12. Dunford, N., Schwartz, J.T.: Linear Operators Part I: General Theory. John Wiley & Sons, New York (1988)
13. Ross, T.J.: Fuzzy Logic With Engineering Application. John Wiley and Sons Ltd., UK (2010)

A On-Line News Documents Clustering Method

Hui Zhang[1], Guo-hui Li[1], and Xin-wen Xu[2]

[1] Department of Engineering, School of Information System and Management,
National University of Defense Technology, Changsha, 410073, China
{zhanghui,guohli}@nudt.edu.cn
[2] College of Basic Education for officers, National University of Defense Technology,
Changsha 410073, China
xinwen_xu@126.com

Abstract. To improve the efficiency and accuracy of on-line news event detection (ONED) method, we select the words that their term frequency (TF) is greater than a threshold to create the vector space model of the news document, and propose a two-stage clustering method for ONED. This method divides the detection process into two stages. In the first stage, the similar documents collected in a certain period of time are clustered into micro-clusters. In the second stage, the micro-clusters are compared with previous event clusters. The experimental results show that the proposed method has fewer computation load, higher computing rate, and less loss of accuracy.

Keywords: On-line news event detection, Vector space model, Two-stage clustering method.

1 Introduction

According to the recent information technology development such as the Internet and electronic documents, a huge number of on-line news documents are delivered over the network every day. they reports thousands of news events happening anywhere in the world day after day. It is very difficult for users to find out required news event reports from such huge documents storage, On-line news event detection (ONED) to detect new event reports in real time from delivered news documents has become an important research area. ONED is the task of identifying the first document discussing a new event in real time, which has not already seen on the stream of news documents. Therefore, ONED systems are very useful for people who need to detect novel information from real-time news stream.

The core problem of ONED is to identify whether two documents are on the same event. Most of state-of-the-art ONED systems compare a new document to all the previous received documents [1, 2], but they are not acceptable in real applications because of the large amount of computation required in the ONED process. In other systems, a document is compared to all the previous news clusters instead of documents. It has been proved that this manner can reduce comparing times significantly, nevertheless, it is less accurate [3, 4]. So, there still exist core problems to be

R. Huang et al. (Eds.): AMT 2012, LNCS 7669, pp. 82–92, 2012.
© Springer-Verlag Berlin Heidelberg 2012

investigated in ONED: how to speed up the detection procedure while no loss the detection accuracy?

Driven by the problem, we propose that: 1) by setting the threshold of TF, we discard the terms that their TF is less than the threshold. It can reduce the dimension of vector space for news document model, and then reduce the computation load of vectors calculation; 2) in order to reduce redundancy matching of similar reports, we proposed a Two-Stage Clustering Method (TSCM) which splits detection process into two stages. In the first stage, the similar news documents collected in a certain period are clustered into micro-clusters, micro-cluster is the set of documents which related to the same news event. In the second stage, every micro-cluster is compared to all previous news event clusters. We implemented a prototype of our system. The evaluation on the TDT4 benchmark shows that the proposed system has fewer computation load, higher computing rate, and less loss of feature information, which significantly improves the efficiency of news event detection without losing accuracy and demonstrates the feasibility of the method.

2   Related Works

In respect of detecting method, Allan [1] and Papka [2] proposed Single-Pass clustering. When a new document was encountered, it was compared with all the previous news reports. If the document did not trigger any reports by exceeding a threshold, it was marked as a new event. Besides, Yang [3] and Lam [4] build up previous news documents clusters, each of which corresponds to a event. In this manner comparisons happen between encountered document and previous clusters. These two methods have a drawbacks that the computational complexity is relatively high, especially the former.

In respect of the representation of document model, most work focus on proposing better methods on comparison of documents and document representation. Brants [5] extended a basic incremental TF-IDF model to include source specific model, good improvements on TDT benchmarks were shown. Stokes [6] utilized a combination of evidence from two distinct representations of a document's content. One of the representations was the usual free text vector, the other made use of lexical chains (created using WordNet) to build another term vector. Then the two representations are combined in a linear fashion. A marginal increase in effectiveness was achieved when the combined representation was used.

In respect of utilization of Named Entity, Giridhar [7] divided document representation into two parts: named entities and non-named entities. Yang [8] gave location named entities four times weight than other terms and named entities. DOREMI [9, 10] research group combined semantic similarities of person names, location names and time together with textual similarity. Zhang Kuo [11], improves Giridhar's work, and discriminated different types of Named Entity and association of each news event class in detail. Fu Yang [12] proposed a rapid detection method which based on the entity recognition techniques. These researches make full use of the feature of named entity, adopt different tactics and improved the performance of the system, but the miss rate and mistake rate of the system were still relatively high.

3   Improvement for Basic Model

In this section, we present the basic ONED model which is similar to what most current systems apply. Then, we propose our new model by extending the basic model. ONED system use news story stream as input, in which stories are strictly time-ordered. Only previously received stories are available when dealing with current story.

We use incremental TF-IDF model for term weight calculation. In a incremental TF-IDF model, the TF in a document is weighted by the inverse document frequency (IDF), document frequency ($df(w)$) is not static but change in real time. The accuracy of ONED will be reduced if ignoring these changes. Therefore, using the incremental TF-IDF ideology which was presented by Brants [5] establish the news story model. The following formulas are used to calculate the weight of feature.

$$weight_t(d, w) = \frac{\log[f(d, w) + 1] \cdot \log \dfrac{(N_t + 1)}{[df_t(w) + 0.5]}}{Z_t(d)} \tag{1}$$

$$Z_t(d) = \sum_w \sqrt{\left\{ \log[f(d, w) + 1] \cdot \log \frac{(N_t + 1)}{[df_t(w) + 0.5]} \right\}^2} \tag{2}$$

$$df_t(w) = df_{t-1}(w) + df_{C_t}(w) \tag{3}$$

Where $f(d, w)$ is the frequency of a term w in a document d, $df_t(w)$ is the number of documents in the collection that contain word w. N_t is the total number of documents at time t, $Z_t(d)$ is a normalization value, $df_{C_t}(w)$ denote the document frequencies in the newly added set of documents C_t.

Increment TF-IDF is based on $f(d, w)$ and $df_t(w)$ to calculate terms weights. The basic idea is that the importance increases proportionally to the number of times a term w appears in the document but is offset by the frequency of the term w in the corpus. Therefore, very low frequency terms w tend to be uninformative, it is not important to distinguish whether compared reports belong to the same event, and can be filtered. We set a threshold θ_d, and only term w having

$$f(d, w) \geq \theta_d \tag{4}$$

is used to establish news documents model, unless not be considered.

By analyzing the feature of network news document from Table 1, we can find that the length of document (DL) is relatively short and the average number of TF is less than 2, where Dataset1 is TDT4 (in Chinese) [15], Dataset2 is getted from reference [18], Dataset3 and Dataset4 are come from Sougou labs [19], so this paper set $\theta_d = 2$. If we set θ_d to 3 or more, the feature space of news document will be too sparse or even zero vector, the feature information is excessively lost, then the accuracy of detection is reduced. By filtering part of terms, the dimension of feature vector space is greatly reduced, and then the similarity calculation of feature vectors is decreased, and the system efficiency is increased obviously.

Table 1. The statistical feature of news documents

	Dataset1	Dataset2	Dataset3	Dataset4
The average for TF	1.549	1.908	1.638	1.747
The average for DL	204.1	427.9	222.4	405.2

4   Two-Stage Clustering Method (TSCM)

4.1   Characteristic Analysis of News Corpus

All kinds of reports from different news sites are delivered over the network immediately once the news events has occurred, so that the similar reports about the same event are very much in a certain period of time. Using the existing methods [1-4] to perform ONED, it is required to measure all similar report to existing reports or event clusters one by one, which cause waste of system computing resources and decrease program efficiency of system. Because each of similar reports is compared to previous document or event clusters, which is similar to repeat matching, so the efficiency of detection is reduced. If the similar reports are clustered into micro-clusters in advance, and then using micro-clusters to match with previous event clusters, the resource consumption brought by the repeat matching will be reduced greatly. Therefore, this paper proposes TSCM to avoid the problem.

4.2   The Model of Two-Stage Clustering Method (TSCM)

The process model of TSCM includes three parts: preprocessing, the generation of micro-clusters, and then the matching of the micro-clusters with the previous event clusters. The process model shows in figure 1.

Preprocessing: it includes word segmentation, stop words filtering and terms frequency statistics. By using of ICTCLAS [13], the word segmentation for news reports was performed, and then the stop words were filtered out by using the stop words list provided in the reference [14], finally, the set of feature for every document was obtained, and the statistics of Terms Frequency are worked out.

The first stage clustering: in a certain period of time t (the setting of t should ensure the requirements for online), the collected news reports were initially clustered into micro-cluster by Single-Pass algorithm [17]. The similar reports about the same event were collected into a micro-cluster. Each of micro-clusters represents by the centroid vector. Initial clustering is characterized by a relatively small number of reports. The amount of information in each report is low. By controlling the threshold λ, the different news cluster can be distinguished basically.

Single-Pass algorithm:

Step1: The documents are processed serially, The representation for the first document becomes the cluster representative of the first cluster;

Step2: Each subsequent document is matched against all cluster representatives existing at its processing time;

Step3: A given document is assigned to one cluster (or more if overlap is allowed) according to some similarity measure;

Step4: When a document is assigned to a cluster, the representative for that cluster is recomputed;

Step5: If a document fails a certain similarity test it becomes the cluster representative of a new cluster.

The second stage clustering: each of micro-cluster formed in the first phase was matched with the detected event clusters one by one in chronological order of reported time of the first report for micro-cluster, by which there is a judgment whether micro-cluster discussed a new event. This stage is characterized by micro-cluster representing the news cluster and matching with previous event clusters. The TSCM method greatly reduced the number of comparing times.

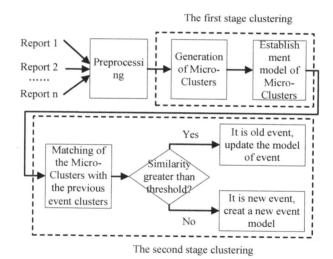

Fig. 1. The process model of two-stage clustering method for ONED

4.3   The Programming of Two-Stage Clustering Method (TSCM)

We assume that $D_t=\{d_1,d_2,...,d_n\}$ represent news corpus collected in time interval t in chronological order of reported time, λ and θ are the threshold in the first and second stage clustering respectively, MC indicates micro-cluster, Cen indicates the centroid of micro-cluster, $MC_s=\{(MC_1,\ Cen_1),(MC_2,\ Cen_2),\ ...,\ (MC_m,\ Cen_m)\}$

indicates the set of micro-clusters, $MaxSim_1$ and $MaxSim_2$ are the maximum similarity for first and second stage respectively, $E=\{e_1,e_2, ...,e_k\}$ indicates the set of event clusters been detected, $D_{New}=\{d_1,d_2,...,d_j\}$ is the set of documents which are the first report of new events. The following is the process design of TSCM.

Step 1: Input news reports set $D_t=\{d_1,d_2,...,d_n\}$, preprocess every news reports collected in D_t, establish the model of news report;

Step 2: Add d_1 to MC_1, feature vector of d_1 as centroid of MC_1, add MC_1 to MC_s;

Step 3: Read next report $d_i(i{\neq}1)$ in chronological order from D_t, calculate the similarities between d_i and every MC in the MC_s, get the $MaxSim_1$ which is the maximum value in all similarities, and label $MC_{MaxSim1}$;

Step 4: $MaxSim_1$ compared with threshold λ, If $MaxSim_1$ do not exceed threshold λ, then the document d_i triggers a new micro-cluster, then create new micro-cluster MC_{New}, add d_i to MC_{New}, feature vector of d_i as the centroid of MC_{New}, finally, add MC_{New} to MC_s; otherwise, the document d_i belong to an old micro-cluster, then add d_i to $MC_{MaxSim1}$, and update the centroid of $MC_{MaxSim1}$;

Step 5: Repeat steps 3 and 4 until all reports in D_t have been compared, output the set of micro-clusters MC_s to step 6;

Step 6: Select a micro-cluster MC_i in chronological order from MC_s, calculate the similarities between MC_i and every event clusters which are from E, get the $MaxSim_2$ which is the maximum value in all similarities, and label $e_{MaxSim2}$;

Step 7: $MaxSim_2$ compared with threshold θ, If $MaxSim_2$ do not exceed threshold θ, then the micro-cluster MC_i triggers a new event, then create new event e_{New}, add MC_i to e_{New} and add the first document of MC_i to D_{New}, centroid vector of MC_i as the centroid of e_{New}, finally, add e_{New} to E; otherwise, the MC_i belong to an old event cluster, then add MC_i to $e_{MaxSim2}$, and update the centroid of $e_{MaxSim2}$;

Step 8: Repeat steps 6 and 7 untill all micro-clusters in MC_s have been compared in turn, output the set of first reports D_{New}.

Step 9: Detection progress finished for this time interval t, goto the step 1 to start next time interval $t+1$.

5   Experimental Results and Evaluation

5.1   Datasets and Evaluation Metric

We used LDC [15] datasets TDT4 (in Chinese) for our experiments. TDT4 have annotated 100 events, only 70 events contain Chinese news reports in datasets. It has 1257 Chinses stories about 70 events that are used for our evaluation, every story only associates with one event. Owing to difference of the continued time and degree of importance, the number of stories is different in different events, ranging from 1 to 140.

We adopt C# as the programming language, and experiments are performed on a PC with Core-i5-2.5GHz CPU and 4GB of main memory. Software environment is Win7 operating system with 64 bit.

TDT Evaluation uses a cost function C_{Det} that combines the probabilities of missing a new story and a false alarm [16]:

$$C_{Det} = C_{Miss} P_{Miss} P_{target} + C_{False} P_{False} P_{non-target} \qquad (5)$$

where C_{Miss} means the cost of missing a new story, P_{Miss} means the probability of missing a new story, and P_{Target} means the probability of seeing a new story in the data; C_{False} means the cost of a false alarm, P_{False} means the probability of a false alarm, and $P_{non-target}$ means the probability of seeing an old story (Referring TDT evaluation [18], $C_{Miss}=1$, $C_{False}=0.1$, $P_{Target}=0.02$, $P_{non-target}=1-P_{Target}$). The cost C_{Det} is normalized such that a perfect system scores 0 and a trivial system, which is the better one of mark all stories as new or old, scores 1:

$$Norm(C_{Det}) = \frac{C_{Det}}{\min(C_{Miss} P_{Miss}, C_{False} P_{False})} \qquad (6)$$

ONED system gives two outputs for each story. The first part is "yes" or "no" indicating whether the story triggers a new event or not. The second part is a score indicating confidence of the first decision. Confidence scores can be used to plot DET curve, i.e., curves that plot false alarm vs. miss probabilities. Minimum normalized cost can be determined if optimal threshold on the score were chosen.

5.2 Experiment Setting

To test the method proposed in this paper, we implemented and tested three systems representing three process design of ONED respectively:

System1 (S1) [1, 2]: System compares a new document to all the previous received documents.

System2 (S2) [3, 4]: System compares a new document to all the previous news document clusters (each cluster corresponds to an old event) instead of documents.

System3 (S3): System divides the ONED process into two phases. In first phase, the similar reports collected in a certain period of time are clustered into microcluster. In second phase, the micro-cluster is compared with previous event clusters.

In the three systems, there are two representation methods for document to establish the model of news stories respectively:

Vector1 (V1): the document are represented with the words which have filtered out stop words.

Vector2 (V2): the document are represented with the words which have filtered out stop words, and their term frequency satisfy formula (4).

5.3 Estimation of Threshold λ for the First Stage Clustering

In the first stage of the TSCM, the set of λ is critical for efficiency and accuracy of detection. The stand or fall of λ directly related to clustering effect in the second stage. If λ is too high, the reports which ought to belong to a micro-cluster are separated, so that the computational cost will increase in the second stage. If λ is too low, the reports which are not supposed to belong to same micro-cluster form into same micro-cluster, which will effect the clustering quality in the second stage.

In this paper, λ is estimated by observing the trends of $Norm(C_{Det})$ curve, The experimental results show in Figure 2. The horizontal axis represents θ which is threshold in the second stage. The longitudinal axis represents $Norm(C_{Det})$. The curve in a two-dimensional coordinate system represents the trends of $Norm(C_{Det})$ according to θ, when λ is set to be a specific value, the difference between the curves represents the trends of $Norm(C_{Det})$ according to λ. The stand or fall of λ is estimated by that: the curve in the figure is closer to the horizontal axis, the detection performance of system is better, it means that the cost for detection error is small. Based on this criterion, λ is set to 0.08 in the experiment.

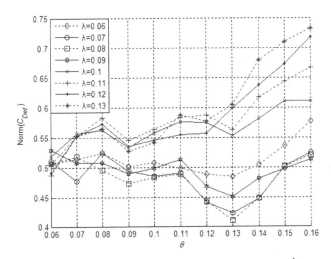

Fig. 2. *Norm (CDet)* curves with different threshold λ

5.4   Efficiency Evaluation and Analysis

System detection efficiency is mainly measured through the time which is consumed by system execution. Efficiency comparison among six groups is shown in table 1. The two metrics about time in the comparison are the system average time consumption and the maximum time consumption where θ takes the different value.

It was obvious that the efficiency of V2 was higher than that of V1. The average efficiency of V2 in S2 and S3 system is as much as 4 times higher in contrast to V1. The efficiency of S1 system is much higher, whose average efficiency increase by nearly 7 times. Theoretically, the efficiency difference between the two models is obviously. By setting the term frequency threshold and filtering out the features with low frequency, the characteristic number of V2 was less than that of V1. In the process of vector similarity computation, the similarity calculation of V2 model will be significantly less than that of V1.

By comparison of different systems, It was obvious that the efficiency of S2 is higher than that of S1, which has been proved in reference [3]. Meanwhile, the efficiency of S3 is higher than that of S2. The reason is that S3 reduced the comparing times in the second stage by using micro-clusters to compare with previous event cluster instead of single document, and then improved the system efficiency. This advantage will be more obvious when there are much more preexisting event clusters and similar reports which collected in a certain period of time.

Table 2. Efficiency comparison with different system and vector

	Average time (s)	Max time (s)
System 1 Vector 1	381	640
System 1 Vector 2	55.4	65.7
System 2 Vector 1	13.7	19.4
System 2 Vector 2	3.39	4.21
System 3 Vector 1	8.69	12.3
System 3 Vector 2	2.03	2.76

5.5   Accuracy Evaluation and Analysis

The accuracy of system is described by plotting DET curves, which is based on the detection results of the experiment. The DET curves were shown in Figure 3 and Figure 4. The figure 3 shows performance of systems in which reports are represented by V1. The figure 4 shows the performance of system in which reports are represented by V2.

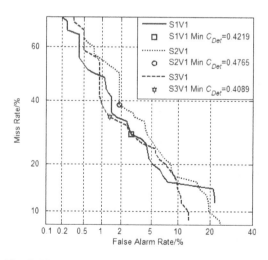

Fig. 3. DET curves for three system represented by V1

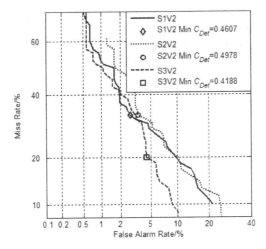

Fig. 4. DET curves for three system represented by V2

In Figure 3 and Figure 4, It shows that the curves of S2 are far away the lower left corner of the coordinate system by comparison with the curves of S1 and S3, so the accuracy of S2 is less than S1 and S3, but it is not very obvious for S1 and S3. By comparison of *Min Norm*(C_{Det}), S1 is better than S2, the S3 is better than S1 in some sort. Meanwhile, The *Min Norm*(C_{Det}) of system represented by V2 is less than system represented by V1 in some sort, so the system using V2 is less loss of accuracy.

6   Conclusion

By setting the threshold of term frequency to filtering out some words which have low term frequency, we select the remained words to build vector space model of news reports, and take advantage of characteristics that the different news site deliver similar reports on the same event in period of time, propose a two-stage clustering method for on-line news event detection. The experimental results show that this method improves efficiency obviously and less loss of accuracy for ONED.

We did not consider the situation that the average of TF equal to 1, and did not make the thorough theoretical analysis for efficiency improving. They will be our work in the future.

Acknowledgements. Thanks to Ting Wang for offering TDT4 dataset, who is a professor in College of Computer, National University of Defense Technology, to Li Jia and Bo-liang Sun for proof reading, and to three anonymous reviewers for their comments.

References

1. Allan, J., Papka, R., Lavrenko, V.: On-line news event detection and tracking. In: Proceedings of the 21st Annual International ACM SIGIR Conference on Research and Development in Information Retrieval, pp. 37–45. ACM Press, New York (1998)

2. Papka, R., Allan, J.: On-line new event detection using single pass clustering TITLE2. Technical Report (1998)
3. Yang, Y., Pierce, T., Carbonell, J.: A study on Retrospective and On-Line Event detection. In: Proceedings of the 21st Annual International ACM SIGIR Conference on Research and Development in Information Retrieval, pp. 28–36. ACM Press, New York (1998)
4. Lam, W., Meng, H., Wong, K., Yen, J.: Using contextual analysis for news event detection. Int'l Journal on Intelligent Systems 16(4), 525–546 (2001)
5. Brants, T., Chen, F., Farahat, A.: A system for new event detection. In: Proceedings of the 26th Annual International ACM SIGIR Conference on Research and Development in Information Retrieval, pp. 330–337. ACM Press, New York (2003)
6. Nieola, S., Joe, C.: Combining semantic and syntactic document classifiers to improve first story detection. In: Proceedings of the 24th Annual International ACM SIGIR Conference on Research and Development in Information Retrieval, pp. 424–425. ACM Press, New York (2001)
7. Kumaran, G., Allan, J.: Text classification and named entities for new event detection. In: Proceedings of the 27th Annual International ACM SIGIR Conference on Research and Development in Information Retrieval, pp. 297–304. ACM Press, New York (2004)
8. Yang, Y., Zhang, J., Carbonell, J., Jin, C.: Topic conditioned Novelty Detection. In: Proceedings of the 8th ACM SIGKDD International Conference, pp. 688–693. ACM Press, New York (2002)
9. Juha, M., Helena, A.M., Marko, S.: Applying Semantic Classes in Event Detection and Tracking. In: Proceedings of International Conference on Natural Language Processing (ICON 2002), pp. 175–183 (2002)
10. Juha, M., Helena, A.M., Marko, S.: Simple Semantics in Topic Detection and Tracking. Information Retrieval 7(3-4), 347–368 (2004)
11. Kuo, Z., Juan-Zi, L., Gang, W.: A new event detection model based on term reweighting. Journal of Software 19(4), 817–828 (2008) (in Chinese)
12. Yan, F., Ming-quan, Z., Xue-song, W.: On-Line Event Detection from Web News Stream. Journal of Software 21(suppl.), 363–372 (2010) (in Chinese)
13. Hua-ping, Z., Qun, L.: Calculation of the Chinese lexical analysis system LCTCLAS. Institute of Computing. Chinese Academy of Sciences (2002) (in Chinese), http://sewm.pku.edu.cn/QA/reference/LCTCLAS/FreeICTCLAS/
14. Xiao-yan, Z.: Research on the Representation Model and Technologies of Link Detection and Tracking on News Topic. National University of Defense Technology (2010) (in Chinese)
15. The linguistic data consortium, http://www.ldc.upenn.edu/
16. NIST. The 2003 Topic Detection and Tracking Task Definition and Evaluation Plan. National Institute of Standards and Technology (NIST) (2003), http://www.itl.nist.gov/iaui/894.01/tests/tdt/tdt2003/evalplan.html
17. Van Rijsbergen, C.J.: Information Retrieval, 2nd edn. Butterworths, London (1979)
18. Rong-lu, L.: Chinese Text Classification Corpus, http://www.nlp.org.cn/docs/docredirect.php?doc_id=281
19. SougouCA corpus, http://www.sogou.com/labs/dl/ca.html

Distributed Protocols for Multi-Agent Coalition Formation: A Negotiation Perspective

Predrag T. Tošić and Carlos Ordonez

University of Houston, 4800 Calhoun Road, Houston, Texas, USA

Abstract. We investigate collaborative multi-agent systems and how they can use simple, scalable negotiation protocols to coordinate in a fully decentralized manner. Specifically, we study multi-agent *distributed coalition formation*. We summarize our past and ongoing research on collaborative coalition formation and describe our original distributed coalition formation algorithm. The present paper focuses on negotiation-based view of coalition formation in collaborative *Multi-Agent Systems* (MAS). While negotiation protocols have been extensively studied in the context of competitive, self-interested agents, we argue that negotiation-based approach may be potentially very useful in the context of collaborative agents, as well – as long as those agents, due to limitations of their sensing and communication abilities, have different views of and preferences over the states of the world. In particular, we show that our coalition formation algorithm has several important, highly desirable properties when viewed as a negotiation protocol.

Keywords: collaborative multi-agent systems, multi-agent coordination, negotiation protocols, coalition formation, distributed consensus.

1 Introduction and Motivation

We study autonomous artificial agents such as softbots, robots, unmanned vehicles and smart sensors, that are fully autonomous and capable of interacting and communicating with each other. We are particularly interested in mechanisms that enable relatively large ensembles (hundreds or even thousands) of such agents to coordinate with each other in a fully decentralized manner, and collaborate in order to jointly complete various tasks. This kind of agents are often referred to as *distributed problem solvers* (DPS) in the Distributed AI literature [2,26,27]. Coalition formation in DPS multi-agent domains is an important coordination and collaboration problem that has been extensively studied by the Multi-Agent Systems (MAS) community [1,2,11,12,13,14]. There are many important collaborative MAS applications where autonomous agents need to form groups, teams or coalitions. The agents may need to form coalitions in order to share resources, jointly complete tasks that exceed the abilities of individual agents, or improve some system-wide performance metric such as the speed of task completion [9,15].

Among various interesting problems in distributed coordination and control of such agent ensembles, we have been extensively studying the problem

R. Huang et al. (Eds.): AMT 2012, LNCS 7669, pp. 93–102, 2012.

of genuinely autonomous, dynamic and fully decentralized coalition formation [19,20,21,22,23]. So far, we have approached coalition formation from two perspectives common in the MAS literature: one is the conceptual multi-agent coordination view of coalition formation, and the other is the practical distributed graph algorithm design. We have integrated these two aspects and proposed a novel, fully distributed, scalable coalition formation graph algorithm named *Maximal Clique based Distributed Coalition Formation* (MCDCF for short) [19,20]. In the present paper, we build on the top of our earlier work on designing, analyzing, simulating and optimizing the MCDCF algorithm, but this time we approach the coalition formation in general, and the MCDCF coalition formation protocol in particular, from a fresh, negotiation-focused perspective.

The rest of this paper is organized as follows. In Section 2, we summarize our MCDCF algorithm for coalition formation among collaborative agents, emphasizing the graph algorithmic and coordination aspects of that algorithm. In Section 3, we first motivate the negotiation view of coalition formation, and make the case for usefulness of such approach even when applied to DPS agents that are *altruistic* "by definition" (as opposed to *self-interested*), as long as those agents are *locally constrained* in terms of their perceptions of and actions in the environment [18,21]. We then cast MCDCF in negotiation terms, and show that it satisfies several highly desirable properties expected from good general-purpose negotiation protocols. Section 4 summarizes our main contributions and outlines the future work.

2   The MCDCF Protocol and Its Main Properties

We address the problem of distributed coalition formation in the following setting. We assume a multi-agent, multi-task dynamic and partially observable environment. The tasks are assumed mutually independent of each other. Each task is of a certain value to an agent. Different tasks may have different values or utilities associated with them; moreover, the utility of a particular task may be differently perceived by different agents. This problem setting is particularly appropriate for many applications involving team robotics and autonomous unmanned vehicles; see, e.g., [5,20,25]. Our agents are assumed to be strictly collaborative, not selfish. The agents have certain capabilities that (may) enable them to service the tasks. Similarly, the tasks have certain resource or capability requirements, so that no agent or coalition of agents whose joint capabilities do not meet a particular task's resource requirements can serve that task [13,20,21]. Agents are assumed capable of communicating, negotiating and making agreements with each other [2,11,14]. Communication is accomplished via exchanging messages. This communication is not free: an agent has to spend time and effort in order to send and receive messages [21]. Our distributed maximal clique based coalition formation algorithm is centered at the idea that, in a *peer-to-peer* (in particular, *leaderless*) MAS, an agent would prefer to form a coalition with those agents that it can communicate with directly, and, moreover, where every member of such a potential coalition can communicate with any other member directly [19]. That is, the preferable coalitions are (maximal) cliques. However, finding a maximal clique in an arbitrary graph is **NP**-hard in the centralized setting [4]. This implies the computational hardness that, in general, each node in the graph faces when trying to determine the maximal clique(s) it belongs to. However, if the degree of a node is sufficiently small, then finding all maximal cliques this node belongs to can be expected to be feasible. If one *a priori* does not know if all the nodes in a given MAS interconnection topology are of small degrees, then one may need to impose additional constraints in order to ensure that the agents are not attempting to solve an intractable problem [20].

The MCDCF algorithm originally introduced in [19,20] is a distributed graph algorithm. The underlying graph captures the communication network topology among the agents, as follows. Each agent is a node in the graph. The necessary requirement for an edge between two nodes to exist is that the two nodes be able to directly communicate with one another. The group broadcast nature of the adopted communication model is primarily due to the application domains that drove the original MCDCF design, namely team robotics and autonomous unmanned vehicles, in particular micro-UAVs [18,21]. The basic idea behind MCDCF is to efficiently and in a fully decentralized manner partition this graph into (preferably, maximal) cliques of nodes. These maximal cliques would usually also need to satisfy some additional criteria in order to form temporary coalitions of desired quality. These coalitions are then maintained until they are no longer preferred by the agents. In an actual MAS application, the MCDCF algorithm may need to be invoked a number of times as a coordination subroutine [19,20].

Agents form coalitions as follows. Each agent (i) first learns of who are its neighbors, then (ii) determines the appropriate candidate coalitions, that the agent hopes are (preferably maximal) cliques that it belongs to, then (iii) evaluates the utility value of each such candidate coalition, measured in terms of the joint resources of all the potential coalition members, then (iv) chooses the most desirable candidate coalition, and, finally, (v) sends this choice to all its neighbors. This basic procedure is then repeated, together with all agents updating their knowledge of (a) what are the preferred coalitions of their neighbors, and (b) what coalitions have been already formed. In a nutshell, MCDCF is a graph algorithm executed in a fully decentralized manner by an ensemble of collaborative, but locally constrained, autonomous agents. Each agent maintains some simple data structures capturing what it knows about other, near-by agents. These data structures include neighborhood lists, candidate coalitions, the current coalition choice (i.e., the current proposal of a coalition), and some auxiliary flags via which the most important aspects of the previous rounds of the protocol are summarized [20,22]. As in any negotiation protocol, what an agent proposes to some nonempty subset of other agents, may or may not get accepted by other agents. In particular, MCDCF can be also viewed as a protocol that solves a variant of *distributed consensus* [6,26]: a coalition proposal becomes an actual, agreed upon coalition only if all members of the proposed coalition agree on that coalition, and moreover they do so at the same time – that is, during the same round of the MCDCF execution. More details on distributed algorithmic aspects on how MCDCF goes through its rounds until each agent has joined some coalition can be found in our prior work [19,20,22,23]. The *negotiation view* of MCDCF is further elaborated upon in the rest of the present paper.

We have rigorously established elsewhere that MCDCF is guaranteed to converge in a finite number of rounds, where an upper bound on that number of rounds can be determined as a function of the total number of agents, N, and the maximum neighborhood size in the underlying communication topology graph [20,21]. When it comes to defining *optimality* of a coalition structure, we have adopted "the bigger, the better" principle – but always under the constraint that each coalition has to be a clique, i.e., that all of its members need to be within a single hop from each other. More details on the graph-theoretic assumptions and the constraint optimization aspects of MCDCF can be found in [19,20,21,22].

3   The Negotiation View of Coalition Formation

Most of the existing literature on multi-agent negotiation assumes *self-interested agents* [6,26,27]. Such agents may still want, in certain situations, to cooperate with each other in various ways, including but not limited to forming teams or coalitions with each other. However, the motivation behind such cooperation is still selfish: so, in the context of coalition formation among self-interested agents, an agent will join a coalition with one or more other agents only if joining that coalition increases the agent's expected individual utility (or some other measure of that agent's individual payoff) [20,26]. Consequently, when it comes to negotiation protocols for coalition formation among self-interested agents, the protocol needs to specify how are the "spoils" of cooperation going to be distributed among the agents. Algorithmic game theory has provided the theoretical framework for systematically addressing these issues [6,26].

We observe, however, that a need for negotiation may arise among non-competing agents, as well – provided that those collaborative, non-selfish agents see the world differently. In particular, if different agents have different local views of the relevant aspects of their environments (such as tasks, resources, other agents, etc.), then it can be expected that these different agents will have different preferences on how would they like the world to be, and what should be the next step (action or sequence of actions) to bring about the most desirable state of the world [27]. These remarks would hold even in purely collaborative, *distributed problem solving* (DPS) settings where agents share a global objective or global utility function to be optimized, and do not have individual utilities per se [18,20]. In other words, the need to negotiate might be due simply to the fact that the agents see the world differently, not necessarily that they are self-interested in the usual sense from economics or game theory. This is of considerable practical importance, since in many collaborative MAS applications involving the real-world software, robotic or other types of autonomous agents, locally bounded sensing and communicating abilities, and hence "world-views", are a rule rather than an exception [21]. In such *locally constrained* MAS settings, it is to be expected that different agents will perceive the world differently, and hence have different preferences over the possible states of the world. Thus, negotiation may be a useful methodology to reach *distributed consensus* among strictly collaborative agents. Examples of distributed applications where negotiation may be the way to go about reaching distributed consensus range from aforementioned team robotics and unmanned vehicle applications [18,25] to decentralized resource sharing in wireless communication networks [8] to enforcing consistency in distributed databases or data warehouses [3].

Consider an application such as a large ensemble of unmanned aerial vehicles (UAVs) deployed on a multi-task mission and spread across a sizable geographic area [5,25]. Assume these UAVs are truly autonomous, and in particular not remotely controlled nor subject to centralized coordination. In such scenario, it is quite likely that different UAV agents will detect different tasks (for example, different regions or points of interest on which to perform surveillance), and hence strive to form different coalitions based on those tasks' estimated values and resource requirements [18]. Moreover, two different UAVs may perceive the same task differently – in terms of the expected utility from completing that task, estimated resource requirements for the task completion, etc. In such scenarios, even though all UAVs work on behalf of the same organization and are "team players" by design, they will still have different local preferences, and, consequently, may want to form different coalitions with their counterparts. The main lesson to be drawn from the general discussion, as well as the UAV example of a collaborative MAS where negotiation may be quite useful, is that sometimes *the differences in agents' perception and preferences*, and not necessarily their "selfishness", drive the need for negotiation. With that in mind, we turn to analyzing the MCDCF algorithm from a negotiation standpoint.

3.1   MCDCF from the Multi-Agent Negotiation Perspective

We outlined the generic problems of MAS coordination and cooperation, and then narrowed down our discussion to the more specific problem of distributed coalition formation. We then summarized our distributed graph algorithm for multi-agent coalition formation, viewed as a coordination subroutine. We subsequently motivated the usefulness and applicability of the negotiation approach to MAS distributed coalition formation – even when the agents aren't self-interested and can in general be expected to share their objectives. After discussing the problem of coalition formation from distributed graph algorithm design and MAS coordination viewpoints, we now discuss that algorithm from a negotiation perspective. We do so at two levels: (i) from the *rules of encounter*, that is, the mechanism design perspective; and (ii) from the perspective of negotiation strategies for each individual agent [27]. We will discuss (i) in detail in the next subsection. To address (ii), we revisit how MCDCF works from a standpoint of a single agent, where, this time, the algorithm is viewed as a negotiation protocol.

We recall that multi-agent negotiation is in general comprised of three major components: the negotiation protocol, the object(s) of negotiation, and the agents' decision making models [6]. A negotiation protocol is a set of rules that govern the interaction among agents, and will be discussed in more detail shortly.

Negotiation objects are a general concept capturing the range of issues over which agents need to reach agreement. In our context, there is only one such *object of negotiation*, namely, who will be forming a coalition with whom. Therefore, our setting as discussed in Section 2 is of the simplest, single object of negotiation nature, and this aspect of negotiation interaction need not be discussed further. Lastly, agents' decision making models have been defined as the decision-making apparatus that the agents employ in order to ensure acting in accordance with the negotiation protocol, so that they ensure satisfying their objectives (which, in our case, is to form coalitions). In general, (i) different agents may have different decision making models, and (ii) each agent's decision making model may depend on the protocol in place, the nature of object(s) of negotiation, and the range of operations that each agent is allowed to perform within the negotiation protocol [6]. In case of MCDCF, agents exchange coalition proposals until an agreement is reached. In particular, an agent engaging in a negotiation protocol based on MCDCF needs to be able to determine (a) whether it should stick to the current coalition proposal or change it to a different candidate coalition, (b) in case of the latter, which among possibly several available candidates for the new coalition proposal to adopt, and lastly (c) whether its coalition proposal has been accepted by the neighboring agents to whom that proposal was sent [22].

From a negotiation perspective, MCDCF is a distributed negotiation protocol that is carried out in several rounds. In each round, every agent, based on the history of the previous rounds and simple internal logic described in detail in [20,21,22], makes a coalition proposal to some nonempty subset of its neighbors. This coalition proposal may either be the same as what the agent proposed in the previous round, or it may be different; in case of the latter, the new coalition proposal is required to be *fresh*, i.e., to be a coalition that this agent has not proposed in any of the prior rounds. Whether an agent that has not joined a coalition yet ought to propose a new coalition or stick to its current choice is determined based on the internal logic that uses the *ChoiceFlag* and *NeighborFlag* information, as well as the proposals (and values of appropriate flags) received from the neighboring agents [19,20,22]. We make an important observation that there is no explicit dependence on the messages (that is, coalition proposals) received from the neighbors during the previous rounds. That is, in each round, based on the proposals (and some additional information, captured by the appropriate flag values) received from the neighbors, an agent appropriately updates its internal state and, at the next round, updates the information about its neighbors. The decision flags (one per agent) are used to keep track, which among an agent's neighbors have already joined a coalition in the prior rounds, and which are still engaged in negotiation. These properties ensure modest memory requirements of each agent: an agent only needs to keep a brief summary of the past rounds, not their detailed history.

In summary, each agent keeps negotiating from one round to the next, until eventually one of the following conditions is met:

– either a coalition proposal made to some of the agent's neighbors in a given round actually coincides with what each of those neighbors is proposing in that same round, in which case the agreement is detected and a new coalition is immediately formed;

– or else, the agent has traversed its list of candidate coalitions and has arrived to the singleton coalition as the only still available option; in that case, the agent recognizes this "exit criterion", notifies the remaining neighbors (if any) of this situation, and forms a singleton coalition, whereupon it changes its *DecisionFlag* to 1 and exits the further negotiation.

In [21], we carefully prove that, under certain assumptions about sparseness of the underlying network topology, the amounts of communication, computation and storage per agent, per round of MCDCF are all reasonably modest. We have also established that the number of rounds is guaranteed to be finite regardless of the underlying topology, and polynomially bounded for sufficiently sparse network topologies.

3.2   The Mechanism Design View of MCDCF

We now turn attention to some desirable properties of negotiation protocols in general, and how well MCDCF viewed as a negotiation protocol for coalition formation measures up with respect to those properties. We first briefly review the basic concepts of the underlying theoretical framework for formulating desirable properties of negotiation protocols; that framework is provided by a sub-area of game theory, called *mechanism design* [6,10,26]. In the context of MAS mechanism design addresses how to define the rules of multi-agent interaction in general, and of negotiation protocols in particular [10]. Specifically, mechanism design for negotiation protocols defines the principles and policies so that, if a given negotiation protocol obeys them, and all the agents involved in negotiation "play by the rules" of the protocol, then certain properties can be guaranteed to hold at the system level, regardless of the exact details of the negotiation, what kind of agreement (if any) is the negotiation going to result in a particular scenario, and the like. Following T. Sandholm in [26], we outline some of the highly desirable properties of a multi-agent negotiation protocol, and then discuss how well our MCDCF algorithm "rates" with respect to those properties.

- *Guaranteed success:* a negotiation algorithm holds this property if it ensures that, eventually (meaning, after finitely many negotiation rounds), an agreement is certain to be reached.
- *Pareto efficiency:* an outcome, O_p, of a negotiation protocol is *Pareto efficient* if there is no other outcome that would make one or more agents better off than in O_p, without making at least one other agent worse off.
- *Individual rationality:* a negotiation protocol is *individually rational* if following the protocol (as opposed to "cheating" or not engaging in it at all) is in the best interest of negotiation participants.
- *Stability:* a negotiation protocol is stable if all agents have an incentive to follow a particular strategy allowed by the protocol; that is, while there may be "legal" ways to deviate from a particular strategy, an agent that would choose such a way to deviate would, in general, be worse off than sticking to

the stable strategy. In game theory, various notions of equilibria have been introduced to capture this notion of stability. The best known such notion is that of Nash equilibrium.

- *Simplicity:* a protocol is simple if each participant who is using the protocol can easily (i.e., computationally tractably) determine its stable or optimal strategy among the allowable strategies.
- *Distribution:* a protocol should be robust and fault-tolerant, esp. with respect to "single points of failure".

The *guaranteed success* property in our case amounts to guaranteed convergence after finitely many rounds. We have established the convergence properties of MCDCF in [20]. Hence, the guaranteed success of MCDCF viewed as a negotiation protocol holds. Similarly, since an agent cannot be worse off than remaining alone, engaging in coalition formation process results in at least as desirable an outcome as not engaging in coalition formation at all, and is therefore, in general, individually rational thing to do. (Discussion on how to properly evaluate, and possibly trade-off, the cost of engaging in MCDCF protocol in hope of joining some nontrivial coalition, versus saving oneself the trouble and the cost and staying on one's own, is beyond our current scope; we just assume, in this paper as well as all our prior work, that the nature of tasks necessitates that agents join coalitions whenever possible.)

More interesting is *Pareto efficiency* of MCDCF. While we have not formally established Pareto optimality in our prior work, the simulations with the original, "baseline" version of MCDCF in practice always resulted in Pareto-optimal final coalition structures. However, the subsequent modifications and optimizations, especially those related to how each agent traverses a lattice of its candidate coalitions [22], while in general ensuring faster convergence once we scaled up our implementation from dozens to several hundreds of agents, also in certain cases may potentially "hurt" the Pareto efficiency. Further modifying the algorithm so as to ensure Pareto efficiency, and then formally establishing that highly desirable property, are the subject of our ongoing work.

When it comes to *simplicity*, the MCDCF protocol is indeed fairly simple, and (unlike most negotiation protocols found in the literature) actually proven to be scalable to hundreds of agents [23,24]; this scalability holds under a single but critically important assumption of *relative sparseness* of the underlying graph [20,21]. We point out that this assumption indeed often holds in practice, and can be imposed by the system designer when it naturally does not hold, as we discuss in detail in [19,20,21]. Our protocol is also very simple in terms of the communication language required for carrying out the negotiation process: the agents need to exchange tuples that encode the current coalition proposal and a few extra bits of auxiliary information (cf. the decision, choice and neighbor flags mentioned earlier [22]). For more details on the exact content of coalition proposal messages that agents exchange, see [21].

Lastly, MCDCF is a fully decentralized, *genuinely distributed* peer-to-peer algorithm; as such, viewed as a negotiation protocol, it is robust to single points of failure and, in particular, satisfies the distribution property above.

4 Summary and Future Work

This paper has three main objectives: it (i) summarizes some aspects of our research on distributed coalition formation, (ii) motivates extending the negotiation paradigm from the traditional, game-theoretic *competitive* multi-agent domains to *collaborative*, distributed problem solving agents that are locally constrained and hence have different local preferences, and lastly (iii) discusses our coalition formation protocol and its properties from a negotiation standpoint. While the original motivation behind MCDCF was admittedly somewhat different, we have argued that our approach to coalition formation actually provides a simple and elegant negotiation protocol that has many desirable properties – and is demonstrably scalable to much larger collaborative MAS than what is commonly found in the existing game theoretic and negotiation-centric MAS literature. This scalability is to a considerable extent due to two particular properties of MCDCF: (1) it is a local algorithm, in that each agent only directly interacts with its immediate, one-hop neighbors; and (2) the messages exchanged during the course of the protocol execution are of a very simple nature.

Our plans for the future work include both short-term, specific problems and a longer-term, broader research agenda. In the short term, we would like to determine whether our coalition formation protocol can be made Pareto efficient via relatively simple modifications to the coalition proposal exchange process. In the longer term, we are interested in further expanding on various distributed, highly scalable negotiation protocols for coordination and cooperation in MAS, and investigating potential benefits of *multi-tiered reinforcement learning* among agents·and agent ensembles for more effective coordination.

Acknowledgments. This work was partially supported by NSF grant IIS 0914861.

References

1. Abdallah, S., Lesser, V.: Organization-Based Cooperative Coalition Formation. In: Proc. IEEE/WIC/ACM Int'l Conf. Intelligent Agent Technology, IAT 2004 (2004)
2. Avouris, N.M., Gasser, L. (eds.): Distributed Artificial Intelligence: Theory and Praxis. Euro. Courses Comp. & Info. Sci., vol. 5. Kluwer Academic Publ. (1992)
3. Garcia-Garcia, J., Ordonez, C.: Consistency-aware evaluation of OLAP queries in replicated data warehouses. In: ACM DOLAP, pp. 73–80 (2009)
4. Garey, M.R., Johnson, D.S.: Computers and Intractability: a Guide to the Theory of NP-completeness. W.H. Freedman & Co., New York (1979)
5. Jang, M.-W., Reddy, S., Tosic, P., Chen, L., Agha, G.: An Actor-based Simulation for Studying UAV Coordination. In: The 16th European Simulation Symposium (ESS 2003), Delft, The Netherlands (2003)
6. Jennings, N.R., Faratin, P., Lomuscio, A.R.: Automated Negotiation: Prospects, Methods and Challenges. In: Group Decision & Negotiation, vol. 10, pp. 199–215. Kluwer (2001)
7. Li, X., Soh, L.K.: Investigating reinforcement learning in multiagent coalition formation. TR WS-04-06, AAAI Workshop Forming and Maintaining Coalitions & Teams in Adaptive MAS (2004)

8. Ma, L., Han, X., Shen, C.C.: Dynamic open spectrum sharing MAC protocol for wireless ad hoc networks. In: Proc. IEEE 1st Int'l Symp. on New Frontiers in Dynamic Spectrum Access Networks (DySPAN 2005), pp. 203–213 (2005)
9. de Oliveira, D.: Towards Joint Learning in Multiagent Systems Through Opportunistic Coordination. PhD Thesis, Univ. Fed. Do Rio Grande Do Sul, Brazil (2007)
10. Parsons, S., Wooldridge, M.: Game Theory and Decision Theory in Multi-Agent Systems. In: Autonomous Agents and Multi-Agent Systems, vol. 5. Kluwer (2002)
11. Sandholm, T.W., Lesser, V.R.: Coalitions among computationally bounded agents. Artificial Intelligence 94, 99–137 (1997)
12. Sandholm, T.W., Larson, K., Andersson, M., Shehory, O., Tohme, F.: Coalition structure generation with worst case guarantees. AI Journal 111, 1–2 (1999)
13. Shehory, O., Kraus, S.: Task allocation via coalition formation among autonomous agents. In: Proceedings of IJCAI 1995, Montreal, Canada, pp. 655–661 (1995)
14. Shehory, O., Sycara, K., Jha, S.: Multi-Agent Coordination Through Coalition Formation. In: Rao, A., Singh, M.P., Wooldridge, M.J. (eds.) ATAL 1997. LNCS (LNAI), vol. 1365, pp. 153–164. Springer, Heidelberg (1998)
15. Shehory, O., Kraus, S.: Methods for task allocation via agent coalition formation. AI Journal 101 (1998)
16. Soh, L.K., Li, X.: An integrated multilevel learning approach to multiagent coalition formation. In: Proc. Int'l Joint Conf. on Artificial Intelligence, IJCAI 2003 (2003)
17. Sun, R.: Meta-Learning Processes in Multi-Agent Systems. In: Zhong, N., Liu, J. (eds.) Intelligent Agent Technology: Research & Development, pp. 210–219. World Scientific, Hong Kong (2001)
18. Tosic, P., Agha, G.: Understanding and modeling agent autonomy in dynamic multi-agent multi-task environments. In: Proc. First European Workshop on Multi-Agent Systems (EUMAS 2003), Oxford, England, UK (2003)
19. Tosic, P., Agha, G.: Maximal Clique Based Distributed Group Formation Algorithm for Autonomous Agent Coalitions. In: Proc. Workshop on Coalitions & Teams, within AAMAS 2004, New York City, New York (2004)
20. Tošić, P.T., Agha, G.: Maximal Clique Based Distributed Coalition Formation for Task Allocation in Large-Scale Multi-agent Systems. In: Ishida, T., Gasser, L., Nakashima, H. (eds.) MMAS 2005. LNCS (LNAI), vol. 3446, pp. 104–120. Springer, Heidelberg (2005)
21. Tosic, P.: Distributed Coalition Formation for Collaborative Multi-Agent Systems, MS thesis, Univ. of Illinois at Urbana-Champaign, Urbana, Illinois, USA (2006)
22. Tosic, P., Ginne, N.: Some Optimizations in Maximal Clique based Distributed Coalition Formation for Collaborative Multi-Agent Systems. In: Proc. IEEE Active Media Technology (AMT 2010), Toronto, Ontario, Canada (2010)
23. Tosic, P., Ginne, N.: Challenges in Distributed Coalition Formation among Collaborative Multi-Agent Systems: An Experimental Case Study on Small-World Networks. In: Proc. Int'l Conf. on Artif. Intelligence (ICAI 2011), Las Vegas, Nevada, USA (2011)
24. Tosic, P.: Distributed Graph-Partitioning based Coalition Formation for Collaborative Multi-Agent Systems: Some Lessons Learned and Challenges Ahead. In: Proc. Int'l Conf. on Artif. Intelligence (ICAI 2012), Las Vegas, Nevada, USA (to appear, 2012)
25. Vig, S., Adams, J.A.: Issues in multi-robot coalition formation. In: Proc. Multi-Robot Systems: From Swarms to Intelligent Automata, vol. 3 (2005)
26. Weiss, G. (ed.): Multiagent Systems: A Modern Approach to Distributed Artificial Intelligence. MIT Press (1999)
27. Wooldridge, M.: An Introduction to Multi-Agent Systems. Wiley (2002)

Multi-Agent Liquidity Risk Management in an Interbank Net Settlement System

Badiâa Hedjazi[1], Mohamed Ahmed-Nacer[2], Samir Aknine[3], and Karima Benatchba[4]

[1] Information Systems Division, CERIST Research Center, 5 Rue des Frères Aissou Ben Aknoun, Algiers, Algeria
[2] Information Systems Laboratory, USTHB University,BP 32 El Alia 16111 Bab Ezzouar, Algiers, Algeria
[3] LIRIS, Université Claude Bernard Lyon,LIRIS UMR 5205INSA de Lyon, Campus de la Doua,Bâtiment Blaise Pascal,20, Avenue Albert Einstein69621 VILLEURBANNE CEDEX France
[4] ESI, National High School of Computer Science, BP 68M OUED SMAR, 16309, El-Harrach, Algiers, Algeria
{badiaa.hedjazi,benatchba}@yahoo.fr, anacer@cerist.dz, samir.aknine@univ-lyon1.fr

Abstract. A net settlement system is a payment system between banks, where a large number of transactions are accumulated, usually waiting until the end of each day to be settled through payment instruments like: wire transfers, direct debits, cheques, ... These systems also provide clearing functions to reduce interbank payments but are sometimes exposed to liquidity risks. Monitoring, and optimizing the interbank exchanges through suitable tools is useful for the proper functioning of these systems. The goal is to add to these systems an intelligent software layer integrated with the existing system for the improvement of transactions processing and consequently avoid deadlock situations, deficiencies and improve system efficiency. We model and develop by multi-agent an intelligent tracking system of the interbank exchanged transactions to optimize payments settlement and minimize liquidity risks.

Keywords: payment system, net settlement system, multi-agent system, liquidity risk, classifier system.

1 Introduction

A net settlement system is a system that processes retail payment instruments: wire transfers, direct debits, cheques, bank cards... These systems also provide clearing functions to reduce the number of interbank payments and therefore cost and consumed time[1]. The payment obligations are divided into two types: High-value payments and low-value payments. Two complementary systems treat these payments: the first is called RTGS (Real Time Gross Settlement) dealing with large amounts and in real time, and the second called net settlement system or DNS (deferred net settlement) or retail payment system which handles retail payments

R. Huang et al. (Eds.): AMT 2012, LNCS 7669, pp. 103–114, 2012.

(cheques, direct debits, bank cards, etc.) on a "deferred net" basis[1].So far, the net settlement system plays only the role of a router in a network formed by the participating banks. Multilateral netting results has a significant reduction in payment flows and liquidity needs, compared to the bilateral netting or the gross settlement, but in the event of the insolvency of a participant, all the underlying transactions settlement of the other participants would be blocked. This creates to non-defaulting participants a *liquidity risk* [2]. Revoking payments (exclusion of failed transactions) when settlement failure, should eliminate this risk. However, risk could remain if the non-defaulting participants do not suspend payments to their defaulting customers [2]. It is necessary to have an additional system to monitor and analyze each received transaction. We model and develop an intelligent tracking system integrated with the net settlement system to minimize the risk of insolvency and liquidity, avoid deadlocks and bypass certain failures by a multilateral optimization of settlement processes through a multi-agent system (MAS) where each participant (bank) is associated with an adaptive agent. In section 2 of this article, we present some optimization models of interbank payment systems. In section 3 we describe our multi-agent balance tracking system. In section 4 we discuss the implementation and our various experiments. In section 5 we conclude with a synthesis of our contributions and give some research perspectives.

2   Related Work

FIFO mechanism is applied by all payment systems, but this mechanism becomes an obstacle for the final high-value payments settlement. Several optimization algorithms have been proposed. The *bilateral optimization* examines the participants in pairs and settles transactions simultaneously for maximum possible value. Güntzer et al [4] proposed *Greedy algorithm* to optimize the netting amount value. This algorithm is optimal for a small number of payments. Renault and Pecceu [5] improved the *Greedy algorithms* and the *Multilateral (*where all participants and all payments are considered simultaneously) *Greedy Las Vegas*. Beyeler et al. [6] proposed the addition of another source of liquidity but this entails additional costs and constraints. All these works have made significant contributions to settlement optimization but their drawback is that they assume that the participating banks behavior remains unchanged and therefore no adaptation or improvement in their decision making process. However, we were inspired by these models, by promoting operations which reduce liquidity risks in the learning model of each bank. If a bank becomes debtor, then we promote operations decreasing the debit value (credit transactions such as cheques if remitting bank or wire transfers if receiving bank).The growing complexity of payment systems requires efficient structures like MAS to their study [7] but this work is devoted to study RTGS systems and not net settlement systems but we were inspired from it to build our system according to multi-agent approach.

3   Multi-Agent Balance Tracking System

Our system is designed for intelligent tracking and processing in an interbank clearing system using a multi-agent system and classifier systems [8] [9] for the reasoning model of the agents associated to banks. It works in collaboration with the net settlement system, to make reliable decisions on transactions and prevent risks.

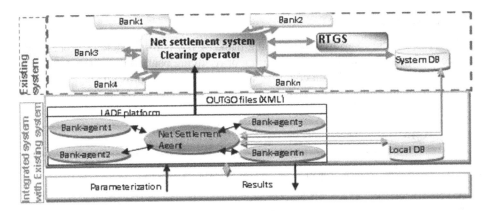

Fig. 1. System representation

Our system is integrated with the existing Net Settlement System. It is composed of a Net Settlement Agent, a set of Bank Agents (BAs) interacting between them, and two databases (DBs) (Fig. 1). *System DB* contains interbank transactions exploited by Net Settlement Agent for extracting data. The BAs exchange messages with each other and with the Net Settlement Agent. It's then a decentralized architecture.

3.1   Net Settlement Agent

Net Settlement Agent manages banks' transactions. This agent is reactive. It reacts to each received payment order and has four functions (Fig. 2): (1) Start or close a day. (2) Extract transactions from *System DB*. (3) Insert and update transactions into *Local DB*. (4) Generate OUTGO files of processed transactions in XML format.

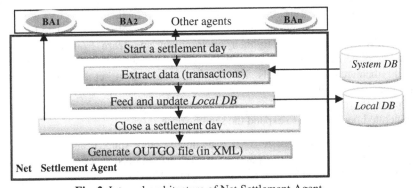

Fig. 2. Internal architecture of Net Settlement Agent

Net Settlement Agent feeds *Local DB* from *System DB* which has a specific format. Hence, the need of a working database for our system, *Local DB*. The Net Settlement agent is reactive. It reacts to each bank payment order reception.

At the beginning of a day, the Net Settlement Agent informs all the BAs of the new day and each start or close of a clearing session. It makes updates by decrementing settlement dates, giving status to balance previous states (creditor or debtor).

3.2   Banks Agents (BAs)

Each BA is associated with a participating bank. A BA is responsible of processing transactions related to it (when the remitting or receiving bank of the transaction corresponds to the BA). ABA is a cognitive agent designed to improve its behavior and profitability by learning and then based on a classifier system (CS) (Fig. 3).

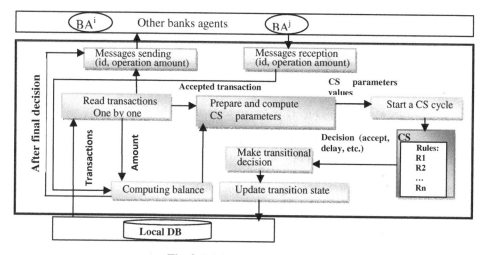

Fig. 3. BA internal architecture

A BA performs the following functions:

(1) Read transactions from Local DB. (2) Prepare CS parameters (Bank State, Bank Type, Rejection Rate, etc.). (3) Start a CS Cycle including, compute Reward and make decision. (4) Compute balance after each transaction and update Local DB.

The BAs gradually adapt to their environment and learn from their past experiences with periodic evaluation of their rules. We choose then classifier systems to build their reasoning models. A classifier system is a set of rules determining agent behavior. It has a mechanism for evaluating its rules by rewarding those who produce more gains. The system starts with a random set of rules; others are generated periodically to expand the search space.

BA Classifier System (CS). A BA CS evaluates bank state, transaction type, bank type, bank threshold, calculates parameters and make decision. Each CS rule consists of 3 parts (**condition**: on 21 bits, **action**: on 2 bits, **fitness**: a real number) (Fig. 4).

Condition: it consists of two parts:

Case: corresponds to encountered cases by a bank during a session.

Parameters: contains seven parameters validating or not the *Case* part.

Action: shows the four possibilities of the agent action (decision):
(1)*Accept*: perform transaction. (2) **Delay**: delay transaction to the next session. (3)*Reject*: Reject transaction but will be processed next day. (4) *Cancel*: Cancel transaction, with the possibility of treatment in the next day.

Fitness: contains the strength of the rule and is a real in the interval [0, 1].

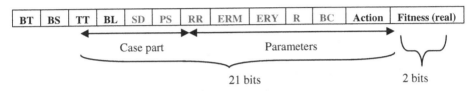

Fig. 4. CS rule representation

BT (Bank type): Coded on 1 bit. If *remitter bank* BT=0 else (*receiver*) BT=1

BS (Bank state): When starting our system all BAs have 'creditor' state. This state changes after interbank exchanges. BS is calculated as in equation (1):

$$BS = C-D \qquad (1)$$

where: $C = \sum$ amounts of credit payment instruments.

$D = \sum$ amounts of debit payment instruments.

BS is coded on 1 bit. If C−D>0 then the bank is *Creditor: 0* else is *Debtor: 1*.

TT (Transaction Type): Instrument type and coded on 4 bits as for example: *Cheque* coded *0000* and *wire transfer* coded *1001*.

BL (Bank limit): Coded on 2 bits (Table 1) and calculated as in (2):

$$BL = (C-D)/X \qquad (2)$$

where: $C = \sum$ amounts of credit payment instruments of the bank.

$D = \sum$ amounts of debit payment instruments of the bank.

$X = 80\%$ of the initial balance of the bank.

We are interested only by negative values of BL (the debtor banks).

Table 1. Binary representation of BL

Interval	Degree	Meaning	Code
[0, 0.4 [	mild	Delaying debit transactions	00
[0.4, 0.7 [	moderate	Sorting transactions to accept (Sort)	01
[0.7, 0.9 [	severe	Sending alarms to participants and Sort	10
>0.9	Very severe	Cancelling debit transactions and Sort	11

SD (Settlement Date): coded on 2 bits where D is the trading day and D1, D2 are the following days (D: *00*, D*1*: *01*, D2: *10*).

PS (Previous State): is the final state of the previous day after balance settlement, and is coded on 1 bit: (*Creditor: 0*; *Debtor: 1*).

RR (Rejection Rate): is the rate of rejected transactions by the number of transactions. It is coded on 2 bits and calculated as in equation (3):

$$RR = (Rejected/ (Processed - Rejected)) \tag{3}$$

where: *Rejected:* number of rejected transactions.

Processed: number of all the transactions to be processed.
If **RR** in [0, 0.4 [then coded 00; If **RR** in [0.4, 0.7[then coded 01;
If **RR** in [0.7, 0.9[then coded 10; if **RR** in [0.9, 1] then coded 11;

ERM (Evolution rate per month): Coded on 2 bits, is amounts evolution rate between 2 consecutive months and calculated as in (4):

$$ERM = ((N-M)/ M)*100 \tag{4}$$

where: N: the value of transactions of the current month.

M: the value of transactions for the previous month.
If **ERM** <0 then coded 00; If **ERM** in [0,50[then coded 01;
If **ERM** in [50,100[then coded 10; if **ERM** >=100 then coded 11;

ERY (Evolution rate per Year): coded on 2 bits; compares amounts of a month of current year with the same of previous year. Is calculated as in (5):

$$ERY = (N/ M)*100 \tag{5}$$

where: N: total amount of all transactions processed in the current month.

M: total amount of processed transactions of same month the previous year.
If **ERY** <0 then coded 00; If **ERY** in [0,50[then coded 01;
If **ERY** in [50,100[then coded 10; if **ERY** >=100 then coded 11;

R (Ratio): Coded on 2 bits, reflecting position of a payment instrument from all payment instruments, and calculated by the formula (6):

$$R= A/ T \tag{6}$$

where : A: the total amount of the transactions by a payment instrument.

T: the total amount of the instruments processed by a bank.
If **Ratio** in [0, 0.4 [then coded 00; If **Ratio** in [0.4, 0.7[then coded 01;
If **Ratio** in [0.7, 0.9[then coded 10; if **Ratio** in [0.9, 1] then coded 11;

BC (Bank category): coded on 2 bits (*Large:00*; *Medium: 01; Small: 10*).

Action: Agent's decision; coded on 2 bits for *Accept, Delay, Reject, Cancel*.
Condition parameters weights depend on Action value (Tables 2, 3, 4, 5):

Table 2. Parameters weights when action is « Accept »

Parameters	Weight	Significance
SD, BC	3	Important
PS, RR, ERM, ERY	2	Moderately important
R	1	Little importance

Table 3. Parameters weights when action is « Delay »

Parameters	Weight	Significance
SD, PS, RR, ERM, ERY, R	2	Moderately important
BC	3	Important

Table 4. Parameters weights when action is « Reject »

Parameters	Weight	Significance
SD, R	1	Little importance
PS, ERM, ERY	2	Moderately important
RR	4	Very important
BC	3	Important

Table 5. Parameters weights when action is « Cancel »

Parameters	Weight	Significance
SD, PS	1	Little importance
RR	4	Very important
ERM, ERY, R	2	Moderately important
BC	3	Important

For the « Case part » of the condition we have the following possibilities:
Where: RC is for receiving bank, RM for remitting, C: Creditor and D: debtor.
We consider: wire transfer as *Debit_instrument*;
card, direct debit, cheque, negotiable instrument as *Credit_instrument*.

1. **If** (BT = *RM* & TT = *Credit_instrument*) *or* (BT = *RC* & TT = *Debit_instrument*)
   then accept.
2. **If** (BT = *RM* & TT = *Debit_instrument* & EB = *C*) *or* (BT = *RC* & TT = *Credit_instrument* & BS = *C*) *then accept.*
3. **If**(BT = *RM* & TT = *Debit_instrument* & EB = *D* & BL є [0,0.4[ ) *or* (BT = *RC* & TT = *Credit_instrument* & BS = *D* & BL є [0, 0,4[ ) *then delay.*
4. **If** (BT = *RM* & TT = *Debit_instrument* & EB = *D* & BL є [0.4, 0.7[ ) *or* (BT = RC & TT = Credit_instrument) & BS = D & BL є [0.4, 0.7[ ) *then reject.*
5. **If** (BT = *RM* & TT = *Debit_instrument* & BS = D & (BL>0.7) *or* (BT = RC & TT = Credit_instrument & BS = D & (BL>0.7) *then* cancel.

Rules Reward: The *Case part* of a classifier is the most important. We associate it with the weight *5/8* of the decision. The *Parameters part* classifier's condition affects less the decision and is associated with the weight *3/8*. If one of the two parts is rewarded then it is multiplied by 1and if it is less rewarded then it is multiplied by *1/4*.We limit the value of the reward (*RW*) in [0, 1] by dividing it by 2.

If action part of C1 (Classifier to evaluate) is « accept »
If (BT=*RM* & TT=*Credit_instrument*) or (BT=*RC* &
TT=*Debit_instrument*)
Then
RW=(5/8+3/8×[(3×SD)+(2×PS)+(2×RR)+(2×(1/ERM))+(2×(1/ERY))+
(3×BC)+(1×R)])/(2×∑weights)
1/ERM and 1/ERY is user to limit the value in [0, 1].
If (BT=*RM* & TT=*Debit_instrument* & BS=*C*) or (BT=*RC* & TT= *Credit_instrument* & BS=*C*)
Then
RW=(5/8+3/8×[(3×SD)+(2×PS)+(2×RR)+(2×(1/ERM))+(2×(1/ERY))+
(3×BC)+ (1×R)])/(2×∑weights)

Else
RW=(5/8×1/4+3/8×[(3×SD)+(2×PS)+(2×RR)+(2×(1/ERM))+(2×(1/ERY))+(3×BC)+(1×R)])/(2×Σweights)

If action part of Cl is « Delay »
If (BT=*RM* & TT=*Debit_instrument* & BS=*D* & BL∈[0,0.4[) or (BT=*RC* & TT=*Credit_instrument* & BS=*D* & BL∈[0, 0.4[)
Then
RW=(5/8+3/8×[(2×SD)+(2×PS)+(2×RR)+(2×(1/ERM))+(2× (1/ERY))   + (3×BC)+ (2×R)])/(2×Σweights)
Else
RW=(5/8×1/4+3/8 × [(2×SD)+(2×PS)+(2×RR)+(2×(1/ERM))+ (2×(1/ERY))+(3×BC)+(2×R)]) /(2×Σweights)

If action part of Cl is « reject »
If (BT=*RM* & TT=*Debit_instrument* & BS=*D* & BL∈[0.4,0.7[) or BT=*RC* & TT=*Credit_instrument*) & BS=*D* & BL∈[0.4, 0.7[)
Then
RW=(5/8+3/8×[(1×SD)+(2×PS)+(4×RR)+(2×(1/ERM))+ (2×(1/ERY))+(3×BC)+(1×R)])/(2×Σweights)
Else
RW=(5/8×1/4+3/8×[(1×SD)+(2×PS)+(4×RR)+(2×(1/ERM))+(2×(1/ERY))+(3×BC)+(1×R)])/(2×Σweights)

If action part of Cl is « Cancel »
If (BT=*RM* & TT=*Debit_instrument* & BS=*D* & (BL> 0.7) or (BT=*RC* & TT=*Credit_instrument*) & BS=*D* & (BL>0.7))
Then   RW=(5/8+3/8×[(1×SD)+(1×PS)+(1×RR)+(2×(1/ERM))+(2×(1/ERY))+(3×BC)+(2×R)])/(2×Σweights)
Else
RW=(5/8×1/4+3/8×[(1×SD)+(1×PS)+(1×RR)+(2×(1/ERM))+(2×(1/ERY))+(3×BC)+(2×R)])/(2×Σweights)

3.3   Transactions Processing and Decision Making

A settlement day is composed of three clearing sessions.

1st session: This session deals with all the new arrival transactions on D-Day and the recovery and processing of already included transactions in *Local DB* of previous days (D-1, etc.) with one of these statements: (1) Transaction's transitional state: *delay*, *reject* or *cancel*. (2) Transitional state *accept*, but acceptance or rejection not confirmed by the receiving bank. (3) Final state *accept*, but their settlement dates not yet reached.

2nd session: Resumes only transactions with *delay* transitional state and processes the new transactions.

3rd session: The same as the 2nd one but updating *Local DB* before closing.

At the end of each session is generated an OUTGO file of processed transactions. Net Settlement Agent extracts all the transactions from *System DB* and copies them into *Local DB*. BA processes each transaction with the remitting bank is this BA.

If the transaction is not yet processed, then process it, generate a transitional decision (state) and update *Local DB*, pending the final state sent by the bank when transitional state is *accept*;

If transaction already processed, the bank final decision is needed. If transaction not rejected, then finished else removed and balance recalculated.

At the end of each clearing session, the system updates the final statement in *Local DB* and generates an OUTGO file in XML format.

Transactions such as *wire transfers* have negative impact on remitting banks than receiving ones. We therefore give decision priority to remitting banks. For the other payment instruments: *card*, *direct debit*, *cheque*, *negotiable instrument*, we give priority to receiving banks. If a *receiving BA* (or *remitting*) processes transactions and results to transitional state *accept* then it informs the *remitting BA* (or *receiving*) to update balances. If the final state is *reject* then the concerned BA updates its balance.

4   Implementation and Experiments

Our system is implemented with JADE multi-agent platform, JAVA, ORACLE DBMS, ART(Artificial Reasoning Toolkit) package for programming CS and DOM for generating OUTGO files. A BA or the Net Settlement Agent can execute several behaviors (specific tasks: ex. *New_settlement_day()*) concurrently (Fig.5).

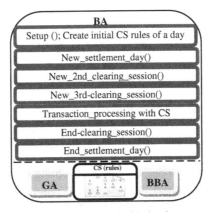

Fig. 5. A BA principal behaviors

At the end of each clearing session the Net Settlement agent generates an OUTGO file in XML format. Each processed transaction contains the decisions of its final or transitional states. These decisions are generated by their corresponding agents. For example the transaction *83* has the transitional decision *accept* and is generated by the *bank agent 3*(BA$_3$) corresponding to the remitting bank number *3*.

The results of our experiments allow us to judge the performance of our system compared to the current system. Our experiments are made with five banks (Bank 1, Bank 2, Bank 3, Bank 4, and Bank 5). The same operations initially with random amounts are considered for both types of experiments. The first type simulates the current system that accepts all operations and the second type that processes transactions by treating them with our multi-agent system. The graph in Fig.6 are calculated without prior treatment, that is to say, all transactions are accepted. Also, Banks often reach the limit value (BL). This puts them in situations of liquidity risks.

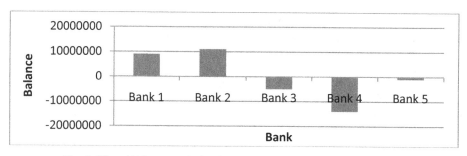

Fig. 6. Virtual balance evolution in the current system for 5 banks / Day

The graph in Fig.7 is generated at the end of a settlement day. It represents the virtual balance evolution achieved in each bank with our system.

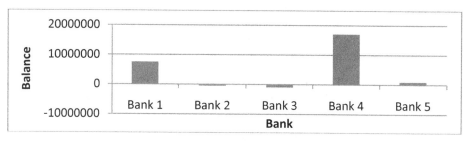

Fig. 7. Virtual balance evolution in our system for 5 banks / Day

The purpose of these two graphs is to compare the effectiveness of the old system compared to ours and to judge the performance of our system. On Fig. 6 and 7 we note that 3 banks (Bank 3, Bank 4, Bank 5) are in high amounts of debtor state (~ -14000000) with the existing system (current) against 2 banks that are in small debtor state amounts (~ -900 000) with our system. By doubling the number of banks in our simulations (10 banks) we noticed that our system is more efficient because all the banks find themselves in a creditor state at the end of the settlement day (Fig. 9). This is explained by the increased number of liquidity sources in the system. This is not the case with the current system even if the number of banks doubles (Fig. 8).

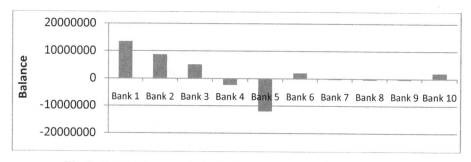

Fig. 8. Virtual balance evolution in the current system for 10 banks / Day

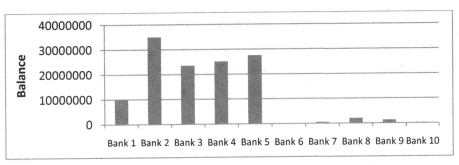

Fig. 9. Virtual balance evolution in our system for 10 banks / Day

Fig. 10 shows the evolution of virtual balances through ten consecutive settlement days. This graph tries to target vulnerabilities in the current system by showing the negative development of some banks, which can cause situations of insolvency, if the problem persists. The system does not matter the number of banks with a negative balance, and gives no warning. Therefore, the balances will evolve, even if it is in debtor state, and this will certainly worsen the financial situation of banks.

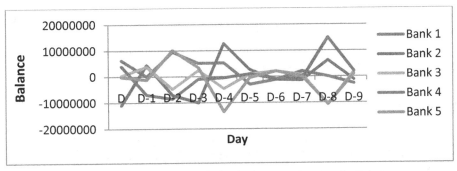

Fig. 10. Virtual balance evolution in the current system for 9 days

Fig. 11 shows the evolution of virtual balances through a period of ten settlement days. This graph shows the improvements made by our system to treatment of balances. In this graph, we note that our system has a positive impact on balances by reducing the margins between negative and positive balances and consequently maintaining stability of the clearing system.

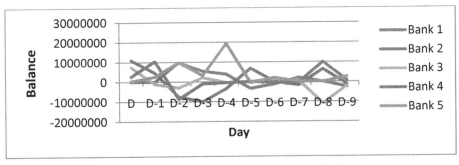

Fig. 11. Virtual balance evolution in our system for 9 days

Our system allows running the entire daily payment processing (three clearing sessions) in each settlement day. The results show that relatively to the existing system, our system allows a significant improvement in the bank balances evolution and significantly minimizes the number of times when banks are in debtor making the payment system less exposed to liquidity risk.

5   Conclusion

Our system is designed as a response to problems posed by the risks of participants failure or insolvency situations in the interbank clearing system, which requires the inversion of the clearing day (removal). This involves the recalculation of clearing balances of the non-defaulting participants. Improving balances management is by a multi-agent decision support system is necessary to reduce liquidity risks. Our current system makes the system more flexible and adaptive by detecting risky transactions and processing them by minimizing liquidity risks. The obtained results show that multi-agent models can be used to better manage the system and resolve payment system major problem which the liquidity risk. Our system fully meets the original goals but some improvements can make our system more flexible such as (1) creating direct interface with the RTGS system in order to manipulate actual balances and (2) adding predicting agents that probe the history data automatically and intelligently to forecast the future evolution of the system.

References

1. Taibi, A.: BDL Système de télécompensation, rapport de banque de la BDL (Mai 2006)
2. Banque de France, Revue de la stabilité financière (3) (Novembre 2003)
3. Atos Euronext/Diamis, Rapport de Conception. Principes Fonctionnels pour la mise en œuvre dansle système d'information, Groupe Atos Origin. Version 2.0. 18. (Juin 2004)
4. Güntzer, M., Jungnickel, D., Leclerc, M.: Efficient algorithms for the clearing of interbank payments. European Journal of Operational Research 106, 212–219 (1998)
5. Renault, F., Pecceu, J.-B.: From PNS to TARGET2: the cost of FIFO in RTGS payment system. In: Leinonen, H. (ed.) Simulation Studies of Liquidity Needs, Risks and Efficiency in Payment Networks. Proceedings from the Bank of Finland Payment and Settlement System Seminars 2005–2006, 320 p. (2006) ISBN 978-952-462-360-5
6. Beyeler, W., Bech, M., Glass, R., Soramäki, K.: Congestion and Cascades in Payment Systems. Physica A: Statistical Mechanics and its Applications 384(2), 693–718 (2007)
7. Galbiati, M., Soramaki, K.: An agent-based model of payment systems. Journal of Economic Dynamics and Control 35(6), 859–875 (2011)
8. Livolant, E.: Apprentissage Multi-Agents par Systèmes de Classeurs, Université de Caen, Laboratoire GREYC, DEA Intelligence Artificielle et Algorithmique (Septembre 2003)
9. Holland, J.H.: Genetic algorithms and classifier systems: Foundations and future directions. In: Refenstette, J.J. (ed.) Proceedings of the Second International Conference on Genetic Algorithms and Their Applications, pp. 82–89. Lawrence Erlbaum Associates, Hillsdale (1987)

An Enhanced Mechanism for Agent Capability Reuse

Hao Lan Zhang[1], Jiming Liu[2], Chaoyi Pang[1,3], and Xingsen Li[1]

[1] NIT, Zhejiang University, Ningbo, Zhejiang Province, China
[2] Hong Kong Baptist University, Kowloon Tong, Hong Kong
[3] The Australian E-health Research Centre, CSIRO, Australia
{haolan.zhang,lixs}@nit.zju.edu.cn, jiming@comp.hkbu.edu.hk,
chaoyi.pang@csiro.au

Abstract. Existing research work on intelligent agents is facing the challenge of extending agent capabilities. One of major difficulties of applying intelligent agents in various software platforms lies in that generalized agent design models may not fit with specified organizational applications. This problem dramatically hampers the practical applications of intelligent agents. In this paper we point out several aspects that might affect the development of agent-based technology; and further address the information reuse and unification issues through suggesting an enhanced agent capability reuse mechanism, which is able to provide an efficient process for agent capability reuse. The proposed design can fulfill the needs for a dynamic agent environment.

Keywords: Agent Capability Reuse, Agent Matching, Multi-agent Systems.

1   Introduction

The intelligent agent (IA) concept has been around for several decades. Many researchers consider the actor model suggested by [1] in 1973 as the early model of today's intelligent agents. The first World Wide Web (WWW) was made available in 1989 by Tim Berners-Lee [2]. The total number of Internet users in the world is 1,966,514,816 by 31 June 2010 [3]. Although we have insufficient resources for conducting a survey on the usage of agent-based applications around the world, it is very clear that users of agent-based applications are far less than WWW users. The reasons for causing this situation in the IA filed are quite complicated and controversial. In this paper we would like to point out two factors among others that hinder the development of agent-based technology.

The first factor is the lack of a unified/standardized framework for different agent applications. Unlike agent-based technologies, all Web-based browsers can recognize HTML (or later XML) as a commonly recognizable language for web pages. Therefore, Web-based technology can be widely accepted by different organizations and users as their browsers support the same language. This improves the efficiency of information sharing and exchanging among various organizations and users. In contrast, researchers and users in the current IA field have not agreed on a universal platform or theory for agent design, communication and cooperation.

R. Huang et al. (Eds.): AMT 2012, LNCS 7669, pp. 115–123, 2012.

The second factor is the inflexibility of adapting to the dynamic environment. In theory, agents are designed with self-learning capabilities; however current machine learning methods can only partially accommodate the needs for perceiving and reacting in a dynamic environment [4, 5]. Compared with agents, Web 2.0 has been extended for supporting many state-of-the-art technologies such as JavaScript, Adobe Flash, XML, and various programming languages. These mechanisms can be easily embedded to HTML Web pages and can fulfill the demands of different users. Thus, Web-based systems have been applied extensively to various applications including online analytical systems, online social networks, e-commerce systems, etc.

The abovementioned two factors reflect that agent design should be extensible, flexible and reusable for supporting various applications and adapting to the dynamic environment. In this paper we suggest a novel agent design, namely the *Domain Specific Component* (DSC) design, which can improve agents' reusability and flexibility. The DSC design utilises a number of components to perform domain-specific tasks. Each component is responsible for a particular task and can be easily plugged into an agent. The DSC design allows an agent to perform multiple tasks and easily extend its functionalities through plugging in new DSC components. DSC components are assembled by a container, which is standardised and supports components reuse among different agents. Thus, the DSC design offers a flexible and unified framework for agent design.

2   Related Work

The DSC design extends the mechanisms described in the Generalized Partial Global Planning (GPGP) framework and further suggests an efficient structure for agent design. GPGP agent architecture [6], shown in Fig. 1, offers a domain-independent framework for agent cooperation and coordination [7].

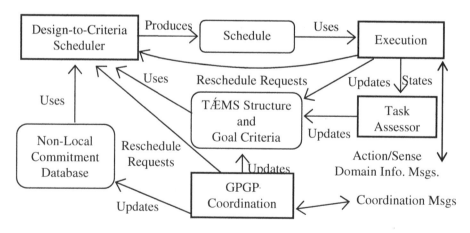

Fig. 1. GPGP agent architecture [7]

Compared with GPGP framework, the DSC-based agent architecture design focuses on individual agent functionalities (represented by DSC items) and allows the reuse of DSC items by other agents.

The research on agent-based capability design has been primarily focusing on the development of agent capability description language and information reuse. The major mechanisms for describing agent capabilities include the Language for Advertisement and Request for Knowledge Sharing (LARKS) [8], Agent Capability Description Language (ACDL) [9], and Interface Communication Language (ICL) [10].

ACDL is introduced to maximize the reuse of agent capabilities over new application domains. It is based on the Knowledge Modeling Framework (KMF). The LARKS allows agents to advertise their capabilities for both syntactic and semantic matching processes. LARKS-based agents are able to use application domain knowledge in any advertisement and request [8].

Modeling of component-based systems is still regarded as a largely unresolved problem in many object-oriented systems according to [11]. The emergence of intelligent agents is helpful to solve the problems in object-oriented systems since agents can be specified on a conceptual level instead of an implementation level [14]. Moreover, agents have a strong adaptability and a self-learning capability, which make the component integration process in agent-based systems much efficient than in object-oriented systems. The design principles for building component-based agents have been described in [14], which provide several preliminary mechanisms to enhance the reusability of agent design. The 'agent specific task' component described in their study is similar to the DSC concept. However, it does not provide more concrete mechanisms about the design procedures and the performance of the design.

Previous research on developing an open and comprehensive agent structure has addressed some fundamental issues including the reusability issue for agent design [8, 10, 12, 13]. The reusability issue for agent design is also related to the other issues, such as agent matchmaking, service advertisement, standardization and learning, etc. The previous work is avail of drawing the basic guidelines for designing DSC-based agents.

3   DSC Usage Centre – A Unified Framework

3.1   A Slot Container – DSC

Each agent has a DSC usage centre, which provides information resources for agents. This component distinguishes the DSC-based agents from existing middle agents or agent facilitators that mainly play as the role of an agent coordinator. The DSC usage centre upgrades the agents' functionalities through acquiring information from the external environment and sending the acquired information in DSC formats back to the centre. The DSC usage centre manages the unused and active DSC items for agents to update their capabilities. Each DSC item is executable and has standardized and predefined inputs and outputs.

Domain Specific Components (DSC)	Item No
Importing external CAD Models	S1
Connecting with Rule-base X	S2
Reasoning CAD item based on X	S3
Idle	S4
...	...

Fig. 2. The slot design of a DSC usage centre

Each DSC usage centre in an agent is a slot container, which offers numerous slots for containing the specializing domain components as shown in Fig 2. Each specializing domain component possesses some special capabilities. Each domain component can plug into a DSC slot to perform specific tasks as an item, for example a domain-component can connect the agent knowledge base to a stock market database to acquire useful knowledge for the agent.

3.2   DSC-Based Agent Design

The DSC-based agent design adopts Beliefs-Desire-Intentions (BDI) model for agent reasoning as agent capabilities can be formalized in a framework [15]. The DSC-based agent design mainly focuses on agents' capabilities, which is reflected by the 'DSC' and 'agent capability registry' components.

Based on the DSC usage centre, the DSC-based agent design can be constructed as shown in the following figure.

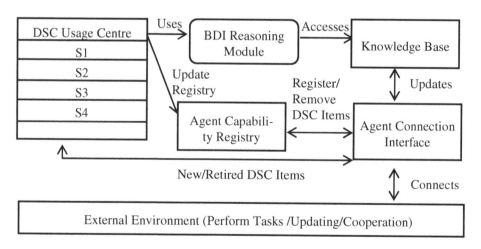

Fig. 3. DSC-based Agent Design

The DSC-based agent design consists of four major components including the DSC usage centre, BDI reasoning module, agent capability registry and agent knowledge base. These four components enhance the flexibility and extensibility of agent capability reuse.

4   Improving Efficiency of DSC Component Reuse

In the previous work, a novel bisection-based structure has been suggested to improve the reusability of the DSC-based agent design. This structure divides DSC into two sections: the front-end section, i.e. *FS*, stores the most recent visited items used by the DSC centre; the back-end section, i.e. *BS*, stores all the DSC-items. This structure adopts a recent-visit-calculation process, which consists of two steps [16].

- Step 1: Calculating the recent visit factors in *Front Section* within time period $[t_a, t_b]$. The total number of visits to a DSC item is calculated as:

$$n = \sum_{i=a}^{b} V(t_i), \text{ and } V(t_i) = 1. \tag{1}$$

where a, b denote the starting point and the ending point of the time period respectively; n is the total number of visits to a DSC item within the time period $[t_a, t_b]$; t_a is the starting point; t_i is the time point that the DSC item is visited within the time period. If a DSC item is visited at time t_i within $[t_a, t_b]$, then this DSC item is the ith visit to the DSC item and $1 \leq i \leq n$. All t_i and t_b are the converted values, which are subtracted by t_a and convert into seconds or a user-defined time scale.

Therefore, a DSC item's recent visit factor can be calculated as the following:

$$E_i = \sum_{i=1}^{n} \left(V(t_i) + \frac{t_b - t_i}{t_i} \times V(t_i) \right)^{\log_n \left(\frac{t_b}{t_i} \times n \right)}$$

$$= \sum_{i=1}^{n} \left(\frac{t_b}{t_i} \times V(t_i) \right)^{(\log_n t_b - \log_n t_i + 1)} \tag{2}$$

where E_i denotes the recent-visit-factor value of a DSC item at time t_i, which indicates a DSC item's usage within the period $[t_a, t_b]$; n is the total number of visits to the DSC and $1 \leq n$; the exponent in Eq. 2, i.e. $(\log_n t_b - \log_n t_i + 1)$, is to enlarge the recent-visit- factor value. If t_i is more recent then its E_i value is greater, meanwhile *log* function is used to limit the exponent value. Eq 2 indicates that: E_i increases when t_i increases. That means: if a visit to the DSC item is more recent then its E_i is greater. The DSC-based system developers or users can define the time period (i.e. $[t_a, t_b]$), which is configurable, to initialise and update the DSC items in *FS*.

- Step 2: Calculating the size of *Front Section*.

It is essential to determine an optimised size of *FS* since it could affect the matching efficiency of the DSC structure. A large FS will decrease the matching speed and a small FS could increase the miss-matching rate. The size of *FS* is based on the target system's DSC item usage frequency; it should also take the miss-rate into consideration. A dynamic alteration method has been used to determine *FS*'s size. An empty *FS* extracts the maximum number of the DSC items with top E_i values from *BS* within *FS*'s predefined size limit. After a period ($[t_0, t_s]$) of running (initial running period), some DSC items in *FS* will be replaced by the items in *BS* and some will be removed from *FS* because of low-usage-efficiency (refer to *B* section). The total number of DSC items in *BS* is calculated as *BS* size, i.e. *A2*. The initial step of this process is set a target miss-rate value and according to the miss-rate value and *A2* to calculate the preliminary size of *A1*.

In the dynamic alteration process, *A1* is affected by E_i values and the miss-rate of searching DSC items in *FS*. The following equitation describes the calculation process.

$$\frac{\sum\limits_{x=1}^{A1} E_x}{\sum\limits_{y=1}^{A2-A1} E_y} \approx \frac{A1}{(A2-A1)} \times \frac{1}{ms} \implies A1 \approx \frac{ms \times \sum\limits_{x=1}^{A1} E_x \times A2}{\sum\limits_{y=1}^{A2-A1} E_y + \sum\limits_{x=1}^{A1} E_x \times ms} \tag{3}$$

where *ms* denotes the miss-rate of searching DSC items in *FS* within the alteration period, and it is always ≤ 1; E_x and E_y denote the E_i values of the DSC items in *A1* and (*A2-A1*) sections, respectively.

The above two steps explain the matching process of the DSC usage centre. The next section will provide the experimental results to testify the efficiency of the bisection process.

5   Experimental Results

The DSC-based agent capability design can offer a flexible and reusable mechanism to agent-based systems. The bisection process of the DSC structure improves the efficiency of the item reuse process. A set of experiments have been conducted to evaluate the DSC-based agent capability design, particularly for the DSC item matching process. The detailed experimental design can be found in [16, 17], due to the space limitation this paper would not give the detailed description about the experimental settings and configurations.

In each experiment, a request is generated randomly based on a knowledge base, which holds around 103 capability descriptions. The DSC usage centre is also generated based on a knowledge base of an agent-based industrial application [16, 17, 18]. The experiments consist of two categories, which are the Functionality Redundancy

Calculation (FRC) experiments and the sets without using the FRC mechanism are called the NFRC sets.

Table 1. Success Rate Comparison of NFRC and FRC mechanisms

DSC item number	Redundant time	Optimized time	NFRC Success rate	FRC Success rate
6	15	12	0.11	0.11
9	18	24	0.29	0.33
12	23	68	0.20	0.29
15	30	31	0.34	0.39
18	40	139	0.42	0.55
21	50	347	0.39	0.46
24	63	169	0.52	0.65
27	80	219	0.61	0.64

Table 1 shows the matching success rates of DSC items based on FRC and NFRC mechanisms. The results show that the requests matching process in FRC is more efficient than NFRC since the all FRC-based success rates based on different request number are greater than the NFRC-based success rates. The improvement on success rates for a single agent is not enormously large based on FRC.

To further evaluate the performance of FRC, we altered the DSC item number (i.e. the DSC slot number) in a DSC usage centre to observe its impact on the success rates of request matching based on 200 requests and remain the other parameters un-changed that are: 39 relationship descriptions and 103 capability descriptions. The results are shown in Fig. 4.

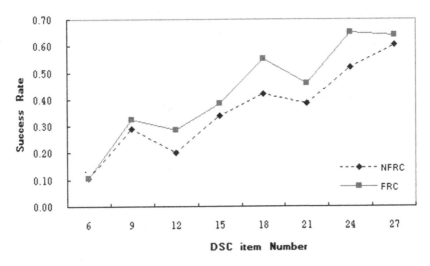

Fig. 4. Success rate comparison based on NFRC and FRC (200 requests)

6   Conclusion

An enhanced mechanism for agent capability reuse has been introduced in this paper. This mechanism adopts a Domain-Specific-Component (DSC) structure to enhance the reusability of an agent-based system.

The DSC design mechanism adopts several traditional methods including the cache model, three-type relationship, Cartesian product, and Chebyshev's theorem. These traditional methods are endowed a new meaning when they are applied to the DSC design for agent-based systems.

The empirical results prove that the FRC method adopted in the DSC-based design enhances matching efficiency and reusability of DSC items.

Acknowledgement. This project is funded by the Zhejiang Philosophy and Social Science Project Fund (Grant No. 11JCSH03YB), Overseas Scholar Science and Technology Excellent Project Funding (Ministry of Human Resources and Social Security of China, 2011), Ningbo Nature Science Fund (Grant No. 2012A610060, 2012A610025) and Ningbo Soft Science Fund (Grant No. 2012A10050).

References

1. Hewitt, C., Bishop, P., Steiger, T.: A Universal Modular Actor Formalism for Artificial Intelligence. In: Proc. of IJCAI, pp. 235–245 (1973)
2. Cailliau, R., Gillies, J.: How the Web Was Born: The Story of the World Wide Web. Oxford University Press (2000)
3. Miniwatts Marketing Group: World Internet Users and Population Stats (2010), http://Internetworldstats.com
4. Dietterich, T.G.: Machine-Learning Research: Four Current Directions. AI Magazine, 97–136 (1997)
5. Mitchell, T.M.: The Discipline of Machine Learning. Technical Report (CMUML-06-108). Carnegie Mellon University (2006)
6. Vidal, J.M.: Fundamentals of Multiagent Systems - with NetLogo Examples, Joes M. Vidal Copyright (2010)
7. Lesser, V., Decker, K., et al.: Evolution of the GPGP/TÆMS Domain-Independent Coordination Framework. Autonomous Agents and Multi-Agent Systems 9, 87–143 (2004)
8. Sycara, K., Klusch, M., Widoff, S.: Dynamic Service Matchmaking among Agents in Open Information Environments. SIGMOD Record 28(1), 47–53 (1999)
9. Gomez, M., Plaza, E.: Extending matchmaking to maximize capability reuse. In: Proc. of AAMAS 2004, New York (2004)
10. Martin, D.L., Cheyer, A.J., Moran, D.B.: The open agent architecture: A framework for building distributed software systems. Applied Artificial Intelligence 13(1-2), 91–128 (1996)
11. Vitharana, P., Zahedi, F., Jain, H.: Design, retrieval, and assembly. Communications of the ACM 46(11), 97–102 (2003)
12. Bradshaw, J.M., Dutfield, S., Benoit, P., Woolley, J.D.: KAoS: Toward an industrial-strength open agent architecture. In: Software Agents, pp. 375–418 (1997)

13. Jennings, N.R.: An agent-based approach for building complex software systems. Communications of the ACM 44(4) (2001)
14. Brazier, F.M.T., Jonker, C.M., Treur, J.: Principles of Component- Based Design of Intelligent Agents. Data & Knowledge Engineering 41(1), 1–27 (2002)
15. Padgham, L., Lambrix, P.: Agent Capabilities: Extending BDI Theory. In: Proc. of the 17th AAAI, pp. 68–73. AAAI Press & The MIT Press (2000)
16. Zhang, H.L., Zeng, W., Van der Velden, C.: A Reusable Agent Design Pattern with Flexibility and Extensibility. Computer Science and Information Systems 8(2), 1229–1250 (2011)
17. Zhang, H.L., Leung, C.H.C., Raikundalia, G.K.: Topological analysis of AOCD-based agent networks and experimental results. Journal of Computer and System Sciences 74, 255–278 (2008)
18. Zhang, H.L., Van der Velden, C., Yu, X., Jones, T., Fieldhouse, I.: Developing A Rule Engine for Automated Feature Recognition from CAD Models. In: Proc. of the 35th Annual Conference of the IEEE Industrial Electronics Society (IECON), Porto, Portugal, pp. 3925–3930. IEEE Press (2009)

A Bayesian Network Approach to Investigating User-Robot Personality Matching

Jungsik Hwang[1], Kun Chang Lee[2,*], and Jaeyeol Jeong[1]

[1] Department of Interaction Science, Sungkyunkwan University, South Korea
{jungsik.hwang,jae10123}@gmail.com
[2] Professor at Business School
WCU Professor at Department of Interaction Science
Sungkyunkwan University, Seoul 110-745, South Korea
kunchanglee@gmail.com

Abstract. Personality analysis has been an important topic in both psychology and human-robot interaction (HRI). The main theme of this paper is to explore the relationship between individuals' personality traits and their tactile interaction patterns with a robot. A sociable robot, Pleo, was used in the experiment. The tactile interaction patterns of the participants with the robot were video-recorded and analyzed. Bayesian network (BN) classifiers such as NBN (naïve BN), TAN (tree-augmented BN), and GBN (general BN) were used to examine the causal relationship between personality traits and touch patterns. The analysis showed that individuals' personality traits could be inferred based on their tactile interaction patterns with a robot. What-if and goal-seeking analysis using GBN confirmed this result. The findings of this paper are promising and its implications are discussed.

Keywords: Personality, human-robot interaction, tactile communication, Bayesian network classifier.

1   Introduction

Sociable robots or socially interactive robots are of particular interest in the field of human-robot interaction (HRI). Humans need social interaction with others, and it is widely known that people like to communicate with sociable robots. Sociable robots have been used for a variety of purposes, including education and therapy. Although there is no universally accepted definition of sociable robots, previous research [1] has suggested that one important characteristic of sociable robots is the ability to engage in human-like social interactions, such as expressing or perceiving emotions and manifesting personality and character. Because social interaction is important for sociable robots, many studies have investigated several factors, such as emotions and personality, which might have an impact on social interactions between people and robots.

---

* Corresponding author.

R. Huang et al. (Eds.): AMT 2012, LNCS 7669, pp. 124–133, 2012.

1.1   Personality in HRI

One of the emerging issues in the field of HRI is the effect of personality on the quality of an interaction. Because personality plays a crucial role in human social interactions, it is also a key element in HRI [2]. Personality can be defined as the set of distinctive qualities or characteristics that distinguish individuals. The "Big Five Inventory" is one of the most widely accepted taxonomies of personality traits [3, 4]. It describes human personality in terms of five traits: extraversion, agreeableness, conscientiousness, neuroticism, and openness.

There are two competing rules regarding the effect of personality on social interactions: the similarity attraction rule and the complementary attraction rule. The similarity attraction rule holds that people prefer to interact with those who are similar to themselves. In contrast, the complementary attraction rule holds that people tend to be attracted to those who are complementary to themselves. Several studies in the field of human-robot interaction have demonstrated either the similarity attraction rule [2] or the complementary attraction rule [4, 5].

Even though both attraction rules have been demonstrated in HRI, it is obvious that identifying the personality traits of the user is the first step in improving the quality of interactions between humans and robots. Therefore, the present paper focuses on inferring people's personality traits based primarily on their tactile interaction with a robot. In particular, we examined the relationship between touch patterns and personality traits by using the Bayesian network (BN) classifiers, with an emphasis on the general Bayesian network (GBN) classifier because the GBN can be used to identify causal relationships between variables [6].

2   Related Works

2.1   Identifying Personality Traits

Several studies in the fields of human-computer interaction and human robot interaction have investigated methods for detecting a user's personality. In [7], recognition of specific personality traits (extraversion and locus of control) in social interactions was examined. Acoustic and visual features were used to classify people's personality traits. In [3], researchers demonstrated that computer users' personality traits could be measured through keyboard and mouse usage. In particular, a subtrait of extraversion turned out to be significantly correlated to the standard deviation of the average time between events.

The main focus of this paper is to identify people's personality traits based on their tactile communication with sociable robots. Animal-like robots are especially good for people to interact with because they have physical bodies and behave autonomously [8, 9]. People's interaction with animal-like robots is mainly through tactile communication. Touch can convey a wide variety of communicative intents, especially in relationships between humans and animals.

2.2   Tactile Communication

The role of touch as a form of communication in human social interactions has been investigated in several studies [10, 11]. For example, the effect of individual differences on tactile behaviors was examined in [11]. The result revealed that several personality factors, such as dominance and surgency, could be used to distinguish tactile-expressive people from less tactile-expressive people.

In the field of HRI, the effect of touch has been also investigated. As social affective touch plays a crucial role for companion animals, it is reasonable to assume that social affective touch would also play an important role for companion robots. Salter et al. [12] investigated the role of touch in HRI. They observed how children played with a mobile robot, and then they collected the robot's infrared sensor readings. The study found that the children's personality types were conveyed by their behavior during their interaction with the robot. Yohanan and MacLean [13] also investigated the role of touch in human-robot interaction. They implemented a robot called the Haptic Creature, which interacted with people mainly through the modality of touch. The researchers examined how people used different kinds of touch to express specific emotions when communicating with the robot. They found that human emotions and intents were revealed through different touches.

The main objective of this paper is to investigate the relationship between people's personalities and their tactile interaction patterns with a robot. In particular, we examined whether the types of tactile interaction between a person and a robot could reveal an individual's personality traits. However, such tactile interaction tends to encompass a wide variety of factors pertaining to both the people and the robot, indicating that we need to consider causal relationships among the relevant factors. For this reason, this paper adopts the GBN as the main vehicle for extracting causal relationships among a set of relevant factors and performing simulations with them to understand the chain links of cause and effect with respect to specific conditions. To the best of our knowledge, this paper is the first to explore the relationship between an individual's personality traits and tactile interaction patterns using BN classifiers.

3   Experiment

3.1   Apparatus

A dinosaur robot from Innvo Labs, Pleo [14], was used in the experiment. Pleo is an interactive robot, and it has been successfully used in the field of human-robot interaction to study human behavior [9]. Pleo contains several actuators and touch sensors inside. The touch sensors embedded on Pleo's head, chin, shoulder, rear, and legs enable Pleo to interact with people by touch. Previous research [2, 4, 5] has shown that the artificial agent's personality has an impact on social interaction. Therefore, we manipulated Pleo's personality to manifest different personality traits (either extraversion or introversion) prior to the experiment. The analysis showed that the manipulation was successful, which means that the participants were able to distinguish two different personalities in Pleo.

3.2   Participants

A total of 31 university students (11 females and 20 males) were recruited for the experiment. The participants' average age was 24.03 with a standard deviation of 2.18. Most of the participants reported that they had no previous experience with Pleo or other types of companion robots. The participants were randomly assigned to either extraverted or introverted Pleo.

3.3   Measurements

Personality. Among the five dimensions of the "Big Five Inventory" [15], we focused on the extraversion dimension because it has been reported to be particularly important in HRI [4]. We used 8 Wiggins [16] personality adjective items in Likert scale format to measure each participant's degree of extraversion. These items have been successfully used to measure personality traits in past studies [4, 5]. The items (α = 0.87) used in the experiment were "cheerful," "enthusiastic," "extravert," "bold," "introvert" (reversely coded), "inward" (reversely coded), "shy" (reversely coded), and "quiet" (reversely coded). The participants were asked to indicate their responses on a scale from 1 (strongly disagree) to 5 (strongly agree).

Tactile Interaction Patterns. To capture the tactile interaction patterns of the participants with Pleo, the participants' interactions with Pleo were video-recorded. The experimenter left the room during the video-recording to leave the participant alone with Pleo so that the participant could interact with Pleo naturally. In the analysis phase, the video-recorded tactile interaction patterns were coded into two items (touched location and type of touch) by two researchers working independently. The touched locations and types of touch were coded along with the time that the touch occurred. Table 1 shows the coded items for each element. We assumed that the touched locations and types of touch would convey different meanings or intentions, and thus they would reveal individuals' personality traits.

Table 1. The coded items and their elements

Items	Elements
Touched locations	head, mouth, jaw, neck, back, tail, belly, front legs, rear legs, other parts
Types of touch	tickle, pat, slap, pick up, hold, shake, poke, pull and push, other types

Fig. 1. The experimental setting

3.4   Procedure

The participants were asked to sign an informed consent upon arrival at the experiment room. Then, the participants were given the questionnaire, which was designed to measure their personality traits and collect demographic information. After they completed the questionnaire, participants were shown tutorial PowerPoint slides. The slides provided a brief introduction to Pleo, describing its shape and function. Detailed information, such as the location of the sensors, was not given to the participants because such information could hinder natural interactions and limit the patterns of tactile interaction. The participants also watched two short video clips: one with Pleo acting alone and one with Pleo playing with a person. After the participants watched the entire tutorial, Pleo was given to them. The participants were asked to play with Pleo for 5 minutes as freely as they wanted. The experimenter started the video-recording and the timer and left the room during the recording. After 5 minutes of free play with Pleo, the participants were asked additional questions. Fig. 1 illustrates the experimental setting.

3.5   Analysis Tool

We used WEKA (Waikato Environment for Knowledge Analysis) [17] to model a Bayesian network to classify personality traits based on tactile interactions with the robot. WEKA provides a toolbox of various classification techniques, such as Bayesian networks and neural networks. In the present paper, we used the Bayesian network classifier to investigate the causal relationships between variables. A Bayesian network is a probabilistic graphical model that illustrates conditional dependencies between variables. In a Bayesian network, each node and arc represents a variable and a probabilistic dependency, respectively. A Bayesian network can be used as a classifier that gives the posterior probability distribution of the class node given the values of the other variables. In addition, the Bayesian network structure is able to represent causal interrelationships among the nodes [6].

4   Results

4.1   Data Preparation

As in previous studies [4, 5], this paper focused on extraversion versus introversion. Therefore, 7 participants showing neither extraversion nor introversion were removed from the original pool of 31 participants to avoid statistical bias. Subsequently, data obtained from 24 participants was included in the dataset. Two researchers each independently coded and analyzed the video-recorded data of the 24 participants. Consequently, 48 records were used as the dataset in the analysis. The dataset was then reprocessed into 5 different datasets, each representing a duration of interaction. Measurement of the 5 interaction durations began at the first touch and ended 1 minute, 2 minutes, 3 minutes, 4 minutes, and 5 minutes later.

The current research examined 24 nodes in the Bayesian network. There were 11 nodes representing the number of touches to the following robot body parts: head, mouth, jaw, neck, back, tail, belly, front legs, rear legs, and other parts of the body. There were 9 nodes representing the types of touches: tickle, pat, slap, pick up, hold, shake, poke, pull and push, and other types of touch. In addition, there were nodes representing the total number of touches, the number of touched locations, the number of kinds of touch, Pleo's personality, and the participant's personality. The nodes were discretized into an optimal number of equal-width bins by using the leave-one-out method provided in WEKA.

4.2   Study 1: Classification Performance

Several algorithms implemented in WEKA were used for Bayesian network structure learning. In study 1, we compared three different types of Bayesian network: the general Bayesian network (GBN), tree-augmented naïve Bayesian network (TAN), and naïve Bayesian network (NBN). The GBN is a kind of unrestricted Bayesian network classifier; it shows the causal relationship between all nodes. The TAN is an extended version of the NBN where the nodes form the shape of a tree. The NBN is the simplest structure; in it, the class node is the parent of all the other nodes. In addition, the NBN does not consider causal relationships between the child nodes. The Bayesian network learning algorithm (HillClimber) showing the best performance on the dataset was found empirically. According to [18], k-fold cross-validation elicits approximately stable estimates independent of k (number of folds). Therefore, 7-fold cross-validation was performed on the datasets. Table 2 shows the classification performances of each Bayesian network according to the different datasets (duration of interaction).

Table 2. The classification performance of each network

	Duration of Interaction				
	1 minute	2 minutes	3 minutes	4 minutes	5 minutes
GBN	77.08%	52.08%	54.17%	58.33%	52.08%
TAN	64.58%	45.83%	54.17%	52.08%	52.08%
NBN	60.42%	56.25%	54.17%	62.5%	62.5%

The performance of the NBN was used as a standard against which the performances of other algorithms could be compared. As can be seen in Table 2, the GBN elicited the best classification performance (77.08%) of the three networks in the first minute of interaction.

4.3   Study 2: What-If and Goal-Seeking Analysis

Study 1 confirmed that individuals' personality traits could be successfully classified by the GBN. The later analysis showed that among the 23 nodes, there were 4 nodes related to individuals' personality traits (class node): the number of pick-ups, the number of rear-leg touches, the number of mouth touches, and Pleo's personality (either extraverted or introverted). Fig. 2 shows the causal relationship between these nodes. These 4 nodes were found to have a direct relationship to people's personality traits. In addition, the number of rear-leg touches was associated with Pleo's personality. The structure of the network was automatically learned using the HillClimber algorithm.

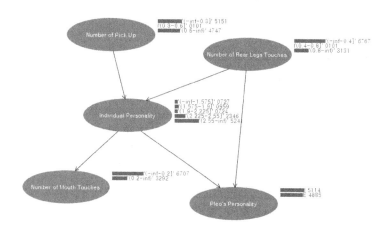

Fig. 2. The GBN with individual personality as a target node

Based on the model depicted above, we conducted what-if analysis and goal-seeking analysis by testing two scenarios.

Scenario 1 (What-If Analysis): If a person picked up extraverted Pleo a few times, touched its rear legs often, and touched its mouth rarely, then how extraverted would that person be?

Originally, the prior probability of a high degree of extraversion had the largest probability. However, when other factors were set to the values indicated by scenario 1, the prior probability of a low degree of extraversion (i.e., introversion) became the highest probability. This means that a person showing the tactile interaction patterns described in scenario 1 would probably have an introverted personality. Fig. 3 illustrates the result of the what-if analysis.

Scenario 2 (Goal-Seeking Analysis): What kinds of tactile interaction patterns would an extraverted person exhibit?

When the individual's personality was set to the highest level (i.e., extraverted), the probability distribution of the other 4 variables changed. Most notably, the number of rear-leg touches decreased, but the number of pick-ups increased. This implies that an extraverted person would probably pick up Pleo more often and barely touch its rear legs compared to an introverted user. Fig. 4 illustrates the result of the goal-seeking analysis.

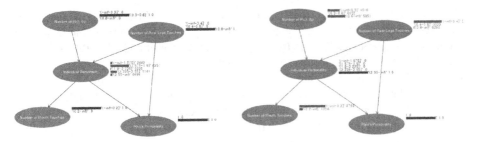

Fig. 3. The result of what-if analysis   **Fig. 4.** The result of goal-seeking analysis

5   Discussion

The result from study 1 revealed that the personality traits could be classified by the tactile interaction patterns with the robot. Among the three different BN classifiers, the GBN showed the highest classification performance (77.08%) in the 1-minute interaction period. Previous studies [7, 19] stressing the importance of a "thin slice" approach for the detection of personality traits supported our experiment's result. Previous studies [7, 19] demonstrated that interaction during a short period of time can be used to classify personality traits more clearly than interaction over a longer period of time. Through our experiments, we also found that that the classification performance using the GBN was highest when the data from the first 1 minute of interaction were used. Moreover, the classification performances drastically decreased afterwards, as shown in Table 2. The video analysis revealed that the most of participants (both extraverted and introverted) showed the greatest interest in Pleo during the first one minute and actively interacted with it during that time. However, the participants seemed to get bored as time went by, and the tactile interaction patterns of the participants became monotonous, resulting in a low classification performance. Thus, we assumed that the individuals' true personality traits were not conveyed through touch in the later part of the interaction.

In study 2, we first found that there were 4 variables that were directly related to the individuals' personality traits. Second, the what-if and goal-seeking analyses revealed the causal relationships among these variables. For example, the analyses showed that introverted people tended to touch Pleo's mouth more than extraverted people but that extraverted people tended to pick Pleo up more often than introverted people. This supported the finding of study 1 that people's personality traits could be inferred from the types of tactile interactions they engaged in with the robot.

6   Conclusion and Future Work

We examined whether people's tactile interactions with a robot could convey their personality traits. We found that an individual's degree of extraversion could be inferred from the person's touch patterns, such as touched locations (e.g., mouth and rear legs) and types of touch (e.g., tickling and picking up). In addition, the GBN was employed to investigate possible causal relationships between the target node and the rest of the relevant nodes that are believed to affect the target node. For the purpose of comparison, the GBN performance was compared with other BN classifiers such as NBN and TAN. We found that the GBN has great potential in terms of prediction accuracy and as a main source of providing causal relationships with which researchers can analyze user-robot personality matching tasks more effectively. One contribution of this paper is that the GBN was first adopted to classify individuals' personality traits based on their tactile interaction patterns. As in previous studies [1, 4, 5], the experiment results revealed that not only the robot's personality but also the individual's personality affected the satisfaction level of the interaction. Therefore, identifying people's personality traits is essential to achieving satisfactory interactions. Our results show the possibility of discovering users' personality traits based on tactile interactions. By sensing user personality traits through touch, a robot could adapt its behavior to increase its attractiveness or the satisfaction of the interaction.

This paper has limitations that can probably be overcome in future research. First, the shape of a robot might have an impact on tactile interaction patterns. For example, the way that people interact with animals is different from the way that people interact with other people. Therefore, it is assumed that people might exhibit different tactile interaction patterns when they are interacting with humanoid robots or machine-like robots. Thus, robot shape could be considered as a moderating factor in future research. Second, certain tactile interaction patterns were not considered in the present research. In addition to the touched locations and the types of touch, other tactile interaction patterns, such as positive and negative touch, might reflect individuals' personality traits. In particular, the video analysis revealed that some participants often kissed Pleo, indicating a highly positive attitude toward the robot. Hence, different types of tactile interaction patterns need to be considered in future research.

Acknowledgements. This study was supported by a grant from the World-Class University program (R31-2008-000-10062-0) of the Korean Ministry of Education, Science and Technology via the National Research Foundation, and also supported by the ubiquitous Computing and Network (UCN) Project, the Ministry of Knowledge and Economy (MKE) Knowledge and Economy Frontier R&D Program in Korea as a result of UCN's subproject 11C3-T2-20S.

References

1. Breazeal, C.: Social interactions in HRI: the robot view. IEEE Transactions on Systems, Man, and Cybernetics, Part C: Applications and Reviews 34, 181–186 (2004)
2. Tapus, A., Țăpuș, C., Matarić, M.: User—robot personality matching and assistive robot behavior adaptation for post-stroke rehabilitation therapy. Intelligent Service Robotics 1, 169–183 (2008)
3. Khan, I.A., Brinkman, W.-P., Fine, N., Hierons, R.M.: Measuring personality from keyboard and mouse use. In: Proceedings of the 15th European Conference on Cognitive Ergonomics: The Ergonomics of Cool Interaction, pp. 1–8. ACM, Portugal (2008)

4. Lee, K.M., Peng, W., Jin, S.-A., Yan, C.: Can Robots Manifest Personality?: An Empirical Test of Personality Recognition, Social Responses, and Social Presence in Human–Robot Interaction. Journal of Communication 56, 754–772 (2006)

5. Isbister, K., Nass, C.: Consistency of personality in interactive characters: verbal cues, non-verbal cues, and user characteristics. International Journal of Human-Computer Studies 53, 251–267 (2000)

6. Madden, M.G.: On the classification performance of TAN and general Bayesian networks. Knowledge Based System 22, 489–495 (2009)

7. Pianesi, F., Mana, N., Cappelletti, A., Lepri, B., Zancanaro, M.: Multimodal recognition of personality traits in social interactions. In: Proceedings of the 10th International Conference on Multimodal Interfaces, pp. 53–60. ACM, Greece (2008)

8. Goris, K., Saldien, J., Vanderborght, B., Lefeber, D.: Probo, an Intelligent Huggable Robot for HRI Studies with Children. In: Chugo, D. (ed.) Human-Robot Interaction. InTech (2010) ISBN: 978-953-307-051-3

9. Kim, E.S., Leyzberg, D., Tsui, K.M., Scassellati, B.: How people talk when teaching a robot. In: Proceedings of the 4th ACM/IEEE International Conference on Human Robot Interaction, pp. 23–30. ACM, La Jolla (2009)

10. Malphurs, J.E., Raag, T., Field, T., Pickens, J., Pelaez-Nogueras, M.: Touch by Intrusive and Withdrawn Mothers with Depressive Symptoms. Early Development and Parenting 5, 111–115 (1996)

11. Deethardt, J.F., Hines, D.G.: Tactile communication and personality differences. Journal of Nonverbal Behavior 8, 143–156 (1983)

12. Salter, T., Dautenhahn, K., Boekhorst, R.T.: Learning about natural human–robot interaction styles. Robotics and Autonomous Systems 54, 127–134 (2006)

13. Yohanan, S., MacLean, K.: The Role of Affective Touch in Human-Robot Interaction: Human Intent and Expectations in Touching the Haptic Creature. International Journal of Social Robotics 4, 163–180 (2012)

14. PLEOworld (July 28, 2012), http://www.pleoworld.com

15. John, O., Srivastava, S.: The Big Five trait taxonomy: History, measurement, and theoretical perspectives. In: Pervin, L., John, O. (eds.) Handbook of Personality: Theory and Research, pp. 102–138. Guilford Press (1999)

16. Wiggins, J.S.: A psychological taxonomy of trait-descriptive terms: The interpersonal domain. Journal of Personality and Social Psychology 37, 395–412 (1979)

17. Hall, M., Frank, E., Holmes, G., Pfahringer, B., Reutemann, P., Witten, I.H.: The WEKA data mining software: an update. SIGKDD Explor. Newsl. 11, 10–18 (2009)

18. Kohavi, R.: A study of cross-validation and bootstrap for accuracy estimation and model selection. In: Proceedings of the 14th International Joint Conference on Artificial Intelligence, vol. 2, pp. 1137–1143. Morgan Kaufmann, Canada (1995)

19. Ambady, N., Rosenthal, R.: Thin slices of expressive behavior as predictors of interpersonal consequences: A meta-analysis. Psychological Bulletin 111, 256–274 (1992)

Modelling Multi-Criteria Decision Making Ability of Agents in Agent-Based Rice Pest Risk Assessment Model

Vinh Gia Nhi Nguyen[1], Hiep Xuan Huynh[2], and Alexis Drogoul[1]

[1] UMI 209 UMMISCO, IRD/UPMC, Institut de recherche pour le développement
Bondy, France
{nngvinh,alexis.drogoul}@gmail.com
[2] DREAM Team, School of Information and Communication Technology,
Cantho Univesity, Vietnam
hxhiep@ctu.edu.vn

Abstract. This paper aims to introduce an agent-based multi-criteria assessment model that support stakeholders to make multi-criteria decisions for rice pest management, this study designs agents with dynamic ability of multi-criteria decision making from information of other agents so that rice pest risk maps are built. A case study is carried out on risk assessment of brown plant hoppers in the Mekong Delta region, simulation results show that rice pest risk index aggregated from criteria and agent-based models are useful tools to support rice pest management.

Keywords: Multi-criteria decision making (MCDM), Agent-based model, ordered weighted averaging (OWA) operator, Rice pest risk (RPR) index, GAMA platform, Brown plant hopper (BPH).

1   Introduction

Rice production is almost affected by rice pest factor because rice pests are not only attack rice but also carry viruses that cause rice disease. There are more than 100 species of pests in rice fields [11] and these rice pests have large impacts on crop losses when their density in rice fields is high. Food and agriculture organization (FAO) has strongly recommended rice producing countries to use integrated pest management (IPM) programs for crop protection [4]. The Vietnamese government applied IPM programs from 1992 [10], IPM strategies requires stakeholders to make rice production and crop protection plans while reducing rice pest population development and minimizing uses of pesticides. Beside advantages of IPM programs, farmers are still facing risks such as weather conditions, flood, potential outbreaks of rice pests and rice diseases, etc. Rice pest population dynamics depend on factors such as rice varieties, rice growth stages, weather conditions (temperature, humidity, rainfall, wind), natural enemies, rice transplanting calendar, fertilizer, insecticide and other environmental indicators. If rice areas have suitable environmental conditions for rice pest residence, migrant rice pest groups will invade into and rice pest density in those areas will be high. Rice pest risk index in this paper is defined as potential

R. Huang et al. (Eds.): AMT 2012, LNCS 7669, pp. 134–144, 2012.

risk level that one rice area is infested by one kind of rice pests, stakeholders can use this index to know which rice areas have potential high pest risk level for preventing purpose. However, building a rice pest risk index have to consider factors/criteria that affect to rice pest life cycle and these factors/criteria need to be aggregated to a single value so that this value can be used to support multi-criteria decision making process in pest risk assessment models. This study proposes a method for building rice pest risk index and designs an agent-based multi-criteria assessment model (called AMCA model) that enable agents like Decision Maker to determine rice pest risk index. The remainder of this paper is outlined as follows: Section 2 summarizes concepts of OWA operators, multi-criteria decision making process and agent-based models in the Mekong Delta region. Method to build rice pest risk index and AMCA model are presented in section 3, an application of assessing risks of Brown plant hopper is described in section 4 and conclusion is pointed out in section 5.

2   Related Work

In practices, multi-criteria decision making process has to be coped with preferences of decision makers and a large collection of information sources, thus computer based decision support systems play an important role in transferring knowledge of experts and opinions of decision makers to MCDM process. In rice pest risk management, choosing criteria related to rice pest life cycle and understanding relative influences between these criteria are key answers to help decision makers assess overall rice pest risk state of regions, multi-criteria decision making process requires aggregating criteria from information sources to create a single value that reflects common meaning of criteria. There are methods that support for unifying criteria [6][17], ordered weighted averaging (OWA) operators are developed by Yager [18] and applied widely in economic and environmental applications, an OWA operator of dimension n is a mapping F: $R^n \rightarrow$ R and has an associated weight vector $W = (w_1, w_2, ..., w_n)^T$ such that it satisfies:

$$w_1 + w_2 + ... + w_n = 1, \text{ with } w_i \in [0,1] \text{ and } i = \overline{1,n} \tag{1}$$

$$F(c_1, c_2, ..., c_n) = \sum_{j=1}^{n} w_j \, b_j \tag{2}$$

Where b_j is the jth largest of values c_i, the function F computes aggregated value from arguments: $c_1, c_2, ..., c_n$. OWA operators have properties of Max, Min and arithmetic mean operators that depend on special forms of weight vector W [18], OWA operators also satisfy properties: commutativity, monotonicity and idempotency [17]. Two measurement factors are used to assess selection of weight vector W:

$$Disp(W) = -\sum_{i=1}^{n} w_i \ln w_i \tag{3}$$

$$Orness(W) = \frac{1}{(n-1)} \sum_{i=1}^{n} (n-i) w_i \tag{4}$$

Dispersion factor (*Disp*) measures degree of information in arguments that is used in aggregation process. Orness factor characterizes andlike or orlike aggregation, this value implies type of aggregation and is in the interval [0,1]. Yager [18] based on variety of problems to describe different methods to obtain weights, Liu [6] classified OWA determination methods into categories: optimization criteria methods, sample learning methods, function based methods, argument dependent methods and preference methods. This paper uses probability density function method [14] to find weights because of its simplicity and flexibility to reach desired degree of Dispersion and Orness [6][14]. In the Mekong Delta-Vietnam, agent-based models has been applied in studies of Brown plant hopper (BPH) that is one kind of the most harmful pests of rice [5]. Studies in [12][15] focused on determining BPH density (individuals/m2) on large scale and predicting the target rice areas that BPH groups can invade into. Agent-based model [7] built two sub-models at field scale: BPH growth model and BPH invasion model on GIS environment, impacts of weather conditions such as temperature, humidity, rainfall and wind were also considered in this model. The MSMEKO model [8] explored possible interactions among agents: decision maker, BPH, rice, weather and spatial objects from GIS data (province, district, commune), this model also used a upscaling method by averaging to assess rice area infested by brown plant hoppers from field to region scales so that the model could help stakeholders in planning agricultural strategy. An extension of the MSMEKO model [9] that used agricultural indicator and OWA operator to determine subjective evaluation values of districts so that stakeholders could know which districts in a province are most influenced by the indicator. AMCA model proposed in this study can help stakeholders understand overall rice pest risk states through *Decision Maker* agents.

3   Agent-Based Multi-Criteria Assessment (AMCA) Model for Rice Pest Management

The AMCA model acts as virtual laboratory to assess rice pest risk levels at different spatial scales (commune, district, province and region). To detect rice areas with high rice pest risk, stakeholders consider criteria/conditions that most impact to pest life

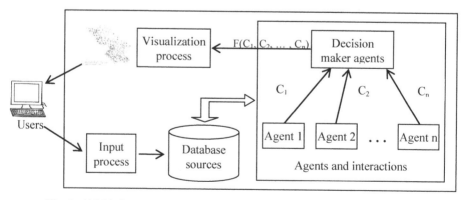

Fig. 1. AMCA framework for evaluating rice pest risk by using OWA operator

cycle to find out suitable habitats for rice pests so that multi-criteria decision making process will focus on these rice areas. Fig.1 shows AMCA framework for rice pest risk evaluation.

Using agent-based model as a tool to help imitate and foresee behaviors of objects in real world [2], in rice production problems, agents can be farmer, decision maker, rice, land use, rice pest, weather, light trap, etc. Interactions among agents have to be clarified. This study equips *Decision Maker* agents an ability to make multi-criteria decisions by aggregating information of other agents. From agents and their interactions, experts can modify flexibly attributes and decision rules of agents, stakeholders can simulate rice pest management strategies to understand emergent phenomena and adjust their plans for purposes of sustainable balance between rice production and natural environment. *Decision Maker* agent at the corresponding scale will aggregate criteria from outputs of other agents within geographic domain that this *Decision Maker* agent manages.

Multi-criteria decision making (MCDM) process in AMCA model as follows:

Step 1. Select criteria that influence rice pest risk events and are available at decision making scale. Standardize criteria so that a comparison between criteria can be made.

Step 2. Determine weights of criteria by using probability density functions [14].

Step 3. Calculate rice pest risk index of each region and visualize rice pest risk GIS map at desired scale.

Step 4. Participate stakeholders and experts in discussion and choosing rice regions with potential high rice pest risk.

Criteria can be selected from opinions of experts, existing rice pest studies and from statistical methods like regression and correlation analysis. A set of criteria is built to help stakeholders determine rice pest risk index, re-addressing factors in step 1 can be repeated for adjusting purpose. The fact is that spatially explicit quantitative data of rice pests are available at observation scale and normally are lacked at region scale, thus stakeholders assess regional rice pest risk by using regional criteria that are derived from statistics and upscaling information models [1][8]. Data of criteria can be entered or be outputs of sub-models inside agents, e.g. outputs of weather forecasting model inside weather agent.

Normal probability density function method [14] is used to calculate weights of OWA, the basic idea is that weight vector W can be considered as a random variable with normal distribution because weight values satisfy conditions in Eq. (1) and (2). The normal distribution has probability density function with mean μ and standard deviation σ:

$$f(x) = \frac{1}{\sigma\sqrt{2\pi}} e^{-\left((x-\mu)^2/(2\sigma^2)\right)}, -\infty < x < +\infty \tag{5}$$

Let n be number of criteria to be aggregated, OWA weights are calculated by:

$$w_i' = \frac{1}{\sigma_n\sqrt{2\pi}} e^{-\left((i-\mu_n)^2/(2\sigma_n^2)\right)}, with \quad i = \overline{1,n} \tag{6}$$

Where μ_n and σ_n are mean and standard deviation of a set of values $\{1, 2, ..., n\}$:

$$\mu_n = \frac{1}{2}(1+n) = \lambda(1+n) \qquad\qquad \sigma_n = \sqrt{\frac{1}{n}\sum_{i=1}^{n}(i-\mu_n)^2} \qquad (7)$$

Normalization of Eq. (6) is proceeded to obtain weight vector W:

$$w_i = \frac{w_i'}{\sum_{j=1}^{n} w_j'}, \quad i = \overline{1,n} \qquad (8)$$

The parameter λ=0.5 will make criteria that is near positions of minimum and maximum values after ordering have low weight values. If $\lambda \in [0,0.5]$, criteria with high values will get high weight values (positively skewed distribution). If $\lambda \in [0.5,1]$, values of weights have distribution like regular increasing monotone function.

In practices, criteria normally are measured in different units (e.g. rice pest density is count number, temperature is in ^0C, ...), thus criteria values need to be standardized by using a common transformation so that experts can compare between criteria and rice pest risk index is computed from standardized values. In this paper, values of criteria are transformed in ranges from 0 to 1, i.e. criteria with value 0 represent no effect on rice pest risk and criteria with value 1 represent the highest effect. AMCA model is implemented on GAMA platform [3], users can enter data of criteria, land use and administrative GIS maps. The model will output resultant rice pest risk GIS maps so that users can select rice areas with the highest rice pest risk index.

4   A Case Study of Risk Assessment of Brown Plant Hopper

MSMEKO model in [8] designed agents and interactions among them to support stakeholders in controlling brown plant hopper (BPH). AMCA model will be integrated into MSMEKO model for purpose of BPH risk assessment at administrative scale (commune, district, province), *Decision Maker* agents at each scale have an action called *pestRiskAssessing()* to aggregate criteria from other agents. From studies in [5][7][13], there are criteria that most affect to BPH risks: temperature, humidity,

Table 1. Transformed criteria values from real data

Criteria	Real value	Qualitative values	Converted values
Temperature	25^0C-30^0C	High	3
	$<25^0C$ or $>30^0C$	Low	1
Humidity	80%-86%	High	3
	<80% or >86%	Low	1
Light trap data (individuals)	>=3000	High	3
	>=1500 and <3000	Medium	2
	<1500	Low	1

rainfall, light trap data, BPH susceptible rice variety, rice transplanting time, natural enemies, pesticide treatments, nitrogen fertilizer applications. Stakeholders need to aggregate these criteria to a single value that characterizes rice pest risk state of one region so that stakeholders have a general view of rice pest risk states between regions before making decisions.

In this study, simulation is experimented on available criteria at district scale: temperature, humidity, light trap data, proportion of BPH susceptible rice area. In addition, some criteria are divided into interval ranges because of suitable levels for BPH risk, a qualitative measurement scale of BPH risk levels (high, medium, low) is used to transform interval ranges of criteria into this scale, then transformed risk levels are converted to corresponding numerical values (3,2,1). This transformation allows users convert interval ranges of criteria to a single utility score [16], table 1 shows an example of converted criteria values, proportion of BPH susceptible rice area is numerical value (min: 0% and max: 100%). To standardize all criteria before aggregation, a linear scaling function [16] is used to map values of criteria into the interval [0,1]:

$$x_i^{new} = \frac{(x_i - x_{min})}{(x_{max} - x_{min})}.NewR \qquad (9)$$

Where x_i^{new} is standardized value, x_i is value of criteria i, x_{min} and x_{max} are the minimum and maximum values of criteria i, $NewR$ is new standardized range ($NewR$=1 for the interval [0,1]). Table 2 presents an example of standardizing criteria:

Table 2. A case of standardized criteria

Criteria	Observed value	Converted values	Standardized value
Temperature	28^{0}C	3	1
Humidity	83%	3	1
Light trap data (individuals)	2100	2	0.5
Proportion of BPH susceptible rice area	65%		0.65

Rice pest risk (RPR) index of each region is determined as follows:

1. Sort values of standardized criteria in descending order:
   For example, standardized values in table 2 after sorting: b= (1, 1, 0.65, 0.5)
2. Compute weight values when parameters n and λ are known:
   In this study, parameter λ=0.1 is chosen because stakeholders need to find rice areas with the highest rice pest risk index. From Eq. (6), (7) and (8), when n=4, weight values w_i are: w_1 =0.369, w_2 =0.305, w_3 =0.208, w_4 =0.118. Orness and Dispersion are computed from Eq. (3) and (4): $Disp(W)$= 1.309, $Orness(W)$=0.642.
3. From Eq. (2), RPR index is calculated:
   RPR = 0.369 * 1 + 0.305 * 1 + 0.208 * 0.65 + 0.118 * 0.5 = 0.868

When *Decision Maker* agent at province scale sends a message to require *Decision Maker* agents at district scale determine rice pest risk index. Action *pestRiskAssessing()* of *Decision Maker* agent at district scale performs aggregation algorithm:

```
declaration action pestRiskAssessing()
    {
        collect data of criteria from other agents;
        standardize criteria criteria to the interval [0,1];
        calculate RPR index;
        return RPR index;
    }
foreach agent is in [Decision_Maker agents at district scale] do
        do action pestRiskAssessing();
        visualize RPR index of each district in GIS maps;
```

Table 3. Weekly raw data of criteria at district scale

District code	District Name	Temperature	Humidity	BPH data in light trap	Proportion of BPH susceptible rice area (%)
103153	TX.CaoLanh	28.50	81.70	37	7.36
103154	CaoLanh	27.63	80.07	4123	7.00
103155	ChauThanh	24.30	81.85	12	6.08
103156	HongNgu	27.05	79.55	7066	27.35
103157	LaiVung	27.50	76.85	719	4.50
103158	LapVo	27.53	79.90	12928	12.63
103159	SaDec	30.00	74.50	152	2.70
103160	TamNong	27.50	77.45	1479	53.24
103161	TanHong	29.00	81.40	491	65.00
103162	ThanhBinh	27.85	78.40	881	76.76
103163	ThapMuoi	27.17	81.43	326	41.61

Table 4. Standardized values of criteria

District code	Temperature	Humidity	BPH data in light trap	Proportion of BPH susceptible rice area (%)
103153	1.00	1.00	0.00	0.07
103154	1.00	1.00	1.00	0.07
103155	0.00	1.00	0.00	0.06
103156	1.00	0.00	1.00	0.27

Table 4. *(Continued)*

103157	1.00	0.00	0.00	0.05
103158	1.00	0.00	1.00	0.13
103159	1.00	0.00	0.00	0.03
103160	1.00	0.00	0.00	0.53
103161	1.00	1.00	0.00	0.65
103162	1.00	0.00	0.00	0.77
103163	1.00	1.00	0.00	0.42

An application of risk assessment of brown plant hopper (BPH) is experimented on 11 districts of *Dong Thap* province in the Mekong Delta region, Vietnam. A system of 24 light traps is setup in this province to catch BPH adults and support stakeholders to predict new BPH invasions. To determine BPH risk levels at district scale for the next week, criteria: temperature, humidity, light trap data and proportion of BPH susceptible rice area are used to calculate BPH risk index of each district. Data of temperature and humidity are forecast average values at district scale in the next week. Proportion of BPH susceptible rice area and light trap data at district scale are computed from averaging values of these criteria in communes that belongs to each district, these values are updated weekly. Table 3 describes weekly raw data of criteria in 11 districts of *Dong Thap* province in winter-spring rice crop. Criteria with interval ranges are transformed by using rules in table 1 and then all criteria are standardized by using linear scaling function in Eq. (9). Table 4 shows standardized values of criteria.

Table 5. Results of RPR index in districts

District code	District Name	b_1 0.369	b_2 0.305	b_3 0.208	b_4 0.118	RPR index
103153	TX.CaoLanh	1.00	1.00	0.07	0.00	0.69
103154	CaoLanh	1.00	1.00	1.00	0.07	0.89
103155	ChauThanh	1.00	0.06	0.00	0.00	0.39
103156	HongNgu	1.00	1.00	0.27	0.00	0.73
103157	LaiVung	1.00	0.05	0.00	0.00	0.38
103158	LapVo	1.00	1.00	0.13	0.00	0.70
103159	SaDec	1.00	0.03	0.00	0.00	0.38
103160	TamNong	1.00	0.53	0.00	0.00	0.53
103161	TanHong	1.00	1.00	0.65	0.00	0.81
103162	ThanhBinh	1.00	0.77	0.00	0.00	0.60
103163	ThapMuoi	1.00	1.00	0.42	0.00	0.76

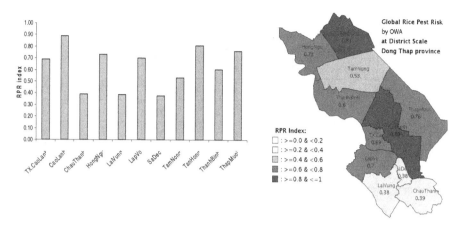

Fig. 2. RPR index chart of districts

Fig. 3. Rice pest risk GIS map of districts

Table 5 presents results of RPR index at district scale, values b_j in Eq. (2) are values at each district row in table 5 after sorting standardized values of criteria (table 4) in descending order, values of weights are in the second row of the table.

Results of rice pest risk index are visualized in chart (Fig.2) and GIS map (Fig.3). In Fig.2, *Cao Lanh* district has the highest rice pest risk index and districts such as *Tan Hong, Thap Muoi, Hong Ngu* have also rather high rice pest risk index in comparison with other districts. From the results, stakeholders will take into account these districts when making rice crop protection plans in next weeks. Because values of criteria are changed over time, RPR index dynamics is also changed. RPR index could be determined at other scales (commune, province) and number of criteria can be different at each scale. This paper aims to explain method of building rice pest risk index in detail, a simple example of BPH risk assessment considers 4 criteria, results of simulation show that rice pest risk index along with experience of stakeholders will be useful for rice pest management.

5   Conclusion

This paper proposes an agent-based multi-criteria assessment model that can be a tool to support stakeholders in assessing global rice pest risk state of system, *Decision Maker* agents use OWA based method and available criteria that are resulted from other agents to determine rice pest risk index of regions. AMCA model is combined with other components of MSMEKO model to help stakeholders evaluate strategies of controlling rice pests. In practices, criteria have relationships with each other, OWA based method does not consider interactions among criteria, thus, choosing method that takes into account interactions between criteria needs to be done in the next step.

Acknowledgments. Authors have many thanks to staffs in Dong Thap plant protection department who supported and provided data of Brown plant hoppers and GIS in Dong Thap province. We appreciate TRIG project for giving a research chance. This publication was made possible through support provided by the IRD-DSF.

References

1. Bierkens, F.P., Finke, P.A., de Willigen, P.: Upscaling and downscaling methods for environmental research. Kluwer Academic Publishers (2000) ISBN-10: 0792363396
2. Drogoul, A., Vanbergue, D., Meurisse, T.: Multi-agent Based Simulation: Where Are the Agents? In: Sichman, J.S., Bousquet, F., Davidsson, P. (eds.) MABS 2002. LNCS (LNAI), vol. 2581, pp. 1–15. Springer, Heidelberg (2003),
   http://dx.doi.org/10.1007/3-540-36483-8_1
3. Drogoul, A., et al.: Gama simulation platform,
   http://code.google.com/p/gama-platform/
4. Food and Agriculture Organization,
   http://www.fao.org/agriculture/crops/
   core-themes/theme/pests/ipm/en/
5. Heong, K.L., Hardy, B.: Planthoppers: New threats to the sustainability of intensive rice production systems in Asia. International Rice Research Institute, Los Baños (2009) ISBN-13: 9789712202513
6. Liu, X.: A Review of the OWA Determination Methods: Classification and Some Extensions. In: Yager, R.R., Kacprzyk, J., Beliakov, G. (eds.) Recent Developments in the Ordered Weighted Averaging Operators: Theory and Practice. STUDFUZZ, vol. 265, pp. 49–90. Springer, Heidelberg (2011), http://dx.doi.org/10.1007/978-3-642-17910-5_4
7. Nguyen, V.G.N., Huynh, H.X., Vo, T.T., Drogoul, A.: On weather affecting to brown plant hopper invasion using an agent-based model. In: Proceedings of the International Conference on Management of Emergent Digital EcoSystems, MEDES 2011, pp. 150–157. ACM Press (2011), http://doi.acm.org/10.1145/2077489.2077517
8. Nguyen, V.G.N., Huynh, H.X., Drogoul, A.: Assessing Rice Area Infested by Brown Plant Hopper Using Agent-Based and Dynamically Upscaling Approach. In: Pan, J.-S., Chen, S.-M., Nguyen, N.T. (eds.) ACIIDS 2012, Part I. LNCS, vol. 7196, pp. 43–52. Springer, Heidelberg (2012), http://dx.doi.org/10.1007/978-3-642-28487-8_5
9. Nguyen, V.G.N., Huynh, H.X., Drogoul, A.: Upscaling and Assessing Information of Agriculture Indicators in Agent-Based Assessment Model from Field to Region Scale. To be appeared in Proceedings of the International Conference on Knowledge and Systems Engineering, KSE 2012. IEEE Computer Society (2012)
10. Ministry of Agriculture and Rural development (MARD) (2012) (in Vietnam),
    http://www.agroviet.gov.vn/Pages/home.aspx
11. Norton, G., Heong, K., Johnson, D., Savary, S.: Rice Pest Management: Issues and Opportunities. In: Rice in the Global Economy: Strategic Research and Policy Issues for Food Security, pp. 297–332. International Rice Research Rice Institute, Los Baños (2010)
12. Phan, C.H., Huynh, H.X., Drogoul, A.: An agent-based approach to the simulation of brown plant hopper invasions (BPH) in the the Mekong Delta. In: Proceedings of the Computing and Communication Technologies, Research, Innovation, and Vision for the Future, pp. 227–232. IEEE Press (2010),
    http://dx.doi.org/10.1109/RIVF.2010.5633134
13. Quang, T.C., Minh, V.Q., Duc, T.T., Tam, T.T.: Application of GIS in predicting medium term of rice infected by Brown plant hopper - A Case study in Dong Thap province. Scientific Journal of Can Tho University 17a, 103–109 (2011)

14. Sadiq, R., Tesfamariam, S.: Probability density functions based weights for ordered weighted averaging (OWA) operators: An example of water quality indices. European Journal of Operational Research 182, 1350–1368 (2007),
http://dx.doi.org/10.1016/j.ejor.2006.09.041
15. Truong, V.X., Huynh, H.X., Le, M.N., Drogoul, A.: Modeling the brown plant hoppers surveillance network using agent-based model: application for the Mekong Delta region. In: Proceedings of the Symposium on Information and Communication Technology 2011, pp. 127–136. ACM Press (2011),
http://doi.acm.org/10.1145/2069216.2069243
16. Voogd, H.: Multicriteria Evaluation for Urban and Regional Planning. Pion Ltd., London (1983)
17. Yager, R.R.: On ordered weighted averaging aggregation operators in multi-criteria decision making. IEEE Transactions on Systems, Man and Cybernetics 18, 183–190 (1988),
http://dx.doi.org/10.1109/21.87068
18. Yager, R.R.: Families of OWA operators. Fuzzy Sets and Systems 59, 125–148 (1993),
http://dx.doi.org/10.1016/0165-01149390194-M

Agent Based Assistance for Electric Vehicles
An Evaluation

Marco Lützenberger, Jan Keiser, Nils Masuch, and Sahin Albayrak

DAI-Labor, Technische Universität Berlin
marco.luetzenberger@dai-labor.de

Abstract. Even before car manufacturers start offering series-produced electric vehicles in a large scale, expectations in the electric powertrain are considerably high. Prospective business perspectives are additionally driven by the so called *Vehicle-to-Grid* technology, which allows electric vehicles to not only procure electric energy, but also to feed energy back into the grid network. However, by using Vehicle-to-Grid, energy literally degenerates into an article of merchandise and becomes of interest to several stakeholders. We have developed a multi-agent system, which embraces this exact view and maximises the interest of several stakeholders in using Vehicle-to-Grid capable electric vehicles. The purpose of this paper is to describe the evaluation of our assistance system and to present collected evaluation results.

1 Introduction

In the next few years, the face of the German road network will be subject to one of the most substantial changes compared to the changes occurred in the last decades. Slowly but steadily, major car manufacturers have been selling their first generation electric vehicles (EVs) in the German automobile market. Efforts in selling electric vehicles are further facilitated by the German government which defined the objective to ensure no less than one million EVs on German roads by the year 2020 [1].

However, this ambitious goal of the federal government is not easily accomplished! Even today, there are still many unsolved problems associated with EVs, for example, consider the limited ranges of EVs and also the insufficient charging infrastructure. The frequency of charging for EVs require significantly more times than the frequency of refuelling conventional petro-fuelled cars. These are only obvious problems!

The additional one million electric consumers will impact the energy grid network. Although the million EVs will not significantly affect the grid network's stability because of the allowable capacity of the German energy grid, the negative effects will increase to urban and industrial electricity needs. Based on the target of the federal government [1], we estimate roughly 200.000 EV registrations –and respectively additional electric consumers– for the capital region of Berlin [2]. Based on previous work [2] we can state, that an addition of 200.000 EVs will compromise Berlin's energy grid significantly. The main reason for this

R. Huang et al. (Eds.): AMT 2012, LNCS 7669, pp. 145–154, 2012.

negative impact is that the grid network's structure can only allow few EVs to be charged simultaneously at the same low voltage systems.

The bottom line is that when it comes to electric mobility, there are still many problems and constraints which have to be considered. These problems and constraints can also be understood as interests of the involved stakeholders. For example, a driver would be mainly interested in unrestricted mobility and in low energy prices. Another example would be from the perspective of an energy supplier. The energy supplier is interested in not compromising his grid infrastructure and to preferably use CO_2 efficient energy from renewable energy sources (for the reason of legitimate requirements) in order to charge EVs.

Recently, we have developed a comprehensive assistance system which optimises the utilisation of vehicle-to-grid (V2G) capable EVs for several stakeholder groups. They are: the driver, the vehicle manufacturer, the charging infrastructure provider and the distribution system operator (DSO).

In our previous publications, we have described the concept [3] and provided implementation details [4,5] of our assistance system. We further proposed several business concepts [6]. In this paper we focus on the evaluation of our assistance system.

The remainder of this paper is structured as follows: In the following section (see Section 2), we will give a short introduction of our assistance system. Then, we present the simulation-based evaluation results of our assistance system (see Section 3). After presenting evaluation results, we discuss the efficiency of our assistance system and wrap up with a conclusion (see Section 4).

2   Concept

In this section, we provide a brief description of the implemented assistance system. The interested reader who wants to learn more about our assistance system can obtain our previous publications [3,4,5,6].

In order to facilitate the effective utilisation of V2G capable EVs (in the interest of all stakeholders), we developed an anticipatory approach in which software agents negotiate a charging and discharging profile for a given amount of time.

This profile reflects the interests of the involved stakeholders and complies with several constraints (e.g. mobility of the driver, restrictions of the energy grid, renewable energies, etc.). For the implementation of an anticipatory system, we required information on the expected behaviour of the driver and his EV. We demanded participating drivers to maintain a personal calendar from which we derived the expected usage pattern. For this purpose, we equipped drivers with a Microsoft Exchange calendar and implemented a smart phone 'app', which allowed mobile users to enter his/her activity patterns on-the-fly. Different facets of the smart phone app are illustrated in Figure 1.

Although it was our intention to affect drivers as little as possible, we required the information of the drivers schedule of the day and the expected usage of his EV in order to optimise charging and feeding (e.g., sending energy back to

Fig. 1. Home screen, calendar and vehicle view of the smart phone application

grid network from EV's battery) schedule that will impact all the stakeholders involved.

We use this information not only to calculate an expected energy requirement curve for the battery of the electric vehicle, but also to identify potential charging or feeding intervals during in the driver's the day. In order to finalise charging and feeding schedule for the day, we compare the potential intervals with a so-called priority curve. We obtain this curve from the DSO, and it represents the share wind energy in the overall electricity production. An example priority curve is shown in Figure 2. While high values indicate an increased availability of wind energy and an interest of the stakeholder for charging processes, low values indicate a low availability of wind energy and an interest for feeding energy back to the energy grid network.

Our assistance system accounts for the driver's preferences. It is possible to specify an individual level of risk tolerance. This tolerance factor determines the driver's willingness to exploit profitable charging and feeding intervals on the expenses of his mobility. Further, drivers are able to specify a comfortability factor, which determines the tolerated distance from a charging station to an appointment location.

In addition to the information on the user, we take infrastructural information into account. For example, our system is aware of detailed characteristics of the charging infrastructure, including the location of charging stations and the available charging rates. Time-dependent information, such as the current availability of charging stations and local load management constraints are supported as well. We retrieve the above mentioned information from a web-service interface provided by the DSO.

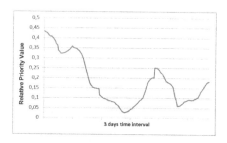

Fig. 2. Share wind energy in the overall electricity production over a three-day interval in the north-eastern region of Germany

In the following we will proceed by presenting our evaluation approach. We evaluated our assistance system in two different ways. First, the main evaluation was performed in a three-week field experiment for which we deployed our application on real hardware components (e.g., charging stations, vehicles, cell phones, etc.). We selected test drivers to evaluate the reliability and suitability of our system. Results of our field test are presented in previous publications [2,5].

In this paper, however, we focus on presenting results of a computer simulation. We developed a simulation framework [7] to draw large-scale implications on future eco systems. As a matter of fact, we initially used our simulation framework to ensure the applications' reliability for the field test. Based on the data we collected from our field tests, we were able to adjusted input parameters (e.g. consumption and charging characteristics, vehicle ranges, etc.) to the simulation framework in order for us to perform large-scale, simulation-based evaluations.

In the following we describe our simulation-based evaluation in more detail.

3   Evaluation

The purpose of our simulation was to measure the efficiency of our assistance system. In order to do so, we compared the effects of the assistance system between drivers which apply our assistance system and those who do not. We defined two different vehicle control patterns: (1) unplanned vehicles and (2) W2V2G (Wind-to-Vehicle-to-Grid) vehicles. We used the patterns to define the behaviour of the simulated vehicles. To have a better understanding of the particular characteristics of each behaviour pattern, we describe the two different types below:

Unplanned Vehicles have no knowledge of the priority signal or the local load management signal which is broadcasted by the DSO. We defined their behaviour in compliance with a driver conceptualisation [8,9]. Unplanned vehicles follow a given activity pattern in the form of several appointments during the day. Whenever an unplanned vehicle arrives at a car park, the current state-of-charge (SOC) is compared to the minimal value tolerated by the driver. In case the SOC is not rated as 'sufficient', the vehicle is charged. For the current rating, unplanned vehicles always select the minimum from the vehicle's and the charging station's capacity.

W2V2G Vehicles apply the driver assistance system. As opposed to unplanned vehicles, charging (and also feeding) intervals are added to the given activity pattern of the drivers. The assistance system is aware of the priority curve and local load management signals, charging intervals are optimised with respect to both, including the selected current rate. W2V2G Vehicles follow their calendar specifications in an exact manner.

For the simulation, we selected three future scenarios, in which we respectively configured the charging infrastructure according to expected numbers [2]. In more detail, we defined six different types of charging stations and configured both, their overall amount as well as their mixture from scenario to scenario. We present details on the charging infrastructure in Table 1.

Table 1. Charging station types that were used for the simulation scenarios 2015, 2020 and 2030, including their charging- and V2G specifications (given in ampere), as well as their absolute number and percentage distribution for each scenario

	Charging	V2G	Mixture		
			2015	2020	2030
Type I	64A	64A	0%	0%	10%
Type II	64A	–	32%	32%	40%
Type III	32A	32A	6%	6%	0%
Type IV	32A	–	32%	32%	0%
Type V	16A	16A	14%	14%	10%
Type VI	16A	–	16%	16%	40%
Absolute number of charging stations			27.900	167.000	200.000

We also defined three different driver types for each future scenario, namely commuters, field workers and a delivery service schema. For the specification of each driver type we used a widely accepted mobility study [10] from which we derived characteristic values for the frequency in which vehicles are used and also for their daily mileage. As such, we defined a typical commuter to travel roughly 100km per day, a field worker to drive around 200km per day, and a deliverer with a daily mileage of approximately 400km.

For each year (2015, 2020 and 2030), we respectively simulated the tree driver types twice. In the first simulation we used our W2V2G assistance to control the behaviour of the vehicles and in the second simulation we used the unplanned control pattern that we have described above.

We evaluated the performance of our assistance system in two general aspects. First, we were interested in the efficiency of our assistance system to increase the utilisation of wind energy. Secondly, we sought to determine in how far our assistance system is able to facilitate the mobility of their drivers. In the following present our simulation results and explain on how we measured the performance of our assistance to contribute to both above mentioned aspects.

3.1   Utilisation of Wind Energy

Whenever it comes to the utilisation of wind energy (or renewable energy in general), the so called 'simultaneity factor' plays an important role.

The simultaneity factor can be understood as the average share of wind energy of all intervals during which electric power was drawn from the grid network. We determined the simultaneity factor by averaging the shares of wind energy during which vehicles were charged. Our simulation results are illustrated in Table 2.

Table 2. Utilisation of renewable energy without considering the current which is applied for charging intervals. The values indicate the average share of renewable energy which was used to charge the simulated vehicles.

	2015		2020		2030	
	Unplanned	W2V2G	Unplanned	W2V2G	Unplanned	W2V2G
Commuter	15.28%	16.00%	15.03%	15.59%	14.28%	15.98%
Field Worker	14.90%	16.83%	15.31%	16.81%	15.19%	16.61%
Delivery Service	15.31%	16.82%	15.19%	17.44%	15.84%	17.41%

Differences between the W2V2G planning system and unplanned drivers are comparably small. We explain this phenomenon with the one-dimensional nature of the simultaneity factor, which takes the average share of wind energy into account, but neglects the intensity of the current that is used to charge vehicles.

To emphasise the importance of the applied current, we performed a second analysis in which we again determined the simultaneity factor, again. Yet, instead of calculating the average value, we weighted the charging intervals according to the applied current. The results for each simulation scenario are illustrated in Table 3.

Compared to the unweighted simultaneity factor, the efficiency of the W2V2G assistance becomes more obvious. As an example, the difference between commuters that were using the assistance system to those who were not, is almost 10%. With an increased daily mileage, both factors literally converge, until –in the 2030 delivery service scenario– the difference between assisted- and unassisted drivers decreases to a little over 2%.

We explain this convergence with the fading optimisation options for the assistance system, which is a direct result to the increased usage of delivery

Table 3. Utilisation of renewable energy under consideration of the applied current. The values indicate the average share of renewable energy which was used to charge the simulated vehicles.

	2015		2020		2030	
	Unplanned	W2V2G	Unplanned	W2V2G	Unplanned	W2V2G
Commuter	12.80%	22.70%	12.36%	22.73%	12.39%	21.38%
Field Worker	15.24%	18.26%	15.43%	18.45%	14.75%	18.04%
Delivery Service	15.02%	17.49%	15.28%	17.82%	15.09%	17.69%

vehicles. An increased usage as well as shorter idle times aggravate the utilisation of periods with a high share of wind energy and cause the performances of the W2V2G and the unplanned control pattern to converge.

For a given driver type, there were only little differences between the results of varying charging infrastructures. We were not able to identify any connection, here.

3.2   Mobility Issues

As a second aspect it was our intention to measure the efficiency of our assistance system to increase the driver's mobility. In order to provide reliable numbers, we analysed our simulation results and counted vehicles that had to cancel at least one scheduled trip due to an insufficient state of charge. The results are illustrated in Table 4.

Table 4. Share of electric vehicles that were affected in their mobility as a result to an insufficient state of charge

	2015		2020		2030	
	Unplanned	W2V2G	Unplanned	W2V2G	Unplanned	W2V2G
Commuter	60.00%	10.00%	53.30%	10.00%	59.30%	07.69%
Field Worker	96.70%	03.33%	93.30%	10.00%	92.30%	13.19%
Delivery Service	100.00%	36.67%	100.00%	48.33%	100.00%	42.86%

The results emphasise the capability of our W2V2G assistance system. While for the commuters and for the field workers the amount of affected W2V2G vehicles remains relatively constant, there is a significant increase of affected vehicles between unplanned commuters and unplanned field workers. This number further increases in the delivery service scenario, where each unplanned vehicle was somehow affected. The W2V2G planning system, however, was able to roughly halve this number.

As the total number of affected vehicles may be a misleading indicator for the degree of mobility, we performed a second analysis in which we determined the total amount of all scheduled trips and compared this number to the amount of trips which actually failed due to an insufficient state of charge of the respective vehicle. The results of this second analysis are illustrated in Table 5.

Table 5. Share trips that failed as a result to an insufficient state of charge

	2015		2020		2030	
	Unplanned	W2V2G	Unplanned	W2V2G	Unplanned	W2V2G
Commuter	02.21%	00.78%	02.00%	00.60%	02.30%	00.47%
Field Worker	04.90%	00.13%	04.55%	00.39%	04.80%	00.51%
Delivery Service	08.68%	01.49%	08.33%	01.95%	08.71%	02.39%

The results underline the efficiency of the W2V2G assistance system to increase the mobility of their drivers. While W2V2G commuters were affected in less than one percent of their scheduled trips, unplanned commuters had to deal with almost three times as many affections. While there were almost no differences between the amount of affections of planned commuters and planned field workers, the number almost doubled between unplanned commuters- and field workers.

In the case of delivery drivers the numbers increased in both cases, nevertheless, the W2V2G planning system was able to lower the number of failures by the factor of 4.

4   Conclusion

In this paper we presented a simulation based evaluation of a recently developed assistance system. This assistance system optimises the utilisation of vehicle-to-grid (V2G) capable electric vehicles with respect to the intentions of several stakeholders.

For this work, we put the main focus on two stakeholder, namely the driver and the distribution system operator (DSO). While drivers are generally interested in an optimised degree of mobility, we understood the interests of the DSO to increase the utilisation renewable energy[1] and to account for structural deficiencies of the low voltage grid.

Indirectly, we account for the interests of other stakeholder as well (though, not as obviously as for driver and for DSO). As an example, for the calculation and for the execution of charging processes we use charging profiles which are recommended by the vehicle manufacturer in order to increase the battery's

---

[1] In fact we focused on wind energy only. However, an additional integration of other types is possible as well.

lifetime. We also account for the interests of the charging infrastructure provider by maximising the utilisation of his charging stations and –as in the case with the DSO– by ensuring a proper operation without jeopardising his low voltage energy grid.

In order to evaluate our approach we defined several future scenarios. We respectively simulated three different driver types for each future scenario and compared the results of drivers which apply the assistance system to those who do not. The simulation results underlined the efficiency of our assistance system in all respects. We were able to increase the utilisation of wind energy by almost 10%. This value decreases with the utilisation and the daily mileage of the vehicles. We explain this phenomenon with the fading alternatives for the assistance system.

We were also able to substantiate the capability of our assistance system to increase the mobility of their drivers. Assisted vehicles encountered significantly less mobility restrictions than unassisted ones. Especially for field workers, the effects of the assistance becomes apparent, as mobility restrictions were reduced by up to 90%.

To conclude — based on our evaluation we can state, that the W2V2G algorithm is able to facilitate a more effective utilisation of renewable energy sources. Further, we demonstrated that our application is able to account not only for individuals but to support the interests of many stakeholders, namely the drivers, the vehicle manufacturer, the charging infrastructure provider and the DSO.

Based on our evaluation we can argue that a respective system is able to contribute to the domain of electric mobility.

References

1. Nationale Plattform Elektromobilität (NPE), Zweiter Bericht der Nationalen Plattform Elektromobilität (May 2011)
2. Eckhardt, C.F., Lindwedel, E., Gödderz, K., Maempel, V., Schwarz, M., Hufnagl, C., Nissen, G., Schulte, U., Jin, D., Stöhr, T., Brosius, O., Heise, J., Buchholz, S., Weber, A., Hajesch, M., Pfab, X., Amiri, R., Schwaiger, M., Brennan, R., Keil, M., Kaluza, S., Krammer, J., Esch, F., Albayrak, S., Keiser, J., Masuch, N., Lützenberger, M., Trollmann, F., Geithner, T., Freund, D., Krems, J., Bär, N., Bühler, F., Eberth, J., Henning, M., Kämpfe, B., Mair, C., Westermann, D., Agsten, M., Schlegel, S.: Steigerung der Effektivität und Effizienz der Applikationen Wind-to-Vehicle (W2V) sowie Vehicle-to-Grid (V2G) inklusive Ladeinfrastruktur. Vattenfall Europe, BMW AG, TU Berlin, TU Chemnitz, TU Ilmenau, Schlussbericht (2011)
3. Keiser, J., Glass, J., Masuch, N., Lützenberger, M., Albayrak, S.: A distributed multi-operator W2V2G management approach. In: Proceedings of the 2nd IEEE International Conference on Smart Grid Communications, Brussels, Belgium, pp. 291–296. IEEE (October 2011)
4. Masuch, N., Keiser, J., Lützenberger, M., Albayrak, S.: Wind power-aware vehicle-to-grid algorithms for sustainable ev energy management systems. In: Proceedings of the IEEE International Electric Vehicle Conference, Greenville, SC, USA, pp. 1–7. IEEE (March 2012)

5. Keiser, J., Lützenberger, M., Masuch, N.: Agents cut emissions – On how a multi-agent system contributes to a more sustainable energy consumption. Procedia Computer Science 10, 866–873 (2012)
6. Masuch, N., Lützenberger, M., Ahrndt, S., Hessler, A., Albayrak, S.: A context-aware mobile accessible electric vehicle management system. In: Proceedings of the Federated Conference on Computer Science and Information Systems, Szczecin, Poland, pp. 305–312 (September 2011)
7. Lützenberger, M., Masuch, N., Hirsch, B., Heßler, A., Albayrak, S.: Predicting future(e-)traffic. In: Balsamo, S., Marin, A. (eds.) Proceedings of the 9th Industrial Simulation Conference, Venice, Italy, Eurosis, pp. 169–176. EUROSIS-ITI (June 2011)
8. Lützenberger, M., Masuch, N., Hirsch, B., Ahrndt, S., Heßler, A., Albayrak, S.: The BDI driver in a service city (extended abstract). In: Tumer, K., Yolum, P., Sonenberg, L., Stone, P. (eds.) Proceedings of the 10th International Joint Conference on Autonomous Agents and Multiagent Systems, Taipei, Taiwan, pp. 1257–1258 (May 2011)
9. Lützenberger, M., Ahrndt, S., Hirsch, B., Masuch, N., Heßler, A., Albayrak, S.: Reconsider your strategy – an agent-based conceptualisation of compensatory driver behaviour. In: Proceedings of the 15th Intelligent Transportation Systems Conference, Anchorage, AK, USA (to appear, 2012)
10. Mobilität in Deutschland 2008. Bonn and Berlin, Germany: Federal Ministry of Transport, Building and Urban Development (2010),
http://www.mobilitaet-in-deutschland.de/
pdf/MiD2008_Abschlussbericht_I.pdf

Event Calculus-Based Adaptive Services Composition Policy for AmI Systems

Huibing Zhang[1,2], Jingwei Zhang[1,2,*], Ya Zhou[2], and Junyan Qian[1]

[1] Guangxi KeyLaboratory of Trusted Software, Guilin University of Electronic Technology,
Guilin, Guangxi, China
[2] Research Center on Data Science and Social Computing,
Guilin University of Electronic Technology, Guilin, Guangxi, China
zhanghuibing@guet.edu.cn, gtzjw@hotmail.com

Abstract. Services composition technology which is used in Ambient Intelligence should have the features of context-aware, partial order and concurrent. It should adaptive to the dynamic change of user preference and context. To meet these requirements, the paper puts forward an adaptive dynamic services composition framework and its implementation mechanism based on event calculus. It studies the basic principles and technologies for descripting domain services, context information and domain rules based on event calculus. On the basis, it details the composition planning mechanism. At last, a prototype system, intelligence application control, is built to verify the effectiveness and availability of the services composition policy.

Keywords: services composition, event calculus, context-aware.

1 Introduction

Ambient Intelligence (AmI) is a "user-centered" system which should provide kinds of services for users adaptively [1]. However, both system resource with feature of heterogeneous, dynamic and distribution and user requirements with feature of personalize, so it's difficult to meet the AmI's requirements [2]. This problem can be solved effectively by using the service-oriented computing (SOC) technology. Any accessible resource in AmI, such as program, sensor, device, can be modeled as Web service by using SOC, so the device-oriented physical space is transmitted into service-oriented information space [3-4]. The user requirement can be meet by compositing some services: according to the dynamic requirement and context, a composite service is created instance and discomposed it after completing the task [5].

Compare with other services composition application system, AmI is more emphasis on context-aware and personalization which should dynamic adjust its behavior to adaptive the special context and user preferences. At the same time, there are lots of concurrent operations and event-triggered activities. The system behavior is partial order in time dimension.So, the services composition framework for AmI should have

---

* Corresponding author.

R. Huang et al. (Eds.): AMT 2012, LNCS 7669, pp. 155–164, 2012.
© Springer-Verlag Berlin Heidelberg 2012

the following features: ①It should have the good ability of context modeling which can model the run-time context, preferences and states constraint; ②It should have the good ability of reasoning which can describe states transition and time constraints; ③It should have the good ability of converting the Web service(OWL-S) into other forms which can be used to automatic services composition flexible and efficient. In order to meet these requirements, the paper puts forward an event calculus-based context-aware service composition framework and its implementation methods. It uses event calculus (EC)as formalize logic and adductive logic programming as reasoning technology to implement context-aware services composition [6-7].

2   Instruction of Event Calculus and Related Works

First-order predicate logic-based event calculus is suitable for description and analysis event-based time-varying domain [8-10]. As a programming framework, event calculus includes ontology, predicates and axioms [11]. The ontology contains events (actions), fluents and time-points which define the basic concepts, attributes, relationships and constraints. The fluents can represent anything which value or state may change with time, such as emotion, room temperature, lamp state (open/close), and so on. The actions can represent any operations which can lead to the change of world states. It can initiates, terminates or releases the values of fluents. Event calculus predicates define the ontology properties and its relationships. The 9 predicates and their meaning are shown in Table 1.

Table 1. Predicates of the Event Calculus

	Formula	Meaning
Base predicates	$Initiates(\alpha, \beta, \tau)$	Fluent β holds after action α at time τ
	$Terminates(\alpha, \beta, \tau)$	Fluent β does not hold after action α at time τ
	$Releases(\alpha, \beta, \tau)$	Fluent β is not subject to the common sense law of inertia after action α at time τ
	$Initially_P(\beta)$	Fluent β holds from time 0
	$Initially_N(\beta)$	Fluent β does not hold from time 0
	$Happens(\alpha, \tau_1, \tau_2)$	Action α starts at time τ_1 and ends at time τ_2
	$HoldsAt(\beta, \tau)$	Fluent β holds at time τ
Auxiliary predicates	$Clipped\ (\tau_1, \beta, \tau_2)$	Fluent β is terminated between times τ_1 and τ_2
	$Declipped(\tau_1, \beta, \tau_2)$	Fluent β is initiated between times τ_1 and τ_2

Shanahan pointed that Partial order planning = event calculus + abduction.Let Γ be a goal, let Σ be a domain description, let$\Delta_0$be an initial situation, and let Ω be the conjunction of a pair of uniqueness-of-names axioms for theactions and fluents mentioned inΣ, and let ψbe a finite conjunction of stateconstraints. A plan for Γ is a narrativeΔsuch that:

$$CIRC[\Sigma; Initiates, Terminates, Releases] \wedge CIRC[\Delta_0 \wedge \Delta; Happens] \wedge \psi \wedge EC \wedge \Omega \models \Gamma$$

Event calculus-based services composition planning is an open study domain. Ozorhan and Okutan introduced event calculus into service composition and implemented 4 types of service composition: Monolithic, Interleaved, Templatebased and Staged [12-13]. They detailed the methods which can transform Web service (OWL-S) into event calculus axioms (ECA). The article [14-15] proposed a method for formal verification of Web service composition by using event calculus. [16] adopted event calculus and workflow to realized services composition. Ishikawa and Chen researched the event calculus-based services composition from other aspects [17-18]. The above mentioned studies achieved good results at Web service modeling and verification based on event calculus. But, as far as we know, there are no study focusing on the AmI's complex context information modeling based on event calculus and using it in AmI system.

3   EC Based Services Composition Framework for AmI

According to the AmI system's special demands, the paper proposed a context-aware dynamic service composition framework based on event calculus. It includes 6 modules, as shown in Figure 1.

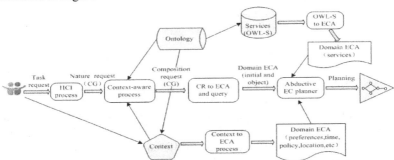

Fig. 1. Event calculus-based context-aware services composition framework

Natural human computer interaction and context-aware processing module implement intelligent interaction and context fusion. The former provides natural langue analysis based on conceptual graph. The latter combines context information into user basic requirement and gets a complete, definite requirement. Figure 2 denotes the processing of requirement "open TV". The module, CR to ECA, submits final requirement to planner.

Fig. 2. Conceptual Graph of user's requirement

3.1   OWL-S to ECA

All of the resources deployed in AmI system modeled as Web services can be described as a 4-tuple:*WSFunctional=(Input, Output, Precondition, Effect)*. Every service in service space can be mapped as an event in event space: Service invoking corresponds to event happens. Service's input parameters correspond to known parameters of event. Service's precondition corresponds to precondition of event happen. Service's output and effect correspond to fluents which are generated by event happen. The details can be seen inarticles [12-13].

3.2   Context to ECA

Context information is rich and it is the key factor for operating action in AmI system. Moreover, context is dynamically changing: ① All kinds of resource (services) evolve independently, and they may be unavailable momentarily. A better service may substitute the old one or new services are published. ② As user movies, available services are differences. ③ Sometime-related context information changes with the time. So it is important to get, model, analyze and use context information in AmI environment. The paper focuses on the rules of each kind of context and their modeling based on event calculus axioms (ECA).

1. Location based service

Location information is an important context and it determines available service. Firstly, location determines the environment and spatiotemporal. Therefore, the location changes often leading to other related context change. Secondly, many services directly related to location. There are always different services at different locations. Last, location can influence system's operating action. For one requirement, the system may have different response because of different location. So it needs to update the context and domain services according to the current location. For example, TV available in the parlor or computer available in the study can be expressed follows:

axiom(initially(have(TV)),[holds_at(at(parlor),t)]).
axiom(initially(have(computer)), [holds_at(at(study),t)]).

2. Data input

In many case, context can be used as data input. System can transmit the data into service's input implicitly and user transparently. So it can effective reduce user inputs, improve user experience and reduce mistakes. For example, a user wants to browse a road map from current location to railway station. He only needs to input the phrase: view the map to train station. The system gets user's current location and train station location automatically, and then takes this information as input data. The context information which acts as input data is converted to known or initial state, and is added to event calculus domain axioms in form of fact clauses. It can be expressed as:

axiom(initially(at(person, location,)),[]).
axiom(initially(at(trainStation, location)),[]).

3. User preference and intelligent reasoning

The personalized of AmI system is mainly presented as user preference. Here, the preference refers to anticipatory affect or behavior disposition under special scenario. So it has significant individual difference and closely related to context. In order to meet user's special preference, AmI system dynamically adjusts its behaviors according to sensed context. Therefore, user preference is a key factor for intelligent reasoning which influences operating action. In event calculus based service planning, user preference can be acted as parameters or precondition of action. For example, a user like TV sports news after going home. It can be expressed as:

axiom(initially(known(channnelsportsNews),
[holds_at(at(parlor),t),holds_at(time(cur_time),t)]).
axiom(initially(known(channnel cartoon),
holds_at(at(parlor),t),holds_at(time(cur_time),t)]).
axiom(initiates(e_Open(Device,TVID,sportsNewts),
On(TVID,sportsNews),t),[holds_at(at(parlor),t),holds_at(time(cur_time),t),holds_at(known
(channnelsportsNews),t),
holds_at(known(channelcartoon),t),holds_at(neg(others(son)),t)]).

At last, user preference also can be acted as implicit requirement which expects some objects in a certain states. The implicit and explicit requirement consist a complete task requirement.

4. Trigger events

Some context can trigger related events which make AmI system adapt to environment. For instance, temperature or humidity can trigger events (actions) of air-conditioning in AmI space: While the temperature exceeded high_temperature, the air-conditioning opens refrigeration. While the humidity exceeded high_humidity, the air-conditioning opens dehumidification. It can be expressed as follows:

axiom(initiates(turnOnCool(x),CoolOn(x),t),
[holds_at(temperature(cur_temperature),t)，holds_at(higher
(current_temperature,high_temperature))，holds_at(neg(CoolOn(x)),t)]).
axiom(initiates(turnOnDehumidification (x),
DehumidificationOn(x),t),[holds_at(humidity
(cur_high_humidity),t)，holds_at(higher(cur_high_humidity ,
high_humidity)),holds_at(neg(Dehumidification (x)),t)]).

5. State constraints

State constraints are a kind of constraints relationship that similar to rules of object's fluents in domain. It reflects the mutual restriction among fluents during system running. In event calculus, state constraints are expressed as predicates holds_at and ¬holds_at. For instance, people will feel comfortable while the temperature and humidity maintained at a value. It can be expressed as follows:

axiom(holds_at(Comfortable,t),[holds_at(temperatureValue,t),holds_at(humidityValue,t)])

6. Precondition

Only meet all preconditions, the service can run and get right results. Context can provide some preconditions. In event calculus, it uses holds_at to express preconditions. For instance, TV should close if there is no human around it. It can be expressed as follow:

axiom(initiates(turnOff(TV),Off(TV),t), [holds_at(neg(at(personID,person_location)),t)]).

3.3   Abductive EC Planning

As practical execution module of service planning, abductive EC planner uses Prolog as its logic programming platform[1], event calculus and abductive theory prover (ATP) as its reasoning logic. It can deal with kinds of special requirement, such as concurrent, trigger, partial-order, continuous variation and so on, in AmI. Event calculus based service composition planning can be expressed as 4-tuple (*SEA, ECA, ConA, Q*), where *SEA i*s domain event calculus axioms which are converted from domain services. *SEA* is used to describe domain state altering or information transformation. *ECA* is domain independent axioms. It is the core of event calculus meta interpreter. *ConA* is domain axioms which get from context. It acts as initial states or state constrains. *Q* is target states which contain some query clauses and represent user requirement.

SEA, ECA and *ConA*constitute the event calculus based logic reasoning knowledge base*K*. Each clause in *K*represents an action, fact or rule. Using*Q* as the starting point of reasoning, it can get an event sequence or fail to exit by match, resolution, backtracking. The event sequence represents a service composition planning.

4   Experiments and Analysis

We apply the event calculus based service planning to our test-bed AmI-Space [1] and use intelligence application control (IAC) to test availability of the service planning. The physical deployment of IAC is shown in figure 3. Ten services are used in IAC:

TurnOnTV(TVID, TVOff,TVOn); TurnOffTV(TVID, TVOn, TVOff);ChannelSelection((TVID, Channel), TVOn, Play on the selection channel); TurnOnLight(LightID, LightOff, LightOn); TurnOffLight(LightID, LightOn, LightOff); TurnOnAircondition(AirconditionID, AirConditonOff, AirConditoinOn); TemperatureSet((AirconditionID, Temperature, CurrentTemperature), ,AirConditionOn, AirConditoin On temperature);TurnOffAircondition(AirconditionID, AirConditonOn, AirConditinOff); OpenCurtain(CurtainID, CurtainClose, CurtainOpen); CloseCurtain(CurtainID, ,CurtainOpen, CurtainClose).

4.1   Scenario and Experiments

Event calculus based service composition planning in IAC can be illustrated by following scenario.

---

[1] http://www.swi-prolog.org

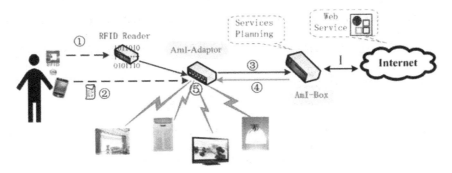

Fig. 3. Architecture of IAC

When comes home, Kevin sits on the sofa and lightly speaks out "open TV" by using his smart phone. Then TV opens automatically and selects the sports news. Curtain closes lightly and gentle light illuminate the parlor. The gentle breeze blows out from air-conditioning. ...

The working process of this scenario as follows:

At the services planning side, it needs to convert domain services into domain axioms, as shown in Figure 4. All of the domain event calculus axioms, domain facts and rules composea complete domain axiom.

axiom(initiates(channelselection(Tvid,Channel),tvplay(Tvid, Channel),T),[holds_at(tvon(Tvid),T)]).
axiom(initiates(turnontv(Tvid), tvon(Tvid), T), [holds_at(tvoff(Tvid),T)]).
axiom(initiates(turnofftv(Tvid),tvoff(Tvid),T), [holds_at(tvon(Tvid),T)]).
axiom(initiates(turnonlight(Lightid),lighton(Lightid),T),[holds_at(lightoff(Lightid),T)]).
axiom(initiates(turnofflight(Lightid),lightoff(Lightid),T),[holds_at(lighton(Lightid),T)]).
axiom(initiates(turnonaircondition(Airconditionid), airconditionon(Airconditionid),T),[holds_at(airconditionoff(Airconditionid),T)]).
axiom(initiates(temperatureset(Airconditionid,Preferencetemperature), temperature(Airconditionid,Preferencetemperature, Currenttemperature),T), holds_at(airconditionon(Airconditionid),T), diff(Preferencetemperature,Currenttemperature)]).
axiom(initiates(closecurtain(Curtainid),curtainclose (Curtainid),T),[holds_at(curtainopen(Curtainid),T)]).
axiom(initiates(opencurtain(Curtainid),curtainopen (Curtainid),T),[holds_at(curtainclose(Curtainid),T)]).
axiom(initiates(turnoffaircondition(Airconditionid), airconditinoff(Airconditionid),T),[holds_at(airconditonon(Airconditionid),T)]).

Fig. 4. Services described by event calculus

At services requester side:

1. When Kevin enters AmI-Space, the system can get complete context information, such as time, location, preference, and so on. All these information are transmitted to the AmI-Box for modeling, analysis and storage by AmI-Adaptor.

2. Some of context information, for example TV or lamp's state and its attribute value are expressed as initial states or state constraints. The following event calculus axioms represent this information.

axiom(initially(curtainopen(Curtainid)),[]).

axiom(initially(tvoff(Tvid)),[]).

axiom(initially(lightoff(Lightid)),[]).

axiom(initially(airconditionoff(Airconditionid)),[]).

3. User speaks out his request "watch TV", then speech- to- text system converts the voice to text and sends to the CGGenerator in AmI-Box.

4. CGGenerator converts the user requirement to conceptual graph, and then generates a complete service request conceptual graph by using domain ontology and context information, as shown in figure 2.

5. Convert the conceptual graph based service request to event calculus query clause which is submitted to service planner. The figure 2.b is expressed as :

?-abdemo([holds_at(tvplay(01,sportsnews),t)],R).

6. In the AmI environment, preferences contain user's requirements which are the expected target states. So some preference information should be expressed as query of event calculus.

?-abdemo([holds_at(curtainclose(02),t),holds_at(temperature(04,22,29),t),
holds_at(lighton(03),t)],R).

7. Combining query of step 5) and step 6), we can get a complete service request query which are submitted to planner. When planner receives the query, it returns an event sequences. Each event corresponds to a service. The system will reach at target state after executing each service according to the time sequence ordered in event sequence.

4.2   Results and Analysis

The planner returns a partial-order event sequences after it receives the query. Figure 5 shows the query and its planning results.

Query: *?-abdemo([holds_at(tvplay(01,sportsnews),t),holds_at(curtainclose(02),t),*
holds_at(temperature(04,22,29),t), holds_at(lighton(03),t)],R).

Planning Results:

R=[[happens(turnonlight(lightid),t6,t6), happens(turnonaircondition(airconditionid), t5, t5),
happens(temperatureset(airconditionid, preferencetemperature),t4,t4),
happens(closecurtain(curtainid),t3,t3),happens(turnontv(tvid),t2,t2),
happens(channelselection(tvid, channel), t1, t1)], [before(t6, t), before(t5, t4),
before(t4, t), before(t3, t), before(t2, t1), before(t1, t)]] .

Fig. 5. Query and planning results

The planning results include 6 atomic services. The time sequence among these services is shown in Figure 6. It has 4 concurrent processes: process P2 and P3 are both single atomic services. Process P1 and P4 are both include two atomic services

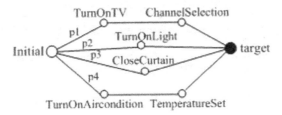

Fig. 6. Services composition corresponding to the planning results

which are sequence structure. From the planning results and its planning procedure we can see that event-based service composition planning method can meet AmI's demands: ① It suitable for three control logic: concurrent, sequence and partial-order. ②It suitable both event logic based and state constraints services composition. Many input/output match based services composition algorithms can only suitable for information transmit service composition. But there are many states altering in AmI environment, the services often only have effects instead of output. Hence, it needs the event logic based services composition. ③ Services composition procedure can naturally reflect user preferences and context constraints.

5   Conclusions

In this paper, we put forward a context-aware automatic service composition framework based on event calculus. It can improve the intelligence of the AmI system and reduce the complexity for AmI developers. And by using event calculus, the services composition mechanism can meet concurrent, partial-order, event state logic. Event calculus based context modeling technology can provide context-aware adaptive services composition. At last, it shows that the planning technology is available and effective by the IAC system. In the future, we will focus on the mechanism of domain axiom automatic generating technology. Another question is time efficiency of event calculus based reasoning.

Acknowledgments. The authors would like to thank the Foundation of Guangxi Key Laboratory of Trusted Software (kx201214, kx201203, kx201114),Nature Science Foundation of Guangxi (No. 2012GXNSFBA053171), National Nature Science Foundation of China (No.61063002,No.61063038),and the Education Department of Guangxi (No. 201010LX154)for their support incurrent research.

References

1. Chen, R., Hou, Y., Huang, Z., He, J.: Modeling the ambient intelligence application system: concept, software, data, and network. IEEE Transactions on Systems, Man, and Cybernetics, Part C: Applications and Reviews 39(3), 299–314 (2009)

2. Lindenberg, J., Pasman, W., Kranenborg, K., et al.: Improving service matching and selection inubiquitous computing environments: a user study. Personal Ubiquitous Computing 11(1), 59–68 (2006)
3. Benazzouz, Y., Sabouret, N., Chikhaoui, B.: Dynamic Service Composition in Ambient Intelligence Environment. In: 2009 IEEE International Conference on Services Computing, pp. 411–418 (2009)
4. Martin, D., Burstein, M., et al.: OWL-S: Semantic Markup for Web Services (2010), http://www.w3.org/Submission/2004/SUBM-OWL-S-20041122/.5.10
5. Medjahed, B., Bouguettaya, A., Elmagarmid, A.: Composing web services on the semantic web. The VLDB Journal 12(4), 333–351 (2003)
6. Shanahan, M.: An abductive event calculus planner. The Journal of Logic Programming 44, 207–239 (2000)
7. Shanahan, M.: The Event Calculus Explained (June 10, 2011), http://www.doc.ic.ac.uk/~mpsha/ECExplained.pdf
8. Kowalski, R.A., Sergot, M.J.: A Logic-Based Calculus of Events. New Generation Computing 4, 67–95 (1986)
9. Shanahan, M.P.: Solving the Frame Problem: A Mathematical Investigation of the Common Sense Law of Inertia. MIT Press (1997)
10. Barros, L.N.D., Santos, P.E.: The nature of knowledge in an abductive event calculus planner. In: Proceedings of the 12th European Workshop on Knowledge Acquisition, Modeling and Management, pp. 328–343. Springer, London (2000)
11. Shanahan, M.: The Event Calculus Explained. In: Veloso, M.M., Wooldridge, M.J. (eds.) Artificial Intelligence Today. LNCS (LNAI), vol. 1600, pp. 409–430. Springer, Heidelberg (1999)
12. Okutan, C., Cicekli, N.K.: A monolithic approach to automated composition of semantic web services with the Event Calculus. Knowledge-Based Systems 23, 440–454 (2010)
13. Ozorhan, E.K., Kuban, E.K., Cicekli, N.K.: Automated composition of web services with the abductive event calculus. Information Sciences 180(19), 3589–3613 (2010)
14. Rouached, M., Perrin, O., Godart, C.: Towards Formal Verification of Web Service Composition. In: Dustdar, S., Fiadeiro, J.L., Sheth, A.P. (eds.) BPM 2006. LNCS, vol. 4102, pp. 257–273. Springer, Heidelberg (2006)
15. Rouached, M., Godart, C.: An event based model for web service coordination. In: Second International Conference on Web Information Systems and Technologies, pp. 81–88 (2006)
16. Aydın, O., Cicekli, N.K., Cicekli, I.: Automated Web Services Composition with the Event Calculus. In: Artikis, A., O'Hare, G.M.P., Stathis, K., Vouros, G. (eds.) ESAW 2007. LNCS (LNAI), vol. 4995, pp. 142–157. Springer, Heidelberg (2008)
17. Ishikawa, F., Yoshioka, N., Honiden, S.: Developing consistent contractual policies in service composition. In: Proceedings of the Second IEEE Asia-Pacific Service Computing Conference, pp. 527–534 (2007)
18. Chen, L., Yang, X.: Applying AI planning to semantic web services for workflow generation. In: The First International Conference on Semantics, Knowledge and Grid, pp. 323–325 (2005)

QoS- and Resource-Aware Service Composition and Adaptation

Quoc Bao Vo and Minyi Li

Faculty of Information & Communication Technologies
Swinburne University of Technology, Australia
{bvo,myli}@swin.edu.au

Abstract. Flexible and adaptive quality-of-service (QoS) is desirable for distributed real-time applications, such as e-commerce, or multimedia applications. The objective of this research is to dynamically instantiate composite services by effectively utilising the collective capabilities of the resources to deliver distributed applications. Related to this objective are the problems of: (1) predicting system and network resources utilisation as well as the user's changing requirements on the provided services, and (2) finding optimal execution plans for a service that meet end-to-end quality requirements and adapting the available resources in accordance to the changing situation. This paper presents a framework for adaptive QoS and resource management in provisioning composite services. We also develop distributed algorithms for finding the multi-constrained optimal execution plan to enable delivery of QoS-assured composite services.

1 Introduction

The Internet infrastructure mainly supports *best-effort connectivity* service. Thus, guaranteeing quality-of-service (QoS) for high-demanding applications that operate in the Internet environment has remained a challenge. The issue arise from the heterogeneity of this environment. Services and contents are typically delivered via multiple network domains (i.e., communication systems owned by various administrative entities). Also, the user could request for the service from a wide variety of access devices, such as PCs, personal digital assistants (PDAs) and cell phones, or next-generation set-top boxes (STBs), with different configurations. Finally, while the best-effort networks might be sufficient for some services such as e-mail and those based on the File Transfer Protocol (FTP), other services such as multimedia and e-commerce applications require better QoS guarantees.

In order to overcome the above issues, a number of technologies and architectures have been introduced. Different forms of *overlay networks* have been developed to provide attractive service provisioning solutions, including service and content overlay networks [1,3]. Services computing on the other hand enables distributed applications, possibly provided by multiple providers, to be discovered, aggregated and deployed. Service Oriented Architecture (SOA) [7,2] further offers a layer of transparency by abstracting the available resources as services to be used by users and other applications. Finally, various middleware services have also been introduced to ensure adaptability and seamless integration of different services. Given various types of services involved

R. Huang et al. (Eds.): AMT 2012, LNCS 7669, pp. 165–175, 2012.

in such an application scenario, cross-layer service composition with QoS assurances has become necessary to cost-effectively create new and seamless Internet services. Thus, an important problem is to provide a framework to enable efficient cross-layer service composition with QoS assurances.

In this paper, we present a QoS and Resource management (QoSRM) model to address the above challenges. Our cross-layer approach operates at all levels of service delivery, from the user's requirements to the resource constraints. To take advantage of the underlying service overlay network, we introduce several efficient algorithms that calculate the optimal execution plan for a given user request of composite service. An execution plan specifies the service instances for a composite service. Given the resource constraints on the service gateways that host the service instances of different instantiations and on the networks connecting between the gateways, our approach enables the selection of execution plans that (i) satisfy the user's QoS requirements, and (ii) optimise the service provider's objectives (such as low costs, good load balancing). Moreover, thanks to a QoS prediction component, our framework is able to adapt to changing user's requirements and evolving conditions during runtime. Due to space limitation, in-depth experiments to demonstrate the effectiveness of our approach as well as the efficiency of the proposed algorithms are omitted (details can be found in the extended version at: http://www.ict.swin.edu.au/personal/bvo/papers/AMT-full.pdf).

2   Motivation and System Model

2.1   Motivating Scenario

We illustrate the problem tackled in this paper by an example scenario in the domain of distributed multimedia applications. Fig. 1 shows a composite service to allow a user to stream multimedia content to her access device. In this scenario, the user can request to watch live sport events, such as the Wimbledon tennis tournament, on her access device (e.g., a PDA, a mobile phone, or a HDTV). First, the user connects to a multimedia portal which authenticates her before a secure session is established to allows her accessing to the subscribed or pay-per-view contents. The user then chooses and requests the content she wants to view (e.g., a tennis match). The content will need to be acquired from a content provider. Then the streaming process begins. While watching the tennis match, the user can also follow another match which is played concurrently by requesting highlights of the other match. Whenever highlights of another match are requested, the service provider will try to acquire them (possibly from other content providers) before merging the highlight content to the current media stream to enable a picture-in-picture view of the highlights.

As shown in Fig. 1, the delivery of the above abstract service process requires services to be instantiated by actual services residing at service gateways V_i. Thus, the the actual services are executed at service gateways and data are transmitted between gateways. This is where the QoS of the services will be realised. In order to guarantee the QoS of the services, most service providers will need a mechanism to ensure that information and data be processed and transmitted in a timely manner with guaranteed reliability and availability. In the domain of multimedia applications, this is usually

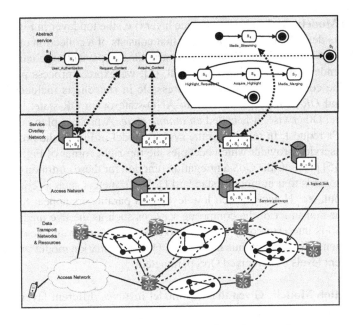

Fig. 1. A composite service delivery framework

achieved through a *service overlay network* (SON). The idea is to have multiple instances of an abstract services distributed among a number of service gateways which
are connected to each other via different network domains with various QoS guarantees.
The service provider, being aware of the constraints on the resources and the network,
will select an *execution plan* for a given composite service that ensures that the Service
Level Agreement (SLA) with the service consumer is satisfied and the system performance is optimised while the cost is minimised. The service gateways not only host the
service instances, but also are equipped with middleware services to perform backend
data processing and act as relay nodes in a SON.

2.2 System Model

In this section, we present the overall system model (see Fig. 2) that realises QoS &
Resource management (QoSRM) in a SOA infrastructure.

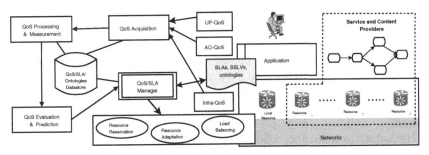

Fig. 2. QoSRM architecture

Composition Model: At the abstract service layer (i.e., the top layer in Fig. 1), a composite service is described by a *process plan* that consists of a collection of component services connected by different composition structures representing control-flow and data-flow dependencies. Following Zeng et al. [8], we express process plans as statecharts [4]. The composition structures expressible in statecharts include: *Sequence, AND-states,* and *OR-states.* If a state is an AND-state or an OR-state, it is called a *compound state.* Otherwise, it is called an *atomic state.* A process plan is designed to fulfilled a user's request. In relation to the system model in Fig. 2, the process plan is defined by the service provider which occupies the top-right corner of Fig. 2.

Jaeger et al [5] presents the QoS aggregation function for these common composition patterns with regards to a number of standard QoS parameters such as cost, execution time, throughput, etc. However, such low-level QoS parameters appear to be of little relevance to the consumer of the composite services such as the live media streaming application who is interested only in the video and audio quality and how smooth the viewing experience is. We will thus introduce a QoS-Information model that takes into account different levels of perceived QoS parameters.

QoS-Information Model: Given that our model involves different layers of service delivery, in order to support QoS & Resource management, measured QoS information as well as other context and resource information needs to be obtained. We distinguish between:

- the *user-perceived QoS* information (*UP-QoS*),
- the *application-oriented QoS* information (*AO-QoS*), and
- the *infrastructure-level QoS* information (*Infra-QoS*).

For example, in the context of the distributed multimedia applications, the UP-QoS specifies measurements over attributes such as video and audio quality, the responsiveness of the media portal and application, and the service cost. The AO-QoS provides application-specific attributes such as video resolution, frame rate, bit rate, security level, etc. On the other hand, the Infra-QoS mainly concerns with the attributes at the Data Transport Networks and Resources layer. Thus, the Infra-QoS is normally specified by parameters such as bandwidth, packet loss, packet delay, jitter, CPU load level, memory usage level, and data storage speed and remaining capacity.

Service Level Agreements (SLAs), Services and Systems Logical Views (SSLVs), and Service Ontologies: The Qos/SLA Manager is the QoSRM core component and takes input from a number of sources:

SLAs: The QoS/SLA Manager needs to know the SLAs between the **composite service provider (CSP)** and various parties: (i) Between the service provider and the user, there is a service contract called the *user contract* (UC); (ii) The SLAs between application-level service providers are normally application specific; and (iii) The QoS guarantees by infrastructure-level providers include specifications on computing resources.

It's important to note that, while there is a close correspondence between UC, application-level SLAs, and infrastructure-level QoS guarantees and UP-QoS, AO-QoS, and Infra-QoS information, respectively, they are different. The QoS information at various layers is the actually perceived QoS during the delivery of the composite service. On the other hand, the contracts and SLAs specify the commitment a service provider makes to its respective service consumer regarding the QoS of the provided service. For instance, while there is no item specifying the viewing experience in a user contract, a user can complain to the service provider that he has had a very choppy viewing experience during the last two live streaming sessions, indicating a low QoS of the service. Similarly, a network provider may have guaranteed with the multimedia provider that the end-to-end bandwidth between two particular service gateways V_i and V_j through his network would be at least 40Mbps, but monitoring traffic between V_i and V_j during the last 24 hours reveals that the actual average bandwidth was only 30Mbps, indicating a breach of QoS guarantee.

SSLVs: The QoS/SLA Manager needs to have the complete logical views of the services and systems which might involve in the delivery of the composite service. In particular, it needs to know the process plan which specifies the composition (i.e., which component services will need to be executed and in which order).

The QoS/SLA Manager also needs to know the underlying service overlay network which enables the end-to-end delivery of the composite service through an instantiation of the abstract services (specified in the process plan) in a specific execution plan. The abstract representation for this purpose typically consists of (1) mappings between the abstract services and the set of service gateways, and (2) information about the logical link between the service gateways. Formally,

Definition 1. Given a set of abstract services $S = \{s_i\}_{i=1}^n$ and a set of service gateways $V = \{V_j\}_{j=1}^m$ that the CSP has control over (i.e., either owned by the CSP or contracted by the CSP from a third-party provider). When there is an instance of the service s_i located at the gateway V_j, this instance is denoted by s_i^j. When there is a network link between two service gateways V_j and V_k, this network link is denoted by ℓ_{jk}.

Service Ontologies and QoS-Mapping: As we are dealing with the cross-layer QoS guarantee problem, one critical issue with our framework is to map between different quality criteria and attributes. However, such mappings are application specific and thus beyond the scope of this paper. In the rest of the paper, we assume that the QoS-mappings between layers are given. More precisely, given a set of constraints over the set of QoS variables at a higher layer, the function that maps these constraints to a set of constraints or requirements over the QoS variables at the lower layer is assumed to be given. In the rest of the paper, given a set V of variables, we denote by $\text{CSTR}(V)$ the set of constraints over the variables of V. Hence, the QoS guarantees over a set of QoS-attributes Q is just a subset of $\text{CSTR}(Q)$.

Definition 2. Given a set of user profiles $\mathcal{UP}$ and the following sets: (1) UQ of UP-QoS attributes, (2) AQ of AO-QoS attributes, and (3) IQ of Infra-QoS attributes.

- The function $\Phi^{UP \to AO}$ maps from $\mathcal{UP} \cup 2^{\text{CSTR}(UQ)}$ to $2^{\text{CSTR}(AQ)}$. That is, the function $\Phi^{UP \to AO}$ maps a user profile from $\mathcal{UP}$ and a subset of $2^{\text{CSTR}(UQ)}$ to a subset of $2^{\text{CSTR}(AQ)}$.[1]

---

[1] We denote by 2^S the power set of the set S.

– Similarly, the function $\Phi^{AO \rightarrow Infra}$ maps from $2^{\text{CSTR}(\mathbf{AQ})}$ to $2^{\text{CSTR}(\mathbf{IQ})}$.

QoS-Acquisition, Measurement and Prediction Model: The QoS/SLA Manager needs to acquire and aggregate QoS information from different sources (e.g., from the user perceived service performance, from the application, and from the network operators and datacentre providers). The information will further be processed and measured using the service ontologies accessible from the QoSRM datastore. Finally, a QoS Evaluation and Prediction component projects relevant information and informs the QoS/SLA Manager about potential SLA violations and about the requirements on the system resources. Further details about these components will be presented later in this paper.

QoS&SLA Management: Based on the current QoS information and predictions as well as the resource needs to fulfill the QoS guarantees, the QoS/SLA Manager will need to perform actions to effectively reserve or allocate resources. The objective of such decisions is to ensure that the SLAs with the service consumers be satisfied while trying to, e.g., minimise cost.

3 QoS- and Resource-Aware Service Composition and Adaptation

As discussed above, the major objective of our framework is to provide both QoS assurance under multiple QoS constraints, and optimal resource utilisation and system performance at the resource and network layer. The consumer initiates a composite service request via a well-known service portal and specifies her desired services and QoS requirements. Then the *composite service provider* (CSP) composes and instantiates a qualified execution path for the requested process plan. We assume that the CSP has some control over the resources of the underlying overlay network. That is, the CSP either owns some service gateways or have Service Level Agreements (SLAs) with the owners of the service gateways and the network operators who provide the network links between the service gateways of the overlay network. On the other hand, the CSP has an SLA contract with the consumer of the composite service.

In order to fulfill the SLA with the service consumer and through the QoSRM, the CSP performs the following two tasks: (1) *service instantiation*, and (2) *dynamic service adaptation*. The service instantiation algorithm chooses and composes the service instances, based on their SLA contracts and current performance to fulfill the QoS guarantees specified in the SLA with the user. The dynamic service adaptation is used during runtime when service outages or SLA violation occur. The algorithm find an alternative execution path to quickly recover the composite service delivery.

Formally, we define the resource-aware service routing and instantiation problem as follows. First, we map a given process plan to a set of *execution graphs*. Under the assumption that we will consider process plans which are acyclic, it's possible to represent execution graphs as directed acyclic graphs (DAGs) as follows.[2]

Definition 3. (Execution graph) Given an acyclic process plan *PP* represented as a statechart, an *execution graph EG* of *PP* is a DAG in which each node represents an atomic state of *PP* and obtained as follows:

---

[2] Zeng et al. [8] presents a semantics-preservation technique to "unfold" a statechart which contains cycles into an acyclic statechart.

- The source t_1 of EG is the initial state of PP.
- Given a DAG containing k nodes $\{t_1, \ldots, t_k\}$ which represents a partially constructed execution graph, a state σ in PP can be added into this DAG as follows.
    - If σ is an AND-state, it will be converted to an execution graph EG_σ whose
      source represents the initial states of all sub-statecharts in this AND-state, and
      sink represents the final states of all sub-statecharts in this AND-state. The sink
      of EG_σ is called the end-node of σ. Furthermore, if σ is the direct successor
      of a state whose end-node is t_i, where $1 \leq i \leq k$, then the DAG contains a
      directed edge from t_i to the source node of EG_σ.
    - If σ is an OR-state, one of the alternative branches of σ is selected to embed
      into the execution graph EG, and all states along other alternative branches of
      σ will be discarded during the construction of EG. The selected branch of σ
      will be converted to an execution graph EG_σ whose source represents the first
      state, and sink represents the last state along that branch. Again, the sink of
      EG_σ is called the end-node of σ. Furthermore, if σ is the direct successor of a
      state whose end-node is t_i, where $1 \leq i \leq k$, then the DAG contains a directed
      edge from t_i to the source node of EG_σ.
    - If σ is an atomic state, it is converted into a node t_{k+1} of the execution graph
      EG. t_{k+1} is also the end-node of σ. Furthermore, if σ is the direct successor
      of a state whose end-node is t_i, where $1 \leq i \leq k$, then the DAG contains a
      directed edge from t_i to t_{k+1}.
- Each node which corresponds to an atomic state in the original process plan is
  labeled by its corresponding abstract service.

If a statechart contains conditional branchings (i.e., OR-states), it has multiple execution
graphs. Each execution graph represents a sequence of tasks to complete the execution
of the given composite service. For example, Fig. 3 illustrate an execution graph of the
process plan presented in the top tier of Fig. 1 which depicts the situation where there
has been no highlight requested.

Note that nodes which are
not labeled by one of the abstract services from the process
plan correspond to the initial
and final states in the original
statechart and are only dummy
services which will not need to
be instantiated and executed.

Fig. 3. An execution graph

3.1   Composite Service Instantiation

We will now consider the problem of dynamically instantiating a given process plan
for a composite service from existing service instances. A service instance s_i is a self-
contained application unit providing a certain functionality (e.g., *User_Authentication*).
The service instantiation problem includes the problem of choosing an execution graph
EG among possibly multiple execution graphs of the given process plan, and the problem of finding a path at the infrastructure layer to enable the execution of the selected

graph. The latter problem corresponds to (1) mapping each node t labeled by the abstract service s_i in EG to a service gateway V_j that hosts a service instance s_i^j of the abstract service s_i, and (2) mapping each directed edge which connects two nodes t_1 and t_2 in EG to an overlay path at the overlay data routing layer.[3] It should be noted that the data can be routed between a pair of service gateways V_i and V_j through multiple routes (that possibly involve the relay service at some intermediate gateways) implies that the service link between V_i and V_j can be instantiated by multiple network links (possibly with different QoS guarantees).

We now consider the example shown in in Figure 1. For the content (located at server V_4) to be streamed to a user's device which has access to gateway V_3, the data could be transmitted via the direct network link between V_4 and V_3. However, when this network link has become highly congested, the data would be routed through service gateway V_5 (acting as a relay node) before reaching V_3 to ensure that the QoS guarantee are not violated even though this may incur additional costs.

Given an execution graph EG, an instantiation $\mathcal{I}$ maps each node of EG that is labeled by an abstract service s_i to a service instance s_i^j and maps each edge in EG that is labeled by a service link to a network link. Then the *composite service instantiation problem* (CSIP) can be described as follows. Given a user contract UC and a set of SLAs $\mathcal{SLA}$ between the CSP and the component service providers as well as the underlying network operators, construct an instantiation $\mathcal{I}$ to satisfy UC while optimising certain objectives (such as minimising cost). We will now present two algorithms to address this problem.

(i) **Tree search algorithm** (TA_Search): The TA_Search algorithm generates possible instance paths in the order of the nodes in the execution graph EG. At each nodes of the execution graph, only the feasible and non-dominated paths will be added to the possible instance paths (See Algorithm 1).

Algorithm 1. TA_Search(EG)

$Paths \leftarrow \emptyset, t \leftarrow EG.source.next()$;
while $t \neq EG.sink$ **do**
    if serviceNode(t) **then**
        Path_Update($Paths, t.allInstances$);
    else if AND-sourceNode(t) **then**
        for all branches EG_σ^i in EG_σ that connecting t **do**
            S_σ^i=TA_Search(EG_σ^i);
        end for
        Path_Update($Paths, S_\sigma^1 \times S_\sigma^2 \times ...S_\sigma^n$); **where** n is the number of branches
        $t \leftarrow EG_\sigma^i.sink$
    end if
    $t \leftarrow t.next()$
end while
return $Paths$

---

[3] An overlay path is essentially a sequence of service gateways such that between two consecutive gateways in the path the QoS of the network connection can be monitored and managed.

Algorithm 2. Path_Update(P, S)

if $P = \emptyset$ then
  $P \leftarrow S$
else
  for all paths p in P do
    for all atomic service instances or AND-state composite instances s in S do
      for all network link sets L linking the current gateways in p to the 1st gateways in s
      do
        if feasible(p,L,s) && !dominated(p,L,s) then
          Expand(p,L,s)
        end if
      end for
    end for
  end for
end if

(ii) Distributed search procedure (DIS_Search): The DIS_Search acts on a similar principle of the TA_Search algorithm. Nevertheless, the process of DIS_Search begins with the CSP broadcasting to all the gateways which host the first service in the execution graph. Each gateway that receives the request message will then execute the function described in Algorithm 3. The process continues until the CSP receives the messages from the gateways which hosts the last abstract service in EG and then the CSP decides on the optimal path. Note that at each stage, some of the dominated paths will be deleted in the next stage by the gateways providing the next services, the reduction of search space is smaller than that of tree search (because gateways can only have a local view of the possible instantiations). Thus, the DIS_Search procedure is slower than TA_Search. The biggest advantage of distributed search is that the search process can be conducted in a decentralised manner. It allows the service gateways to communicate and search for optimal paths by themselves rather than disclose all information to the CSP. This is particularly important when service gateways are owned by different service providers.

Algorithm 3. DIS_Search(P, t)

Path_Update(P, $this$.instance(t)); $t \leftarrow t$.next();
if EG-EndNode(t) then
  msg(CSP, P)
else if serviceNode(t) then
  broadcast(t.$allGateways$,P,t)
else if AND-sourceNode(t) then
  for all nodes t' connecting to t do
    broadcast(t'.$allGateways$,P,t')
  end for
else
  broadcast(t.next().$allGateways$,P,t.next())
end if

3.2   QoS Measurement and Prediction

Our approach to QoS measurement and prediction is based upon the mechanisms of (a) performance modeling and measurement, and (b) risk analysis. Several analytic models for predicting the performance of computing and network systems have been developed based on the theory of Queueing Networks (QNs) [6]. A QN has a certain number of queues. Each of which represents either a physical component (such as a processor, an I/O device, or a network domain) or a logical component such as software resources (i.e., threads, database locks, critical sections). A standard analytic QN model then takes input parameters which specify the workload intensity of the system (i.e., how many requests are present in the system, or the arrival rate) and the service demand (of a class of request of a given workload), and calculates the performance metrics (such as the response time) of the measured system under the given workload.

4   Related Work and Conclusion

Most QoS-assurance mechanisms [8] are built to work with web services which operate under the luxury of being relatively stable, with static information about servers' locations and resource availability. However, for most real-world distributed applications, this assumption could significantly affect the service QoS guarantees. By introducing an QoS and Resource management model that operates under dynamic resource constraints, we are making important progress towards solving this problem.

QoS assurance mechanisms for service overlay networks have been investigated by Duan et al [1]. Gu and Nahrstedt [3] present a fully decentralised service composition framework for delivering distributed multimedia service with statistical QoS assurances. The algorithm presented in their paper based on a *probing* mechanism is quite similar to our distributed search for optimal service instantiations. Their paper however does not take into account a number of important composition structures such as conditional branch, exclusive OR, and conditional loop.

In this paper we present a model for resource-aware service delivery with QoS assurance. The proposed model consists of a cross-layer architecture for a Qos and Resource Management framework and algorithms to enable optimal service selection and instantiation. Moreover, the QoS measurement and prediction components are also introduced to the system performance to be measured and forecasted so that potential QoS violations could be avoided. We further discussed how the proposed framework can also handle QoS-assured runtime adaptation.

Acknowledgments. We thank the **Australian Research Council** for partially supporting this research (grants DP0987380 and DP110103671).

References

1. Duan, Z., Zhang, Z.-L., Hou, Y.T.: Service overlay networks: SLAs, QoS, and bandwidth provisioning. IEEE/ACM Trans. Netw. 11(6), 870–883 (2003)

2. Erl, T.: Service-Oriented Architecture: Concepts, Technology, and Design. Prentice Hall, New York (2005)
3. Gu, X., Nahrstedt, K.: Distributed multimedia service composition with statistical QoS assurances. IEEE Transactions on Multimedia 8(1), 141–151 (2006)
4. Harel, D., Naamad, A.: The statemate semantics of statecharts. ACM Trans. Softw. Eng. Methodol. 5(4), 293–333 (1996)
5. Jaeger, M.C., Rojec-Goldmann, G., Mühl, G.: QoS Aggregation for Web Service Composition using Workflow Patterns. In: EDOC, pp. 149–159 (2004)
6. Menascé, D.A., Almeida, V.A.F., Dowdy, L.W.: Performance by Design: Computer Capacity Planning by Example. Prentice Hall International (2004)
7. Papazoglou, M.P., Georgakopoulos, D.: Service-oriented computing: Introduction. Commun. ACM 46(10), 24–28 (2003)
8. Zeng, L., Benatallah, B., Ngu, A.H.H., Dumas, M., Kalagnanam, J., Chang, H.: QoS-Aware Middleware for Web Services Composition. IEEE Trans. Software Eng. 30(5), 311–327 (2004)

An Event-Driven Energy Efficient Framework for Wearable Health-Monitoring System

Na Li, YiBin Hou, and ZhangQin Huang

Embedded Software and Systems Institute, Beijing University of Technology, Beijing, China
nana439@emails.bjut.edu.cn, {ybhou,zhuang}@bjut.edu.cn

Abstract. Wearable health-monitoring system requires keeping the balance between energy consuming and user's real-time service demands with continuous sensing, wireless communication and processing. In this paper, we present a design framework for an Energy Efficient Wearable Health-monitoring System (EEWHMS). EEWHMS uses smartphone as a central unit to process data from wearable sensors, with event-driven energy management strategy to save energy. State transitions and continuous being still of user is used to adjust duty cycle of the system. We present the design, implementation, and evaluation of EEWHMS, which collects user's information with accelerometer and physiological sensors, and sends it to an Android phone by Bluetooth. According to evaluation of power of Bluetooth chip, CPU load of smartphone and response time when emergency happened, the system demonstrates its capability to keep balance between real time and long term sensing in an energy-efficient manner.

Keywords: energy efficient, event-driven, state transition of human.

1 Introduction

Wearable health-monitoring systems (WHMS) have garnered a lot of attention from research community and the industry during the last decade as it is pointed out by the numerous and yearly increasing corresponding research and development efforts [1- 2]. A key challenge of these applications is processing raw data from multiple sensors to get user's context aware information and obtaining higher level service on battery-operated device in real time. Such applications are emerging, e.g.: AMON [3], Smart Vest [4], WEALTHY [5].

WHMS may comprise various types of miniature sensors, wearable or even implantable. It requires extracting more meaningful user's information in real time so that physiological/activity-related, location-aware services, natural user interfaces can be more adaptive to the user's personal health condition. The information is got from the wearable or built-in sensors, such as: biosensors, accelerometer, GPS, microphone, etc. Based on the hardware configuration, research prototypes of WHMS are divided into several categories: 1) custom-designed platform; 2) smart textiles; 3) wireless sensor motes; 4) Bluetooth-enabled biosensors and smart-phones [6]. System in this paper combines category 2) and 4). Related information of context data is concerned, including activity, physiological data and location.

R. Huang et al. (Eds.): AMT 2012, LNCS 7669, pp. 176–185, 2012.
© Springer-Verlag Berlin Heidelberg 2012

Power consumption (and battery size) appears to be perhaps the biggest technical issue in current these wearable computing applications [6], so careful energy management is required to keep the balance between battery capacity of system and real time responsiveness with continuous data collection. Overall, energy consumption in system can be divided into three domains: sensing, wireless communication and data processing [7]. In most mobile sensing applications, energy savings can be achieved by shutting down some sensors and changing sensor duty cycles, which are stovepipe solutions that attempt to address this challenge but in a very application specific way [8].

In this paper, we present the design, implementation, and evaluation of EEWHMS, an energy efficient wearable health-monitor system that employs an event-driven energy management strategy. We divided work mode of wireless transmission unit (WTU) into two classes: active and inactive. The active mode includes searching, connecting, transmitting; while inactive mode means WTU on sleep state. Classifiers on smartphone are used to recognize user's state (activity, physiology and location information) in real time. User's state transitions and continuous being still state are respectively regarded as wake-up event and sleep event to trigger changes of duty cycle of WTU. Based on types of sensors, event model is built. The benefits of our energy management strategy are following: first, it is generic, without dependence on special hardware in the system; second, event-driven model makes the system responds in real time to the user's state transition and emergency; third, it is flexible to add/remove decision rules in the management unit for XML descriptor used.

The remainder of the paper has the following structure. Section 2 describes related work of energy efficient strategy for wearable computing system, especially using mobile phone as processing unit. Section 3 introduces the framework of EEWHMS and event-driven strategy. Section 4 provides our prototype implementation. Section 5 presents an evaluation of the system in terms of response time, device lifetime and CPU load in mobile phone. Section 6 makes a discussion. And section 7 concludes.

2   Related Works

There have been some researches done in the area of reducing energy consumption for the similar application: continuous sensing on mobile phone. Similar researches have been done in wireless sensor network [9-10], but difference between them is that the former uses a smartphone as a processing unit, which is battery-driven and also used as communication tool in user's daily life. The problem of energy management of these systems has been well-explored in the literature. For example, Hong Lu utilized changes of user's mobility and behavior pattern's to trigger some sensors to power hungry processing in his Jigsaw continuous sensing engine [8], which turned down GPS when its accelerometer detected that the user was on a still state to make the system in an energy-efficient manner. Yi Wang used hierarchical sensor management strategy and powered only a minimum set of sensors with appropriate sensor duty cycles to improve device battery life [9]. The system can not respond rapidly to user's emergency because it adopted periodic sensing strategy. Bodhi Priyantha et al. developed LittleRock [12], which added a dedicated low-power microcontroller to the phone and some process of sensor data are done on it to decide when to wake up the main processor. This method needs an extra core, which is not likely to be adopted by phone manufacturers. Event-driven power-saving method is investigated by Shih et. al [13]. Their work focused on reducing the idle power that a device consumes in a "standby" mode. The device will be powered on only when there is an incoming or outgoing call or when the user needs to use the phone. While these works only focus on the user's information accurate, robust of system and system's energy efficient [14-16], how to balance real time and energy efficiency is less taken into account; moreover, they used build-in sensors on smartphone, so information about user is limited and it is difficult to add external sensors to the system.

In our system design, we build on these relevant reaches and advance EEWHMS. In this paper, switching duty cycle of WTU is used to save energy, which is triggered by the event-driven strategy.

3   Method

WHMS in this paper is comprised of sensors, device processor, wireless transmitting unit, smartphone, background service unit. The device processor is responsible for some straightforward work, for example: collecting raw sensor data, packaging, comparison and computing. The first four are battery-operated.

In order to achieve balance between energy efficient and real time, we employ the framework of EEWHMS, core component of it is the event-driven energy management strategy, which runs on processing unit and controls duty cycle of WTU. The architecture of system is shown in figure 1.

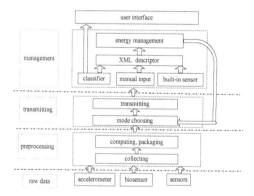

Fig. 1. Architecture of EEWHMS

3.1   Event Model

Event-driven strategy provides a method for dealing with complex event streams [17]. In this paper, state transition and continuous being still are regarded as events, for example: motion state to still state, normal to heart trouble, normal to manual alarm, and so on. Events are divided into three classes: activity, physiological info and location. Each corresponds to different queues and classifiers on smartphone. Activity event is the key factor to reflect user's state. We build event set as following:

- Event element: state transition, being still, and emergency. Emergency includes falling, physiological exception and manual input alarm.
- Event type: wake-up event and inactive event. The former one is to make WTU into active state, while the latter one is to notice it into sleep state.
- Starting time: time when event arrives at the corresponding event queue.
- Relations between events: priority principle is based on event type. Wake-up event is prior to inactive event, but there is no priority between wake-up events.
- Structure of an event: <ID, event type, starting timestamp >.

Event-driven strategy is achieved by specifying an XML-format event descriptor as input, which is flexibly to add new state. Following code is a sample of events descriptor and the rules definition. It can be seen that energy management strategy does not only focus on current state, but also on the previous state as context. For example, although user is on still state, it needs to check if there is a falling or emergency happened before, then different measures are taken.

```xml
<?xml version="1.0" encoding="UTF-8" standalone="yes"?>
<eventList>
  <event>
    <id>1</id>
    <type> inactive </type>
    <possible_consequences>
      <consequence>
        <condition>After_walk</condition>
        <operation> inactive </operation>
      </consequence>
      <consequence>
        <condition>After_fall</condition>
        <operation>active</operation>
      </consequence>
    </possible_consequences>
  </event>
  <event>
    <id>2</id>
    ...
  </event>
</eventList>
```

3.2   Event-Driven Strategy

Each event is processed by corresponding event processing agent. Classifier agent (CA) reports classification results of activity and physiology to event agent (EA), which is responsible for storing and managing events. The energy management agent (EMA) gets events from EA and decides whether changing WTU duty cycle. This process is shown as figure 2.

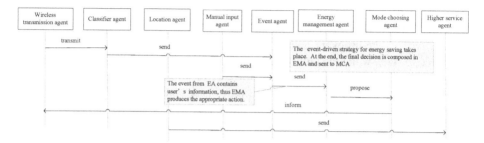

Fig. 2. Interacting process between agents under event-driven strategy

Assuming a scenario, user has sit for a period of time, and CA recognizes user's still state and sends to activity event queue. If there is no state transition coming and the previous activity is not falling, then EA assign event type "inactive event" and send to EMA. EMA checks if there is a "wake-up event" to confirm that user is on "safe being still", and proposes MCA to extend duty cycle of WTU. If there is an emergency, location agent (LA) is triggered and sends location information to higher service agent (HSA).

3.3 Adaptive Duty Cycle of WTU

WTU adapts its duty cycles according to user's state. Here, we adopt a simple adaptive technique based on duration of being still. Duty cycle of transmission unit will become longer when user is still for a period of time, except emergency happens. The pseudo-code is summarized as following:

```
sleep_fraction =1;
onStillState:
   sleep_fraction += n_{s1};//increasing sleep fraction
if (sleep_fraction>N)  sleep_fraction=N;//Max fraction
   Sleep(sleep_fraction);
```

4   Experimental Trails and Results

4.1 Experimental Setup

In this section, we will describe an implementation of wearable computing system using EEWHMS framework and evaluate its performance in terms of real time and energy efficient. All these experiments were done in lab environment. In our previous work [18-19], the algorithm for steady state recognition and falling detection by accelerometer was proposed and run on smartphone.

External sensors (accelerometer ADXL345, heart rate sensor HK-08A, temperature sensor DS18B20) and built-in GPS sensor and base station ID on smartphone are used. Raw sensor data is transmitted to smartphone by Bluetooth and smartphone connects with remote server by Wi-Fi. Blue Core 5 chip is adopted, which has logic units (DSP and VM): DSP collects raw data and sends it to VM, which packages and transmits it to smartphone. Android smartphone (CPU: dual-core, 1.6GHZ; memory: 1GB; vision of Android: 4.0) is used as a processing unit. User interface on it is shown in figure 3 (b). External sensors connect to Bluetooth unit by wires; they are powered by the same battery (Li-rechargeable battery, 3.2V). We integrated these devices into a vest, as shown in figure 3 (a). Accelerometer and temperature are near the center-of-gravity position, electrodes of heart rate sensor are put on the hear position, Bluetooth model is on the right pocket position; smartphone are put into outside pocket. Subjects including 6 males and 4 females wore the vest doing daily activities. We focus on falling without recovery, so subject fell and kept still for a period of time. There is manual alarm button on the screen of smartphone, so subject buttoned it when he simulated emergency happening.

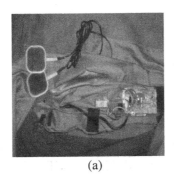

(a)   (b)

Fig. 3. (a) Wearable sensors and WTU in vest (b) Interface on the smartphone

Sample rate of each sensor is shown in table 1 and DSP collects raw data at 20 HZ. In active state, smartphone receives package about every 200ms. According to datasheet of chips used in experiment, power of them under different supply voltage is shown in table 1 (excluding external consumptions on peripheral circuit):

Table 1. Power of some devices under different supply voltage

Sensor/wireless transmission	voltage	power on different states		
	Voltage(V)	Active(mW)	Sleep(uW)	Sampling rate(mW)
Accelerometer	2.5	0.035	1.3225	0. 035 (100HZ)
Temperature	5	7.5	5	3.52 (1 HZ)
Heart Rate	3.3	2.64	1.71	2.64 (60~120 HZ)
Bluetooth	3.3	660	49.5	0.0495~660

We mainly make Bluetooth sleep to save energy because we find that power of it is significantly higher than that of other units. When WTU is in sleep state, DSP works as usual but won't disturb VM. In practice, when MCA detects user on still state for 10 seconds without any emergence events, it will message VM by Bluetooth to enter inactive mode. DSP stops sending data to VM and simply compares the new samples of accelerometer with current data to check if the user is still being static. The condition of being still is measured by expression (1):

$$(|a_x(t)-a_x(t-1)|<0.1g)\ \&\ \&(|a_y(t)-a_y(t-1)|<0.1g)\ \&\ \&(|a_z(t)-a_z(t-1)|<0.1g) \qquad (1)$$

If user keeps being still, DSP just messages VM periodically to transmit data. Different thresholds are also used to infer exceptions on temperature and heart rate, if there is abnormal data, DSP will wake VM immediately.

4.2   Experimental Results

We evaluate the performance of system from three aspects: power of Bluetooth, CPU load of smartphone, and response time to change of user's state.

Power of Bluetooth model is shown in table 2 (all these operations finished on DC regulated power supply, voltage is 3.2V).

Table 2. Power of Bluetooth model under different states

State	power of Bluetooth model (mw)
Bluetooth inquiring and scanning (no sensor connected)	192~320
Connecting finishes but no data being transmitted	224
Data being transmitted (temperature and accelerometer)	256
Data being transmitted(temperature, accelerometer, heart rate)	256
Sleep	192

Data on table 2 includes energy consumption of peripheral circuit. Bluetooth model on sleep state can save 25% energy than it on active state.

Workload of CPU of smartphone is shown in figure 5. In experiment, there was no incoming call or other applications running on smartphone, screen of smartphone was kept awake. In figure 5(a), smartphone processes data every 200ms; in figure 5 (b), energy management strategy is adopted, but a fixed duty cycle of WTU, 1 second, is used; in figure 5 (c), adaptive duty cycle of WTU is adopted, sleep fraction is 1 second, but duty cycle will not exceed to 10 seconds.

DSP can wake up VM immediately when detecting abnormal data, so response time of system is not slowdown comparing that without energy efficient strategy. Average response time is about 5304 ms when falling happened (user is walking before falling), here, average transmitting and unpacking time of 100 sets data is 5300ms, and average running time of fall detection algorithm is 4ms.

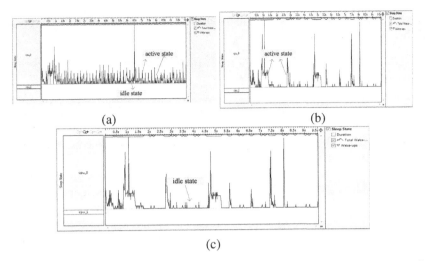

(a)   (b)

(c)

Fig. 4. CPU workload of smartphone :   (a) without energy efficient strategy   (b) with fixed duty cycle when user on still state   (c) with adaptive duty cycle when user on still state

5   Discussion

EEWHMS achieves energy efficiency with event-driven approach. The results in section 4 summarize the performance of system from energy consumption and real time. This section reports a more in-depth analysis of results.

In this paper, we only save power when being still, but human activities tends to have durations on the order of ten seconds or more. Some activities, such as: cycling, upstairs/downstairs, last considerably longer and successive data is similar [8]. This redundancy could be eliminated by similarity detector. But if this similarity detection was processed on DSP/VM in Bluetooth model, it is beyond the ability of them. Moreover, if more sensors are in the system, each sensor has its own similarity detection algorithm, which will be a large workload. So choosing apposite similarity detection algorithm should be taken into account.

Thresholds are set manually based on experimentations, e.g.: time of continuous being still, but there is some randomness on human activity, so there may be a situation that WTU change its duty cycle repeatedly. Some sensors may reduce this instability, for example, microphone can be used to get environmental information to infer user's state, e.g.: talking with others, attending the lecture, which is more accurate for changing duty cycle of units.

Alternate wireless communication, e.g.: zigbee, IrDA, MICS, are more lower-power technologies. IrDA and MICS are for short-range exchange of data over infra-red light (1-2m) and in support of low-rate data of medical devices. They have not been utilized widely by researches due to the lack of commercially available solutions. But the 802.15.6 IEEE Task Group is planning the development of a communication standard optimized for low power device and operation on, in or around the human body [20]. If these wireless communication technologies were used, energy of WHMS can be saved more.

6   Conclusion

WHMS can provide rich information about user and environment for higher layer application and services. However, the energy consumption by sensors and transmission unit, coupled with limited battery capacities, make it infeasible to be continuously running for long time.

In this paper, we presented the design, implementation and evaluation of the energy efficient wearable health-monitoring system (EEWHMS). The core component of it is an event-driven energy management strategy, which adjusts the duty cycle to improve device battery life according to state transition and emergency detected by classifier. Experimental results show that EEWHMS is able to provide a high level of real time and more than 30% gain on device lifetime.

Future work may involve testing other wireless transmission, e.g.: zigbee and development of similarity detection to reduce processing replicate data.

References

1. Lymperis, A., Dittmar, A.: Advanced Wearable Health Systems and Applications, Research and Development Efforts in the European Union. IEEE Eng. Med. Biol. Mag. 26, 29–33 (2007)
2. Pantelopoulos, A., Bourbakis, N.G.: Prognosis-a Wearable Health-Monitoring System for People at Risk: Methodology and Modeling. IEEE Transactions on Information Technology in Biomedicine 14, 613–621 (2010)
3. Anliker, U., Ward, J.A., Lukowicz, P., Troster, G.: AMON: a Wearable Multiparameter Medical Monitoring and Alert System. IEEE Transactions on Information Technology in Biomedicine 8, 415–426 (2004)
4. Pandian, P.S., Mohanavelu, K., Safeer, K.P., Kotresh, T.M., Shakunthala, D.T., Gopal, P., Padaki, V.C.: Smart Vest: Wearable Multi-parameter Remote Physiological Monitoring System. Med. Eng. Phys. 30, 466–477 (2008)
5. Lymperis, A., Paradiso, R.: Smart and Interactive Textile Enabling Wearable Personal Applications: R&D State of the Art and Future Challenges. In: Proc. 30th Ann. Int. IEEE EMBS Conf., pp. 5270–5273 (2008)
6. Pantelopoulos, A., Bourbakis, N.G.: A Survey on Wearable Sensor-based Systems for Health Monitoring and Prognosis. IEEE Transactions on Systems, Man, and Cybernetics-Part C: Applications and Reviews 40, 1–12 (2010)
7. Latre, B., Braem, B., Moerman, I., Blondia, C., Demeester, P.: A Survey on Wireless Body Area Network. Wireless Networks 17, 1–18 (2011)
8. Lu, H., Yang, J., Liu, Z., Lane, N.D., Choudhury, T., Campbell, A.T.: The Jigsaw Continuous Sensing Engine for Mobile Phone Applications. In: The 8th ACM Conference on Embedded Networked Sensor System (SenSys 2010), pp. 71–84 (2010)
9. Sundaresan, S., Koren, I., Koren, Z., Krishna, C.M.: Georgia Tech.: Event-driven Adaptive Duty-cycling in Sensor Networks. International Journal of Sensor Networks 6, 89–100 (2009)

10. Huang, Y., Yang, Y., Cao, J., Ma, X., Tao, X., Lu, J.: Runtime Detection of the Concurrency Property in Asynchronous Pervasive Computing Environments. IEEE Transactions on Parallel and Distributed Systems 23(4), 744–750 (2012)
11. Wang, Y., Lin, J., Annavaram, M.: A Framework of Energy Efficient Mobile Sensing for Automatic User State Recognition. In: Proceeding of the 7th International Conference on Mobile Systems, Applications, and Services (MobiSys 2009), pp. 1–14 (2009)
12. Priyantha, B., Lymberopoulos, D., Liu, J.: LittleRock: Enabling Energy-efficient Continuous Sensing on Mobile Phones. Pervasive Computing 10, 12–15 (2010)
13. Shih, E., Bahl, P., Sinclair, M.J.: Wake on Wireless: an Event Driven Energy Saving Strategy for Battery Operated Devices. In: Proceeding of the 8th Annual International Conference on Mobile Computing and Networking (MobiCom 2002), pp. 106–171 (2002)
14. Krause, A., Ihmig, M., Rankin, E.: Trading off Prediction Accuracy and Power Consumption for Context-aware Wearable Computing. In: Proceeding of the Ninth IEEE International Symposium on Wearable Computers, pp. 20–26 (2005)
15. Miluzzo, E., Lane, N., Fodor, K., Peterson, R., Lu, H., Musolesi, M., Eisenman, S., Zheng, X., Campbell, A.: Sensing Meets Mobile Social Networks: The Design, Implementation and Evaluation of the CenceMe Application. In: The 10th ACM Conference on Embedded Networked Sensor System (SenSys 2010), pp. 337–350 (2008)
16. Aghera, P., Rosing, T.B., Fang, D., Patrick, K.: Energy Management in Wireless Healthcare Systems. In: 2009 International Conference on Information Processing in Sensor Networks (IPSN 2009), pp. 363–364 (2009)
17. Dunkel, J., Fernandez, A., Ortiz, R., Ossowski, S.: Energy-driven Architecture for Decision Support in Traffic Management Systems. Expert Systems with Applications 38, 6530–6539 (2011)
18. Li, N., Hou, Y., Huang, Z.: A Real-time Algorithm based on Triaxial Accelerometer for the Detection of Human Activity State. Proceeding of the 6th International Conference on Body Area Networks (BODYNETS 2011), pp. 103–106 (2011)
19. Hou, Y., Li, N., Huang, Z.: Triaxial Accelerometer-based Real Time Fall Event Detection. In: International Conference on Information Society, i-Society 2012 (2012)
20. Yuce, M.R., Ng, S.W.P., Myo, N.L., Lee, C.H., Khan, J.Y., Liu, W.: A MICS Band Wireless Body Sensor Network. In: IEEE Wireless Communications and Networking Conference (WCNC 2007), pp. 2473–2478 (2007)

Learning Style Model for e-Learning Systems

Mohamed Hamada

Graduate School of Computer Science
The University of Aizu,
Aizuwakamatsu, Fukushima, Japan
hamada@u-aizu.ac.jp

Abstract. E-learning is the use of digital devices and networks to transfer skills and knowledge to learners. E-learning applications and tools include active media, multimedia, web-based learning, computer-based learning, virtual education tools, etc. In such rich learning environment learners need a method to specify a clear learning path that matches with their learning preferences. Learning style model can provide a basis for such methods. In this paper we study and implement a web-based learning style model that can help learners to find a learning path in an e-learning system.

1   Introduction

E-Learning environments have established a tradition in technology systems for individual learning. To better utilize such systems in the learning process, learners have to be aware of their learning preferences. A learning style model can help learners to identify their learning preferences. It also supports to adopt suitable learning materials to enhance learner's learning process.

On the other hand, teachers can gain by knowing their students' learning preferences. From the teacher's point of view, if they figure out their students' learning preferences, they can adjust their teaching style and adopt suitable materials to best fit with the students' preferences.

If there is a mismatch between a learner's learning style and the way learning materials are presented, students are more likely to lose motivation to study.

Integration of learning style into e-learning systems can lead to an intelligent and adaptive learning system that can adjust the content in order to ensure faster and better performance in the learning process.

So far, several learning style models have been developed (e.g., [3], [5], [7], and [8]) for the realization of the learning preferences of learners. Among these models, that of Felder and Silverman [3] is simpler and easier to implement through a Web-based quiz system, as in [10]. The model classifies learners into four axes: active versus reflective, sensing versus intuitive, visual versus verbal, and sequential versus global. Active learners gain information through a learning-by-doing style, while reflective learners gain information by thinking about it. Sensing learners tend to learn facts through their senses, while intuitive learners prefer discovering possibilities and relationships. Visual learners prefer images, diagrams, tables, movies, and demos,

R. Huang et al. (Eds.): AMT 2012, LNCS 7669, pp. 186–195, 2012.

while verbal learners prefer written and spoken words. Sequential learners gain understanding from details and logical sequential steps, while global learners tend to learn a whole concept in large jumps.

We found that there are three week points in Felder-Silverman "Learning Style Index" (LSI) model:

1. LSI allows students to choose between two alternatives. However, in real life, not everything is black or white. Hence freedom has to be given to learners to choose among several alternatives in a fuzzy-like system.
2. LSI did not support social/emotional learning dimension which has important effect on the learning preferences of learners and the learning process as a whole.
3. The current implementation of LSI supports individual learners only. It means that teachers can only get the learning preferences of individual students separately. Therefore teachers cannot get collective learning preferences of all students (or a group of students) in their class.

Based on this observation and our additional enhancement to LSI, our paper has the following five contributions.

1. Extending the LSI computational system to allow a multi-choice among several alternatives in a fuzzy-like system.
2. Extending LSI to a new social/emotional dimension to reflect its impact on the students' learning preferences in an e-learning environment.
3. A new web-based implementation that is connected to an SQL database server was introduced. Our implementation allows teachers to analyze the LSI for a group of learners and help them to obtain a wider bird-view of the learning preferences of their students. Our implementation can also provide graphical representations for the students' learning preferences in addition to the ability to distinguish between male and female learners. This allows us to obtain a deeper understanding of the effect of gender differences on the learning process.
4. E-learning systems are widely used and rapidly increasing. The integration of a learning style index in an intelligent and adaptive e-learning system is useful to help e-learners to navigate through different available learning materials. As a case study we show the integration of our enhanced learning style index into an intelligent and adaptive e-learning system for automata and theory of computation.

The rest of the paper is organized as follows. Section two covers background in learning style model. Our study on the learning style model and the extension of existing systems will be introduced in section three. Section four covers our web-based implementation of the enhanced learning style model. Section five describes the integration of our implemented learning style index into an e-learning system for the theory of computation topics. We conclude the paper in Section six.

2   Learning Style Model

Among the existing learning styles' models we chose Felder-Silverman model for the following reasons: it is widely known and applicable, it can describe learning styles in more details than other models, and its reliability and revalidation have already been tested. Felder-Silverman Learning Style Model can classify learners according to a scale of four dimensions; processing, perception, input and understanding, see Table 1. Each of these dimensions consists of contrastive attributes listed below.

Table 1. Learning and Teaching Styles

Learning Style		Teaching Style	
Process	Active	Student participation	Active
	Reflective		Passive
Perception	Sensory	Content	Concrete
	Intuitive		Abstract
Input	Visual	Presentation	Visual
	Verbal		Verbal
Understanding	Sequential	Understanding	Sequential
	Global		Global

- Active/Reflective learners

Active learners understand and memorize better what they went through. They tend to retain and understand information best by doing something active. "Let's try it out and see how it works" is an active learner's phrase. Reflective learners understand better what they quietly think about it first. "Let's think it through first" is the reflective learner's phrase.

- Sensory/Intuitive learners

Sensory learners tend to be interested in facts and solve the problem with well-established methods. It means that they dislike solving problem with complicated methods. In addition, they tend to be patient with details and good at memorizing facts and doing hands-on work. Intuitive learners tend to discover relationship to actual world and possibilities and be good at finding new concepts and understand abstract contents better. Besides, they can solve problem faster than sensory learners and often make mistakes.

- Visual/Verbal learners

Visual learners are good at remembering what they see; pictures, graphs, charts, movies, and demonstrations. Verbal learners tend to understand information best by progressing verbal instruction or using words from instructor. In addition, if they talk with someone to study, they can understand more effectively. They prefer to study in a group.

- Sequential/Global learners

Sequential learners prefer presentations that proceed step by step and are good at understanding the relationship to things they have already studied. They tend to solve problems logically and piece by piece. It means that they figure out things like heaping up books. Global learners tend to learn in large jumps, absorbing material almost randomly without seeing connections and the suddenly "getting it." In sum, they understand many small parts and then suddenly they have a brainwave and find out the large part. Once they have grasped the big picture, they may be able to solve complex problems quickly or put things together in novel ways.

The Index of Learning Style (LSI), based on Felder-Silverman Learning Style model, is an outline questionnaire for identifying learning styles. The ILS consists of a total of 44 questions for afore-mentioned four dimensions, and each dimension has 11 questions. These preferences are expressed with values between +11 to -11 and one problem has 1 or -1 (minus 1). For example, if you answer a question related to "active/reflective" attributes and your answer has an active preference, +1 is added to the score, whereas 1 is subtracted from the score if you answer the question with a reflective preference. That is, the degree of preference for each of dimension is just the algebraic sum of all values of the answers to the eleven questions.

$$i_{DIM} = \sum_{i=1}^{11} q_i^{DIM} \tag{1}$$

$DIM = \{A/R, S/I, V/V, S/G\}$ refers to each of the four dimensions (A/R for Active/Reflective, S/I for Sensory/Intuitive, V/V for Visual/Verbal, and S/G for Sequential/Global). The vector of indexes $I=\{i_{A/R}, i_{S/I}, i_{V/V}, i_{S/G}\}$ describes attributes in each dimension. $Q=\{q_1, q_2, \ldots, q_{11}\}$ is the sum of questions belonging to each dimension, and each q_i indicates the contribution given by i-th question within the eleven questions for each DIM to detect preference. 1 or -1 is substituted into q_i. Results are divided into three groups, according to points (Figure 1). If your score is between 3 to -3, you are categorized into "well balanced". If your score is between -5 and -7, or between 5 and 7, you are classified into "moderate preference". If your score is between -9 and -11 or between 9 and 11, you are grouped into "strong preference".

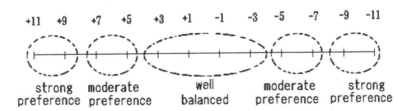

Fig. 1. LSI Evaluation System

Reliability of LSI system was established in western style educational institutes because the western style culture allows a clear-cut "yes/no" answers for queries. On contrary, the reliability of LSI is not clear in Asian educational institutes because Asian culture (especially Japanese) tend to give unclear fuzzy answers for queries. Hence, in order to be able to study the reliability of LSI in Asian educational institutes, it is necessary to extend the traditional "yes/no" style for answers to a new fuzzy-like system with index of five levels. This extension will be introduced in the next section.

3   Enhanced Learning Style Index (ELSI)

In this section, our new model is proposed. It extends the Felder-Silverman LSI model in two ways: a fuzzy-like evaluation system, and a new social/emotional dimension.

3.1   Fuzzy-Like Evaluation System

Our model is based on answers of an ascending risk scale of 1 to 5 (Figure 2). Our new assessment system extends the traditional Felder-Silverman model as follows.

$$i_{DIM} = \sum_{i=1}^{11} q_i^{DIM-} - \sum_{i=1}^{11} q_i^{DIM+}$$

(2)

DIM, Q and I are same as in section 3. q_i^{DIM+} and q_i^{DIM-} are the attributes which represents the contrast of each dimension. The Felder-Silverman model only has 1 or -1 value assigned to each dimension when the learner answers a question. Our new model has five different values assigned for each question. Depends on the choice of the learner on the 5 scale values of the question's answer, q_i^{DIM+} and q_i^{DIM-} will take one of the values: 1, 0.75, 0.5, 0.25, or 0 (positive or negative). If you choose the first option in q_i, the value +1 will be assigned to q_i^{DIM+} and 0 assigned to q_i^{DIM-}. When you choose the second option in q_i, the value +0.75 will be assigned to q_i^{DIM+} and +0.25 will be substituted to q_i^{DIM-}. When you choose the third (middle) option in q_i, the value +0.5 will be assigned to q_i^{DIM+} and +0.5 to q_i^{DIM-}. When you choose the forth option in q_i, the value +0.25 will be assigned to q_i^{DIM+} and +0.75 to q_i^{DIM-}. Finally, when you choose the fifth option in q_i, the value 0 will be assigned to q_i^{DIM+} and +1 to q_i^{DIM-}. Summation of the values assigned to the attribute q_i^{DIM+} will be calculated and summation of the values assigned to the attribute q_i^{DIM+} will also be calculated. Then subtraction between the two calculated values of the two attributes will be the result of one's learning preference. For example, suppose that the first choice is closest to "active" and fifth choice is closest to "reflective." If you choose the first option in this question, +1 point will be added to the attribute of "active". If you choose the second option, +0.75 will be added to the attribute of "active" and also +0.25 will be added to the attribute of "reflective." Likewise, if you choose the third option, +0.5 will be added to "active" and +0.5 to "reflective" and so on. Then the result of the learning preference in the active/reflective dimension will be calculated by subtracting the total value assigned to "reflective" from that assigned to "active".

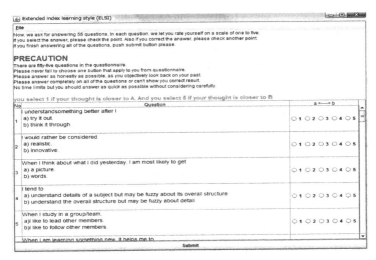

Fig. 2. Enhanced LSI Five Options System

Following the change in the point allocation system, we changed the degrees of preference (Figure 3). If your score is between 11 and 7.5, or between -11 and -7.5, you are categorized into "strong preference." If your score is between 7.5 and 3.5, or between -7.5 and -3.5, you are classified into "moderate preference." If your score is between 3.5 and 2, or between -3.5 and -2, you are grouped into "some preference." If your score is between -2 and 2, you fall into "well balanced".

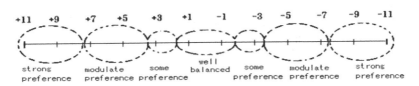

Fig. 3. Enhanced LSI Evaluation System

Several works was done to analyze the Felder-Silverman LSI model for example [9, 11, 12, 13], but none of them considered the extension of the evaluation system as we did.

3.2   Social/Emotional Dimension

Social emotional learning (SEL) is a process for helping people to develop the fundamental skills for achieving an effective life. SEL teaches the skills we all need to handle ourselves, our relationships, and our work, effectively and ethically. SEL holds the next five key:

- Self-awareness: assessing one's feelings, interests, values, and strengths,
- Self-management: regulating one's emotions to handle stress, control impulses, and persevere in overcoming obstacles,
- Social awareness: is able to take the perspective of and empathize with others,
- Relationship skills: establishing and maintaining healthy and rewarding relationships based on cooperation,
- Responsible decision-making: making decisions based on consideration of ethical standards, safety concerns, appropriate social norms, respect for others, and likely consequences of various actions.

These skills include recognizing and managing emotions, developing caring and concern for others, establishing positive relationships, making responsible decisions, and handling challenging situations constructively and ethically. SEL is a framework for school improvement.

Teaching SEL skills helps create and maintain safe, caring learning environments. Social and emotional skills are implemented in a coordinated manner, school wide, from preschool through high school. Lessons are reinforced in the classroom, during out-of-school activities, and at home. Educators receive ongoing professional development in SEL. Families and schools work together to promote students' social, emotional, and academic success.

We extended the Felder-Silverman LSI model by adding a new "realistic" dimension that concerns with the effect of emotion and social learning styles. To this extent we added a new set of eleven questions to the quiz system of LSI for this new dimension.

Table 2 summarizes the new realistic (social/emotional) dimension, where the main attributes for both categories are the following:

- Social learners prefer reading books, discussions, social interaction, recognized and valued, and they may need repetition for detail,
- Emotional learners are affected by their emotion.

Table 2. Realistic (social/emotional) dimension

Realistic Learner	
Social	Emotional
Social learners tend to be big picture people; concepts are more interesting than details.	Emotions can affect the learning process, in both a positive and negative way.
They are motivated by relationships and care a great deal about what others think of them.	When a learner experiences positive emotions, the learning process can be enhanced.
They make more effort to attract people's attention.	When a learner experiences negative emotions, the learning process can be disabled.
As a result, they are vulnerable to criticism. They also prefer cooperation rather than completion.	

4   Web-Based Implementation

We developed a web-based application with a web server, application server and a database for a better analysis of the learning style. Advantages of our developed system are as follows.

1. Easy to use through its user-friendly interface
2. Easy to integrate into E-learning systems. As we will explain in section seven.
3. Easy to find and analyze the learning style of a group of learners. This enables the teachers to have a bird's view of the learning preferences of all students in the class.
4. Easy to access and use anytime anywhere.

The overview of our system is shown in Fig. 4. The system consists of the following components: a user-friendly graphical interface, a web-server, an application server, and a database module.

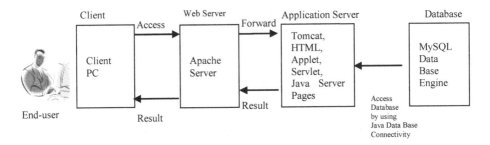

Fig. 4. Extended ILS System Outline

The learning preferences computational module of our system, resides in the application server, uses the new calculation model described in section four. This calculation model will provide more detailed information about the learning style of a learner.

The learner can access the system through the user interface: that is a java applet runs on a web browser (Fig. 2). The learner then fills in all the answers of the quiz system and then submits the answers to the Apache [1] web server through the client PC. The web server will then passes it to the Tomcat [14] application server. The application server will run the computation module of the system to compute the learning preferences of the user. The application server will send the result back to the user through the Apache web server and the Clint PC. A copy of the result will also be stored in the MySql database which is connected to the application server through the "Java DataBase Connector" JDBC. JDBC provides methods for querying and updating data in a database. Our system also provides functions that maintain statistics with the user gender distinguished. This helps educators to analyze the learning styles of their group students (even with gender taken into consideration) and then prepare suitable teaching materials and adapt their teaching style accordingly.

5   ELSI Integration into e-Learning Systems

Compared to traditional learning systems, E-learning [2] provides a more comfortable learning environment, where learners can learn at their convenient. E-learning systems are widely used and rapidly increasing. Hamada [4] provides an e-learning system for automata theory and theory of computation based on Java2D technology [6] (Fig. 5). Hamada's e-learning system for automata and theory of computation is composed of several components: an animated (movie-like) welcome component, a hypertext introduction to the theory of computation topics, a finite state machine (FSM) simulator, a Turing machine (TM) simulator, a self-assessment component, a chatting component for supporting online collaborative learning, and other three components showing automata visual examples such as video player, rice cocker, and tennis game. Novice automata learners find it difficult to grasp these comprehensive materials that were designed to meet all kinds of learning preferences. Learners do not know where they should start.

Fig. 5. Automata e-Learning System Interface

To facilitate their learning process, we extend Hamada's e-learning system by adding a new component (Fig. 6) for learning style. This new component enables the user to find his/her learning preferences and hence chose the suitable components form the rich automata e-learning system.

ELSI component integrated into the e-learning

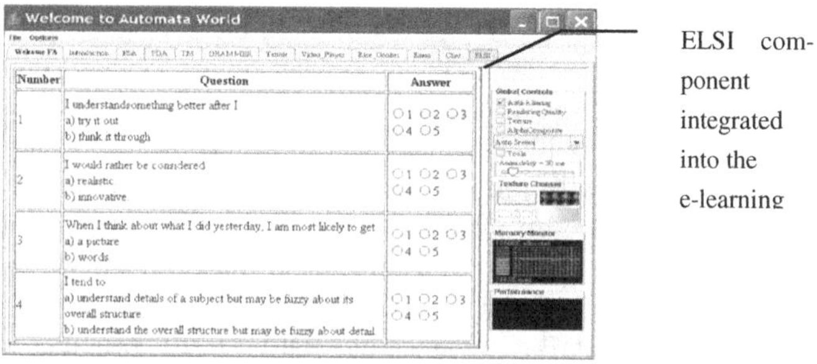

Fig. 6. ELSI Integrated into Automata e-Learning System

The integration of our enhanced learning style system into Hamada's automata e-learning system needs access to the source code. Since both systems are written in Java, there was no compatibility problem in the integration process. Technical details of the integration are omitted here due to lack of space.

6   Conclusion

In this research, we developed an enhanced version of a learning style index taking into consideration culture differences. We implemented our model in a way that allows learners to easily check their learning preferences. Moreover teachers can have a wider bird-view on their students' learning preferences. To this extent our implementation utilizes several useful tools such as: web-based interface, java applets, Apache web server, Tomcat application server, MySQL database, and JDBC connector.

To show the flexibility and usefulness of our implemented system we integrated it into an e-learning system that is based on Java2D technology and contains intensive set of learning materials to support all kind of learners. By this integration we believe that the e-learning system will be more effective since novice learners can now grasp their way into the system easily. However this is just a starting, a follow-up and evaluation of the integration is necessary and that what we plan to do in our future research. We also plan to study the reliability of the model.

References

1. The Apache Software Foundation (2006), http://www.apache.org
2. The Advanced Distance Learning group (2007), http://www.adlnet.gov
3. Felder, R., Silverman, L.: Learning and teaching styles in engineering education. Engineering Education 78(7), 674–681 (1988)
4. Hamada, M.: An Integrated Virtual Environment for Active and Collaborative e-Learning in Theory of Computation. IEEE Transactions on Learning Technologies 1(2), 1–14 (2008)
5. Herrmann, N.: The Creative Brain. Brain Books, Lake Lure (1990)
6. Java2D of Sun Microsystems (2007), http://www.sun.com
7. Kolb, D.: Experiential Learning: Experience as the Source of Learning and Development. Prentice-Hall, Englewood Cliffs (1984)
8. Myers, I.: Gifts Differing. Consulting Psychologists Press, Palo Alto (1980)
9. Ahmad, N.B.H., Shamsuddn, S.M.H.: Mapping Student Learning Characteristics Into Integrated Felder Silverman Learning style Using Rough Sets Classifier, July 3-4. University Technology Malaysia (2007)
10. Soloman, B., Felder, R.: Index of Learning Style Questionnaire, http://www.engr.ncsu.edu/learningstyle/ilsweb.html
11. Viola, S.R., Graf, S., Kinshuk, Leo, T.: Analysis of Felder-Silverman Index of Learning Styles by a Data-driven Statistical Approach. In: Proceedings of the Eighth IEEE International Symposium on Multimedia, pp. 959–964 (December 2006)
12. Silvia, R.V., Sabine, G., Kinshuk, Tommaso, L.: Investigating relationships within the index of learning styles; a data driven approach. Journal of Interactive Technology and Smart Education 4(2), 7–18 (2007)
13. Soloman, B., Felder, R.: Index of learning style questionnaire (2009), http://www.engr.ncsu.edu/learningstyle/ilsweb.html (accessed December 20, 2011)
14. Tomcat (2012), http://tomcat.apche.org (accessed March 15, 2012)

A Trajectory-Based Recommender System for Tourism

Ranieri Baraglia[1], Claudio Frattari[2], Cristina Ioana Muntean[3],
Franco Maria Nardini[1], and Fabrizio Silvestri[1]

[1] ISTI–CNR, Pisa, Italy
{name.surname}@isti.cnr.it
[2] University of Pisa, Pisa, Italy
claudiofrat2002@hotmail.it
[3] Babes-Bolyai University, Cluj-Napoca, Romania
cristina.muntean@econ.ubbcluj.ro

Abstract. Recommendation systems provide focused information to users on a set of objects belonging to a specific domain. The proposed recommender system provides personalized suggestions about touristic points of interest. The system generates recommendations, consisting of touristic places, according to the current position of a tourist and previously collected data describing tourist movements in a touristic location/city. The touristic sites correspond to a set of points of interest identified a priori. We propose several metrics to evaluate both the spatial coverage of the dataset and the quality of recommendations produced. We assess our system on two datasets: a real and a synthetic one. Results show that our solution is a viable one.

Keywords: tourist recommender systems, trajectory pattern mining.

1 Introduction

Recommendation systems deal with providing focused information to users that can likely be of their interest in a set of objects belonging to a specific domain (music, movies, books, etc.). Such systems are common in search engines and social networks, as well as in any situation where a focused suggestion might be of value. This work presents a recommender system that allows to provide personalized information about locations of potential interest to a tourist. The system generates suggestions consisting of touristic places, according to the current position of the tourist that is visiting a city and a history of previous visiting patterns from other users. For the selection of tourist sites, the system uses a set of points of interest (PoI) identified a priori. The system is structured into two modules: one operating offline and one operating online with respect to the current visit of a tourist. The offline one is used to create the knowledge model, that is then used for calculating suggestions. It is executed periodically when new GPS data are available for updating the knowledge model. The online one uses information from the current visit path of a tourist and the knowledge model to calculate a list of suggestions as possible next locations to visit.

R. Huang et al. (Eds.): AMT 2012, LNCS 7669, pp. 196–205, 2012.

2   Related Work

In [11], authors perform travel recommendations by mining multiple users' GPS traces. They model multiple users location histories with a tree-based hierarchical graph. Based on the graph, authors propose a HITS-based inference model, which regards the access of an individual in a location as a directed link from the user to that location. The recommendation process uses a collaborative filtering model that infers users' interests in an unvisited location based on her location histories and those of others. Results show that the HITS-based inference model outperforms baseline approaches like rank-by-count and rank-by-frequency.

Monreale *et al.* propose WhereNext, a method aimed at predicting with a certain accuracy the next location of a moving object [9]. The prediction uses previously extracted movement patterns named Trajectory Patterns, which are a concise representation of behaviors of moving objects as sequences of regions frequently visited with typical travel times. A decision tree, named T-pattern Tree [4], is built by using Trajectory Patterns and used as predictor of the next location of a new trajectory, finding on the tree the best matching path.

Kurashima *et al.* use the list of locations visited in the past by a user to incrementally build new trajectories that maximize their likelihood in the mixed topic-Markov model [6]. The best k routes satisfying a maximum time and distance constraints are returned.

Other works on geographical recommender systems use the concept of Point of Interest (PoI). In [3] a mobility-aware recommendation system, called PILGRIM, is proposed. It uses the location of a user to filter recommended links. The authors build models relating resources to their spatial usage pattern, used to calculate a preference metric when the current user is asking for resources of interest.

Lucchese *et al.* propose a novel random walk-based algorithm for the interactive generation of personalized recommendations of touristic places of interest based on the knowledge mined from photo albums and Wikipedia [8].

Our solution blends together user friendly approaches in the above mentioned literature, in order to produce exhaustive tourist recommendations. We expand trajectory mining from the simple use of GPS coordinates to the use of significant/relevant PoIs able to bring value to users according their interest. While WhereNext is able to produce significant predictions, we intent to deliver a broader solution: a recommendation system able to assist and offer not only the next step [9], but a list of suggestions from which the user can choose from.

3   The Proposed System

As Figure 1 shows, the architecture of our recommender system has two main modules: offline and online. The first one aims to create the knowledge model, which is the basis for computing suggestions. Its execution takes place when new data is available for updating the knowledge model. The online module uses the current user information and the knowledge model in order to produce a list of suggestions.

Building the Knowledge Model: The data processed by the offline module consists of a dataset of trajectories representing the movements of users in a certain period of time, as detected by their GPS devices, and a set of PoIs with their coordinates. The trajectories initially have the following format: $T = < (x_1, y_1, t_1), (x_2, y_2, t_2), \ldots, (x_n, y_n, t_n) >$, where (x, y) are the coordinates on the Cartesian plane, and t is the timestamp. To facilitate the trajectory mining process, we define the following distance function: $N : \Re^2 \to P(\Re^2)$, describing whether the points of a trajectory are related to a PoI. In order to do this, the Cartesian plane is divided into regions so that if a point falls in the region of a PoI, then this point is assigned to that region. This function allows us to simplify the process of mining, which manages trajectories expressed as: $T'' = < (A, t_1), (B., t_2), \ldots, (C, t_n) >$, where capital letters represent regions and t timestamps, instead of trajectories expressed as coordinates. It reduces the cardinality of the set of elements on which the knowledge model is computed. To divide the plane into regions [2] we exploited the QuadTree technique, which consists of a class of hierarchical data structures that have the common characteristic of recursive division of space [10]. The adopted QuadTree-based algorithm divides the plane into regions of the same shape, but of different sizes.

Regions can be very small where there are several close PoIs. Their size depends on the PoIs distribution, and especially on the distance from one PoI to another. The QuadTree technique is very efficient, it performs data comparison and data insertion with a complexity equal to $O(logn)$, where n are the number of PoIs in input. In the worse cases, corresponding to poor PoIs distributions, thus unbalanced trees, it can run in $O(n)$.

For constructing the knowledge model we first transform the dataset of trajectories of regions according to the QuadTree division of space. This allows us to identify points that fall in a specific region with identifier id. Thus the trajectories are represented as: $A \underset{\alpha}{\to} B$, where A and B are two regions and α is the estimated time needed to move from A to B. Then, to build the knowledge model the T-Pattern Tree is used. Frequent trajectories are trajectories with a support greater than a threshold value σ. The T-pattern Tree is built incrementally and

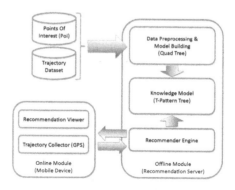

Fig. 1. Architecture of the proposed recommender system

trajectories having a prefix in common are overlapped on the tree to avoid unnecessary branches duplication. Each node is identified by a tuple $< id, region, support >$, where id is the node identifier, $region$ is the concerned region and $support$ is the sum of the supports of the various trajectories that have $region$ in that position. Each edge has associated a tuple $< [t_{min}, t_{max}], P(A \to B)] >$, where $[t_{min}, t_{max}]$ represents the estimated time needed to pass from a parent region to a child one, and $P(A \to B)$ indicates the probability of moving from region A to region B. The probability value is computed as the trajectories support value of a child node divided by the trajectories support value of its parent node. The T-pattern Tree can thus be modeled as a Markov chain and is used to find the similarity between trajectories frequently traveled and the current-analyzed one.

Computing Suggestions: User locations are obtained from GPS systems and sent to the offline module whenever a new position is detected. The recent movements of a user are used to set up the *current trajectory*, which is then compared with each practicable T-pattern Tree path. For each pattern match, a score, called *Punctual Score*, is calculated by assigning a value to each node of the current trajectory, which is then compared with those contained in T-pattern Tree. As in [9], the use of this score is designed to measure the reachability of a node r by a trajectory T which has already reached the parent node $r-1$. The punctual score indicates the matching of a node compared with the trajectory taken into consideration. The comparison may lead to three different cases for the calculation of the punctual score: 1) The current region is equal to the current node, and reached within the expected time. The punctual score is equal to the support associated to the node; 2) The current region is equal to the current node, but not reached within the maximum expected time. The punctual score is computed as: $node.support/\beta * d_t$, where β is a constant and d_t is the distance in time between the end interval and the time when the current region is reached; 3) The current region is not equal to that of the current node, the punctual score is computed as: $node.support/(\beta * d_t + \alpha * d_s)$, where α is a constant. Moreover, we specify the distance time tolerance th_t and distance space tolerance th_s as the maximum value that d_t and d_s can assume; when they exceed the specified values the punctual score of the current node is set equal to 0. The total pattern tree score $PathScore$ is computed as follows. Given a trajectory tr, a path $P = [p_1, p_2, ..., p_k]$ and a Punctual Score $PScore_k$ defined on each $p_k \in P$, the three indexes are computed as: 1) $AvgScore(tr, P) = \frac{\sum_1^n PScore_k}{n}$; 2) $SumScore(tr, P) = \sum_1^n PScore_k$; 3) $MaxScore(tr, P) = max\{PScore_1, \ldots, PScore_k\}$. AvgScore generalizes the concept of similarity by averaging the distance between the actual trajectory and the T-pattern Tree's pattern; SumScore is based on the concept of depth, the highest score is assigned to the longest path that intersects the actual trajectory; MaxScore considers only the node with the highest score based on the fact that if a trajectory has a good match with a node, it will probably have a good overall match. In our tests we use the SumScore method and assign the highest score to the longest path that intersects the trajectory. To carry out suggestions, candidates with higher PathScore and their children are returned, indicating the next regions that one can get from the current position. Once the regions are found, we look for the associated PoIs, which are suggested to a tourist. For our solution, we create a knowledge model for the category of PoIs that represent tourist sites. Then, by default the user will receive suggestions on tourist places that are updated according to his moves. Moreover, by using the information associated to the edges we can also provide, as a function of the available means of transport, an approximate time and cost needed to reach a suggested place. There are three cases where the system fails to make a prediction: 1) The current trajectory is longer than each T-pattern Tree's trajectory; 2) The current trajectory is spatially too distant from any trajectory on the T-pattern Tree; 3) The current trajectory is temporally too far away from any trajectory on the T-pattern Tree. These events are directly dependent on the quality of the T-Pattern used. It is therefore necessary to evaluate a priori the predictive power of the set of T-Pattern used as the quality of the predictions depends on the spatial and temporal characteristics of this set. The T-pattern are sequences of spatial regions of different sizes and intervals of time. The sizes of these regions are the key to being able to produce the predictions of good quality. For example, considering only T-pattern covering a small portion of the total space, can not be processed predictions reliable.

4   Evaluation

We measure the effectiveness and the efficiency of the proposed solution by using two trajectory sets: *synthetic* and *real*, and a set of predefined PoIs. Moreover, the performance results obtained by the proposed solution were compared with those obtained by a greedy solution that carries out a list of suggestions made up of regions closer to the current location.

The synthetic dataset was created using a trajectory generator for a specific geographic area. It takes as input a dataset of PoIs, which are combined in sequences that form trajectories. The set of PoIs adopted during the tests include all the most important (monumental/artistic) PoIs in Florence, generated using information from Wikipedia. Through the Wikipedia API, we are able to retrieve the spatial coordinates of a list of PoIs in Florence.

By setting the parameters like the distance threshold d and the number of trajectories to build, it is possible to customize the dataset. This flexible mechanism allows us to generate a dataset of a predetermined size. Accordingly, for building a trajectory the following steps are required: i) we randomly select the starting point from 1022 PoIs extracted from Wikipedia, ii) we identify the starting point neighbors from the set of PoIs closer than d, iii) we rank the neighboring points by means of a function that minimizes the distance between the candidate PoI and the one currently analyzed; we select the next step in the trajectory according to the score associated to each candidate PoI - the higher the score, the higher the probability of selection, and iv) we terminate the trajectory building when the desired length is reached. The decision to model the interest for a PoI by the distance was influenced by the good results shown in [7]. The resulting synthetic dataset contains 20000 trajectories.

The real dataset is made up of data coming from Flickr. The trajectories are built using the photos submitted by users. A photo may have additional data such as the time it was taken and the geospatial coordinates of the object depicted. We considered the subject in the photos as potential real PoIs.

We built the dataset from information relating to photos taken in Florence from January 2004 to January 2010. With the data obtained it was possible to build daily trajectories for each user. The spatial coordinates associated with an individual photo may not exactly coincide with the PoI photographed. In order to build trajectories a data structure called R-Tree [5] was used which permits us to assign a rectangular area to each PoI belonging to the set of PoIs. If the coordinates of a PoI extracted from the dataset of photos fall into one of these rectangular areas, the representative corresponding PoI is assigned to the trajectory. Building the real dataset consists of three phases: 1) Build the R-Tree. Each node represents a geographical area and its children represent sub-areas. 2) Extract information from photos: user id, date and spatial coordinates. 3) Extract the PoI. For each pair of spatial coordinates contained in the list of PoIs from Flickr, we verify whether it is contained in one of the R-Tree's leaf nodes. If so, the PoI identifying that bounding box is added to the user's trajectory.

Evaluating the Quality of the Trajectory Set: The quality of the trajectory set is a key element for building the knowledge model with which the recommendations are computed. Therefore, it is important to understand in advance whether a set is valid for the effective evaluation of suggestions. To this end, in [9] the authors have proposed a method to establish a correspondence between the accuracy and the value of the support for a set of association rules. In our case, the ability to make accurate predictions also depends on the spatial characteristics of a set of T-pattern, not only on support. We refer to it as *Coverage*. The following indexes were adopted by us for assessing the Coverage:

- **SpatialCoverage** (SC) measures the fraction of the total space covered by the trajectory set as: $SpatialCoverage = \frac{\cup_{Tp \in Tp_{set}} Space(Tp)}{TotalSpace}$, where $Space(Tp)$ is the function that assigns to each T-pattern a portion of the plane that it fails to cover. $TotalSpace$ is the total space where tourists move around;

- **DataCoverage** (DC) defines the fraction of trajectories that go over the support value. This is computed as: $DataCoverage = \frac{|T| - Tpset}{|T|}$, where T is the trajectory set and $Tpset$ is the number of extracted trajectories which satisfy the support value.

- **RegionSeparation** (RS) measures the prediction accuracy as function of the prediction granularity. It is computed as: $RegionSeparation = \frac{MinimalRegion}{AVG_{r \in Tp \in Tp_{set}}}$, where $MinimalRegion$ is the minimum spatial granularity corresponding to a PoI within the considered space and $AVG_{r \in Tp \in Tp_{set}}$ is the average size of regions belonging to the trajectory set.

- **Rate** correlates all above three metrics as follows:
$Rate = SpatialCoverage \cdot DataCoverage \cdot RegionSeparation$

As can be seen from Table 1, as the number of PoIs per region increases, the Rate value decreases. Even if the SpatialCoverage index increases, the other two indexes RegionSeparation and DataCoverage decrease. In fact, the higher values for Rate, i.e 0.06 for the real trajectory set and 0.08 for the synthetic trajectory set, are obtained when there are five PoIs per region. This is because RegionSeparation, and consequently Rate, rewards the correspondence between PoIs and regions. Almost identical Rate values were obtained for the same test with synthetical sets 5000 and 10000 trajectories. It shows that RegionSeparation is independent of the size of the trajectory set. The value of the support used to conduct the test is equal to 1.

Evaluating the Effectiveness and Efficiency: To evaluate the effectiveness of suggestions we adopted an empirical approach that estimates the percentage of errors in making recommendations using a test set [9]. The set of samples is divided into two disjoint subsets, a *training set* used to build the knowledge model (90%) and a *test set* (10%) used for evaluation. Each trajectory in the test set is iteratively divided into two parts: the first part represents the current trajectory on which we want to receive recommendations, and the second part is used for comparisons with the suggested regions. Initially, the current trajectory is represented by the first region of the analyzed trajectory and the remaining regions are used for comparison. A trajectory is divided in this way until the second part contains a single region. The tests to evaluate the efficacy of the proposed solution were conducted by computing a list of 10 regions as a suggestion. For evaluating the effectiveness, we adopted the following metrics:

- **Prediction Rate** (PR) is the percentage of trajectories for which the system is able to make a prediction.
- **Accuracy** (A) is the percentage of trajectories for which the system returns a list of suggestions containing the region that, in the test set, immediately follows the last region of the current trajectory.
- **Modified Accuracy** (MA) refines Accuracy. In [7] is shown that people tend to minimize the distance between locations. Accordingly, a tourist in a region may move to another PoI in the same region or in a different one. The region where the tourist is located is added to the suggestions.
- **Average Error** (AE) is the average error percentage computed for each trajectory. A trajectory of n regions is divided $n - 1$ times, and $n - 1$ comparisons are made. The result of each comparison is true, if the list of

Table 1. Coverage value for the real and synthetic trajectory sets

PoIs Region	Real Dataset				Synthetic Dataset			
	SC(%)	RS(%)	DC(%)	Rate	SC(%)	RS(%)	DC(%)	Rate
5	0,70	0,12	0,75	0,06	0,91	0,92	0,71	0,08
10	0,67	0,06	0,75	0,03	0,96	0,05	0,64	0,03
20	0,86	0,03	0,69	0,02	0,97	0,03	0,61	0,02
30	0,83	0,02	0,66	0,01	1	0,02	0,06	0,01

suggestions contains the next region, and false otherwise. Let a the number of false comparisons, the error rate for the related trajectory is equal to $a/n - 1$.

- **Omega** (Ω) measures the immediate utility of the generated suggestions [1]. It is computed as:

$$\Omega = \sum_{i=1}^{Ns} \frac{\sum_{k=1}^{n_k} [p_k \in \{S_i^{1,k} \cap R_i^{k+1,n_k}\}] \frac{f(k)}{n_k}}{N_s} \quad (1)$$

where N_s is the number of trajectories of the test set, n_k is the number of regions in the current trajectory, $f(k)$ is a function assigning a weight to a suggested region and p_k is a functions that returns 1 for a correct prediction and 0 on the contrary.

Table 2 shows the values of the metrics evaluated to measure the efficacy of the computed suggestions. They were computed by varying the number of PoIs per region and using a knowledge model built with the value of the support equal to 1. The number of PoIs in a region significantly affects the effectiveness of the system. Increasing the number of PoIs per region, the regions become larger and the prediction becomes easier because the knowledge model needs fewer examples to correctly predict the regions. Accordingly, the probability that a region is correctly suggested increases. The best value for MA and A is achieved with 30 PoIs per region, reaching a maximum value of 80.19% for MA, 66.21% for A, and 56.49% for PR. As can be seen from Table 2, AE decreases as the number of PoIs in the regions increases, leading to a higher accuracy in suggestions. The values for Ω indicate that the immediate utility of suggestions varies little by changing the number of PoIs per region. PR computed on the synthetic set is greater than the one on the real set.

Table 3 shows the performance evaluated by varying the support value σ, used to generate the knowledge model. Increasing σ, the percentage of trajectories correctly predicted decreases, revealing the non monotonic property of the support. Also, the number of trajectories used to build the knowledge model decreases and the model loses part of its predictive power. Even if for small values of the support PR is high enough, A never increases above 50%, instead decreases progressively. The support value that ensures the best A is equal to 1 for both the real and synthetic trajectory sets.

Concerning the real trajectory set, better results are obtained for $\sigma = 4$, when PR is equal to 36.18%. A for high support values is related to a small

Table 2. Effectiveness by varying the number of POIs per region

PoIs Region	Real Dataset					Synthetic Dataset				
	PR(%)	A(%)	MA(%)	AE(%)	Ω	PR(%)	A(%)	MA(%)	AE(%)	Ω
5	48,24	30,45	35,00	74,35	5,74	100	49,52	55,32	46,12	4,53
10	48,24	47,27	57,95	63,29	4,59	100	47,46	56,32	49,40	4,37
20	56,47	59,80	71,65	48,10	5,31	100	67,55	70,34	30,15	5,36
30	56,49	66,21	80,19	39,39	4,15	100	81,36	83,01	16,15	4,84

percentage of predicted trajectories, less than half of the available trajectories. In fact, a peak (67.27%) of the MA index correspond to only 36.18% of the total trajectories. Moreover, MA w.r.t A increases of about the 10%, it means that there are at least 10% of users move to a PoI within the same current region.

The AE values varies between 63.29% and 76.80%. It is important to note that this measure of error is only related to the percentage of suggested trajectories. Ω has a similar value for the first three tests, and then decreases to 1.31, showing that the suggestions related to the support value equal to 10 have the highest immediate utility. Ω is referred only to trajectories that have a correct prediction, so for the support value equal to 10, Ω refers only to 26.32% of the suggested trajectories.

Concerning the synthetic set, PR is always equal to 100%, A has a trend inversely proportional w.r.t. the support. The synthetic dataset maintains a PR greater than the one obtained on the real dataset.

The system efficiency was evaluated measuring the average elapsed time to compute a list of suggestions on a trajectory. This time depends highly on the cardinality of T-pattern Tree. The bigger and deeper the tree, the more the execution time grows and the cost of prediction rises. Tests were conducted by considering 100, 150, 200, and 300 requests per minute. In such tests the proposed system demonstrate to be able to respond quickly, with an average elapsed time of 1500 ms. For 100 requests per minute the average response time remains constant below 500 ms, then slightly increasing in the case of 150 requests per minute. Reasonable values are obtained also in the case of 200 and 300 requests per minute, with average response time values between 1500 and 2000 ms.

We present a comparison of the proposed solution versus a greedy solution, Nearest. Due to the scarcity of recommendation systems available and low availability of datasets used in other articles, we developed a simple recommender called Nearest, which returns a list of suggestions containing regions closest to the current region of the tourist, thus not adopting any process of mining.

Both methods, ours and Nearest, are evaluated on the synthetic dataset and best performance values for parameters, namely σ equal to 1 and number of PoIs per region equal to 30. In the case of our system we obtain a PR of 100%, an A of 81.36% and a MA of 83.01%, while Nearest obtained a PR of 100%, an A of 60.40% and a MA of 63.40%. Due to the fact that tests were conducted on

Table 3. Results varying the support and 10 POIs per region

σ	Real Dataset					Synthetic Dataset				
	PR(%)	A (%)	MA (%)	AE (%)	Ω	PR (%)	A (%)	MA (%)	AE (%)	Ω
1	48,24	47,27	57,95	63,29	4,59	100	47,46	53,32	49,40	4,37
2	48,24	43,18	53,41	64,68	4,57	100	42,70	48,15	52,63	3,91
4	36,18	51,51	67,27	63,82	4,18	100	37,76	42,30	57,42	3,50
6	36,18	45,45	61,82	71,32	2,71	100	33,74	38,41	61,10	3,11
8	36,18	36,36	53,94	76,80	1,93	100	30,92	35,21	63,75	2,73
10	36,18	26,36	46,06	63,82	1,31	100	28,80	33,16	67,65	2,27

synthetic data sets, PR for both system is 100%. However a clear performance of our system in respect with Nearest can be acknowledged from an increased A and MA.

5 Conclusions

We proposed a recommendation system that can assist a tourist visiting a city. It is able to generate suggestions of potential PoIs, depending on the current position of a tourist, and a set of trajectories describing the paths previously made by other tourists. The best effectiveness is achieved when the support σ is equal to 1 and when the number of PoIs per region is equal to 30. We also evaluated our proposed system against a simple baseline solution, which produces a suggestion list of regions closer to the tourist's current position. Results show that our solution clearly outperforms that one. We also proved that the response time enables it to be used interactively.

Our research has been funded by the POR-FESR 2007-2013 No 63748 (VIS-ITO Tuscany) project and POSDRU/88/1.5/S/60185 (Investing in people!).

References

1. Baraglia, R., Silvestri, F.: An online recommender system for large web sites. In: Proc. of IEEE/WIC/ACM WI 2004. IEEE Computer Society, Washington, DC (2004)
2. Baraglia, R., Frattari, C., Muntean, C.I., Nardini, F.M., Silvestri, F.: Rectour: a recommender system for tourists. In: Proceedings of the First Workshop on Tourism Facilities (TF 2012) Colocated with the 2012 IEEE/WIC/ACM International Conference on Web Intelligence (2012)
3. Brunato, M., Battiti, R.: Pilgrim: A location broker and mobility-aware recommendation system. In: Proc. of PerCom, pp. 265–272. IEEE (2003)
4. Giannotti, F., Nanni, M., Pinelli, F., Pedreschi, D.: Trajectory pattern mining. In: Berkhin, P., Caruana, R., Wu, X. (eds.) KDD, pp. 330–339. ACM (2007)
5. Guttman, A.: R-trees: a dynamic index structure for spatial searching. SIGMOD Rec. 14(2), 47–57 (1984)
6. Kurashima, T., Iwata, T., Irie, G., Fujimura, K.: Travel route recommendation using geotags in photo sharing sites. In: Proc. CIKM, pp. 579–588. ACM (2010)
7. Lee, K., Hong, S., Kim, S.J., Rhee, I., Chong, S.: Slaw: A new mobility model for human walks. In: Proc. IEEE INFOCOM, pp. 855–863. IEEE (2009)
8. Lucchese, C., Perego, R., Silvestri, F., Vahabi, H., Venturini, R.: How Random Walks Can Help Tourism. In: Baeza-Yates, R., de Vries, A.P., Zaragoza, H., Cambazoglu, B.B., Murdock, V., Lempel, R., Silvestri, F. (eds.) ECIR 2012. LNCS, vol. 7224, pp. 195–206. Springer, Heidelberg (2012)
9. Monreale, A., Pinelli, F., Trasarti, R., Giannotti, F.: Wherenext: a location predictor on trajectory pattern mining. In: Proc. of KDD, pp. 637–646. ACM (2009)
10. Samet, H.: Hierarchical Spatial Data Structures. In: Buchmann, A., Smith, T.R., Wang, Y.-F., Günther, O. (eds.) SSD 1989. LNCS, vol. 409, pp. 191–212. Springer, Heidelberg (1990)
11. Zheng, Y., Xie, X.: Learning travel recommendations from user-generated gps traces. ACM TIST 2(1), 2 (2011)

An Adaptive Method for the Tag-Rating-Based Recommender System

Xi Yuan and Jia-jin Huang

International WIC Institute, Beijing University of Technology, China
yuanxi.gloria@yahoo.com.cn, hjj@emails.bjut.edu.cn

Abstract. In this paper, we propose an adaptive method for recommender system based on users' preference to items represented by the ratings of users. This method defines a term-association matrix to describe the relation between tags and items properties. A gradient descent method is employed to compute the association matrix. The association matrix is then used to implement the two kinds of recommendation, namely, tag recommendation and items properties recommendation.

1 Introduction

In Web 2.0, the overload information makes Web users difficult to find useful Web information. Recommender systems provide available methods to help users be free from large data by predicting and recommending information in which Web users may be interested.

Collaborative filtering (CF) recommendation is one of the most successful techniques. CF recommends items among people with similar tastes [2,3]. Besides, content-based recommendation (BCR) can provide recommendations by encoding users' preferences from textual information [6]. Hybrid recommendation [8,12] combines CF and BCR into a single integrated model [4].

The advent of Web 2.0 brings a new form of user-centric method, folksonomy [10], which allows users not only to tag items for their own characters with user-defined words but also to upload items to express their opinions. The tag-based recommender system has attracted more attention recently.

How to create the Web user interest model is the key issue of the tag-based recommender systems. The main directions for the research can be divided into matrix-based methods, clustering-based methods and graph-based methods. In matrix-based methods, Xu et al [14] proposed latent semantic analysis (LSA) to compute the included-angle cosine between tags and items by using tag-item matrix. Xu et al [15] further introduced higher-order singular value decomposition (HOSVD) to improve recommendation quality and stability, which finds association between users, items and tags by combining them into a framework. With respect to clustering-based methods, Reyn et al [9] constructed a scenario-based CF model based on the similarity of tags, which recommends items by abstracting tags to the user vectors and counting users similarity. Jonathan et al [5] applied the TF-IDF formula to cluster tags by a hierarchical method, and

R. Huang et al. (Eds.): AMT 2012, LNCS 7669, pp. 206–214, 2012.

constructed the Web user interest model by users' interests on items. As for graph-based methods, Andreas et al [1] proposed a modified PageRank algorithm, namely FolkRank, which consider the connection between users, tags and items to an undirected graph. Guan et al [17] proposed a framework based on graph Laplacian to model interrelated multi-type objects.

Although the social tag is especially useful for both searching and organizing items, many studies argue that not all the tags benefit recommendation [16] because the unrestricted nature of the tagging function is liberating [7]. In tagging systems, tags with free style will interfere the analysis of the structure and users' behaviors.

Currently, ratings have been regarded as an effective and simple form of recommendation. Tags represent users' interests and preferences in more detail, but the tags with free style also interfere the analysis of user behaviors and system structure. For these reasons, we construct the Web interest model by combining tags with ratings, and construct a tag-rating-based term-association matrix for high-efficiency recommendation.

2 The Web User Interest Model of the Tag-Rating-Based Recommender System

2.1 A User-Item-Tag-Rating Fourfold Graph and the Vector Space Model

In the tag-rating-based recommender system, the recommender system consists of users, items, tags and ratings [18]. Users express their preferences by tagging items with high ratings.

Let $D = (U, I, T, R)$ denote the four parts. The component $U = \{u_1, u_2, ..., u_k\}$ is the set of users, each user $u_i(1 \leq i \leq k)$ is modeled as a vector $\mathbf{u_i} = (\mu_1, \mu_2, ...\mu_m)$ over the set of tags $T = \{t_1, t_2, ..., t_m\}$, and each tag $t_s(1 \leq s \leq m)$ is represented by the weight $\mu_s(1 \leq s \leq m)$ of $\mathbf{u_i}$; $I = \{i_1, i_2, ..., i_p\}$ is the set of items, each item $i_j(1 \leq j \leq p)$ is modeled as a vector $i_j = (\nu_1, \nu_2, ..., \nu_n)$ over the set of item properties $G = \{g_1, g_2, ..., g_n\}$, and each property $g_t(1 \leq t \leq n)$ is represented by the weight $\nu_t(1 \leq t \leq n)$ of $\mathbf{i_j}$; $R = \{r_{11}, r_{12}, ..., r_{ij}, ..., r_{kp}\}$ is the set of ratings, each r_{ij} presented the rating of item i_j by user u_i. An example is shown in Fig. 1.

In this paper, the rating relation is used to describe users' preference to items. Each rating relation is measured by ratings given by users. In the next section, we will describe the definition of rating relation and use it to construct the Web user interest model.

2.2 The Definition of Rating Relation

In this section, we will introduce the definition of rating relation in two kinds of recommendation which can be used to measure user preference to items in the tag-rating-based recommender system.

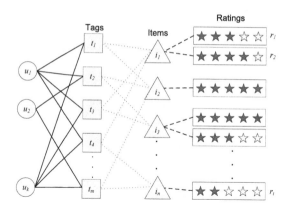

Fig. 1. The illustration of a user-tag-item-rating fourfold graph

The (u, i, t, r) is defined to a user-item-tag-rating ($UITR$) quaternion. Given a tag, for $i, i' \in I$ and $u, u' \in U$. The rating relation is defined as follows:

$$(u, i, t) \succ (u', i', t') \Leftrightarrow \text{The rating of item } i \text{ with } t \text{ by user } u \text{ is higher than } i'$$
$$\text{with } t' \text{ by user } u'. \tag{1}$$

In the tag-rating-based recommender system model, the components rely on 2-dimensional projections of the user-tag-item (UTI) matrix, which reduce the dimensionality of the data but sacrifice its informational content. We produce an item-user (IU) matrix projection and a user-item (UI) matrix projection. We define them to be binary, which the users and the items vectors are indicated by whether or not the user has annotated the tag and the item has classified to the property. The weights $\mu_s (1 \le s \le m)$ and $\nu_t (1 \le t \le n)$ have only two values, 1 or 0. Thus, we propose two kinds of recommendation: UI recommendation and IU recommendation which recommend items properties and tags respectively.

So, we separate the Eq. (1) for two kinds of recommendation, and define the individual rating relation as follows:

For UI recommendation, the rating by users on particular item and tag can be specified by individual rating relation as follows:

$$u \succ_{i,t} u' \Leftrightarrow \text{The rating of item } i \text{ with tag } t \text{ by user } u \text{ is higher than } u'.$$
$$\Leftrightarrow (u, i, t) \succ (u', i, t)$$

$$\tag{2}$$

For IU recommendation, the rating of items on particular user and tag can be specified by individual rating relation as follows:

$$i \succ_{i,t} i' \Leftrightarrow \text{The rating of item } i \text{ with tag } t \text{ by user } u \text{ is higher than } i'.$$
$$\Leftrightarrow (u, i, t) \succ (u, i', t)$$

$$\tag{3}$$

The subscript of Eq. (2) and Eq. (3) emphasized the fact that the rating relation is defined to particular item, user and tag.

2.3 The Web User Interest Model of the Tag-Rating-Based Recommendation System

In this section, based on the algorithm proposed by Wong et al [13] combined with the social tag, we propose a Web interest model by constructing IU term-association matrix and UI term-association matrix. In order to be short, we only introduce the construction of IU term-association matrix, the construction of UI term-association matrix is similar.

We construct an IU term-association matrix to describe the relation between users and items by a bilinear function:

$$g(i, u) = \sum_{i=1}^{n} \sum_{j=1}^{m} \nu_i a_{ij} \mu_j = \mathbf{i A u^T}$$

(4)

where a_{ij} measures the strength of association between item property g_i and tag t_j, and $\mathbf{A} = (a_{ij})$ is the IU term-association matrix, which is not necessarily a symmetric matrix, rows and columns of $\mathbf{A}$ is determined by the dimension of the item properties and the tags. The construct matrix $\mathbf{A}$ need to satisfy the condition: for any $UITR$ quaternion $(u, i, t, r), (u, i', t, r') \in UITR$,

$$i \succ_{u,t} i' \Rightarrow \mathbf{i A u^T} >_t \mathbf{i' A u^T}$$

(5)

We called $\mathbf{w} = \mathbf{i} - \mathbf{i'}$ is a difference vector, so the condition $\mathbf{i A u^T} >_t \mathbf{i' A u^T}$ can be written as $\mathbf{w A u^T}$, the set W consisting of item rating vector teams defined by:

$$W = \{(\mathbf{w}, \mathbf{u}) | \mathbf{w} = \mathbf{i} - \mathbf{i'}, i \succ_{u,t} i'\}$$

(6)

Obviously, the problem of finding the IU term-association matrix $\mathbf{A}$ satisfying Eq. (5) is reduced to a problem of finding a solution matrix to satisfying condition as follows:

$$\mathbf{w A u^T} >_t 0, (\mathbf{w}, \mathbf{u}) \in W$$

(7)

Borrowing the algorithm in [13], we will introduce the procedure of calculating the IU term-association matrix.

(i) We start with an initial matrix $\mathbf{A}^{(0)}$ and let $k = 0$, usually matrix $\mathbf{A}^{(0)}$ is an unit matrix.

(ii) If $\mathbf{w A u^T} >_t 0$, we say that matrix $\mathbf{A}$ correctly describes the rating relationship $i \succ_{u,t} i'$, so we defined:

$$\Gamma(\mathbf{A}^{(k)}) = \{(\mathbf{w}, \mathbf{u}) | (\mathbf{w}, \mathbf{u}) \in W \wedge \mathbf{w A}^{(k)} \mathbf{u^T} \leq 0\} \subseteq W$$

(8)

If $\Gamma(\mathbf{A}^{(k)}) = \Phi$, terminate the procedure.

(iii) $\Gamma(\mathbf{A}^{(\mathbf{k})})$ is the term-association matrix in the $(k+1)th$ iteration, which is obtained by the gradient descent method, and calculated by:

$$\mathbf{A}^{(k+1)} = \mathbf{A}^{(k)} + [\sum_{(\mathbf{w},\mathbf{u}) \in \Gamma(\mathbf{A}^{(\mathbf{k})})} \mathbf{w}]^{\mathbf{T}} \mathbf{u} \qquad (9)$$

(iv) Let $k = k + 1$, go back to (ii).

If the solution matrix exists, we can find it by the algorithm above. But in practice it is difficult to find the solution matrix $\mathbf{A}$ satisfying the termination condition (ii), because the increased complexity of the data increase the complexity of the matrix. Thus, we find the solution matrix $\mathbf{A}$ by calculating the accuracy of the recommendation when $k = 0, 1, 2, 3$ and the term-association matrix is $\mathbf{A}^1, \mathbf{A}^2, \mathbf{A}^3$ respectively.

This algorithm provides a systematic method to construct term-association, and needn't to introduction any particular parameters. For the UI recommendation, all we need is to exchange the place of $\mathbf{w}$ and $\mathbf{u}$ and use $\mathbf{n} = \mathbf{u} - \mathbf{u}'$ instead of $\mathbf{w} = \mathbf{i} - \mathbf{i}'$. Tags can be recommended by the IU algorithm above in the IU recommendation and for the UI recommendation, item properties could be recommended by the UI term-association matrix.

2.4   The Network Configuration of the Tag-Rating-Based Recommender System

By constructing the Web interest model above, tags can be recommended by finding IU term-association matrix $\mathbf{A}$. In this section, we describe the network configuration of the tag-rating-based recommender system and give the algorithm of calculating the IU term-association matrix $\mathbf{A}$.

Figure 2 shows the network configuration. Two directions of the network represent the processing of the UI recommendation and the IU recommendation, which term-association matrixes calculated by $g(u, i) = \mathbf{uAi}^{\mathbf{T}}$ and $g(i, u) = \mathbf{iAu}^{\mathbf{T}}$, respectively.

For the IU recommendation, the input layer is represented by an item node i. The node i connects property node g_i by an individual weight μ_i. The output layer consists of a node which pools the user terms with individual weight $\beta_j = 1$. For the UI recommendation, the input layer is represented by a user node u, which connects tag node t_j by an individual weight. The output layer consists of a node which pools the item terms with individual weight $\alpha_i = 1$. The weight between item property g_i and tag t_j is represented by a_{ij} and a_{ji}, which is the element of the matrix $\mathbf{A}$.

3   Experimental Evaluation

In this section we describe the methods used to gather and pre-process our datasets and our evaluation metrics. The algorithm above is used as two kinds of recommendation. This section mainly present the recommendation results.

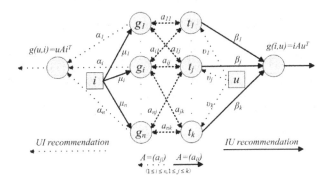

Fig. 2. The tag-rating-based recommender system network configuration

3.1   Dataset and Data Pre-processing

Our experiments are conducted by using the *MovieLens* (*Version*1.0 (*May*2011)) dataset which are gathered from the corresponding *MovieLens* Web site. Users are allowed to rating and tag movies on *MovieLens* Web site. The dataset contains users, movies, tags and ratings, which need to remove rarely-occurring information to reduce the noise in the data.

The data of users, movies, tags, ratings are gathered into one data which the movies are more than 20 for each user, and are expressed by (u, i, t, r) quaternion. The data is resulted in 29694 annotations with 185 users, 4238 movies and 6570 tags, and each movie can be classified into several genres from 20. The set of tags $T = \{t_1, t_2, ..., t_m\}$ and the set of item properties $G = \{g_1, g_2, ..., g_n\}$ are respectively represented by the tags and the movie genres in the dataset, which weight are indicated by whether or not the movie has classified to the genre and the user has annotated the tag. Most important, the rating vector teams of IU and UI recommendation are need to be extracted from the (u, i, t, r) quaternion to Adata dataset and Bdata dataset. If $r > r'$, the rating relations are $i \succ_{u,t} i'$ and $u \succ_{i,t} u'$.

In the experiment, the training and test sets are created by 10-fold cross-validation method [11]. Firstly, each dataset is randomly split into 10 mutu-ally exclusive subsets of approximately equal size. Each time the test set T_k is clustered by two randomly subsets and the corresponding training set TR_k is clustered by the other eight subsets, 10 times $(1 \leq k \leq 10)$ in all. Finally, we get 10 test sets $(T_1, T_2, ..., T_{10})$ and 10 training set $(TR_1, TR_2, ..., TR_{10})$ for each dataset.

For IU recommendation, the top 20% tags that calculated by $\mathbf{i} \times \mathbf{A} = \mathbf{u}$ are chosen to be the $finalTags$, and the tags that tagged for m_i in the training set are chosen to be the $targetTags$. The $finalTags$ and the $targetTags$ have eliminated duplicate entries. The accuracy of i_i for each time is gotten by: for $1 \leq k \leq 10$,

$$Acc_k(i_i) = \frac{NUM(finalTags \cap targetTags)}{NUM(targetTags)} \tag{10}$$

Table 1. The accuracy of the IU recommendation

items	i_{25}	i_{785}	i_{1219}	i_{2076}	i_{2730}	i_{3676}	i_{4194}	i_{5390}	i_{6258}	i_{6669}
$k = 1$	57.38	59.26	51.55	66.67	57.81	60.14	52.78	57.78	53.84	57.41
$k = 2$	60.00	63.09	53.13	70.00	60.95	63.54	54.73	60.29	57.16	60.00
$k = 3$	65.10	66.71	60.00	70.83	66.16	67.27	61.80	65.97	63.09	65.87
$k = 4$	68.49	69.20	63.37	72.39	68.65	69.63	64.60	68.57	65.85	68.49
$k = 5$	68.89	69.75	63.90	73.57	69.10	69.96	65.68	69.07	66.46	69.02
$k = 6$	62.38	63.05	57.38	68.78	62.67	63.09	58.33	62.50	59.73	62.47

The average accuracy of i_i is gotten by Eq. (11):

$$Ave(i_i) = \frac{\sum_{k=1}^{N} Acc_k(i_i)}{N}, (1 \leq N \leq 10) \tag{11}$$

For UI recommendation, the top 20% genres that calculated by $\mathbf{u} \times \mathbf{A} = \mathbf{i}$ are chosen to be the $finalGenres$, and the genres that classified to the movies tagged by u_j are chosen to be the $targetGenres$. Noticeable, the $finalGenres$ and the $targetGenres$ have eliminated duplicate entries. The accuracy of u_j for each time is gotten by: for $1 \leq k \leq 10$,

$$Acc_k(u_j) = \frac{NUM(finalGenres \cap targetGenres)}{NUM(targetGenres)} \tag{12}$$

The average accuracy of u_j is gotten by Eq. (13):

$$Ave(u_j) = \frac{\sum_{k=1}^{N} Acc_k(u_j)}{N}, (1 \leq N \leq 10) \tag{13}$$

3.2   Results and Discussion

The IU recommendation recommend tags for each movie by IU term-association matrix $\mathbf{A}$, which is calculated by Eq. (8). However, in practice, it is difficult to find the solution matrix, so we let $k = 1, 2, ...$ find the result matrix $\mathbf{A}$ with best recommendation effect. The average accuracies of selected movie are shown in Table 1. It's easy to find that the curve of $k = 5$ has the best accuracy for

Table 2. The accuracy of the UI recommendation

users	u_{2643}	u_{8787}	u_{12265}	u_{19923}	u_{23172}	u_{38662}	u_{45290}	u_{51954}	u_{68228}	u_{71331}
$k = 1$	40.40	39.48	28.92	14.95	39.51	16.67	36.88	25.16	22.86	18.43
$k = 2$	50.00	41.67	32.20	17.50	41.67	20.00	40.40	28.72	26.45	20.98
$k = 3$	77.07	65.70	47.79	30.48	66.02	32.22	59.85	42.78	40.40	34.32
$k = 4$	79.54	71.56	56.89	42.42	74.07	46.10	69.06	53.11	50.95	47.41
$k = 5$	78.53	68.57	54.04	41.67	71.90	43.40	66.67	51.58	50.00	45.05

average 60%, which means that the recommendation is the best when $\mathbf{A}^5$ is the result matrix.

In this paper, tags are freely assigned by users, which results in much noise of tags. The next task is using the clustering methods and anti-spam technique to reduce the noise and provide high-quality recommendations.

Similar to the IU recommendation, finding the result matrix by setting $k = 1, 2, ...$, the UI recommendation gives a best average accuracy of 50% when $k = 4$. The average accuracies of selected user are shown in Table 2.

4   Conclusion

In this paper, we have constructed the Web user interest model and the network configuration of the tag-rating-based recommendation system by combining the rating relation with user-item-tag vector teams. We provide two kinds of recommendation strategies, namely the IU recommendation and the UI recommendation, which can be used to provide tags recommendation and item properties recommendation respectively. The relation between items and users is represented by a term-association matrix $\mathbf{A}$. Given items, the IU recommendation calculates weighs of tags and selects the top 20% tags to recommend. Given users, the UI recommendation calculates weighs of item properties and selects the top 20% item properties to recommend. The results show that the algorithms are effective. The next task is constructing multi-layer Web interest model for diversity recommendation and to improve the recommendation qualities.

Acknowledgements. This work is supported by Beijing Natural Science Foundation (4102007), the CAS/SAFEA International Partnership Program for Creative Research Teams, the China Postdoctoral Science Foundation Funded Project (2012M510298), Projected by Beijing Postdoctoral Research Foundation (2012ZZ-04), and the doctor foundation of Beijing University of Technology (X0002020201101).

References

1. Hotho, A., Jäschke, R., Schmitz, C., Stumme, G.: Information Retrieval in Folksonomies: Search and Ranking. In: Sure, Y., Domingue, J. (eds.) ESWC 2006. LNCS, vol. 4011, pp. 411–426. Springer, Heidelberg (2006)
2. David, G., David, N., Brain, M.O., Douglas, T.: Using Collaborative Filtering to Weave an Information Tapestry. Communication of the ACM-Special Issue on Information Filtering 35(12), 61–70 (1992)
3. John, S.B., David, H., Carl, K.: Empirical Analysis of Predictive Algorithms for Collaborative Filtering. In: Proceedings of the Fourteenth Conference on Uncertainty in Artificial Intelligence (UAI 1998), pp. 43–52 (1998)
4. Jonathan, G., Thomas, S., Bamshad, M.: Resource Recommendation in Social Annotation Systems: A Linear-Weighted Hybrid Approach. Journal of Computer and System Sciences 78(4), 1160–1174 (2012)

5. Gemmell, J., Shepitsen, A., Mobasher, B., Burke, R.: Personalizing Navigation in Folksonomies Using Hierarchical Tag Clustering. In: Song, I.-Y., Eder, J., Nguyen, T.M. (eds.) DaWaK 2008. LNCS, vol. 5182, pp. 196–205. Springer, Heidelberg (2008)

6. Marko, B., Yoav, S.: Content-Based Collaborative Recommendation. Communications of the ACM 40(3), 66–72 (1997)

7. Mathes, A.: Folksonomies-Cooperative Classification and Communication Through Shared Metadata. Computer Mediated Communication (2004)

8. Qing, L.: Clustering Approach for Hybrid Recommender System. In: Proceedings of the 2003 IEEE/WIC International Conference on Web Intelligence (WI 2003), pp. 33–38 (2003)

9. Reyn, N., Shinsuke, N., Jun, M., Shunsuke, U.: Tag-based Contextual Collaborative Filtering. IAENG International Journal of Computer Science 34(2), 214–219 (2007)

10. Jäschke, R., Marinho, L., Hotho, A., Schmidt-Thieme, L., Stumme, G.: Tag Recommendations in Folksonomies. In: Kok, J.N., Koronacki, J., Lopez de Mantaras, R., Matwin, S., Mladenič, D., Skowron, A. (eds.) PKDD 2007. LNCS (LNAI), vol. 4702, pp. 506–514. Springer, Heidelberg (2007)

11. Ron, K.: A Study of Cross-Validation and Bootstrap for Accuracy Estimation and Model Selection. In: Proceedings of the 14th International Joint Conference on Articial Intelligence (IJCAI 1995), pp. 1137–1145 (1995)

12. Robin, B.: Hybrid Web Recommender Systems. The Adaptive Web, 377–408 (2007)

13. Wong, S.K.M., Cai, Y.J., Yao, Y.Y.: Computation of Term Associations by a Neural Network. In: Proceedings of the 16th Annual International ACM SIGIR Conference on Research and Development in Information Retrieval (SIGIR 1993), pp. 107–115 (1993)

14. Xu, Y.F., Zhang, L.: Personalized Information Service Based on Social Bookmarking. Implementing Strategies and Sharing Experiences. In: Proceedings of the 8th International Conference on Asian Digital Libraries, pp. 475–476 (2005)

15. Xu, Y., Zhang, L., Liu, W.: Cubic Analysis of Social Bookmarking for Personalized Recommendation. In: Zhou, X., Li, J., Shen, H.T., Kitsuregawa, M., Zhang, Y. (eds.) APWeb 2006. LNCS, vol. 3841, pp. 733–738. Springer, Heidelberg (2006)

16. Kerstin, B., Claudiu, S., Wolfgang, N., Raluca, P.: Can all tags be used for search? In: Proceedings of the 17th ACM Conference on Information and Knowledge Management (CIKM 2008), pp. 193–202 (2008)

17. Guan, Z.Y., Bu, J.J., Mei, Q.Z., Wang, C.: Personalized Tag Recommenation Using Graph-based Ranking on Multi-Type Inerrelated Objects. In: Proceedings of the 32nd International ACM SIGIR Conference on Research and Development in Information Retrieval (SIGIR 2009), pp. 540–547 (2009)

18. Zhang, Z.K., Zhou, T., Zhang, Y.C.: Tag-Aware Recommender Systems: A State-of-the-Art Survey. Journal of Computer Science and Technology 26(5), 767–777 (2011)

Comparative Study of Joint Decision-Making on Two Visual Cognition Systems Using Combinatorial Fusion

Amy Batallones[1], Cameron McMunn-Coffran[2], Kilby Sanchez[1],
Brian Mott[1], and D. Frank Hsu[3]

Laboratory for Informatics and Data Mining, Department of Computer and Information
Science, Fordham University, New York, NY 10023, USA.
{abatallones,kisanchez,bmott}@fordham.edu,
cameron@dsm.fordham.edu, hsu@trill.cis.fordham.edu

Abstract. In processing multimedia technologies or decision-making in visual
cognition systems, combination by both simple average and weighted average
are used. In this paper, we extend each visual cognition system to a scoring system using Combinatorial Fusion Analysis (CFA). We investigate the performance of the combined system in terms of individual system's performance and
confidence. Twelve experiments are conducted and our main results are: (a)
The combined systems perform better only if the two individual systems are
relatively good, and (b) overall, rank combination is better than score combination. In addition, we compare the three types of averages: simple average M_1,
weighted average M_2 using σ, and weighted average M_3 using σ^2, where σ is related to confidence of each system. Our results exhibit a novel way to better
make joint decisions in visual cognition using Combinatorial Fusion.

Keywords: Combinatorial Fusion Analysis (CFA), decision-making, visual
cognition, multiple scoring systems (MSS), score function, rank function, score
combination, rank combination, rank-score characteristics (RSC) function.

1 Introduction

Research into decision-making has been of growing interest in recent decades. Of
note are full studies conducted using human subjects, which have yielded considerable amounts of data. Combining multiple aspects of vision alone [7, 18], or even
coupled with other senses [5-7,13], visual sensory perception plays a vital role in the
environmental interpretation, decision-making, and determinations of human beings.

Previous research into the interactive decision-making of people, specifically based
on visual perception, has been conducted by many groups including Bahrami et al[1],
Kepecs et al[11], and Ernst and Banks[2]. By way of a number of trials, including
varying degrees of noise, feedback, and communication, Bahrami plotted the data
against four predictive models: Coin-Flip(CF), Behavioral Feedback(BF), Weighted
Confidence Sharing(WCS), and Direct Signal Sharing(DSS). Of these four models,
Bahrami concludes that the WCS model is the only one that can be fit over the empirical data. He finds that communication between the pairs often aids in the accuracy of

R. Huang et al. (Eds.): AMT 2012, LNCS 7669, pp. 215–225, 2012.

the decision-making and can significantly improve the overall performance of the pair.

Marc O. Ernst expands on the concept of weighted confidence sharing[4] between pairs by proposing a hypothetical soccer match during which two referees determine whether the ball falls behind a goal line. Ernst agrees with the proposal Bahrami has given, that simply taking the style approach of Behavioral Feedback, or even a Coin-Flip, omits information which could lead to an optimal joint decision between the pair. Though Ernst gives an indication that the WCS model can lead to a beneficial joint determination, he concludes that there are improvements to be made to this approach. With Ernst's scenario, Bahrami's WCS model can be applied as the distance of the individual's decision(d_i) divided by the spread of the confidence distribution (σ), which is d_i/σ_i. A modified version of WCS(which closely resembles DSS) using sigma-square can produce a more accurate estimate through the joint opinion, which is represented as d_i/σ_i^2. In a validation of Bahrami, Ernst also notes that joint decision-making comes with a cost when individuals with dissimilar judgments attempt to come to a consensus in such a manner. Bahrami and Ernst set forth very different experimental methods, but their aim is very much the same: an algorithm is devised for optimal decision-making between two individuals, all based on their visual sensory input.

The fusion of multiple scoring systems (MSS) (see Hsu et al [9, 10, 19]) via Combinatorial Fusion Analysis has been used in a wide variety of research areas ([9, 10], [12, 14, 16, 19]). Combined systems are better than individual systems when the scoring systems individually are both diverse and perform well ([8-10], [12, 14], [19]). In this paper, we recorded the decisions of pairs of humans in a modified version of the soccer match proposed by Ernst [3, 4]. The research presented herein is an expansion on our previous investigations into the same domain [15, 16]. In this paper, we focused particularly on the individual's confidence in his or her decision and the effect it had on joint decision-making. We investigate how Combinatorial Fusion Analysis can optimize the joint decision-making process. In Section 2, we review the concept of multiple scoring systems in the Combinatorial Fusion Analysis (CFA) framework and treat each visual cognition system as a scoring system. In Section 3, we describe the experiments, which consist of twelve pairs of human observers, and discuss the result, often applying the Combinatorial Fusion Analysis. Section 4 summarizes the results and offers future work.

2   Combining Visual Cognition Systems

2.1   Multiple Scoring Systems (MSS)

When an individual needs to make a decision based on visual input, he or she often considers a variety of multiple choices. These choices may be either explicit or implicit, and several methods have been presented to combine these scoring systems ([1], [2], [4], [11], [13]). In this paper, we use the CFA framework [9-10] to combine scoring systems in order to optimize their joint decision.

Let D be a set of candidates with $|D|=n$. Let $N=[1,n]$ be the set of integers from 1 to n and R be the set of real numbers. In CFA, the scoring system A consists of a score function s_A and a rank function r_A on the set D of possible n positions. Hsu et al. [9-10] defined a new characteristic function for scoring systems called Rank-Score Characteristic (RSC) function as: $f_A : N \rightarrow R$,

$$f_A(i) = (s_A \circ r_A^{-1})(i) = s_A(r_A^{-1}(i)) \tag{1}$$

Two methods of combination are used for a set of p scoring systems $A_1, A_2, \ldots, A_p$ on the set D of locations. One is score combination (SC):

$$s_{sc}(d) = (\sum_{i=1}^{p} s_{Ai}(d)) / p \tag{2}$$

The other is rank combination (RC):

$$s_{rc}(d) = (\sum_{i=1}^{p} r_{Ai}(d)) / p \tag{3}$$

where d is in D, s_A and r_A are score function and rank function from D to R and N respectively.

2.2   Visual Cognition System as a Scoring System

Combinatorial fusion is used to combine the decisions of two individuals receiving the same visual input. In this case, the two different scoring systems in CFA are the two individuals' separate decisions on where he or she perceived a target has landed on a plane. Each spot on the landing plane can be considered a candidate that which an individual can score, and the two individuals serve as the scoring systems, p and q.

(A)   Each test subject was asked a radius measurement of confidence about his or her decision, which allows for a weighted evaluation of the visual area. We use this radius r to calculate the spread of the distribution around the perceived landing point of the target, calling it σ. In this paper, we use:

$$\sigma = 0.5r \tag{4}$$

This allows us to create normal distribution probability curves for each participant, labeled P and Q. The σ values are used to determine the positions of the combined means and denoted as M_i, such that $m_i = d(M_i, A)$, where A is the actual site. P, Q, and A exist in a two dimensional space as x- and y- coordinates. Three formulas are used to calculate the mean of P and Q:

Average mean, defined as:

$$M_1 = (P[x,y] + Q[x,y]) / 2 \tag{5}$$

σ mean, defined as:

$$M_2 = (P[x,y]/\sigma_1 + Q[x,y]/\sigma_2) / (1/\sigma_1 + 1/\sigma_2) \tag{6}$$

and σ^2 mean, defined as:

$$M_3 = (P[x,y]/\sigma_1{}^2 + Q[x,y]/\sigma_2{}^2) / (1/\sigma_1{}^2 + 1/\sigma_2{}^2) \qquad (7)$$

These three different combined means fall somewhere in between points P and Q and M_i is determined as a coordinate.

(B)   The range of confidence σ extends beyond the scope of line PQ, so the scope of the observation area to either side of P and Q is widened. The upper and lower bounds of the extension (P' and Q') are appended to P and Q respectively using a percentage of the longer of the two distances, PM_i or M_iQ. Hence this is the middle point of P'Q', and d(P,Q) is the distance between P and Q   (Fig. 1).

Fig. 1. Diagram of the layout of intervals used to organize the data in each experiment

The length of the line segment P'Q' is then partitioned into 127 intervals d_i, i = 1, 2, …, 127 with each interval length d(P',Q')/127. Each of the 127 intervals is scored by multiple scoring systems using the CFA framework. Our two individuals serve as the distinct scoring systems p and q who perceive the locations of P and Q, respectively, as their decided visual spot.   In this paper, all three combined means are used to observe the effect it has on CFA performance. This total length between P' and Q' is divided into 127 distinct intervals, with the center interval containing the M_i being used (Fig. 1). We refer to this extended space with P'Q', and A as extreme points and the 127 intervals as the common visual space for the two visual cognition systems p and q.

(C)   An arrangement of intervals (d_1, d_2, …, d_{127}) and their corresponding scores are generated for each participant's observation, or scoring systems p and q, by constructing a normal distribution curve based on the coordinates of where a subject believed the target landed, P and Q respectively.   Normal distribution is used:

$$Y = (1/(\sigma * \sqrt{(2\pi)})) * e^{(-(x - \mu)**2)/(2*\sigma**2)} \qquad (8)$$

where x is a normal random variable, μ is the mean, and σ is the standard deviation. In theory, a normal distribution curve infinitely spans, therefore our two scoring systems p and q create overlapping distributions that span the entire visual plane.   Each of the 127 intervals d_1, …, d_{127} has a score by p and a score by q.   These scores can be combined using score or using rank in the CFA framework.   Multiple scoring systems described in section 2.1 are used to obtain rank and score combinations, respectively marked as C and D. In this paper, we observed the effect that interval mapping on different M_i had on the performance of C and D.

2.3   Combining Two Visual Cognition Systems

For each of the 127 intervals d_1, …, d_{127}, the score value of p and q are combined, and then the rank value of p and q are combined. Score combination of systems p and q is labeled as C, and rank combination of systems p and q is labeled as D. The score function of the combination by score C is defined as:

$$s_C(d_i) = [s_p(d_i) + s_q(d_i)] / 2 \qquad (9)$$

The score function of the combination by rank D is defined as:

$$s_D(d_i) = [r_p(d_i) + r_q(d_i)] / 2 \qquad (10)$$

Each of the score functions, $s_C(d_i)$ and $s_D(d_i)$, are sorted in descending order to obtain the rank function of the score combination, $r_C(d_i)$, and the rank function of the rank combination, $r_D(d_i)$. Each interval d_i is ranked. CFA considers the top ranked intervals in C and D as the optimal points and these points are used for evaluation. The performance of the seven points (P, Q, M_1, M_2, M_3, C_1, and D_1) is determined by each points' numerical distance from the target A, the shortest distance being the highest performing point (Fig 2).

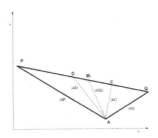

Fig. 2. Layout of M_1, C, and D in relation to P, Q, and A. The distances between the 5 estimated points and the actual site are noted on each line connecting the point to the target.

3   Experiments

3.1   Data Sets

The target that participants observed was a construct of metal plates, nuts, and a bolt sizing at 1.5 by 1.5 inches. This target was designed to be heavy enough to be thrown far distances, small enough to be hidden once on the ground, and of irregular shape to limit travel once in the grass. A confidence measuring tool was constructed using a wooden pole with a foam core x- and y- axes of 36 by 36 inches. Participants consisted of pairs chosen at random, although the individuals were already acquainted before the experiment. The participants were situated 40 feet from a marked plane of 250 by 250 inches and stood 10 feet apart from one another. There were five coordinators on site—one coordinator stood with the first participant, the second coordinator stood with the other, the third coordinator to the side of the participants, and the fourth and fifth coordinators on either side of the marked plane.

The third coordinator threw the target into the plane from next to where the participants stood. Then the participants independently and simultaneously directed the two pre-designated coordinators to where they believed the target landed and a small marker was placed on the ground at each spot. The simultaneous and independent determination of landing sites works to minimize the amount of time taken to mark the participants' initial decisions and to minimize the effect of one participant's decision on the other's decision. Then the confidence tool was taken to the plane and each participant was asked his or her radius of confidence around the spot he or she perceived the target landed. This was done by sliding a pointer stick along the x- axis of the tool until the participant expressed his or her confidence radius and then drawing the circle in the air to confirm. Finally, the x- and y- coordinates for the three points (P, Q, and A) were recorded. For each experiment, eight numerical values were obtained: the two x-coordinates of P and Q, the two y-coordinates of P and Q, the two confidence values for P and Q, and the x- and y- coordinates for actual A. The participants were interviewed for information including gender, height, eyesight, and other factors that may influence visual perception. This process was repeated for twelve experiments: four female-female pairs, four male-male pairs, and four male-female pairs (Table 1).

3.2   Results and Analysis

The line segment PQ is obtained using the decision of Participant p, marked as P, and the decision of Participant q, marked as Q. The radii of confidence are used to calculate the two σ values to locate the coordinates of points $M_1, M_2,$ and M_3 along the extended P'Q'. In order to combine and compare the two visual decision systems of p and q, identical data points must be evaluated by the different systems. In our current study, we increased the number of intervals from 9 intervals in our previous study [15] to 127 and in each case we use M_i, i = 1,2,3.

When P'Q' has been sectioned into the 127 intervals mapped according to M_i, the intervals are scored according to the normal distribution curves of P and Q using the standard deviation σ_P and σ_Q respectively.   Both systems assume the set of common interval midpoints $d_1, d_2, d_3,..., d_{127}$. Each scoring system, p and q, consists of a score function, as explained in Section 2.1. We define score functions $s_P(d_i)$ and $s_Q(d_i)$ that map each interval, d_i, to a score in systems P and Q, respectively.   The rank function of each system maps each element d_i to a positive integer in N, where N = $\{x \mid 1 \leq x \leq 127\}$. We obtained the rank functions $r_P(d_i)$ and $r_Q(d_i)$ by sorting $s_p(d_i)$ and $s_Q(d_i)$ in descending order and assigning a rank value from 1 to 127 to each interval respectively.

Table 1. Coordinates of P, Q, and A and confidence radius of P and Q for the 12 experiments

	X	Y	Radius		X	Y	Radius
	Exp. 1				**Exp. 7**		
Male (P)	111.5	134.5	11.5	Female (P)	22	190.5	7
Male (Q)	78.5	105	16	Male (Q)	17	227.75	6
Actual (A)	94	124		Actual (A)	14.75	195	
	Exp. 2				**Exp. 8**		
Female (P)	23.5	56	7	Male (P)	98.75	57	12.5
Female (Q)	112	96.75	21.5	Male (Q)	71.25	25.5	12
Actual (A)	28.5	43		Actual (A)	16.5	1	
	Exp. 3				**Exp. 9**		
Female (P)	105	134.25	21	Female (P)	205.5	15	17
Male (Q)	78.5	87.75	22	Female (Q)	204	21.5	6.5
Actual (A)	39.5	119		Actual (A)	203	26	

Table 1. *(Continued)*

Exp. 4				Exp. 10			
Female (P)	229.25	151.5	14	Female (P)	100.5	4.5	19.5
Male (Q)	256	162.5	15.5	Female (Q)	172	25.25	6
Actual (A)	216.25	149.75		Actual (A)	127	9.5	
Exp. 5				**Exp. 11**			
Female (P)	125.5	13.5	0.5	Male (P)	236.25	43	4
Female (Q)	112.75	57.25	3	Male (Q)	234	72.75	4.5
Actual (A)	113.75	46		Actual (A)	229	51.5	
Exp. 6				**Exp. 12**			
Female (P)	184.5	108.25	21.5	Male (P)	98.5	-75.5	10
Male (Q)	164.5	249.75	12	Male (Q)	99	30	12
Actual (A)	173.25	212.5		Actual (A)	96	4	

Using the formula defined in Section 2.2 and the score and rank combinations C and D, respectively defined in Section 2.3, the three combined means M_1, M_2, and M_3 are obtained. The distances from each of the seven points (P, Q, M_1, M_2, M_3, C, and D) to actual target A are computed. The points are ranked by performance from 1 to 7 (Table 2). The point with the shortest distance from the target is considered the best.

In 5 experiments, M_1 performed the best out of the three midpoints M_1, M_2, and M_3. In 6 experiments, M_3 performed the best, and in 1 experiment, M_2 performed the best (Table 3). For all 11 experiments where either M_1 or M_3 performed the best, if the better performing individual was more confident, M_3 was the highest performing midpoint (Experiments 2, 4, 6, 8, 9, and 11), while if the worse performing individual was more confident, M_1 was the highest performing midpoint (Experiments 3, 5, 7, 10, and 12). In every one of these 11 experiments, rank combination D performed either equally as well or better when the better midpoint was used to map intervals than when the alternate midpoint was used. When using this discretionary mapping procedure, 8 of the D rank scores outperformed the D rank scores of the worse performing midpoint (Experiments 2, 3, 5, 7, 8, 9, 10, and 12), and 3 were of equal performance (4, 6, and 11). None performed more poorly (Table 4).

Out of the 12 cases, when mapping the intervals by using M_1 for all experiments, as was done in our previous paper [16], D ranks 1st in 4 cases (Experiments 3, 5, 7, and 12). When using the discretionary mapping procedure, D ranks 1st in 7 cases (2, 3, 5, 7, 8, 9, and 12) (Table 4). In 23 out of 24 cases of experiments ran on M_1 and M_3, C performed worse than D.

Table 2. Performance ranking of P, Q, M_1, M_2, M_3, C, D_1, and D_2 when using Combinatorial fusion on scoring system P and Q for each experiment (E). 127 intervals are used using statistical mean: (a) M_1, (b) M_2, and (c) M_3.

(2)(a) Common visual space using mean M_1												
	E1	E2	E3	E4	E5	E6	E7	E8	E9	E10	E11	E12
P	6	1	7	1	7	7	2	7	7	3	6	7
Q	7	7	2	7	2	6	7	1	1	7	7	2
M_1	2	6	3	6	3	3	3	6	6	1	3	3
M_2	1	5	4	5	4	2	4	5	5	4	2	4
M_3	3	4	5	4	5	1	5	4	4	5	1	5
C_1	4	2	6	3	6	4	6	2	3	6	4	6
D_1	-	2	1	2	1	4	-	2	2	2	-	-
D_2	4	-	-	-	-	-	1	-	-	-	4	1

Table 2. *(Continued)*

	E1	E2	E3	E4	E5	E6	E7	E8	E9	E10	E11	E12
(2)(b) Common visual space using mean M_2												
P	5	3	7	2	6	7	1	7	7	2	6	6
Q	7	7	1	7	1	6	6	3	1	6	7	1
M_1	2	6	2	6	2	3	2	6	6	1	3	2
M_2	1	5	3	5	3	2	3	5	5	3	2	3
M_3	3	4	4	4	4	1	4	4	4	4	1	4
C_1	6	1	5	3	7	4	7	1	3	7	4	7
D_1	4	1	6	1	5	4	-	1	2	5	4	-
D_2	-	-	-	-	-	-	5	-	-	-	-	5

	E1	E2	E3	E4	E5	E6	E7	E8	E9	E10	E11	E12
(2)(c) Common visual space using mean M_3												
P	6	3	7	1	7	7	1	7	7	2	5	7
Q	7	7	1	7	1	6	7	3	2	7	7	1
M_1	2	6	2	6	2	3	2	6	6	1	3	2
M_2	1	5	3	5	3	2	3	5	5	3	2	3
M_3	3	4	4	4	4	1	4	4	4	4	1	4
C_1	4	1	5	3	5	4	5	1	3	5	6	5
D_1	4	1	6	2	5	4	5	1	1	5	4	5
D_2	-	-	-	-	-	-	-	-	-	-	-	-

Table 3. Performance comparison among M_1, M_2, and M_3 for the 12 experiments

Mean	Experiments											
	1	2	3	4	5	6	7	8	9	10	11	12
M_1	2	3	1	3	1	3	1	3	3	1	3	1
M_2	1	2	2	2	2	2	2	2	2	2	2	2
M_3	3	1	3	1	3	1	3	1	1	3	1	3

Table 4. Comparison between performance and confidence radius of (P, Q), performance of M_i, and performance ranking of D when using M_1 and M_3

Experiment Number	Performance of P and Q	Confidence Radius (P, Q)	Best M_i	Performance of D Using M_1	Performance of D Using M_3
1	(20.41, 24.52)	(11.5, 16)	M_2	4	4
2	(13.93, 99.30)	(7, 21.5)	M_3	2	1
3	(67.25, 49.98)	(21, 22)	M_1	1	6
4	(13.12, 41.74)	(14, 15.5)	M_3	2	2
5	(34.56, 11.29)	(0.5, 3)	M_1	1	5
6	(104.86, 38.26)	(21.5, 12)	M_3	4	4
7	(8.53, 32.83)	(7, 6)	M_1	1	5
8	(99.50, 59.98)	(12.5, 12)	M_3	2	1
9	(11.28, 4.61)	(17, 6.5)	M_3	2	1
10	(26.97, 47.68)	(19.5, 6)	M_1	2	5
11	(11.17, 21.83)	(4, 4.5)	M_3	4	4
12	(79.54, 26.17)	(10, 12)	M_1	1	5

3.3   Discussion

Rank combination D gives relatively good results when we examined individual systems P, Q, and systems of combination C, D, and M_i across all 12 experiments. In addition, for 11 out of 12 experiments where either M_1 or M_3 were the best performing midpoints, mapping intervals around the better performing midpoint increases D's average performance, as opposed to if it were always mapped according to one M_i. Out of these 11 experiments, D's average performance when intervals are mapped around the better performing M_i outranked the alternate mapping scheme in 8 instances and performed the same in 3 instances. There were no instances where D performed worse on the better M_i mapping scheme.

In contrast to the previous experiment [15-16], this method of interval mapping takes into account the performance of M_i, and thus the reliability of the individuals' confidence. This can possibly serve useful in experiments where the record of an individual's performance is known. If an individual is confident and known to perform well in visual cognitive decisions, mapping intervals around M_3 may be the best scheme. If an individual is confident but is known to perform poorly or if his performance is unknown, mapping around M_1 may be the best scheme. Further analysis on more experiments is required to investigate the significance of the difference between mapping schemes.

The 3 instances where P and Q were the worst performing points, D also performed worse than all three midpoints. This may signify that CFA is better only if P and Q are relatively good. D performed poorly on both M_1 and M_3 in Experiments 1, 6, and 11 where P and Q were worse than all midpoints. D also performed poorly on M_3 in Experiment 3, 5, 7, and 10 where one P or Q is ranked 1st or 2nd and the other is ranked 7th—however, in all these cases D performs well on M_1, ranking at either 1st or 2nd. This also signifies that mapping around just M_1 performs more consistently well than just M_3. Further analysis must be performed to investigate the effect the performance of P and Q has on the performance of CFA.

Based on this data and data from the previous experiment, combination by rank on M_1 appears to be the most consistent. This could indicate that combination by rank is particularly sensitive to extracting the meaningful characteristics of a best-performing individual decision. In almost all of our experiments, combination by rank based on M_1 outperformed M_1 itself, showing that a weighted average is a preferred method of combining decisions. In addition, our analysis demonstrates that combinatorial fusion is a useful vehicle for driving weighted combination. Rank and score combination, while seemingly similar in theory actually produce very different results when applied to real data. While this dissimilarity is somewhat surprising and will require further investigation, it is interesting to note the dramatic difference in performance between rank and score combination.

The dissimilarity and inaccuracy of individual decisions is the main factor in determining when two people should or should not combine their decisions. Our current research signifies that D has the potential to perform best only if P and Q are relatively good.

4   Conclusion and Future Work

We applied Combinatorial Fusion Analysis to augment the joint decision-making process. The CFA framework is advantageous for the study of joint decision-making based on visual cognitive systems. Our current analysis further explores ways to optimize this joint decision-making. Our work provides a fresh observation and consideration of previously established cognition models. As seen in other domains, rank and score combination of multiple scoring systems is useful when the scoring systems both perform well and are diverse ([8], [17], [19]).

In the future, we must also consider computing the rank-score function, f_P and f_Q as well as study the cognitive diversity [10] between two systems p and q, $d(f_P, f_Q)$. Further study on the physical relationship between points P, Q, and A will also be conducted. The sample size of 12 experiments is still small and we will be continually conducting trials and modifying experiment methodology as needed. Our results also further signify that P and Q performance must both be relatively good in order for the rank combination D to perform well.

We will continue to add more trials to the data pool and use the CFA framework to analyze different aspects of the data and how it affects decision-making, like gender or occupation. We look to conduct trials where more people are added to each experiment as well. Our research has demonstrated that CFA can serve as a useful tool in understanding how to derive the best decision from a pair of individually made decisions. CFA demonstrates much flexibility in combining multiple visual cognition systems. It also can work well in other situations where multiple scoring systems are applied.

Acknowledgements. A. Batallones acknowledges her gratitude to the Clare Boothe Luce Program and Fordham University for support in Summer, 2012.

References

1. Bahrami, B., Olsen, K., Latham, P., Roepstroff, A., Rees, G., Frith, C.: Optimally interacting minds. Science 329(5995), 1081–1085 (2010)
2. Ernst, M.O., Banks, M.S.: Humans integrate visual and haptic information in a statistically optimal fashion. Nature 415, 429–433 (2002)
3. Ernst, M.O.: Learning to integrate arbitrary signals from vision and touch. Journal of Vision 7(5), 7, 1–14 (2007)
4. Ernst, M.O.: Decisions made better. Science 330(6010), 1477 (2010)
5. Gepshtein, S., Burge, J., Ernst, O., Banks, S.: The combination of vision and touch depends on spatial proximity. J. Vis. 5(11), 1013–1023 (2009)
6. Gold, J.I., Shadlen, N.: The neural basis of decision making. Annual Review of Neuroscience 30, 535–574 (2007)
7. Hillis, J.M., Ernst, M.O., Banks, M.S., Landy, M.S.: Combining sensory information: mandatory fusion within, but not between, senses. Science 298(5598), 1627–1630 (2002)
8. Hsu, D.F., Taksa, I.: Comparing rank and score combination methods for data fusion in information retrieval. Information Retrieval 8(3), 449–480 (2005)
9. Hsu, D.F., Chung, Y.S., Kristal, B.S.: Combinatorial Fusion Analysis: methods and practice of combining multiple scoring systems. In: Hsu, H.H. (ed.) Advanced Data Mining Technologies in Bioinformatics. Idea Group Inc. (2006)
10. Hsu, D.F., Kristal, B.S., Schweikert, C.: Rank-Score Characteristics (RSC) function and cognitive diversity. Brain Informatics, 42–54 (2010)
11. Kepecs, A., Uchida, N., Zariwala, H., Mainen, Z.: Neural correlates, computation and behavioural impact of decision confidence. Nature 455, 227–231 (2008)
12. Lin, K.-L., Lin, C.-Y., Huang, C.-D., Chang, H.-M., Yang, C.-Y., Lin, C.-T., Tang, C.Y., Hsu, D.F.: Feature selection and combination criteria for improving accuracy in protein structure prediction. IEEE Transactions on NanoBioscience 6(2), 186–196 (2007)

13. Lunghi, C., Binda, P., Morrone, C.: Touch disambiguates rivalrous perception at early stages of visual analysis. Current Biology 20(4), R143–R144 (2010)
14. Lyons, D.M., Hsu, D.F.: Combining multiple scoring systems for target tracking using rank–score characteristics. Information Fusion 10(2), 124–136 (2009)
15. McMunn-Coffran, C., Paolercio, E., Liu, H., Tsai, R., Hsu, D.F.: Joint decision making in visual cognition using Combinatorial Fusion Analysis. In: Conference Proceedings of the IEEE International Conference on Cognitive Informatics and Cognitive Computing, pp. 254–261 (August 2011)
16. McMunn-Coffran, C., Paolercio, E., Fei, Y., Hsu, D.F.: Combining visual cognition systems for joint decision making using Combinatorial Fusion. In: Proceedings of the 11th IEEE International Conference on Cognition Informatics and Cognition Computing, pp. 313–322 (2012)
17. Ng, K.B., Kantor, P.B.: Predicting the effectiveness of naive data fusion on the basis of system characteristics. J. Am. Soc. Inform. Sci. 51(12), 1177–1189 (2000)
18. Tong, F., Meng, M., Blake, R.: Neural basis of binocular rivalry. Trends in Cognitive Sciences 10(11), 502–511 (2006)
19. Yang, J.M., Chen, Y.F., Shen, T.W., Kristal, B.S., Hsu, D.F.: Consensus scoring for improving enrichment in virtual screening. Journal of Chemical Information and Modeling 45, 1134–1146 (2005)

Research on Touch as a Means of Interaction in Digital Art

Mingwei Zhang[1] and Guoyong Dai[2]

[1] Camberwell College of Arts, University of the Arts London, UK, SE5 8UF
[2] Zhejiang Shuren University, Hangzhou, China, 310014

Abstract. This paper explores touch as a medium of creative expression in digital art. Important to this work is haptic interaction which takes place via a physical interface. Touch is not simply a tactile sensation, but also a way to explore an environment by combining sensory information such as pressure and temperature. Therefore an artist's creation can be understood by the viewer through the sensory information received. The main focus of the paper is on the sense of touch and it will help our understanding to investigate the following four artworks and corresponding technologies. The selected works include both visual and audio interactions. *Touch Me* concentrates on the sensory experience perceived through the skin, while *A-Volve, Colorful Touch Palette* and *FuSA2 Touch Display* are concerned with the virtual hand in technological systems. I believe that multi-touch, which enables a user to interact with more than one finger, will become the most important tool for experiencing sensory digital art in the future.

Keywords: touch, haptic interaction, information technology, digital experiment.

1 Introduction

The sense of touch is perhaps a little undervalued as far as the development of human-computer systems is concerned. Through the haptic interface, the user can not only see objects on the screen, but can also touch and manipulate them to produce a more realistic sense of immersion. 'The more life-like and immersive graphical displays become, the more tempting it is to explore the data-like objects in the real world by touching as well as seeing and hearing' (Hansen,1998).

Touch plays an irreplaceable role in interaction. In recent years, many artists have taken a keen interest in haptic technologies which interact with the viewer. The viewer can now use gloves covered with sensors that not only simulate sensation, but also allow the manipulation of distant objects.

As the abundance of haptic technology within our environment has increased, the role played by haptic interaction via the physical interface within art will increasingly continue to influence the works of artists in the new millennium. Many artists are interested in research conducted by high-tech laboratories. For example, an experiment using LCD and photoelasticity with a soft- touch panel in Japan enabled the user

R. Huang et al. (Eds.): AMT 2012, LNCS 7669, pp. 226–236, 2012.

to manipulate 3D shapes and experience tactile feedback using a gel which created soft sensations. Another experiment enabled a user to interact with more than one finger, and I believe that this 'multi-touch interaction' will become the most important tool for experiencing sensory interaction. The dominance of technology seems to be a fundamental aspect of much of today's art.

In order to explore how haptic technology influences artists and how to comprehend touch as a medium to experience the virtual world or as a metaphor for culture, I will investigate four examples of new technology which use touch to experience the virtual world. *Colorful Touch Palette* and *FuSA2 Touch Display* which are concerned with the virtual hand in technological systems were developed in high-tech laboratories, whilst sensors were used in the creation of the other two artworks; *A-Volve* which uses an interactive interface, and *Touch Me* which concentrates on the sensory experience perceived through the skin.

2   Research Context

2.1   *A-Volve* - Interacting with Artificial Life Creatures

It is the spectators, through their interaction, who give form to the artistic process. Our installations are not, therefore, static, predefined and predictable, but are more like the visible manifestation of a living process. (Sommerer & Mignonneau , p.146)

Fig. 1. A-Volve by Scommerer and Mignonneau, 1994

This interactive installation created by Christa Sommerer and Laurent Mignonneau in 1994 is a human haptic interaction which reflects the relationship between life and the artificial environment by the touching of a unique pond: a horizontal blue screen which creates an illusion of three-dimensionality. Visitors interact with virtual organisms in the pond and create small artificial creatures by drawing the shapes with a sensor pencil. The visitor can experience the interaction of the human hand with an artificial creature. These creatures are sensitive to the visitor's hand and react to its movements. When two creatures mate they create a third creature which inherits some of their characteristics, and when a creature becomes hungry i.e. when it runs out of energy, it will prey on the nearest creature to it in order to replenish its energy stores.

With built-in survival instincts, a creature will evolve to become as fast as possible in order to avoid becoming prey. Thus, as an interactive installation, A-Volve provides a clear link between aesthetics and evolution.

This was the first time Sammerer and Mignonneau successfully employed complex technology in their work, using touch as a medium to present their concept whilst inviting the visitor to immerse himself into a virtual world. A-Volve not only expresses the artists' concept of the interaction between the real and the virtual world, but also expresses how the artificial creatures react and interact within their own artificial environment. In this way, A-Volve blurs the borders between the real and the virtual worlds.

Virtual reality(VR) is now commonly used of any space created by or accessible through computer, ranging from the 3D world of a game to the Internet as an alternate virtual reality constructed by a vast networked communication space. The original meaning of VR, however, referred to a reality that fully immersed its users in a three-dimensional world generated by a computer and allowed them an interaction with the virtual objects that comprise that world. (Paul, 2008, p.125)

In A-Volve not only do the creatures interact with each other as described above,but they also interact with the visitors, thus making the system truly interactive for the general public. Mignonneau used a camera detection system and interface technology to detect when the visitors put their hands into the pool of water. The creatures respond to any pressure exerted by the hand by moving away at first, but after catching them they will relax, gradually stop swimming and you can actually hold them in your hand. As the visitors become more interested in their creations they begin to see each other and communicate with each other through the simulated evolutionary system. Thus, A-Volve is not only concerned with the interaction between the artificial creatures and visitor's behavior but also encourages the visitors to communicate with each other. It was to artworks such as A-Volve which express artificial life using digital technologies that Licklider (1960, p.140) was referring to when he said, 'Man-computer Symbiosis enable man and computer to cooperate in making decision and controlling complex situations' Many digital artworks express artificial life using digital technologies to specified variables and the possibility of behaviors such as fleeing and attacking. A-Volve is clearly an example of this.

Steven Johnson said new technology transforms the way we create and communicate, and he predicted that new types of interfaces would alter the style of our conversations, prose and thoughts in the future and he rightfully foresaw that interface designs would be strongly linked to artistic innovation as they reach out into the applications of our daily lives. (Sommerer et al., 2008, p.9)

2.2   *Touch Me* -Skin as an Analogue System

I am fascinated by the dermis as a material, sensor, contact organ and medium for non-verbal communications. While our lives are increasingly surrounded with artificial intelligence and digital networks I am interested in the directness of the skin - skin as an analogue system. [...] The *Touch Me* installation invites visitors to create

Fig. 2. Touch *me* by Zane Berzina, 2006

visual responses through touch and metamorphoses from one state into another. (Berzina, 2008, p.147)

This kind of installation which uses a skin interface as a medium which serves as an intelligent natural material. Just like natural skin, sk-interfaces respond to pressure, heat, and light. The name sk-interface stands for skin interface and is meant to emphasise the state of 'inbetweenness'. 'while skin and interface can be readily identified as solid words in their own right, tempting us to make the immediate, literal association of 'skin as interface', it is the title's hyphen that demands our attention'(Jens Hauser, 2008, p.6). Sk-interfaces are one of the many tools used by artists to comment on the evolution of our skin.

Zane Berzina aims to translate skin technology and explore the relationship between skin and the nervous system into his work. Our skin, being clearly linked to the nervous system, acts as a mirror or physical interface which reflects our emotions. In his installation *Touch Me*, Zane chose to use a wall, quite an ordinary object, to express this concept. This sensory installation invites visitors to interact with the wall through the medium of touch. The wall interacts with the body temperature of the visitors by becoming a darker shade of color, leaving various marks such as hand prints on its surface. These marks gradually fade away as the heat dissipates, just as sensations we receive via the interaction of our skin with various physical phenomena take time to disappear.

Tactile sensation is one of various psychological states of our skin. The sociologist Georg Simmel(1998) said 'touch is the confirmatory sense which collects information and confirms data received by the other senses and therefore is our actual sense of reality'.

According to Ingrid Graz, the skin (which is the largest organ in the body covering an area of two square meters) is not only a sensing organ but also directly reacts to the sensed impressions. Various research groups have done work on transducers, which mimic the properties and functionalities of the human skin. 'Virtual Reality technology allows a transgression of boundaries between male/female, human/machine, time/space. The self becomes situated beyond the skin' (Stelarc, p.251)

Just as our skin provides the body with a living interface for interaction with the outside environment, so sk-interface can provide the artist with an interface on which he explore the ever-narrowing boundaries between art and science in modern culture.

2.3   *Colorful Touch Palette* -Creating Novel Tactile Painting

Colorful Touch Palette is a novel interactive painting interface that provides a rich tactile sensation. Users can select various tactile textures, paint the textures on the canvas, and experience the tactile sensation of paint with the electro tactile simulation. The interface could be used to design spatial tactile patterns for surface prototyping or to support innovations in artistic tactile paints. Compared to a conventional multi-touch interface, this provides tactile feedback.

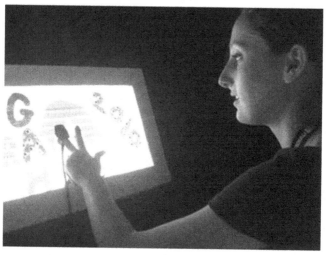

Fig. 3. Colorful Touch Palette by Yuki Hirobe, Shinobu Kuroki, Katsunari Sato, Takumi Yoshida, Kouta Minamizawa, Susumu Tachi, 2010

This installation, which was developed by Yuki Hirobe, Shinobu Kuroki, Katsunari Sato, Takumi Yoshida, Kouta Minamizawa, Susumu Tachi, gives the painter a feeling of digital colors through haptic feedback. The system consists of a 'Tactual Cap', a 'Touch Palette' and a 'Texture Canvas'. The 'Tactual Cap' is a cap-shaped device which fits onto the user's finger and provides the tactile feedback. The user blends colors selected from the 'Touch Palette' on to the 'Texture Canvas' using the cap. The user experiences shades of color as degrees of weight, providing the user with tactile feedback. The system can be potentially used to create novel tactile paintings. In this approach, the relationship between digital colors as a virtual material and the sense of touch is explored. Colorful Touch Palette is designed to bring corporeality into the field of digital art. This new art form enhances the experience and enjoyment of the digital painting process for the user to a level which approaches that of real painting. It allows the user to perceive corporeally the virtual attributes of the digital color via the force feedback brush and canvas. This technology will undoubtedly help digital painters realise the importance of touch in creative art and give them new inspiration.

It is hoped that in time the technology will develop so that the painter will also be able to experience the viscosity of the paint used. Although this installation seems to be solely concerned with the haptic information via digital painting, if it were developed so that it could remember, store, and represent the actions of people, it could become a system for sharing and experiencing other people's feeling through digital Art. It may even be expanded to include audio sensation in future work.

According to Victor Grillo(p.118), digital painting has been a permanent feature in fields of art, and represents   technology used in the artistic process. The Colorful Touch Palette not only allows artists to connect with the digital nature of the works produced, but also allows them to rediscover the physical, tactile dimension of the painting process.

2.4   *FuSA2 Touch Display* - A Tactile Display in Texture

This display, which uses plastic optical fiber(POF) bundles to give the visitor tactile sensations and visual feedback ,was developed by Human Interface Engineering Lab at Osaka University. The system uses POF bundles which resemble soft pliable hairs creating a soft tactile multi-colored surface which can be manipulated and appreciated by the human hand, the visual display and multi-touch input detection technique is a simple construction. 'A technique for creating a touch-sensitive input device is proposed which allows multiple, simultaneous users to interact in an intuitive fashion' (Dietz & Leigh, 2001). The techniques developed in the creation of this display will surely provide new potential possibilities for multi-touch.

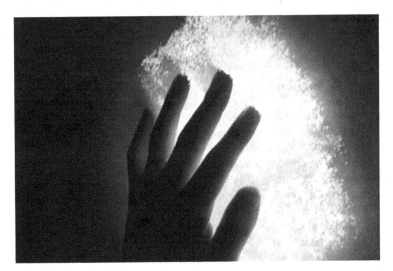

Fig. 4. FuSA2 Touch Display by Human Interface Engineering Lab at Osaka University, 2010

The purpose of this experiment is to supply direct communication between people and the environment by using a POF interface. The act of stroking hair or fur is a direct means of communicating warm emotions such as love empathy and comfort. For example, people unconsciously stroke dogs and cats giving them feelings of warmth and comfort. The direct interaction between the skin and fur plays an important role in this communication. The *FuSA2 Touch Display* gives the visitor a soft tactile feedback from a visually furry, haptic interface.

The possibilities of virtual realities, it appears, are as limitless as the possibilities of reality. It provides a human interface that disappears- a doorway to other world. (Fisher, p.117)

The human-computer interface is not only important for the interaction between participants and artworks, but also for the communication they provide between the user and the computer system. It should not only help users acquire an accurate model of the computer system, but also be able to accommodate different types of users. With advanced sensors and displays the amount of information communicated by the human-computer interface, between the user and the virtual environment has dramatically increased.

The *FuSA2 Touch Display* with its tactile furry texture communicates intuitively, appealing to human sensibilities. In order to provide the information presented on the surface of the furry texture, the display uses the physical properties of optical fibre. An infra-red camera on the surface is used to detect when it is touched and a light is emitted. The image which appears reacts to the movements of the hand. Techniques like this, based on the characteristics of the furry surface, could be developed to provide various applications for high-resolution interface interaction.

With increased overlaps between technologies, materials, academic research and science it is inevitable that artists find themselves experimenting on the edges of new technologies and practice. (Stubbs, 2008, p4)

Experiments not that dissimilar to one mentioned above actually led to Digital Art itself, nowadays technology is evolving all around us and is part of our everyday lives. So it is no wonder that artists make use of these new materials and processes in their creative work.

2.5   Natural User Interface

Digital Artists have always used new technologies and interfaces to fit the conceptual content for their interactive artworks. The natural user interface (NUI) enables the visitor to use relatively natural motions or gestures to manipulate the on-screen content. For example, in a presentation at TED in 2006, Jefferson Han showed that Multi-touch-sensing enables the user to interact with more than one finger on a computer interface. During the presentation, he expressed his interest in exploring the creative applications of NUI technology. He gave a simple example drawing a figure on the screen which he was able to manipulate just like puppet using his finger. 'Multi-touch-sensing was designed to allow non-techies to do masterful things while allowing power users to be even more virtuosic' Han (2006) says.

Today's haptic interfaces provide only a bare approximation of a human's sensitivity for touch. In distinguishing between today's force-feedback devices and the tactile feedback interfaces needed to make virtual objects feel authentic. (Hansen, 1998)

According to Hansen, a haptic interface provides the sensations across the skin to feel the virtual images, and represents movement over three degrees of freedom. In this way, a tactile display transmits an object's tactile sensation to the user making it feel natural. There is a wide range of NUI being used and developed at the moment including Perceptive Pixel, Microsoft Surface, 3D immersive Touch, Xbox Kinect and Dragon Naturally Speaking. As Steve said 'You don't have to look very far to realize that technology is becoming more natural and intuitive'. On a typical day, millions of people use touch technology on their phones, at the ATM, at the grocery store and in their cars.

The emergence of the new mass cultures of digital and virtual representation have enabled new enquiry into the significance of the role of the body in culture, and more particularly the significance of the hand in making culture. (Pajaczkowska, 2010)

As mentioned above, Colorful Touch Palette is a good example of a Multi-touch interface. This experiment provides a visual design system in a tactile setting. This system uses high-density electro tactile simulation to match the characteristics of the human activities and could enable users to feel a variety of tactile sensation via their fingertip. In this way, a digital artist can not only use Multi-touch-sensing to create digital art, but also as a source of inspiration. In this way both 'Touch Palette' and 'Texture Canvas' can be handles just like real tools and materials. By contrast, FuSA2 Touch Display uses POF bundles as a Multi-touch interface to give the visitor direct tactile sensations and visual feedback.

Colorful Touch Palette and FuSA2 Touch Display are just two of a growing number of systems being developed in high-tech laboratories, which blur the borders between the real world and the virtual reality. Computer and other special systems which provide simulation through tactile sensation are increasingly being used to simulate alternate worlds as users prefer to interact with this genre of installation. As Marshall McLuhan explained ' all natural material or reality to be transformed into a system' and he believed that it is possible to translate and transform any given material reality into another. So it is no wonder that artists make use of multimedia art which involves a substantial amount of technical supports or media and do not rely solely on traditional fields of art. Marshall McLuhan said 'the hybrid or the meeting of two media is a moment of truth and revelation from which new form is born'. Therefore, it could be said that the techniques developed in the creation of this system will provide new potential possibilities for digital art. A technology system generates information from the virtual world via an interface and channels this information to your senses which could include the sense of touch.

Technology today does not simply imply a physical implement, a 'machine', mechanical or electronic, but a systematic, disciplined, collaborative approach to a chosen objective. There is a new technology that Daniel Bell has called 'intellectual technology'; this is what artists must accept and understand. The medium, in this case technology, is not in itself the message; it becomes a message when it is in a vital dialogue with out most authentic contemporary needs [...]. (Kepes, 1968, p.268)

Hence, the importance of artistic creativity in the development of new forms of interactions and communications can never be over-emphasised.

2.6   Non-linear Interaction

According to Sammerer and Mignonneau, an interactive installation is not a finished product. Although it is initially created by the artist, it then takes on a life of its own; a dynamic process which also interacts with its environment. They feel that this process should evolve in a non-linear fashion allowing the visitor to continually venture into the unknown. 'nonlinear interaction: interaction should be easy to understand but also rich so that the visitor is able to continuously discover different levels of interaction experiences'((Sommerer & Mignonneau, 2008, P. 254).

Thus, it is important that the tactile experience of feeling the interface be an integral part of work. The cases of *A-Volve* and *Touch Me* illustrate this. *A-Volve* as tactile art evolves in a non-linear rather than in a systematic fashion. *Touch Me* invites visitors to interact with the wall using their hands creating a non-systematic dynamic process. Roland Penrose said 'It has often been said that art is a means of communication with the artist himself who is also an inventor eager to discover new methods of improving and facilitating his own techniques'.

It can be seen from *A-Volve* and *Touch Me* that the artist's primary consideration is no longer concerned with the materials used in their creations i.e wood, stone, metal etc. Increasingly, the disembodied image of digital art is evoked, into which the visitors can physically and sensorially immerse himself. The digital artist exploits the available technology in his creation as we have discussed.

3   Conclusion

Unlike conventional art representation, this paper has explored a new digital art representation which uses touch to connect the artists and visitors. From the examples above, we've seen that touch can be in the form of haptic interaction, human skin interface or tactile sensations. These interfaces give the visitor a tactile feedback which blurs the borders between the real and the virtual world.

The digital artist is at the fore-front of modern technology, incorporating it into their creative work where appropriate technology exists and using it as a source of inspiration to extend the boundaries of their own creativity; the results of this artistic exploration seemingly developing into a virtuous cycle of inspiration with the relationship between digital art and science appearing almost symbiotic in nature. On reviewing *Art and Electronic Media*, it would appear that the relationship between digital art and technology is built on an ad hoc basis and the various collaborations involving particular artists and technological laboratories are provisional in nature.

There is no doubt that touch as a medium has become increasingly important in digital artwork; particularly in those installations where the visitor is able to receive messages via the interaction with artificial realities. These man-machine and man-man communications not only promote the new aesthetics based on real-time interaction, but also the integration of technology and art which provides the visitor with a new genre of tactile sensations.

Haptic technology as an artistic medium is a unique form of digital art which communicates the artist's concept to a computer system. Digital Artists, far from being overawed by the complexities of haptic technology, appear to revel in the opportunities it affords for creative expression in their artwork. With the haptic device allowing the artist to have direct contact with a virtual object that produces real-time sound or image, it would seem highly likely that touch is set to become an integral part of the interface culture.

References

1. Berzina, Z.: Re-thinking Touch. In: Hauser, J. (ed.) Sk-interfaces: Exploding Borders- Creating Membranes in Art, Technology and Society, 1st edn. Liverpool University and FACT, Liverpool (2008)
2. Bryant, A., Pollock, G.: Digital and Other Virtualities: Renegotiating the Image. I.B.Tauris & Co Ltd., London (2010)
3. Colson, R.: The Fundamentals of Digital Art. AVA Publishing SA, Switzerland (2007)
4. Fisher, S.: Interface. In: Rockwood, C. (ed.) Digital and Video Art. Florence de Meredieu, France (2005)
5. Grillo, V.: The computer and it double: painting machines. In: Rockwood, C. (ed.) Digital and Video Art. Florence de Meredieu, France (2005)
6. Hansen, M.: New Philosophy for New Media, 1st edn. The MIT Press, Massachusetts (2006)
7. Hauser, J.: Sk-interfaces: Exploding Borders- Creating Membranes in Art, Technology and Society, 1st edn. Liverpool University and FACT, Liverpool (2008)
8. Hillner, M.: Virtual Typography. Thames & Hudson Ltd., London (2009)
9. Kepes, G.: Art &Technology-Conversation. In: Shanken, E. (ed.) Art and Electronic Media, 2nd edn. Phaidon Press Inc., London (2008)
10. Leopoldseder, H.: CyberArt 2010: International Compendium-Prix Ars Electronica, 1st edn. Hatje Cantz, Austria (2010)
11. Dietz, Leigh, D.: Diamond Touch: a multi-user touch technology (2001)
12. Paul, C.: Digital Art. Thames&Hudson Ltd., London (2008)
13. Sommerer, C., Mignonneau, L., King, D.: Interface Cultures- Artistic Aspect of Interaction. Rutgers University, North America (2008)
14. Russett, R.: HYPERANIMATION: Digital images and Virtual Worlds. John Libbey Publishing Ltd., United kingdom (2009)
15. Stelarc: The Body is Obsolete. In: Shanken, E. (ed.) Art and Electronic Media, 2nd edn. Phaidon Press Inc., London (2008)
16. Stubbs, M.: Welcome sk-interfaces to FACF. In: Hauser, J. (ed.) Sk-interfaces: Exploding Borders- Creating Membranes in Art, Technology and Society, 1st edn. Liverpool University and FACT, Liverpool (2008)
17. Sommerer, C., Mignonneau, L.: Designing Interfaces for Interactive Artworks. In: Shanken, E. (ed.) Art and Electronic Media, 2nd edn. Phaidon Press Inc., London (2008)
18. Sommerer, C., Mignonneau, L.: Christa Sommerer and Laurent Mignonneau. In: Rockwood, C. (ed.) Digital and Video Art. Florence de Meredieu, France (2005)
19. Wood, A.: Digital Encounters, 1st edn. Routledge, London (2007)

20. Ballme, S.: Microsoft Plans a Natural Interface Future Full of Gestures, Touchscreens, and Haptics (2010),
    http://www.fastcompany.com/1720964/microsoft-plans-a-natural-interface-future-finally-understands-windows-isnt-touchscreen-frie (accessed March 28, 2011)
21. Hansen, M.: It Just Feels Right (1998),
    http://www.sensable.com/documents/documents/CGW_ItJustFeelsRight.pdf (accessed March 28, 2011)
22. Han, J.: The Future of Interfaces is Multi-Touch (2006),
    http://www.multitouchexpert.com/tag/jeff-han/ (accessed March 28, 2011)
23. Human Interface Engineering Lab: FuSA2 Touch Display, http://www-human.ist.osaka-u.ac.jp/fusa2/ (accessed March 28, 2011); Hirobe, Y., Kuroki, S., Sato, K., Yoshida,T., Minamizawa, K., Tachi, S.: Colorful Touch Palette (2010),
    http://tachilab.org/modules/projects/CTP.html (accessed March 28, 2011)

Perceptual Image Hashing with Histogram of Color Vector Angles

Zhenjun Tang[1], Yumin Dai[1], Xianquan Zhang[1], and Shichao Zhang[1,2,*]

[1] Department of Computer Science, Guangxi Normal University, Guilin 541004, P.R. China
[2] Faculty of Information Technology, University of Technology, Sydney, NSW 2007, Australia
zhangsc@gxnu.edu.cn

Abstract. Image hashing is an emerging technology for the need of, such as image authentication, digital watermarking, image copy detection and image indexing in multimedia processing, which derives a content-based compact representation, called image hash, from an input image. In this paper we study a robust image hashing algorithm with histogram of color vector angles. Specifically, the input image is first converted to a normalized image by interpolation and low-pass filtering. Color vector angles are then calculated. Thirdly, the histogram is extracted for those angles in the inscribed circle of the normalized image. Finally, the histogram is compressed to form a compact hash. We conduct experiments for evaluating the proposed hashing, and show that the proposed hashing is robust against normal digital operations, such as JPEG compression, watermarking embedding, scaling, rotation, brightness adjustment, contrast adjustment, gamma correction, and Gaussian low-pass filtering. Receiver operating characteristics (ROC) curve comparisons indicate that our hashing performs much better than three representative methods in classification between perceptual robustness and discriminative capability.

Keywords: Perceptual hashing, image hashing, image authentication, color vector angle, color histogram.

1 Introduction

As today's digital images are easy to copy, edit and redistribute, content security of the digital images has become an actual and challenging issue. An efficient way for content protection is digital watermarking [1], which embeds a signal, called watermark, into the image and achieves authentication by verifying integrity of the extracted watermark. But modification during the embedding procedure will inevitably degrade visual quality of the image. This weakness makes digital watermarking unsuitable for those applications with highly demanding quality, such as medical images. Another alternative technology for data authentication is cryptographic hash functions, e.g., MD5 and SHA-1, which extract a short string called authentication code from the input message. Conventional cryptographic hash functions do not

---

* Corresponding author.

R. Huang et al. (Eds.): AMT 2012, LNCS 7669, pp. 237–246, 2012.
© Springer-Verlag Berlin Heidelberg 2012

modify the input data and then preserve its digital representation. However, these algorithms are sensitive to bit-level change. One bit difference will lead to a completely different string. This property makes it applicable to text or data file verification, but unsuitable for digital images. In practice, digital images will undergo normal digital operations, which change the digital representation of images, but keep their visual appearances unchanged. Therefore, image authentication code should be a visual content-based string. In this work, we study the issue of image hashing, an emerging multimedia technology which can totally preserve image quality and achieve image authentication.

Image hashing is a technology for deriving a content-based compact representation, called image hash, from an input image, and has been widely used in many applications, such as image retrieval, image authentication, digital watermarking, image copy detection, image indexing, and multimedia forensics. In general, image hash function must satisfy two basic properties as follows. (1) Perceptual robustness: Visually identical images have the same or very similar hashes no matter what their digital representations are. In other words, image hashing should be robust against normal digital operations, such as image compression and geometric transforms. (2) Discriminative capability: Different images have different image hashes. It means that hash distance between different images should be large enough.

Many researchers have devoted themselves to developing image hashing algorithms in last decade. For example, Venkatesan et al. [2] used wavelet coefficient statistics to construct image hashes. Lin and Chang [3] designed image authentication system using robust hashing. Lefebvre et al. [4] pioneered the use of Radon transform (RT) to construct robust hash. Swaminathan et al. [5] used discrete Fourier transform coefficients to produce image hashes. Kozat et al. [6] proposed to calculate hashes using singular value decompositions (SVDs). Monga and Mihcak [7] are the first of applying non-negative matrix factorization (NMF) to derive image hashing. This NMF-based hashing is resilient to geometric attacks. Ou et al. [8] applied the RT to the input image, randomly selected 40 projections to perform 1-D discrete cosine transform (DCT), and took the first AC coefficient of each projection to construct hash. Tang et al. [9] used structural features to design image hashing and proposed a similarity metric for tamper detection.

Although many image hashing algorithms have been successfully developed for real applications, there are still some limitations. For example, most of the existing hashing algorithms [4, 7, 9] consider only the gray images. For color images, they use luminance components in YCbCr color space for representation. As hue and saturation information are discarded, discriminations existing in color images are not fully exploited. On the other hand, many hashing algorithms [2, 3] are sensitive to rotation. Some algorithms [4, 6, 8] are resilient to rotation, but their discriminative capabilities are not desirable. In this work, we propose a robust image hashing algorithm based on histogram of color vector angles. Our algorithm can overcome the above mentioned weaknesses due to the following strategies. (1) The color vector angle is calculated by fully exploiting all components of the color pixels, and therefore makes our hashing discriminative. (2) Since histogram is a global invariant statistics, our hashing is resistant to image rotation. Our experiments demonstrate that the proposed algorithm is robust against normal digital operations including image rotation with arbitrary angle, and has good discriminative capability.

The rest of this paper is organized as follows. Section 2 describes the proposed image hashing. Section 3 and Section 4 present the experimental results and performance comparisons, respectively. Conclusions are finally drawn in Section 5.

2   Proposed Image Hashing

As shown in Fig. 1, our image hashing is generated in four phases. The input image is preprocessed to produce a normalized image in first phase. Color vector angles of the normalized image are then extracted. The third phase is a procedure of computing the histogram of those color vector angles in a circle region. In last phase, DCT is exploited to compress histogram and significant coefficients are selected as image hash. The proposed hashing is illustrated as follows.

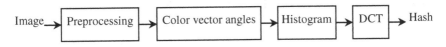

Fig. 1. Block diagram of our image hashing

2.1   Preprocessing Phase

In this phase, the input image is converted to a normalized size $M \times M$ by bi-cubic interpolation. This operation is to make our hash resistant to those images with different resolutions. After resizing, the image is then blurred by a Gaussian low-pass filter, which can be achieved by a convolution mask. Let $T_{\text{Gaussian}}(i, j)$ be the element in the i-th row and the j-th column of the convolution mask. It can be obtained with Formula (1) as follows.

$$T_{\text{Gaussian}}(i, j) = \frac{T^{(1)}(i, j)}{\sum_i \sum_j T^{(1)}(i, j)} \tag{1}$$

where $T^{(1)}(i, j)$ is defined as

$$T^{(1)}(i, j) = e^{\frac{-(i^2 + j^2)}{2\sigma^2}} \tag{2}$$

in which σ is the standard deviation of all elements in the convolution mask. The filtering manipulation is useful to alleviating influences of minor modifications on the final hash, such as noise contamination.

2.2   Color Vector Angle Extraction

A color image can be depicted by its hue, saturation and luminance, where the hue represents color appearance, the saturation describes the amount of white contained in the color, and the luminance, also called intensity, is an indicator of brightness. Clearly, image features only extracted from luminance component are ineffective in measuring color change. To overcome this weakness, we exploit color vector angles to generate robust hash. Color vector angle is insensitive to intensity variations, but sensitive to the differences of hue and saturation when changing visual appearance of an image [10]. It has been successfully used in edge detection [10] and image retrieval [11]. Comparing with the Euclidean distance in RGB color space, color vector angle is more effective in evaluating perceptual differences between two colors. To illustrate this, we take Fig. 2 (adopted from [11]) as an example, where the color pair (C1, C2) is more similar than (C3, C4). The Euclidean distance between C1 and C2 is the same with that between C3 and C4, but the vector angle of (C1, C2) is much smaller than that of (C3, C4). The advantage of color vector angle is attributed to its sensitiveness to hue differences. This means that color vector angle can capture perceptual color difference. Let $P_1=[R_1, G_1, B_1]^T$ and $P_2=[R_2, G_2, B_2]^T$ be two colors in RGB color space, where R_1 and R_2, G_1 and G_2, B_1 and B_2, are their red, green, and blue components, respectively. The angle θ between P_1 and P_2 can be calculated with Formula (3) as follows.

$$\theta = \arcsin\left(1 - \frac{(P_1^T P_2)^2}{P_1^T P_1 P_2^T P_2}\right)^{1/2} \tag{3}$$

To reduce computation, we use the sine value of θ for representation in our hashing.

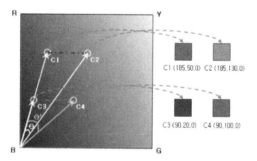

Fig. 2. Comparisons between the Euclidean distances and color vector angles of two color pairs

As the calculation of color vector angle requires two colors, we set a reference color $P_{\text{ref}}=[R_{\text{ref}}, G_{\text{ref}}, B_{\text{ref}}]^T$, where R_{ref}, G_{ref} and B_{ref} are the means of red, green and blue components of all pixels in the normalized image. With the reference color P_{ref}, we can calculate $\sin\theta_{i,j}$ for the pixel $P_{i,j}$ in the i-th row and j-th column ($1 \le i, j \le M$). And then, we can obtain the matrix $\mathbf{A}$ of color vector angles of the input image as follows.

$$A = \begin{bmatrix} \sin\theta_{1,1} & \sin\theta_{1,2} & \cdots & \sin\theta_{1,M} \\ \sin\theta_{2,1} & \sin\theta_{2,2} & \cdots & \sin\theta_{2,M} \\ \cdots & \cdots & \cdots & \cdots \\ \sin\theta_{M,1} & \sin\theta_{M,2} & \cdots & \sin\theta_{M,M} \end{bmatrix} \tag{4}$$

2.3 Histogram Calculation

Since rotation manipulation often takes image center as origin of coordinates, those pixels in the inscribed circle of an image are kept unchanged. Therefore, we calculate histogram of those color vector angles in the inscribed circle of A and use it to represent the input image. To do so, we find those color vector angles for histogram computation, whose coordinates (x, y) satisfy Formula (5) as follows.

$$(x - x_c)^2 + (y - y_c)^2 \le r^2 \tag{5}$$

where (x_c, y_c) is the coordinates of image center and $r=M/2$ is the radius of the inscribed circle. If M is an even number, $x_c=y_c=M/2+0.5$. Otherwise, $x_c=y_c=M/2$. As the result of $\sin\theta$ is a real number in the interval $[0, 1]$, we quantize the interval with a step size 0.005 and then obtain $N=201$ discrete values, i.e., 0.000, 0.005, 0.010, ..., 1.000. Let $f(t)$ be the value of the $(t+1)$-th bin of the histogram, where $t=0,1,...,N-1$. Consequently, the total number of the histogram bins is N.

2.4 Compression Phase

To make the hash as short as possible, we exploit one dimensional DCT to compress the histogram of color vector angels. This is done as follows.

$$C(u) = a(u) \sum_{t=0}^{N-1} f(t) \cos\left[\frac{(2t+1)u\pi}{2N}\right] \quad u = 0,1,...,N-1 \tag{6}$$

where $a(u)$ is determined by

$$a(u) = \begin{cases} \sqrt{1/N}, & u = 0 \\ \sqrt{2/N}, & u = 1,2,...,N-1 \end{cases} \tag{7}$$

Then, we use the first n AC coefficients to form the hash $\mathbf{h}$:

$$h(i)=C(i) \tag{8}$$

where $h(i)$ is the i-th element of $\mathbf{h}$, and $i=1, 2, ..., n$.

2.5   Hash Similarity

To measure the similarity between a pair of image hashes $\mathbf{h}_1$ and $\mathbf{h}_2$, we take L_2 norm as similarity metric, which is defined as follows.

$$d(\mathbf{h}_1, \mathbf{h}_2) = \sqrt{\sum_{i=1}^{n} [h_1(i) - h_2(i)]^2} \qquad (9)$$

where $h_1(i)$ and $h_2(i)$ are the i-th elements of $\mathbf{h}_1$ and $\mathbf{h}_2$, respectively. The more similar the images of the input hashes, the smaller the L_2 norm. If the L_2 norm is smaller than a pre-defined threshold T_d, the two images are considered as visually identical images. Otherwise, they are different images.

3   Experimental Results

We conduct experiments for illustrating the performances of our hashing. Comparisons with the existing algorithms will be given in Section 4. In the following experiments, all images are resized to 512×512, and blurred by a 3×3 Gaussian low-pass mask with a unit standard deviation. Total number of the used AC coefficients is 30. In other words, $M=512$, $\sigma=1$ and $n=30$.

3.1   Perceptual Robustness

Five standard color images sized 512×512 are taken as test images and attacked by different digital operations, which are achieved by Photoshop, MATLAB and Stir-Mark 4.0 [12]. Detailed operations and their parameter values are summarized in Table 1. As rotation will expand the sizes of the processed images, only the 361×361 central parts of the original and the processed images are taken for hash generation. After the digital operations, each image has 60 attacked versions.

Table 1. The used operations and their parameter values

Tool	Operation	Description	Parameter value
Photoshop	Brightness adjustment	Photoshop's scale	10, 20, −10, −20
Photoshop	Contrast adjustment	Photoshop's scale	10, 20, −10, −20
MATLAB	Gamma correction	γ	0.75, 0.9, 1.1, 1.25
MATLAB	3×3 Gaussian low-pass filtering	Standard deviation	0.3, 0.4, ..., 1.0
StirMark	JPEG compression	Quality factor	30, 40, ..., 100
StirMark	Watermark embedding	Strength	10, 20, ..., 100
StirMark	Scaling	Ratio	0.5, 0.75, 0.9, 1.1, 1.5, 2.0
StirMark	Rotation	Angle in degree	1,2,5,10,15,30,45,90, −1,−2,−5,−10,−15, −30, −45, −90

Extract image hashes of the original and the attacked images, and calculate their similarities by the Formula (9). The maximum, minimum and mean L_2 norms of each operation and its standard deviation are presented in Table 2. It is observed that most L_2 norms of the test operations are all smaller than 14000 except a few results. This means that we can choose T_d=14000 as a threshold to resist most of the above operations. In this case, 2.67% attacked images are falsely judged as different images. If T_d reaches 16000, there are only 1.33% attacked images considered as different images.

Table 2. Maximum, minimum, and mean L_2 norms of different operations and their standard deviations

Operation	Max.	Min.	Mean	Standard deviation
Brightness adjustment	12869	2510	6176	3023.2
Contrast adjustment	16498	3927	8805	3914.1
Gamma correction	13102	2855	6622	3313.6
3×3 Gaussian low-pass filtering	880	2	397	268.5
JPEG compression	11948	3094	6459	2534.2
Watermark embedding	14421	751	5772	3449.3
Scaling	10237	3219	5771	2418.3
Rotation	19708	1802	6939	3933.0

3.2   Discriminative Capability

To construct a color image database for discrimination test, we downloaded 67 images from the picture channel of the SOHU.com, captured 33 images by digital

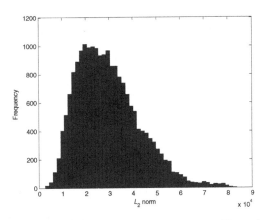

Fig. 3. L_2 norm distribution between hashes of different images

cameras, and took 100 images from the Ground Truth Database [13]. These image sizes range from 256×256 to 2048×1536. Calculate image hashes of the 200 color images and compute L_2 norm between each pair of hashes. Thus, 19900 results are available. Fig. 3 is the distribution of the results, where the x-axis is the L_2 norm and the y-axis represents the frequency of L_2 norm. It is observed that the maximum, minimum, and mean distances are 83665, 2883, and 29583, respectively, and the standard deviation is 13118.4. When T_d=14000, 9.37% different images are falsely judged as similar images. If T_d=16000, there are 14.37% different images considered as visually identical images.

4   Performance Comparisons

In this section, we compare our algorithm with some notable hashing methods, i.e., the SVD-SVD hashing [6], the NMF-NMF-SQ hashing [7], and the RT-DCT hashing [8]. To make fair comparisons, the same images used in Section 3 are also adopted to validate their perceptual robustness and discriminative capability. Since the input images of these algorithms [6, 7, 8] are gray images, we convert the RGB color images into YCbCr color space and take their luminance components for hash generation. The parameters used in the SVD-SVD hashing are as follows: the first number of overlapping rectangles is 100, rectangle size is 64×64, the second number of overlapping rectangles is 20 and the rectangle size is 40×40. For the NMF-NMF-SQ hashing, the used parameter values are: sub-image number is 80, height and width of sub-images are 64, rank of the first NMF is 2 and rank of the second NMF is 1. The SVD-SVD hashing and the NMF-NMF-SQ hashing both use L_2 norm as distance, while the RT-DCT hashing takes normalized Hamming distance as metric.

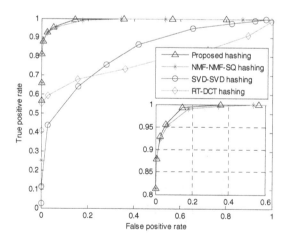

Fig. 4. ROC curve comparisons among different hashing methods

Exploit the assessed methods to generate image hashes and calculate their hash distances between each pair of images. ROC graph [14] is then exploited for comparing classification performances. We choose thresholds for each hashing, calculate the true positive rate (TPR) and false positive rate (FPR), and arrive at the ROC curves. Fig. 4 is the ROC graph comparisons among our hashing and the above three methods. We observe that ROC performances of our hashing are slightly better than those of the NMF-NMF-SQ hashing, and both the algorithms are much better than other hashing methods. For example, when FPR is near 0, TPRs of our hashing and the NMF-NMF-SQ hashing are both 0.82 and those of the SVD-SVD hashing and RT-DCT hashing are about 0.11 and 0.55, respectively. Similarity, when TPR reaches 1.0, optimal FPR of our hashing is about 0.36, and those of the NMF-NMF-SQ hashing, the SVD-SVD hashing and the RT-DCT hashing is approximately 0.53, 0.95 and 0.98.

5   Conclusions

In this paper, we have proposed a robust image hashing algorithm based on the histogram of color vector angles. As color vector angle is effective in measuring hue and saturation differences, our hashing has a good discriminative capability. Since pixels in the inscribed circle are not changed by rotation, the histograms of their color vector angles are almost the same before and after rotation, which makes our hashing resistant to image rotation with arbitrary angle. Experiments have demonstrated that our hashing is robust against normal digital operations, such as JPEG compression, watermarking embedding, scaling, rotation, brightness adjustment, contrast adjustment, gamma correction and Gaussian low-pass filtering. ROC curve comparisons between our hashing and three notable algorithms are carried out, and the results have shown that our hashing has better performances than the compared algorithms.

Acknowledgements. This work is supported in part by the Australian Research Council (ARC) under large grant DP0985456; the China "1000-Plan" National Distinguished Professorship; the China 863 Program under grant 2012AA011005; the China 973 Program under grant 2013CB329404; the Natural Science Foundation of China under grants 61165009, 60963008, 61170131; the Guangxi Natural Science Foundation under grants 2012GXNSFGA060004, 2012GXNSFBA053166, 2011GXNSFD018026, 0832104; the Guangxi "Bagui" Teams for Innovation and Research; the Project of the Education Administration of Guangxi under grant 200911MS55; and the Scientific Research and Technological Development Program of Guangxi under grant 10123005–8.

References

1. Qin, C., Chang, C.C., Chen, P.Y.: Self-embedding fragile watermarking with restoration capability based on adaptive bit allocation mechanism. Signal Processing 92, 1137–1150 (2012)
2. Venkatesan, R., Koon, S.-M., Jakubowski, M.H., Moulin, P.: Robust image hashing. In: 7th IEEE International Conference on Image Processing, pp. 664–666. IEEE Press, New York (2000)

3. Lin, C.Y., Chang, S.F.: A robust image authentication system distinguishing JPEG compression from malicious manipulation. IEEE Transactions on Circuits System and Video Technology 11, 153–168 (2001)
4. Lefebvre, F., Macq, B., Legat, J.-D.: RASH: Radon soft hash algorithm. In: 11th European Signal Processing Conference, pp. 299–302 (2002)
5. Swaminathan, A., Mao, Y., Wu, M.: Robust and secure image hashing. IEEE Transactions on Information Forensics and Security 1, 215–230 (2006)
6. Kozat, S.S., Venkatesan, R., Mihcak, M.K.: Robust perceptual image hashing via matrix invariants. In: 11th IEEE International Conference on Image Processing, pp. 3443–3446. IEEE Press, New York (2004)
7. Monga, V., Mihcak, M.K.: Robust and secure image hashing via non-negative matrix factorizations. IEEE Transactions on Information Forensics and Security 2, 376–390 (2007)
8. Ou, Y., Rhee, K.H.: A key-dependent secure image hashing scheme by using Radon transform. In: IEEE International Symposium on Intelligent Signal Processing and Communication Systems, pp. 595–598. IEEE Press, New York (2009)
9. Tang, Z., Wang, S., Zhang, X., Wei, W.: Structural feature-based image hashing and similarity metric for tampering detection. Fundamenta Informaticae 106, 75–91 (2011)
10. Dony, R.D., Wesolkowski, S.: Edge detection on color images using RGB vector angles. In: IEEE Canadian Conference on Electrical and Computer Engineering, vol. 2, pp. 687–692. IEEE Press, New York (1999)
11. Kim, N.W., Kim, T.Y., Choi, J.S.: Edge-Based Spatial Descriptor for Content-Based Image Retrieval. In: Leow, W.-K., Lew, M., Chua, T.-S., Ma, W.-Y., Chaisorn, L., Bakker, E.M. (eds.) CIVR 2005. LNCS, vol. 3568, pp. 454–464. Springer, Heidelberg (2005)
12. Petitcolas, F.A.P.: Watermarking schemes evaluation. IEEE Signal Processing Magazine 17, 58–64 (2000)
13. Ground Truth Database,
http://www.cs.washington.edu/research/
imagedatabase/groundtruth/
14. Fawcett, T.: An introduction to ROC analysis. Pattern Recognition Letters 27, 861–874 (2006)

Data Hiding Method Based on Local Image Features

Xianquan Zhang[1], Zhenjun Tang[1], Tao Liang[1], Shichao Zhang[1,2,*],
Yingjun Zhu[1], and Yonghai Sun[1]

[1] Department of Computer Science, Guangxi Normal University, Guilin 541004, P.R. China
[2] Faculty of Information Technology, University of Technology, Sydney, NSW 2007, Australia
zhangsc@gxnu.edu.cn

Abstract. We propose a data hiding method with high embedding capacity and good fidelity. It is done by block classification, data embedding, and pixel adjustment. The image blocks are firstly divided into three categories: smooth block, edge block and textural block. Different secret bits are then adaptively embedded into different blocks in terms of the image types by using least-significant-bit (LSB) substitution. After data embedding, the changed pixels are adjusted to minimize distortion. Many experiments are conducted to validate the effectiveness of the proposed method.

Keywords: Data hiding, least-significant-bit (LSB) substitution, RS steganalysis, embedding capacity, wavelet transform.

1   Introduction

Nowadays, people can easily transmit message to others by the Internet. Since the Internet is open to anyone, malicious attackers are easy to intercept and monitor the transmitted data by using powerful tools. Thus, protection of the transmitted data is in demand. Data encryption [1] is a conventional method for data protection, which converts the private data to the meaningless data. As malicious attackers cannot understand the meaningless data, data protection is achieved. However, as the meaninglessness generally implies importance of the transmitted data, it will attract the attackers' attention and make them try to decrypt these data.

Data hiding, also called information hiding, is a technology of concealing secret data and finds applications in copyright protection, content authentication, image forensics, data binding, secret communications, and so on [2]. It embeds the secret data into a cover image and then produces stego-image. Since there is no obvious visual difference between the cover image and the stego-image, attackers cannot perceive the existence of the secrete data and then the stego-image can be securely transmitted. In general, data hiding techniques should satisfy the basic requirements [2, 3]: (1) Imperceptibility or fidelity: Embedding should not cause significant perceptual change in the stego-images. In other words, stego-images should have good visual quality. (2) Embedding capacity: The number of secret bits embedded into the

---

* Corresponding author.

R. Huang et al. (Eds.): AMT 2012, LNCS 7669, pp. 247–256, 2012.

cover image should be as high as possible. In fact, the above two requirements contradict each other. More secret bits embedded will lead to worse visual quality of the stego-image. It is a challenging task to develop high performance data hiding methods simultaneously satisfying a high embedding capacity and a good visual quality.

Many data hiding techniques have been reported in the literature. One of the common used techniques is to manipulate the least-significant bit (LSB), which directly replaces the LSBs of the cover image with the secret bits. As the number of the used bit-planes increase, the LSB method can embed more secret bits into the cover image. But the quality of the stego-image will be degraded. To improve the visual quality, Chan and Cheng [4] proposed an optimal pixel adjustment process and combined it with the simple LSB substitution to design a data hiding scheme [5]. Wu and Tsai [6] calculated the difference of two consecutive pixels and then embedded secret data by replacing those difference values in a certain range. In another study, Wang and Tsai [7] proposed a data hiding method based on best-block matching and k-means clustering. This method has a high embedding capacity. However, as the secret data must be digital images, applications of this method are limited. Recently, Yang [8] exploited the LSB substitution in combination with the inverted pattern approach to improve image quality. Lee and Tsai [9] proposed a data hiding scheme based on a human vision model with distortion-minimizing capabilities. This method achieves good image quality by block pattern coding and dynamic programming. In another work, Chen et al. [10] detected the edge pixels by the combination of the Canny edge detector and the fuzzy edge detector, and then embedded secret data using LSB substitution.

In this paper, we propose a data hiding method with high embedding capacity and good fidelity. The proposed method achieves high embedding capacity by adaptively exploiting multiple lowest bit-planes to embed secret data, and reaches good fidelity by pixel adjustment. The rest of the paper is organized as follows. Section 2 introduces the block feature and its classification. Section 3 describes the proposed method. Experimental results are presented in Section 4 and conclusions are finally drawn in Section 5.

2   Block Feature and Its Classification

Data hiding in digital image can be viewed as a process of adding a weak signal (secret data) to a strong signal (cover image). If the strength of the weak signal is smaller than a contrast threshold, human vision system (HVS) can not perceive the change. The contrast properties of HVS show that different image regions have different contrast thresholds, which are generally affected by the luminous intensity, texture complexity, and so on. In smooth regions, HVS is sensitive to additive noises. Therefore, the contrast threshold is low and a few secret bits can be embedded. In non-smooth regions, HVS is insensitive to additive noises. Thus, the threshold is high and more secret bits are tolerated. In general, a cover image contains different type regions. To make the embedding capacity as high as possible, we must identify the type of each region before data embedding. To do so, we divide the cover image into non-overlapping blocks and classify blocks into different categories. The block classification is illustrated as follows.

2.1   Classification of Smooth and Non-smooth Blocks

Wavelet transform is an effective method for signal analysis. Its basic thought is multi-resolution decomposition. Fig. 1 is a schematic example of three-level 2-D discrete wavelet transform (DWT). The 2-D DWT can decompose an image into a series of subbands. The LL subband contains most information of the original image and depicts the coarse characteristic. It can be used to approximate the original image. Other subbands, such as HL, LH, and HH subbands, preserve the edge information in different directions. The more edges the image has, the bigger the value of high frequency coefficient is. So we can exploit the wavelet coefficients to classify blocks into two categories: smooth blocks and non-smooth blocks.

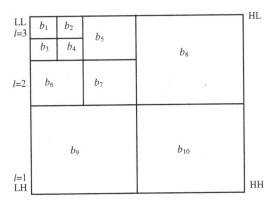

Fig. 1. Schematic example of three-level 2-D DWT

To do so, we divide the cover image into non-overlapping 8×8 blocks, apply three-level 2-D DWT to each block and then obtain 10 subbands, i.e., b_1, b_2,..., b_{10}, as shown in Fig. 1. Let H and W be the height and width of a subband, and $d_t(i, j)$ be a coefficient in the i-th row and the j-th column of the t-th subband (t=1, 2, ..., 10). Thus, the energy of the t-th subband can be defined as

$$e_t = \frac{1}{HW} \sum_{i=1}^{H} \sum_{j=1}^{W} d_t(i, j)^2 \qquad (1)$$

Next, we can derive a ratio between the high-frequent subbands and the low-frequent subband as follows.

$$E = \frac{e_2 + e_3 + \cdots + e_{10}}{e_1 + \varepsilon} \qquad (2)$$

where ε is a small constant to avoid singularity. In fact, the E value is an index of the block texture. The richer the block texture, the bigger the E value. To normalize the E

value, we apply three-level 2-D DWT to the original image and calculate the energy ratio $\overline{E}$ of the whole image. Thus, the normalized value ρ is defined as

$$\rho = \frac{E}{\overline{E}} \tag{3}$$

If the ρ value is greater than a pre-defined threshold u, the block is considered as a smooth block. Otherwise, it is viewed as a non-smooth block.

2.2   Synthetic Image

During the data hiding, we embed different secret bits into different blocks in terms of their ρ values. After embedding, the stego-image is not totally equal to the cover image. Thus, the ρ values of blocks in the stego-image will be inevitably changed. This may lead to false extraction. To avoid this case, the ρ values must be kept unchanged. To do so, we create a synthetic image by using the high bit-planes of the original image, and then identify the block types of the original image by calculating the ρ values of blocks in the synthetic image. This is based on the following considerations. Firstly, the high bit-planes of the original image are unchanged since only the low bit-planes are replaced with the secrete bits. Secondly, the high bit-planes contain the main details of the original image. In practice, the synthetic image can be obtained by replacing the k ($k = 1, 2,\dots , 8$) lowest bit-planes of the original image with 0 and keeping the $(8 - k)$ highest bit-planes unchanged.

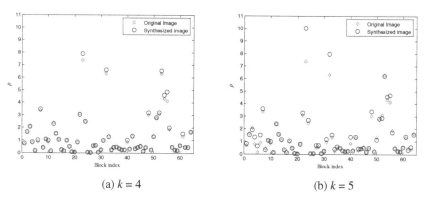

(a) $k = 4$ (b) $k = 5$

Fig. 2. ρ value comparisons between the corresponding blocks of the original and synthesized images

To validate the effectiveness of the synthesized images, many experiments are conducted. A typical example is presented here, where the standard image Peppers sized 256×256 is used again. We convert it to a low resolution version sized 64×64 for visibility, and create the synthesized images with different k values of the low resolution image. Then, we divide the low resolution image and the synthesized images into 8×8 non-overlapping blocks, and calculate their ρ values. The results are listed in Fig. 2, where the abscissas are the block indices and the ordinates are the ρ values. It is observed that there is slight difference between the ρ values when $k \leq 5$. In general, the visual changes between the original and synthesized images are imperceptible when $k \leq 4$. In this case, we can exploit the synthesized image to represent the original image. For some images, the k value can be 5. But $k \geq 6$ is unsuitable.

2.3   Non-smooth Block Classification

Image edge is an important feature for HVS. Additive noises added around the edge are easily captured by HVS. Thus, to make the stego-image good quality, the edges must be preserved. It means that a moderate noise can be added to the region with edges. Obviously, we must judge whether or not a block has edges before data embedding. This can be done as follows. For a non-smooth block, we exploit an edge detection algorithm, such as Prewitt operator, to extract the edge and calculate the number of edge pixels in the block. Then, we compute the ratio r between the numbers of the edge pixels and total pixels. If r is bigger than a pre-defined threshold v, the block is considered as an edge block. Otherwise, it is a textural block. In experiments, the used threshold is $v=12.5\%$. It means that a block is considered as an edge block if 12.5% pixels are edge points.

We now conduct experiments to show the differences between the extracted edges of the original and synthesized images. It is found that the differences are slight when $k \leq 5$ and become remarkable when $k>5$. Fig. 3 is a typical example, where the Prewitt operator is used to detect the edges of the original Peppers and its synthesized images. We observe that the edges of synthesized images with $k \leq 5$ are almost the same with those of the original image. However, when $k=6$, there are a great difference.

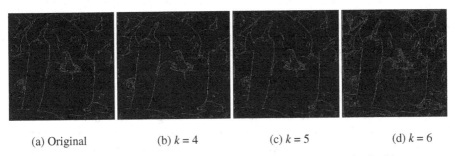

(a) Original   (b) $k = 4$ (c) $k = 5$ (d) $k = 6$

Fig. 3. Edge comparisons between the original Peppers and its synthesized images

2.4   Block Classification

In summary, the image blocks are classified into smooth blocks and non-smooth blocks, and further, the non-smooth blocks are divided into edge blocks and textural blocks. The HVS properties show that blocks of different types can tolerate different secret bits under the premise that visual changes of the blocks are invisible. In other words, HVS is sensitive to additive noises in a smooth block and the number of embedded bits in a smooth block is small. Furthermore, for the edge block and the textural block, HVS is more and more insensitive to additive noises and then more secret bits can be embedded in these regions.

3   Proposed Method

In data embedding, we firstly identify the block type and then adaptively embed secret bits into the blocks. To determine the block type, we create the synthesized images with different values (1, 2,..., k), calculate the ρ values of blocks, and apply edge detection operator to extract image edges. Let the used highest bit be h. To make the stego-image good visual quality, we let $h \leq k - 1$ and then adjust the $(h + 1)$-th bit to minimize pixel distortion. The pixel adjustment is illustrated as follows.

3.1   Pixel Adjustment

To make visual change as small as possible, we adjust the values of the modified pixels as follows. Let the values of the original and modified pixels be x and x'. Thus, their difference is $f=x - x'$. The modified pixel is then adjusted by using the following rules.

$$\text{If } f > 2^{h-1} \text{ and the } (h + 1)\text{-th bit is 0, let the } (h + 1)\text{-th bit be 1.} \tag{4}$$

$$\text{If } f < -2^{h-1} \text{ and the } (h + 1)\text{-th bit is 1, let the } (h + 1)\text{-th bit be 0.} \tag{5}$$

This manipulation can make the pixel change small, leading to a high peak signal-to-noise ratio (PSNR) value between the cover and stego-images.

3.2   Data Embedding

Let the used highest bit-planes in smooth block, edge block and textural block be h_1, h_2 and h_3, respectively. Thus, $h_1 \leq h_2 < h_3$. Since the k-th bit is used for pixel adjustment, $h_3 \leq k-1$. For the applications that the secret bit number is known in advance, we can adjust the parameters, i.e., h_1, h_2, h_3, and u, to find the optimum values leading to good performances in PSNR and anti-analysis. During the data hiding, we select blocks in terms of their ρ values, i.e., from big to small, and then embed secret bits into blocks one by one. This is based on the considerations that the block with a big ρ value has rich texture and moderate noises added to such textural region are invisible.

To evaluate the total performance of the proposed embedding method, we design a metric combining the PSNR and the estimated number of embedded bits obtained by the RS (regular and singular groups) steganalysis [11]. Specifically, we use the weighted PSNR subtract the weighted estimated number of embedded bytes as follows.

$$m = w_1 \times p - w_2 \times n \tag{6}$$

where p and n are the PSNR and the estimated number of embedded bytes respectively, w_1 and w_2 are their weights. The bigger the m value, the better the total performances. This is because a good data hiding method means a big PSNR value and a small estimated number of embedded bytes. Note that p and n are both the normalized values ranging from 0~1, which are done by using their maximum values. One can increase the w_1/w_2 value to make the proposed method emphasis the PSNR/anti-analytical requirement. The detailed steps are as follows.

STEP 1: Set the parameter values, i.e., k, v, w_1, and w_2.

STEP 2: Encrypt the secret bits using a binary chaos sequence and the bitwise exclusive-OR operator.

STEP 3: Create the synthesized images with different values (1, 2, ..., k) and divide them into non-overlapping blocks sized 8×8. Calculate the p values and r values of the blocks, and sort the p values.

STEP 4: As the ranges of h_1, h_2 and h_3 are $[0, k-2]$, $[h_1, k-2]$ and $[h_2+1, k-1]$ respectively, we use an iterative attempt to find optimum parameters for data embedding. For each combination values of h_1, h_2 and h_3, we select the threshold u from big p values to small values. For each u value, we identify the block types, select blocks from big p values to small values, and then embed the encrypted bits into each block from low bit-planes to high bit-planes using the LSB substitution. If all bits are embedded, we adjust the changed pixels using the Formulas (4) and (5), and calculate the PSNR value p and the estimated number n to find the m value by the Formula (6). Thus, we terminate the u value selection, and compute the next parameter combination. Otherwise, we fail to embed all bits. In this case, we choose next u value and embed bits again until all bits are used. If we cannot find the u value to embed all bits, we discard the combination values. Use all parameter combinations to find their m values. The combination values with maximum m are the optimum parameters and the corresponding embedding result is the final stego-image.

Note that the k, u, v, w_1, w_2, h_1, h_2, h_3 and the length of secret bits are used as secret keys, which are shared between the sender and the receiver.

3.3   Data Extraction

Data extraction is an inverse process of the data embedding. Detailed steps are as follows.

STEP 1: Create the synthesized image with the k value and divide it into 8×8 non-overlapping blocks. Calculate the p values and r values of the blocks, and sort the p values.

STEP 2: Determine the type of each block in terms of the u and v values.

STEP 3: Select blocks from big p values to small values, and extract the embedded bits from low bit-planes to high bit-planes in terms of the block types, i.e., the values of h_1, h_2 and h_3.

STEP 4: Decrypt the extracted bits using the exclusive-OR operator and the same chaos sequence, and then the decrypted results are the secret bits.

4   Experimental Results

In the experiments, the PSNR is used as the object metric to measure the visual quality of stego-image. The used parameters are: $k = 4$ and $v=12.5\%$.

4.1   Given Secret Bits

To validate the performance of given secret bits, we exploit standard test images to conduct experiments with different secret bits, and compare our method with Lee and Tsai's method [9]. The typical PSNR comparisons are listed in Table 1, where $w_1=1$ and $w_2=1$, meaning that our method simultaneously considers the PSNR and the estimated number of secret bits. It is found that the PSNR values of our method are all bigger than those of Lee and Tsai's method.

Table 1. PSNR comparisons between the proposed method and Lee and Tsai's method

Cover image	Embedded data (Bytes)	Lee and Tsai's method	Proposed method				
		PSNR(dB)	PSNR(dB)	h_1	h_2	h_3	u
Lena	200	55.40	64.92	1	2	3	25.14
Lena	1200	55.19	57.98	0	0	1	0.80
Airplane	200	54.86	62.96	1	2	3	80.90
Airplane	1200	55.28	56.52	1	2	3	9.54

We exploit the RS steganalysis algorithm to test the anti-analytical abilities of our method and Lee and Tsai's method. Comparison results with standard color images sized 256×256 are shown in Table 2. It is observed that the estimated bytes in the stego-images created by our method are all smaller than those in the stego-images generated by Lee and Tsai's method. This means that our method is more secure than Lee and Tsai's method.

Table 2. Estimated byte comparisons between the proposed method and Lee and Tsai's method

Cover image	Embedded bytes	Estimated bytes of the cover image	Lee and Tsai's method Estimated bytes	Proposed method Estimated bytes	h_1	h_2	h_3	u
Lena	200	380	498	210	0	0	1	6.99
Lena	1200	380	1174	276	0	0	2	3.92
House	200	508	597	500	2	2	3	32.18
House	1000	508	1330	510	2	2	3	32.18
Airplane	200	731	752	697	1	2	3	56.37
Airplane	1200	731	1341	714	1	2	3	21.59

4.2   Capacity Test

To test the embedding capacity of our method, the secret-hiding ratio (SHR) is exploited, which is calculated as

$$SHR = \frac{\text{Number of the secret bits}}{\text{Number of the cover bits}} \times 100\% \qquad (7)$$

In the experiments, standard gray images sized 256×256 such as Airplane, Lena, and Baboon are used as the cover images. Typical results are presented in Table 3. We observe that the PSNR values are still above 40 dB even when SHR reaches 30%. This means that our method has a high embedding capacity and the stego-images have good visual quality. Moreover, for the images with many smooth regions such as Airplane and Lena, most of the estimated bytes decrease when SHR<25%. For textural images such as Baboon, the estimated bytes become small when SHR=15%, but increase as SHR grows.

Table 3. Performances under different SHR values

Cover image	Estimated bytes of the cover image	SHR(%)	h_1	h_2	h_3	u	PSNR (dB)	Estimated bytes of the stego-image
Airplane	427	15	1	1	2	0.1086	49.10	5
		20	1	1	3	0.0512	42.11	38
		25	2	2	3	0.6076	43.78	526
		30	2	2	3	0.0259	40.72	436
Lena	335	15	1	1	2	0.3668	47.44	264
		20	1	2	3	0.4539	43.24	252
		25	2	2	3	0.1069	41.34	80
		30	2	2	3	0.0811	40.67	46
Baboon	2102	15	2	2	3	3.0341	46.51	1421
		20	2	2	3	2.4736	45.20	3957
		25	2	2	3	2.0038	44.14	3500
		30	2	2	3	0.4009	40.96	6161

5   Conclusions

In this paper, we have proposed a data hiding method based on local image content. The proposed method creates synthesized images using the high bit-planes of the cover image, classifies image blocks into three categories based on the synthesized image, and adaptively embeds secret bits in terms of the block types. Many experiments have been conducted to validate the proposed method, and the results have shown that our method can embed more secret data into the cover image and preserve a good visual quality of the stego-image.

Acknowledgements. This work is supported in part by the Australian Research Council (ARC) under large grant DP0985456; the China "1000-Plan" National Distinguished Professorship; the China 863 Program under grant 2012AA011005; the China 973 Program under grant 2013CB329404; the Natural Science Foundation of China under grants 61165009, 60963008, 61170131; the Guangxi Natural Science Foundation under grants 2012GXNSFGA060004, 2012GXNSFBA053166, 2011GXNSFD018026, 0832104; the Guangxi "Bagui" Teams for Innovation and Research; the Project of the Education Administration of Guangxi under grant 200911MS55; and the Scientific Research and Technological Development Program of Guangxi under grant 10123005–8.

References

1. Van De Ville, D., Philips, W., Van de Walle, R., Lemanhieu, I.: Image scrambling without bandwidth expansion. IEEE Trans. Circuits Syst. Video Technol. 14, 892–897 (2004)
2. Moulin, P., Koetter, R.: Data-hiding codes. Proceedings of the IEEE 93, 2083–2126 (2005)
3. Lin, I., Lin, Y., Wang, C.: Hiding data in spatial domain images with distortion tolerance. Computer Standards & Interfaces 31, 458–464 (2009)
4. Chan, C., Cheng, L.M.: Improved hiding data in images by optimal moderately significant-bit replacement. IEE Electron. Lett. 37, 1017–1018 (2001)
5. Chan, C., Cheng, L.M.: Hiding data in images by simple LSB substitution. Pattern Recognition 37, 469–474 (2004)
6. Wu, D., Tsai, W.: A steganographic method for images by pixel-value differencing. Pattern Recognition Letters 24, 1613–1626 (2003)
7. Wang, R., Tsai, Y.: An image-hiding method with high hiding capacity based on best-block matching and k-means clustering. Pattern Recognition 40, 398–409 (2007)
8. Yang, C.: Inverted pattern approach to improve image quality of information hiding by LSB substitution. Pattern Recognition 41, 2674–2683 (2008)
9. Lee, L., Tsai, W.: Data hiding in grayscale images by dynamic programming based on a human visual model. Pattern Recognition 42, 1604–1611 (2009)
10. Chen, W., Chang, C.C., Le, T.: High payload steganography mechanism using hybrid edge detector. Expert Systems with Applications 37, 3292–3301 (2010)
11. Fridrich, J., Goljan, M., Du, R.: Detecting LSB steganography in color and gray-scale images. IEEE Multimedia 8, 22–28 (2001)

Fast Flow Visualization on CUDA Based on Texture Optimization

Ying Tang[1], Zhan Zhou[1], Xiao-Ying Shi[1,2], and Jing Fan[1,*]

[1] School of Computer Science and Technology,
Zhejiang University of Technology, Hangzhou,China
[2] School of Information Engineering,
Zhejiang University of Technology, Hangzhou,China
`{Ytang,fanjing}@zjut.edu.cn, zhanzhanlove.hi@gmail.com,`
`shixiaoying8888@yahoo.com.cn`

Abstract. Flow visualization plays an important role in many scientific visualization applications. It is effective to visualize flow fields with moving textures which vividly capture the properties of flow field through varying texture appearances.Texture-optimization-based (TOB) flow visuliaztion can produce excellent visualization results of flow fields. However, TOB flow visualization without acceleration is time-consuming. In this paper, we propose fast flow visualization based on the accelerated parallel TOB flow visualization which is entirely implemented on CUDA. High performance is achieved since most time-consuming computations are performed in parallel on GPU and data transmission between CPU and GPU are arranged properly. The experimental results show that our TOB flow visualization generates results with fast synthesis speed and high synthesis quality.

Keywords: CUDA, flow visualization, texture optimization, GPU parallel computing.

1 Introduction

Flow visualization is a very important branch of scientific visualization which can be applied to lots of fields, such as automotive design, meteorology, medical imaging and water conservancy.

In recent years, the rapidly developing methods of texture-based flow visualization [1-2], [6-7] not only capture the rich details of a static flow field, but also vividly show the movements of the dynamic flow field. LIC (Line Integral Convolution) [1] and spot noise [2] are two basic algorithms of texture-based flow visualization. Both methods can generate directional features of textures which are suitable for the flow field visualization. However the textures generated by these methods are blurry to some extent. In addition, most of them use random noise texture to visualize the flow field instead of the real-world pictures with rich visual patterns.

Kwatra *et al.* proposed a texture-optimization-based (TOB) flow visualization to solve the problem [3-4]. Flow fields are visualized and animated with the

---

* Corresponding author.

R. Huang et al. (Eds.): AMT 2012, LNCS 7669, pp. 257–267, 2012.
© Springer-Verlag Berlin Heidelberg 2012

user-provided textures, and it produces good visualization effects. But this method involves much iterative optimization which makes it computationally expensive and limits its applicability to visualizations with the requirement of fast speed. Han *et al* [9] and Huang *et al* [19] accelerated the texture optimization by a GPU-based algorithms. Their algorithms are implemented on traditional GPU by specific GPU programming interface involving a lot of tricky GPU operations which makes them difficult and inefficient to realize. In this paper, we propose the CUDA (Compute Unified Device Architecture)-based flow visualization algorithm which is entirely implemented on GPU. CUDA is a parallel computing platform and programming model launched by NVIDIA [14]. It enables dramatic increases in computing performance by harnessing the power of GPU. It has several advantages over traditional general GPUs, such as shared memory, faster downloads and reads back to and from the GPU. Besides, the GPU program with CUDA is easier to understand and implement.

In this paper, our contribution is utilizing the powerful performance of CUDA architecture to implement a parallel algorithm of TOB flow visualization, which greatly increases the synthesis efficiency.

The rest of the paper is organized as follows: A brief overview of the related work is given in Section 2. Section 3 describes the TOB flow visualization in detail. In Section 4, we describe how to use CUDA to realize the parallel algorithm of TOB flow visualization. Experimental results are presented and discussed in Section 5. Section 6 concludes the paper with suggestions for future work.

2   Related Work

Texture-based flow visualization plays an important role in the flow visualization field. Most of previous methods were based on Line Integral Convolution (LIC) [1, 5]. Van Wijk [6] extended LIC-based visualization with texture animation to visualize the flow field. These methods adopt the random noise as visual texture patterns. It can depict rich details of the flow field effectively but is not suitable for visualize more complex flow field which comes from the real-world flow, like water, cloud and fire etc [9].

In recent years, some methods have been proposed to visualize flow field by using the real-world textures. Zhang *et al.* proposed the progressively-variant texture technique which synthesized the textures according to the user-specified vector fields [10]. Taponecco *et al.* presented the steerable texture synthesis technique [11] in which several samples with different rotations and scales are used to generate textures with the same vector filed. A similar approach was proposed by Lefebvre *et al* [12]. And Fisher *et al* .extended this method to three dimension surface flow field [23]. However, all of the above methods were applied only to static flow fields and were not extended to deal with the dynamic flow fields. Using the animated sequence of general textures to visualize the flow on 3D surfaces flow is a rising technique in recent years. It deals with both the static and dynamic flow field, and is more effective than the noise methods. The consecutive animated sequences of textures were generated by warping the texture image according to the surface flow field in [6-7]. However, the textures are made to stretch, compact or twist during the process, especially in the vicinity of singularity (such as sink and source) in flow field. This method would produce the results which appear different from the sample texture. Lefebvre *et al* [4] solved the problem by re-synthesizing the large deformed texture region, which ensured the similarity between the synthesis results and sample texture but did not ensure the consistency between the consecutive frames.

Kwatra *et al* [3] presented a global optimization solution algorithm to address this issue. It added a correction step after warping the former frame to keep the consistency and similarity. The correction step adopted the energy minimization function. This technique is capable to generate good visual appearances, but with slow computation speed. Han *et al* [9] accelerated it by a GPU-based parallel algorithm. However, the algorithm based on traditional GPU involves many programming optimization tricks which make it difficult to implement. In this paper, we study and implement the TOB flow visualization algorithm based on CUDA architecture.

3   Texture Optimization-Based Flow Visualization

The texture sequences are synthesized to keep consistency and similarity, namely the textures not only flow in a coherent fashion (consistency) but also maintain their structural elements (similarity) under the control of flow field. Fig.1 shows two criteria to be satisfied the texture sequences are synthesized: 1) Flow Consistency: the motion of synthesized textures should be in accordance with the flow; 2) Texture Similarity: the visual appearance of target textures should be similar to the source.

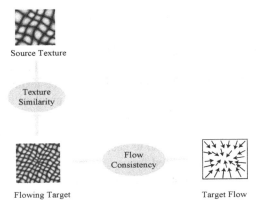

Fig. 1. Two criteria for texture optimization-based flow visualization

Kwatra *et al* [3] proposed that textures are synthesized in a frame-by-frame fashion to visualize the flow and two steps are performed when a new frame is synthesized. At first, the warp step warps the former frame according to the flow field. After the first step, a correction step is adopted to make the warped frame look similar to the exemplar. The two steps are described in the following subsections.

3.1   Warp Step

Let X_{i+1} denote the synthesized texture and X_i denote its former one. Let f_i denote the flow field function for frame X_i.

To get next frame X_{i+1}, X_i is first warped by flow function f_i and get the result $f_i(X_i)$. In order to get the pixel value q of warped frame, we start from q backtracks along with the flow field to find p which means $q=f_i(p)$. Fig.2 shows the result of the warp step.

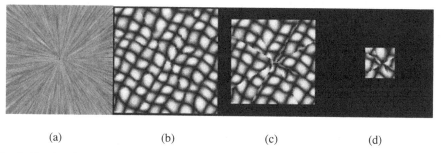

| (a) | (b) | (c) | (d) |

Fig. 2. Result of warp step. From the pictures we can see that warp step can keep the flow consistency but can't keep the texture similarity. Indeed it will disappear at last along the sink flow.

3.2   Correction Step

Fig.2 demonstrates that when only use the warp step the synthesized sequences will not maintain the similarity of the texture just the flow consistency. So after warp step the correction step should be applied.

In correction step the synthesized frame needs to be close to the warped frame to ensure flow consistency and also be similar to the source texture to ensure texture similarity. Correction step is casted as an optimization problem where the energy is defined for maintaining each goal of correction step.

To keep the flow consistency, flow energy is defined to make sure the synthesized frame as close as the warped frame. And to keep the texture similarity, texture energy is defined to make sure the synthesized frame as similar as the source texture. Equation (1) (2) (3) shows the flow energy, texture energy and the energy sum of them.

$$E_f(X) = \| x - w \|^2 \tag{1}$$

$$E_t(X) = \sum_{p \in X^+} \| x_p - z_p \|^2 \tag{2}$$

$$E(X) = \sum_{p \in X^+} \| x_p - z_p \|^2 + \lambda \| x - w \|^2 \tag{3}$$

$E_f(X)$ is computed as the squared differences between the warped frame and the synthesized frame. And $E_t(X)$ is computed as the total of all individual neighborhood energy where individual neighborhood energy refers to the squared difference between the neighborhood of one pixel in the subset of synthesized texture X^+ and its nearest neighborhood in the source texture. X refers to the synthesized texture, X^+ refers to the subset of X , x is the vectorized X, w is the vectorized warped texture, x_p is the neighborhood of p , z_p is the vectorized pixel neighborhood in Z (source texture) who is most similar to x_p,$\lambda$is a relative weighting coefficient.

The problem to visualize flow fields is formulated to minimize the energy $E(X)$. A EM-like algorithm is applied to solve the optimization problem.

EM-like iterative algorithm is divided into two steps: E-step and M-step. In E-step, according to the energy minimum requirement, the output pixels' neighborhoods $\{x_p\}$ are updated while neighborhoods $\{z_p\}$ are unchanged; in M-step, $\{x_p\}$ are fixed while $\{z_p\}$ are updated. E-step and M-step are processed in turn, and the optimized global solution is obtained until it is convergent or reaches the specific iterative number. Table 1 shows the procedure of correction step which are derived from [3].

Table 1. pseudo-code of correction step (Algorithm 1) and TOB flow visualization (Algorihthm 2)

Algorithm 1. Correction Step

$z_p^0 \leftarrow$ random neighborhood in $Z \forall p \in X^+$

for iteration n=0: M **do**

$$x^{n+1} \leftarrow \text{argmin}_x [\sum_{p \in X^+} \| x_p - z_p^n \|^2 + \lambda \| x - w \|^2] // \text{E step}$$

$$z_p^{n+1} \leftarrow \text{argmin}_v [\| x_p - v \|^2 + \lambda \| y - w \|^2]$$

// M-step , v is neighborhood in Z and y is the same as x

//except for neighborhood x_p which is replaced with v

If $z_p^{n+1} = z_p^n \ \forall p \in X^+$ **then**

$x = x^{n+1}$ **break**

end if

end for

Algorithm 2. Texture Optimization-Based Flow Visualization

for i=1:N

$X_{i+1} \leftarrow f_i(X_i)$ //warp step

   for j=1:K   // K is the consecutive neighborhood number

       CorrectionStep(X_{i+1}, f_i) //correction step

   end for

end for

3.3   Flow Visualization

Let $f = (f_1, f_2,..., f_{N-1})$ denote the input flow fields and $(X_1, X_2,..., X_N)$ denote the texture sequences being synthesized, X_1 is the initial frame which can be synthesized by texture optimization [3]. During synthesis of frame $i+1$, $f_i(X_i)$ is the warped texture which is the result of Warp step, then the Correction step is used to synthesize frame $i+1$. Table 1 shows the algorithm of TOB flow visualization.

4   Fast TOB Flow Visualization on CUDA

In [3], the hierarchical tree search is utilized in M-step which makes the EM-like algorithm time-consuming and the least squares solver is used in E-step which makes the result blurry to some extent. To solve the problem, paper [9] incorporated k-coherence [8] into both E-step and M-step of the original EM solver in [3]. And in our implementation, we have used a CUDA-friendly *k*-coherence algorithm [21] to accelerate the flow visualization speed. The warped step and correction step are also implemented on CUDA architecture system.

In CUDA a thread hierarchical model is used which is: " thread" → "thread blocks" → "thread-block grid". Each thread block contains a certain number of threads. A thread block grid can be divided into multiple thread blocks.

Our method can be divided into two phrases: Warp and Correction step. In order to fully utilize CUDA, we design the algorithm deliberately, including data transfer between CPU and GPU, allocating graphic memory appropriately and designing the dimension of both block and grid carefully.

When synthesizing a new frame before Correction step, the warped frame and initial new frame are obtained by warping the former frame. Before warping, the number of blocks is set to be the size of X_{i+1}, and the number of threads in each block is 1, such that each block processes one pixel. X_i and the flow field are bounded with texture memory.

4.1   Correction Step

During correction step, the basic step is divided into E-step and M-step. Table 2 shows correction step which incorporates the parallel k-coherence.

Table 2. Pseudo-code of correction step with *k*-coherence

Algorithm 3. Correction Step with *k*-coherence

$z_p^0 \leftarrow$ random neighborhood in Z $\forall p \in X^+$

for iteration n=0: M **do**

$\qquad x^{n+1} \leftarrow$ arg min $_{x, x(p) \in k(p)}[\sum_{p \in X^+} \| x_p - z_p^n \| + \lambda \| x - w \|^2]$ //E -step

$\qquad z_p^{n+1} \leftarrow$ arg min $_{v, v \in k(p)}[\| x_p - v \|^2 + \lambda \| y - w \|^2]$ // M-step

$\qquad$ **if** $z_p^{n+1} = z_p^n$ $\forall p \in X^+$ **then**

$\qquad\qquad x = x^{n+1}$ **break**

$\qquad$ **end if**

end for

In M-step, the CUDA-friendly k-coherence search is used to solve the neighborhood searching problem. For each q in the synthesizing frame, candidate set $c(q)$ is built by the union of the neighborhood's similar set of q. And then search in $c(q)$ to find the neighborhood that closet to q's neighborhood using an improved parallel reduction algorithm. The searched nearest neighborhood is saved in the corresponding array. Since each output pixel is independent on any other output pixels, the number of blocks is equal to the size of X^+, and the number of threads in one block is equal to the number of candidates. We copy the sample texture, pre-computed similarity set, warped texture ($f_i(X_i)$) and X_{i+1} to texture memory (due to un-changing of them during the whole M step) , use shared memory and registers to complete parallel k-coherence search and copy result back to host, since access to shared memory is much faster than to global memory. In the kernel function of M-step kernel we pre-cache the k-coherence candidate set in shared memory.

The adoption of improved parallel reduction is to avoid time expensing of data transfer between CPU and GPU and improve the efficiency of the whole execution processing. The idea of parallel reduction is similar to the merging sort algorithm. Fig.3(a) shows the procedure of reduction algorithm in one block. In our reduction process, the plus operation is replaced by compare operation. In thread block, each thread processes the data respectively; the former half threads compare the first half data in data set with the second half, and the smaller ones are moved to the first half of data set. After one process, the number that needs to be compared is half down, and all of them are stored in the first half of data set. After certain iteration, the lowest one in data set is the first item .To all blocks and all threads in the blocks this process is parallel and the number we need to compare is constant, so the parallel reduction has constant time complexity. Fig.2 (b) shows the improved parallel reduction.

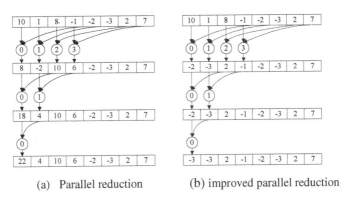

(a) Parallel reduction   (b) improved parallel reduction

Fig. 3. Parallel reduction theory

After M-step, E-step follows. During E-step, in order to minimize the energy function (3), we copy the warped texture ($f_i(X_i)$) to the texture memory. In E-step we also use CUDA-based k-coherence search to find $\{x_p\}$. The number of blocks equals to the size of X_{i+1} and the threads number in each block is equal to the candidate number. The allocated *uchar4* global array *cu_CorrectCoord* has the same size as that of X_{i+1}. Since in E-step the value of $\{z_p\}$ does not change, we copy *cu_MatchCoord*, pre-computed similarity set and $f_i(X_i)$ to texture memory.

The E-step kernel function executes as follow:

1. Declare an *int4* array *distance* and an *uchar2* array *kcoh* in the shared memory. Since when calculating the energy, an average value will be accessed by all threads, a *float4* type variable *ref_color* should be declared in the shared memory. And another *float4* type shared variable *ref_color_con* is needed as the warped color. Then according to the similarity set, compute candidate coordinates and store them in *kcoh*.
2. According to *blockIdx.x* and *blockIdx.y* get the position *p2* of the pixel being synthesized. According to *p2* the average value *ref_color* and the constrained color value *ref_color_con* is calculated. Then to the current thread we get the corresponding candidate *p1*.The energy value among *p1*, *ref_color* and *ref_color_con* is calculated, it is stored in *distance*.
3. Synchronize all operations above. After that we use the improved parallel reduction algorithm to search the candidate point with lowest energy. This candidate point will be stored to corresponding position in *cu_CorrectCoord*.

At last *cu_CorrectCoord* is copied back to host.

5   Results

We have implemented our algorithm with the following experimental hardware configuration: CPU: Intel Core™ 2Duo 2.83 GHz, Memory: 2GB, Graphics card: GeForce 8800 GTX. The algorithm is developed in Visual Studio 2008 with CUDA4.0.

In our CUDA implementation, we use successive neighborhood sizes of 16*16 and 8*8 pixels at each frame, and perform 1~2 iterations for each neighborhood.

We follow the method in [15-16] to define several user-specified flow fields.

Similar to the pre-computation in [9], for *k*-coherence search, the similarity set is pre-computed for each neighborhood. During pre-processing, an improved CUDA-based KNN search [17, 21] is used.

Fig.4 presents the results of our TOB flow visualization on CUDA, which shows that our technique is able to achieve high quality flow visualization by keeping consistency and similarity.

From Table 3, we can see that our method is almost 200+ faster than the original method and 10 times faster than the discrete solver [9]. We utilize the powerful parallel computation capacity of GPU under CUDA to implement the accelerated E-step and M-step, which makes the execution time of TOB flow visualization reduce one order of magnitude compared with previous methods.

Table 3. Synthesis Time for Different Synthesis Schemes

	Flow-guided synthesis (256^2—>256^2) (each frame)
Original texture optimization[3]	20~60s
Discrete texture optimization[9]	>1.2s
Texture optimization-base flow visualization on CUDA	180ms~200ms

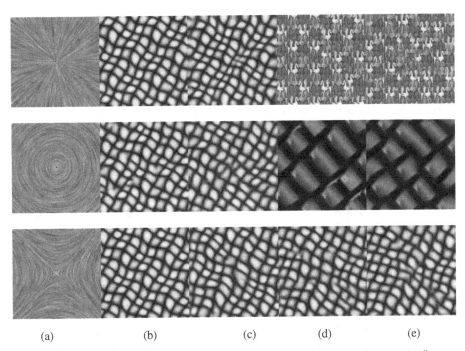

(a)   (b)   (c)   (d)   (e)

Fig. 4. The results of flow visualization. Pictures of column (a) are flow fields. For the first row column (b) gives frame1, and column (c) is frame 20. (d) is frame 1 and (e) is frame 20. The second row is similar to first.   At the last row, from (b) to (e) are frame 1, frame 10, frame 50 and frame 100, respectively.

6   Conclusion

We adopt the *k*-coherence parallel algorithm to the CUDA-based algorithm, which greatly accelerates the Correction step. Experimental results show that our CUDA-based flow visualization algorithm can not only achieve the similarity and consistency goals but also produce texture sequence at high speed.

In our future work, we will continue to study CUDA parallel techniques, and further explore how to use the general texture to achieve efficient flow visualization which can vividly simulate the feature details of the flow and has a higher speed.

Acknowledgements. This work is supported by National Natural Science Foundation of China (61173097, 61003265), Zhejiang Natural Science Foundation of China (Z1090459), Zhejiang Science and Technology Planning Project of China (No. 2010C33046), and Tsinghua–Tencent Joint Laboratory for Internet Innovation Technology.

References

1. Falk, M., Weiskopf, D.: Output-Sensitive 3D Line Integral Convolution. IEEE Transactions on Visualization and Computer Graphics, 820–834 (2008)
2. Falk, M., Seizinger, A., Sadlo, F., Uffinger, M., Weiskopf, D.: Trajectory-Augmented Visualization of Lagrangian Coherent Structures in Unsteady Flow. In: 14th International Symposium on Flow Visualization (2010)
3. Kwatra, V., Essa, I., Bobick, A., Kwatra, N.: Texture Optimization for example-based synthesis. ACM Transactions on Graphic 24(3), 795–802 (2005)
4. Wei, L.Y., Lefebvre, S., Kwatra, V., Turk, G.: State of art in example-based texture synthesis. In: Eurographics 2009, State of the Art Report, EG Association (2009)
5. Cabral, B., Leedoml, C.: Imaging vector fields using line integral convolution. In: Proceedings of ACM SIGGRAPH 1993, p. 263. ACM, New York (1993)
6. VanWijk, J.J.: Imagebased flow visualization. ACM Transactions on Graphics 21(3), 7 (2002)
7. Rasmussen, N., Enright, D., Nguyen, D., Marino, S., Sumner, N., Geiger, W., Hoon, S., Fedkiw, R.: Directable photorealistic liquids. In: 2004 ACM SIGGRAPH. Eurographics Symposium on Computer Animation, pp. 193–202 (2004)
8. Lefebvre, S., Hoppe, H.: Parallel controllable texture synthesis. ACM Transactions on Graphic 24(3), 777–786 (2005)
9. Han, J., Zhou, K., Wei, L., Gong, M., Bao, H., Zhang, X., Guo, B.: Fast example-based surface texture synthesis via discrete optimization. The Visual Computer 22(9), 918–925 (2006)
10. Zhang, J., Zhou, K., Velho, L., Guo, B., Shum, B.Y.: Synthesis of progressively-variant textures on arbitrary surfaces. ACM Transactions on Graphics 22(3), 295–302 (2003)
11. Taponecco, F.: Steerable texturesynthesis. In: Proceedings of Eurographics, pp. 57–60 (2004)
12. Lefebvre, S., Hoppe, H.: Appearance-space texture synthesis. ACM Transactions on Graphics 25(3), 541–548 (2006)
13. Yu, Q., Neyret, F., Bruneton, E., Holzschuch, N.: Scalable real-time animation of rivers. Computer Graphics Forum (Proceedings of Eurographics 2009) 28(2) (March 2009)
14. CUDA Programming Guide ver. 1.0, NVIDIA (2007)
15. Zhang, E., Mischaikow, K., Turk, G.: Vector field design on surfaces. Tech. Rep. 04-16, Gerogia Institute of Technology (2004)
16. Chen, G., Kwatra, V., Wei, L.Y., Hansen, C.D., Zhang, E.: Design of 2D Time-Varying Vector Fields. IEEE Transactions on Visualization and Computer Graphics (2011)
17. Garcia, V., Debreuve, E., Barlaud, M.: Fast k nearest neighbor search using GPU. In: IEEE Computer Society Conference on Computer Vision and Pattern Recognition Workshops, pp. 1–6 (2008)
18. Huang, H., Tong, X., Wang, W.: Accelerated parallel texture optimization. Journals of Computer Science and Technology 22(5), 761–769 (2007)
19. CUDA C Best Practices Guide v4.0, Navidia (2011)

20. Laramee, R.S., Hauser, H., Doleisch, H., Vrolijk, B., Post, F.H., Weiskopf, D.: The State of the Art in Flow Visualization: Dense and Texture-Based Techniques. Proc. Computer Graphics Forum 23(2), 203–221 (2004)
21. Tang, Y., Shi, X., Xiao, T., Fan, J.: An improved image analogy method based on adaptive CUDA-accelerated neighborhood matching framework. Vis. Comput. 28, 743–753 (2012)
22. Van Wijk, J.J.: Spot noise-Texture Synthesis for Data Visualization. Computer Graphics (Proceedings of ACM SIGGRAPH 1991) 25, 309–318 (1991)
23. Fisher, M., Schroder, P., Desbrun, M., Hoppe, H.: Design of tangent vector fields. ACM Transactions on Graphics 26(3), 56:1–56:9 (2007)

A Message Passing Graph Match Algorithm Based on a Generative Graphical Model

Gang Shen and Wei Li

School of Software Engineering, Huazhong University of Science and Technology
Wuhan 430074, China
gang_shen@mail.hust.edu.cn, liwei.silvia@gmail.com

Abstract. In This paper, we present a generative model to measure the graph similarity, assuming that an observed graph is generated from a template by a Markov random field. The potentials of this random process are characterized by two sets of parameters: the attribute expectations specified by the the template graph, and the variances that can be learned by a maximum likelihood estimator from a collection of samples. Once a sample graph is observed, a max-product loopy belief propagation algorithm is applied to approximate the most probable explanation of the template's vertices, mapped to the sample's vertices. As demonstrated by the experiments, compared with other algorithms, the proposed approach performed better for near isomorphic graphs in the typical graph alignment and information retrieval applications.

Keywords: Markov random fields, loopy belief propagation, approximate graph matching.

1 Introduction

Graph is a versatile tool to represent abstract and complex components and their relationships. It has been a universally accepted practice to model objects of interest as graphs in many scientific and engineering fields. Specifically, we consider the application of graphs in the area of the image understanding and pattern recognition: usually images are modeled as graphs consisting of key points depicted by various features as a prerequisite step for further processing. Typical applications of this approach can be found in object recognition, information retrieval, and image alignment, etc([1],[2]).

In the above mentioned applications, it is essential to explore the similarity of two graphs. As a structured object, a graph cannot be simplified as a data point in a metric space, therefore the Euclidean distance based similarity measures and algorithms are no longer applicable. One way to measure the similarity of two graphs is to match these graphs in the first place. As a fundamental problem in computer science, finding the exact match of two graphs is NP hard. Except for a few special classes of graphs [3], people instead look for the scalable approximate solution for large graphs in more efficiently ways. A literature survey on early work in graph match can be found in [2]. As of today, many new techniques have

R. Huang et al. (Eds.): AMT 2012, LNCS 7669, pp. 268–277, 2012.

been introduced to solve the match problem approximately and efficiently with the help of different heuristic([4][6]), approximation([5][10]), or relaxation of the original match problem([9]).

In this paper, we propose a generative model embedding both attribute similarity and structure compatibility of vertices and edges in the generation process. No matter the graphs are directed or undirected, we use a probabilistic graphical model (a Markov random field based on the template graph, Section 2) and the associated parameters to represent the generation process from one graph to the other. Then, for arbitrary graphs, we apply a loopy belief propagation (LBP) algorithm to find the most probable assignment for vertices (Section 3). The intuition is that the maximum probability to generate an observed graph from a template graph gives a measure of similarity between the two graphs, and the most probable assignment represents a match between the two sets of vertices. We designed two experiments to test the effectiveness of the proposed method in Section 4. For near isomorphic graphs (the hotel and house in CMU images), the proposed method outperformed all other non-learning algorithms([13], [14], [15], [16]). When applied to information retrieval (COIL 100 images), the proposed method delivered better performance than the baseline.

As reported, when LBP converges, it gives a good approximation of the real inference[7], and has been applied in many applications including maximum weighted matching problem of graphs[8]. In the experiments of this paper, we found that when the noise level is contained, the proposed method converged to the exact solution with a only few iterations.

2 Markov Random Field Model

We consider the attributed graphs, i.e. the graphs with well-defined vertex attributes and edge weights. Let a graph G be denoted as $G = < V, E >$, where V is a finite set of vertices (nodes), and E is a finite set of edges defined on $V \times V$. For directed attributed graphs (for simplicity, we assume there is no self loop on any vertex), a vertex is associated with an n-dimensional (each dimension may be either numerical or categorical) vector a describing that the attributes of that node, and an edge with a start vertex and an end vertex may also be mapped into an m-dimensional weight vector w. We first remove the direction of the edges to get an undirected version of G (directions may be processed using parameters in the later stage). Now let the vertices of a template G_1 represent a set of random variables and satisfy the Markov property, thus forming a Markov random field. Suppose any arbitrary graph in the collection is created by that template G_1 following a multivariate Gaussian process. When a sample graph $G_2 = < V_2, E_2 >$ is observed, we first estimate a set of potentials for G_1 to generate G_2. Then we try to infer the most probable explanation for vertices in G_1. Finally, the similarity $S(G_1, G_2)$ is taken as a measure of how good the most probable explanation reserves the structure of G_1.

In the usual setup of graph matching, one seeks the solution to the following optimization problem

$$\max[\sum_{i \in V_1} a_i(x_i)\delta_i(x_i) + \sum_{(i,j) \in E_1} w_{ij}(x_i, x_j)\delta_{ij}(x_i, x_j).] \tag{1}$$

where function $a_i(x_i)$ represents the vertex compatibility score of $i \in V_1$ and $x_i \in V_2$, and $w_{ij}(x_i, x_j)$ denotes the edge compatibility score of $(i,j) \in E_1$ and $(x_i, x_j) \in E_2$ respectively, while function $\delta_i(x_i)$ is a indicator taking 1 when node i in graph G_1 is assigned to x_i in G_2 and taking 0 otherwise, and the indicator $\delta_{ij}(x_i, x_j)$ associates an edge in G_1 to an edge in G_2,

$$\delta_{ij}(x_i, x_j) = \begin{cases} 1, & i = x_i, j = x_j, \text{and} (i,j) \in E_1, (x_i, x_j) \in E_2 \\ 0, & \text{otherwise} \end{cases}.$$

Let the samples be generated by the MRF based on the template G_1 following a Gaussian process $\mathcal{N}(\mu, \theta)$. The unary potential is defined as

$$\phi_i(x_i) = \frac{1}{\sqrt{2\pi}\sigma_1} e^{-\frac{(a(x_i)-a(i))' \Lambda_1 (a(x_i)-a(i))}{2\sigma_1^2}},$$

and the binary potential is

$$\psi_{ij}(x_i, x_j) = \frac{1}{\sqrt{2\pi}\sigma_2} e^{-\frac{(w(x_i,x_j)-w(i,j))' \Lambda_2 (w(x_i,x_j)-w(i,j))}{2\sigma_2^2}} \delta_{i,j}(x_i, x_j),$$

where the positive definite matrices Λ_1 and Λ_2 represent a way to mix the different metrics or labels in different dimensions of vertex attributes and edge weights. By adding this indicator function $\delta_{i,j}(x_i, x_j)$, the directional information of the edges in a directed graph is embedded into the edge potential.

It is worth noticing that the attributes and weights used in the potential definitions should vary in the particular applications, so as to better pack the local structural characteristics as well as attributes into the generative model parameters. For example, in our experiments concerning with the CMU images (see Section 4), the 60 dimensional shape context features are used as vertex index vector. After the vertices are obtained by Delaunay triangulation, the edge is measured by the length in terms of pixels (Euclidean distance of two vertices), and this length is taken as the edge weight.

Once a set of samples is observed, we are able to estimate the variances σ_1^* and σ_2^* using the maximum likelihood estimator,

$$\sigma_1^* = \max_{\sigma_1} \prod_{G_2} \prod_{i \in V_1, x_i \in V_2} \phi_i(x_i|\sigma_1),$$

$$\sigma_2^* = \max_{\sigma_2} \prod_{G_2} \prod_{(i,j) \in E_1, (x_i,x_j) \in E_2} \psi_i(x_i, x_j|\sigma_2).$$

Consequently, the joint probability of the random variables on the MRF based on G_1 can be written as

$$P(G_1|\theta) = \prod_{i \in V_1} \phi_i(x_i) \prod_{(i,j) \in E_1} \psi_{ij}(x_i, x_j) \propto \exp[-\sum_{i \in V_1} \frac{(a(x_i) - a(i))' \Lambda_1 (a(x_i) - a(i))}{2\sigma_1^2}$$

$$-\sum_{(i,j) \in E_1} \frac{(w(x_i, x_j) - w(i,j))' \Lambda_2 (w(x_i, x_j) - w(i,j))}{2\sigma_2^2}]. \tag{2}$$

When a sample G_2 is drawn and observed, we may find $\max P(G_1|G_2, \theta)$ by optimizing (2) with the constraints that all vertices in G_1 have to take values in V_2 (non-convex constraints). In case $a_i(x_i) = \phi_i(x_i)$ and $w_{ij}(x_i, x_j) = \psi_{ij}(x_i, x_j)$, we can have the main result of this section expressed as the following theorem.

Theorem: the solution to $\max P(G_1|G_2, \theta)$ is identical to the solution of (1) when the following conditions hold
1) G_1 and G_2 are isomorphic, and
2) if $i \neq j$ then $a(i) \neq a(j)$, or
3) $w(i, j) = w(k, l)$ only if $i = k$ and $j = l$.

The proof of the above theorem relies on the uniqueness of the solution. When there is no guarantee of the unique optimization solution or the two graphs of interest are not but close to isomorphic, we formulate the graph match problem as the maximization of $P(G_1|G_2, \theta)$. After we get the solution to $\max P(G_1|G_2, \theta)$, we define the matching compatibility of a pair of edges in G_1 and G_2 as

$$m_{ij}(x_i^*, x_j^*) = \frac{1}{1 + (w(x_i^*, x_j^*) - w(i,j))' \Lambda_2 (w(x_i^*, x_j^*) - w(i,j))}.$$

However, when $|E_1|$ and $|E_2|$ are small, $m_{ij}(x_i^*, x_j^*)$ is a rather coarse measure for edges. Similarly, the vertex matching compatibility is given below,

$$m_i(x_i^*) = \frac{1}{1 + (a(x_i^*) - a(i))' \Lambda_1 (a(x_i^*) - a(i))}.$$

Finally, we are able to define the graph similarity of G_1 and G_2 as follows

$$S(G_1, G_2) = \frac{\sum_i m_i(x_i^*)}{|V_1|}.$$

Apparently, $m_{ij}(x_i^*, x_j^*) \in (0, 1]$, and $m_i(x_i^*) \in (0, 1]$. When $G_2 = G_1$ and the optimal solution is a perfect match, it follows that $S(G_1, G_2) = 1$.

3   Loopy Belief Propagation Algorithm

As we know, the exact solution of $\max P(G_1|G_2, \theta)$ is also a NP inference problem and we need to approximate the solution in an efficient way. Loopy belief propagation can be used to approximately infer the marginal probability (sum-product) or most probable explanation (max product) of vertices in a probabilistic graphical model with loops. We apply the max product belief propagation to the proposed MRF for the approximate solution of $\max P(G_1|G_2, \theta)$. Denote b_i the belief vector for vertex i, and $M_{i \rightarrow j}$ the message i passes to a neighboring vertex $j \in N_i$, where N_i is the set of i's neighbors in G_1, and let ϕ_i and ψ_{ij} be the potentials represent vertex compatibility and edge compatibility as defined in the previous section, then we have

$$b_i(x_i) \propto \phi_i(x_i) \prod_{j \in N_i} M_{j \rightarrow i}(x_i)$$

$$M_{i \rightarrow j}(x_j) = \max_{x_i}[\phi_i(x_i)\psi_{ij}(x_i, x_j) \prod_{k \in N_i/j} M_{k \rightarrow i}(x_i)].$$

When the size of N_i is big, the message value tends to be close to zero after iterations, leading to the problems of overflowing or accumulated round-up errors. In implementation, we instead use a modified logarithm message passing scheme. This is analogous to a repeated voting process, assuming similar vertices have similar neighbors. Let

$$b_i(x_i) = \phi_i'(x_i) + \sum_{j \in N_i} M_{j \rightarrow i}(x_i) + \eta,$$

$$M_{i \rightarrow j}(x_j) = \max_{x_i}[\phi_i'(x_i) + \psi_{ij}'(x_i, x_j) + \sum_{k \in N_i/j} M_{k \rightarrow i}(x_i)],$$

where η is a normalizer used to prevent overflow, and the logarithm potentials are $\phi_i'(x_i) = \log(\phi_i(x_i))$ and $\psi_{ij}'(x_i, x_j) = \log(\psi_{ij}(x_i, x_j))$.

After the messages get converged (in most cases they do) or have reached a preset threshold for repetitions, we get the final believes which can further be used in linear assignment algorithms (e.g. Hungarian algorithm) to find the vertex assignment. Or, we may simply compute $x_i^* = \arg\max_{x_i} b_i(x_i)$. In our implementation, a sequential message passing is taken for its simplicity,

$$b_i(x_i) = \phi_i'(x_i) + \sum_{j \in N_i} M_{j \rightarrow i}(x_i),$$

$$M_{i \rightarrow j}(x_j) = \max_{x_i}[b_i(x_i) + \psi_{ij}'(x_i, x_j) - M_{j \rightarrow i}(x_i)].$$

Then $b_i(x_i)$ is normalized by $b_i(x_i) \leftarrow b_i(x_i) + \eta$ before it is used for the continuing calculation on other vertices. We note that when overflow is present, messages should also be normalized. In order to avoid saving the edge potentials in a storage of size $O(|V_1||V_2|)$, we only store the vertex potentials and calculate $\psi_{ij}(x_i, x_j)$ on the fly. To test the convergence of the believes, we set a threshold for the error: when b_i is within the ϵ range of its last value b_i', it is considered converged. After all vertices have converged believes, the message passing stops.

Let $m = |V_1|$ and $n = |V_2|$, d_m and d_n be the average numbers of neighboring vertices in G_1 and G_2, C be the number of iterations in the proposed algorithm. It is obvious that the time complexity of the belief propagation algorithms mainly sits in the message calculation. Usually in a single iteration of the belief propagation, one needs to know $d_m \times n$ messages for each of the m vertices. In the proposed algorithm, the calculation of each message $M_{i \to j}(x_j)$ only takes $O(d_n)$ operations, and thus reduces the complexity by $O(d_m)$. In our experiments, this proposed algorithm converges from several to a couple of dozens of iterations, and by constraining C for early stop if not convergedthe total complexity of the proposed algorithm is $O(Cd_m d_n mn)$. In the typical setting of image processing, we often work on sparse graphs (e.g., the triangulated graphs in Section 4), the real complexity is far less than $O(Cm^2 n^2)$. The space complexity of the proposed algorithm comes mainly from the storage of messages, which is $O(m^2 n)$, while can be lowered to $O(\max(m^2, n^2, d_m mn))$ using sparse storage.

4   Experiments and Evaluation

We implemented the proposed algorithm in C++ on a personal computer with a 2GHz Intel Core processor and 2G memory. Experiments were set for two applications: first, we studied the graph alignment problem using CMU images, in particular the house and hotel series (http://vasc.ri.cmu.edu/idb/html/motion/house/index.html); second, we investigated the information retrieval problem using the Columbia object image library (COIL100, http://www.cs.columbia.edu/CAVE). The house and hotel image series in the CMU dataset contain 111 and 101 frames of 3D models taken at different angles. We used the mismatch rate to evaluate our results and compared with some other algorithms. In COIL-100, each of the 100 objects has 72 color pictures taken at different angles with 5 degrees apart. In the second experiment, we took COIL-100(RAG) test set from IAM database (http://www.iam.unibe.ch/fki/databases/iam-graph-database), including 10 randomly selected pictures for every one of the 100 objects. Then we applied the Precision-Recall curves and Precision@N, two extensively used criteria to evaluate our results, and compared with the baseline.

In the experiments, we used the regularized weights and attributes, defined as follows

$$w'(x_i, x_j) = \frac{w(i,j)}{\max |w(x_i, x_j) - w(i,j)|},$$

$$a'(i) = \frac{a(i)}{\max \sqrt{(a(x_i) - a(i))'(a(x_i) - a(i))}}.$$

And we set Λ_1 and Λ_2 to identity matrices to get the potential $\phi_i(x_i)$ and $\psi_{ij}(x_i, x_j)$.

One nontrivial point to make is that when the size of pictures is small, parameters $\theta = \{\sigma_1, \sigma_2\}$ learned by MLE from G_1 and G_2 may not work well. We instead selected from a set of candidate parameters by testing them in a limited portion of the dataset until a good choice is found, then apply them in the rest of the dataset.

In our graph alignment experiments, we used separate parameters for house and hotel data. In the house dataset, we set $1/2\sigma_1^2 = 25$ and $1/2\sigma_2^2 = 200$, while for hotel, we chose $1/2\sigma_1^2 = 100$ and $1/2\sigma_2^2 = 400$. In the information retrieval experiments, for simplicity, we used the same parameters for all object classes (they are all very small represented by graphs). We did a grid search in [1,1000] for $1/2\sigma_1^2$ and $1/2\sigma_2^2$, with a step of 10, until we found a reasonably good candidate.

Fig. 1. Matching results for the frame 2 and frame 92 in the house dataset

The datasets used in graph alignment was from [13], in which each frame of the pictures had 30 manually labeled key points and the adjacency relationships were formed using Delaunay triangulation. In our experiments, we took the edge length in pixels as the weights, and adopted shape context descriptors as the vertex attributes. We used a simple basic shape context without normalization and other processing, thus lack of the rotational invariance and scale invariance in the presence of big noise, because we intended to verify the robustness of the proposed message passing algorithm. We tested the performance of the proposed approach with CMU house and hotel datasets for various baselines (distances in frames). Among the algorithms under comparison were linear assignment([15]), quadratic assignment (graduated assignment[16]), and quadratic normalization 0.0001. We also compared our approach with a spectral matching algorithm, SMAC([14]). Our approach does not require training on examples, but we also implemented the learning algorithms on top of linear and quadratic assignment. Except for SMAC, which was implemented by MatLab code provided in [14], all other implementations were in C++([13]).

Fig. 1 shows matching results of the two pictures in the CMU house datasets with a distance of 90 frames. The large variation in angles gave rise to the difficulty in matching. As we can see, the proposed algorithm outperformed the linear assignment plus learning algorithm, the best of all other ones used in comparison.

While the learning version of linear assignment algorithm made 6 mismatches (thick red lines in the upper part of Fig. 1), the proposed approach presented a perfect match (lower part). Fig. 2 shows the comparisons of different algorithms

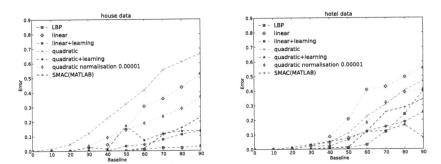

Fig. 2. Matching results for the house and hotel dataset on different baselines

using varying baselines. It can be seen how the approach proposed in this paper (denoted LBP) performed in error (mismatch) rate. In house experiments (left), LBP consistently delivered better results than others when the noise level went up, while SMAC and linear+learning had the similar error rates at relatively lower noise levels. In hotel experiments (right), if the frame distances were with 60, our approach kept low error rates. Even after the noise level went to greater than 70 frames, LBP remained the best among all non-learning algorithms.

In the information retrieval experiments, we assumed that the pictures of an object were related and belonged to a class. In the test set, there were 100 classes, each had 10 samples. We computed the similarities of these 1000 graphs to a query graph, and sorted them by the similarity scores. To evaluate the results, we took the arithmetic average of repeated tests for all query pictures to get the overall retrieval quality. Let the number of classes be C, the number of samples in a class be N, we define the retrieval depth as r, i.e. each retrieval returns the first r items with the highest scores. When we take depth r for CN such queries, the total number of retrieved items is rCN, while the data set has NCN individual samples. The precision at retrieval depth r is defined as

$$precision = \frac{|relevant(r)|}{rCN},$$

where $|relevant(r)|$ represents the number of items belonging to the same class as the query results at depth r. We define recall rate at r as

$$recall = \frac{|relevant(r)|}{CN^2}.$$

Setting r from 1 to CN we got the precision-recall pairs using several distinct parametersthe values of which are shown in Fig. 3. It can be seen that at the

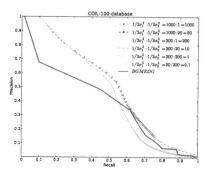

Fig. 3. Precision-recall curves for COIL-100 dataset

same recall rate, LBP had a higher precision than the benchmark BGMEDG for big $1/2\sigma_1^2 : 1/2\sigma_2^2$ ratios (≥ 10). If we took lower $1/2\sigma_1^2 : 1/2\sigma_2^2$ values, the results were not as good, indicating that the structural compatibility was not a major factor differentiating two distinct objects. The vertex attribute was the 64 dimension color histogram in a block, conveying much more information in determining the object classes. LBP would fail to classify correctly if the vertex attributes were not given sufficiently large weight.

5   Conclusions

In this paper, we extended the research in [12] in two directions: first, we use a loopy BP algorithm to approximate optimal solution of the inference problem on arbitrary directed graphs rather than use single path dynamic programming or reduce the graphs to trees; second, we presented a MRF model reflecting both attribute and structural compatibilities of vertices and edges. When applied to experiments, the proposed approach demonstrated good results at the cost of slightly high computational complexity involved in message calculation. The experiments in this paper also implied that for near isomorphic graphs, this approach worked better for graphs with more informative connection relationships (CMU datasets). As an extension, evidence (known matches) and constraining additional connections may be added to the model to derive a dependable, informative and simplified inference ([17]). However, we notice that paths and subsets of nodes are critical in graph matching. As we know, probabilistic graphical models assume the conditional independence of random variables on nodes separated by others, but in graph matching, the role of paths cannot be downplayed[11]. In the current setting, we only considered the local attributes and weights of the graph components, while the substructures surrounding the vertices and relating the edges may give more insights in determining a match. In the future research, we plan to include connectivity information into the graphical model parameters and apply learning mechanisms to better determine the weights and other arguments, in addition to improving BP algorithm's performance.

Acknowledgments. This project is supported in part by NSFC fund 61073095.

References

1. Bunke, H.: Graph Matching: Theoretical Foundations, Algorithms and Applications. In: Proceedings Vision Interface (2000)
2. Conte, D., Foggia, P., Sansone, C.: Thirty Years of Graph Matching in Pattern Recognition. International Journal of Pattern Recognition and Artificial Intelligence 18, 265–298 (2004)
3. Ullmann, J.R.: An Algorithm for Subgraph Isomorphism. Journal of ACM 23(1), 31–42 (1976)
4. Egozi, A., Keller, Y., Guterman, H.: A Probabilistic Approach to Spectral Graph Matching. IEEE Transactions on Pattern Analysis and Machine Intelligence (2012)
5. Bengoetxea, E., Larranaga, P., Bloch, I., et al.: Inexact Graph Matching by Means of Estimation of Distribution Algorithms. Pattern Recognition 35, 2867–2880 (2002)
6. Melnik, S., Garcia-Molina, H., Rahm, E.: Similarity flooding: A versatile graph matching algorithm and its application to schema matching. In: Proceedings of 18th International Conference on Data Engineering, pp. 117–128 (2002)
7. Murphy, K.P., Weiss, Y., Jordan, M.I.: Loopy belief propagation for approximate inference: an empirical study. In: Proceedings of the 15th Conference on Uncertainty in Artificial Intelligence, pp. 485–492 (1999)
8. Sanghavi, S., Malioutov, D., Willsky, A.: Linear Programming Analysis of Loopy Belief Propagation for Weighted Matching. In: Advances in Neural Information Processing Systems, pp. 1281–1288. MIT Press, Cambridge (2007)
9. Zaslavskiy, M., Bach, F., Vert, J.P.: A Path Following Algorithm for the Graph Matching Problem. IEEE Transactions on Pattern Analysis and Machine Intelligence 31(12), 2227–2242 (2009)
10. Yan, X., Zhu, F., Yu, P., Han, J.: Feature-Based Similarity Search in Graph Structures. ACM Transactions on Database Systems 31, 1418–1453 (2006)
11. Tiakas, E., Papadopoulos, A., Manolopoulos, Y.: Graph Node Clustering via Transitive Node Similarity. In: Proceedings of 14th Panhellenic Conference on Informatics, pp. 72–77 (2010)
12. Caelli, T., Caetano, T.S.: Graphical models for graph matching: Approximate models and optimal algorithms. Pattern Recognition Letters 26, 339–346 (2005)
13. Caetano, T.S., McAuley, J.J., Cheng, L., et al.: Learning Graph Matching. IEEE Transactions on Pattern Analysis and Machine Intelligence 31(6), 1048–1058 (2009)
14. Cour, T., Srinivasan, P., Shi, J.: Balanced graph matching. In: Proc. NIPS, pp. 313–320 (2006)
15. Jonker, R., Volgenant, A.: A shortest augmenting path algorithm for dense and sparse linear assignment problems. Computing 38(4), 325–340 (1987)
16. Gold, S., Rangarajan, A.: A graduated assignment algorithm for graph matching. IEEE Transactions on Pattern Analysis and Machine Intelligence 18(4), 377–388 (1996)
17. Bayati, M., Shah, D., Sharma, M.: Max-Product for Maximum Weight Matching: Convergence, Correctness, and LP Duality. IEEE Transactions on Pattern Analysis and Machine Intelligence, 1241–1251 (2008)

Fast Content-Based Retrieval from Online Photo Sharing Sites

Gerald Schaefer and David Edmundson

Department of Computer Science, Loughborough University, Loughborough, U.K.
gerald.schaefer@ieee.org, d.edmundson@lboro.ac.uk

Abstract. Literally billions of images have been uploaded to photo sharing sites since their inception, comprising a staggering wealth of visual information. However, effective tools for querying these collections are rare and keyword based. Since users rarely annotate their images, this approach is only of limited use. Content-based image retrieval (CBIR) extracts features directly from images and bases searches on these features. However, conventional CBIR approaches require a dedicated system that performs feature extraction during photo upload and a database system to store the features, and are hence not available to the average user. In this paper, we present a very fast content-based retrieval method that performs feature extraction on-the-fly during the retrieval process and thus can be employed client-side on images downloaded from photo sharing sites such as Flickr.

Our approach is based on the fact that images uploaded to Flickr are stored in a JPEG format optimised to minimise disk space and bandwidth usage. In particular, we exploit the optimised Huffman compression tables, which are stored in the JPEG headers, as image descriptors. Since, in contrast to other approaches, we thus have to read only a fraction of the image file and similarity calculation is of low complexity, our approach is extremely fast as demonstrated by the bandwidth used to retrieve images from the Flickr photo sharing site. We also show that nevertheless retrieval performance is comparable to CBIR using colour histograms which is at the core of many CBIR systems.

1 Introduction

Visual information is becoming more and more important and at a rapid rate. One of the prime examples where this can be observed is the exponential growth of images uploaded to photo sharing sites such as Flickr[1] where literally billions[2] of images are available. Although this clearly comprises a staggering amount of visual information, retrieval tools are typically rather simplistic and implemented as simple keyword based searches. Since users rarely annotate images [1], searches thus often do not return useful images.

---

[1] http://www.flickr.com
[2] http://blog.flickr.net/en/2011/08/04/6000000000/

R. Huang et al. (Eds.): AMT 2012, LNCS 7669, pp. 278–287, 2012.
© Springer-Verlag Berlin Heidelberg 2012

Content-based image retrieval (CBIR) techniques [2,3,4,5,6] extract image features such as colour or texture features directly from the image data and can hence be employed also when no textual annotations are available. However, conventional CBIR requires the features to be extracted during image upload and thus a dedicated system as well as a database to store the feature data. As such, this approach would not be available to users who want to perform content-based queries from a site such as Flickr.

While a client-side solution is in principle possible, this would require all image data to be downloaded during the retrieval process, then the features to be calculated and finally retrieval based on these features to be performed. Unfortunately, in particular the first step is prohibitive in terms of the time it requires, even with today's bandwidth provisions.

In this paper, we therefore present a very fast method to perform content-based retrieval of images stored on the Flickr photo sharing site. Since our method relies on data contained in the header of JPEG images, only a fraction of the whole image file needs to be downloaded during the retrieval process, leading to a significant speedup in terms of retrieval time. Experimental results confirm that while this enables interactive online retrieval for a database uploaded to Flickr, retrieval accuracy is comparable to classical image retrieval using colour histograms and colour based retrieval in the JPEG compressed domain.

The remainder of the paper is organised as follows. Section 2 briefly introduces CBIR concepts, while Section 3 describes the JPEG image compression algorithm. Section 4 then describes our contribution which allows for very fast JPEG image retrieval from online photo sharing sites. Experimental results are given in Section 5, and Section 6 concludes the paper.

2 Content-Based Image Retrieval

Colour was the first type of feature exploited for CBIR. Swain and Ballard [7] introduced the use of colour histograms, which record the frequencies of colours in the image, to describe images in order to perform object recognition and image retrieval. As similarity measure they introduced histogram intersection which quantifies the overlap between two histograms and can be shown to be equivalent to an L_1 norm. Starting from early CBIR systems such as QBIC [8] or Virage [9], colour histograms or related features have been at the core of many CBIR implementations [2]. Other kinds of features that are commonly used for CBIR include spatial colour, texture and shape features for which various types of algorithms have been suggested in the literature [2,3].

Almost all CBIR techniques operate in the pixel domain; that is they are calculated from pixel values in the images. On the other hand, virtually all images are stored in compressed form, most commonly in JPEG format, to reduce storage and bandwidth requirements. Consequently, for feature calculation the images need to be fully decompressed to first arrive at pixel data leading to a computational overhead during feature generation.

Faster feature extraction is possible using compressed domain CBIR algorithms [10,11] which operate directly on compressed image data and hence require only partial decoding. However, these methods are still not sufficiently fast for online image retrieval, e.g. retrieval from highly dynamic databases from the web such as photo sharing sites, or without access to internal feature databases. Here, pre-calculated features are not available (to the query process), and therefore must be calculated client-side "on-the-fly" during retrieval. As this must be conducted for every image in the dataset, the time taken is significant as the complete image file needs to be read in and partial decompression conducted.

3   JPEG Image Compression

JPEG [12] is not only the most popular image compression technique[3], it has also been adopted as an ISO standard for still picture coding. JPEG compression is based on the discrete cosine transform (DCT), a derivative of the discrete Fourier transform. First, an (RGB) image is usually converted into the YCbCr space. The reason for this is that the human visual system is less sensitive to changes in the chrominance (Cb and Cr) channels than in the luminance (Y) channel. Consequently, the chrominance channels can be downsampled by a factor of 2 without significantly reducing image quality, resulting in a full resolution Y and downsampled Cb and Cr components.

The image is then divided (each colour channel separately) into 8×8 pixel sub-blocks and DCT is applied to each such block. The 2-d DCT for an 8×8 block $f_{xy}, x, y = 0 \ldots 7$ is defined as

$$F_{uv} = \frac{C_u C_v}{4} \sum_{x=0}^{7} \sum_{y=0}^{7} f_{xy} \cos \left(\frac{(2x+1)u\pi}{16} \right) \cos \left(\frac{(2y+1)v\pi}{16} \right) \qquad (1)$$

with $C_u, C_v = 1/\sqrt{2}$ for $u, v = 0$, $C_u, C_v = 1$ otherwise.

DCT has energy compactification close to optimal for most images which means that most of the information is stored in a few, low-frequency, coefficients (due to the nature of images which tend to change slowly over image regions). Of the 64 coefficients, the one with zero frequency (i.e., F_{00}) is termed "DC coefficient" and the other 63 "AC coefficients". The DC term describes the mean of the image block, while the AC coefficients account for the higher frequencies. As the lower frequencies are more important for the image content, higher frequencies can be neglected which is performed through a (lossy) quantisation step that crudely quantises higher frequencies while preserving lower frequencies more accurately.

The AC and DC components of the image are stored in separate streams for each colour channel. While the AC coefficients are zig-zag, run-length and entropy encoded, the DC stream is differentially encoded. That is, rather than storing the actual DC values, the differences between DC values are saved. As

---

[3] Up to 95% of all images on the web are JPEG images [13].

DC values range in $[-1024; 1024]$, the range of possible differences between DC components is $[-2048; 2048]$. The difference values are stored as two components: the first component, known as the "DC code", represents the size of the change in number of bits, while the second component stores the actual difference between the DC blocks.

The DC codes are then entropy encoded, and Huffman coding [14] is employed for this step. A standard Huffman table is provided by the JPEG group and is commonly used to compress images.

4 Fast JPEG Retrieval from Online Photo Sharing Sites

While normally JPEG images need to be fully decompressed to arrive at pixel data and hence enable feature calculation, it is also possible to derive image features directly in the JPEG compressed domain [15,16]. These features are based on the DCT coefficients expressed in Equation (1) and hence still require partial decoding of the file to undo the entropy, run-length and differential coding stages as well as the reversal of the quantisation step. To further reduce the computational load, it is possible to utilise only the DC stream (and hence ignore AC data), e.g. to calculate a colour histogram of the image [17,13]. For online image retrieval this approach is however still not nearly fast enough as the overhead associated with downloading the complete image files is dominant. Consequently, it appears to be impossible to perform content-based image retrieval from photo sharing sites such as Flickr, by applying CBIR methods on a set of images retrieved from there, within a reasonable time frame.

In this paper, we present a method that allows for exactly this kind of image retrieval. Our approach performs extremely fast image retrieval based solely on information contained in the header of JPEG files. Our method also utilises only the DC data of JPEG images which, since the DC term represents the average of an 8×8 block, corresponds to a downsampled version of the image [17]. That retrieval based on subsampled images can give similar results to retrieval based on full resolution pictures has already been demonstrated in [7], and we hence do not necessarily eliminate crucial information when using only the DC stream.

Now looking closer at how DC data is encoded (see Section 3), we see that rather than the DC terms themselves the differences between neighbouring DC terms (that is differences between the averages of neighbouring blocks) are stored. This data is however directly useful for content-based retrieval as has been shown in [18]. This comes from the fact that differences of neighbouring DC coefficients essentially give a description of the image gradient and hence a feature that encapsulates image variance as well as edges and uniform areas (the latter giving 0-differences between neighbouring blocks). Interestingly also, a histogram of these simple differences has been shown [18] to provide better CBIR performance than other, more sophisticated features such as the LBP operator [19] applied to DC terms as in [17].

Speedwise this approach is fast but not significantly so compared to other JPEG compressed domain algorithms [15,16] as it still requires entropy decoding and hence also reading in of the complete image files. We therefore go one step further and present an algorithm that is based solely on data available in the header of JPEG images.

Following the JPEG compression scheme, the DC differences are then entropy coded using Huffman coding [14]. While, as mentioned, standard tables (one for luminance and one for chrominance channels) defined by the JPEG group are typically used for this assignment, the Huffman tables can also be optimised to lead to increased compression in an image adaptive way. For this optimisation, a frequency table is built for each DC code and each code is then assigned a unique prefix code that assigns shorter bit strings to the most frequently occurring DC Codes (i.e. the most occurring DC differences), and longer strings to those less common. This optimisation is commonly employed by major image websites such as Flickr or Google Images[4]. For the former it is performed during photo upload and does not require any user intervention, while all JPEG images that can consequently be downloaded from Flickr contain image adapted Huffman tables.

The Huffman tables contain statistical information about DC difference occurrences in the image and hence provide an approximation of the DC difference histograms that were used in [18]. This indeed is the core of our idea and we therefore employ the optimised Huffman tables directly as image features. Since they are stored in the header of JPEG files, only a fraction of the image file needs to be downloaded for feature extraction. Our methods should hence be extremely fast for online image retrieval which we will seek to demonstrate in Section 5.

To compare the Huffman tables of two JPEG images, we use the length in bits of the prefix code assigned to each DC code directly as a feature. To calculate a distance representing dissimilarity between images, we utilise the L_1 norm between the feature vectors. Thus, for two images I_1 and I_2 with bitlength vectors f_{I_1} and f_{I_2}, the dissimilarity between the images is calculated as

$$d(I_1, I_2) = \sum_{i=1}^{12} |f_{I_1}(i) - f_{I_2}(i)|, \tag{2}$$

where i indicates the DC code of which there are at most 12 in a JPEG Huffman table. In cases where an entry does not exist in the Huffman table, the corresponding bitlength in the feature vector is set to the maximum over all other bitlengths plus 1 to indicate that the corresponding DC code appears even less frequently than all the other ones.

To incorporate both intensity and colour information, distances between luminance and chrominance DC tables are calculated and added to give a combined distance measure.

---

[4] http://images.google.com

5 Experimental Results

In our experiments we performed retrieval of images from the photo sharing site Flickr. As mentioned, all Flickr images contain optimised Huffman tables which we can hence readily exploit for our presented image retrieval algorithm. We uploaded the UCID dataset [20], a database of about 1400 images, to Flickr and wrote an application that, given a Flickr query image, performs online retrieval on the UCID images hosted on Flickr. That is, it performs feature extraction both on the query image and all database images, before calculating distances between query and model images and presenting the images to the user sorted by visual similarity. Our Flickr Browser application is available at http://www-staff.lboro.ac.uk/~cogs/software/flickr-browser/.

For comparison, we also integrated two other methods into our application, one from the pixel domain and one from the compressed domain. For the former we chose colour indexing [7] based on RGB colour histogram due to its popularity and wide spread in CBIR systems. As compressed domain method, we use Jiang et al.'s direct content access algorithm [13] where the authors exploited the fact that the average colour of an 8×8 block can be obtained directly from the DC component of the block. While they also showed that it is possible to extract the average value for 4×4 blocks through simple operators involving the first 3 AC components, they found this to be less effective, and we therefore also use DC based colour histograms calculated in YCbCr colour space (i.e. the colour space JPEG images are usually encoded in).

In Table 1 we give the average bandwidth required (as averages of more than 200 separate queries) for all implemented methods. We chose the bandwidth to give a measure that is independent of network load and connection speed; clearly low bandwidth indicates high retrieval speed while high bandwidth requirements equate to low speed retrieval processes. Since Flickr stores images at multiple resolutions, we give results for *large* images which are 500 pixels along the larger dimension and are hence of similar size to the original uploaded images as well as for *small* images which are thumbnails with a maximal dimension of 100 pixels.

Looking at Table 1, we can see that colour indexing based on *large* Flickr images is clearly prohibitive in terms of required bandwidth as it results in a download of almost 150 MB of image data. Using the *small* images significantly reduces the bandwidth requirement to under 6 MB, however it is also known

Table 1. Average bandwidth requirements of an online query for all CBIR methods

	bandwidth [KB]
colour indexing [7] (large)	146,012
colour indexing [7] (small)	5,751
Jiang *et al.* [13] (large)	146,012
Jiang *et al.* [13] (small)	5,751
Huffman table (large)	1,398

Fig. 1. Sample query on the UCID dataset based on colour histograms (left) and our Huffman method (right). The query image is the image on the top left (and hence also the first ranked retrieval) while the top 20 retrieved images for both methods are shown.

that retrieval based on thumbnails typically does not match that based of the original images.

While Jiang *et al.*'s method operates in the compressed domain of JPEG and calculates colour histograms based on DCT data, we can see that this does not affect the amount of data that needs to be downloaded since the complete image file needs to be read to arrive at the coefficient data. While the feature calculation itself is about 7 times faster than that for colour indexing [15] this speedup is negligible in comparison with the time required for downloading the images.

Finally, we look at the results of our presented method. Since we require only the JPEG headers, the bandwidth requirements are much lower even for the *large* images for which we report the results here. With less than 1.5 MB of data downloaded this corresponds to a reduction of more than 3 orders of magnitude compared to the other methods, and still leading to a more than 4-fold speedup when comparing it to retrieval of the *small* images for the other algorithms. Furthermore, as we utilise the Huffman tables directly as image features, feature calculation itself is not necessary which results in a further speedup compared to other approaches.

That we do not compromise in terms of retrieval accuracy is demonstrated in Figures 1 to 3. Figure 1 gives a retrieval example and comparison between colour indexing (based on *large* images) and our method. While it is clear that the bandwidth required for our algorithm is significantly lower (our application

Fig. 2. Retrieval results of the same query as in Figure 1 using Jiang *et al.*'s method and our approach

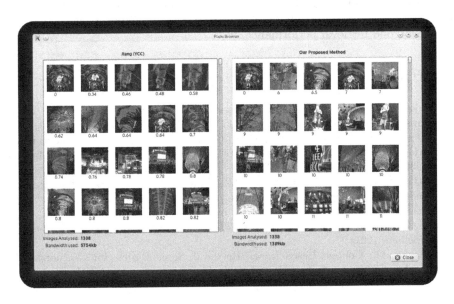

Fig. 3. Another retrieval example using Jiang *et al.*'s method and our Huffman table algorithm

actually performs retrieval for both methods concurrently which results in the grid for our method to be "filled" much faster than for other methods), we still get good retrieval results with both approaches retrieving two matches in the top 20 ranked images.

Figure 2 shows results for the same query using Jiang *et al.*'s method (on *small* images) and our approach. Clearly, for the latter the results are the same as in Figure 1[5], while for the compressed domain colour histogram algorithm none of the top 20 images contain the building of the query image.

Another retrieval example is given in Figure 3, again for Jiang *et al.*'s and our algorithm. Clearly, for the chosen query image a colour-based approach works very well while our approach, based on differences between neighbouring blocks lacks somewhat behind yet still retrieves some relevant images.

6   Conclusions

Using current algorithms, content-based image retrieval in an online fashion from photo sharing sites without access to internal feature databases and hence requiring client-side feature extraction and comparison does not allow for interactive retrieval times since the complete files of all images need to be downloaded during the query process. In this paper, we have presented a fast method for retrieving images from Flickr based on information contained solely in the header of JPEG files. In particular, we exploit image adaptive Huffman tables, which are generated during photo upload to Flickr, as image features and hence require only a fraction of the files to be downloaded during retrieval resulting, as demonstrated, in a significant reduction of bandwidth requirements while still providing good retrieval performance.

References

1. Rodden, K.: Evaluating Similarity-Based Visualisations as Interfaces for Image Browsing. PhD thesis, University of Cambridge Computer Laboratory (2001)
2. Smeulders, A., Worring, M., Santini, S., Gupta, A., Jain, R.: Content-based image retrieval at the end of the early years. IEEE Trans. Pattern Analysis and Machine Intelligence 22, 1249–1380 (2000)
3. Datta, R., Joshi, D., Li, J., Wang, J.Z.: Image retrieval: Ideas, influences, and trends of the new age. ACM Computing Surveys 40, 1–60 (2008)
4. Schaefer, G.: Mining Image Databases by Content. In: Fernandes, A.A.A., Gray, A.J.G., Belhajjame, K. (eds.) BNCOD 2011. LNCS, vol. 7051, pp. 66–67. Springer, Heidelberg (2011)
5. Schaefer, G.: Content-Based Image Retrieval: Some Basics. In: Czachórski, T., Kozielski, S., Stańczyk, U. (eds.) Man-Machine Interactions 2. AISC, vol. 103, pp. 21–29. Springer, Heidelberg (2011)

---

[5] Some equi-distant images appear in a slightly different order. The actual bandwidth results vary to a certain degree due to different sizes of the query images as well as due to other concurrent applications and difficulties in measuring the "true" bandwidth.

6. Schaefer, G.: Content-Based Image Retrieval: Advanced Topics. In: Czachórski, T., Kozielski, S., Stańczyk, U. (eds.) Man-Machine Interactions 2. AISC, vol. 103, pp. 31–37. Springer, Heidelberg (2011)

7. Swain, M., Ballard, D.: Color indexing. Int. Journal of Computer Vision 7, 11–32 (1991)

8. Flickner, M., Sawhney, H., Niblack, W., Ashley, J., Huang, Q., Dom, B., Gorkani, M., Hafner, J., Lee, D., Petkovic, D., Steele, D., Yanker, P.: Query by image and video content: The QBIC system. IEEE Computer 28, 23–32 (1995)

9. Bach, J., Fuller, C., Gupta, A., Hampapur, A., Horowitz, B., Humphrey, R., Jain, R.: The Virage image search engine: An open framework for image management. In: Storage and Retrieval for Image and Video Databases. Proceedings of SPIE, vol. 2670, pp. 76–87 (1996)

10. Mandal, M., Idris, F., Panchanathan, S.: A critical evaluation of image and video indexing techniques in the compressed domain. Image and Vision Computing 17, 513–529 (1999)

11. Schaefer, G.: Content-based retrieval of compressed images. In: International Workshop on Databases, Texts, Specifications and Objects, pp. 175–185 (2010)

12. Wallace, G.: The JPEG still picture compression standard. Communications of the ACM 34, 30–44 (1991)

13. Jiang, J., Armstrong, A., Feng, G.: Direct content access and extraction from JPEG compressed images. Pattern Recognition 35, 1511–2519 (2002)

14. Huffman, D.: A method for the construction of minimum redundancy codes. Proceedings of the Institute of Radio Engineers 40, 1098–1101 (1952)

15. Edmundson, D., Schaefer, G.: Performance comparison of JPEG compressed domain image retrieval techniques. In: IEEE Int. Conference on Signal Processing, Communications and Computing (2012)

16. Edmundson, D., Schaefer, G.: An overview and evaluation of JPEG compressed domain retrieval techniques. In: 54th International Symposium ELMAR (2012)

17. Schaefer, G.: JPEG image retrieval by simple operators. In: 2nd International Workshop on Content-Based Multimedia Indexing, pp. 207–214 (2001)

18. Schaefer, G., Edmundson, D.: DC stream based JPEG compressed domain image retrieval. In: 8th Int. Conference on Active Media Technology (2012)

19. Ojala, T., Pietikäinen, M., Harwood, D.: A comparative study for texture measures with classification based on feature distributions. Pattern Recognition 29, 51–59 (1996)

20. Schaefer, G., Stich, M.: UCID - An Uncompressed Colour Image Database. In: Storage and Retrieval Methods and Applications for Multimedia. Proceedings of SPIE, vol. 5307, pp. 472–480 (2004)

Interactive Exploration of Image Collections on Mobile Devices

Gerald Schaefer[1], Matthew Tallyn[1], Daniel Felton[1], David Edmundson[1], and William Plant[2]

[1] Department of Computer Science, Loughborough University, Loughborough, U.K.
gerald.schaefer@ieee.org
[2] School of Engineering and Applied Science, Aston University, Birmingham, U.K.

Abstract. Image collections are ever growing and hence visual information is becoming more and more important. Moreover, the classical paradigm of taking pictures has changed, first with the spread of digital cameras and, more recently, with mobile devices equipped with integrated cameras. Clearly, these image repositories need to be managed, and tools for effectively and efficiently searching image databases are highly sought after, especially on mobile devices where more and more images are being stored. In this paper, we present an image browsing system for interactive exploration of image collections on mobile devices. Images are arranged so that visually similar images are grouped together while large image repositories become accessible through a hierarchical, browsable tree structure, arranged on a hexagonal lattice. The developed system provides an intuitive and fast interface for navigating through image databases using a variety of touch gestures.

1 Introduction

Due to advances in camera technology and the associated drop of equipment costs, many people nowadays own personal image collections of hundreds to thousands of images. Moreover, a significant proportion of these images is taken with mobile devices rather that bespoke cameras. Obviously these collections need to be managed so that the user is able to retrieve certain images at ease. However, as users typically refrain from annotating their image collections with e.g. keywords or descriptions [1], to automate this is a non-trivial task.

Common tools display images in a one-dimensional linear format where only a limited number of thumbnail images are visible on screen at any one time, thus requiring the user to search back and forth through thumbnail pages to view all images. Obviously, this constitutes a time consuming, impractical and exhaustive way of searching images, especially in larger collections. Furthermore, the order in which the pictures are displayed is based on attributes like file names that often do not reflect the actual image contents and hence cannot be used to speed up the search.

R. Huang et al. (Eds.): AMT 2012, LNCS 7669, pp. 288–296, 2012.

Recently, several approaches have been introduced which provide a more intuitive interface to browsing and navigating through image collections [2,3]. In this paper, we introduce such an image browsing system that allows for interactive exploration of image repositories on mobile devices such as smartphones. Since nowadays a large percentage of images are captured using and stored on such devices, the presented approach fills a highly needed gap of managing these collections. Our approach presents an image database in such a way that visually similar images (images of similar colour) are grouped together while large image collections are managed on a browsable hierarchical tree structure, arranged on a hexagonal lattice. User interaction is performed through various touch gestures leading to an intuitive way of exploring image datasets. An initial user study confirms the usefulness of the developed approach.

2 Image Browsing

Image database navigation systems have been shown to provide an interesting and useful alternative to image retrieval systems [2,3]. The idea here is to provide a visualisation of a complete image collection together with browsing tools for an interactive exploration of the database.

Visualisation methods for image repositories can be grouped into three main categories: mapping-based, clustering-based, and graph-based approaches [3,4]. Mapping-based techniques employ dimensionality reduction techniques to map high-dimensional image feature vectors to a low-dimensional space for visualisation. Typical examples examples use principal component analysis (PCA) [5,6], multi-dimensional scaling (MDS) [7], or non-linear embedding techniques [8] to define a visualisation space onto which to place images. Clustering-based visualisations group visually similar images together, often in a hierarchical manner. Among the systems that employ such an approach are [9,10,11,12,13,14]. In graph-based navigation systems, images are the nodes of a graph structure, while the edges of the graph show relationships between images (e.g., joint keywords or visual similarity). Graph-based systems include [15,16,17,18].

Once a database has been visualised, it should then be possible to browse through the collection in an interactive, intuitive and efficient way [3,19]. We can distinguish between horizontal browsing which works on images of the same visualisation plane, and includes operations such as panning, zooming, magnification and scaling, and vertical browsing which allows navigation to a different level of a hierarchically organised (often clustering-based) visualisation.

3 Mobile Image Browsing

In this paper, we present an effective and efficient tool for exploring and querying large image collections on mobile devices. Our system provides users with a visualisation of images generated solely from the image pixel data, overcoming the reliance on metadata. Image thumbnails are displayed on a hexagonal lattice, based on mutual colour similarity of the images. Users can effortlessly navigate through the visualisation using a variety of browsing operations to effectively explore the image collection. Large image collections are visualised in a hierarchical manner so that images are accessed in different layers, while usage of the visualisation space is maximised through the application of image spreading algorithms.

Arranging visually similar images close together within a visualisation space has been shown to support more efficient image retrieval when compared to retrieval from a random arrangement of images [20]. Similarly, it has been shown that image overlap, often occurring in mapping-based visualisations, leads to user confusion and that a regular arrangement of images supports faster navigation [20].

We follow these principles and arrange images according to their mutual colour content similarity. More precisely, we utilise the median colour in HSV space (which better correlates with human perception than RGB) of an image as an image descriptor, from which a position in the visualisation space is derived using the value V (describing brightness) and hue H descriptors. Hue is the perceptual attribute usually associated with 'colour' and is defined in HSV as [21]

$$H = \cos^{-1} \frac{0.5[(R-G)+(R-B)]}{\sqrt{(R-G)(R-G)+(R-B)(G-B)}} \tag{1}$$

where R, G, and B are red, green and blue pixel values. Hue constitutes an angular attribute, going from red to yellow to green to blue back to red. Each image is thus characterised by its H and V values, which in turn directly define the co-ordinates of the image within the (2-dimensional) visualisation space by mapping H to the x- and V to the y-axis. While the features employed might at first seem simple, average colour descriptors have been shown to work at least as well as higher-dimensional (and computationally more complex) image descriptors, such as colour distributions (histograms), for image database browsing [20].

Following [20], we also organise images on a regular lattice with no image overlap; however in contrast to there we do not employ a grid structure but rather follow [22] and utilise a hexagonal lattice onto which the images are placed. This has the advantage that, when the images are organised in a space-filling arrangement, each row and column of images is visually displaced from its neighbouring rows/columns. This would not be possible using a regular square grid structure where larger visual gaps are needed to delineate images clearly. The space saved as a result of using a hexagonal lattice enables larger or more images to be displayed within the visualisation. In addition, on a hexagonal lattice, the six neighbours of a hexagon are equidistant from the middle cell while on a square architecture, the neighbours at the diagonal are further away than the horizontal and vertical neighbouring cells [23]. Clearly, images are not of hexagonal shape and hence need to be cropped for display. This is done by inscribing a hexagon of maximal size at the centre of the image and cropping the part outside this area.

Only a limited number of images from the database can be displayed simultaneously in any visualisation approach. However, even personal image collections on mobile devices may contain hundreds or thousands of pictures. Akin to the approach in [14], we address this problem by employing a hierarchical approach to image database visualisation and browsing. Construction of the hierarchical data structure is performed in a recursive manner. At the root layer (which in our default implementation consists of 17×14 hexagons) we determine for each image into which of the cells it falls. If a cell contains only a single image, this is the image that is displayed in the hexagon. If more than one image is mapped to a cell, the image whose hue/value co-ordinates are closest to the cell centre is displayed as a representative image for that image cluster. Then at the next level of the hierarchy, this cluster is expanded by subdividing the hexagon into a set of smaller hexagons and performing the above procedure (that is, mapping each image in the cluster to a sub-cell, and - should multiple images still fall in the same cell - descend into the next level of the hierarchy) again. The complete image data structure can thus be regarded as a hierarchical tree where intermediate nodes correspond to image clusters and terminal nodes represent individual images. Equivalently, one can envisage the image database as being visualised as layers of different resolutions, with the user being shown one of these layers and having the ability to move up or down to other layers.

While this hierarchical approach allows effective and efficient access even to large image databases, it is inevitable that only a certain amount of cells will actually be filled with images. Similarly, if two images have very similar image descriptors (i.e. have similar median hue and value co-ordinates), such as two photographs of the same scene, these images are likely to be mapped to the same cell even on lower layers of the data structure. Therefore, we employ two image spreading algorithms in order to maximise the available visualisation space and to minimise the number of generated browsing layers.

To reduce the number of empty cells and thus distribute images within the visualisation more evenly, each empty cell in a layer is located. If a given empty cell has 4 or more neighbouring hexagons that contain images, a relative percentage of these images is moved to the empty cell. The images moved from each of the occupied cells will be those with hue and value co-ordinates closest to the hue and value of the centroid of the empty cell (i.e. those images closest to the borders between the two cells) as illustrated in Figure 1. The overall effect of this approach is that more cells are filled, hence making better use of the visualisation space. We repeat this process three times to arrive at a compromise between filling empty cells and maintaining the original positions of the images within the browsing environment.

In order to prevent the creation of too many layers, we employ a second spreading strategy which is designed to distribute images that fall within the same (parent) cell. In particular, we adopt the 'place', 'bump' and 'double-bump' strategies of [20] illustrated in Figure 2 if, for a particular parent node, fewer than 25% of the cells are filled. For cells that contain multiple images, the idea is to keep the representative image in the cell while moving the other images to a nearby cell with little change to the overall configuration. To do so, we initiate a spiral scan around the cell, checking whether any of its 6 neighbouring cells is

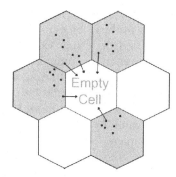

Fig. 1. Five images (represented as black dots) from the 4 neighbouring hexagons (which contain 25 images in total) will be moved to the vacant cell, so that each of the cells contains five images

empty where we can 'place' an image. Should all neighbouring cells be occupied, we move to the next ring (which contains 12 cells) looking for an empty hexagon. If one is found, the image of the closest cell in the first ring is placed there and an image from the original cluster is placed in the first ring cell ('bump'). If necessary, the same principle can be extended to the third ring (which contains 18 hexagons) by moving images from the second to the third and from the first to the second ring before placing an image ('double bump').

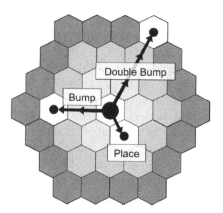

Fig. 2. A graphical representation of our 'place', 'bump' and 'double-bump' spreading strategies

While an effective visualisation paradigm is important for intuitive image database navigation, it is only through interaction with such a system that its true potential and usefulness becomes apparent. To support intuitive and efficient interaction, our system provides a variety of browsing tools and operations, all based on touch gestures of the mobile device. The user may effortlessly pan the current display through a dragging operation similar to those provided e.g. in document editing/viewing applications. Using such a pan operation, the user is able to focus on a different area in the visualisation space. Panning along the horizontal main axis will bring up images of different hues while panning along the vertical axis allows to explore brighter or darker images. As hue is a circular quantity, panning continuously along the main axis will eventually lead back to the starting point. Panning is supported not only on the root layer but on all layers of the visualisation hierarchy. In lower layers, it is possible to move from images belonging to one parent cell to those corresponding to neighbouring parent cells, thus enabling panning along the entire layer.

A continuous zoom operation is also implemented activated by a two-finger pinch gesture. This allows to zoom in on the current display, hence giving the user the possibility to inspect the images contained more closely, or to zoom out to a view where an overview of all images in the layer is possible.

While the above operations allow interaction within one particular layer of the visualisation (i.e. so-called horizontal browsing [19]), it is vertical navigation (that is, navigation from one layer to the next) that allows for intuitive management of large image collections. As detailed earlier, an image cluster is visualised through the use of a representative image. By double-tapping an image, the respective image cluster is expanded allowing the user to delve further into the browsing structure, and to bring up images that were not shown before.

4   Experimental Results

While the system as described above can of course be implemented on literally any mobile device with a touchscreen, our actual implementation is based on Apple's iOS environment[1] and our browser hence runs on Apple's iPhone and iPod Touch devices.

Figures 3 and 4 show two sample "screenshots" of our application in action. Figure 3 shows the root layer of the system running on a database of about 1400 images [24], while Figure 4 displays the application after a vertical browsing operation, i.e. after the user has selected one of the images on the previous layer to be expanded.

As has been shown [25], an objective evaluation of image browsing systems is difficult if not impossible to perform. We therefore performed an initial, user-based study to gather some feedback. For this, we asked a group of eight computer science undergraduate students to participate and installed the application on their smartphones. The users were hence able to browse through their own private image collections using our browsing systems. The initial feedback received during these sessions was very promising since all users commented that they found the developed application very useful.

In order to provide a more quantitative analysis, and on a larger database, we used the MPEG-7 Common Colour Dataset [26] which contains close to 5500

---

[1] http://developer.apple.com/devcenter/ios

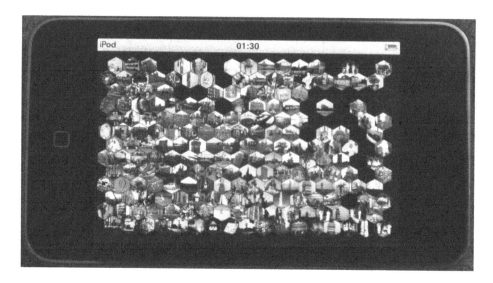

Fig. 3. UCID image database [24] visualised in the image browser on an iPod Touch

images and 10 of its predefined ground truth queries, and count the number of user interactions that are needed to reach, starting from a query image, all its corresponding model images. We compare this to the equivalent number of user interactions that is required to do the same on a common file browsers as well as using the ImageSorter browser [27] which is also based on idea of arranging images by derived content features but on a single visualisation plane.

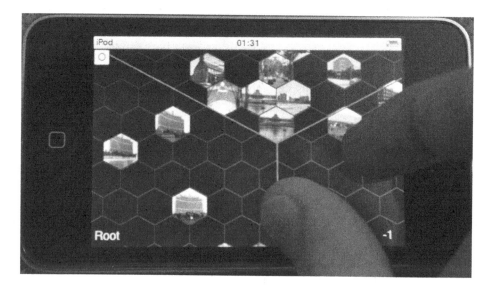

Fig. 4. Image browser after navigating into the next layer of the browsing structure

We found that for our image browser the average number of user interactions was 4.9 compared to 9.4 using ImageSorter and 36.5 using a file browser [22]. This clearly demonstrates the usefulness of our image browsing approach and also shows that employing a hierarchical approach to managing image collections leads to a more effective browsing interface.

5 Conclusions

In this paper, we have presented a content-based image browsing system for image collections stored on mobile devices such as smartphones. Images are described by colour features and arranged so that images with similar features are grouped together in the browsing display. Arrangement is performed on a regular hexagonal lattice which can be browsed through panning and zooming operations using intuitive touch gestures. Larger image collections are accessed through a hierarchical browsing structure akin to clustering where an image cluster can be opened thus revealing similar images not displayed before. An initial user study confirms that our approach provides a useful and intuitive system for managing personal image collections on mobile devices. Clearly, while this is certainly encouraging, a more formal study is necessary and is planned for future research.

References

1. Rodden, K., Wood, K.: How do people manage their digital photographs? In: SIGCHI Conference on Human Factors in Computing Systems, pp. 409–416 (2003)
2. Heesch, D.: A survey of browsing models for content based image retrieval. Multimedia Tools and Applications 40, 261–284 (2008)
3. Plant, W., Schaefer, G.: Visualisation and Browsing of Image Databases. In: Lin, W., Tao, D., Kacprzyk, J., Li, Z., Izquierdo, E., Wang, H. (eds.) Multimedia Analysis, Processing and Communications. SCI, vol. 346, pp. 3–57. Springer, Heidelberg (2011)
4. Plant, W., Schaefer, G.: Visualising image databases. In: IEEE Int. Workshop on Multimedia Signal Processing, pp. 1–6 (2009)
5. Moghaddam, B., Tian, Q., Lesh, N., Shen, C., Huang, T.: Visualization and user-modeling for browsing personal photo libraries. Int. Journal of Computer Vision 56, 109–130 (2004)
6. Keller, I., Meiers, T., Ellerbrock, T., Sikora, T.: Image browsing with PCA-assisted user-interaction. In: IEEE Workshop on Content-Based Access of Image and Video Libraries, pp. 102–108 (2001)
7. Rubner, Y., Guibas, L., Tomasi, C.: The earth mover's distance, multi-dimensional scaling, and color-based image retrieval. In: Image Understanding Workshop, pp. 661–668 (1997)
8. Nguyen, G., Worring, M.: Interactive access to large image collections using similarity-based visualization. Journal of Visual Languages and Computing 19, 203–224 (2008)
9. Krischnamachari, S., Abdel-Mottaleb, M.: Image browsing using hierarchical clustering. In: IEEE Symposium on Computers and Communications, pp. 301–307 (1999)

10. Chen, J., Bouman, C., Dalton, J.: Hierarchical browsing and search of large image databases. IEEE Trans. Image Processing 9, 442–455 (2000)
11. Bartolini, I., Ciaccia, P., Patella, M.: Adaptively browsing image databases with PIBE. Multimedia Tools and Applications 31, 269–286 (2006)
12. Gomi, A., Miyazaki, R., Itoh, T., Li, J.: CAT: A hierarchical image browser using a rectangle packing technique. In: 12th Int. Conference on Information Visualization, pp. 82–87 (2008)
13. Schaefer, G., Ruszala, S.: Image Database Navigation on a Hierarchical MDS Grid. In: Franke, K., Müller, K.-R., Nickolay, B., Schäfer, R. (eds.) DAGM 2006. LNCS, vol. 4174, pp. 304–313. Springer, Heidelberg (2006)
14. Schaefer, G.: A next generation browsing environment for large image repositories. Multimedia Tools and Applications 47, 105–120 (2010)
15. Chen, C., Gagaudakis, G., Rosin, P.: Similarity-based image browsing. In: Int. Conference on Intelligent Information Processing, pp. 206–213 (2000)
16. Heesch, D., Rüger, S.: NNk networks for content-based image retrieval. In: European Conference on Information Retrieval, pp. 253–266 (2004)
17. Dontcheva, M., Agrawala, M., Cohen, M.: Metadata visualization for image browsing. In: 18th Annual ACM Symposium on User Interface Software and Technology (2005)
18. Worring, M., de Rooij, O., van Rijn, T.: Browsing visual collections using graphs. In: Int. Workshop on Multimedia Information Retrieval, pp. 307–312 (2007)
19. Plant, W., Schaefer, G.: Navigation and browsing of image databases. In: Int. Conference on Soft Computing and Pattern Recognition, pp. 750–755 (2009)
20. Rodden, K.: Evaluating Similarity-Based Visualisations as Interfaces for Image Browsing. PhD thesis, University of Cambridge Computer Laboratory (2001)
21. Sangwine, J., Horne, R.: The Colour Image Processing Handbook. Chapman & Hall (1998)
22. Plant, W., Schaefer, G.: Image retrieval on the honeycomb image browser. In: 17th IEEE Int. Conference on Image Processing, pp. 3161–3164 (2010)
23. Sheridan, P., Hintz, T., Alexander, D.: Pseudo-invariant image transformations on a hexagonal lattice. Image and Vision Computing 18, 907–917 (2000)
24. Schaefer, G., Stich, M.: UCID - An Uncompressed Colour Image Database. In: Storage and Retrieval Methods and Applications for Multimedia 2004. Proceedings of SPIE, vol. 5307, pp. 472–480 (2004)
25. Plant, W., Schaefer, G.: Evaluation and benchmarking of image database navigation tools. In: Int. Conference on Image Processing, Computer Vision, and Pattern Recognition, vol. 1, pp. 248–254 (2009)
26. Moving Picture Experts Group: Description of core experiments for MPEG-7 color/texture descriptors. Technical Report ISO/IEC JTC1/SC29/WG11/N2929 (1999)
27. Barthel, K.: Automatic image sorting using MPEG-7 descriptors. In: Workshop on Immersive Communication and Broadcast Systems (2005)

The Derived Kernel Based Recognition Method of Vehicle Type

Zhen Chao Zhang[1], Yuan Yan Tang[1], Chang Song Liu[2], and Xiao Qing Ding[2]

[1] Department of Computer and Information Science,
Faculty of Science and Technology, University of Macau, Macau, China
bravobrava@126.com, yytang@umac.mo
[2] Department of Electronic Engineering, Tsinghua University
State Key Laboratory of Intelligent Technology and Systems Beijing, 100084, China
lcs@tsinghua.edu.cn

Abstract. This paper applies the updated derived kernel algorithm into the vehicle type recognition, which is a heated research area based on the method of pattern recognition and digital image processing because of its significant usage on exit and entry administration, traffic and vehicle control and toll collection. The method of two- layer derived kernel on neural response is involved in extracting useful features from vehicle images, for the method itself has better capacity of decreasing the negative influences from different colors and vehicle speeds, background condition interference and blur noises. Some clustering algorithms are employed on the process of templates construction, and the first nearest neighbor algorithm on pattern classification. Since our method can get rid of the disturbances from similar parts of vehicle images and make the most of the features of representative parts, the vehicle type recognition accuracy reaches above 95% as high.

Keywords: vehicle type recognition, derived kernel, templates construction, neural response, classification.

1   Introduction

Recently since the rapid development of vehicle manufacture and its wide usage on many aspects of daily life, there come quite a number of issues concerned on vehicles. As the result, more and more research topics on vehicle area have become significant and applicable, one of which is vehicle type recognition mostly applied on crimes detection, Electronic Toll Collection management, exit and entry administration, and security control of restricted areas.

There have been different methods used on the vehicle type recognition, such as Support Vector Machine, Perimeter-Based Noise Filter, extension-based recognition model [10], and Scale Invariant Feature Transform [11]. Edge detection method is widely and effectively used, in [1], Sobel operator, typically used to find the approximate absolute gradient magnitude at each point in an input grayscale image, together with edge direction and detection method is proposed. An algorithm based on vertical

R. Huang et al. (Eds.): AMT 2012, LNCS 7669, pp. 297–306, 2012.

edge detection and color alternation in grayscale image, is proposed to locate the vehicle in [2]. In [3], canny edge detector is applied for extracting feature information after the process of dilation.

Neural networks can be used in the area of vehicle type identification, because of their ability to derive patterns from complicated or imprecise data. In [4], BP network learning algorithm is improved by using momentum and genetic algorithm after analyzing the defects of the BP network learning algorithm. A new recognition approach based on Radial Basis Function Neural Networks (RBFNN) is mentioned in [5]. Transform methods, represented by Fast Fourier Transform, Discrete Curvelet Transform and Discrete Wavelet Transforms are used in [7] for image feature extraction and vehicle type recognition. In [6], it recognizes vehicle type in different lighting conditions with the application of wavelet and contourlet as tools for feature extraction, which is robust to illumination and scale variation.

As things stand now, some methods mentioned above show the difficulty in dealing with the vehicle type recognition problem with high standard of accuracy. One of the challenges on that area, as we conclude, is how to handle with the similar parts and the representative parts of the vehicle images. According to our experiment and research, the features of similar parts of vehicle images, such as some backgrounds, vehicle wheels and lamps, are easy to make confusion when recognizing; and the information of representative parts, such as vehicle color, landmark and contour is hard to use for discrimination. And the method we propose, with two layer neural response feature extraction and process, which is based on each useful small image patch from more subtle perspective, can get rid of the disturbance from the similar parts of vehicle images, and make the most of the representative parts, so that the features obtained from vehicle images can be better used for enhancing the recognition accuracy.

2   Derived Kernel Algorithm

In this section, the derived kernel algorithm as an updated feature extraction and similarity measurement method is introduced into the recognition of vehicle type. That algorithm is firstly proposed by [12], under development of [13] [14], and it is encouraged to read them for more specific details and other applications.

2.1   Preliminaries

For the whole process of vehicle type recognition, there are four ingredients needed to define for the preparation of derived kernel part listed as followed:

Nested Patches Definition
The initial definition of the derived kernel is made in the case of the nested patches definition on one vehicle image in R^2. By the definition, that vehicle image is proposed to be made up of three different layers of patches: named u, v and Sq. Sq represents the whole vehicle image domain. The left part of Fig. 1 shows the different nested patch domains on one randomly selected vehicle image, where $u \in v$, $v \in Sq$ and $Sq \in R^2$.

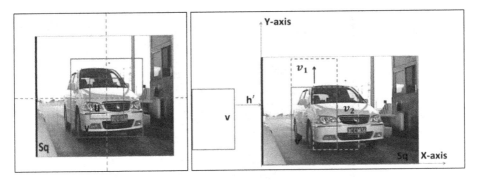

Fig. 1. Nested patches definition & a set of translation v through Sq area

Image Patches Transformation

A set of transformations of a certain vehicle image from a patch to another patch is depicted as the right part of Fig. 1. We define a finite set H_u of transformations that are maps from u to v with h: $u \rightarrow v$, and similarly H_v with h $\nearrow$: $v \rightarrow$ Sq. The transformation can be translation, scaling and rotation. In this paper, considering the actual problems, the sets of transformations are assumed to be finite and limited to only translations for simplicity. As the right part of Fig. 1, we make coordinate patch v transform through the whole image area Sq, and there should come a series of limited coordinate patches named from v_1 to v_n. And on the image we just use v_1 and v_2 as representatives.

Image Function Definition

Assume that we are given an image function on Sq, denoted by $I_{Sq} = \{f: Sq \rightarrow [0, 1]\}$ (where f is the image with size Sq), as well as the image function I_u, I_v defined on sub-patches u, v respectively. As Fig. 2, there come two image patches h'_1 and h'_2 through the Sq area.

Image Function Restriction

Besides the ingredients defined previously, there is an axiom which is used to restrict any vehicle images. And f $\circ$ h: $u \rightarrow$ [0, 1] is in I_u if f $\in I_v$ and h$\in H_u$, similarly f $\circ$ h': $v \rightarrow$ [0, 1] is in I_v if f $\in I_{Sq}$ and h' $\in H_v$. The mathematical expressions are shown as followed:

$$f \circ h : u \rightarrow [0,1] \in I_u, f \in I_v, h \in H_u \tag{1}$$

$$f \circ h' : v \rightarrow [0,1] \in I_v, f \in I_{Sq}, h' \in H_v \tag{2}$$

Fig. 2. Image function definition, where h′$_1$ and h′$_2$ correspond to two different translation (one is upward movement, another is right side movement)

2.2 Neural Response

In the following part, the neural response is shown based on the above described recursive definition which will be on a hierarchy of local kernel. The bottom-up design style is applied on the construction of derived kernel on neural response.

2.2.1 The derived kernel process begins with the normalized, non-negative, reproducing kernel $I_u \times I_u$, which is denoted by $\widehat{K_u}(f, g)$ in formula (3).The value of $\widehat{K_u}(f, g)$ is the normalized result of $K_u(f, g)$, which is inner product operation of image function $f(x)$ and $g(x)$ on patch area u.

$$\widehat{K_u}(f, g) = \frac{K_u(f,g)}{\sqrt{K_u(f,f)K_u(g,g)}} \tag{3}$$

The function f (x) and g (x) on patch area u is belong to I_u, which can be considered as two vehicle images needed to compare for computing the similarity, shown as formula (4). Since the value zero cannot be allowed in the denominator, $K_u(f, f)=0$ is excluded in order to satisfy the normalization.

$$K_u(f, g) = <f(x), g(x)> \tag{4}$$

2.2.2 For the first layer neural response on image function f (x) with one selected template t is denoted as $N_v(f, t)$, which is shown as (5) where f ∈ I_v, h∈ H_u and t ∈ T_u.

$$N_v(f, t) = \max_{h \in H} \widehat{K_u}(f \circ h, t) \tag{5}$$

As the left part of Fig. 3, it shows the whole process of one-layer neural response on image function f. For easy understanding, we use t_1' instead of t', h_1' instead of h'. And f ∘ h_1' can be considered as a certain patch of the vehicle image of size v, if f ∈ I_{Sq} is an image patch of size Sq and h_1' ∈ H_v. $\widehat{K_u}(f \circ h_1', t_1')$ is the result of normalized inner product of f ∘ h_1' and t_1', and $\widehat{K_u}(f \circ h_n', t_n')$ is the result of normalized inner product of f ∘ t_n' and t_n'. $N_v(f, t_1')$ is the best match (the maximum of values from $\widehat{K_u}(f \circ h_1', t_1')$ to $\widehat{K_u}(f \circ h_n', t_1')$) of the template t_1' in the image patch f.

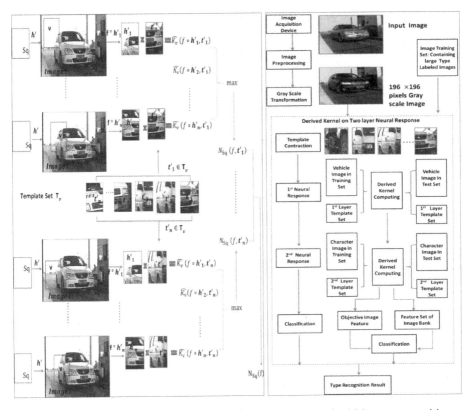

Fig. 3. The process of one layer neural response & the process of vehicle type recognition

The derived kernel of $I_v \times I_v$ is then defined as the formula (7).While f, g $\in$ I_v, can be considered as two vehicle images. The value of $\widehat{K_v}(f, g)$ is obtained by normalizing $K_v(f, g)$ according to (4).

$$N_v(f) = \{N_v(f, t_1), N_v(f, t_2) \ldots N_v(f, t_n)\} \qquad (6)$$

$$(f, g) = <N_v(f), N_v(g)> \qquad (7)$$

For example, for a certain vehicle image, such as image f, after finishing the first neural response, there comes the result set $N_v(f)$, including $N_v(f, t_1), N_v(f, t_2)$ to $N_v(f, t_n)$ shown as (6); so does another vehicle image g, through the first neural response there comes the result set $N_v(g)$.

2.2.3 The process above is repeated by defining the second layer neural response shown as (8)

$$N_{Sq}(f, t') = \max_{h' \in H_v} \widehat{K_v}(f \circ h', t') \qquad (8)$$

While in this case f$\in I_{Sq}$, $t' \in T_v$ and h' $\in H_v$. Consequently, the derived kernel on $I_{Sq} \times I_{Sq}$ is given by (10), where f, g$\in I_{Sq}$. Just as the former procedure, the final derived kernel $\widehat{K_{Sq}}$ is obtained by normalizing K_{sq}.

$$N_{Sq}(f) = \left\{N_{Sq}(f, t_1'), N_{Sq}(f, t_2') \dots N_{Sq}(f, t_n')\right\} \tag{9}$$

$$K_{Sq}(f, g) = < N_{Sq}(f), N_{Sq}(g) > \tag{10}$$

3   Vehicle Type Recognition

In the following section, the whole process of vehicle type recognition method is depicted, including image preprocessing, gray scale transformation, derive kernel on neural response, templates construction and classification.

3.1   Image Preprocessing

Since the original images we obtain are randomly sized, and it would have negative effects on image analysis and recognition process. As the result, we make the size of those entire images changed as 196×196 pixels.

3.2   Gray Scale Transformation

In this part, we transform images from original RGB type into gray scale type. And the common used formula is shown as (11). Considering 32-bit integer operation and avoiding floating point operation, we do some adjustment and apply (12) for gray scale transformation.

$$\text{Gray} = R \times 0.299 + G \times 0.587 + B \times 0.114 \tag{11}$$

$$\text{Gray} = (R \times 299 + G \times 587 + B \times 114 + 500)/1000 \tag{12}$$

3.3   Derived Kernel on Neural Response

In our research, two layers of neural response are employed. As we can see in Fig. 4, the captured smaller image patches h_1 and h_2 (which construct one layer patch set named H_u mentioned in section 2) from two different vehicle images have too many similar parts, so that the extracted information would be not enough for type identification. At the same time, with the assistance of the second layer bigger image patches h'_1 and h'_2 (constructing another layer set named H_v), the acquired feature would include more complete and useful information, including smaller parts of vehicles (such as vehicle wheels and lamps) and bigger parts (such as vehicle gate and general contour).

Specifically speaking, in our research, for each image, we would extract six bigger image patches as the first neural response, on each of which the smaller image patches as the second layer neural response would be chosen through the translation, image function and restriction as we mentioned in Section 3.

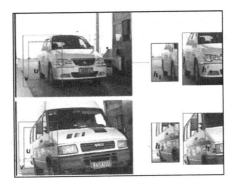

Fig. 4. The function of first layer and second layer templates comparison

3.4   Template Construction

Template construction counts important on the whole process of vehicle type recognition method based on derived kernel, which connect the mathematical model to a real world setting. In our method, we assume T_u and T_v are discrete and finite template sets as (13) and (14). The template sets $T_u \in I_v$, $T_v \in I_{sq}$ restrict the templates selection scope.

$$T_U = \{t_1, t_2, t_3, \dots t_n\} \tag{13}$$

$$T_v = \{t'_1, t'_2, t'_3, \dots t'_n\} \tag{14}$$

Templates can be seen as representative image patches frequently encountered in the daily life, and in our research, they can be a certain number of vehicle image patches frequently appeared and they are used for vehicle type matching and recognition. The construction of template sets is considered as important step, and will finally reflects on the accuracy and time efficiency.

We employ some common and standard 196 ×196 pixels vehicle images which we transform into gray-scale ones. And then from those processed images, we extract some templates with specific size. The first step is to select template sets randomly through image translation and transformation. For each vehicle type (there are three types totally, including truck, van and car), 10 smaller image patches with size u are randomly chosen as the first layer templates and 8 larger image patches with size v as the second layer templates. For higher recognition accuracy, the size of image patches should be determined by a series of testing experiments, which will be shown on the following session IV. Since there are three kinds of vehicle types, and for each kind of types, there are four corresponded images to be used for templates selection. So after random selection, the total number of templates is 120 for the first layer and 96 for the second layer. The specific process to choose those templates through translation and transformation way is depicted as Fig. 1, Fig. 2 and their explanation on section 2.

So far, the basic templates construction is finished with random selection method. For gaining the representing capacity of the template set, some clustering algorithms can be applied and the entire templates can be divided into several dozen of classes. The idea is to extract some templates which locate on the central position of a certain class and which have proper ability representing for the other templates on that same class. Specifically speaking, in our method, the K-means clustering and hierarchical clustering algorithm are employed to partitionally and hierarchically select smaller amount of   representative templates on two derived kernel layers.

The advantage of applying the clustering algorithms into template construction is as followed:

- It reduces the number of involved templates so that the time consumption on template construction and recognition process is diminished to some extent.
- Since some singularities and their disturbance can be avoided by the clustering algorithm, the representing ability of the selected templates is increased, which helps to enhance the final recognition accuracy.

3.5   First Nearest Neighbor Classification

In our research, the first-nearest neighbor classification method is involved, which requires two sets: training set, and testing set. Together with the template set used for the neural response on derived kernel, there are totally three data sets used.

The images on the training set are selected, preprocessed and transformed from the real-world vehicle images; they will be put into the specific type subset according to their own type. The function of the training set is used to help to recognize the type label of the vehicle images on the testing set by comparing and computing the similarity level of extracted features from images on the training set and testing set. For each vehicle type, we will employ 50 images, and there are three types needed to recognize, including car, van and truck. So for the training subset, there are 150 vehicle images.

The idea of the first-nearest neighbor classification is that an unlabeled testing set is given the label from the closest image in training set. Specifically speaking, the images on the training set are used to compare with the object ones about their feature similarity through hierarchical derived kernel. When a certain image in the training set, as an example, has the largest similarity level with the unlabeled object one (which goes beyond the threshold), the label of that image on the training set would be assigned to the object one as the recognized result. The test set, composed of various object vehicle images is used to test the accuracy of our research.

4   Experiment and Result

The experiment of ensuring the size of each layer templates are depicted as followed. Firstly, we just use one layer derived kernel on neural response for experiment and realize that the template size of 146×146 pixels can lead to the highest recognition accuracy (compared with other template size for one layer). And then we employ two layer derived kernel on neural response and fix the size of the first layer template as 146×146 pixels (the size of vehicle images have already been fixed as 196×196 pixels as we mentioned previously), and as the change of the size of the second layer templates, the recognition accuracy fluctuate. We determine the second layer template size, with which the related experiment result shows the highest accuracy. After the series of experiments, we can conclude that the derived kernel with 146 ×146 pixels first layer template size and 96 × 96 pixels second layer template size neural responses have the greatest recognition accuracy. That's because template with that size include more exclusive and representative feature information with less similarity disturbance.

The experiment of vehicle type recognition on the image bank, or the testing set as we mentioned previously, is presented. All the images employed for training, testing and constructing templates are randomly selected from the internet or some image bank, and they are originally RGB types with different size. Unlike other image bank just depicting the frontal part of vehicles, the images we employ describing the vehicle contour with side view can record more vehicle information (Fig. 5 shows some image examples). On the testing set, we totally employ 150 vehicle images, 50 of which depict cars, 50 vans and 50 trucks.

According to the testing experiment, the recognition accuracy on vehicle images is as high as above 95%. Besides, according to the experiment result the recognition accuracy of truck images is highest compared with the one of the others. The reason is that truck images have more exclusively representative feature, which makes the feature similarity between truck and other vehicles more distances. The following table 1 shows the performance comparison of some related approaches, and some of the statistics on that table are based on [1] [8] [9] [11].

Table 1. Comparision of Different Approaches

Source Paper	Approach and Accuracy	
	Approach Accuracy	Main Proposed Approach
[8]	87.3%	Different gradient features
[9]	89%	Scale invariant feature transform
[1]	89.22%	Sobel operation and edge detection
Our Paper	Above 95%	Derived kernel based recognition method

5   Conclusion

In this paper, we have described the derived kernel based vehicle type recognition. The vehicle type can be recognized with high level of accuracy with the assistance of first nearest classification and clustering algorithm, since it can avoid the disturbance from similar parts of vehicles and make the most of the information from the representative vehicle parts. Although we have dealt with the template construction and described the clustering algorithm applied together with random selection method on template construction, the templates selection and the number of templates used are still the critical element influencing the time consumption and recognition accuracy, which will be one open topic for future research.

Acknowledgments. The authors wish to thank the State Key Laboratory of Intelligent Technology and Systems of Tsinghua University for the equipment assistance and image acquisition.

References

1. Wang, W.: Reach on Sobel Operator for Vehicle Recognition. In: International Joint Conference on Artificial Intelligence, JCAI 2009, April 25-26, pp. 448–451 (2009)
2. Mello, C.A.B., Costa, D.C.: A Complete System for Vehicle License Plate Recognition. In: 16th International Conference on Systems, Signals and Image Processing, IWSSIP 2009, June 18-20, pp. 1–4 (2009)
3. Munroe, D.T., Madden, M.G.: Multi-class and single class classification approaches to vehicle Model recognition from images. In: Proc. AICS (2005)
4. Wang, Z., Li, S.: Research and implement for vehicle license plate recognition based on improved BP network. In: 2010 International Conference on Computer and Communication Technologies in Agriculture Engineering (CCTAE), June 12-13, vol. 3, pp. 101–104 (2010)
5. Wang, W.: Vehicle Type Recognition Based on Radial Basis Function Neural Networks. In: International Joint Conference on Artificial Intelligence, JCAI 2009, April 25-26, pp. 444–447 (2009)
6. Arzani, M.M., Jamzad, M.: Car type recognition in highways based on wavelet and contourlet feature extraction. In: 2010 International Conference on Signal and Image Processing (ICSIP), December 15-17, pp. 353–356 (2010)
7. Kazemi, F., Samadi, S., Pourreza, H., Akbarzadeh, M.: Vehicle Recognition Based on Fourier, Wavelet and Curvelet Transforms - a Comparative Study. IJCSNS Int. Jou. of Comp. Sci. and Net. Secu. 7(2) (2007)
8. Petrovis, V.S., Cootes, T.F.: Analysis of features for rigid structure vehicle type recognition. In: Proc. of the British Machine Vision Conference (2004)
9. Dlagnekov, L.: Video-based car surveillance: License plate make and model recognition. Master's Thesis, University of California at San Diego (2005)
10. Yigui, Q., Yuanyuan, Y., Ling, Z., Xing, L.: The Extension-Based Recognition Method of Vehicle Type. In: Third International Conference on Convergence and Hybrid Information Technology, ICCIT 2008, November 11-13, vol. 2 (2008)
11. Iqbal, U., Zamir, S.W., Shahid, M.H., Parwaiz, K., Yasin, M., Sarfraz, M.S.: Image based vehicle type identification. In: 2010 International Conference on Information and Emerging Technologies (ICIET), June 14-16, pp. 1–5 (2010)
12. Smale, S., Poggio, T., Caponnetto, A., Bouvrie, J.: Derived distance: towards a mathematical theory of visual cortex. CBCL Paper. MIT, Cambridge (2007)
13. Zhang, Z.C., Tang, Y.Y.: License plate recognition algorithm based on derived kernel. In: 2012 International Conference on Wavelet Analysis and Pattern Recognition (ICWAPR), July 15-17 (in press, 2012)
14. Smale, S., Rosasco, L., Bouvrie, J., Caponnetto, A., Poggio, T.: Mathematics of the neural response. Found. Comput. Math. 10, 67–91 (2010)

An Approach to Define Flexible Structural Constraints in XQuery

Emanuele Panzeri and Gabriella Pasi

University of Milano-Bicocca
Viale Sarca 336, 20126 Milano, Italy
{panzeri,pasi}@disco.unimib.it

Abstract. This paper presents a formal definition of an extension of the XQuery Full-Text language: the proposed extension consists in adding two new flexible axes, named **below** and **near**, which express structural constraints that can be specified by the user. Both constraints are evaluated in an approximate way with respect to a considered path, and their evaluation produces a path relevance score for each retrieved element. The formal syntax and the semantics of the two new axis are presented and discussed.

1 Introduction

The increasing number of huge collections of highly structured XML documents has stimulated in recent years a wealth of research aimed at improving XML querying to both increase query languages expressiveness, and to provide an approximate matching of queries with the consequent ranking of the retrieved elements [3,18]. The first XML query languages were designed based on a data-centric view of XML repositories to allow an efficient access to complex data structures; these languages were finalized to the specification of structural constraints as well as content-related constraints (specification of exact values for XML element contents) in a Data Base style: the results produced by such constraints evaluation is a set of relevant elements. Later, several proposals appeared based on a document-centric view of XML repositories; such approaches have been classified by the information retrieval (IR) community as content-only search (CO) and content and structure search (CAS) [14]. CO approaches were mainly aimed at allowing the specification of keyword based queries in an IR style, where query evaluation produces a ranking of the retrieved XML elements [5]. CAS approaches were defined to allow the formulation of constraints on both documents content and structure [14]. CAS approaches that were based on the syntax of XPath [15], constituted a first attempt to merge the IR and DB search paradigms. Since then, the importance of merging the IR and the DB search perspectives has been widely recognized, and it has recently culminated in the W3C standard XQuery Full-Text (XQ-FT) [17] extension. The evaluation of XQ-FT queries produces a set of weighted elements, where scoring is based on a keyword based matching in textual elements. The problem of providing a ranking of XML

R. Huang et al. (Eds.): AMT 2012, LNCS 7669, pp. 307–317, 2012.
© Springer-Verlag Berlin Heidelberg 2012

elements retrieved by a query based on both content and structural constraints has been addressed in [3], where a query relaxation technique that provides an approximate structural matching was introduced.

None of the above approaches allows users to directly specify the structural relaxations via ad-hoc predicates with a score computation. More recently, in [6], an approach to structural relaxation in XPath via new user specified constraints is proposed; a RDBMS is extended to evaluate relaxed structure matching. However the query evaluation does not provide any ranking of the retrieved fragments. As outlined in [18] and [6], querying highly structured databases or document repositories via structured query models (as XQuery is) forces the users to be well aware of the underlying structure, which is not trivial. In the above cases, users could benefit of a query language that allows a direct specification of flexible structural constraints that easily allow to require the relative position of important nodes, independently of an exact knowledge of the underlying structure(s). To achieve this aim, in this paper, we propose a formal extension of XQuery Full-Text, where two new flexible structural axes, specified by the predicates `below` and `near` are defined. The work reported in this paper was originated by a previous research where a flexible extension of the XQuery language was advocated and informally sketched in [9,7]. This is the first work that proposes the full syntax and semantics of the formal extension.

The proposed extension allows to obtain: (1) a ranking based on content predicates evaluation only (as in the original XQ-FT), (2) a ranking based on the flexible structural constraints evaluation (based on our proposal), or (3) a ranking based on a linear combination of the two above scores, which the user may also specify via the `order-by` clause, as it will be explained in the paper.

In summary, the main contributions of this paper are: (1) to define a formal extension of XQ-FT with two new flexible axes, thus allowing users to explicitly specify their tolerance to an approximate structural matching, while not forcing them to be aware of all the possible structural variations of the data/document structure; (2) to define an ad-hoc approximate matching of the flexible structural constraints thus allowing both a ranking only based on approximate structure matching, and a ranking based on a combination of content predicates and the new flexible structural predicates (while preserving a ranking based only on content predicates).

The paper is organized as follows: Section 2 reviews the research work related to introducing flexibility in XML query evaluation. Section 3 presents the proposed extension of the XQuery Full-Text language with the new flexible structural constraints: both the syntax and the semantics of the new constraints are formalized as well as some usage examples. Section 4 concludes the paper.

2   Related Work

As outlined in Section 1, several approaches to introduce some flexibility in XML retrieval have been proposed in last years, by both the database and the IR communities [8,10,12,16] In IR, the approaches to inquiry XML documents have been classified as content-only search (CO), and both content and structure search (CAS). CAS approaches (proposed in the IR research context) consider both document content and structure in query formulation and evaluation. CAS approaches include: TopX [13], NEXI [15], TeXQuery [1], and FleXPath [4].

Most CAS approaches, like TopX [13], do not consider the content-related and the structural constraints equally important; in fact, they employ a two stage evaluation strategy by which the evaluation of content predicates is first performed (as done with CO queries), and then the obtained results are analysed, and eventually filtered out from the final result set, based on the structural constraints satisfaction. Amer-Yahia et al. [2] define some relaxations in XML structure and content querying such as the introduction of generalized data-types, the adoption of edit distances on paths, and some operations to modify the structure such as delete node, insert intermediate nodes or rename nodes. The aim of these approaches is to modify the structure specified in the query in order to relax the selection of a candidate fragment during a query evaluation. The language NEXI [15] was defined to propose a common language for CAS approaches: it is a *reduced* version of XPath where the only supported axis are the *descendant* and *self* axis. Another important XML query language is TeXQuery [1] which provides a set of full-text search features, such as Boolean connectives, phrase matching, proximity distance. TexQuery is the precursor of the XQuery Full-Text[17] language extension of single path pattern queries or more complex twig pattern queries.

FleXPath [4] is the first approach proposing an approximate matching of structural query constraints by a formalization of relaxations in the evaluation of the structure specified in the queries; it constitutes the first algebraic framework for spanning relaxations. This approach has been further developed in [3], where path scoring is formalized as an approximation of twig scoring by loosing correlations between query nodes in score computation. After the two above seminal contributions, subsequent research has addressed the problem of approximate structural matching of XML data [18]. It should be noted that all previous approaches introduce flexibility in the evaluation process of conventional queries, and therefore these approaches do not allow users to specify flexible constraints that explicitly require the application of an approximate structural matching providing fragment scores distinct from scores produced by the keyword-based evaluation. This means that the user has no way to distinguish between structural constraints in the query the evaluation of which has to produce a set of fragments, and flexible structural constraints the evaluation of which has to produce weighted fragments, with structural fragment scores distinct from content related fragment scores (usually computed by CAS approaches, XQ-FT included).

The novelty of our approach is to rely on users' specification of flexible structural constraints which require an approximate matching of XML nodes. It is worth noticing that also in the DB community a formalization of flexible structural queries was proposed in [16] and [6]; the authors defined in fact two new axes to introduce flexibility in XML structural matching. However, the above approaches neither compute a relevance score for each matched fragment, nor offer users the possibility to combine content and structural scores in a user-defined fragment ranking as proposed in this work.

3   The Proposed Extension

In this section, we introduce the proposed extension that integrates the new predicates `below` and `near` with the XQ-FT syntax. Like the `descendant` axis, also the constraint `below` is specified as a flexible axis of a path expression: its evaluation is aimed at identifying elements (called *target nodes*) that are direct descendants of a node (called the *context node*). However, differently from the `descendant` evaluation, the `below` constraint computes a numeric score for each retrieved node. The constraint `near` is specified as a flexible axis of a path expression; it allows to identify XML elements (*target nodes*) connected to the *context node* through *any path*. Also by this axis, for each retrieved fragment a score is computed.

3.1   Flexible Structural Constraints

The main innovative characteristics of the `below` and `near` constraints are: (i) they are user-specified, and (ii) their evaluation produces a weighted set of nodes. A node weight is computed based on the node closeness to the ideal paths identified by the flexible constraints. We outline that ranking based on user specified structural constraints is a new approach to XML querying, and that, to our knowledge, no XML query engine provides this feature. The proposed approach allows to obtain a node ranking either based on structural constraints evaluation only, or based on content constraints evaluation only, or on a combination of them, which may also be decided by a user. It is important to notice that the integration of the `below` and `near` axes in the XQuery syntax allows to specify them in *any* XQuery predicate, as explained in Section 3.2.

The flexible axes can also be used in conjunction with positional predicates: as the matched elements are returned in decreasing order of relevance, the positional predicates are referred to the rank of the fragments. The following definitions, as well as the provided examples, make use of the *unabbreviated* language form where each axis is explicitly specified. For example, the query *//book/description* is written as */descendant :: book/child :: description*.

The Constraint "Below": The evaluation of the constraint `below` extends the XQuery `descendant` axis evaluation by computing, for each retrieved node, a *path relevance score* that is inversely proportional to the path distance from the ideal path identified by the `below` constraint. The ideal path is the one where the target node is a direct descendant (direct child) of the context node.

An example of the `below` evaluation is graphically sketched in Fig. 1(a): the query "/descendant:: person/below::name" is evaluated against the XML

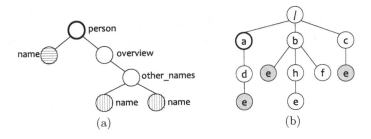

Fig. 1. (a) Graphical representation of the `below` constraint evaluation for the query: `/descendant::person/below::name` (b) The `near` constraint evaluation for the query: "`a/near(3)::e`"

document[1] fragment shown in the figure. The node labeled **person** is the *context* node for the **below** axis evaluation, while all nodes labeled **name** are the *target* nodes. The different filling of the **name** nodes indicates the ranking produced by the below constraint evaluation: the **name** element filled with horizontal lines represents the ideal element, due to the fact that it is the direct descendant of the **person** element. The nodes filled with vertical lines, instead, do not have a direct child relationship with the *context* node, and thus their score is proportional to their distance from the **person** element. Based on the example in Fig. 1(a) the elements retrieved in decreasing order of relevance estimate, are: (i) the node **person/name**; and (ii) the nodes **person/name/other_names/name** filled with vertical lines. An important observation is that, although the XQ-FT standard constraints could allow to formulate complex queries with a behavior similar to the one associated with **below**, the use of explicit flexible constraints is clearly more user-oriented and better complies with the XQ-FT scoring mechanism.

The Constraint "Near": The **near** constraint requires to find target nodes that are "in the neighborhoods" of the context node, in all directions (not only on the descendant axis, but also with respect to siblings and ancestors). A parameter can be specified with the **near** constraint to indicate the maximum distance between the context node and the target node: nodes reachable with more than n arcs form the context node will be excluded from the retrieved elements. The parameter allows users to control the **near** evaluation by avoiding to search in the whole XML graph for matching target nodes. The syntax of the **near** constraint is: `/near(n)::label` (based on the considered application, it could make sense to specify a default value for n). The **near** constraint evaluation computes a relevance score for each matching node: this score is inversely proportional to the distance of the target node from the context node (by considering that the ideal node is directly connected to the context node). The evaluation function of the **near** constraint is formally defined in Section 3.5.

---

[1] In the example the document is taken from the INEX DataCentric collection created from the IMDD movie database.

As an example, in Fig. 1(b) the evaluation of the query "a/near(3)::e" is shown: the node labeled a with bold border is the *context* node, while the e nodes with the filled background are the nodes matched by the example query. Note that the e node with path /b/h/e will not be retrieved because its distance from the context node is more than 3 arcs. To summarize the example in Fig. 1(b) the nodes matched by the query "a/near(3)::e", in decreasing order of relevance estimate, are: (i) node /a/d/e; and (ii) nodes /b/e and /c/e.

3.2 Syntax

The extended grammar (based on the Core XPath grammar as defined in [11]) that includes the below and near constraints is expressed in EBNF (Extended Backus–Naur Form) as follows:

```
locpath ::= '/' locpath | locpath '/' locpath | locpath '|' locpath|locstep
locstep ::= axis '::' t | locstep '[' bexpr ']'
axis     ::= xpathAxis | axisNear | axisBelow
xpathAxis::= 'self' | 'child' | 'parent' | ...
axisNear ::= 'near' | 'near(' number ')'
axisBelow::= 'below'
```

where locpath is the start symbol, named the *location path*; bexpr represents a boolean *filter expression* used as a filter for location paths; t denotes tag labels of document nodes; axis denotes the axis relations in the XPath language (xpathAxis represents all the standard axes, not fully listed for sake of readability) as well as the new flexible axis near and below. As outlined in Section 3.1 the below constraint produces the same *node-set* of the XPath axis descendant; however the evaluation of the below constraint associates a score with each node in the retrieved *node-set*. To manage such numeric scores, an optional Score Variable named score-structure has been introduced, in addition to the XQ-FT score variable, in the for FLWOR[2] clause. We extend the XQuery for clause, defined in [17], as follows:

```
ForClause    ::= "for" "$" VarName TypeDeclaration? PositionalVar?
  FTScoreVar? StructScoreVar? "in" ExprSingle ("," "$" VarName
  TypeDeclaration? PositionalVar? FTScoreVar? "in" ExprSingle)*
FTScoreVar   ::= "score" "$" VarName
StructScoreVar::= "score-structure" "$" VarName
```

where Varname is a variable name; TypeDeclaration is a variable type declaration; and ExprSingle is the actual query for node selection as defined in the XQuery language. An example of XQuery expression including both the Full-Text and below axis evaluation is the following:

---

[2] The FLWOR acronym stands for For-Let-Where- OrderBy-Return that represents the ability of the XQuery language to support selection and iterations over XML elements.

```
for $item score $scoreFT score-structure $scoreS in
  person/below:name[text() contains text "brad"] order by $scoreS
  return <i scS='{$scoreS}' scFT='{$scoreFT}'>$item</i>
```

By this query the user declares her/his interest in all nodes labeled **name** that contain the text *brad*; such nodes must have a **person** node as ancestor. The resulting **name** nodes are then ranked based on their distance from the context node **person** (as obtained by the **below** axis evaluation), and stored in the **$scoreS** variable. The results are then returned from the XQuery expression evaluation in an XML form where both structural and Full-Text scores are stored as well as the **name** node textual contents. An example of obtained results is the following:

1)   `<i scS="1" scFT="1">Brad</i>`
2)   `<i scS="0.3" scFT="0.62">Brad Winsley</i>`

A linear combination of the two scores can be provided by the user to obtain an overall ranking score. Further details on the **below** axis evaluation and scoring computation will be given in Section 3.5.

3.3   Semantics

In this section, we present the formal definition of the semantics of the proposed flexible structural constraints. An important observation here is that a key feature of the new structural constraints is the computation of a relevance score, that should be formally defined in the constraint semantics. However, in this paper, we comply to the formal definition of the XQ-FT scores semantics, where to justify the computation of scores in the evaluation of XQ-FT expressions, we introduce second-order functions. The produced scores are then used in **FLWOR** clauses to assign scoring-variables values. For further details, refer to [17, Chapter 4.4]. We extend then the semantic function S of Core XPath (as defined in [11]) by adding the **below** and the **near** semantics to define the *node-sets* retrieved by the axis constraint evaluation. In Equation (1), the semantics of the axis evaluation is provided as a reference:

$$S[\![\chi :: t]\!](N_0) := \chi(N_0) \cap T(t) \tag{1}$$

where the axis relation function $\chi : 2^{dom} \to 2^{dom}$ is defined as $\chi(N_0) = \{x \mid \exists x_0 \in X_0 : x_0 \chi x\}$ (thus overloading the χ relation name), N_0 is a set of context nodes, and a query π evaluates as $S[\![\pi]\!](N_0)$. It should be noticed that in the presented EBNF, the flexible constraints **below** and **near** can be nested without any limit in any query.

The "Below" Semantics: As previously outlined the definitions provided in this section are finalized to identify the set of nodes retrieved by the new axes evaluation. In this sense, the formal semantics of the constraint **below** is the same formal semantics of the **descendant** constraint, as both of them identify the same set of nodes, i.e., those having node t as a descendant of context node

N_0. In other words, both of them allow to match all the descending nodes of the context node n_0 with a given label t. The only and important difference between the **below** and the **descendant** constraints is the computation of *path relevance scores* that are produced by the **below** constraint evaluation: for each fragment matching the descendant constraint (and thus returned in the set of retrieved fragments) a score is computed, as it will be described at an operational level in section 3.5. Based on this assumption, we may then assert that:

$$S[\![below :: t]\!](N_0) = S[\![descendant :: t]\!](N_0) \qquad (2)$$

As defined in Section 3.2, the **below** constraint can be inserted in any query with an unlimited nesting. The relevance score of a retrieved fragment is computed as an aggregation of all the nested **below** axes evaluation. To achieve this aim, we use the function *min* as an aggregation operator as we require that **all** constraints be satisfied. We will further address in a future work the important issue of the selection of alternative aggregation operators.

The "Near" Semantics: The **near** axis allows to define a maximum distance x that acts as a threshold on the number of arcs between the context node and the target node; nodes the distance of which is more then x arcs are filtered out from the possible results. Following the CoreXPath semantics introduced before, we define here below the **near** constraint semantics; also in this case (as already outlined for the **below** constraint) the score computation is formally defined as in [17]. The semantics is:

$$S[\![near(x) :: t]\!](N_0) := Near(N_0, x) \cap T(t) \qquad (3)$$

where $Near(N_0, x)$ is the function that returns all nodes with a maximum distance of x from the set of nodes in N_0. As an example we present the $near(3)$ semantics. In this example the constraint **near(3)** specifies a maximum distance of 3 nodes between the context node and the target node. Here below all the matching paths of **/near(3)::l** are listed.

$$S[\![/near(3) :: l]\!](n_0) := S[\![/child :: l]\!](n_0) \quad (4)$$
$$\cup S[\![/child :: */child :: l]\!](n_0) \cup S[\![/child :: */child :: */child :: l]\!](n_0)$$
$$\cup S[\![/parent :: l]\!](n_0) \cup S[\![/parent :: */parent :: l]\!](n_0)$$
$$\cup S[\![/parent :: */parent :: */parent :: l]\!](n_0) \cup S[\![/parent :: */child :: l]\!](n_0)$$
$$\cup S[\![/parent :: */child :: */child :: l]\!](n_0) \cup S[\![/parent :: */parent :: */child :: l]\!](n_0)$$

For each matching path represented in (4) a *path relevance score* is computed.

3.4   Flexible Constraints and Query Branching

An important issue related to the flexible axis evaluation concerns the aggregation of queries involving multiple flexible constraints and branching in fragment selection. The flexible constraints, as described in Section 3.1, allow to associate with each node involved in the flexible part of the query a relevance score in the range $[0, 1]$, which is used to compute a ranking of the selected fragments. While in flat queries this can be done without any difficulties, a particular observation should be made for branching queries where the selection node that needs to be ranked appears in a different branch than the one/ones that use the flexible constraints, thus obtaining a path relevance score.

Let us consider the complex query `person[descendant::act/near(4)::title[contains(.,'gran torino')]]/child::name` and the XML document fragment shown in Fig. 2(b). Let us suppose that the user is interested in finding the names of people involved in the movie entitled "Gran Torino." The user interest is mainly in, but not limited to, people who acted in such movie: by using the constraint **near** the user requires also to find people who worked as director, producer, etc. (even if with a lower structural relevance). In Fig. 2(a), the tree representation of the query is shown: the underlined *name* element identifies the selection node; the edges between two nodes identify the *axes*-constraints (the label identifies the specified axis, i.e., `child` and `near`). Dotted lines represent filtering functions, in this example the `contains` function. The element **person** is also called *branching point*.

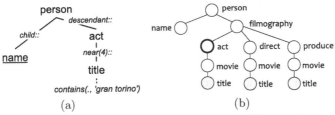

Fig. 2. (a)The tree-representation of the query `person[descendant::act/near(4)::title[contains(.,'gran torino')]]/child::name` (b) the XML fragment graph

From the above example is clear how the evaluation of the right branch of Fig. 2(a) can produce a set of elements for a single person (i.e., *Clint Eastwood* was involved in the movie as the main actor, the director and the producer). In this case, each retrieved fragment has a score associated, and it is not clear how the final score should be computed and assigned to the branching point to allow a ranking of the elements matched in the left branch of the example.

As a first reasonable solution to address this situation, we assign to the branching point element (in our example the **person** element), a score which is the maximum value among those obtained by evaluating the right branches. Although the choice of applying the `max()` aggregation is quite natural to obtain an optimistic aggregation, other aggregation schemes will be investigated.

3.5   The Proposed Approximate Evaluation

In this section, the evaluation functions of the new axes **near** and **below** are defined; each function computes the *path relevance score* of a document path with respect to the query path. Each score is in the interval $[0, 1]$ where 1 represents a full satisfaction of the axis constraint evaluation, while values less than 1 are assigned to target nodes *far* from the context node. Nodes with a path relevance score of zero will not be retrieved as they are not relevant to the user query.

As previously outlined the notion of path *closeness* is related to the concept of node distance (intended as the number of arcs connecting two nodes). Both the `near` and the `below` evaluation functions are based on a count of the of arcs between the context node and the target node to compute the path relevance score; however alternative evaluation functions can be easily implemented.

The "Below" Constraint Evaluation Function: As previously stated, the `below` axis produces the same *node-set* result as the XPath `descendant` axis; however, for each node, a score is computed based on the *distance* between the context node and the target node. The path relevance score for the `below` axis evaluation, with a context node c and a target node t, can be computed as :

$$w_{below}(c,t) = \frac{1}{|desc\_arcs(c,t)|}. \tag{5}$$

Where $desc\_arcs(c,t)$ is a function that, given two XML nodes c and t (where t must be a descendant of node c), returns the set of descending arcs from c to t. The score computed for the `below` axis is inversely proportional to the distance of the nodes c and t, thus giving to nodes *closer* to the context node a higher score than the one given to nodes *far* from the context node.

The "Near" Constraint Evaluation Function: As explained in Section 3.3, the `near` axis evaluation allows to retrieve nodes that are *close* to the given context node in every path direction; in the `near` axis evaluation the maximum allowed distance that can occur between the two nodes is taken into account.

In Equation (6), the evaluation function used to compute the *path relevance score* based on the `near` axis is defined, where c is the context node, t is the current target node, l is the maximum allowed distance and $arcs(c,e)$ is a function that returns the set of arcs in the shortest path between c and t.

$$w_{near}(c,t,l) = \begin{cases} \frac{1}{|arcs(c,t)|} & \text{if } |arcs(c,t)| \leq l \\ 0 & \text{else.} \end{cases} \tag{6}$$

Like the `below` scoring, score is inversely proportional to the distance of the context node from the target node. The function assign higher values if the target node is *closer* to the context node, while the relevance score decreases to zero as the distance increases.

4   Conclusions and Future Work

In this work, a new approach to XML querying has been presented: two new flexible axes, named `below` and `near`, have been introduced into the XQuery Full-Text language, to give users the possibility of specifying flexible structural constraints. For each axis evaluation, a *path relevance score* is computed to produce a ranked list of elements. Each retrieved XML element is evaluated based on the notion of *path closeness* between the considered element and the ideal element specified in the query. In this work, the syntax, the semantics and the axes evaluation functions have been defined, as well as an initial analysis of the branching issue for the proposed axis evaluation. Further research will address the branching issue for the `below` and the `near` axes evaluation, as well as the definition of meaningful aggregation strategies for such cases.

Ongoing works are being conducted related to the implementation of the two new axes inside a XML Query engine, able to handle both the XQuery Full-Text specification and the flexible structural constraints `below` and `near`.

References

1. Amer-Yahia, S., Botev, C., Shanmugasundaram, J.: TeXQuery: A Full-Text Search Extension to XQuery. In: WWW 2004, pp. 583–594. ACM (2004)
2. Amer-Yahia, S., Cho, S., Srivastava, D.: Tree Pattern Relaxation. In: Jensen, C.S., Jeffery, K., Pokorný, J., Šaltenis, S., Bertino, E., Böhm, K., Jarke, M. (eds.) EDBT 2002. LNCS, vol. 2287, pp. 496–513. Springer, Heidelberg (2002)
3. Amer-Yahia, S., Koudas, N., Marian, A., Srivastava, D., Toman, D.: Structure and Content Scoring for XML. In: VLDB 2005, pp. 361–372 (2005)
4. Amer-Yahia, S., Lakshmanan, L.V.S., Pandit, S.: FleXPath: flexible structure and full-text querying for XML. In: SIGMOD 2004, pp. 83–94 (2004)
5. Amer-Yahia, S., Lalmas, M.: XML search: languages, INEX and scoring. In: SIGMOD 2006, pp. 16–23 (2006)
6. Bhowmick, S.S., Dyreson, C., Leonardi, E., Ng, Z.: Towards non-directional Xpath evaluation in a RDBMS. In: CIKM 2009, pp. 1501–1504 (2009)
7. Campi, A., Damiani, E., Guinea, S., Marrara, S., Pasi, G., Spoletini, P.: A fuzzy extension of the xpath query language. J. Intell. Inf. Syst., 285–305 (2009)
8. Chinenyanga, T.T., Kushmerick, N.: An expressive and efficient language for XML Information Retrieval. JASIST 53, 438–453 (2002)
9. Damiani, E., Marrara, S., Pasi, G.: A flexible extension of XPath to improve XML querying. In: SIGIR 2008, pp. 849–850 (2008)
10. Fuhr, N., Großjohann, K.: XIRQL: A Query Language for Information Retrieval in XML Documents. In: SIGIR, pp. 172–180 (2001)
11. Gottlob, G., Koch, C., Pichler, R.: Efficient algorithms for processing xpath queries. ACM Trans. Database Syst. 30, 444–491 (2005)
12. Theobald, A., Weikum, G.: Adding Relevance to XML. In: Suciu, D., Vossen, G. (eds.) WebDB 2000. LNCS, vol. 1997, pp. 105–124. Springer, Heidelberg (2001)
13. Theobald, M., Schenkel, R., Weikum, G.: TopX and XXL at INEX 2005. In: Fuhr, N., Lalmas, M., Malik, S., Kazai, G. (eds.) INEX 2005. LNCS, vol. 3977, pp. 282–295. Springer, Heidelberg (2006)
14. Trotman, A., Lalmas, M.: The Interpretation of CAS. In: Fuhr, N., Lalmas, M., Malik, S., Kazai, G. (eds.) INEX 2005. LNCS, vol. 3977, pp. 58–71. Springer, Heidelberg (2006)
15. Trotman, A., Sigurbjörnsson, B.: Narrowed Extended XPath I (NEXI). In: Fuhr, N., Lalmas, M., Malik, S., Szlávik, Z. (eds.) INEX 2004. LNCS, vol. 3493, pp. 16–40. Springer, Heidelberg (2005)
16. Truong, B.Q., Bhowmick, S.S., Dyreson, C.: SINBAD: Towards Structure-Independent Querying of Common Neighbors in XML Databases. In: Lee, S.-g., Peng, Z., Zhou, X., Moon, Y.-S., Unland, R., Yoo, J. (eds.) DASFAA 2012, Part I. LNCS, vol. 7238, pp. 156–171. Springer, Heidelberg (2012)
17. W3C. XQuery/XPath FullText (March 2011), http://www.w3.org/TR/xpath-full-text-10
18. Yu, C., Jagadish, H.V.: Querying Complex Structured Databases. In: VLDB 2007, pp. 1010–1021 (2007)

DC Stream Based JPEG Compressed Domain Image Retrieval

Gerald Schaefer and David Edmundson

Department of Computer Science, Loughborough University, Loughborough, U.K.

Abstract. The vast majority of images are stored in compressed JPEG format. When performing content-based image retrieval, faster feature extraction is possible when calculating them directly in the compressed domain, avoiding full decompression of the images. Algorithms that operate in this way calculate image features based on DCT coefficients and hence still require partial decoding of the image to arrive at these. In this paper, we introduce a JPEG compressed domain retrieval algorithm that is based not directly on DCT coefficients but on differences of these, which are readily available in a JPEG compression stream. In particular, we utilise solely the DC stream of JPEG files and make direct use of the fact that DC terms are differentially coded. We build histograms of these differences and utilise them as image features, thus eliminating the need to undo the differential coding as in other methods. In combination with a colour histogram, also extracted from DC data, we show our approach to give (to our knowledge) the best retrieval accuracy of a JPEG domain retrieval algorithm, outperforming other compressed domain methods and reaching a performance close to that of the best performing MPEG-7 descriptor.

1 Introduction

With literally billions of images at our disposal, visual information retrieval is of high importance. The vast majority of images is stored in compressed form, most commonly in JPEG format[1]. While most content-based image retrieval (CBIR) algorithms, which extract image features for retrieval purposes [2,3,4], operate on pixel data and hence require full decompression of the images, compressed-domain algorithms [5] provide faster feature extraction since they need only partial image decoding.

In this paper, we present a novel JPEG compressed domain algorithm, which, in contrast to previous approaches, does not need to undo the differential coding stage in JPEG. Rather, we make use of the differentially coded DC data directly and employ it as an image feature in form of a histogram. We show that this descriptor, which captures image gradient and texture information, works well for image retrieval and that in combination with a colour histogram, which is also extracted from DC data, excellent retrieval performance is achieved as demonstrated by an extensive set of experiments. Impressively, our approach is shown to outperform all JPEG CBIR algorithms to date.

---

[1] Up to 95% of all images on the web are JPEG images [1].

R. Huang et al. (Eds.): AMT 2012, LNCS 7669, pp. 318–327, 2012.

2   JPEG Image Compression

JPEG [6] is not only the most popular image compression technique, it has also been adopted as an ISO standard for still picture coding. JPEG compression is based on the discrete cosine transform (DCT), a derivative of the discrete Fourier transform. First, an (RGB) image is usually converted into the YCbCr space. The reason for this is that the human visual system is less sensitive to changes in the chrominance (Cb and Cr) channels than in the luminance (Y) channel. Consequently, the chrominance channels can be downsampled by a factor of 2 without significantly reducing image quality, resulting in a full resolution Y and downsampled Cb and Cr components.

The image is then divided (each colour channel separately) into 8×8 pixel sub-blocks and DCT applied to each such block. The 2-d DCT for an 8×8 block $f_{xy}, x, y = 0 \ldots 7$ is defined as

$$F_{uv} = \frac{C_u C_v}{4} \sum_{x=0}^{7} \sum_{y=0}^{7} f_{xy} \cos\left(\frac{(2x+1)u\pi}{16}\right) \cos\left(\frac{(2y+1)v\pi}{16}\right) \qquad (1)$$

with $C_u, C_v = 1/\sqrt{2}$ for $u, v = 0$, $C_u, C_v = 1$ otherwise.

DCT has energy compactification close to optimal for most images which means that most of the information is stored in a few, low-frequency, coefficients (due to the nature of images which tend to change slowly over image regions). Of the 64 coefficients, the one with zero frequency (i.e., F_{00}) is termed "DC coefficient" and the other 63 "AC coefficients". The DC term describes the mean of the image block, while the AC coefficients account for the higher frequencies. As the lower frequencies are more important for the image content, higher frequencies can be neglected which is performed through a (lossy) quantisation step that crudely quantises higher frequencies while preserving lower frequencies more accurately.

The AC and DC components of the image are stored in separate streams for each colour channel. While the AC coefficients are zig-zag, run-length and entropy encoded, the DC stream is differentially encoded and also entropy coded.

3   JPEG Compressed Domain CBIR Algorithms

While normally JPEG images need to be fully decompressed to arrive at pixel data and hence enable feature calculation, it is also possible to derive image features directly in the JPEG compressed domain [7,8]. These features are based on the DCT coefficients expressed in Equation (1) and hence still require partial decoding of the file to undo the entropy, run-length and differential coding stages as well as the reversal of the quantisation step.

One of the earliest attempts of JPEG image retrieval is that of Shneier and Abdel-Mottaleb [9] who divide an image into a set of windows for which keys are constructed and then paired for comparison, resulting in a bitstream as feature vectors. Lay and Guan [10] investigated the use of low energy JPEG coefficients, and selected the top four coefficients whose sums were stored in histogram form. Schaefer [11] proposed a JPEG retrieval method that uses both colour and texture descriptors extracted from only the DC terms. A colour histogram [12] is constructed from chromaticity DC data, while a texture descriptor is obtained by applying the LBP operator [13] on the DC data of the luminance (Y) channel and generating a histogram of LBP descriptors. Jiang, Armstrong and Feng [1] also exploited the fact that the average colour of an 8×8 block can be obtained directly from the DC component of the block, and thus calculate a colour histogram in a fashion similar to [11] yet based on all three colour (YCbCr) channels. Feng and Jiang [14] analysed the distribution of moments (means from DC and variances from AC data), and thus extract a rough estimation of texture information which is then encoded in histogram form. Chang *et al.* [15] built a bitstream based on differences between DC values in scanned entropy data, while AC information is employed, based on the first 9 AC coefficients and the 8 differences between them. Eom and Choe [16] presented an edge histogram detector that resembles the MPEG-7 non-homogeneous texture descriptor [17], while the algorithm by Lu, Li and Burkhardt [18] extracts colour (histogram), texture (energy histogram) and basic edge information.

4   DC Stream Based JPEG Image Retrieval

In this paper, we present an effective and efficient JPEG compressed domain retrieval algorithm. Similar to other methods [11,1] we exploit only the DC coefficients, i.e. F_{00} in Eq. (1), which essentially corresponds to a downsampled version of the image. That this does not necessarily lead to compromised retrieval accuracy was shown in [12] where it was found that image retrieval on subsampled images and on their original counterpart led to similar retrieval performance.

4.1   DC Difference Histogram

All the JPEG retrieval algorithms discussed in Section 3 require partial decoding of the image to arrive at DCT data. That is, first the entropy coding needs to be reversed, followed by the runlength and zig-zag coding for the AC stream and differential coding for the DC stream, while finally the quantisation step is undone. Our approach in this paper, does not require the last two steps. Since it is based solely on DC data we operate directly on the DC differentially encoded data stream. To the best of our knowledge, our approach is the first method that does not require to decompress the differentially encoded DC stream.

In fact, the core of our idea is that the differentially encoded DC data is directly useful for image retrieval. Since DC values characterise the average of image blocks, the differences between them describe important information about the image gradient, image texture and about edges in the image as well as about uniform image areas (where the differences would be 0 or close to it). We therefore propose to encode the DC difference data in form of a histogram over the whole image. Since there are three colour channels (Y, Cb, and Cr), we build a histogram for each of the channels. However, these histograms need to be carefully defined to be useful for image retrieval. Since the DC values are in the range $[-1024; 1024]$ the possible differences between the DC components are in $[-2048; 2048]$. However, lower differences are statistically much more likely than higher ones. We therefore define the boundaries of the histogram bins to be $\pm(2^m)$, $m = \{0, 1, 2, ..., 11\}$ with an additional bin for 0. That is, the first bin of the histogram encodes 0 differences, the second bin absolute differences of 1, the third absolute differences in $[2; 3]$, the fourth in $[4; 7]$, ..., the last bin absolute differences in $[1024; 2048]$.

We normalise the histograms (to sum to 1) and compare them using an L_1 norm. To integrate the three histograms, we add the calculated distances yet weigh them 0.6:0.2:0.2 for Y:Cb:Cr into a single measure to put more emphasis on the luminance channel which is dominantly affected by image texture.

4.2   Combination with DC Colour Histogram

While our new DC difference descriptor provides a simple description of image texture, it has been shown that by combining such a feature with a colour descriptor improved retrieval performance can be achieved [11]. We therefore follow this and augment our DC difference histogram with a colour histogram. This again can be efficiently calculated in the JPEG compressed domain as has been shown in [11,1]. We make use again only of the DC coefficients for this, since it has also been shown that while AC coefficients can be employed to calculate a more accurate histogram, this might actually lead to a decrease in retrieval performance [1]. Our colour histogram is calculated as a 3-dimensional uniform $8 \times 8 \times 8$ histogram in YCbCr colour space (so that no colour conversions are necessary), and again we use the L_1 norm on normalised histograms for distance calculation.

In order to allow emphasising either the colour or the texture component, we utilise a weighted average of the two distances. That is, if d_t is the distance calculated for the DC difference histograms and d_c that for the colour histograms the combined distance is established as $d = \alpha d_c + (1 - \alpha)d_t$ with $\alpha \in [0; 1]$.

5   Experimental Results

We structure our experiments in three parts. We first aim to show that our novel method, that directly exploits the differentially coded DC stream, provides a useful image retrieval technique on its own. We will then combine it with DC based colour histograms to increase retrieval accuracy and will seek a good weighting between the colour and texture terms. We will compare this method with a series of pixel domain MPEG-7 descriptors [17] and the current state-of-the-art of compressed domain JPEG methods [7], both in terms of retrieval accuracy and computational complexity. Finally, we will seek to show that our technique scales well by running experiments on a larger dataset.

To start with, we are interested in the retrieval performance of our novel DC difference descriptor from Section 4.1, both in terms of accuracy as well as speed. For this, we employed the UCID dataset [19], which consists of 1338 uncompressed images of which 262 are query images for which matching ground truth images are defined, and compress it using JPEG to a typical quality setting of q=85. The ground truth allows to assess the performance of any CBIR algorithm, and we chose the suggested modified average match percentile (AMP),

defined as the average of $MP_Q = \frac{100}{S_Q} \sum_{i=1}^{S_Q} \frac{N - R_i}{N - i}$ over the whole dataset as

performance measure, where N is the number of images in the dataset, S_Q the number of model images for the given query, and R_i is the rank in which the i-th model image was retrieved.

The results are listed in Table 1 which gives the retrieval performance (in terms of AMP) on the UCID dataset, the total number of CPU instructions (in millions) required to extract the features from all 1338 images in the database, and the total number of CPU instructions (in millions) required to make a query against the dataset (i.e., to extract features for a query image and compare the derived features against those of the other 1337 feature vectors from the database). All algorithms were implemented and compiled with GCC (ver. 4.5.2), and the experiments were run on an Intel Core 2 Duo 2.66GHz CPU under the Linux operating system.

For comparison, we also implemented the texture retrieval method from [11]. This approach is also based solely on DC coefficients and applies the LBP texture operator [13] on the DC terms of the luminance channel. LBP (local binary patterns) provide a descriptor of a pixel and its immediate (8-)neighbourhood which are aggregated over the whole image in form of a histogram. Looking at Table 1, we can see that this results in a respectable AMP of just over 83% on the UCID dataset. However, we can also see that our method, though in principle simpler, clearly outperforms this, giving an AMP of about 88.5%. Looking at the number of instructions required for feature extraction, it can be seen that our method is faster although it operates on all three colour channels, whereas [11] uses only the luminance channel. This clearly demonstrates that our method allows for faster feature calculation since it does not require to decompress the differentially coded DC data. Feature calculation is also faster, by about one third, which stems from the fact that our descriptor contains 36 histogram bins compared to the 256 of [11].

Table 1. Results of our proposed descriptor vs. the DC LBP descriptor from [11]

	retrieval accuracy [AMP]	feature extraction	image retrieval
		[10^6 CPU instructions]	
DC LBP texture [11]	83.04	5,431.66	6.64
DC difference histogram	88.53	5,109.33	4.29

Table 2. Retrieval performance on UCID dataset for the combined colour and texture approach

α	retrieval accruacy [AMP]
0.0	89.31
0.1	89.62
0.2	90.23
0.3	90.82
0.4	91.35
0.5	91.78
0.6	92.04
0.7	92.08
0.8	91.69
0.9	90.58
1.0	88.53

Next, we combine our DC difference histogram with a DC colour histogram as detailed in Section 4.2. As mentioned there, we can assign different weights α to put more emphasis on either the colour or texture component on the combined algorithm. We performed retrieval on the UCID database, setting $\alpha = \{0.0, 0.1, 0.2, ..., 1.0\}$, and give the results, again in terms of average match percentile, in Table 2.

In Table 2, clearly the results at the extremes, i.e. with α set to 0 or 1 represent the performance of each subcomponent on its own with $\alpha = 1$ the result already given in Table 1 for our method, while $\alpha = 0$ is the performance of the DC colour histogram on its own. We can further see that the results for these two cases are also the lowest ones in Table 2 which confirms that a combination of these complimentary features is indeed useful. The best results were achieved with $\alpha = 0.7$ which puts about twice as much emphasis on the colour histogram than on the DC difference one, and it is this setting that we will use in the remainder of our experiments.

To enable an objective comparison of our method with the literature, we implemented all eight JPEG compressed domain algorithms discussed in Section 3 which represent the current state-of-the-art in this field [7] as well as four popular pixel domain algorithms, namely colour indexing [12] and three of the MPEG-7 descriptors (Scalable colour, Colour layout, and Edge histogram [17]). We ran all 13 algorithms on the UCID benchmark database and recorded the retrieval performance as well as the computational complexity of each method. The results are given in Table 3 which is laid out in the same fashion as Table 1.

From Table 3, it is evident that our proposed approach works extremely well and gives an average match percentile of just above 92%. In fact, out of all compressed domain algorithms it gives the best retrieval performance and is therefore (to the best of our knowledge) the best performing JPEG CBIR algorithm on this dataset to date. The only algorithm that gives even better accuracy is the MPEG-7 Scalable colour descriptor which (again to our knowledge) is the algorithm that gives the highest retrieval performance on the UCID database. On the other hand, we can see that the Scalable colour algorithm is also by far the

Table 3. Results of our retrieval experiments on the UCID dataset

	retrieval accuracy [AMP]	feature extraction [10^6 CPU instructions]	image retrieval
Colour histogram [12]	90.95	38,257.99	35.40
MPEG-7 Scalable colour [17]	92.97	164,483.13	128.72
MPEG-7 Colour layout [17]	89.75	59,003.76	45.58
MPEG-7 Edge histogram [17]	88.79	50,510.43	41.88
Shneier and Abdel-Mottaleb [9]	76.00	7,897.27	9.14
Lay and Guan [10]	91.97	5,744.54	6.29
Schaefer [11]	90.35	5,760.45	9.96
Jiang et al. [1]	89.31	5,186.35	10.68
Feng and Jiang [14]	84.71	8,685.18	9.79
Chang et al. [15]	88.33	5,420.10	7.18
Eom and Choe [16]	87.08	7,937.75	18.54
Lu et al. [18]	90.82	6,549.43	14.55
proposed approach ($\alpha = 0.7$)	92.08	5,528.18	11.24

slowest of all methods as it calculates a colour histogram based on a sliding window approach. The other pixel domain methods are faster but are still relatively slow. In comparison, all JPEG methods, including the one we propose in this paper, are considerably faster for both feature extraction and image retrieval.

In order to see how well our method scales with respect to database size, we merged the UCID and MIR Flickr datasets [20]. The latter consists of 25,000 photos taken from the image sharing site Flickr[2]. We compress the UCID images to match the quality setting of the MIR images, and retain the UCID defined query set, hence ending up with a database of 26,338 images and 262 queries. Retrieval results on this dataset are given in Table 4.

From there we can see that our approach does indeed scale well and also gives the best performance of all JPEG retrieval algorithms on this larger dataset with an AMP of 95.3%. Only the pixel domain Scalable colour descriptor outperforms this.

The good retrieval performance is illustrated in a retrieval example from the combined UCID and MIR database shown in Figure 1. In there, we show one of the query images from the UCID set together with the top 8 ranked images retrieved (from the UCID+MIR database) by the LBP descriptor of [11], our proposed DC difference descriptor, colour indexing [12], the MPEG-7 Scalable colour descriptor [17], the compressed domain methods by Lay and Guan [10] and Schaefer [11] and our proposed (combined) algorithm. We can see that the LBP descriptor on DC data on its own performs quite poorly here, failing to retrieve a single match in the top 8 images, while our novel DC difference algorithm returns three. (Pixel domain) colour histograms perform quite well retrieving six images, the same as the MPEG-7 Scalable colour descriptor although for the latter these are also ranked in the top 6. Of the compressed domain approaches, Lay and Guan's method gives quite good performance, returning four matches

---

[2] http://www.flickr.com

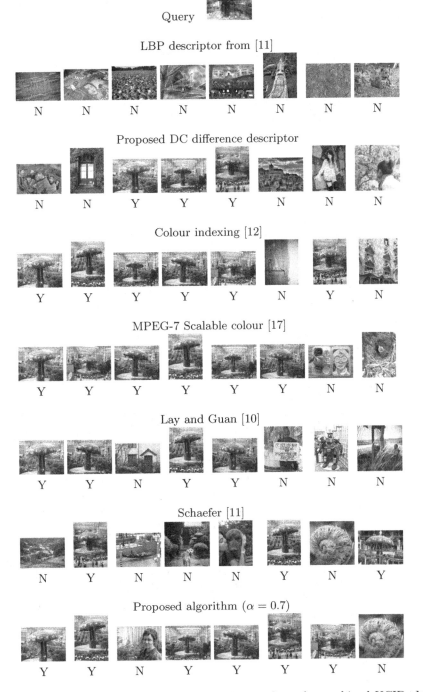

Fig. 1. Retrieval example, showing the top 8 images from the combined UCID+MIR dataset retrieved by various algorithms. Y indicates correct retrieved image, N otherwise.

Table 4. Results of our retrieval experiments on the combined UCID and MIR dataset. The algorithm by Chang *et al.* can only be applied to images of the same size and hence cannot be run on the combined UCID+MIR database.

	retrieval accuracy [AMP]
Colour histogram [12]	94.14
MPEG-7 Scalable colour [17]	96.28
MPEG-7 Colour layout [17]	92.64
MPEG-7 Edge histogram [17]	91.70
Shneier & Abdel-Mottaleb [9]	88.64
Lay & Guan [10]	95.19
Schaefer [11]	93.38
Jiang *et al.* [1]	94.40
Feng and Jiang [14]	79.44
Chang *et al.* [15]	N/A
Eom and Choe [16]	87.44
Lu *et al.* [18]	93.95
proposed approach ($\alpha = 0.7$)	95.30

in the top 8 retrieved images, while Schaefer's algorithm performs less well with three matches. Finally, we can see that our proposed approach, combining DC difference histograms with DC based colour histograms, provides clearly better retrieval accuracy, retrieving 6 correct images in the top 8.

Overall, our extensive experiments clearly demonstrate that our proposed algorithm provides excellent retrieval performance, outperforming all JPEG compressed domain methods to date, and almost matching the results of the best pixel domain algorithms, while allowing fast feature calculation and comparison.

6 Conclusions

In this paper, we have introduced an effective and efficient JPEG compressed domain retrieval algorithm. Similar to other approaches, our technique exploits JPEG DC coefficients, yet in contrast to them, it does not require to undo the differential encoding stage of JPEG. Rather, we directly exploit these differences which provide information about image gradient and texture and sum them into a histogram descriptor. Improved retrieval performance can be achieved by augmenting this approach with a DC based colour histogram. Extensive experiments have confirmed that this algorithm provides excellent retrieval performance, outperforming all JPEG CBIR algorithms to date, while allowing fast feature calculation and comparison, hence making it a very useful choice for generic content-based image retrieval.

References

1. Jiang, J., Armstrong, A., Feng, G.: Direct content access and extraction from JPEG compressed images. Pattern Recognition 35, 1511–2519 (2002)

2. Smeulders, A., Worring, M., Santini, S., Gupta, A., Jain, R.: Content-based image retrieval at the end of the early years. IEEE Trans. Pattern Analysis and Machine Intelligence 22, 1249–1380 (2000)
3. Datta, R., Joshi, D., Li, J., Wang, J.Z.: Image retrieval: Ideas, influences, and trends of the new age. ACM Computing Surveys 40, 1–60 (2008)
4. Schaefer, G.: Mining Image Databases by Content. In: Fernandes, A.A.A., Gray, A.J.G., Belhajjame, K. (eds.) BNCOD 2011. LNCS, vol. 7051, pp. 66–67. Springer, Heidelberg (2011)
5. Mandal, M., Idris, F., Panchanathan, S.: A critical evaluation of image and video indexing techniques in the compressed domain. Image and Vision Computing 17, 513–529 (1999)
6. Wallace, G.: The JPEG still picture compression standard. Communications of the ACM 34, 30–44 (1991)
7. Edmundson, D., Schaefer, G.: Performance comparison of JPEG compressed domain image retrieval techniques. In: IEEE Int. Conference on Signal Processing, Communications and Computing (2012)
8. Edmundson, D., Schaefer, G.: An overview and evaluation of JPEG compressed domain retrieval techniques. In: 54th International Symposium ELMAR (2012)
9. Shneier, M., Abdel-Mottaleb, M.: Exploiting the jpeg compression scheme for image retrieval. IEEE Trans. Pattern Analysis and Machine Intelligence 18, 849–853 (1996)
10. Lay, J.A., Guan, L.: Image retrieval based on energy histograms of the low frequency DCT coefficients. In: IEEE Int. Conference on Acoustics, Speech and Signal Processing, vol. 6, pp. 3009–3012 (1999)
11. Schaefer, G.: JPEG image retrieval by simple operators. In: 2nd International Workshop on Content-Based Multimedia Indexing, pp. 207–214 (2001)
12. Swain, M., Ballard, D.: Color indexing. Int. Journal of Computer Vision 7, 11–32 (1991)
13. Ojala, T., Pietikäinen, M., Harwood, D.: A comparative study for texture measures with classification based on feature distributions. Pattern Recognition 29, 51–59 (1996)
14. Feng, G., Jiang, J.: JPEG compressed image retrieval via statistical features. Pattern Recognition 36, 977–985 (2003)
15. Chang, C., Chuang, J., Hu, Y.: Retrieving digital images from a JPEG compressed image database. Image and Vision Computing 22, 471–484 (2004)
16. Eom, M., Choe, Y.: Fast extraction of edge histogram in DCT domain based on MPEG7. In: Int. Conference on Enformatika, Systems Sciences and Engineering (2005)
17. Sikora, T.: The MPEG-7 visual standard for content description - an overview. IEEE Trans. Circuits and Systems for Video Technology 11, 696–702 (2001)
18. Lu, Z., Li, S., Burkhardt, H.: A content-based image retrieval scheme in JPEG compressed domain. Int. Journal of Innovative Computing, Information and Control 2, 831–839 (2006)
19. Schaefer, G., Stich, M.: UCID - An Uncompressed Colour Image Database. In: Storage and Retrieval Methods and Applications for Multimedia 2004. Proceedings of SPIE, vol. 5307, pp. 472–480 (2004)
20. Huiskes, M., Lew, M.: The MIR Flickr retrieval evaluation. In: ACM Int. Conference on Multimedia Information Retrieval (2008)

A Comparative Study of Community Structure Based Node Scores for Network Immunization

Yuu Yamada and Tetsuya Yoshida

Graduate School of Information Science and Technology,
Hokkaido University
N-14 W-9, Sapporo 060-0814, Japan
{yamayuu,yoshida}@meme.hokudai.ac.jp

Abstract. Network immunization has often been conducted by removing nodes with large network centrality so that the whole network can be fragmented into smaller subgraphs. Since contamination (e.g., virus) is propagated among subgraphs (communities) along links in a network, besides centrality, utilization of community structure seems effective for immunization. We have proposed community structure based node scores in terms of a vector representation of nodes in a network. In this paper we report a comparative study of our node scores over both synthetic and real-world networks. The characteristics of the node scores are clarified through the visualization of networks. Extensive experiments are conducted to compare the node scores with other centrality based immunization strategies. The results are encouraging and indicate that the node scores are promising.

1 Introduction

Contamination (e.g., virus) is usually propagated among subgraphs (communities) along links in a network. For preventing the spread of contamination over the whole network, it is necessary to remove (or, vaccinate) contaminated nodes. Since contamination is propagated among communities in a network, for effective network immunization, it is important to identify nodes which play the role of intermediating or connecting communities.

Most previous work on network analysis considers the community structure of a network in terms of links in a network (e.g., graph cut) [7]; however, we consider it in terms of nodes in a network, and proposed community structure based node scores for network immunization [11]. Based on a quality measure of communities for node partitioning [4], a vector representation of nodes in a network is constructed, and the community structure in terms of the distribution of node vectors is utilized for calculating node scores. Two types of node score are proposed based on the direction and the norm of the constructed node vectors.

In this paper we report a comparative study of our node scores over both synthetic and real-world networks. The characteristics of node scores are clarified through network visualization, and they are compared with other centrality

R. Huang et al. (Eds.): AMT 2012, LNCS 7669, pp. 328–337, 2012.

based immunization strategies. Comparison with other centrality based immunization strategies shows that our node scores are promising, since they can exploit the community structure of a network without relying on the externally supplied community labels of nodes.

Section 2 explains network immunization and centralities. Section 3 describes our community structure based node scores. Section 4 reports a comparative study and discusses the results. Section 5 summarizes our contributions.

2 Network Immunization

2.1 Preliminaries

We use a bold italic lowercase letter to denote a vector, and a bold normal uppercase letter to denote a matrix. $\mathbf{X}_{ij}$ stands for the element in a matrix $\mathbf{X}$, and $\mathbf{X}^T$ stands for the transposition of $\mathbf{X}$. $\mathbf{1}_n \in \mathbb{R}^n$ stands for a vector where each element is 1.

Let n stands for the number of nodes in a network G, and m stands for the number of links in G [1]. Since most social networks are represented as undirected graph without self-loops [6], we focus on this type of networks in this paper.

The connectivity of a network is represented as a square matrix $\mathbf{A} \in \{0,1\}^{n \times n}$, which is called an adjacency matrix. $\mathbf{A}_{ij} = 1$ if the pair of vertices (i,j) is connected; otherwise, 0. For an undirected graph without self-loops, its adjacency matrix $\mathbf{A}$ is symmetric and its diagonal elements are set to zeros.

2.2 Network Immunization

Epidemics (e.g, virus) are often propagated through the interaction between nodes (e.g., individuals, computers) in a network. If a contaminated node interacts with other nodes, contamination can spread over the whole network. In order to protect the nodes in the network as much as possible, it is necessary to dis-

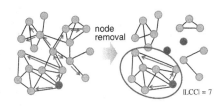

Fig. 1. Network immunization

connect (or, remove) the contaminated node so that the major part of the network. For instance, the largest connected component (LCC) of a network can be prevented from the contamination by removing several nodes (see Fig. 1).

2.3 Network Centrality

Various notions of "network centrality" have been studied in social network analysis [6,8]. Since nodes with many links can be considered as a hub in a network, the degree (number of links) of a node is called **degree centrality**. On the other hand, **betweenness centrality** focuses on the shortest path along which

---

[1] We also call a network as a graph, a node as a vertex, and a link as an edge.

information is propagated over a network. By enumerating the shortest paths between each pair of nodes, betweenness centrality of a node is defined as the number of shortest paths which go through the node.

Similar to the famous Page Rank, eigenvector centrality utilizes the leading eigenvector of the adjacency matrix $\mathbf{A}$ of a network, and each element (value) of the eigenvector is considered as the score of the corresponding node. Based on the approximate calculation of eigenvector centrality via perturbation analysis, another centrality (called dynamical importance) was also proposed in [9].

By assuming that community labels of nodes in a network can be provided, perturbation analysis of node centrality is utilized for exploiting the relation among communities in [5]. However, although various methods have been proposed for community discovery from networks [8,10], it is still difficult to identify the true community labels.

3   Community Structure Based Node Scores

Our node scores consider the community structure in terms of nodes, not links as in most previous approaches [7]. A vector representation of nodes is constructed based on modularity to reflect the community structure of a network.

3.1   Node Vectors Based on Community Structure

Modularity has been utilized as a standard for community discovery in network analysis[4]. It was shown that the maximization of modularity can be sought by finding the eigenvector for the largest eigenvalue of the following matrix [7]:

$$\mathbf{B} = \mathbf{A} - \mathbf{P} \tag{1}$$

where $\mathbf{P} = \mathbf{k}\mathbf{k}^T/2m$ for $\mathbf{k} = \mathbf{A}\mathbf{1}_n$ ($\mathbf{k}$ is the degree vector).

By utilizing several eigenvectors of $\mathbf{B}$ in eq.(1) with several largest positive eigenvalues, the modularity matrix $\mathbf{B}$ can be approximately decomposed as:

$$\mathbf{B} \simeq \mathbf{U}\mathbf{\Lambda}\mathbf{U}^T \tag{2}$$

where $\mathbf{U} = [\mathbf{u}_1, \cdots, \mathbf{u}_q]$ are the eigenvectors of $\mathbf{B}$ with the descending order of eigenvalues, and $\mathbf{\Lambda}$ is the diagonal matrix with the corresponding eigenvalues. Based on eq.(2), the following data representation was proposed [7]:

$$\mathbf{X} = \mathbf{U}\mathbf{\Lambda}^{1/2} \tag{3}$$

In our approach, the i-th row of $\mathbf{X}$ in eq.(3) is used as the vector representation of the i-th node, and is called as a node vector.

3.2   Inverse Vector Density

A node score was proposed in terms of the mutual angle between node vectors in eq.(3) [11]. The number of "near-by" node vectors is utilized for identifying

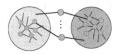

Fig. 2. Synthetic network (CL_*_*_)

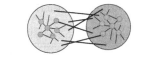

Fig. 3. Synthetic network (CL_*)

border nodes. This node score is defined as:

$$\mathbf{D} = diag(\|\boldsymbol{x}_1\|, \dots, \|\boldsymbol{x}_n\|) \tag{4}$$

$$\mathbf{X}_1 = \mathbf{D}^{-1}\mathbf{X} \tag{5}$$

$$\boldsymbol{\Theta} = \cos^{-1}(\mathbf{X}_1\mathbf{X}_1{}^T) \tag{6}$$

$$f(\boldsymbol{\Theta}_{ij}, \theta) = \begin{cases} 1 & (\boldsymbol{\Theta}_{ij} < \theta) \\ 0 & (otherwise) \end{cases} \tag{7}$$

$$ivd(\boldsymbol{x}_i) = \frac{1}{\sum_j^n f(\boldsymbol{\Theta}_{ij}, \theta)} \tag{8}$$

where $\mathbf{D}$ in eq.(4) is a diagonal matrix with elements $\|\boldsymbol{x}_1\|, \dots, \|\boldsymbol{x}_n\|$, and θ is a threshold. The value of $ivd(\cdot)$ in eq.(8) corresponds to the score.

The function f in eq.(7) checks if the angle θ_{ij} in eq.(6) is less than the specified threshold θ. Finally, since border nodes have relatively small number of near-by node vectors, the node score is calculated by taking the inverse of the number of near-by vectors. This node score is called IVD (inverse vector density).

3.3   Community Centrality Based Inverse Vector Density

Removal of hub nodes, which act as mediators of information diffusion over the network, also seems effective for network immunization. However, the node score of a hub node gets rather small with IVD. One of the reasons is that, the direction of each node vector is utilized in IVD, but its norm is not yet utilized.

The square norm of a node vector was regarded to what extent the node is central to a community [7], and was named as community centrality: $cc(\boldsymbol{x}_i) = \boldsymbol{x}_i^T \boldsymbol{x}_i$. This was reflected on IVD, and another node score was proposed as [11]:

$$ccivd(\boldsymbol{x}_i) = cc(\boldsymbol{x}_i) \times ivd(\boldsymbol{x}_i) \tag{9}$$

This is called CCIVD (Community Centrality based Inverse Vector Density).

4   Evaluations

4.1   Experimental Settings

Networks Extensive experiments were conducted over both synthetic and real-world networks. Utilized networks are shown in Table 1 and Table 2.

Table 1. Synthetic networks

dataset	#nodes	#links (ave.)
CL_2_5_1	105	708
CL_3_5_1	165	1064
CL_2	100	355.7
CL_3	150	548.1
CL_4	200	744.3
CL_5	250	931.3

Table 2. Real-world networks

dataset	#nodes	#links
karate	34	78
dolphins	62	159
lesmis	77	254
polbooks	105	441
netscience	379	914
celegansneural	297	2148

When constructing synthetic networks, each component (community) was generated using Barabási-Albert (BA) model [1] by setting the degree distribution $p(k) \propto k^{-3}$, where k denotes the degree of a node. The initial degree was set to 4 in order to generate sparse networks. After generating communities, they were connected either through five intermediate nodes (as illustrated in Fig. 2) or directly with links (Fig. 3). Since synthetic networks are generated based on random networks, we constructed 10 networks for each type and report the average result.

As real world networks, we used three networks in Table 2, which are available as GML (graph markup language) format [2].

Quality Measures. Following the quality measure in [5], the relative size S of the largest connected component (LCC) in a network was measured against the node occupation probability p. After removing some nodes from a network with n nodes, these are calculated as:

$$S = \frac{|LCC|}{n}, \quad p = \frac{\#remaining\ nodes}{n} \tag{10}$$

where $|LCC|$ is the number of nodes in LCC. The smaller S is, the better a immunization strategy of networks is, since it can prevent the spreading of contamination over the whole network (see Fig. 1).

Compared Methods. For comparison, network immunization based on various concepts of network centrality in Section 2.3 were compared. The node with the maximum centrality was repeatedly selected and removed in each method:

D : degree centrality
B : betweenness centrality
EVC : eigenvector centrality
CC : community centrality
ModC : meta-graph based centrality [5]

In our methods, parameters (q and θ) were set based on preliminary experiments.

---

[2] http://www-personal.umich.edu/~mejn/netdata/

Fig. 4. Visualization result (CL_2_5_1)   **Fig. 5.** Visualization result (CL_5)

Immunization Strategies Network immunization was conducted by removing the node with the maximal node score (e.g., centrality). The following strategies were evaluated for the calculation of node scores:

single : scores were calculated only once with respect to the whole network.
recalc : scores were re-calculated when a node is removed from a network.

The strategy recalc can utilize up-to-date node scores even after some nodes are removed, in compensation for the additional computational cost. Since ModC requires re-calculation of centrality, it was not evaluated for single strategy.

4.2 Visualization of Node Scores

In order to verify that our node scores can identify nodes which connect communities in a network, we compared the node scores and the score with **B** (betweenness centrality). The size of each node is depicted proportional to its node score in each method. The results with single strategy are shown in Fig. 4(CL_2_5_1), Fig. 5(CL_5), and

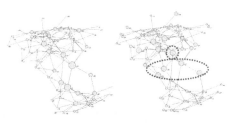

Fig. 6. Visualization result (dolphins)

Fig. 6 (dolphins). The left-hand side in these figures corresponds to B (betweenness centrality), and the right-hand side to IVD (the result of CCIVD was similar to IVD and thus not shown here).

By comparing the visualized networks, we can see that IVD could identify nodes which connect communities (encompassed with dotted red circles the figures). Since the visualized networks with IVD is similar to those with B, which is known to be effective for immunization despite its rather large time complexity, the node score can be said as effective for identifying intermediating nodes among communities.

4.3 Results of Synthetic Networks

Results of synthetic networks (average of 10 runs) are shown in Fig. 7 and Fig. 8. The horizontal axis is the node occupation probability p, and the vertical one is the relative size S of LCC in eq.(10). In the legend, gray lines with "x" are

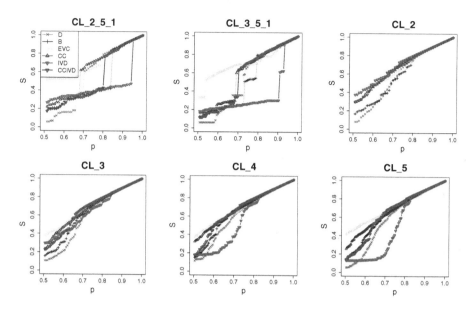

Fig. 7. Results on synthetic networks (single)

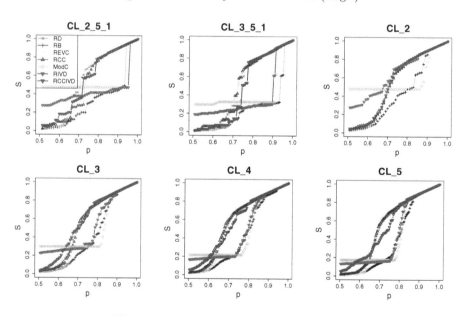

Fig. 8. Results of synthetic networks (recalc)

for D, black lines with "+" for B, yellow lines with squares for EVC, blue lines with upper triangles for CC, and water blue lines with diamonds for ModC. The proposed node scores are shown with lower triangle (IVD (green lines) and CCIVD (red lines)) in Fig. 7. For recalc strategy (Fig. 8), prefix "R" is put on the method name (except for ModC).

As shown in Fig. 7 and Fig. 8, IVD effectively immunized the networks in both single and recalc strategies. Especially, it showed the best performance for single strategy (rapid decrease of S with respect to p), and showed similar result with RB for recalc strategy around $p \geq 0.8$. On the other hand, unfortunately, CCIVD did not outperform B for single strategy, but it showed good performance for recalc strategy when p gets small (i.e., after large number of nodes are removed from a network). Compared with ModC, which also utilizes the community structure of a network (but requires community labels), RIVD outperformed ModC for all p, and RCCIVD showed better performance when p gets small.

4.4 Results of Real-World Networks

Results of real world networks in Table 2 are shown in Fig. 9 and Fig. 10. Both IVD and CCIVD showed almost equivalent performance with B for single strategy. For recalc strategy, the performance of RCCIVD was similar to RB (especially for karate and lesmis), and CCIVD outperformed IVD for real-world networks. This would be because the removal of hub nodes (with large community centrality) is effective for immunization of real-world networks. As in synthetic networks, the performance of ModC saturated after a network was divided into disconnected components. On the other hand, with the proposed methods, the value of S continued to fall even after a network was divided into disconnected components.

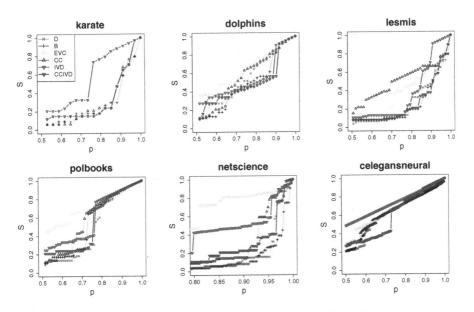

Fig. 9. Results of real-world networks (single)

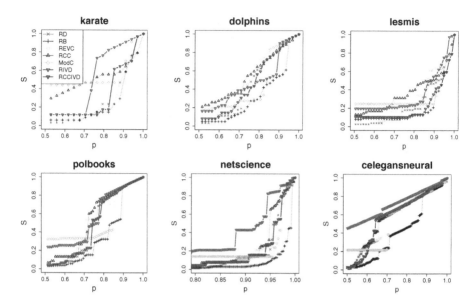

Fig. 10. Results of real-world networks (recalc)

4.5   Discussions

Our node scores (IVD and CCIVD) showed comparable performance with B (be-tweenness centrality), which is known to be effective for network immunization, in most networks. As in RB, the performance of these methods improved with recalc strategy, albeit this strategy requires much more computational effort. In addition, CCIVD showed better performance than CC, which is solely based on the norm of node vectors. This indicates the effectiveness of IVD for reflecting the community structure of a network in terms of the distribution of node vectors.

As shown in [5], utilization of the community structure of a network is effective for network immunization. However, finding community labels of nodes with maximum modularity is NP-complete [3]. The proposed approach can exploit the community structure of a network in terms of the distribution of node vectors, without relying on the externally supplied community labels of nodes.

5   Concluding Remarks

This paper reported a comparative study of community structure based node scores over both synthetic and real-world networks. Since contamination is prop-agated among groups of nodes (communities) through intermediating nodes in a networks, such nodes are identified based on the community structure of a network *without* requiring community labels of nodes. The characteristics of the proposed node vectors was analyzed. Extensive experiments were conducted to compare the node scores with other centrality based immunization strategies. The results are encouraging, and indicate that the node scores are promising.

Immediate future work includes more in-depth analysis of the node vectors and their relations. Especially, we plan to conduct the analysis of kernel density in terms of the histogram of node vectors for determining the appropriate parameter (e.g., θ) in our approach.

Acknowledgements. This work is partially supported by the grant-in-aid for scientific research (No. 24300049) funded by MEXT, Japan, Casio Science Promotion Foundation, and Toyota Physical & Chemical Research Institute.

References

1. Barabási, A.-L., Albert, R.: Emergence of scaling in random networks. Science 286, 509–512 (1999)
2. Brandes, U.: A faster algorithm for betweenness centrality. Journal of Mathematical Sociology, 163–177 (2001)
3. Brandes, U., Delling, D., Gaertler, M., Görke, R., Hoefer, M., Nikoloski, Z., Wagner, D.: On modularity clustering. IEEE Transactions on Knowledge and Data Engineering 20(2), 172–188 (2008)
4. Clauset, A., Newman, M.E.J., Moore, C.: Finding community structure in very large networks. Physical Review E 70(6), 066111 (2004)
5. Masuda, N.: Immunization of networks with community structure. New Journal of Physics 11, 123018 (2011), doi:10.1088/1367-2630/11/12/123018
6. Mika, P.: Social Networks and the Semantic Web. Springer (2007)
7. Newman, M.: Finding community structure using the eigenvectors of matrices. Physical Review E 76(3), 036104(2006)
8. Newman, M.: Networks: An Introduction. Oxford University Press (2010)
9. Restrepo, J.G., Ott, E., Hunt, B.R.: Characterizing the dynamical importance of network nodes and links. Physical Review Letters 97, 094102 (2006)
10. Yoshida, T.: Toward finding hidden communities based on user profile. Journal of Intelligent Information Systems (2011) (accepted)
11. Yoshida, T., Yamada, Y.: Community Structure Based Node Scores for Network Immunization. In: Anthony, P., Ishizuka, M., Lukose, D. (eds.) PRICAI 2012. LNCS, vol. 7458, pp. 899–902. Springer, Heidelberg (2012)

Semantic Precision and Recall for Evaluating Incoherent Ontology Mappings

Qiu Ji[1], Zhiqiang Gao[1], Zhisheng Huang[2], and Man Zhu[1]

[1] School of Computer Science and Engineering, Southeast University, Nanjing, China
{jiqiu,zqgao,mzhu}@seu.edu.cn
[2] Department of Computer Science, Vrije University Amsterdam, The Netherlands
huang@cs.vu.nl

Abstract. Ontology mapping plays an important role in the Semantic Web, which generates correspondences between different ontologies. Usually, precision and recall are used to evaluate the performance of a mapping method. However, they do not take into account of the semantics of the mapping. Thus, semantic precision and recall are proposed to resolve the restricted set-theoretic foundation of precision and recall. But the semantic measures do not consider the incoherence in a mapping which causes some trivialization problems. In this paper, we propose semantic measures for evaluating *incoherent ontology mappings*. Specifically, a general definition of semantic measures is given based on a set of formal definitions capturing reasoning with incoherent mappings. Then we develop a concrete approach to reasoning with incoherent mappings, which results in some specific semantic measures. Finally, we conduct experiments on the data set of conference track provided by OAEI[1].

1 Introduction

Ontology mapping plays an important role in the Semantic Web to solve heterogeneous problems between semantically presented data sources. So far, many ontology mapping algorithms have been developed (see surveys in [17,2,18]). With the increasing number of ontology mapping algorithms, some evaluation measures have been used to compare them, such as precision and recall [3]. These measures compare a mapping provided by a mapping algorithm with a reference mapping on syntactic level without considering the semantics of the mapping. Here, a reference mapping is usually created by domain experts and is reliable. To deal with this problem, Euzenat proposes the semantic precision and recall in [5], where a mapping is evaluated in a semantic way.

Recently, there is an increasing interest in dealing with logical contradictions caused by mappings (see [15] and [16] for examples). In some of these work, an ontology mapping is translated into description logic (DL) axioms. This treatment of an ontology mapping is useful for ontology integration [11], mapping

---

[1] OAEI indicates ontology alignment evaluation initiative, which is a platform to evaluate various ontology mapping systems.

R. Huang et al. (Eds.): AMT 2012, LNCS 7669, pp. 338–347, 2012.
© Springer-Verlag Berlin Heidelberg 2012

revision [16], mapping debugging [15] and mapping evaluation [12,14]. In these scenarios, *incoherence*[2] of a mapping is usually an unavoidable problem. As reported in [6] about the results in OAEI, more than 80% analyzed mappings are incoherent for most of the mapping systems.

Although incoherence is different from inconsistency[3] as an incoherent ontology can be consistent, we also encounter some trivialization problems because of the explosive problem of inconsistency (i.e. everything can be inferred from an inconsistent ontology): first, an unsatisfiable concept is a sub-concept of any concept; second, an unsatisfiable concept is equivalent to any other unsatisfiable concept. Thus, when defining a semantic measure for an incoherent mapping, we cannot apply standard DL semantics as it will result in counter-intuitive results.

In order to overcome the trivialization problems, we provide a set of formal definitions capturing reasoning with incoherent mappings. Based on these definitions, we then give a general definition of semantic precision and recall. To instantiate the general semantic measures, one concrete approach to reasoning with incoherent mappings is proposed. The key benefit of our incoherent-tolerant approach is that, every correspondence which is inferred by using our approach can be explained in a meaningful way. That is to say, for each such correspondence c, we can find a coherent subset from the merged ontology to infer c. This coherent subset can be served as a justification of c from the merged ontology.

2   Preliminaries

We introduce the notions of an unsatisfiable concept and incoherence in a DL-based ontology defined in [9]. More details about Description Logics (DLs for short) can be found in the DL handbook [1].

Definition 1. *(Unsatisfiable Concept)* *A named concept C in an ontology O is unsatisfiable iff for each model $\mathcal{I}$ of O, $C^{\mathcal{I}} = \emptyset$. Otherwise C is satisfiable.*

This definition means that a named concept is unsatisfiable in an ontology O iff the concept is interpreted as an empty set by all models of O.

Definition 2. *(Incoherent Ontology)* *An ontology O is incoherent iff there exists at least one unsatisfiable concept in O. Otherwise O is coherent.*

Given two ontologies O_1 and O_2, we can define correspondences.

Definition 3. *(Correspondence)[7]* *Let O_1 and O_2 be two ontologies, Q be a function that defines sets of mappable elements $Q(O_1)$ and $Q(O_2)$. A correspondence is a 4-tuple $\langle e, e', r, \alpha \rangle$ such that $e \in Q(O_1)$ and $e' \in Q(O_2)$, r is a semantic relation, and $\alpha \in [0,1]$ is a confidence value. A mapping consists of a set of correspondences.*

---

[2] A mapping is incoherent if there is a concept in the merged ontology which is interpreted as an empty set and no such concepts exist in the single ontologies.

[3] A mapping is inconsistent if the merged ontology is inconsistent and two single ontologies are consistent. An ontology is inconsistent iff. there is no model in the ontology.

There is no restriction on function Q and semantic relation r. The mappable elements $Q(O)$ could be concepts, object properties, data properties and individuals. As for the semantic relations, we mainly consider the equivalence relation and subsumption relation. A mapping is a set of correspondences whose elements are mappable. The correspondences can be divided into two categories:

Definition 4. *(Complex/Non-Complex Correspondences)[8] A correspondence $c = \langle e, e', r, \alpha \rangle$ is non-complex if both e and e' are atomic concept names or property names in the corresponding aligned ontologies. Otherwise c is complex.*

In this paper, we only consider those non-complex correspondences to focus on the explanation of our semantic measures. Note that, most of the existing ontology mapping algorithms or systems generate non-complex mappings [8].

Definition 5. *(Mapping Semantics)[13] Given a mapping $\mathcal{M}$ between ontologies O_1 and O_2. A correspondence $\langle e, e', r, \alpha \rangle \in \mathcal{M}$ can be converted to a DL axiom in this way: $t(\langle e, e', r, \alpha \rangle) = er e'$, where t is a translation function. $t(\mathcal{M}) = \{t(c) : c \in \mathcal{M}\}$.*

Take a correspondence $c = \langle SocialEvent, Document, \sqsubseteq, 1.0 \rangle$ as an example. We have $t(c) = SocialEvent \sqsubseteq Document$. For convenience, $O_1 \cup_{\mathcal{M}} O_2$ is used to indicate the merged ontology $O_1 \cup O_2 \cup t(\mathcal{M})$.

Definition 6. *(Incoherent mapping)[13] Assume we have two ontologies O_1 and O_2 and a mapping $\mathcal{M}$ between them. $\mathcal{M}$ is incoherent with O_1 and O_2 iff there exists a concept C in O_i with $i \in \{1, 2\}$ such that C is satisfiable in O_i and unsatisfiable in $O_1 \cup_{\mathcal{M}} O_2$. Otherwise, $\mathcal{M}$ is coherent.*

Definition 6 is given based on a consistent merged ontology. The incoherence of a mapping most occurs when matching ontologies on terminological level.

3   Formal Definitions

3.1   Reasoning with Incoherent Mappings

To cope with the trivialization problems in a mapping, we define the meaningfulness of an incoherency reasoner which is inspired by the notion of meaningfulness of an inconsistency reasoner given in [10]. An inconsistency reasoner is one which can reason with inconsistent ontologies, without relying on a repair of the inconsistencies in the ontologies. Namely, an inconsistency reasoner can return meaningful answers, without suffering from the explosive problem in the classical reasoning (i.e, any statement is a consequence of an inconsistent knowledge base). Similarly we define an incoherency reasoner as one which can reason with incoherent ontologies without relying on a repair of the incoherency in the ontologies. That is, an incoherency reasoner can return meaningful answers, without suffering from the trivialization problems caused by unsatisfiable concepts in the classical reasoning.

Definition 7. *(Meaningfulness) Assume we have two ontologies O_1 and O_2 and a mapping $\mathcal{M}$ between them. For a non-complex correspondence $c = \langle e, e', r, \alpha \rangle$, an answer provided by an incoherency reasoner is meaningful iff the following condition holds:*

$$\Sigma \Vdash t(c) \Rightarrow$$
$$(\exists \Sigma' \subseteq \Sigma)(\Sigma' \not\models e \sqsubseteq \bot \ \text{and} \ \Sigma' \not\models e' \sqsubseteq \bot \ \text{and} \ \Sigma' \models t(c)),$$

where $\Sigma = O_1 \cup_{\mathcal{M}} O_2$. An incoherency reasoner is regarded as meaningful iff all of the answers are meaningful.

In this definition, t is a translation function (see Definition 5). This definition indicates that, for a non-complex correspondence c, if $t(c)$ can be inferred from an incoherent set Σ then there exists a subset Σ' of Σ such that Σ' is coherent w.r.t. c and Σ' can infer $t(c)$ using classical reasoning. Here, we say Σ' is coherent w.r.t. c, if e and e' are satisfiable in Σ'. Definition 7 implies that an incoherency reasoner performs as a classical reasoner if Σ is coherent w.r.t. c.

In Definition 7, we only ensure that the signatures appearing in c are satisfiable in Σ'. This means there may exist some other unsatisfiable entities in Σ' and thus Σ' is still incoherent. Since our aim is to avoid the trivial inference for c, we do not care about other unsatisfiable concepts. In our view, it is reasonable to have other unsatisfiable entities in Σ' as repairing a mapping is not our aim.

As completeness and soundness are two main properties of a reasoner, we analyze them for our reasoner according to the definitions given in [10].

Definition 8. *(Local completeness) For a non-complex correspondence c, an incoherency reasoner is locally complete w.r.t. a coherent subset Σ' w.r.t. c of Σ ($\Sigma = O_1 \cup_{\mathcal{M}} O_2$) iff the following condition is satisfied:*

$$\Sigma' \models t(c) \Rightarrow \Sigma \Vdash t(c).$$

Namely, if $t(c)$ can be inferred from the given coherent subset Σ' w.r.t. c by applying a standard DL reasoner, it should be also inferred by Σ by using the incoherency reasoner.

Definition 9. *(Local soundness) For a non-complex correspondence c, an answer to a query $\Sigma \Vdash t(c)$ is regarded as locally sound w.r.t. a subset $\Sigma' \subseteq \Sigma$ ($\Sigma = O_1 \cup_{\mathcal{M}} O_2$) which is coherent w.r.t. c, iff the following condition is satisfied:*

$$\Sigma \Vdash t(c) \Rightarrow \Sigma' \models t(c).$$

That is, if $t(c)$ can be inferred by using our reasoner, it should be implied by the given coherent subset Σ' w.r.t. c by applying a standard DL reasoner.

3.2   Semantic Precision and Recall

With the introduction of the definitions that an incoherency reasoner should fulfill, we can adapt the required notations in [5] to define our semantic measures.

Definition 10. (α-**Consequence of a mapping**) *Given a mapping $\mathcal{M}$ between two ontologies O_1 and O_2, a correspondence c is an α-consequence of O_1, O_2 and $\mathcal{M}$ iff we have*

$$O_1 \cup_\mathcal{M} O_2 \Vdash t(c).$$

This also can be written as $\mathcal{M} \Vdash_{O_1,O_2} c$.

Definition 11. (**Closure of a mapping**) *Given a mapping $\mathcal{M}$ between two ontologies O_1 and O_2, the closure of the mapping, denoted as $Cn^{Inco}(\mathcal{M})$, can be defined as the following:*

$$Cn^{Inco}(\mathcal{M}) = \{c | \mathcal{M} \Vdash_{O_1,O_2} c\}.$$

That is, the closure of a mapping is the set of all α-consequences.

Definition 12. (**Semantic precision and recall**) *Given a reference mapping $\mathcal{R}$, the semantic precision of some mapping $\mathcal{M}$ is defined as below:*

$$P_{sem}^{Inco}(\mathcal{M}, \mathcal{R}) = \frac{|\mathcal{M} \cap Cn^{Inco}(\mathcal{R})|}{|\mathcal{M}|} = \frac{|\{c \in \mathcal{M} : \mathcal{R} \Vdash_{O_1,O_2} c\}|}{|\mathcal{M}|}.$$

The semantic recall is defined by

$$R_{sem}^{Inco}(\mathcal{M}, \mathcal{R}) = \frac{|Cn^{Inco}(\mathcal{M}) \cap \mathcal{R}|}{|\mathcal{R}|} = \frac{|\{c \in \mathcal{R} : \mathcal{M} \Vdash_{O_1,O_2} c\}|}{|\mathcal{R}|}.$$

Now we discuss some properties of the semantic measures defined in [5]. First, it is obvious that P_{sem}^{Inco} and R_{sem}^{Inco} satisfy *positiveness, completeness-maximality* and *correctness-maximality*. The property of positiveness means the values of semantic measures are no less than zero. Completeness-maximality is a property to indicate that the value of the recall is 1 if all correspondences in $\mathcal{R}$ can be inferred by $\mathcal{M}$. Similarly, the property of correctness-maximality shows that we know that the value of the precision is 1 if all correspondences in $\mathcal{M}$ can be inferred by $\mathcal{R}$. Second, our semantic measures may not satisfy the property *boundedness*. This property means the values of semantic measures should be no less than those values obtained by using standard measures without considering semantics. Take $R_{sem}^{Inco}(\mathcal{M}, \mathcal{R})$ as an example. If $\mathcal{M}$ is incoherent, a correspondence $c \in \mathcal{M}$ may not be included in $Cn^{Inco}(\mathcal{M})$ if we fail to find a coherent subset Σ' w.r.t. c such that $\Sigma' \models t(c)$. That is, we may not always have $\mathcal{M} \subseteq Cn^{Inco}(\mathcal{M})$ and thus the boundness w.r.t. R_{sem}^{Inco} may not hold.

4   Concrete Approach to Reasoning with Incoherent Mappings

From Section 3 we can see that reasoning with incoherent mappings is vital for our definitions of semantic measures. Here, we propose a novel approach to reasoning with an incoherent mapping based on selection functions. This

---

Algorithm 1: The algorithm to compute $\mathcal{M} \Vdash_{O_1,O_2} c$

Data: Two ontologies O_1 and O_2, a mapping $\mathcal{M}$ between the two ontologies, and a non-complex correspondence $c = \langle e_1, e_2, r, \alpha \rangle$.

Result: Boolean value

```
 1 begin
 2 │   Σ ← O₁ ∪ℳ O₂;
 3 │   Σ' ← ∅;
 4 │   k ← 0;
 5 │   if t(c) ∈ Σ then
 6 │   └   return true
 7 │   while s_k(Σ, t(c)) ≠ ∅ do
 8 │   │   Σ' ← Σ' ∪ s_k(Σ, t(c));
 9 │   │   if Σ' ⊨ e₁ ⊑⊥ or Σ' ⊨ e₂ ⊑⊥ then
10 │   │   │   Σ' ← Σ' \ s_k(Σ, t(c));
11 │   │   │   for ax ∈ s_k(Σ, t(c)) do
12 │   │   │   │   Σ' ← Σ' ∪ {ax};
13 │   │   │   │   if Σ' ⊨ e₁ ⊑⊥ or Σ' ⊨ e₂ ⊑⊥ then
14 │   │   │   │   └   Σ' ← Σ' \ {ax};
15 │   │   if Σ' ⊨ t(c) then
16 │   │   └   return true
17 │   └   k ← k + 1;
18 │   return false
19 end
```

---

approach (see Algorithm 1) is proposed based on the linear extension strategy in [10] which uses selection functions. Unlike the approach given in [10], ours deals with incoherence instead of inconsistency. Besides, two heuristic strategies have been exploited to improve the efficiency.

Algorithm 1 takes ontologies O_1 and O_2, a mapping $\mathcal{M}$ between them and a correspondence c as inputs and outputs whether c can be inferred by $\mathcal{M}$ based on a relevance-directed selection function s. An axiom is directly relevant to another axiom if they share at least one signature[4]. An axiom is directly relevant to a set S of axioms if this axiom is directly relevant to an axiom in S. $s_k(\Sigma, t(c))$ indicates a set of axioms which are directly relevant to $s_{k-1}(\Sigma, t(c))$ $(k > 0)$ and $s_0(\Sigma, t(c))$ includes those axioms which are directly relevant to $t(c)$.

In the approach, we first check whether c is explicitly included in $\mathcal{M}$ (see Line 5) to improve the reasoning efficiency. If not, the approach iterates on the sets of selected axioms and terminates if the current set of selected axioms s_k is empty (see Line 7) or if $t(c)$ can be inferred (see Line 15). For each iteration, if adding the current $s_k(\Sigma, t(c))$ to Σ' makes the modified Σ' incoherent w.r.t. c, then we check the axioms in $s_k(\Sigma, t(c))$ one by one. In this way, those relevant axioms are kept for reasoning as many as possible.

It is easy to check that the incoherency reasoner based on Algorithm 1 satisfies meaningfulness, local soundness and local completeness. Besides, the semantic measures defined by this approach satisfy the property of boundedness. This is because for an incoherent mapping $\mathcal{M}$, any $c \in \mathcal{M}$ can be inferred by $Cn(\mathcal{M})$ and thus we have $\mathcal{M} \subseteq Cn(\mathcal{M})$. So the boundedness property is satisfied.

---

[4] A signature means an atomic concept name, property name or individual name.

5   Experimental Evaluation

Our measures have been implemented using OWL API 3.0.0 and the standard reasoning tasks are performed using Pellet[5]. Our experiments were performed on a laptop with 2.13 GHz Intel(R) Core(TM) i3 CPU and 2.00 GB of RAM using Windows 7. Sun's Java 1.6.0 was used for Java-based tools and the maximum heap space was set to 1GB. When evaluating a mapping by applying a specific evaluation method, the timeout is set to 30 minutes.

In this section, we compare our measures (marked as "SF-based approach") with those without considering semantics (marked as "No semantics") and the semantic measures [5] defined by a standard DL reasoner (marked as "DL semantics"). We also compare various mapping systems by applying different measures.

5.1   Data set

We use the data set in the conference track provided by OAEI 2009. It consists of a set of expressive ontologies, the reference mappings and the mappings generated by the mapping systems which have participated in the contest (see [6] for more details). For our tests, we choose those ontology pairs whose corresponding reference mappings are available. In this way, 16 out of 21 reference mappings are selected which involve 16 ontology pairs and 7 individual ontologies. Among these selected individual ontologies, ontology edas has 624 axioms and the sizes of other ontologies vary from 116 to 354. The ontology pairs are listed as followings: 1: cmt-conference, 2: cmt-confOf, 3: cmt-edas, 4: cmt-ekaw, 5: cmt-iasted, 6: cmt-sigkdd, 7: confOf-edas, 8: confOf-ekaw, 9: confOf-iasted, 10: confOf-sigkdd, 11: edas-ekaw, 12: edas-iasted, 13: edas-sigkdd, 14: ekaw-iasted, 15: ekaw-sigkdd, 16: iasted-sigkdd. The participated systems provide mapping results for all of the selected ontology pairs. All participated systems can be seen in Figure 1.

Notes: (1) We do not consider the correspondences between a data property and an object property as we pay more attention to the explanation of our evaluation approaches. (2) The individuals in ontologies iasted and edas have been removed since they cause inconsistency in some merged ontologies and we currently only focus on dealing with incoherence. (3) Those correspondences whose confidence values are no less than 0.2 are regarded as correct. (4) Both of the systems AgrMaker and aroma generate 93.75% incoherent mappings. As for system ASMOV, only 6.25% incoherent mappings are generated. For others, all of the generated mappings between the selected ontology pairs are incoherent.

5.2   Comparison of Various Evaluation Methods

To compare various evaluation measures, we choose the mapping results generated by the system kosimap. Because this system provides more interesting results which contain more correspondences associated with distinct confidence values.

---

[5] http://clarkparsia.com/pellet/

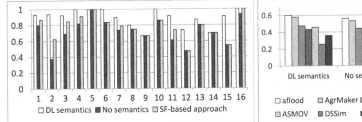
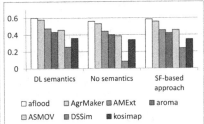

Fig. 1. The left figure shows the performance of various evaluation measures w.r.t. the recall over the mappings generated by the system kosimap. The figure on the right shows the comparison of different mapping systems w.r.t. the average f-measures.

The figure on the left in Figure 1 shows the performance of various evaluation methods to compute the recalls. From the figure, we first observe that the standard recall (i.e. "No semantics") always returns the lowest value. This can be explained by the fact that the semantic recalls defined by DL semantics and our approach satisfy the boundedness property. Second, the values returned by the semantic recall defined by DL semantics are no less than those returned by that defined by our approach. It is because DL semantics suffers from the trivialization problems, thus many meaningless correspondences can be inferred. Finally, we can observe that the difference between the values returned by the two semantic recalls is obvious. The average value of the difference is 0.15. Besides, the largest one is 0.36 for the mapping between ekaw and sigkdd. It shows us that many meaningless correspondences have been inferred by the approach using DL semantics when computing semantic recalls.

Similarly, we can analyze the results with respect to the precisions. First, we can see that the same values are returned by the semantic precisions for all reference mappings except the mapping between ontologies edas and iasted which is incoherent. Second, the values returned by the semantic precisions are no less than the value returned by the standard precision as the semantic precisions also satisfy the boundedness property.

5.3   Comparison of Mapping Systems

To compare different mapping systems using our measures, all selected ontology pairs except the pairs edas-iasted and iasted-sigkdd are considered as our approach fails to return recall values for some mappings between the two pairs within the time limit. The figure on the right in Figure 1 shows the comparison of various mapping systems w.r.t. the average f-measure. Here, f-measure is a harmonic mean (i.e. $\frac{2*precision*recall}{precision+recall}$) of precision and recall.

Obviously, different semantic measures produce similar values for a specific system. Take system aflood as an example. The average values are 0.602 and 0.585 returned by the f-measures defined by DL semantics and ours respectively. This is because the precisions always produce the same values and the difference among the recall values is not very big although it is obvious.

We can also see that different approaches to compute f-measures rank the systems differently. This is mainly caused by the positions of the systems AMExt, aroma and ASMOV. It maybe because that the percentage of implied correspondences in a mapping generated by ASMOV is higher than that generated by aroma. Thus the ordering of the two systems is changed when the evaluation approaches vary from the standard approach to the approach using DL semantics.

6   Related Work

Many measures have been proposed to evaluate ontology mappings in the literature. Here, we focus on those measures related to precision and recall.

As the standard measures are sensitive to the syntactics of a mapping, a general framework has been proposed in [4]. The framework are instantiated by three different measures considering some particular aspects of mapping utility based on the proximity of correspondence sets. The weakness of these measures is that they do not take the semantics of the ontologies into account.

Due to the restricted set-theoretic foundation of the traditional precision and recall, Euzenat defined semantic measures in [5], where a first-order model theoretic semantics is adopted. The semantic measures are proposed using the deductive closure bounded by a finite set. However, their measures are not tailored to evaluate incoherent mapping.

In [8], a simplified version of semantic precision and recall has been proposed. This restricted version defines the semantics of mappings according to a translation into a logical theory in [13]. Comparing their work, we use the same mapping semantics but different formulas to compute precision and recall. Another difference is that the measures defined in [8] do not deal with incoherence.

In [12], the proposed quality measures consider the number of unsatisfiable concepts and the effort to repair an incoherent mapping separately. The main difference between this work with ours is that, we consider the logic implication to see whether a correspondence can be inferred or not, where every correspondence inferred by using our approach can be explained in a meaningful way.

7   Conclusion and Future Work

In this paper, we proposed some novel semantic precision and recall for evaluating incoherent ontology mappings. We first provided a set of formal definitions which capture reasoning with incoherent mappings. Based on these definitions, we gave a general definition of semantic precision and recall. Then we proposed a concrete incoherence-tolerant reasoning approach based on selection functions, which results in a pair of specific semantic precision and recall. Finally, our experimental results showed that our measures are promising to evaluate incoherent mappings which ensures each correspondence inferred by using our approach can be explained in a meaningful way. This guarantee has been implemented by finding a coherent subset w.r.t. a correspondence using our approach.

In the future, we will extend our work to evaluate inconsistent ontology mappings and develop new methods to improve the efficiency of our approach. Besides, more data sets will be considered.

Acknowledgements. We gratefully acknowledge funding from the National Science Foundation of China under grants 60873153, 60803061, and 61170165.

References

1. Baader, F., Calvanese, D., McGuinness, D.L., Nardi, D., Patel-Schneider, P.F. (eds.): The Description Logic Handbook: Theory, Implementation, and Applications. Cambridge University Press (2003)
2. Bellahsene, Z., Bonifati, A., Rahm, E. (eds.): Schema Matching and Mapping. Springer (2011)
3. Do, H.-H., Melnik, S., Rahm, E.: Comparison of Schema Matching Evaluations. In: Chaudhri, A.B., Jeckle, M., Rahm, E., Unland, R. (eds.) NODe-WS 2002. LNCS, vol. 2593, pp. 221–237. Springer, Heidelberg (2003)
4. Ehrig, M., Euzenat, J.: Relaxed precision and recall for ontology matching. In: K-CAP Integrating Ontologies Workshop, Banff, Alberta, Canada (2005)
5. Euzenat, J.: Semantic precision and recall for ontology alignment evaluation. In: IJCAI, pp. 348–353 (2007)
6. Euzenat, J., et al.: Results of the ontology alignment evaluation initiative 2009. In: OM, pp. 15–24 (2009)
7. Euzenat, J., Shvaiko, P.: Ontology Matching. Springer, Heidelberg (2007)
8. Fleischhacker, D., Stuckenschmidt, H.: A practical implementation of semantic precision and recall. In: CISIS, pp. 986–991 (2010)
9. Flouris, G., Huang, Z., Pan, J.Z., Plexousakis, D., Wache, H.: Inconsistencies, negations and changes in ontologies. In: AAAI, pp. 1295–1300 (2006)
10. Huang, Z., van Harmelen, F., ten Teije, A.: Reasoning with inconsistent ontologies. In: IJCAI, pp. 454–459. Morgan Kaufmann (2005)
11. Jiménez-Ruiz, E., Grau, B.C., Horrocks, I., Berlanga, R.: Ontology Integration Using Mappings: Towards Getting the Right Logical Consequences. In: Aroyo, L., Traverso, P., Ciravegna, F., Cimiano, P., Heath, T., Hyvönen, E., Mizoguchi, R., Oren, E., Sabou, M., Simperl, E. (eds.) ESWC 2009. LNCS, vol. 5554, pp. 173–187. Springer, Heidelberg (2009)
12. Meilicke, C., Stuckenschmidt, H.: Incoherence as a basis for measuring the quality of ontology mappings. In: OM, pp. 1–12 (2008)
13. Meilicke, C., Stuckenschmidt, H.: An Efficient Method for Computing Alignment Diagnoses. In: Polleres, A., Swift, T. (eds.) RR 2009. LNCS, vol. 5837, pp. 182–196. Springer, Heidelberg (2009)
14. Meilicke, C., Stuckenschmidt, H., Šváb-Zamazal, O.: A Reasoning-Based Support Tool for Ontology Mapping Evaluation. In: Aroyo, L., Traverso, P., Ciravegna, F., Cimiano, P., Heath, T., Hyvönen, E., Mizoguchi, R., Oren, E., Sabou, M., Simperl, E. (eds.) ESWC 2009. LNCS, vol. 5554, pp. 878–882. Springer, Heidelberg (2009)
15. Meilicke, C., Völker, J., Stuckenschmidt, H.: Learning Disjointness for Debugging Mappings between Lightweight Ontologies. In: Gangemi, A., Euzenat, J. (eds.) EKAW 2008. LNCS (LNAI), vol. 5268, pp. 93–108. Springer, Heidelberg (2008)
16. Qi, G., Ji, Q., Haase, P.: A Conflict-Based Operator for Mapping Revision. In: Bernstein, A., Karger, D.R., Heath, T., Feigenbaum, L., Maynard, D., Motta, E., Thirunarayan, K. (eds.) ISWC 2009. LNCS, vol. 5823, pp. 521–536. Springer, Heidelberg (2009)
17. Shvaiko, P., Euzenat, J.: A Survey of Schema-Based Matching Approaches. In: Spaccapietra, S. (ed.) Journal on Data Semantics IV. LNCS, vol. 3730, pp. 146–171. Springer, Heidelberg (2005)
18. Shvaiko, P., Euzenat, J.: Ontology matching: state of the art and future challenges. IEEE Transactions on Knowledge and Data Engineering (2012)

A Self-organization Method for Reorganizing Resources in a Distributed Network

Xiaolong Guo and Jiajin Huang

International WIC Institute, Beijing University of Technology,
Beijing, 100124, China
{gxlvip,hjj}@emails.bjut.edu.cn

Abstract. Using bio-inspired agents to reorganize resource has been adopted to address the distributed resource optimization issue in distributed networks. This paper presents a self-organization algorithm to reorganize resources by the use of autonomous agents which can exchange their resources each other. Agents are equipped with two operations (pull and push operation) and three behaviors (best selection, better selection and random selection). And at every moment, agents probabilistically choose a behavior to perform. Experimental results indicate that the strategy has a positive influence on system performance.

1 Introduction

A distributed network (e.g. P2P networks) consists of a number of connected nodes while there is no central controller in the distributed network. So how to find the resource descriptors in the distributed network becomes a key problem because those resources of the network need be obtained and accessed effectively [1]. To discovery a number of useful descriptors in a short time, a solution is to put similar resource descriptors together effectively in restricted regions by the use of a self-organization method.

How to solve these problems in distributed environments is also a key problem in Web Intelligence [2,3]. By using bio-inspired algorithms to reorganize resource has been adopted in distributed networks. The So-grid system employs several mobile agents to reorganize resources. In So-grid, the agent travel the Grid through P2P interconnections. While the agent is traveling, it picks up the resource descriptor if the resource descriptor is different from others in the available region and it drops down the resource descriptors at appropriate nodes [1]. As same as mentioned in [4], agents in the So-grid system may make a large amount of random movements before they pick up or drop down resource descriptors, therefore some nodes may not be visited. In this case, these descriptors in non-visited nodes may not be placed at suitable nodes. The work of [5] uses an agent-view to connect the agent society. An agent-view contains agents whose contents are similar. However, these agents exchange their all contents in their agent-views with each other, which results in much more communication costs. In order to overcome the shortcoming of the above work, this paper presents

R. Huang et al. (Eds.): AMT 2012, LNCS 7669, pp. 348–356, 2012.

a method to reorganize descriptors by using agents that reside on the nodes. These agents can (1) select descriptors which are not suitable to stay the node and then push these descriptors to other nodes until that a suitable node can keep these descriptors; (2) decide whether they keep these descriptors or not according to their local environments when they receive descriptors pushed by other agents. This paper is partly inspired by the work of [4], who proposed an ant clustering algorithm in which data are placed in a 2-Dimension Grid and an agent represents a data object. However, our work is different from the application domain of [4]. Our work faces a large-scale, dynamic and distributed network environment. So we need a method to be adapted to the network environment.

In [4], agents can be clustered dynamically by collaboration based on local information in a 2-Dimension Grid. In other words, moving behaviors of agents have autonomy features. An autonomous agent system includes the environment of agents, agent's profiles and the behavior rules of agent [6,7]. In this paper, we design a kind of agent which resides on the node of a network. An agent's local environment is the set of descriptors in a restricted region. Agents push or pull the resource descriptors to put similar resource descriptors together dynamically. By the certification of experiments, the method is valid in networks with different structures, different scales, and dynamically changing numbers of nodes.

2   The Formulation of the Self-organization Strategy

According to [6,7,8], an agent essentially has the following properties: it can live in an environment; it can sense its local environment since there is no central controller and it has some behaviors driven by certain purposes.

A network is represented as graph $G = <V, E>$, where $V = \{v_1, v_2, \cdots v_N\}$ is a set of nodes and $L = \{(v_i, v_j) | 1 \leq i, j \leq N, i \neq j, v_i, v_j \in V\}$ is a set of edges. $N = |V|$ represents the total number of nodes in a network, and (v_i, v_j) indicates that an edge exists between node v_i and v_j. As a result, v_i and v_j are so-called neighbor nodes with each other.

The profile of node v is defined by $v = \{nodeID, \Gamma, D\}$, where $nodeID$ denotes the identifier of a node, Γ represents total neighbors of node v ($v.\Gamma = \{v_i | (v, v_i) \in L\}$), D represents the set of descriptors in node v.

The local environment of agent a_i is defined as E_l, which is a set of resource descriptors both in the node resided by agent a_i and the node's neighbors. At time t, the local environment of agent a_i residing on node v_i is $E_l(a_i(t)) = \{d | d \in \bigcup v.D, v \in v_i.\Gamma \bigcup \{v_i\}\}$. Corresponding to the agent's local environment, we will set the entire network G as an agent's living environment E_g. It is impossible for agents to know about the situation of E_g in distributed environments. Therefore the local environment is most important for an agent. Each agent can only sense their own local environments and the local environments E_l could be better if there are more same class of descriptors. Each agent's local environment E_l is not static, which can be modified by other agents' behaviors which are defined by the profile of agent a_i.

The profile of agent a_i is $\{ID, nodeID, behaviors, rules\}$, where ID denotes the identifier of the agent $a_i.nodeID$ represents the identifier of a node resided by agent $a_i.behaviors$ includes the pull and push behaviors. According to the local environment, an agent decides whether it pushes resource descriptors from the node resided by it to other nodes or pull the suitable resource descriptor to the node resided by it from other nodes. Thus, $rules$ are used to decide the two behaviors. By these behaviors, each agent changes their local environments. Agents interact with each other in the changing local environments.

Borrowing the concept of fitness measure in [1,9], we also let agent a decide whether it will pull or push the descriptor d at time t based on a fitness function $f_a(d, t)$ given by

$$f_a(d, t) = \frac{|E_l(a(t))|}{R_d}, \tag{1}$$

where R_d is the number of the descriptors of the whole class in the local environment $E_l(a(t))$ of agent a at time t, and $|E_l(a(t))|$ is the overall number of descriptors that are accumulated in the local environment at time t. From Equ. (1), we can see the more there are the same class of descriptors d in the local environment of agent a, the more the descriptor d is suitable in the node resided by agent a.

At time t, agent a selects descriptor $d \in a(t).nodeID.D$ with least fitness $f_a(d, t)$ and then evaluates push probability function to decide whether or not to push descriptor d out of the resided node. The push probability is shown by

$$p_{push}^a(d, t) = (\frac{k_1}{k_1 + f_a(d, t)})^2, d \in a(t).nodeID.D. \tag{2}$$

When f is much lower than the constant k_1 (i.e. there are no descriptors as same class as descriptor d in the local environment of agent a), the push probability is higher. As more descriptors as the same class as descriptor d are accumulated in the local environment of agent a, the fitness of the descriptor d increases. With the increase of the fitness (specially f is much higher than k_1), the value of the push probability for descriptor d becomes lower. So at time t, given a random value r, agent a does not select descriptor d when $p_{push}^a(d, t) < r$; agent a finds the node is not suitable to descriptor d when $p_{push}^a(d, t) > r$ and then will push descriptor d out of the node. One of three push behaviors based on the primitive behaviors of autonomous agents are probabilistically performed [6,7]. (a) agent a pushes the descriptor d to the node selected randomly from $a.nodeID.\Gamma$. (b) agent a pushes the descriptor d to the node with the most f for the descriptor from $a.nodeID.\Gamma$. (c) agent a randomly selects two nodes from $a.nodeID.\Gamma$, and then push the descriptor to the node with more f for the descriptor d between the two nodes.

When agent a receives a descriptor d from agent a' at time t, the agent evaluates pull probability function in order to decide whether it will pull descriptor d to the node resided by it or not. The pull probability is shown by

$$p_{pull}^a(d, t) = (\frac{f_a(d, t)}{k_2 + f_a(d, t)})^2, d \in a(t).nodeID.D. \tag{3}$$

As opposite to the push probability, the more same descriptors are accumulated in the agent's local environment, the higher the pull probability for this class of descriptors becomes. When $p_{pull}^a(d,t) < r$, the agent does not receive the descriptor d. When $p_{pull}^a(d,t) > r$, the agent receives the descriptor d. Let $D_{pull}^a(t)$ be the set of descriptors pulled by agent a from others nodes at time t. And at the next time $t+1$, we have $a(t+1).nodeID.D = a(t).nodeID.D \bigcup D_{pull}^a(t)$. By these behaviors, each agent's local environment are changed. At the next time, agents decide their behaviors in new local environments. From this perspective, we can see agents interact with each other in the changing local environment.

The measures of push and pull probability function are similar to the pick and drop probability function shown in [1,9]. However, agents in [1,9] can move in the enviroment, where agents in this paper need not move in the network and each agent resides in a node of the network. The push and pull operations could be embedded in the push-pull protocol of P2P [10] to implement a real-world application. According to the algorithm framework in [1,4,9], we summarize our algorithm as Algorithm 1.

Algorithm 1. A Self-organization Algorithm for Reorganizing Resources

---

01. Initialize the parameters
02. FOR t=1 to T
03. FOR each agent a DO
04. $a(t).nodeID.D = a(t-1).nodeID.D \bigcup D_{pull}^a(t-1)$
05. $D_{pull}^a(t) = \emptyset$
06.   IF agent a receives the non-acceptance message from other agents
07.   Keep descriptors pushed by agent a at time t-1
08.   END IF
09.   Select a descriptor d in $a.nodeID.D$ with least fitness $f_a(d,t)$
10.   Compute the push probability $p_{push}^a(d,t)$
11. r=random($[0,1]$)
12.   IF $p_{push}^a(d,t) > r$
13.   Probabilistically determine a primitive behavior to select a node in $a(t).nodeID.\Gamma$
14.   Push descriptor d to the node
15.   END IF
16.   FOR each descriptor d' pushed by other agents
17.   Compute the pull probability $p_{pull}^a(d',t)$
18. r'=random($[0,1]$)
19.   IF $p_{pull}^a(d',t) > r'$
20. $D_{pull}^a(t) = D_{pull}^a(t) \bigcup \{d'\}$
21.   ELSE
22.   Send the non-acceptance message to the agent pushing descriptor d'
23.   END IF
24.   END FOR
25. END FOR
26. END FOR

---

3 Experiment

To prove the effectiveness of the proposed algorithm, we set a series of experiments to answer the following questions. Can the strategy remain effective in different networks? The networks include networks generated by different network generators, networks with different numbers of networks, and networks with the dynamically growing scale. Firstly, we get a network topology by the use of a network generator, store a certain number of descriptors on nodes, and then assign each agent in each node. We use three kinds of complex networks which are listed in the Table 1. The GLP [11] and WS [12] networks are used to test the effectiveness in networks generated by different network generators and networks with different numbers of networks. The active network is used to test the effectiveness networks with the dynamically growing scale. Then, we let agents operate the resource descriptors by the push or pull operation as shown in Algorithm 1.

Table 1. The structures of networks

Network	Parameter
GLP	Power law network with $\alpha=2$
WS	Small World network (the neighbor connection is 5)
Active	Dynamic network (increasing 20 nodes at a moment)

The effectiveness of the strategy is evaluated trough a spatial entropy function. For agent a, let $fr_a(i)$ denote the percent of descriptors of class C_i in the local environment of agent a and Nc denote the number of classes. The local entropy $En_l(a)$ [1] is given by

$$En_l(a) = \frac{-\sum_{i=1}^{N_c} fr_a(i) \lg fr_a(i)}{\lg N_c}. \tag{4}$$

Based on the local entropy, the overall entropy Eo is defined as the average of En_l of all agents. According to the well-known Shannon's formula, the more minimal is the Eo, the more effective is the strategy. In each experiment, we run the algorithm ten times and the shown results are an average value of the ten results.

The experiment about the effectiveness in different kinds of networks is done in WS and GLP networks. In each kinds of network, the numbers of nodes are 1000 and 2000 respectively. The results are shown in Fig. 1. Fig. 1(a) shows that the value of overall entropy decreases from 0.96 to 0.4 after 1000 cycle times (T=1000) and then tends to a stable state in the WS network with 1000 nodes; the value of overall entropy is stable at 0.6 in the GLP network with 1000 nodes. Fig. 1(b) shows the similar result in networks with 2000 nodes. From Fig. 1,

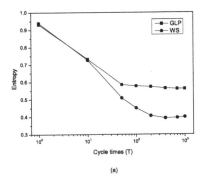
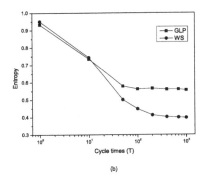

Fig. 1. Results in two kinds of networks. (a) results in networks with 1000 nodes; (b) results in networks with 2000 nodes.

we can see that the presented algorithm is useful for the WS network and the ability of clustering is better in the WS network.

The results of the experiment in the WS networks with different number of nodes are shown in Fig. 2. The number of nodes is from 100 to 5000. Fig. 2(a) shows that the value of overall entropy becomes stable with the increasing of the cycle times. And from Fig. 2(b), we know that the results of experiment could be better if the network contains more nodes.

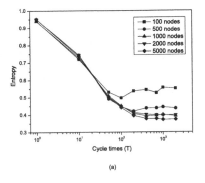
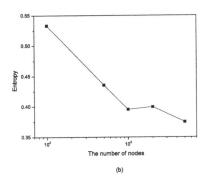

Fig. 2. Results in a kind of network with different numbers of nodes

The results of the experiment in the active network are shown in Fig. 3. The active network is a WS-based network with dynamically increasing 20 nodes at a moment. From Fig. 3, we can see that the value of overall entropy is about 0.4 at the stable state which is as same as the result in WS networks shown in Fig. 1 and Fig. 2. These results of experiments tell us that the ability of the clustering is similar whatever the network is dynamic or not.

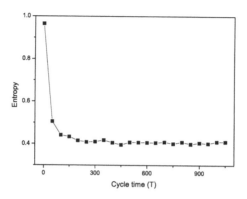

Fig. 3. Results in a dynamic network

The experimental results of relevant parameters are shown in Fig. 4. The network is a WS network with 1000 nodes. Fig. 4 shows that the value of entropy is lower and the result is better when the parameter k_1 is bigger than the parameter k_2.

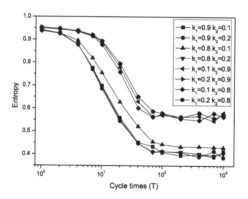

Fig. 4. Results with different combinations of k_1 and k_2 parameters

Fig. 5 shows how the random selection affects the result. We set the ratio of random selection is 0%, 5%, 10%, and 15% respectively. From Fig. 5, we can see that the random selection behavior affects the value of overall entropy and the stability of the entropy curve. When the percent of the random selection is lower, the overall entropy value is lower. But the entropy value is a little higher if there is lack of the random selection, as the strategy maybe stuck in a local optima without random selection [7].

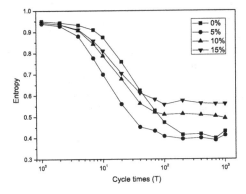

Fig. 5. Results with different assignment of random selection

4   Conclusions

This paper presents a self-organization algorithms for the resource reorganization in a decentralized network which is no central control. Inspired by the work of [9], two different kinds of agent behaviors are defined as push and pull operation. These behaviors change agents' local environments and agents exchange descriptors with each other. Simulated experiential results showed that the algorithm is effective in the controlled propagation and reorganization of information. In this primary method, we only transplant the similar data cluster method of [4] into the network environment. The future work will focus on how to balance data distribution in nodes and how to design effective positive feedback [7] and self-learning [8] principles for agents to speed up the reorganization.

Acknowledgements. This work is supported by Beijing Natural Science Foundation (4102007), the CAS/SAFEA International Partnership Program for Creative Research Teams, the China Postdoctoral Science Foundation Funded Project (2012M510298), Projected by Beijing Postdoctoral Research Foundation (2012ZZ-04), and the doctor foundation of Beijing University of Technology (X0002020201101).

References

1. Forestiero, A., Mastroianni, C., Spezzano, G.: So-Grid: a Self-organizing Grid Featuring Bio-inspired Algorithms. ACM Transactions on Autonomous and Adaptive System 3(2), 1–37 (2008)
2. Zhong, N., Liu, J.M., Yao, Y.Y.: In Search of the Wisdom Web. IEEE Computer 35(11), 27–31 (2002)
3. Zhong, N., Liu, J.M., Yao, Y.Y.: Envisioning Intelligent Information Technologies (iIT) From the Stand-Point of Web Intelligence (WI). Communications of the ACM 50(3), 89–94 (2007)

4. Xu, X.H., Chen, L., He, P.: A Novel Ant Clustering Algorithm Based on Cellular Automata. Web Intelligence and Agent Systems 5(1), 1–14 (2007)

5. Zhang, H.Z., Croft, W.B., Levine, B.N., Lesser, V.R.: A Multi-agent Approach for Peer-to-peer Based Information Retrieval System. In: Proceeding of the 3rd International Joint Conference on Autonomous Agents and Multiagent Systems (AAMAS 2004), pp. 456–463 (2004)

6. Gao, C., Liu, J.M., Zhong, N.: Network Immunization with Distributed Autonomy-oriented Entities. IEEE Transactions on Parallel Distributed Systems 22(7), 1222–1229 (2011)

7. Liu, J.M., Jin, X.L., Tsui, K.C.: Autonomy Oriented Computing (AOC): from Problem Solving to Complex System Modeling. Kluwer Academic Publishers (2005)

8. Liu, J., Zhong, W.C., Jiao, L.C.: A Multiagent Evolutionary Algorithm for Combinatorial Optimization Problems. IEEE Transactions on Systems, Man, and Cybernetics - Part B: Cybernetics 40(1), 229–240 (2010)

9. Bonabeau, E., Dorigo, M., Theraulaz, G.: Swarm Intelligence: from Natural to Artificial Systems. Oxford University Press (1999)

10. Gueret, C.: Nature-Inspired Dissemination of Information in P2P Networks. In: Abraham, A., et al. (eds.) Computational Social Network Analysis, pp. 267–290 (2010)

11. Bu, T., Towsley, D.: On Distinguishing between Internet Power Law Topology Generators. In: Proceeding of 2002 IEEE International Conference on Computer Communications (INFOCOM 2002), pp. 638–647 (2002)

12. Watts, D.J., Strogatz, S.H.: Collecive dynamics of 'small-world' networks. Nature 393, 440–442 (1998)

Extracting Property Semantics from Japanese Wikipedia

Susumu Tamagawa[1], Takeshi Morita[2], and Takahira Yamaguchi[1]

[1] Keio University
{s_tamagawa,yamaguti}@ae.keio.ac.jp
[2] Aoyama Gakuin University

Abstract. Here is discussed how to build up ontology with many properties from Japanese Wikipedia. The ontology includes is-a relationship (rdfs:subClassOf), class-instance relationship (rdf:type) and synonym relation (skos:altLabel) moreover it includes property relations and types. Property relations are triples, property domain (rdfs:domain) and property range (rdfs:range). Property types are object (owl:ObjectProperty), data (owl:DatatypeProperty), symmetric (owl:SymmetricProperty), transitive (owl:TransitiveProperty), functional (owl:FunctionalProperty) and inverse functional (owl:InverseFunctionalProperty).

Keywords: Wikipedia, ontology, property definition, ontology learning.

1 Introduction

It is useful to build up large-scale ontologies for information searching and data integration. Among the large-scale ontologies are WordNet [1] and Cyc [2]. However, it takes many costs to build these ontologies by hands. Moreover, the manual ontology engineering process makes many bugs come up and maintenance and update difficult. For these reasons, more attention comes to build (semi) automatic ontologies on research, ontology learning.

Wikipedia, the Web-based open encyclopedia, becomes popular as new information resources [4]. Since Wikipedia has rich vocabulary, good updatability, and semistructuredness, there is differences between Wikipedia and ontologies when compared with free text. Thus, ontology learning from Wikipedia becomes popular. In addition more attention comes to publish and share RDF databases that describe instances in recent year. This is called Linked Open Data (LOD). It is easy to search information by linking instances mutually and we can use them easily for a lot of our applications and services. In English, DBpedia [5] is widely used as a hub on LOD. However DBpedia has less Japanese information by problems between languages and There is no hub that takes the place of DBpedia in Japan.

We proposed a large-scale and general-purpose ontology (called JWO:Japanese Wikipedia Ontology below) learning methods to construct relations (Is-a relationship, Class-Instance relationship, property domain, synonyms and instance relations with properties) using the Japanese Wikipedia as resources [3]. However, there are some problems in our ontology compared with OWL specification. For

R. Huang et al. (Eds.): AMT 2012, LNCS 7669, pp. 357–368, 2012.

example, we didn't define property ranges and types. So in this paper, we extract property names (that are including following types: object, data, symmetric, transitive, functional and inverse functional) and relations (that are triples, property domains and property ranges). After that, we add recent techniques to proposed techniques [3] and construct a large-scale and general-purpose ontology including class scheme hierarchy and many properties.

This paper is structured as follows: We introduce related works about deriving ontology from Wikipedia in Section 2. In Section 3 we explain about JWO and its property definition and the detailed extraction techniques to Wikipedia. In Section 4 we show the result of experiment that we actually applied the extraction techniques to Wikipedia. In Section 5 we show overall view and characteristics of the Wikipedia Ontology. Finally we present conclusion of this paper and our future work.

2   Related Work

Auer et al.'s DBpedia [5] constructed a large information database to extract RDF from semi-structured information resources of Wikipedia. They used information resource such as Infobox, external link, categories the article belongs to. However, properties and classes are built manually and there are only 170 classes and 720 properties. In addition, many properties from Infobox have not been integrated in ontology because properties by hand and Infobox properties are separating.

YAGO2 [6] was enhancing of the knowledge base of YAGO. It was not only enhancing WordNet but also spatially and temporally enhancing by using Wikipedia and GeoNames. They defined relations such as wasBornOnDate and isLocatedIn and related them and instances. However, they built these properties by hands and these domains and ranges ware built by hands too.

Fei Wu & Daniel Weld [7] built up a rich ontology by combining Wikipedia Infobox with WordNet using statistical-relational learning. They mapped Infobox classes (Template name) to WordNet nodes and they mapped attributes between related classes allowing property inheritance. This was a rich ontology but they constructed properties using only Wikipedia Infobox structure and they didn't examine property types.

3   Property Definition in Japanese Wikipedia Ontology

We proposed JWO learning methods to construct relations using the Japanese Wikipedia as resources [3]. Figure 1 shows JWO overview. We constructed Is-a relationships by matching the character string related to Wikipedia categories, matching category name and infobox template name and Scraping a table of contents headings, Class-Instance relationships by scraping listing pages, properties and triples (Note, in this paper, triple means the three sets of "instance, property, a value of propertyff") by modeling infoboxes, property domains by matching category name and class names and synonyms by scraping redirect links. In this session, we explain new methods and some proposed methods about property definition. Property definition consists following relationships and types. Note, they enclosed in parentheses are vocabularies (classes and properties) defined by OWL[1], RDFS[2], each of which corresponds to extracted relationships.

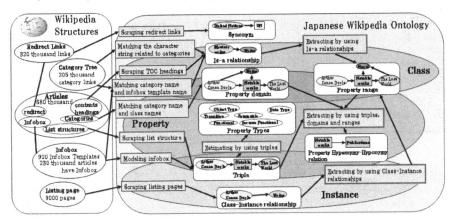

Fig. 1. Japanese Wikipedia Ontology Overview

A. Property Name and Triple
B. Property Domain (rdfs:domain)
C. Property Range (rdfs:range)
D. Property Hypernymy-Hyponymy Relationship(rdfs:subPropertyOf)
E. Property Types
    E1. Object Type (owl:ObjectProperty)
    E2. Data Type (owl:DatatypeProperty)
    E3. Symmetric Type (owl:SymmetricProperty)
    E4. Transitive Type (owl:TransitiveProperty)
    E5. Functional Type (owl:FunctionalProperty)
    E5. Inverse Functional Type (owl:InverseFunctionalProperty)

3.1 Extracting Property Names and Triples

Property names and triples in JWO are built by following two methods.

1. Extracting by modeling infobox
2. Extracting by scraping list structure in articles

Extracting Property Names and Triples by Modeling Infobox. This method was recently proposed by us in [3]. The three sets that have an infobox, "article name, item and its value" can also be seen as three other sets, "instance, property, and a value of property". Thus, the triple can be extracted from the infobox by scraping and using this structure. Further it is possible to

---

[1] OWL: http://www.w3.org/TR/owl-ref/
[2] RDFS: http://www.w3.org/TR/rdf-schema/

classify these property into owl:ObjectProperty and owl:DatatypeProperty as a
property type by using modeling rules for 40 infobox templates (They can cover
around 70% articles including infobox). Actually, the type of properties from
infobox templates are defined as owl:ObjectProperty, when values of the properties are resources. On the other hand, the type of the properties are defined
as owl:DatatypeProperty, when values of the properties are literals.

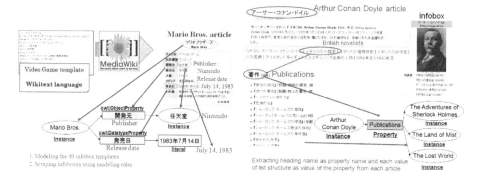

Fig. 2. Extracting by modeling infobox   **Fig. 3.** Extracting by scraping list
structure

**Extracting Property Names and Triples by Scraping List Structure in
Articles** A lot of Wikipedia articles have list structures. This method extracts
triples from list structures as "article name - heading name - each value". At that
time, we check categories that each article belongs and collect a lot of heading
names that appear in each category. As a result, it becomes possible to extract
the category that the article belongs as a property domain. The procedure of
this method is shown following a - d.

a. Extract categories and headings from each article of Wikipedia dump data.
b. Check occurrence rate of heading name from each category using (a).
c. Remove a few occurrence rate (5 or less) heading names from (b).
d. Extract heading name as property name and each value of list structure as
   value of the property from each article.

Figure 3 shows the example. In this case, we extract a heading name "Publications"
as a property and each value of list structure as property values in an article like
"Arthur Conan Doyle" (e.g. "The Land of Mist", "The Lost World"). A value of
property like "The Lost World" can be extracted by modeling Infobox triple too
but a value of property "The Land of Mist" can be extracted newly by this method.

3.2   Extracting Property Domain

This method was recently proposed by us in [3]. A subject of Infobox discussed
in 3.1 was an article name as an instance. For this reason, checking a category to
which an article that is a subject belongs makes it possible to define a domain

of a property. Thus, Infobox template name is extracted as a domain of each property. In Figure 4, categories that article "Ruby" belongs have a chance of defining as a domain of a property " Developer".

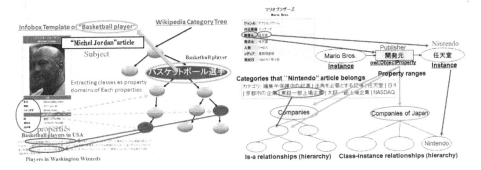

Fig. 4. Extracting property domain   **Fig. 5.** Extracting property range

3.3   Extracting Property Range

It was comparatively easy to define property domains because an instance name as a subject of a property corresponds to an article name in a triple and Infobox template name of an article can be considered to be a property domain. However, about property ranges, an instance name as a object of a property can't conclude to an article name, so it is difficult to define property ranges of all properties like property domain. Thus, for property ranges, we use following two methods.

1. Extracting by using Class-Instance relationships
2. Extracting by using Is-a relationships

It is often linked if a word has already had an article in Wikipedia and an article name corresponds to an instance name. So we matches an object of property triple (instance) extracted in 3.1 and instances of Class-Instance relationships in JWO. After that, we extract classes as property ranges that the instance belongs. Next, for extracting property ranges that can't be extracted by a previous technique, as well as a previous technique, we matches categories that an article (which is same name as an object of property triple) belong and classes of Is-a relationships in JWO. After that, we extract classes as property ranges. Figure 5 shows an example. In this case, a value of "Publisher" property "Nintendo" belongs to "Companies of Japan" class in Class-Instance relationships, so we extract "Companies in Japan" as a property range. In addition, we extract matched classes by matching categories that "Nintendo" article belongs and Is-a relationships.

3.4   Extracting Property Hypernymy-Hyponymy Relationship

The outline of the article is a feature of the infobox. Then, we try extracting hypernymy-hyponymy relationships by using this feature from extracted property names in 3.1. First, we matches values of extracted property names by modeling Infobox triple and by scraping list structure in Wikipedia articles in each subject of triples (instances). Then, when at least one matched value exists, we extract the extracted property name by scraping list structure in Wikipedia articles as an upper property candidate of the extracted property name by modeling Infobox triple. Next, we match domains and ranges in these candidate properties, then, when at least one matched properties exists, we extract these properties as a property hypernymy-hyponymy relationship.

Figure 6 show an example. In this case, a subject of triple "Arthur Conan Doyle" has a property "Publications" and its values "The Land of Mist, The Lost World and so on " and a property " Notable works" and its values "The Lost World, A Study in Scarlet and so on". Thus, we get "Publications" as an upper property candidate and " Notable works" as a lower property candidate. After that, when we match domains and ranges of these properties, we know that these properties have same domains like "Novelist" and same ranges like "Novel". Thus, we extract " Publications'- Notable works" as a property hypernymy-hyponymy relationship.

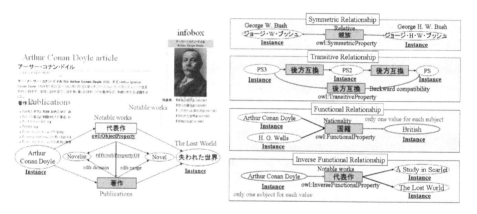

Fig. 6.   Extracting   property hype-hyponymy relation

Fig. 7. Estimating property types

3.5   Estimating Property Types

Extracted properties by modeling Infobox triple in 3.1 have already classified Object type and Data type. In this method, We try estimating symmetric, transitive, functional and inverse functional property type using extracted triples in 3.1. Note, symmetric property is a property for which holds that if "X-P-Y" is a triple of P, then "Y-P-X" is also an triple of P, transitive property is a property which means that if "X-P-Y" is a triple of P, and "Y-P-Z" is also triple of P, then we can infer "X-P-Z" is also a triple of P, functional property is a property which can have only one value "Y" for each instance "X" and inverse functional property is a property which can have only one instance "X" for each value "Y".

We try estimating symmetric property. We take a subject X and a value Y of each property P and if a triple "Y-P-X" of P exist, then we extract the property P as symmetric property candidate α. In addition, we try estimating with the ratio of all triples of P to extracted triples as symmetric relationships candidate. In figure 7, a property "Relative" holds the symmetric relation triples " George W. Bush-Relative- George H. W. Bush" and " George H. W. Bush-Relative- George W. Bush". Similarly we try estimating transitive, functional and inverse functional property by querying triples.

4   Experimental Results and Observations

Now we discuss the extracted results and observations from Wikipedia dump data in January 2012 using methods in 3.

4.1   Results and Observations of Property Names and Triples

Results and Observations by Modeling Infobox We extracted a total of 10,407 property names and 2,958,029 triples by using the method 3.1. We could convert Wikipedia article names including infobox into 221,318 instances as a subject of property triples. For example, we could extract "address" and "genre" property as object property and "birthPlace" and "Founded" property as datatype property.

Results and Observations by Scraping List Structure in Wikipedia Articles We extracted a total of 4,510 property names and 3,508,716 triples by using the method 3.1. We could convert Wikipedia article names into 274,874 instances as a subject of property triples. For example, "Staff", "Cast", "TV drama" and "Movie" property could be extracted. These properties can't be extracted by modeling infobox triple. We extracted 1,000 samples from the parent population of 3,508,716 triples and determined their truthfulness and falseness. We used following expression (1) as a formula for 95% confidence interval. In the formula (1), n represents the number of samples, N represents population, and $\hat{p}$ represents the estimated amount which is the number of accuracy samples divided by total number of samples. The results had an accuracy rate of 92.5±1.63%. Many errors were scraping errors and they were extracted when a lot of information was described in each line of list structures. For example, "Track listings" property can be seen in many Wikipedia album or single articles of singers but list structures in these articles has often described information such as Writers, Released Date and Length besides the Track listings. Thus, Writers, Released Date and Length was extracted as values of "Track listings" property. Fortunately, we think that it is possible to remove these errors to add more detailed rules of list structure with each category that articles belong.

$$[\quad \hat{p}-1.96\sqrt{\left(1-\tfrac{n}{N}\right)\tfrac{\hat{p}(1-\hat{p})}{n-1}} \quad , \quad \hat{p}+1.96\sqrt{\left(1-\tfrac{n}{N}\right)\tfrac{\hat{p}(1-\hat{p})}{n-1}} \quad] \qquad (1)$$

Table 1 shows the number of properties, triples, instances as a subject and accuracies of triples by using both methods in 3.1. We can extract 14,498 properties and 6,219,669 triples by using two methods (duplications are excluded). the accuracy rate of all 6,219,669 triples is 94.3±1.44%. This shows that the accuracy rate fells a little compared with modeling Infobox triple method but triples increases by around as much as 2 times. In addition, we could convert Wikipedia article names into 378,782 instances as a subject of property triples including 157,464 articles which didn't have infobox.

Table 1. The number of properties, triples, instances as a subject and accuracies of triples by both methods

Method	#Properties	#Triples	#Instances@	accuracy
By modeling infobox triples	10,407	2,958,029	221,318	95.2±1.33%
By scraping list structure	4,510	3,508,716	274,874	92.5±1.63%
Both methods	14,498	6,219,669	378,782	94.3±1.44%

Table 2. Example of extracted property domain and range

Property	Property Domain	#instance
Staff	TV program	26,251
Cast	drama	21,140
Staff	drama	10,871
Institution	Station	10,088

Property	Property Range	#Instance
TV Animation	Japanese Comics	23,195
Cast	Japanese Actor	20,633
TV Animation	Animation at midnight	15,956
Cast	Living people	44,766
Performer	Living People	22,175
Movie	Japanese Movie	16,370

4.2   Results and Observations of Property Domain

We could extract 12,613 property domains for 14,498 extracted properties in 4.1. Extracted properties by modeling Infobox triple has infobox template as domains, thus we could define property domain of all properties by modeling Infobox triple. In addition, we could define property domain of 2,206 properties by scraping list structure in Wikipedia articles. The accuracy rate was 94.8±1.22%. Table 2 shows examples of property domain and range. When a property has plurals property domains, those a common upper concept classes isn't defined. For example, "Staff" property has "TV program" and "TV drama" as property domain and it has a lot of classes as property domain besides them (e.g. " Radio program", "Baseball team"). In this case, we have to integrate a common upper concept classes.

4.3   Results and Observations of Property Range

First, we could extract 4,813 property ranges and 51,233 relationships for 14,498 extracted properties by using Class-Instance relationship method in 3.3. The accuracy rate was 84.3±2.24%. Many ranges are specialized in domain (e.g.in table 2 "Animation at midnight" range of "TVAnimation" property, "Japanese Actor" range of "cast" property). For example of errors, "Native dress" as property range of "nationality" property was extracted. This is an error caused because Class-Instance relationship is wrong. Many errors by this method are same reason, so we think that the accuracy by this method goes up by improving the accuracy of Class-Instance relationship, too.

Next, we could extract 3,641 property ranges and 35,946 relationships for 14,498 extracted properties by using Is-a relationship method in 3.3. The accuracy rate was 90.2±1.82%. A lot of ranges with high utilization rates include instances that are person who consists of actor such as a object of "cast" property, that concern train stations such as a object of "entrepreneur" property and that are countries such as a "Nationality". Thus, many ranges are related class like country, person, and trains. For example of errores, "Article on mathematics" was extracted as a range of "victory Frequency" property or "numberOfAccommodations" property. This is a error caused because of modeling insufficient, normally the property type becomes owl:DatatypeProperty, thus the range becomes literal (rdfs:Literal). In addition, "Living People" was extracted as a range of "Track listings" property. This is an error caused due to the scraping rule being inadequate.

Table 3. Example of property hypernymy-hyponymy relationship

Upper Property	Lower Property	#Appearance
Cast	Performer	2,082
Staff	DirectedBy	1,919
Staff	screenplayBy	1,514
Gods in shrine	Main God in shrine	237
Related company	Major subsidiaries	227

4.4 Results and Observations of Property Hypernymy-Hyponymy Relationship

We could extract 1,387 property hypernymy-hyponymy relationships by using a method in 4.4. We determined all relationships truthfulness and falseness by hand. Thus the accuracy was 57.5%. Then, we counted articles that relationship appears in each relationship and measured the relation between the number of appearance and the accuracy. As a result, number of appearance n is inversely proportional to the number of hypernymy-hyponymy relationships, but the accuracy is the highest when appearance n is 18 or more, and the accuracy is 75.7%. Table 3 shows examples of property hypernymy-hyponymy relationship.

There are a lot of property hypernymy-hyponymy relationships concerning televisions and movies like "cast" and "staff". And there are a lot of relations including the word "related" like "relatedCompany-majorSubsidiaries", too. For example of errors, the most incorrect are relations that upper and lower property name are the same meaning like "majorShareholder-mainShareholder". This is because separate property names were extracted because of synonymous problem.

4.5   Results and Observations of Property Types

We tried estimating by using methods in 3.5 with 14,498 extreacted properties and 6,219,669 extracted triples.

We could extract 10,927 symmetric relation triples and 415 symmetric property candidates. We determined all 415 properties truthfulness and falseness by hand. Thus the accuracy was 45.1%. "similarCommendationBadge" property had the most inclusion rate. There are a lot of properties by this method including words like "adjoin", "connect", and "related" (e.g. "adjoiningConstellation", "connectedRoad" and " relatedSchool"). But this method could extract such as "partner" and "relative".

Next, we could extract 340 transitive relation triples excluding symmetric relation and 210 transitive property candidates. However, we couldn't find transitive property in these candidates. As one of the reason of such a result, we think there is covering problem in Wikipedia. When we extract by this method, it is necessary to extract at least three triples that become the transitive relations. Thus, it is necessary to cover the information by Infobox or list structure as the same property name in the Wikipedia articles. For extracting transitive property, we have to refine property names, integrate the same one, and try extracting other method that use a part of nostructuredness information in articles.

And we could extract 185,700 functional relation triples and 2,267 functional property candidates. Thus the accuracy was 54.3%. For example, There were " Throws and Bats" property and "Administrative divisions" property. They are functional relations and their properties have only one instance as a value of property. Similarly, we could extract 47,295 inverse functional relation triples and 3,670 inverse functional property candidates and the accuracy was only 22.4%. As this reason, there is a problem of mark difference or synonym of property name like "Main work" and "Notable work" property. For example of inverse functional property, There were "Main belonging artist" property and "Collection listing" property. The most errors of both functional and inverse functional properties were properties which have values as literals. Type of these properties has to owl:DatatypeProperty (e.g. "Number of total games" as functional property,).

5   Evaluation and Observations of Japanese Wikipedia Ontology

5.1   Overall View of Property in Japanese Wikipedia Ontology

Table 4 shows accuracy, the number of properties, the number of triples of each property type in JWO. Table 5 shows the numbers and the 95% accuracy of triples, property domains, property ranges and property hypernymy-hyponymy relationships. JWO has 14,498 property and 6,219,669 triples. These properties have domains and ranges. The total number of relationships is over 6 million, indicating that a very large-scale ontology with many properties was built through ontology learning.

Table 4. Acuracy, the number of properties and triples of each property type in JWO

Types	#Property	#Acuracy	#Triple
All types	14,498	-	6,219,669
Datatype	204	-	982,771
Object	260	-	1,466,480
Symmetric	415	45.1%	21,854
Transitive	210	0%	1,020
Functional	2,267	54.3%	185,700
InverseFunctional	3,670	22.4%	47,295

Table 5. the number of relationships and accuracies in JWO Property

Relationship	#Relationships	Accuracy
All triples of property	6,219,669	94.3±1.44%
By modeling infobox	2,958,029	95.2±1.33%
By scraping list structure	3,508,716	92.5±1.63%
Property Domain	78,616	94.8±1.22%
Property Range	81,165	84.3±2.24%
Property Hyper-Hyponymy Relationship	1,387	57.5%

Table 6. Example of comparing Japanese Wikipedia Ontology to DBpedia

Concept	DBpedia		Japanese Wikipedia Ontology	
	Property	Examples of value	Property	Examples of value
	Genre*	Short story	Genre*	Short story
Person	notable works	Rashomon (1915)	notable works*	Rashomon
Ryunosuke	children	Hiroshi.A(eldest son)	child*	Hiroshi.A
Akutagawa	birth date+	1892-03-01	birthDate+	March 1, 1892
(Writer)	other	6 prop. 6 triples	other	7 prop. 63 triples
Inanimate	alt maxi	130m	Altitude	max:130m
Object	maire	Bertrand Delanoe	maire	Bertrand Delanoe
Paris(place)	other	22 prop. 69 triples	other	3 prop. 17 triples

5.2 Feature and Utility of Japanese Wikipedia Ontology

We further compared the Wikipedia Ontology properties to DBpedia [5] properties, which are representative, preexisting, RDF database. When comparing it, we used a Japanese data set of DBpedia. Table 6 shows some examples of compared properties. As for the number of extracted triples by our method, JWO (6,219,669) has around more 2.5 million triples than DBpedia (3,811,024). However, as for the number of extracted properties, our ontology (14,498) was only about 1,000 properties more than DBpedia (13,429). For this reason, DBpedia has an original property name like "wikiPageUsesTemplate". In addition, DBpedia extracts the triple directly from the Wikipedia infobox. Thus, a lot of property names are extracted in English or by abbreviation, so DBpedia doesn't specialize in Japanese. In addition, JWO can be 2.4 times also in the number of instances as a subject compared with DBpedia. This shows more article names are converted to an instance as a subject of property. In table 6, As for DBpedia, it doesn't specialize in Japanese so we can't understand meaning of some properties. DBpedia has original property names and detailed data type properties in place domain like "yearSun". Instead in JWO, there are detailed object type properties (e.g. "Publications" and "Family" in person domain, "sports" in place domain). In addition, "Family" property, "Child" property and " Consort" property in JWO are hymernymy-hyponymy relationship. This is a feature of JWO. Moreover, Wikipedia Ontology has detailed synonyms and class hierarchy compared with DBpedia because of specializing in Japanese. Thus, it is used be as a hub of Japanese Linked Open Data.

6   Conclusion

In this paper, we proposed and evaluated a method of building a large-scale and general-purpose Japanese ontology with many properties using the Japanese Wikipedia as resources. Through this study, not only were we able to show that Wikipedia is valuable resources for ontology learning for is-a relationships, but also showed its value in extracting other relationships.

The total numbers of articles in Wikipedia Japan as of January 2012 are over 750 thousands and it is still growing rapidly. Ontology becomes more popular for end users by Linked Open Data and the necessity of a general ontology construction has risen. However, ontology constructed from Wikipedia is useful because of including lower concepts like properties and instances. We provided JWO and search support tool using Linked Open Data as ways of utilizing JWO, WiLD : Wikipedia Linked Data Application, in our project page [8].

References

1. Bond, F., Isahara, H., Fujita, S., Uchimoto, K., Kuribayashi, T., Kanzaki, K.: Enhancing the Japanese WordNet. In: The 7th Workshop on Asian Language Resources, in Conjunction with ACL-IJCNLP (2009)
2. Lenat, D.B., Guha, R.V.: Building Large Knowledge Based Systems. Addison-Wesley (1990)
3. Tamagawa, S., Sakurai, S., Tejima, T., Morita, T., Izumi, N., Yamaguchi, T.: Learning a Large Scale of Ontology from Japanese Wikipedia. In: 2010 IEEE/WIC/ACM International Conference on Web Intelligence, pp. 279–286 (2010)
4. Nakayama, K., Hara, T., Nishio, S.: Wikipedia Mining for an Association Web Thesaurus Construction. In: Benatallah, B., Casati, F., Georgakopoulos, D., Bartolini, C., Sadiq, W., Godart, C. (eds.) WISE 2007. LNCS, vol. 4831, pp. 322–334. Springer, Heidelberg (2007)
5. Auer, S., Bizer, C., Kobilarov, G., Lehmann, J., Cyganiak, R., Ives, Z.G.: DBpedia: A Nucleus for a Web of Open Data. In: Aberer, K., Choi, K.-S., Noy, N., Allemang, D., Lee, K.-I., Nixon, L.J.B., Golbeck, J., Mika, P., Maynard, D., Mizoguchi, R., Schreiber, G., Cudré-Mauroux, P. (eds.) ISWC/ASWC 2007. LNCS, vol. 4825, pp. 722–735. Springer, Heidelberg (2007)
6. Hoffart, J., Suchanek, F., Berberich, K., Weikum, G.: YAGO2: A Spatially and Temporally Enhanced Knowledge Base from Wikipedia. Research Report MPI-I-2010-5-007, Max-Planck-Institut für Informatik (2010)
7. Wu, F., Weld, D.S.: Automatically Refining the Wikipedia Infobox Ontology. In: International World Wide Web Conference 2008, pp. 634–644 (2008)
8. Wikipedia Ontology Project, http://www.wikipediaontology.org

An Ontology Based Privacy Protection Model for Third-Party Platform

Haojun Yu[1], Yuqing Sun[1,*], and Jinyan Hu[2]

[1] School of Computer Science and Technology, Shandong University, Jinan250101, China
fengzhengjie2005@163.com, sun_yuqing@sdu.edu.cn
[2] School of Economics, Shandong University, Jinan 250100, China
hwx@sdu.edu.cn

Abstract. A third-party trading platform is a web based system that provides services for sellers and buyers. On such platform, users are required to provide personal information to ensure the authenticity and undeniability of a transaction. In this paper, we propose an ontology based privacy protection model for third-party platform, which allows buyers and sellers to define privacy policies according to their preferences and converts policies into ontology based forms. We introduce the property of *good* sellers who require the minimal information from buyers while satisfying other trading requirements. The proposed policy matching algorithm finds such sellers as candidates for a buyer request. A practical example is given to illustrate our model.

Keywords: third-party platform, privacy policy, ontology.

1   Introduction

A third-party trading platform is a web based system that provides business related services for sellers and buyers. According to a complete business process, services on such platform can be divided into three categories: pre-business services such as information publication and contract negotiation, in-process services such as trade confirmation and payment, post-business services such as logistics, etc. During this process, both buyers and sellers are required to release their personal information so as to ensure the authenticity and undeniability of a transaction, such as name, address, and intentions, etc. Such information includes identity information and business information, which involves buyer's privacy and may face various attacks [12]. Therefore, it is important to protect personal information against misuse.

Privacy protection on third-party platforms has attracted much attention from academia, government and business. Many professionals have devoted to this topic and a lot progresses have been made. From the view of law, the Online Data Privacy Regulation by the European Commission[6]makes many restrictions on web service providers. For example, a provider is only allowed to collect and use buyers' personal

---

* Corresponding author.

R. Huang et al. (Eds.): AMT 2012, LNCS 7669, pp. 369–376, 2012.
© Springer-Verlag Berlin Heidelberg 2012

information with their permissions. The regulation also considers buyers' preference as privacy policies, which state the ways a party gathers, uses, and releases personal information it collects. Privacy policies [9] should inform the buyer what specific information is collected, and whether it is kept confidential, shared with partners, or sold to other firms. These regulations provide a legal basis for privacy protection.

In this paper, we propose an ontology based privacy protection model for third-party trading platform according to the current Data Privacy Regulations. In our model, both buyers and sellers are allowed to express their privacy preferences as policies and then the policies can be compared based on domain ontologies. We introduce the property of *good* sellers who require the minimal information from buyers while satisfying other trading requirements. And then we propose a policy matching algorithm to automatically select such sellers. We implement our model using ontology editor Protégé and schema matching tool COMA++, and illustrate our model with a practical example. The rest of the paper is organized as follows. In Section 2, we present the ontology based privacy protection model. In Section 3, we propose the policy parsing and matching method. Then we discuss how the model could be implemented in Section4. In Section 5, we conclude the paper.

2   Related Work

Considering web based privacy protection, many models are proposed. Tunner et al. propose a privacy protection model for web based services[7].It introduces the concept of agents on behalf of buyers to negotiate with sellers on how much personal information should be released. In their model, sellers need to state why they collect personal information and should provide multiple elective information choices for buyers to choose. Bonatti et al.[3]show that the competition between sellers can reduce their information requests through a game-theoretic approach. They present a privacy protection mechanism based on Vickrey auction. Because the auction mechanism is truthful, the proposed mechanism induces sellers to ask for the exactly necessary information from buyers to deliver their service effectively and securely. However, these methods do not provide a way for buyers to define their privacy policies according to their preferences.

Some research works study how to implement the privacy policy interaction between users and systems. Garcia et al. [6] use semantic web technology to present an ontology-based privacy policy definition model. Buyers define some allowed constraints on the operations of their personal information. Hacker et al. [5] propose privacy ontology to enable buyers to understand the content on web so that buyers can match their preferences to the seller's policy and decide whether to use the given service. Gaoet al. [8] propose an ontology based approach for privacy protection in different application scenarios. They specify privacy policies in a semantic way and abstract the policies as trust attributes for privacy ontology, which would be used in evaluating trust in different applications.Hu et al. [13] propose a semantic legal policy definition model, which is in compliance with the laws. Then the policies are enforced automatically at the super-peer to enable Law-as-a-Service. The above methods only consider the case of exact policy match, but ignore the case that sellers and buyers may express the same object in different words.

From the above analysis, we find that quite a few shortcomings exist in current privacy protection and should be improved. First is the privacy preferences definition. Since people have various considerations on their personal information [4] and the required information in different business are not the same, it is necessary to allow buyers to define privacy policy according to their own preferences. Second is the automatic matching requirement. For the same information, sellers and buyers may express their understanding in different ways, such as home address and family address. So, there should be a semantic-based method of policy matching so as to get rid of ambiguity. Third point is the minimization principle. Consider the case that there are multiple candidate sellers with different requirements on buyers' personal information, the third-party platform should choose the *good* sellers who require minimal information while satisfying buyers' preferences.

3   The Ontology Based Privacy Protection Model

3.1   Basic Concepts

According to the latest Online Data Privacy Regulation [6], privacy policies on a third-party platform should include three aspects: the personal privacy items, such as name, contact information, the purpose of sellers' using information, such as credit card for purchase, and the process mode of personal information, such as the ID numbers must be deleted after transactions. We introduce the following basic notions.

Definition 1 (Item). An item is an attribute of a buyer and its associated sensitivity. Let PRI_ITM be a set of all attributes in a given system. An item is denoted as a pair of $t=<a, s>$, where $a \in PRI\_ITM$, $s \in [0..1]$, denotes an attribute and its sensitivity.

The attribute sensitivity shows how an attribute is important for a buyer. For a given item $t=<a, s>$, we use the notion $t.a$ and $t.s$ to denote an attribute a and its sensitivity s. The higher sensitivity is, the more important the attribute is for the buyer.

Privacy policy for third party platform involves three aspects. Privacy information is denoted as a finite set of items that a buyer releases in a transaction. Purpose is a finite set of ways that how sellers use privacy information. Process mode is denoted as a set of ways how sellers process privacy information after a transaction. Let INFO, PURPOSE, and PROCESS respectively denote the set of all privacy information, purposes of information, and process modes of information in a given system.

Definition 2 (Privacy Rule, PR for short). A privacy rule specifies the purpose and process mode of a set of privacy items. It is defined as a triple PR= (PIn, Pur, Pro), where $PIn \subseteq INFO$, is a set of privacy items, $Pur \subseteq PURPOSE$ is a set of purposes, and $Pro \subseteq PROCESS$ is a set of process modes.

For example, the privacy rule *PR*= ({*<name, 0.4>*, *<telephone, 0.6>*}, {*notification*}, {*retain*}) declares that the name and telephone are allowed to use in a transaction for the purpose of notification. The sensitivities of these attributes are 0.4 and 0.6, respectively. After the transaction, the seller should retain this information.

Definition 3 (Privacy Policy, PP for short). A privacy policy is a set of rules specified by a buyer or a seller before a transaction, which specifies how their information could be used in the periods of this transaction (e.g. registration, payment, logistics). It is denoted as PP= {PR_1, PR_2...PR_k}, where k is an integer, PR_i, i∈[1..k], PR_i∈PR, is a privacy rule. Let PSET be the set of all privacy policies used in a system.

3.2   The Ontology Based Privacy Protection Model

In this section, we describe the proposed ontology based privacy protection model in details. Our model (see Figure 1) contains three aspects.1) *Policy Definition* allows buyers and sellers to define privacy policies according to their preferences, which are based on the proposed definition.2) *Semantic Parse* converts policies into ontology-basedforms.3) *Policy Match* enforces an automatic policy matching to find the good service for buyers according to the policy matching algorithm.

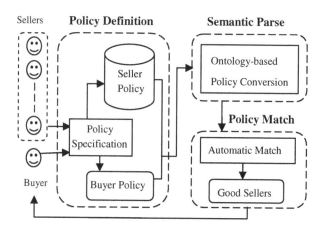

Fig. 1. An Ontology Based Privacy Protection Model

The details of a privacy policy matching process are given below. First, before transactions, sellers publish their services on a third-party trading platform with the specifications of privacy policies, namely the required information for buyers. These policies are stored in the seller policy repository. When a buyer invokes a transaction, he/she defines the privacy policies according to his/her preferences. Then, the trading system on the third-party platform semantically parses sellers' policies and the buyer's policy by converting them into ontology forms. Finally, according to policy matching, the platform finds the *good* seller candidate policy for the buyer and returns policy and the related services to the buyer. Policy matching is a group of matches between a buyer's policy and a set of sellers' policies, and is used to find the *good* seller policy. We present the policy matching problem as follows.

Definition 4 (Policy Matching Problem, PMP for short). Given a buyer privacy policy *pol*, a set of sellers privacy policies p_{set}= {pol_1, pol_2...pol_k}, where k is an integer, $pol_i \in PSET$, $i \in [1..k]$. The policy matching problem is to find the policy in p_{set} satisfying one of the following constraints:

1) *exact match*: pol_i exactly matches pol if and only if the following formula holds

$pol.\ PIn \supseteq pol_i.\ PIn \wedge pol.\ Pur = pol_i.\ Pur \wedge pol.Pro = pol_i.\ Pro \wedge minSizeof(pol_i.PIn) \wedge$
$minimal(pol_i.s)$, where $pol_i.s = \sum_{t_j \in pol_i.PIn} t_j.s$, $t_j.s$ is the sensitivity defines by the user, $pol_i.s$ is the sum of all sensitivities in the policy pol_i.

2) *fuzzy match*: pol_i fuzzy matches pol if and only if the following formula holds

$!(pol.PIn \supseteq pol_i.PIn \wedge pol.Pur = pol_i.Pur \wedge pol.Pro = pol_i.Pro) \wedge maximum(pol_i.d)$,

where

$pol_i.\ d = \sum_{t_j \in pol_i.PIn, t_k \in pol.PIn} similarity(t_j.a, t_k.a)$, is the *fuzzy match degree* of a seller policy pol_i.

Exact match refers to the case that a seller's required attributes, the purpose and process mode of this information satisfies the buyer's privacy preference. For policies satisfying *exact match*, we find the *good* seller policy with the minimal information requirement, i.e. minimal size of $pol_i.\ PIn$, and the minimal sum of sensitivities. If there is not any seller's policy *exact match* with the buyer's policy, we calculate the *fuzzy match degree* for all seller policies. To find which seller policy satisfies the buyer's preference as much as possible, we give top priority to the attributes with higher similarities. So we calculate the sum of similarities and find the *good* seller policy with the maximum *fuzzy match degree*. So, the above constraints are suitable for different cases for the buyer to choose the *good* seller.

4   Ontology-Based Policy Parsing and Matching

In this section, we parse privacy policies to eliminate semantic ambiguity. Ontology represents [11] knowledge as a set of concepts and the relationships between those concepts. OWL is the web ontology language[1], which provides abundant semantic expression and supports reasoning between concepts. By means of converting policies into OWL, the system can realize ontology-based policy expression and policy matching, and then it can find the good policy.

We propose a policy matching algorithm *Policy Match* to solve the PMP problem. We introduce a threshold α to judge if the corresponding elements in two policies refer to the same object. If the *similarity* between two elements is greater than α, it can be considered that they express the same object. The value of α is set by buyers. In this paper, we suppose the value of α be 0.8 by experience. According to the definition of *exact match*, if the *similarity* for each pair of element in two policies is greater than α, policy pol and pol_i are called *exact match*. Otherwise, we calculate the *fuzzy match degree*. We introduce a function *match()* to judge if a seller policy pol_i satisfies the buyer policy pol.

$$match(pol_i) = \begin{cases} true & \text{if } similarity(pol_i.element, pol.element) > \alpha, \forall pol_i.element \\ \sum_{t_j \in pol_i.PIn, t_k \in pol.PIn} similarity(t_j.a, t_k.a) & \text{otherwise} \end{cases}$$

If $match(pol_i)$=true, it means the policy satisfies the buyer policy pol, then put the seller policy pol_i in a set p'. After getting the values of $match()$ of all policies in the set p_{set}, we use the algorithm $Policy\ Match$ to acquire the $good$ seller policy. If the set p' is not empty, we find the $good$ policy with the minimal information requirement and the minimal sum of sensitivities. If p' is null, we choose the $good$ policy with the maximum $fuzzy\ match\ degree$. The algorithm $Policy\ Match$ is given below:

Algorithm *PolicyMatch(pol, p_{set}, match)*

```
For each pol, in p_set
  If (match(pol,)==true)
    Put pol, in p';
If (p'!=Φ)
  Sort p' by size of PIn in descending order;
  Choose the minimal ones, and put in p_0;
  For each pol, in p_0
pol_i.s=∑_{t_j∈pol_i.PIn} t_j.s ;
  Sort p_0 by pol,.s in descending order;
  Choose the minimal one, and put in p_1;
  Return p_1;
Else
  Sort p_set by match(pol,);
  Choose the maximum one, and put in p_2;
  Return p_2;
```

We would illustrate our method with a practical example in next section.

5   System Implementation and an Illustrative Example

We implement our model in an environment with Intel 2 cores, 2.4GHz CPU, 2GB of memory and 320GB disk. We choose the ontology editor Protégé to convert privacy policies into ontology based expressions, and choose COMA++ as the schema matching tool for policy match. The selected data set PRI_ITM= {register name, gender, age, name, e-mail, company phone, home address, ID number}, the set PURPOSE= {inform, delivery, verify, refund}, and the set PROCESS= {disclosure, retain, delete} to present all available data in the system.

In our model, we use OWL to define each *element* in a policy as a class and define *sensitivities* of attributes as data property [10]. Then according to operating on data properties, we get the sum of sensitivities to realize the algorithm *Policy Match*. COMA ++ [2] is a schema and ontology matching tool. Figure 2 describes how COMA++ works. The left figure shows the *similarity* of the corresponding nodes. Experiences show that if the *similarity* is larger than 0.8, this pair of nodes can be considered as the same, i.e. the two nodes express the same object. The right part of the figure 2 shows the whole policy match in COMA++.

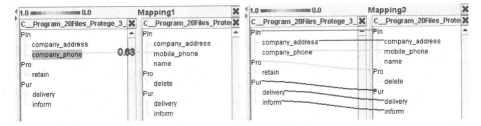

Fig. 2. the Policy Match in COMA++

Example. Suppose that a buyer privacy policy is pol = ({<companyphone,0.3>, <companyaddress,0.3>,<mobilephone,0.4>}, {inform, delivery}, {delete}), and the policies of candidate sellers are pset={pol_x, pol_y, pol_z, pol_w},where pol_x=({<company adress,1>,<companyphone,1>},{delivery,inform},{retain}),pol_y=({<officeaddress,1>, <mobilephone,1>},   {delivery,inform},{delete}), pol_z=({<companyaddress,1>, <officephone,1>},{delivery,inform},{delete}), pol_w=({<officeaddress,1>, <companyphone,1>,<mobliephone,1>},{delivery, inform},{delete}).

After policy parsing and matching, the algorithm returns the policy p_1=pol_z and its related service information to the buyer. The analysis of the results is as follows. In the phase of exact match, due to similarity (<pol.delete, pol_x.retain>) =0.65<α, the policy polx doesn't satisfy the buyer's preference, so it is removed. After exact match, we can get the policy set p'= {pol_y, pol_z, pol_w}. Next, based on algorithm Policy Match, due to the size of pol_w. PIn=3>2(the number of the least information need is 2) and pol_y.s=0.7>0.6(the lowest sum of sensitivities is 0.6), so the policies pol_y, pol_w are removed. At last, the algorithm returns the seller policy p_1=pol_z to the buyer. And this result is in accordance with our intuitive understanding.

6   Conclusions

In this paper, we present an ontology-based policy definition and matching model for privacy protection on third-party trading platform. Our model allows buyers and sellers to define privacy policy according to their preferences and then converts policies into ontology based expression forms. We introduce the concept good property of sellers who require the minimal information from buyers while satisfying other trading requirements. We propose a privacy policy matching algorithm to find such seller for a user request. Some experiments are performed to verify our model and an illustrative example is given. In the future, we would improve the policy matching algorithm and conduct more experiments to get an exact value of the threshold in the algorithm.

Acknowledgment. Part of this work is supported by the National Natural Science Foundation of China (61173140), the Science Foundation of Shandong Province(Y2008G28), and the Independent Innovation Foundation of Shandong University (2010JC010).

References

1. Denker, G., Kagal, L., Finin, T.: Security in Semantic Web using OWL. Information Security Technical Report 10, 51–58 (2005)
2. Aumueller, D., Do, H.H., Massmann, S., Rahm, E.: Schema and Ontology Matching with COMA++. In: Proceedings of the 5th ACM SIGMOD International Conference on Management of Data, New York, pp. 906–908 (2005)
3. Bonatti, P.A., Faella, M., Galdi, C., Sauro, L.: Towards a Mechanism for Incentivating Privacy. In: Atluri, V., Diaz, C. (eds.) ESORICS 2011. LNCS, vol. 6879, pp. 472–488. Springer, Heidelberg (2011)
4. Fenz, S.: An Ontology-and Bayesian-based Approach for Determining Threat Probabilities. In: Proceedings of the 6th ACM Symposium on Information, Computer and Communications Security, New York, pp. 344–354 (2005)
5. Hecker, M., Dillon, T.S., Chang, E.: Privacy Ontology Support for E-Commerce. In: Proceeding of IEEE Computer Society 2008, Internet Computing, pp. 54–61 (2008)
6. Garcia, D.Z., Toledo, M.B.: A Web Service Privacy Framework Based on a Policy Approach Enhanced with Ontologies. In: Proceedings of the 11th IEEE International Conference on Computational Science and Engineering, San Paulo, pp. 209–214 (2008)
7. Tumer, A., Dogac, A., Toroslu, I.H.: A Semantic-Based User Privacy Protection Framework for Web Services. In: Mobasher, B., Anand, S.S. (eds.) ITWP 2003. LNCS (LNAI), vol. 3169, pp. 289–305. Springer, Heidelberg (2005)
8. Gao, F., He, J., Peng, S.: An Approach for Privacy Protection Based-on Ontology. In: Proceedings of 2010 Second International Conference on Networks Security, Wireless Communications and Trusted Computing, Wuhan, Hubei, pp. 397–400 (2010)
9. Carminati, B., Ferrari, E., Heatherly, R.: A Semantic Web Based Framework for Social Network Access Control. In: Proceedings of the 14th ACM Symposium on Access Control Models and Technologies, New York, pp. 177–186 (2009)
10. Masoumzadeh, A., Joshi, J.: OSNAC: An Ontology-Based Access Control Model for Social Networking Systems. In: Proceedings of IEEE International Conference on Social Computing, Minneapolis, MN, pp. 751–759 (2010)
11. Qian, J.A., Jiang, X.H., Sun, T.F.: Privacy Ontology-based Personalized Access Control Model. Information Security and Communications Privacy 2, 67–73 (2011)
12. Lan, L.H., Ju, S.G., Liu, S.C.: Survey of study on privacy preserving data publishing. Application Research of Computers 27, 2822–2827 (2010)
13. Hu, Y.J., Wu, W.N., Cheng, D.R.: Towards law-aware semantic cloud policies with exceptions for data integration and protection. In: Proceedings of the 2nd International Conference on Web Intelligence, Mining and Semantics (2012)

Evaluating Ontology-Based User Profiles

Silvia Calegari and Gabriella Pasi

Department of Informatics, Systems and Communication (DISCo),
University of Milano-Bicocca,
v.le Sarca 336/14, 20126 Milano (Italy)
{calegari,pasi}@disco.unimib.it

Abstract. User profiles play an important role in any process of personalization as they represent the user's interests and preferences. Only if a user profile faithfully represents the information related to a user a system may rely on it. This paper shortly presents a comparative evaluation between two distinct approaches that analyze textual documents for defining user profiles based on the usage of the YAGO general purpose ontology. The performed evaluations compare the two approaches both by the robust index measure and their efficiency.

Keywords: User profile, Ontology, Evaluation.

1 Introduction

The issue of personalization is becoming more and more important in various research domains. In fact, there is an increasing need to define personalized systems that tailor their outcomes to the users' context, to the aim of better complying to their expectations. A user profile plays a key role for the definition of personalized systems; it models several knowledge dimensions related to a user, such as his/her personal data, background knowledge, topical preferences, etc. The knowledge represented in a user profile is analyzed and then used to improve the standard behaviour of the considered system. A personalized system works well if the knowledge stored into a user profile represents at best the information related to a user.

In the literature several formal representations of user profiles have been proposed, such as sets of weighted keywords, semantic networks or hierarchies of concepts [4]. Ontologies have been recently considered as a valuable support to express a more structured and complete knowledge representation of user profiles. In fact, they allow to enrich the expressiveness of the information represented in a profile by using formal languages like RDFS or OWL. The existing models to build user profiles based on ontologies are mainly focused on approaches either relying on data mining techniques [6,10] or adopting external reference knowledge [3,8] to capture the meaning of the user's preferences.

In this paper our attention is on strategies that make use of ontologies (like external reference knowledge) to build user profiles. In particular, our approach

R. Huang et al. (Eds.): AMT 2012, LNCS 7669, pp. 377–386, 2012.

allows to generate ontological user profiles based on the general purpose ontology YAGO [9]. YAGO consists of several million of entities and facts, where a fact is a triple of two entities and the relation between them. In YAGO 99 relations have been defined. In this work, two strategies able to extract the meaningful entities and facts from YAGO are considered. The first strategy [1] is able to disambiguate the YAGO information acquired during the knowledge extraction process by combining the user's local knowledge (i.e., user's documents), and the user's global information (i.e., the YAGO ontology). The second strategy makes use of the query2YAGO query processor [5] that is aimed to search for YAGO facts according to a specific syntax. In this paper, we have extended query2YAGO in order to define a new methodology that allows to navigate and extract information from YAGO related to the user topical interests.

The comparative evaluations of the two above strategies include both qualitative evaluations and efficiency evaluations; the qualitative evaluations allow to analyze the obtained user profiles in terms of amount of noisy information gathered by the considered extraction process. The efficiency evaluations are finalised at testing the time required by each of the two strategies to build the user profiles.

The paper is organized as follows: Section 2 shortly introduces the considered strategies for the ontological user profiles definition, whereas in Section 3 the evaluations are presented that compare the effectiveness of the two methods. In Section 4 some conclusions are stated.

2   Building Ontological User Profiles

In this section the two considered strategies that make use of the YAGO ontology to express user's preferences in a semantically meaningful way are described. Both of them take in input a set of documents representing the user's preferences as well as the YAGO knowledge-base, and they generate in output a user profile constituted by the meaningful portions of YAGO related to the contents of the provided documents. Figure 1 shows an overview of the process undertaken by both strategies. The set of documents that are representative of the user's interests is indexed by a standard procedure. The output of the indexing procedure is a set of weighted keywords where the weights are computed by applying one of the classic weighting functions (e.g., the standard normalized Tf-Idf [7]). We call the selected weighted keywords the *interest-terms*. Then the two methods analyze the set of interest-terms to extract the YAGO sub-graphs as shown in Figure 1; the methods differ in the extraction process. Finally, the obtained YAGO knowledge portions are formally represented into the ontological language RDFS[1].

In Subsection 2.1 a short explanation of the YAGO ontology is provided, whereas in Subsections 2.2 and 2.3 the two methodologies for the user profile extraction are described.

---

[1] http://www.w3.org/TR/PR-rdf-schema

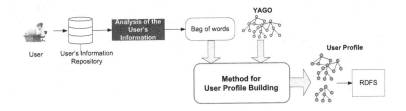

Fig. 1. Overview of the general process to build a user profile

2.1   The YAGO Ontology in a Nutshell

YAGO is one of the largest knowledge bases actually available, and it is composed of entities and facts. Currently, the YAGO knowledge base contains about 1.95 million entities and 19 million facts. The YAGO model has been defined as an extension of RDFS as explained in [9] where entities represent the objects in a world knowledge-base model. YAGO's authors have defined entities as *abstract ontological objects*; more specifically entities may be literals, or words, or classes, or relations, or fact identifiers. Entities that are neither fact identifiers nor relations are defined as *common entities*. Common entities that are not classes are called *individuals*. Entities constitute `arguments` of a fact, and a single fact is a triple constituted by two entities and the relation name linking them. An example of a fact is *(Mia Farrow, isMarriedTo, Frank Sinatra)*, with the meaning that *Mia Farrow has been married to Frank Sinatra*. Moreover, with each fact a *fact identifier* is associated to link, for example, URL information with the knowledge of an other fact. If the fact *(Mia Farrow, isMarriedTo, Frank Sinatra)* has the identifier #1, then it is possible to generate a new fact as *(#1, foundIn, http://en.wikipedia.org/wiki/Mia_Farrow)* to have more information on *Mia Farrow* and her marriage. Based on these notions, the YAGO model $\mathcal{M}_{YAGO}$ is defined as:

Definition 1 $\mathcal{M}_{YAGO} = \langle E, \mathcal{F} \rangle$, *where* $E = \mathcal{I} \cup \mathcal{C} \cup \mathcal{R}$ *where* $\mathcal{I}$ *is the set of fact identifiers,* $\mathcal{C}$ *is the set of common entities,* $\mathcal{R}$ *is the set of relation names, and* $\mathcal{F}$ *is the set of YAGO facts.*

Based on the application of the two strategies described in Sections 2.2 and 2.3, a user profile is defined as a subset of YAGO. Formally, a user profile is defined as $\mathcal{UP} = \langle E_{\mathcal{UP}}, \mathcal{F}_{\mathcal{UP}} \rangle$ such that $\mathcal{UP} \subseteq \mathcal{M}_{YAGO}$.

2.2   Building the User Profile: The First Strategy.

Figure 2 shows the main phases of the technique proposed in [1] to automatically build user profiles by using YAGO. Here below, a short explanation of each phase is reported.

***Common Entities Identification* Phase.** The objective is to discover the set of the YAGO common entities that are related to the set of interest-terms $\mathcal{IT}$

by string containment. Let *int* be an interest-term and *c* be a common entity, then *int* can be equal to *c* or it can be contained into *c*.

Common Entities Disambiguation Phase. The objective is to reduce the noisy information gathered by the previous phase. This means to eliminate the YAGO common entities not related to the user topical interests by considering two types of information: local knowledge and global knowledge, respectively. The local knowledge on *c* determines its importance with respect to the set of interest-terms. Then the local knowledge weight of a common entity *c*, w_{LK}^c, is computed as an average of the weights that are associated with the interest-terms related to the common entity *c*. Instead, the global knowledge on *c* allows to explore its possible interpretations in YAGO. To this aim, we have considered the YAGO facts where common entities are linked by the *Type* (or *instance-of*) relation. The global knowledge weight of a common entity *c*, w_{GK}^c, is computed on the basis of how many other common entities share the same YAGO knowledge linked by the *Type* relation.

To associate an overall score with a common entity *c*, a linear combination of its weights w_{LK}^c and w_{GK}^c is applied as $w_c = \alpha * w_{LK}^c + (1 - \alpha) * w_{GK}^c$, where $0 \le \alpha \le 1$. The parameter α has been set to the value 0.6 in order to give a slightly higher importance to the local knowledge. A threshold value *t* is used to individuate the common entities that are representative of the user's interests.

Rules for the Knowledge Extraction Phase. The objective is to extract the YAGO facts containing the set of common entities obtained as output of the previous phase by the definition of four specific rules. The YAGO entities constitute the *arguments* of a fact where *arg1* identifies the YAGO entities that appear as first argument of a fact, whereas *arg2* identifies the YAGO entities that appear as second argument of a fact. The four rules are: rule 1) only *arg1* carries information useful to a common entity identification, rule 2) only *arg2* carries information useful to a common entity identification, rule 3) both *arg1* and *arg2* together carry information useful for common entities, rule 4) either *arg1* or *arg2* may contain information useful to a common entity identification. Each of the 99 YAGO relations has been manually associated with the right rule.

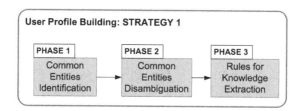

Fig. 2. Overview of the process to build a user profile

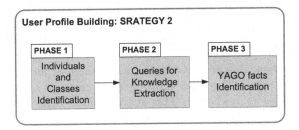

Fig. 3. Overview of the process to build a user profile

2.3   Building the User Profile: The Second Strategy

The second strategy we consider is proposed for the first time in this paper; it is able to semi-automatically extract portions of YAGO related to the set of interest-terms by using a simple query processor named *query2YAGO* [9]. The *query2YAGO* Java application has been defined by the developers of the Max-Plank Institute in order to search YAGO facts according to a specific query language. This query processor has been integrated in the NAGA semantic search engine [5] that can operate on knowledge-bases organized as graphs (like YAGO) in order to search the sub-graphs that match the user's query.

We have extended *query2YAGO* in order to search the YAGO sub-graphs related to the set of interest-terms as described in this section. Before providing the explanation of the phases devoted to the definition of the user profile, it is necessary to give a short explanation of the *query2YAGO* syntax as a technical documentation of the language is not available.

The syntax of a generic query is: $Q = e_1 \ r \ e_2$, where $e_1, e_2 \in \mathcal{C} \cup \mathcal{I}$ and $r \in \mathcal{R}$, with the meaning of searching for facts containing e_1 as the first argument of a fact and e_2 as the second argument of a fact linked by the relation name r. During the query evaluation, the system automatically expands each argument (both e_1 and e_2) with its possible semantic interpretations in YAGO by analyzing the *Means* relation. For example, if $e_1 = guitar$ then the following query is implicitly defined: $Q = guitar \ means \ ?y$, where $?y$ individuates all the *meanings* of *guitar* in YAGO. Then, the output is the following set of common entities related to *guitar*: $E_{guitar} = \{ Guitar\_album, Guitar\_song, Matt\_Murphy\_(blues\_guitarist), wordnet\_\_guitarist \}$. Each common entity in E_{guitar} will substitute the original e_1 indicated by the user during the formulation of his/her query (the same happens for e_2). This way, if $|E_{e_1}| = M$ and $|E_{e_2}| = N$ then $M \times N$ queries will be defined and evaluated in order to extract all the YAGO facts related to e_1 and e_2 linked by the relation name r. When a user is interested in finding all the YAGO portions related to e_1 and e_2 without any constraint on the relation names linking them, it is possible to define the following query: $Q = e_1 \ ? \ e_2$, where the character ? indicates all the YAGO relation names. Here below, an explanation of all the phases for building the user profile is reported.

***Individuals and Classes Identification* Phase.** From the set of interest-terms $\mathcal{IT}$, the user is asked to identify two sub-sets: (1) sub-set of individuals, and (2) sub-set of classes, respectively. To separate the interest-terms allows to formulate specific queries in *query2YAGO* to constrain the navigation of the YAGO knowledge-base in a proper way. Formally, the user judges $\mathcal{IT}$ to identify the sub-set of individuals $\mathcal{I}_{\mathcal{IT}}$ and the sub-set of classes $\mathcal{C}_{\mathcal{IT}}$ such that $\mathcal{IT} = \mathcal{I}_{\mathcal{IT}} \cup \mathcal{C}_{\mathcal{IT}}$ and $(\mathcal{I}_{\mathcal{IT}} \cap \mathcal{C}_{\mathcal{IT}}) = \emptyset$. This is the unique phase that requires the intervention of a user in the whole user profile building process.

***Queries for Knowledge Extraction* Phase.** The objective is to extract the YAGO sub-graphs related to the two sub-sets (i.e., $\mathcal{I}_{\mathcal{IT}}$ e $\mathcal{C}_{\mathcal{IT}}$) identified in the previous phase. To this aim, the *query2YAGO* query processor has been extended in order to manage four types of queries. These queries can be logically divided into two groups: (1) the first two types of queries allow to extract the YAGO knowledge where facts contain relation names that directly link individuals and classes in $\mathcal{IT}$, and (2) the last two types of queries allow to select additional facts related to individuals and classes in $\mathcal{IT}$ by exploring the YAGO knowledge-base. An explanation of the four types of queries is given here below:

First Type: it selects the YAGO facts where a direct association between an individual $i_j \in \mathcal{I}_{\mathcal{IT}}$ and a class $c_k \in \mathcal{C}_{\mathcal{IT}}$ exists. At this phase, two types of queries have to be defined in order to seek for individuals (or classes) appearing either as the first argument of a fact or as the second argument of a fact. Thus, two types of queries are defined as $Q_1 = i_j ? c_k$ and $Q_1 = c_k ? i_j$.

Second Type: it selects the YAGO facts containing two individuals $i_j, i_s \in \mathcal{I}_{\mathcal{IT}}$ where $i_j \neq i_s$. The query is defined as $Q_1 = i_j ? i_s$.

Third Type: it selects the YAGO facts where an individual $i_j \in \mathcal{I}_{\mathcal{IT}}$ shares the same knowledge of two classes $c_k, c_h \in \mathcal{C}_{\mathcal{IT}}$ where $c_k \neq c_h$. The aim is to extract from YAGO additional information related to individuals and classes in $\mathcal{IT}$. To this aim a sequence of the following queries: $Q_3 = i_j ? ?x; c_k ? ?x; c_h ? ?x$ is defined.

At this step, the main problem is to disambiguate the YAGO information obtained during the evaluation of the above sequence of queries. In fact, as previously reported, each interest-term (like an individual or a class) is automatically expanded with its possible meanings in YAGO by the *Means* relation. So that, if a user is interested in the sport tennis and *Agassi* $\in \mathcal{I}_{\mathcal{IT}}$, then it can be substituted with the two following YAGO individuals i.e., *Andre Agassi* and *Carlos Agassi*, but only the first one can satisfy the user's interests as, in this example, the user has preferences on the *tennis* topic. In order to select only facts related to the user's interests and preferences, this type of query searches YAGO facts where the knowledge associated with an individual is the same also for the two classes. We consider two classes because the use of a unique class is not sufficient for extracting non relevant facts; in fact, as it happens for an individual, the YAGO-interpretations of a class can be also ambiguous. By adopting the knowledge of an individual plus the knowledge of two classes can reduce the possibility to discover noisy information from YAGO.

Fourth Type: it selects the YAGO facts where three classes $c_w, c_k, c_h \in C_{IT}$ with $c_w \neq c_k \neq c_h$ share the same knowledge. To this aim the following queries are generated: $Q_4 = c_w$? $?x$; c_k ? $?x$; c_h ? $?x$.
The idea underlying this type of query is the same of the previous one; the objective is to extract additional knowledge from YAGO related to the classes in C_{IT}. We consider three classes in order to minimize the number of non-relevant YAGO facts with respect to the user's expectations.

YAGO facts Identification **Phase.** The outputs of the execution of the four types of queries are sub-portions of the YAGO knowledge related to the analyzed set of interest-terms IT. Unfortunately, the output provided by *query2YAGO* is not conform to the YAGO fact syntax (i.e., a YAGO fact is a triple between two common entities and the relation name linking them). For example, if *Andre Agassi* $\in I_{IT}$ and *player* $\in C_{IT}$, then the result of the following query $Q_1 = $ "*Andre Agassi*" ? *player* is RESULT="?=type, ?"*Andre Agassi*"=Andre_Agassi, ?player=wordnet_player". Thus, a parsing is needed in order to redefine, for example, the previous result in the standard YAGO fact syntax i.e., *(Andre Agassi, Type, wordnet player)*. To reconduct the set of results obtained by the *Queries for Knowledge Extraction* phase in the YAGO triples allows to convert them in the ontological language RDFS by using predefined scripts. The *YAGO facts Identification* step is then divided into two sub-phases: (1) to analyze each result obtained by the previous phase in order to modify it according to the YAGO fact syntax, and (2) to remove duplicated YAGO facts from the set of facts produced by sub-phase (1).

3   Evaluations

Evaluations have been performed to compare the strategies of Section 2 to both assess the quality of the knowledge gathered in the user profiles, and analyze their behaviour in terms of efficiency. Qualitative evaluations test the quality of the information stored in the user profiles in terms of ambiguous knowledge gathered.

To evaluate the two strategies, we have asked ten users to collect the documents that are more representative of their interests. The ten main user's collections (made up of 100 documents each) are representative of the following topics: *architecture, astronomy, botany, cuisine, health and fitness, literature, music, tennis, travel* and *wine*. The topics selected by the users are related to a broad spectrum of knowledge, and this allows to test the effectiveness of the two methodologies in different areas of interest.

The ten users had also a significant role during the qualitative evaluations; they have acted as assessors by evaluating the quality of the twenty sets of common entities defined in the user profiles obtained with the two applied methodologies on the ten topics. In fact, for each set of common entities obtained by the application of the two strategies of Section 2, each user has expressed a judgement on each common entity by classifying it into two distinct groups: a set of

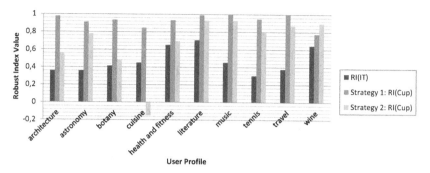

Fig. 4. RI metric evaluations for all the user profiles

positive common entities, and a set of uncorrelated common entities (i.e., not in line with the user's interests).

In Subsection 3.1 the qualitative evaluations are presented, whereas in Subsection 3.2 the efficiency evaluations are reported.

3.1 Qualitative Evaluations

To perform a qualitative evaluation of a user profile, we have adopted a simple measure called the robustness index (RI) [2]. In this paper, we adapt the RI metric to assess the quality of the set of common entities $\mathcal{C}_{\mathcal{U}}$ obtained as the outcome of the presented knowledge extraction processes. In fact, the extracted set of common entities $\mathcal{C}_{\mathcal{U}}$ may contain some noisy information, the amount of which we want to evaluate. We assume then that $\mathcal{C}_{\mathcal{U}}$ consists of two subsets, $\mathcal{C}_{\mathcal{U}}^{+}$ and $\mathcal{C}_{\mathcal{U}}^{-}$, $\mathcal{C}_{\mathcal{U}} = \mathcal{C}_{\mathcal{U}}^{+} \cup \mathcal{C}_{\mathcal{U}}^{-}$, where $\mathcal{C}_{\mathcal{U}}^{+}$ identifies the positive common entities, and $\mathcal{C}_{\mathcal{U}}^{-}$ identifies the uncorrelated common entities. A common entity is identified as positive when it is semantically correlated to the considered user topical interest; on the contrary, a common entity is identified as uncorrelated when it is out of topic with respect to the considered user interest. The RI metric is defined as $RI(\mathcal{C}_{\mathcal{U}}) = \frac{|\mathcal{C}_{\mathcal{U}}^{+}| - |\mathcal{C}_{\mathcal{U}}^{-}|}{|\mathcal{C}_{\mathcal{U}}|}$, where $(|\mathcal{C}_{\mathcal{U}}^{+}| + |\mathcal{C}_{\mathcal{U}}^{-}|) = |\mathcal{C}_{\mathcal{U}}|$ and $-1 \leq RI(\mathcal{C}_{\mathcal{U}}) \leq 1$. Clearly, $\mathcal{C}_{\mathcal{U}} = 1$ if all common entities are classified as positive, while $\mathcal{C}_{\mathcal{U}} = -1$ if all common entities are classified as uncorrelated.

Experiments. As the input of the proposed methodologies is constituted by a set of interest-terms related to the user topical interests, our intuition is that if the number of uncorrelated interest-terms in $\mathcal{IT}$ is high, it can be more difficult to extract from $\mathcal{M}_{YAGO}$ the knowledge related to the right user topical interests. Figure 4 reports the values for $RI(\mathcal{IT})$, and $RI(\mathcal{C}_{\mathcal{U}})$ for both the two presented strategies over all user profiles. High RI values (i.e., closer to 1) are obtained when a few uncorrelated information is acquired by the considered knowledge extraction strategy. If there is a high number of uncorrelated interest-terms, it is plausible to assume that also a high number of uncorrelated common entities will be obtained by the extraction process. Both the two strategies for user profile building allow to obtain positive RI values by analyzing the set $\mathcal{IT}$. This

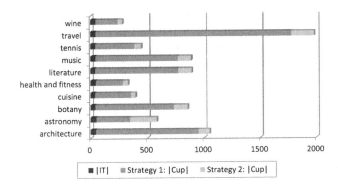

Fig. 5. Analysis of the number of common entities

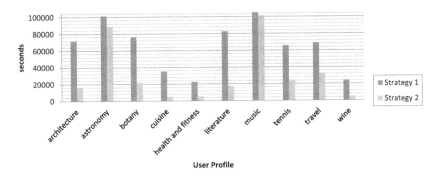

Fig. 6. Efficiency evaluations

indicates that they are able to select non-ambiguous knowledge in spite of the presence of noisy knowledge as starting point of their knowledge extraction process. By comparing the behaviour of the two strategies, it emerges that the first strategy outperforms the second one in all the user profiles but the user topical interest on *wine*. This means that the *Strategy 1* allows to better disambiguate the YAGO knowledge portions during the several phases of the process thanks to the analysis of the user local information that acts as an additional indicator of the knowledge from YAGO that is related to the user's interests. The good results obtained by *Strategy 1* in terms of robust index are also reinforced by analyzing the amount of common entities inserted in each user profile, as shown in Figure 5. In fact, not only *Strategy 1* extracts less uncorrelated knowledge from YAGO than *Strategy 2*, but it allows to obtain more information related to each user topical interest.

3.2 Efficiency Evaluations

The two considered strategies are now analyzed in order to consider the execution time necessary to obtain the YAGO sub-graphs. Figure 6 shows that for most user profiles *Strategy 2* outperforms *Strategy 1* (i.e., less execution time is used).

Similar timings are instead obtained for the user topical interests on *astronomy* and *music*; this happens because in YAGO some topics cover a huge amount of facts with respect to other topics (like *cuisine*), and then it is possible that the two strategies could manage the same amount of YAGO information before extracting the relevant one.

In general, the process of knowledge extraction of *Strategy 2* works faster than *Strategy 1*, but it exhibits a worst behaviour in terms of the robust index if compared with *Strategy 1*.

4   Conclusions and Future Works

In this paper both qualitative and efficiency evaluations have been presented to compare two distinct approaches that make use of the YAGO ontology to extract and represent user profiles from a set of textual documents representing users' interests. The evaluations have outlined some interesting results: the first considered strategy is better in selecting non ambiguous information, whereas the second strategy is more efficient than the first one.

In future works we will also compare the ontological user profiles defined by the strategies analysed in this paper with other methodologies presented in the literature that are able to build user profiles represented as ontologies.

References

1. Calegari, S., Pasi, G.: Personal ontologies: generation of user profiles based on the YAGO ontology. Information Processing & Management (2012) (in Printing)
2. Carpineto, C., Romano, G.: A Survey of Automatic Query Expansion in Information Retrieval. ACM Comput. Surv. 44(1), 1:1–1:50 (2012)
3. Daoud, M., Tamine, L., Boughanem, M.: A Personalized Graph-Based Document Ranking Model Using a Semantic User Profile. In: De Bra, P., Kobsa, A., Chin, D. (eds.) UMAP 2010. LNCS, vol. 6075, pp. 171–182. Springer, Heidelberg (2010)
4. Gauch, S., Speretta, M., Chandramouli, A., Micarelli, A.: User Profiles for Personalized Information Access. In: Brusilovsky, P., Kobsa, A., Nejdl, W. (eds.) Adaptive Web. LNCS, vol. 4321, pp. 54–89. Springer, Heidelberg (2007)
5. Kasneci, G., Suchanek, F.M., Ifrim, G., Ramanath, M., Weikum, G.: NAGA: Searching and ranking knowledge. In: ICDE, pp. 953–962. IEEE (2008)
6. Li, Y., Zhong, N.: Mining ontology for automatically acquiring web user information needs. IEEE Trans. Knowl. Data Eng. 18(4), 554–568 (2006)
7. Robertson, S.: Understanding inverse document frequency: on theoretical arguments for idf. Journal of Documentation 60(5), 503–520 (2004)
8. Speretta, M., Gauch, S.: Miology: A web application for organizing personal domain ontologies. In: International Conference on Information, Process, and Knowledge Management, eKNOW 2009, pp. 159–161 (February 2009)
9. Suchanek, F.M., Kasneci, G., Weikum, G.: Yago: A large ontology from wikipedia and wordnet. Journal of Web Semantic 6(3), 203–217 (2008)
10. Tao, X., Li, Y., Zhong, N.: A personalized ontology model for web information gathering. IEEE Trans. on Knowl. and Data Eng. 23, 496–511 (2011)

Semantic Information with Type Theory of Acyclic Recursion

Roussanka Loukanova

Independent Research, Uppsala, Sweden
rloukanova@gmail.com

Abstract. The paper provides introduction into the language, semantics, and the reduction calculus of acyclic recursion. The expressiveness of the theory for conveying semantic information, relevant for applications in active media technologies, is demonstrated by representing naming information.

1   Background of L_{ar}^λ Algorithmic Intensionality

With his work on formal languages of recursion, Yiannis Moschovakis initiated development of a new approach to the mathematical notion of algorithm, beginning with work on the mathematics of algorithms in 1994, see [4]. One of the most exciting potentials of the approach is for applications to computational semantics of artificial and natural languages (NLs). In particular, the theory of acyclic recursion L_{ar}^λ in [5] models the concepts of meaning and synonymy. The semantics of each meaningful language expression is provided by its algorithmic meaning, called *referential intension*, which is the algorithm for computing the denotation of the expressions in the domains of the semantic models. For initial applications of L_{ar}^λ to computational syntax-semantics interface in Constraint-Based Lexicalized Grammar (CBLG) of human language (HL), see [3]. The formal system L_{ar}^λ is a higher-order type theory, which is a proper extension of Gallin's TY$_2$, see [1], and thus, of Montague's Intensional Logic (IL), see [6]. The type theoretic system L_{ar}^λ extends Gallin's TY$_2$, at the level of the formal language and its semantics, by using two kinds of variables (*recursion variables*, called alternatively *locations*, and *pure variables*) and by formation of an additional set of *recursion terms*. The recursion terms are formed by using a designated recursion operator, which is denoted by the constant where and used in infix notation.

In the first part of this paper, we give the formal definitions of the syntax and the denotational semantics of L_{ar}^λ. We introduce the intensional semantics of L_{ar}^λ with some intuitions behind it. Then we introduce the rules of the reduction calculus of the type theory L_{ar}^λ, which is its central part, and some key theoretical results that are essential for the algorithmic meanings. All notions are supplemented by examples from English language.

In the rest of the paper, we present renderings of specific kinds of NL expressions that contain proper names, into terms of L_{ar}^λ, which present patterns with potentials for extended work.

R. Huang et al. (Eds.): AMT 2012, LNCS 7669, pp. 387–398, 2012.

2   Introduction to the Type Theory L_{ar}^{λ}

2.1   Types of L_{ar}^{λ}

The set *Types* is the smallest set defined recursively as follows (using a widespread notation in computer science):

$$\tau :\equiv e \mid t \mid s \mid (\tau_1 \to \tau_2) \qquad\qquad (Types)$$

The type e is associated with primitive objects (entities called individuals) of the domain, as well as with the terms of L_{ar}^{λ} denoting individuals. The type s is for states consisting of various context information such as a possible world (a situation), a time moment, a space location, and a speaker; t is the type of the truth values.

2.2   Syntax of L_{ar}^{λ}

Vocabulary of L_{ar}^{λ}: The vocabulary of L_{ar}^{λ} consists of pairwise disjoint sets:

- *Constants:* $K = \bigcup_{\tau \in Types} K_\tau$,
  where, for each $\tau \in Types$, $K_\tau = \{\, c_0, c_1, \ldots, c_{k_\tau}\,\}$ is a finite set.
- *Pure variables:* $PureVars = \bigcup_{\tau \in Types} PureVars_\tau$,
  where, for each $\tau \in Types$, $PureVars_\tau = \{\, v_0, v_1, \ldots\,\}$.
- *Recursion variables (locations):*
  $RecVars = \bigcup_{\tau \in Types} RecVars_\tau$,
  where, for each $\tau \in Types$, $RecVars_\tau = \{\, p_0, p_1, \ldots\,\}$.

The Terms of L_{ar}^{λ}: The recursive rules for the set of L_{ar}^{λ} terms can be expressed by using a notational variant of BNF, with the assumed types given as superscripts:

$$A :\equiv c^\tau : \tau \mid x^\tau : \tau \mid B^{(\sigma \to \tau)}(C^\sigma) : \tau \mid \lambda v^\sigma (B^\tau) : (\sigma \to \tau)$$
$$\mid A_0^\sigma \text{ where } \{\, p_1^{\sigma_1} := A_1^{\sigma_1}, \ldots, p_n^{\sigma_n} := A_n^{\sigma_n}\,\} : \sigma$$

where $\{\, p_1 := A_1, \ldots, p_n := A_n\,\}$ is a set of assignments that satisfies the acyclicity condition defined as follows:

Acyclic System of Assignments: For any terms $A_1 : \sigma_1, \ldots, A_n : \sigma_n$, and locations $p_1 : \sigma_1, \ldots, p_n : \sigma_n$ (where $n \geq 0$, and for all i, j, if $i \neq j$, $1 \leq i, j \leq n$, then $p_i \neq p_j$), the system of assignments $\{\, p_1 := A_1, \ldots, p_n := A_n\,\}$ is called *acyclic* iff the following *acyclicity condition* condition is fulfilled: there is a ranking function $\mathsf{rank} : \{\, p_1, \ldots, p_n\,\} \longrightarrow \mathbb{N}$ such that, for all $p_i, p_j \in \{\, p_1, \ldots, p_n\,\}$, if p_j occurs free in A_i, then $\mathsf{rank}(p_j) < \mathsf{rank}(p_i)$.

Intuitively, an acyclic system $\{\, p_1 := A_1, \ldots, p_n := A_n\,\}$ defines recursive computations of the values to be assigned to the locations $p_1, \ldots, p_n$, which close-off after a finite number of steps; ranking $\mathsf{rank}(p_j) < \mathsf{rank}(p_i)$ means that the value A_i assigned to p_i, may depend on the values of the free occurrences of the location p_j in A_i, and of all other free occurrences of locations with lower rank than

p_j. The formal syntax of L_{ar}^λ allows only recursive terms with acyclic systems of assignments. The FLR, see [4], allows cyclicity, but is an untyped system. Typed language of full recursion L_{lr}^λ has the formal syntax of L_{ar}^λ without the acyclicity requirement. The classes of languages of recursion (FLR, L_{lr}^λ, and L_{ar}^λ) have two semantic layers: denotational semantics and referential intensions. The recursive terms of L_{ar}^λ are essential for encoding two-fold semantic information.

2.3   Denotational Semantics of L_{ar}^λ

Definition 1. *An L_{ar}^λ structure is a tuple $\mathfrak{A} = \langle \mathbb{T}, \mathcal{I}, \mathrm{den} \rangle$, where $\mathbb{T}$ is a set, called a* frame, *of sets (or classes, depending on the area of application, with relevant technical details) $\mathbb{T} = \{\, \mathbb{T}_\sigma \mid \sigma \in \textit{Types} \,\}$, and the following conditions (S1)–(S4) are satisfied:*

(S1) $\mathbb{T}_e \neq \varnothing$ *is a nonempty set (class) of entities called* individuals, $\mathbb{T}_t = \{\, 0, 1, er \,\} \subseteq \mathbb{T}_e$ *is the set of the* truth values, $\mathbb{T}_s \neq \varnothing$ *is a nonempty set of objects called* states.

(S2) $\mathbb{T}_{(\tau_1 \to \tau_2)} = \{\, p \mid p \colon \mathbb{T}_{\tau_1} \longrightarrow \mathbb{T}_{\tau_2} \,\}$.

(S3) $\mathcal{I}$ *is a function $\mathcal{I} \colon K \longrightarrow \mathbb{T}$ (called the* interpretation function *of $\mathfrak{A}$) such that for every constant $\mathsf{c} \in K_\tau$, $\mathcal{I}(\mathsf{c}) = c$ for some $c \in \mathbb{T}_\tau$.*

(S4) *Given that the set G of all variable assignments is $G = \{\, g \mid g \colon \textit{PureVars} \cup \textit{RecVars} \longrightarrow \mathbb{T}$ and $g(x) \in \mathbb{T}_\sigma$, for every $x \colon \sigma \,\}$, the denotation function* den $\colon \textit{Terms} \longrightarrow \{\, f \mid f \colon G \longrightarrow \mathbb{T} \,\}$ *is defined, for each $g \in G$, by recursion on the structure of the terms:*

(D1) $\mathrm{den}(x)(g) = g(x)$; $\mathrm{den}(\mathsf{c})(g) = \mathcal{I}(\mathsf{c})$;

(D2) $\mathrm{den}(A(B))(g) = \mathrm{den}(A)(g)(\mathrm{den}(B)(g))$;

(D3) $\mathrm{den}(\lambda(x)(B))(g) \colon \mathbb{T}_\tau \longrightarrow \mathbb{T}_\sigma$, *where $x \colon \tau$ and $B \colon \sigma$, is the function such that, for every $t \in \mathbb{T}_\tau$, $[\mathrm{den}(\lambda x(B))(g)](t) = \mathrm{den}(B)(g\{x := t\})$;*

(D4)

$$\mathrm{den}(A_0 \text{ where } \{\, p_1 := A_1, \ldots, p_n := A_n \,\})(g) \tag{1a}$$

$$= \mathrm{den}(A_0)(g\{\, p_1 := \overline{p}_1, \ldots, p_n := \overline{p}_n \,\}) \tag{1b}$$

where for all $i \in \{1, \ldots, n\}$, the values $\overline{p}_i \in \mathbb{T}_{\tau_i}$ are defined by recursion on $\mathrm{rank}(p_i)$, *so that:*

$$\overline{p_i} = \mathrm{den}(A_i)(g\{\, p_{k_1} := \overline{p}_{k_1}, \ldots, p_{k_m} := \overline{p}_{k_m} \,\}), \tag{2}$$

where $p_{k_1}, \ldots, p_{k_m}$ are the recursion variables $p_j \in \{p_1, \ldots, p_n\}$ such that $\mathrm{rank}(p_j) < \mathrm{rank}(p_i)$.

2.4   Intensional Semantics

The notion of intension in the languages of recursion covers the most essential, computational aspect of the concept of meaning. The *referential intension*, $\int(A)$, of a meaningful term A is the tuple of functions (a recursor) that is defined by the denotations $\mathsf{den}(A_i)$ $(i \in \{0, \ldots n\})$ of the parts (i.e., the head sub-term A_0 and of the terms $A_1, \ldots, A_n$ in the system of assignments) of its canonical form $\mathsf{cf}(A) \equiv A_0$ where $\{ p_1 := A_1, \ldots, p_n := A_n \}$. Intuitively, for each meaningful term A, the intension of A, $\int(A)$, is the *algorithm* for computing its denotation $\mathsf{den}(A)$. Two meaningful expressions are synonymous iff their referential intensions are naturally isomorphic, i.e., they are the same algorithm. Thus, the algorithmic meaning of a meaningful term (i.e., its sense) is the information about how to compute its denotation step-by-step: a meaningful term has sense by carrying instructions within its structure, which are revealed by its canonical form, for acquiring what they denote in a model. The canonical form $\mathsf{cf}(A)$ of a meaningful term A encodes its intension, i.e., the algorithm for computing its denotation, via: (1) the basic instructions (facts), which consist of $\{ p_1 := A_1, \ldots, p_n := A_n \}$ and the head term A_0, which are needed for computing the denotation $\mathsf{den}(A)$, and (2) a terminating rank order of the recursive steps that compute each $\mathsf{den}(A_i)$, for $i \in \{0, \ldots, n\}$, for incremental computation of the denotation $\mathsf{den}(A) = \mathsf{den}(A_0)$.

Thus, the languages of recursion offer a formalisation of central computational aspects of Frege's distinction between sense and denotation, with two semantic "levels": Referential Intensions (Algorithms) and Denotations.

$$\underbrace{\text{NL Syntax} \Longrightarrow L_{lr}^{\lambda} \Longrightarrow \text{Referential Intensions (Algorithms)} \Longrightarrow \text{Denotations}}_{\text{Computational Semantics}}$$

The reduction calculus of the type theory L_{ar}^{λ} is effective and has strong mathematical properties. The calculus of the intensional (i.e., algorithmic) synonymy in the type theory L_{ar}^{λ} has a restricted β-reduction rule, which contributes to the high expressiveness of the language of L_{ar}^{λ}. There are arguments for restricted β-reduction with respect to preserving the intensional meanings (i.e., the algorithmic senses) of the L_{ar}^{λ}-terms, which comply with meanings of natural language expressions, but also of L_{ar}^{λ}-terms that are type theoretic constructs, see [5] and [2]

Characteristically, ambiguities of various kinds and underspecified expressions are abundant in NL. There are ongoing developments in computational linguistics for representation of ambiguities, underspecification, and related phenomena. Most theories use various meta-level variables and techniques. The reduction calculus of the type theory L_{ar}^{λ} has unique potentials for expressing underspecified semantic phenomena, at its object level. This is presented by rendering the sentence (3a) to the L_{ar}^{λ} recursion term (3b):

$$\text{John loves his wife, and Mary hates him.} \xrightarrow{\text{render}} \tag{3a}$$

$$[p_1 \ \& \ p_2] \text{ where } \{p_1 := L(h_1)(j), \ j := john, \tag{3b}$$
$$L := \lambda(x)love(w(x)), \ w := \lambda(x)wife(x),$$
$$p_2 := H(h_2)(m), \ m := mary, H := \lambda(x)hate(x)\}$$

By taking the free variables h_1 and h_2 in (3b) to be recursion variables, (3b) represents underspecified, abstract linguistic meaning of the sentence (3a). If left unbound by assignment, h_1 and h_2 can be interpreted as deictic pronouns obtaining their referents by speaker's references (e.g., modeled by a variable assignment). Alternative readings of (3a) can be obtained by extending the system of assignments in (3b), respectively by: (1) $h_1 := j$ and/or $h_2 := j$; (2) $h_2 := h_1$ and h_1 is free (note that in this case, h_2 is bound).

2.5 Some Key Theoretical Features of L_{ar}^{λ}

The type theory of recursion L_{ar}^{λ} (and L_{lr}^{λ}) introduces a new approach to the mathematical notion of algorithm, which is open for theoretical developments. The reduction calculus L_{ar}^{λ} (and L_{lr}^{λ}) is effective and has potentials for varieties of applications. For self-containment of the paper, this section presents the reduction system of L_{ar}^{λ} and some of the major results (without proofs) that are essential for algorithmic semantics of the language of L_{ar}^{λ} (and L_{lr}^{λ}). The reduction rules define a reduction relation between terms, $\Rightarrow \subseteq Terms \times Terms$. The *congruence* relation, denoted by $\equiv_c$, is the smallest relation between L_{ar}^{λ}-terms ($A \equiv_c B$) that is reflexive, symmetric, transitive, and closed under: term formation rules; renaming of bound variables, without causing variable collisions; and re-ordering of the assignment equations within the acyclic recursion systems of the recursion terms. The set of *reduction rules* of L_{ar}^{λ} is as follows:

Congruence (cong)
    If $A \equiv_c B$, then $A \Rightarrow B$
Transitivity (trans)
    If $A \Rightarrow B$ and $B \Rightarrow C$, then $A \Rightarrow C$
Compositionality
    If $A \Rightarrow A'$ and $B \Rightarrow B'$, (rep1)
    then $A(B) \Rightarrow A'(B')$
    If $A \Rightarrow B$, then $\lambda u\,(A) \Rightarrow \lambda u\,(B)$ (rep2)
    If $A_i \Rightarrow B_i$, for $i = 0, \ldots, n$, then (rep3)
    A_0 where $\{\, p_1 := A_1, \ldots, p_n := A_n \,\}$
    $\Rightarrow B_0$ where $\{\, p_1 := B_1, \ldots, p_n := B_n \,\}$
The head rule (head)
    $\left(A_0 \text{ where } \{\, \vec{p} := \vec{A} \,\} \right)$ where $\{\, \vec{q} := \vec{B} \,\}$
    $\Rightarrow A_0$ where $\{\, \vec{p} := \vec{A},\ \vec{q} := \vec{B} \,\}$
    given that no p_i occurs free in any B_j
    for $i = 1, \ldots, n,\ j = 1, \ldots, m$
The Bekič-Scott rule (B-S)
    A_0 where $\{ p := (B_0 \text{ where } \{\, \vec{q} := \vec{B} \,\}),\ \vec{p} := \vec{A} \}$
    $\Rightarrow A_0$ where $\{ p := B_0,\ \vec{q} := \vec{B},\ \vec{p} := \vec{A} \}$
    given that no q_i occurs free in any A_j
    for $i = 1, \ldots, n,\ j = 1, \ldots, m$
The recursion-application rule (recap)
    $(A_0 \text{ where } \{\, \vec{p} := \vec{A} \,\})(B)$
    $\Rightarrow A_0(B)$ where $\{\, \vec{p} := \vec{A} \,\}$
    given that no p_i occurs free in B for $i = 1, \ldots, n$

The application rule   (ap)

   $A(B) \Rightarrow A(p)$ where $\{p := B\}$

   given that B is a proper term and p is a fresh location

The λ-rule   (λ)

   $\lambda u \, (A_0$ where $\{\, p_1 := A_1, \ldots, p_n := A_n \,\})$

   $\Rightarrow \lambda u \, A'_0$ where $\{\, p'_1 := \lambda u \, A'_1, \ldots, p'_n := \lambda u \, A'_n \,\}$,

   where for all $i = 1, \ldots, n$, p'_i is a fresh location and A'_i is the result of the
   substitution $A'_i := A_i \{p_1 := p'_1(u), \ldots, p_n := p'_n(u)\}$.

Two major results are essential to the algorithmic senses of the terms:

Theorem 1 (Canonical Form Theorem). *Existence and uniqueness of the canonical forms: For each term A, there is a unique, up to congruence, irreducible term denoted by $\mathsf{cf}(A)$, such that:*

1. $\mathsf{cf}(A) \equiv A$ *or* $\mathsf{cf}(A) \equiv A_0$ where $\{\, p_1 := A_1, \ldots, p_n := A_n \,\}$
2. $A \Rightarrow \mathsf{cf}(A)$
3. *if $A \Rightarrow B$ and B is irreducible, then $B \equiv_c \mathsf{cf}(A)$, i.e., $\mathsf{cf}(A)$ is the unique up to congruence irreducible term to which A can be reduced.*

The canonical forms have a distinguished feature that is part of their computational (algorithmic) role: they provide algorithmic patterns of semantic computations. The more general terms provide algorithmic patterns that consist of sub-terms with components that are recursion variables; the most basic assignments of recursion variables (of lowest ranks) provide the specific basic data that feeds-up the general computational patterns. The more general terms and sub-terms classify language expressions with respect to their semantics and determine the algorithms for computing the denotations of the expressions.

Theorem 2 (Referential Synonymy Theorem). *Two terms A, B are referentially synonymous, $A \approx B$, iff there are explicit, irreducible terms (of appropriate types), $A_0, A_1, \ldots, A_n, B_0, B_1, \ldots, B_n$, $n \geq 0$, such that:*

1. $A \Rightarrow_{\mathsf{cf}} A_0$ where $\{\, p_1 := A_1, \ldots, p_n := A_n \,\}$,
2. $B \Rightarrow_{\mathsf{cf}} B_0$ where $\{\, p_1 := B_1, \ldots, p_n := B_n \,\}$,
3. $\models A_i = B_i$, *for all $i = 0, \ldots, n$, i.e., $\mathsf{den}(A_i)(g) = \mathsf{den}(B_i)(g)$ for all variable assignments g.*

Intuitively, $A \approx B$ iff (a) in case when A and B have algorithmic senses (i.e., A and B are proper), their denotations (which are equal) are computed by the same algorithm; (b) otherwise (i.e., A and B are immediate), A and B have the same denotations.

3   Proper Names

We assume that proper names, like John, Mary, etc., by default are rendered to L^λ_{ar} constants of type $\widetilde{\mathsf{e}}$, e.g.:

$$\text{Mary} \xrightarrow{\text{render}} mary : \tilde{e} \tag{4}$$

Syntactic co-occurrences with other expressions may invoke transformation of the names to expressions of appropriate types (so called type-shifting), e.g., by the type-driven coordination operator in L_{ar}^λ, see [5]. A name like Mary may be represented by the term $\lambda PP(mary)$ or its canonical form, $\lambda PP(mary) \Rightarrow_{cf} \lambda PP(m)$ where $\{m := mary\}$.

3.1 Proper Names and Co-denotations

The examples (5a)–(5e) are based on an example from [5] and are included here to highlight the special role of the constants in L_{ar}^λ. Different occurrences of the same sub-expression, in a larger expression, are rendered in L_{ar}^λ by using different recursion variables (locations). These expressions and the corresponding recursion variables can co-denote the same object, but are not computationally equivalent, i.e, they are not referentially synonymous. For example, two different occurrences of the proper name Mary in the sentence Mary likes Mary can co-denote the same object, as in (5a) and (5d). In (5a), the explicit and irreducible L_{ar}^λ term (a constant) $mary$ is assigned to different recursion variables m_1 and m_2, which co-denote the same object den($mary$) computed twice. Such a sentence would most likely be used with different stress over the name Mary (e.g., by uttering it differently), without necessarily denoting the same individual, which is represented by additional indexes in (5b) and (5d). In L_{ar}^λ, this represents co-denotation between the terms in (5a)–(5e), in case den($mary$) = den($mary_1$) = den($mary_2$), without co-reference between them.

$$\text{Mary likes Mary.} \xrightarrow{\text{render}} like(m_1, m_2) \text{ where } \{m_1 := mary, \tag{5a}$$
$$m_2 := mary\}$$

$$\text{Mary}_1 \text{ likes Mary}_2. \xrightarrow{\text{render}} like(m_1, m_2) \text{ where } \{m_1 := mary_1, \tag{5b}$$
$$m_2 := mary_2\}$$

$$\text{Mary likes herself.} \xrightarrow{\text{render}} like(m, m) \text{ where } \{m := mary\} \tag{5c}$$

$$\text{Mary likes Mary}_2. \xrightarrow{\text{render}} like(m_1, m_2) \text{ where } \{m_1 := dere(mary), \tag{5d}$$
$$m_2 := mary\}$$

$$\text{Mary likes herself.} \xrightarrow{\text{render}} like(m, m) \text{ where } \{m := dere(mary)\} \tag{5e}$$

Antecedent–anaphora relations can be represented in L_{ar}^λ via co-indexing and is summarised by (6b)-(6c); for more details see [2]. The terms (6b) and (6c) represent strict, *reflexive anaphora* via two kinds of co-indexing. Typical case of reflexive anaphora are the reflexive pronouns, like "herself", "himself", etc., in human languages (HL). Reflexives can be regulated by a HL grammar constraints, by lexical and phrasal syntactic rules that may vary depending on specific languages. Such a grammar can be provided with syntax-semantics interface rules that define a rendering procedure for translating HL phrases into L_{ar}^λ terms. The grammar and its rules for the render procedure can be supplemented with co-indexing operators that take care for the two kinds of co-indexing, λ-co-indexing that equates pure variables, as in (6b), and *ar-co-indexing* that equates recursion variables, as in (6c). co-indexing arguments of terms denoting relations (typically, relation constants), as in options (6b)-(6c), and ruling out (6d).

The antecedent-anaphora relation in (6c) is expressed by ar-co-indexing, which identifies arguments with recursion variables, and thus represents strict antecedent–anaphora relation between syntactic arguments, which is semantically reflexive.

Mary likes herself. (6a)

$$\xrightarrow{\text{render}} \lambda x \, like(x,x)(m) \text{ where } \{m := mary\} \qquad \text{(via } \lambda\text{-co-index)} \quad \text{(6b)}$$

$$\approx like(m,m) \text{ where } \{m := mary\} \qquad \text{(via ar-co-index)} \quad \text{(6c)}$$

$$\not\approx like(m_1, m_2) \text{ where } \{m_1 := mary, \, m_2 := m_1\} \quad \text{(inappropriate)} \quad \text{(6d)}$$

4 Naming Predicate

Names as Generalized Quantifiers with Naming Property: Names are quite naturally interpreted referentially to denote individuals in various contexts, e.g., by agents such as speakers, listeners, and other users that can be either humans or computerised entities. Names, e.g., in sentences like the above (5a)–(5e), and in more common ones, like "Mary likes John", typically are used to denote individuals, which may vary from state to state, or, as in (5d)–(5e), to denote rigidly the same individual across states.

In many cases, the information about a naming act, while important, is available "silently", as resource information, not as a direct part of the denotation. However, there are cases, when a name like "Mary" is a component of expressions in contexts that require naming information in the denotation. In such cases, given that the name occupies a syntactic position that is consistent with the type $(\tilde{e} \to \tilde{t}) \to \tilde{t}$ for quantifiers, the name can be "lifted up" to this type by rendering it to a L_{ar}^λ term as in (7b). The term in (7b) represents the name as a combination of referential denotation, expressed by the L_{ar}^λ constant *mary*, and naming information. The values of the denotation of the constant *mary* are assigned to the recursion variable m, which, additionally, has to satisfy the constraint expressed by the sub-term $named(m)(\text{Mary})(p)$. The constraint $named(m)(\text{Mary})(p)$ approximately stands for "She$_p$ named her$_m$ Mary". In this constraint, one of the arguments of the constant *named* is filled up by the entity Mary, which is the name itself[1]. The argument for the one who does the naming is filled by a free recursion variable p, the values of which are *underspecified* by

---

[1] In the examples in this paper, the name is a single word, but the technique can be generalized to more complex identity expressions.

the term (7b). The term (7b) denotes the set of all properties that Mary has, in addition to the property of being named Mary by an unspecified individual represented by the free recursion variable p.

$$\text{Mary} \xrightarrow{\text{render}} \tag{7a}$$

$$\lambda P\big(P(m) \ \& \ n\big) \text{ where} \{m := mary, \ n := named(m)(\text{Mary})(p)\} \tag{7b}$$

4.1   Predicative Naming

The name "Mary" may occur in an NP expression like a Mary without reference to any specific individual. E.g., the NP a Mary can occupy a predicative position denoting the property of being named Mary and we can render it to a L_{ar}^{λ} term of type $(\widetilde{e} \to \widetilde{t})$ denoting the set of all individuals named Mary:

$$\text{a Mary} \approx \text{named Mary} \xrightarrow{\text{render}} \tag{8a}$$

$$\lambda x\big[named(x)(\text{Mary})(p)\big] \tag{8b}$$

$$\Rightarrow \ \lambda x\big[named(x)(N)(p) \text{ where } \{N := \text{Mary}\}\big] \tag{8c}$$

$$\Rightarrow_{cf} \ \lambda x\big[named(x)(N'(x))(p)\big] \text{ where } \{N' := \lambda x\text{Mary}\} \tag{8d}$$

$$\Rightarrow_{\eta} \ \lambda x\big[named(x)(N)(p)\big] \text{ where } \{N := \text{Mary}\} \qquad \text{by } (\eta\text{-Red}) \tag{8e}$$

The λ-rule is a key computational means, which preserves correct argument structure. It presents the general possibilities for argument links and fill-ups, which sometimes, in special cases like (8c), creates unnecessary λ-abstractions. This is straighten by the η-Reduction Rule (η-Red), as in (8e), introduced in [3].

In the expression a Mary, the name Mary occupies a syntactic position for the syntactic category of nominal expressions, such as common nouns requiring a specifier and typically rendered into terms of type $(\widetilde{e} \to \widetilde{t})$. This suggests that the occurrence of the word Mary in the expression a Mary, in (8a), refers to a naming action of type $(\widetilde{e} \to \widetilde{t})$, by approximating property denoted by the VP "p gives the name Mary to _", or "p names _ Mary", without explicit reference to the individual giving the name and applicable to the named individuals. The specifier word a functions as syntax-semantics interface in producing the expression a Mary, which has a predicative interpretation similar to the passive participle named Mary.

In order the terms to have occurrences of names that denote themselves, the formal language L_{ar}^{λ} has to include such names as constants. Also, the names have to be elements of the domain frame for objects of type $\widetilde{e}$.

In the sentence (9a), the verb be, called the copula, takes a predicative complement a Mary. The syntax-semantics interface role of the verb be is to pass the rendering of the subject NP as the argument of the rendering of the predicative NP complement a Mary. Thus the sentence (9a) is rendered to the term (9b).

Mary is a Mary $\approx$ Mary is named Mary $\xrightarrow{\text{render}}$ $\qquad$ (9a)

$$\Big(\lambda x\big[named(x)(\mathsf{Mary})(p)\big]\Big)(mary) \qquad (9b)$$

$$\Rightarrow \Big(\lambda x\big[named(x)(\mathsf{Mary})(p)\big]\Big)(m) \text{ where } \{m := mary\}$$
by (ap) $\qquad$ (9c)

$$\Rightarrow \Big(\lambda x\big[[named(x)(N)(p)] \text{ where } \{N := \mathsf{Mary}\}\big]\Big)(m) \text{ where } \{m := mary\}$$
by (ap), (recap), (rep2), (rep1), (rep3) $\qquad$ (9d)

$$\Rightarrow \Big(\lambda x\big[named(x)(N'(x))(p)\big] \text{ where } \{N' := \lambda x\mathsf{Mary}\}\Big)(m) \text{ where } \{m := mary\}$$
by (λ-rule), (rep1), (rep3) $\qquad$ (9e)

$$\Rightarrow \Big(\big[\lambda x\big[named(x)(N'(x))(p)\big]\big](m) \text{ where } \{N' := \lambda x\mathsf{Mary}\}\Big) \text{ where } \{m := mary\}$$
by (recap), (rep3) $\qquad$ (9f)

$$\Rightarrow_{\mathsf{cf}} \lambda x\big[named(x)(N'(x))(p)\big](m) \text{ where } \{N' := \lambda x\mathsf{Mary}, m := mary\}$$
by (head) $\qquad$ (9g)

$$\Rightarrow_\eta \lambda x\big[named(x)(N)(p)\big](m) \text{ where } \{N := \mathsf{Mary}, m := mary\}$$
by (η-Red) $\qquad$ (9h)

5   Names as Referential Descriptions

Let denote the set of well-formed NL expressions, determined by some computational grammar, by the abbreviation NL-exprs. We assume that the grammar is equipped by syntax-semantics interface, e.g., see [3], which associates well-formed NL expressions with L^λ_{ar}-terms, i.e., the grammar defines a render function (or relation): render: NL-exprs $\to Terms(L^\lambda_{ar})$.

For expository purposes, here we assume a simplified version of the states (the details of elaborations depend on the area of applications). Semantic frames include the set $\mathbb{T}_s$ of states as having components of utterances. For the purposes of this paper, we set:

$$\mathbb{T}_s = \mathbb{T}_w \times \mathbb{T}_{\mathsf{time}} \times \mathbb{T}_{\mathsf{space}} \times \mathbb{T}_{\mathsf{user}} \times \mathbb{T}_{\mathsf{UserRefs}} \qquad (10)$$

where $\mathbb{T}_{world} \neq \varnothing$ is a nonempty set of possible worlds, taken as primitive objects; $\mathbb{T}_{\mathsf{time}} \subseteq \mathbb{T}_{\mathsf{e}}$ is a set of time moments; $\mathbb{T}_{\mathsf{space}} \subseteq \mathbb{T}_{\mathsf{e}}$ is a set of space locations; $\mathbb{T}_{\mathsf{space}} \subseteq \mathbb{T}_{\mathsf{e}}$ is a set of *users* of utterance states, who are usually *speakers*, but not always; $\mathbb{T}_{\mathsf{UserRefs}}$ is a set of functions, called *users' references* (which is an alternative of the technical term *speakers' references*):

$$\delta: \mathsf{NL\text{-}exprs} \longrightarrow \mathbb{T}, \quad \text{such that, for every } E \in \mathsf{NL\text{-}exprs} \qquad (11a)$$

$$\delta(E) = \mathsf{den}\big(\mathsf{render}(E)\big) \qquad (11b)$$

We assume that the language L_{ar}^λ has constants time, space, user, UserRefs, denoting elements, correspondingly, of the sets $\mathbb{T}_{time}$, $\mathbb{T}_{space}$, $\mathbb{T}_{user}$, $\mathbb{T}_{UserRefs}$.

$$\text{den}\left(the(p)(\text{ResSt}(s))\right) = \tag{12}$$

$$\begin{cases} \text{the unique } y \in \mathbb{T}_e \text{ such that} \\ \text{den}(p)(b \mapsto y)(\text{den}(\text{ResSt}(s))) = 1, & \text{if such unique } y \text{ exists,} \\ er, & \text{otherwise.} \end{cases}$$

where $\text{den}(\text{ResSt}) \colon \mathbb{T}_s \to \mathbb{T}_s$, is a L_{ar}^λ-constant[2] ResSt : (s $\to$ s) for a function, which maps states s_i into states s_j.

Some possibilities for rendering naming expressions, with the simple proper names as a special case, are as follows:

Verified referential usage of a name:

$$\text{John} \xrightarrow{\text{render}} \text{dere}\,(d)\,(r) \text{ where } \{\, d := the(p), \tag{13a}$$
$$p := \lambda x\big(j = R(x)\big), \tag{13b}$$
$$R := \lambda x\big(refer(x)(n)(u)\big), \tag{13c}$$
$$u := \text{user}(r),\ r := \text{ResSt}(s), \tag{13d}$$
$$j := john,\ n := \text{John}\,\} \tag{13e}$$

where $john : K_{\widetilde{e}}$ is a L_{ar}^λ-constant, and *refer* is a predicate constant such that $refer(x)(n)(u)$ stands for "u, i.e. user(r), refers to x with the naming expression n". The sub-term (13b) requires, i.e., verifies, that the user refers with the name John to the individual denoted by the constant *john*. In case that the user u refers with the name John to an individual k, i.e., $den\big(\lambda x\big(refer(x)(n)(u)\big)(b \mapsto k)\big)(\text{den}(r)) = 1$, but $j \neq k$, the term (13a)-(13e) evaluates to er.

Unverified referential usage of a name:

$$\text{John} \xrightarrow{\text{render}} \text{dere}\,(d)\,(r) \text{ where } \{\, d := the(p), \tag{14a}$$
$$p := R, \tag{14b}$$
$$R := \lambda x\big(refer(x)(n)(u)\big), \tag{14c}$$
$$u := \text{user}(r),\ r := \text{ResSt}(s), \tag{14d}$$
$$j := john,\ n := \text{John}\,\} \tag{14e}$$

The term (14a)-(14e) does not require, i.e., it does not verify, that the user refers with the name John to the individual denoted by the constant *john*.

---

[2] A more elaborated option would be the function den(ResSt) to depend on the user(s_i) (i.e., user's references) in the state s_i: ResSt$(s_i) \equiv$ UResSt(user(s_i)).

6   Future Work

More research is needed for investigating the various kinds of names, their classification and co-occurrences with other language phenomena, especially in the context of interactive interfaces. I.e., it is interesting to investigate interactive transitions from unverified semantic information, e.g., the referential usage of naming information, represented by the term (14a)-(14e), to either verifying or refuting it, depending on contextual information.

Names participate in various kinds of larger expressions whose analysis is very promising for new theoretical development in computational semantics and for various applications. This is important in active media technologies, where naming information is abundant, and often requires verification and interactive updates. From computational point, the analysis of the names and other semantic phenomena would be better by using some computational grammar for rendering NL expressions in L_{ar}^λ terms. This can be done by using constraint-based lexicalized grammar (CBLG) approaches, in general, by covering variety of grammar systems. It is very interesting to see the rendering defined with Milt-lingual grammatical framework, GF in particular.

Elaboration of the domain frames of L_{ar}^λ is important for better coverage of semantic phenomena and context dependency. A primary future development is the domain of the states, e.g., by using a suitable version of situation theory, which provides elaborate factual information about contexts, states, and events in specific application areas.

References

1. Gallin, D.: Intensional and Higher-Order Modal Logic. North-Holland (1975)
2. Loukanova, R.: Reference, co-reference and antecedent-anaphora in the type theory of acyclic recursion. In: Bel-Enguix, G., Jiménez-López, M.D. (eds.) Bio-Inspired Models for Natural and Formal Languages, pp. 81–102. Cambridge Scholars Publishing (2011)
3. Loukanova, R.: Semantics with the language of acyclic recursion in constraint-based grammar. In: Bel-Enguix, G., Jiménez-López, M.D. (eds.) Bio-Inspired Models for Natural and Formal Languages, pp. 103–134. Cambridge Scholars Publishing (2011)
4. Moschovakis, Y.N.: Sense and denotation as algorithm and value. In: Oikkonen, J., Vaananen, J. (eds.) Lecture Notes in Logic, vol. 2, pp. 210–249. Springer (1994)
5. Moschovakis, Y.N.: A logical calculus of meaning and synonymy. Linguistics and Philosophy 29, 27–89 (2006)
6. Thomason, R.H. (ed.): Formal Philosophy: Selected Papers of Richard Montague. Yale University Press, New Haven (1974); Edited, with an introduction by R.H. Thomason

Pyxis: An Active Replication Approach for Enhancing Social Media Services

Sen Li[1,2], Beihong Jin[1], Yuwei Yang[1,2], and Wenjing Fang[1]

[1] Institute of Software, Chinese Academy of Sciences, Beijing 100190, China
[2] Graduate University of Chinese Academy of Sciences, Beijing 100190, China

Abstract. To support the rapid increase of social media, social media services are demanded to be efficient, highly available and scalable. The paper provides an active replication approach Pyxis to enhance social media services. Following Pyxis, either the newly-built social media services or the rebuilt ones can tolerate site failures and network partition failures, and thus achieve better user experience by providing continuous services on the basis of keeping required causal consistency. The experimental results show that Pyxis can adapt to a large number of concurrent users, and recover from failures in acceptable intervals.

Keywords: Social Media Services, Active Replication, Fault Tolerance, Availability, Consistency.

1 Introduction

The media is a platform to propagate information. The Internet is referred to as the fourth-generation media after newspapers, radios, and televisions. The rapid development of Web 2.0 creates social media. In comparison with traditional media, social media emphasizes its interaction with users. Social media permits users to publish and propagate information online, and to accept information selectively rather than passively and indiscriminately. Nowadays, the typical social media includes social networking services, microblogs, wikis, and VoD (Video-on-Demand) websites.

From the perspective of software, social media depends on social media services, i.e., a group of Web 2.0 applications, which allow the creation and exchange of user-generated contents. In this paper, we focus on the availability of social media services so as to improve user experience.

Social media services can employ data replication to enhance their performance, availability and scalability, and further user experience. For example, in a microblogging system, microblogs can be replicated to multiple sites. As thus, a user's request can be randomly dispatched to one of the sites so that the system throughput can be improved. On the other hand, the microblogging system can normally respond to users' requests even if some site fails. Another example comes from VoD services. Facing a large amount of requests of accessing the same video, a conventional implementation method is to divide the whole video into multiple blocks, each with

R. Huang et al. (Eds.): AMT 2012, LNCS 7669, pp. 399–410, 2012.

equal play duration. And then, every block is replicated to multiple sites and an index is provided for copies of blocks. Whenever a user makes a request for accessing a block, the request will be forwarded to the location of the corresponding block by retrieving the index file. Such method can improve the performance of watching videos. Meanwhile, users still can watch videos smoothly in the case of site failures.

However, using replication may lead to data inconsistency. Currently, eventual consistency is the consistency offered by many distributed storage services which adopt the replication technology. This kind of consistency means that all the replicas will be consistent eventually, in other words, stale data may be returned to the users. We claim that social media needs stronger consistency than the eventual consistency. Here, we think that it is not suitable for social media to decide whether to reach consistency by the elapsing time, and the consistency required by social media should be decided by the execution order of operations on replicated data. In detail, if there is a causal relation between two operations in an application, then we hope the two operations should be executed in a cause-first order at all replicated sites. Such consistency requirement is called causal consistency. For example, a microblogging system needs to sort the messages according to the fact that a user posts a message and then some other users comment on the message, that is, operation "posting a message" and operation "commenting on the message" are required to satisfy causal consistency. Similarly, operation "adding a follower on a microblogger" and operation "removing the follower on the microblogger" can be viewed to form a pair of causal operations. Moreover, if a microblogger uploads some pictures in his shared album, then his followers can receive the prompt message about the uploading of the pictures. The interested followers may view these pictures. Here, operation "viewing the picture" obviously relies on operation "uploading a picture". If a microblogging system with multiple replicated sites only provides the eventual consistency, then the followers may not necessarily view the newly-uploaded pictures immediately. At least, the causal consistency is supported so that the followers can be guaranteed to view the newest pictures.

On the design of distributed data services, only two out of consistency, availability and partition tolerance are supposed to be guaranteed simultaneously, which is called CAP conjecture [1]. In general, for Internet applications, it is necessary to support partition tolerance. However, while the network is available, the tradeoff between consistency and performance is regarded as more important [2].

Social media services (SMSs in short) are a kind of distributed data services, so they need failure detecting and failover, moreover, they are required to provide the consistency implied by themselves and provide better user experience for users. While keeping such SMSs in mind, this paper presents an active replication approach named Pyxis. Pyxis takes advantage of replication and achieves the above goals for SMSs.

The rest of this paper is organized as follows. Section II introduces the related work. Section III gives the overview of Pyxis. Section IV describes the failover mechanism in Pyxis. Section V evaluates Pyxis by experiments. The last section concludes this paper.

2   Related Work

SMSs may be constructed from scratch such as Twitter, Tencent weibo, or implemented by utilizing existing distributed storage systems. For example, Windows Azure Storage (WAS) [4], a scalable cloud storage system, is used for applications which provide SMSs inside Microsoft. In particular, WAS provides serializability by utilizing Paxos protocol. Therefore, we investigate data replication technology and distributed storage systems which adopt replication.

In terms of data replication, it is an enabling technology for distributed services, where a group of replica managers (RMs in short) manage data replicas according to the following design choices: the working mode of RMs (as peers or master-slaves), the update propagation policy, the consistency guarantee method and the failure processing mechanism.

If a replication method works on the master-slaves mode, i.e., each replica is designated a primary RM responsible for coordinating write operations, then it is called the passive replication method. This method is simple but has limited availability. However, in any active replication method where RMs work as peers, multiple RMs are permitted to directly carry out write operations on replicas. In these active replication methods, the topology among RMs, the communication protocols for update propagation and the concurrency control over operations on replicas have a significant impact on their availability and consistency. For example, employing reliable totally-ordered multicast can implement sequential consistency, even in presence of fail-stop of replica servers. However, epidemic propagation can improve availability on the basis of offering causal consistency [3].

As far as the distributed storage systems are concerned, replication has been adopted by many key-value storage systems. Take Amazon Dynamo [5] and Apache Cassandra [6] as examples. In Dynamo, data are partitioned and replicated using a consistent hashing ring algorithm, whose versions can be distinguished by the vector clocks of data. Dynamo introduces sloppy quorum and a gossip-based protocol to maintain the consistency among replicas during updates, and provides eventual consistency. Cassandra uses the partition and replication methods similar to Dynamo, but it provides tunable consistency from eventual consistency to sequential consistency by configuring different numbers of replicas during a read/write operation.

Now, some researches focus on improving the consistency in distributed storage systems by decreasing the degree of inconsistency or providing stronger consistency than eventual consistency. For example, scalable causal consistency is the goal of COPS which is a geo-replicated wide-area key-value storage system [9]. However, [7] targets for reducing the number of stale reads. It gives an intelligent estimation model to estimate the rate of stale reads, and then automatically adjusts the number of replicas involved in read operations so that the stale data returned to users can be below the tolerable level of an application. [8] treats the number of update conflicts as the metrics of inconsistency. Based on such inconsistency definition, it presents a replica management method based on the probability of update conflict, which ensures two levels of consistency for applications, i.e., session consistency and serializability.

Compared with the existing work, Pyxis has the following features:

- Pyxis provides a set of replication techniques for SMSs, which can keep causal consistency needed by SMSs. By applying Pyxis, existing SMSs can be upgraded, or a new SMS can be built.
- Pyxis can support a lot of concurrent users by propagating the updates through LRP (Least Recently Push) messages in an optimistic way.
- Pyxis provides failure detecting and recovery, and the experimental data verify that the failover time is acceptable.

3 Overview of Pyxis

Pyxis specifies a set of APIs. Therefore, a new SMS (Social Media Service) can be built on the basis of Pyxis, or, as shown in Fig. 1, an existing SMS can be upgraded by taking advantage of Pyxis APIs, where any access to data from SMS clients needs to be transformed through Pyxis APIs.

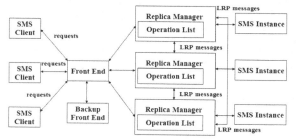

Fig. 1. Enhancing an SMS by Pyxis

The APIs provided in Pyxis are shown in Fig. 2. For convenience, all the replicated objects which belong to SMS components are called "SMS component objects" for short. In Pyxis APIs, *Register()* is used to specify that the operations (add/update/delete) to all the replicated objects managed by SMS components should satisfy casual consistency. *Add()* is used to create SMS component objects and assign their values. *Update()* and *Query()* are for updating and querying the values of SMS component objects,

void Register(path:string, op:int)
boolean Add(obj:string, val:string)
boolean Update(obj:string, val:string)
boolean Delete(obj:string)
string Query(obj:string)
string Invoke(op:string, type:OP_CODE)

Fig. 2. APIs in Pyxis

respectively. *Delete()* is for deleting SMS component objects. *Invoke()* can invoke the operations of SMS component objects.

Most importantly, Pyxis sets guidelines for Front End and a group of RMs (short for Replica Managers). Front End is supposed to connect with all the clients and RMs. After receiving requests from clients, Front End is responsible for designating appropriate RMs to deal with these requests and returning results to corresponding clients. Backup Front End is introduced to ensure that client requests can still be processed even if Front End fails to work, thus improving the availability. Meanwhile, Pyxis provides the methods of update propagation and failure processing for RMs.

Regarding update propagation, Pyixs generates unique identifiers for update operations to replicated data using multi-dimensional timestamp technique.

Multi-dimensional timestamp t is a vector in the form of $(t1, t2,..., tn)$, where ti is a *non-negative integer* and n is the number of RMs. Let t and s be two multi-dimensional timestamps. The definition of the relation $\leq$ between t and s is: $t \leq s \equiv (t_1 \leq s_1 \wedge ... t_n \leq s_n)$. The relation $\leq$ satisfies reflexivity, anti-symmetry and transitivity, so it is a partial-order relation. Each RM maintains a multi-dimensional timestamp rm_ts, and its component rm_tsi records the number of update operations received from RMi. Each RM also maintains a list of multi-dimensional timestamps ts_table, and the *i-th* component $ts\_table(i)$ is the latest timestamp rm_ts of RMi. In this way, Pyxis can sort the operations according to the partial-order relations ($\leq$) between the timestamps. Compared with quorum-based replication strategies, Pyxis tends to be highly available and scalable.

Each RM keeps an operation list *op_list* including three types of records: update records, ACK records, and reconfiguration records. An update record u contains the following components: *op, update_id, prev, cid, time* and *rm_id*, where *u.op* gives the content of the update operation, *u.update_id* is a unique multi-dimensional timestamp of the client update request assigned by an RM, *u.prev* is a collection of multi-dimensional timestamps of update operations on which u depends, *u.cid* is a unique identifier of the client update request assigned by Front End, *u.time* represents the time when the update record is written into *op_list*, and *u.rm_id* is a unique identifier of the RM generating the update record. An ACK record contains the following components: *ack_id, cid, time* and *rm_id*, where *a.ack_id* is a unique multi-dimensional timestamp of the client ACK message assigned by an RM, the value of *a.cid* is the same as the *cid* of the corresponding update request, *a.time* records the time when the ACK record is added into *op_list*, and *a.rm_id* is a unique identifier of the RM generating the ACK record. The details of reconfiguration records will be introduced in Section 4 (see 4.3).

In Pyxis, update propagation adopts TSAE (short for Time-Stamped Anti-Entropy) protocol[10] to ensure that the operations received from other RMs are continuous and sorted. In details, update record messages between RMs are propagated in a push style. Each RM periodically sends LRP messages to the RM which is least recently pushed.

An LRP message m consists of *log, ts* and *id*, where *m.log* represents a list of update records, *m.ts* is the *rm_ts* of the RM sending LRP message, and *m.id* is the index of the RM in *ts_table*.

Before an RM sends an LRP message m to another RM r, it generates the update record list *m.log* according to r's timestamp maintained in its *ts_table*. For example, if RMi is going to send an LRP message m to RMj, RMi needs to orderly insert the update records of *op_list* whose timestamps are not less than *ts_table(j)* into *m.log*, and set *m.ts* to *rm_ts* and *m.id* to its index i. After an RM receives LRP messages from other RMs, it will update the corresponding components of its *ts_table*.

On the other hand, RM failure and communication interruption may affect the execution of the operations submitted by clients. When RM failure or network partition happens, query/update requests may be prevented, which is due to the fact that the query/update requests depend on some update operations submitted before. For instance, after an RM r accomplishes an update operation u from a certain client c, RM r returns a "successful execution" message to client c. However, just before the update record u being propagated to another RM, network partition happens, which causes that RM r disconnects with the other RMs, Front End and the backup Front End. Therefore, Front End has to send subsequent query/update requests from client c to other RMs, but these requests depending on u cannot be executed until r recovers from failure. To solve the problems mentioned above, Pyxis provides the algorithms of failure detection, failure recovery and reconfiguration.

4   Failover Mechanism

4.1   Failure Detection

Each RM periodically sends heartbeat messages to the other RMs. When an RM recognizes that it has not received heartbeat messages from a certain RM r, it supposes that RM r fails to work or network partition happens, and starts a reconfiguration procedure. If a certain RM cannot connect to most of the other RMs which means that the RM is in a smaller network partition, the RM infers its failure, and stops receiving and processing update and query requests.

In particular, when a certain RM finds that the Coordinator RM, an RM with the highest ID who acts as the coordinator in reconfiguration, fails to work, it starts an election of Coordinator RM using Bully Algorithm.

4.2   Failure Recovery

In order to guarantee the recovery of a failed RM, each RM not only stores an operation list in its memory, but also maintains a log file in its persistent storage. Before new updates and ACK records are inserted into the operation list, they should be written into the log file. The records in both the operation list and the log file are in the same format.

Moreover, each RM executes a snapshot operation in a fixed period, and records its own state in a state file. When a snapshot operation is in execution, RM r stops processing all the newly-received messages including LRP messages, query and update requests. Then RM r informs the corresponding SMS instance to carry out the persistence procedure, and waits until it gets the notification from the SMS instance, indicating the persistence procedure has been finished. After that, r appends its state information to the state file. The state information consists of val_ts and cid_list, where val_ts is a multi-dimensional timestamp representing the number of the executed update operations, and cid_list is a collection of the executed update operations' $cids$. At last, r stores the operation list into a new log file and deletes the old one, and this operation is called log recycle.

When an RM begins to restart after failure, it waits for the corresponding SMS instance to recover to the state when the last snapshot is taken, and then reads the state and log files to recover its own state. To be specific, the steps of RM recovery are as follows:

- It reads its own state information from the state file;
- It reads all the records from the log file, and appends them to the operation list merging the *uid* of each record into its *rm_ts*;
- It communicates with the other RMs and updates its *ts_table* with the *rm_tses* achieved from the other RMs.

4.3   Reconfiguration Algorithm

An RM is supposed to be unavailable and needs to be reconfigured when it is in the following situations: a) the RM fails to work in a relatively long period; b) network partition leads some RM to disconnect to most of the other RMs.

Pyxis accomplishes deleting failed RMs and adding new RMs through reconfiguration algorithm, and allows users to decide the values of configuration options in a configuration file of each RM. As for RM *r*, its configuration options contain *DEL_OLD_RM*, *AUTO_REJOIN*, *rm_id* of RM *r*, a list of *rm_ids* of all the RMs, and a list of all the RMs' IP addresses. *DEL_OLD_RM* is set to be true if an RM is to be deleted directly after its failure, or false if it is to be added again. *AUTO_REJOIN* is set to be true if an RM is to be added automatically after being deleted, or false if it is supposed to be added manually by application managers later.

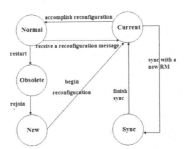

Fig. 3. State Transition in an RM

To support the reconfiguration algorithm, Pyxis defines five different states for RMs, including Normal, Current, Obsolete, Sync, and New.

- **Normal:** A Normal RM is not in the reconfiguration procedure, or it is preparing for the recovery of another RM (see Phrase 1 of the reconfiguration procedure)
- **Current:** A Current RM is in the reconfiguration procedure, and belongs to the new configuration.
- **Obsolete:** An Obsolete RM is in the reconfiguration procedure, and belongs to the old configuration but not the new one.
- **Sync:** When the Coordinator RM receives a "rejoin" message from a certain RM, it turns to Sync state.
- **New:** An RM in New state can be a newly-deployed RM or a rejoining one after deleted. The New RM has not begun the reconfiguration procedure.

Fig. 3 shows the state transition diagram for an RM. After a Normal RM finds that another RM fails and receives a reconfiguration message from the Coordinator RM, it turns to Current state. When a Current RM accomplishes reconfiguration operations, it returns to Normal. If a Normal RM fails to work and restarts again, it turns to Obsolete state. When an Obsolete RM rejoins, it turns to New state. A New RM becomes Current after receiving the reconfiguration message from the Coordinator RM. When the Coordinator RM in Current state receives a "sync" message from the failed RM which has restarted, it begins to synchronize its log file to the restarting RM and turns to Sync state. After accomplishing synchronization, the Sync RM returns to Current.

In support of reconfiguration algorithm, configuration objects are introduced to record RMs' configuration information, including the number of RMs, the array of *rm_ids,* and the version number which identifies the running configuration. Each RM in the reconfiguration procedure needs to keep two configuration objects, that is, *old_config* and *current_config*.

Moreover, an RM *r* generates a reconfiguration record when executing reconfiguration operations. The reconfiguration record is written into the operation list of RM *r*, and it is used to separate the update records of the operation list into two parts which respectively belong to the old configuration and the new one. The reconfiguration record contains old and new configuration information, *rm_id* of *r* and a multi-dimensional timestamp *uid* of this record.

The following part is going to explain the processing procedure of reconfiguration algorithm in detail. In order to ensure that the system can still process client requests before failover, we prescribe that in the period after some RM fails and before it restarts, the system still uses the old version number, indicating that the failed RM does not receive any client requests and the other RMs are normal in this period. Every reconfiguration operation leads to the increase of the version number by one. There are three phrases of the reconfiguration procedure, and the specific steps are described as follows:

Phrase 1: After detecting a failed RM or network partition, Normal RMs will stop log recycle, and elect Coordinator RM if necessary. The specific steps are as follows:

1. The RM which has detected a failed RM or network partition starts an election of Coordinator RM among available RMs using Bully Algorithm if necessary.
2. The Coordinator RM generates a new configuration object named *new_config*, inserts it into a failure message *f* and sends the message *f* to all the Normal RMs.
3. The RMs receiving the failure message *f* stop recycling log files so that the records deleted from the operation lists can be saved and synchronized to newly-joining RMs. Meanwhile, when they are processing LRP messages, once they propagate the update records to all the RMs in the new configuration *new_config*, they can remove the records from their operation lists, thus ensuring that all the operation lists in memory can be recycled to prevent memory from leaking.

Phrase 2: After the failed RM has restarted and all the update records of the old version have been propagated to all the Normal RMs, the Normal RMs will delete the failed RM from their configurations.

1. After recovery, the failed RM obtains the other RMs' addresses in the original process group from the configuration file, reconnects to these RMs using TCP, and sends its heartbeat messages.

2. After receiving reconfiguration messages, normal RMs turn to Current state: they stop processing new client requests but not for LRP messages, and execute reconfiguration operations. Each Current RM only sends its LRP messages to the other Current RMs. If a Current RM receives an LRP message including a reconfiguration record of some RM r, the Current RM sends an ACK message to RM r. When a Current RM receives ACK messages from all the other Current RMs, it can remove the operation records of the old version from the operation list. When the failed RM which has restarted receives the reconfiguration message, it executes reconfiguration operations and turns to Obsolete state. Note that Front End is not going to send any client requests to an Obsolete RM, and the Obsolete RM sends LRP messages to Current RMs and only processes ACK messages.

3. Once a Current RM has received all the operation records of the old version, and sent all of its old-version operation records to all the other Current RMs, it can remove the old configuration information and return to Normal. If an Obsolete RM has sent all of its old-version operation records to all the Current RMs, and the value of DEL_OLD_RM is set to be true, the Obsolete RM removes the SMS instance and exit, otherwise, Phrase 3 is carried out.

4. If the value of DEL_OLD_RM is true, RM continues to recycle log files when it returns to Normal state; otherwise, RM continues to recycle log files when Phrase 3 of the reconfiguration is accomplished.

Phrase 3: A New RM rejoins. A New RM can be a failed RM which has restarted. The detailed steps are in below:

1. If the value of $AUTO\_REJOIN$ is true, the removed RM sends a "rejoin" message which contains its rm_ts (as $sync\_ts$) to the Coordinator RM after Phrase 2; otherwise, it waits for the "rejoin" command from application managers.

2. The Coordinator RM turns to Sync state after receiving the "rejoin" message. The Coordinator RM starts a sync thread, and the sync thread is responsible for reading the operation records whose timestamps are not less than $sync\_ts$ from the log file and sending them to the New RM. Considering that the whole number of the operation records to be read from the log file might be quite large, the record number read each time is set to be at most MAX_NUM. These read operation records are inserted into sync messages which are sent to the New RM. Receiving the sync messages, the New RM appends the operation records to its operation list, correspondingly updates its rm_ts, and then replies messages including updated rm_ts (as a new $sync\_ts$) to the Coordinator RM. The Coordinator RM compares the received $sync\_ts$ with its own rm_ts, and if the difference (i.e., the number of operation records unsent to the new RM) between them is less than MAX_DIFF, it stops receiving and processing client requests and LRP messages. The Coordinator RM will not execute reconfiguration operations or continue receiving and processing LRP messages, until all the rest of operation records have been synchronized to the New RM.

3. The coordinator RM generates new configuration and reconfiguration messages, and broadcasts the reconfiguration message to all the RMs.

4. After receiving the reconfiguration message, each RM turns to Current state: it stops processing query and update requests, changes the recycle strategy when processing LRP messages, and executes reconfiguration operations. The new

recycle strategy is that RM is not going to recycle the operation records until they are propagated to all the RMs in the new reconfiguration.

5. When the New RM has received all the operation records of the old version, it returns to Normal state.

5   Evaluation

We conduct experiments to evaluate the performance and availability of Pyxis in terms of throughput, response time and failover time. Here, failover time refers to the time of successful execution of a causal operation Z submitted by a client when the corresponding RM or SMS instance breaks down (marked as X, which will be restarted) and operations which Z depends on have been successfully executed before X breaks down. Therefore, throughput and respond time reflect the performance of an upgraded SMS, and failover time measures availability of an upgraded SMS.

Our experimental environment includes two machines connected by a Gigabit Ethernet switch. Two servers are Dell PowerEdge T610, each with two 4-core Intel E5506 processors and 12GB memory. All servers run XenServer 5.5, and in each XenServer, one or more CentOS Linux 5.4 virtual machines (VMs) are installed. The basic configuration of a VM is 1-core processor and 1GB memory. In our experiments, every RM and the corresponding SMS instance (a microblogging system prototype) are deployed in one VM. Clients and a Front End are deployed in one VM.

In the first group of experiments, we observe how the number of concurrent clients impacts on the performance of an SMS. In experiments, LRP message interval is set to 3 seconds. Every client sends causal write requests in a rate of 1 per second. We record the throughput and response time of the microblogging system. The results are shown in Fig. 4-5.

From the experimental results, we see that both the throughput and the response time have a direct proportion with the number of clients while the LRP message interval and the number of RMs are invariable; and the increase of RMs' number has little effect on the throughput and response time while the number of client threads remains the same.

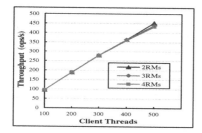 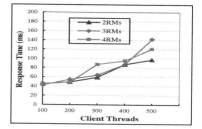

Fig. 4. Throughput vs. No. of client threads   **Fig. 5.** Response time vs. No. of client threads

In the second group of experiments, LRP message interval varies from 1 second to 10 seconds, however, the number of concurrent clients is set to 300 and every client also sends causal write requests in a rate of 1 per second. Fig.6-7 show the experimental results.

From the experimental results, we can see that there is a positive relation between LRP message intervals and response time, that is, shortening LRP message intervals leads to the decrease of response time, but throughput is relatively stable; and the number of RMs has little efforts on the throughput and response time while the LRP message interval is fixed.

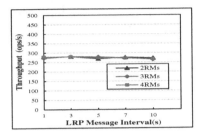
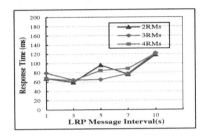

Fig. 6. Throughput vs. LRP message interval   **Fig. 7.** Response time vs. LRP message interval

In Pyxis, by default client's timeout value is set to 20 seconds. RM broadcasts heartbeat messages every 3 seconds, if an RM's heartbeat message has not been received for more than 30 seconds, then the RM is regarded as broken down by other RMs. We control an RM to be killed, and then restarted after 30 seconds in order to ensure that the other RMs could detect the failure. In the third group of experiments, LRP message interval is set to 3 seconds. The number of clients varies from 100 to 500. The results are shown in Fig. 8.

The experimental results show that more RMs lead to longer failover time while LRP message interval remains the same. Due to our algorithm, when an RM switches to Current state, *reconfiguration record* must be sent to all other RMs, and this RM could never recover to Normal state until receiving all the other RMs' *reconfiguration records*. As a result, when the number of RMs increases, failover time increases.

Another important point is that the number of clients has little effect on failover time. The reason is that when an RM e.g. *r* breaks down, the worst case is that update requests which have arrived at RM *r* in last 3 seconds fail to be sent to other RMs. After restarting, RM *r* will propagate these unsent update requests which can be executed immediately on other RMs.

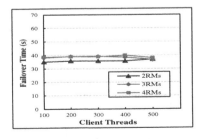
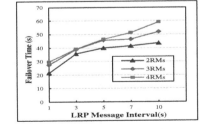

Fig. 8. Failover time vs. No. of client threads   **Fig. 9.** Failover time vs. LRP message interval

In the fourth group of experiments, the number of client threads is set to 300. LRP message interval varies from 1 second to 10 seconds. The experimental results are shown in Fig. 9.

From the data in Fig. 9, we note that 1) When the number of client threads is fixed, more RMs lead to longer failover time; 2) Longer LRP message interval leads to longer failover time while RMs' number remains the same; 3)When the LRP message interval is 1 second, failover time is less than the failure detecting time. This is because the RM has sent part of client operations to other RMs before its breakdown.

6   Conclusions

Social media services require high availability. As social media's influence expands, failures on social media services may lead to heavy even unexpected losses. By employing replication technology, Pyxis can enhance the availability of social media services. Pyxis enables social media services to recover within an acceptable duration and continue to provide normal functionality.

Acknowledgments. This work was supported by the National Natural Science Foundation of China under Grant 60970027 and Grant 91124001.

References

1. Gilbert, S., Lynch, N.: Brewer's conjecture and the feasibility of consistent available partition-tolerant web services. ACM SIGACT News (2002)
2. Brewer, E.: Cap twelve years later: How the "rules" have changed. Computer 45(2), 23–29 (2012)
3. Ladin, R., Liskov, B., Ghemawat, S.: Providing High Availability Using Lazy Replication. ACM Transactions on Computer Systems 10(4), 360–391 (1992)
4. Calder, B., Wang, J., Ogus, A., et al.: Windows azure storage: a highly available cloud storage service with strong consistency. In: 23rd ACM Symposium on Operating Systems Principles, pp. 143–157. ACM, New York (2011)
5. DeCandia, G., Hastorun, D., Jampani, M., Kakulapati, G., Lakshman, A., Pilchin, A., Sivasubramanian, S., Vosshall, P., Vogels, W.: Dynamo: Amazon's Highly Available Key-Value Store. In: The 21st ACM SIGOPS Symposium on Operating Systems Principles, pp. 205–220. ACM, New York (2007)
6. Lakshman, A., Malik, P.: Cassandra: A Decentralized Structured Storage System. ACM SIGOPS Operating Systems Review 44, 35–40 (2010)
7. Chihoub, H., Ibrahim, S., Antoniu, G., Perez, M.S.: Harmony: Towards Automated Self-Adaptive Consistency in Cloud Storage. In: IEEE Cluster 2012, Beijing, China (2012)
8. Kraska, T., Hentschel, M., Alonso, G., Kossmann, D.: Consistency Rationing in the Cloud: Pay only when it Matters. Proc. VLDB Endowment 2, 253–264 (2009)
9. Lloyd, W., Freedman, M.J., Kaminsky, M., Andersen, D.G.: Don't Settle for Eventual: Scalable Causal Consistency for Wide-Area Storage with COPS. In: ACM SOSP 2011, Cascais, Portugal, October 23-26 (2011)
10. Petersen, K., Spreitzer, M.J., Terry, D.B., Theimer, M.M., Demers, A.J.: Flexible Update Propagation for Weakly Consistent Replication. In: Proceedings of the Sixteenth ACM Symposium on Operating Systems Principles, SOSP 1997 (December 1997)

Social Network Analysis of Virtual Worlds

Gregory Stafford, Hiep Luong, John Gauch, Susan Gauch, and Joshua Eno

University of Arkansas, Fayetteville, AR 72701
{gstaffor,hluong,jgauch,sgauch,jeno}@uark.edu

Abstract. As 3D environments become both more prevalent and more fragmented, studying how users are connected via their avatars and how they benefit from the virtual world community has become a significant area of research. An in depth analysis of the virtual world social networks is necessary to evaluate its worlds, to understand the impact of avatar social networks on the virtual worlds, and to improve future online social networks. Our current efforts are focused on building and exploring the social network aspects of virtual worlds. In this paper we evaluate the Second Life social network we have created and compare it to other social networking sites found on the web. Experimental results with data crawled from Second Life virtual worlds demonstrate that our approach was able to build a representative network of avatars in virtual world from the sample data. The analysis comparison between virtual world social networks and others in flat web allows us to gauge measures that better explore the relationship between locations linked by multiple users and their avatars. Using this comparison, we can also determine if techniques of personalization search and content recommendation are feasible for virtual world environments.

Keywords: Virtual Worlds, Social Network Analysis, Web Crawling, Second Life.

1   Introduction

Virtual world environments allow users to navigate through and interact with online content in a 3D space using avatars as virtual representations. According to Virtual Worlds Review[1], virtual worlds such as Second Life, Open Simulator, and Active Worlds, are growing both in popularity and in numbers. They have received significant attention from the public at large, from businesses and other organizations, and from scholars in disciplines as diverse as law, sociology, psychology, math, and, more recently, information systems [12]. Providing users with an easy way to locate online content quickly and efficiently has been a large area of interest. Several social networking web sites on 'The Flat Web' including Facebook, Flickr, LinkedIn, MySpace, Orkut and YouTube collect user information, user preferences, and online activities to better predict what web content will be of interest to users in the future. Virtual world environments have the same need to quickly and easily locate content relevant to their

---

[1] http://www.virtualworldsreview.com

R. Huang et al. (Eds.): AMT 2012, LNCS 7669, pp. 411–422, 2012.

interests, especially when content is presented with a full 3D environment that can be quite overwhelming to new users. Therefore, understanding how avatars behave when they connect to social networking virtual communities such as Second Life (see Figure 1) creates opportunities for better interface design, richer studies of social interactions, and improved design of content distribution systems.

In this work, we present an analysis of user workloads in both virtual worlds and online social networks. Comparing the social networks of virtual worlds to those found on the web not only shows whether or not current relevant content search is viable on the virtual world platform but may also elucidate how users behave differently in an immersive virtual environment compared to social networking sites. In addition, an analysis of the virtual world social network can give researchers insight to the behaviors of users when immersed in a virtual environment. Additionally if a proper social network can be constructed, then techniques developed for the web that bring relevant content to users and generate user preference models may be implemented on this new platform. Bringing these technologies to virtual worlds will enhance this already expanding community and helping new users find interesting content in an initially overwhelming environment.

In this paper, we begin with a summary of related work on social networking in virtual worlds. Then, we introduce how we have built a social network of avatars and present measures to compare the social network of virtual worlds to those on the current flat web. In the next section, we report our experimental results and evaluation of this comparison. The final section presents conclusions and discusses our ongoing and future work in this area.

Fig. 1. Social interaction in Second Life virtual world

2   Related Work

There has been tremendous interest in online social networks such as Facebook, Flickr, LinkedIn, MySpace, Orkut and YouTube. Boyd [2] provides a history of these online social networks and describes how typical interactions within a social network take place. They also discuss the complications that arise when the real world and online social networking overlap, causing social disruptions in many cases.

Mislove [13] applied quantitative measures based on network analysis to study the underlying social networks of Flickr, LinkedIn, Orkut and YouTube. In another study of Facebook, Nazir [14] deployed three applications (*The Got Love, the Fighter's Club* and *Hugged*) and captured the activities of users to better understand the social interactions of 3-4 million Facebook users. Analysis of these online social networks may provide insight to the integrity of the virtual world social network formed using our methods. The variety of networks included in this research allows for good analysis of how exactly our social network behaves.

An important issue in social network analysis is dealing with link definition that relates two entities the network. Adamic [1] has identified community structure from overlapping memberships in groups of people. While the various link definitions portrayed different entity behavior separately, Matsuo [11] found that experts and authorities are easily located while trustworthiness and other important factors are more visible when this information is integrated.

How virtual environments reflect or effect behaviors in the real world is another active area of research. In recent work by Doodson [4], users accessed Second Life and were given a survey to answer questions based on their interactions within the virtual world and in the real world. The survey concluded that personalities portrayed in the two worlds varied greatly. Similarly, Lewis et al. [10] examined a Facebook social network of campus individuals, and found that the lives of the students differed significantly than their lives represented in their Facebook profiles.

When sampling a large scale network it is important to understand the various ways a network can be sampled and the effect the sampling technique has on the structural properties and distributions of the social network being evaluated. Leskovec [9] performed statistical comparisons between a number of sampling techniques and found that comparing the degree distributions of the sample data set versus the full data set is an effective way to evaluate sample data set. They also found that a sample size of 15% of the entire data set is sufficiently representative in many applications. Gjoka [7] perform a similar comparison of sampling techniques against a uniform sampling of the same network. The conclusion is that two variations of the random walk sampling method are most similar to the uniform sample structurally. These techniques, where applicable, provide an effective method to generate a representative sample.

3   Analysis of Virtual Worlds

In this section, we will present how we collected data from Second Life virtual world and built a social network of avatars from this data. Then, we introduce several basic and advanced measures that are used to evaluate social networks.

3.1   Crawling Data in Virtual Worlds

To collect data for this study, we have chosen to mimic common web crawlers and make use the intelligent virtual world crawler that we had developed for a previous project [5][6]. Our crawler is able to move semi-autonomously, either following a set of waypoints provided to them or by following other avatars. Using this crawler we were able to extract from Second Life a listing of groups and avatars that have sub-scribed to them. An avatar may list publicly the groups in which it is subscribed. With each group having its own profile, the group then lists each member that has made the subscription public. Our crawler supports searching for groups and locations using the existing Second Life search service and can request descriptive information for par-cels, avatars, groups, avatars, and objects. All extracted information is stored in a collection of indexed tables in a MySQL database, and used to create our virtual world social network.

3.2   Building a Virtual World Social Network

Our current efforts are focused on building and exploring the social network aspects of virtual worlds. Given the data we were able to collect using our crawler; we represent avatars as nodes in our network, and use undirected edges between nodes to represent relationships between avatars. The weights assigned to edges correspond to the number of Second Life groups the two avatars share in common. Larger weights correspond to stronger inferred relationships between the pair of avatars. Edges with weight zero (no groups in common) are excluded from the graph.

   We have developed a sparse graph representation to store our virtual world social network in memory and perform a range of graph analysis algorithms. Our network consists of 1,628,532 unique avatars that we crawled from the Second Life server. To add edges to the graph, we processed the list of all 310,702 Second Life groups, and for each groups we incremented the weight of the edges between all possible pairs of avatars in the group. As group size gets larger, this leads to an explosion in the num-ber of edges added to the graph. For example, a group with 100 avatars will produce 4,950 edges, while a group with 1000 avatars will yield 499,500 edges. Our largest group of 28,328 avatars would produce a staggering 401,223,628 edges.

   Including all of the edges from all Second Life groups would exceed our graph software capacity, so we have to represent the social network with a subset of edges. Since random edge sampling has been shown to perform poorly [9], we have chosen to exclude all groups above a specified maximum size when constructing the social network graph. If we assume that the strength of avatar relationships is inversely related to the group size, excluding large groups will have the effect of ignoring weak edges between avatars when creating the social network graph.

3.3   Social Network Analysis

In this section, we describe the graph metrics we will use to characterize our virtual world social network and compare our experimental observations to prior work in social network analysis.

— *Degree:* The degree of a node in an undirected graph is the number of adjacent edges. Since edges represent the number of groups shared by two avatars, the degree of a node is equal to the number of relationships for each avatar.
— *Fraction of Avatars:* The number of avatars in the sample used for analysis compared to the total number of avatars in the virtual world.
— *Average Node Degree:* The average of all of the avatar's degrees in the sample.
— *Number of Groups:* The number of unique Second Life groups that are represented within the sample used for analysis.
— *Average Groups Per Users:* The average number of groups per user within a sample. Groups exceeding a predefined size limit are not considered in this average.
— *Average Path Length.* This is the average of the shortest paths between all pairs of nodes in the network. Due to the size and complexity of our networks we took a random sample of one hundred nodes and calculated the shortest paths from these random nodes to all other nodes to estimate the average path length.
— *Radius:* The radius of a graph is defined as the minimum eccentricity of the graph. Given a node n, the eccentricity of n is the maximum shortest path between n and any other node [8]. For our networks we calculated the radius of the largest graph component. We also took one hundred random seed nodes and calculated the eccentricity of each of them. The minimum eccentricity of these seed nodes gave us our radius measurement for our social network.
— *Diameter.* While not simply twice the radius, the diameter metric does find the shortest path between the edges of the network. The diameter is the maximum of the eccentricities across all nodes. Similar to the radius, we take a random sample of one hundred seed nodes and calculated the eccentricities between a seed node and all other nodes in the network.
— *Joint Degree Distributions.* The joint degree distribution displays the rate at which nodes of a certain degree connect to nodes of another degree. This degree distribution is represented as 2D matrix of the node degrees, the coordinate value, or shade of gray at a pixel, represents the strength of connection between nodes of that degree.
— *K-Nearest Neighbor Distributions.* We define kNN (k-Nearest Neighbors) degree distribution to represent the tendency of nodes of a certain degree to connect to higher or lower degree nodes. A value of a point on the scatter plot attributed with a given degree is the average degrees of all nodes connected to a node of this degree.

4   Comparison of Social Networks

4.1   Data Collection

Using our Second Life crawler [6] we simulated the movements of an avatar and collected and stored all information that was sent from the Second Life server to the graphical client. A summary of the data we collected over several months is shown in Table 1. Although this is not as complete as the original data stored on Second Life servers, we feel this data collection provides large representative cross section of regions/avatars/groups that exist within Second Life. The next section will present in detail how we extract a representative sample of this data to build a social network of avatars within this virtual world.

Table 1. Summary of data collection

Total Regions Crawled	20,901
Total Unique Avatars	1,628,532
Total Unique Groups	310,702
Total Group Subscriptions	9,534,064

4.2   Experiments

A consideration when setting a maximum group size limit to be included within the social network is the real relationship between avatars in the same group. When groups of large size are added to the social network it increases the strength of existing links between avatars and adds links between several pairs of avatars. In order to keep this fact in consideration, we have taken various samples of our social network for comparison between them and those found in other studies. Traditionally samples would be taken at a 25%, 50%, 75%, and 100% but due to the power law distribution of the group sizes [3] we believe it necessary to take sample of 70th, 80th, and 90th percentiles while including a 99th percentile sample as our maximum. This decision was made based on the significantly small changes between basic measures of the generated networks. Figure 2 represents the percentages of groups included in a sample given a maximum group size.

Table 2 shows that ninety percent of groups have a membership of 41 members or less. This suggests a behavior of group members wanting to keep groups within a manageable size or being unable to find membership for their group. Larger group sizes are most likely public groups or large popular companies and the power law behavior is not surprising. As the group size increases, the amount of edges added to the graph is exponential. Additionally, this affects computation time of most of the metrics as many rely on finding all pairs shortest past for a given node. Given these facts we have chosen to specify a maximum group size of 457 in order to represent 99% of the groups included in the crawled data while avoiding high edge groups in the top one percent.

Table 2. Values of the Group Distribution and the corresponding percentiles

Group size	2	5	10	13	18	41	457	28328
Percentile	25%	50%	70%	75%	80%	90%	99%	100%

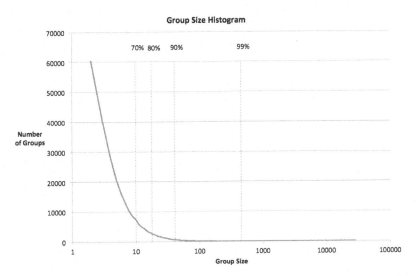

Fig. 2. Distribution of virtual world group sizes with different main percentiles marked

4.3   Evaluation of Social Networks

Using the data collected above, we have defined two social networks of avatars. The first social network (*All_Edges*) uses all edges between avatars. The second social network uses all edges with edge weights greater than one (*All_Except_1_Edge*). The resulting graph is a network of avatars that share two groups or more in common, which is a stronger connection than avatars that share only one group in common.

In Table 3 we show a side-by-side analysis of our two social networks and their samples based on percentile. These results are compared with statistics observed in online social networking reported by Mislove [13]. The YouTube and Live Journal networks are directed networks. The edges in these networks have a direction from one node to the other and involve several other analysis techniques. The Orkut social network is an undirected network that also contains a large amount of edges. Our network has approximately half the number of nodes (avatars) while the number of edges is still larger than other networks. This suggests that the removal of edges with a weight of one is justified in constructing a social network based on implicit link definitions.

More complex statistics of our network are presented in Table 4 alongside online social networking site data [13]. The average path length of the virtual world social networks does not deviate significantly from online social networks either. Our social network composed of edges of all weight (*All_Edges*) has a slightly lower average path length due to the amount of edges. The virtual world social network composed of edges with a weight greater than one (*All_Except_1_Edge*) strengthens the relation represented by a link, but adversely increases the average path it would take to get from one node to the other. Similar cases are found when comparing the radius and diameter of the graphs found in virtual worlds and those found on the web including Flickr, Live Journal, and YouTube. It suggests that even though edges of weight one can be removed the furthest an avatar is from another in the largest component of our virtual world social network is a shorter distance than those found in online social networks.

Table 3. Comparison of basic measures of social networks

| Percentile of Groups | Virtual World Social Network | | | | | | | | Online Social Networks [13] | | |
| | All Edges | | | | All Except 1 Edge | | | | YouTube | Live Journal | Orkut |
	70th	80th	90th	99th	70th	80th	90th	99th			
Number of Avatars/Users	447,825	576,991	760,445	1,273,002	97,396	153,369	239,006	604,861	1,157,827	5,284,457	3,072,441
Fraction of Users	27.4%	35.4%	46.6%	78.1%	6.0%	9.4%	14.7%	37.1%	Unknown	95.4%	11.3%
Number of Links	1,819,062	4,582,699	14,975,625	315,455,515	109,732	250,017	647,953	10,070,845	4,945,382	77,402,652	223,534,301
Avg. Node Degree	8.12	15.88	39.39	495.61	2.25	3.26	5.42	33.30	4.29	16.97	106.1
Number of Groups	21,9058	251,102	279,778	307,602	21,9058	251,102	279,778	307,602	30,087	7,489,073	8,730,859
Avg. Groups Per User	1.96	2.30	2.78	4.28	1.96	2.30	2.78	4.28	0.25	21.25	106.44

Table 4. Structural properties of the social networks

| Graph | VW Social Network | | Online Social Networks [13] | | | |
	All_Edges (99%)	All_Except_1_ Edge (99%)	Flickr	Live Journal	YouTube	Orkut
Avg Path Length	2.96	4.89	5.67	5.88	5.10	4.25
Radius	7	11	13	12	13	6
Diameter	9	14	27	20	21	9

In Table 5 we compare the social network found in our virtual world to two studies involving Facebook applications and a variety of other online social sites [14]. The 99th percentiles of both versions of our network are presented and have similar statistics of the online social networks. The number of components is quite large in the network with edges of a weight of two or more. This is expected, as the removal of edges with a weight of one will disconnect the graph substantially.

Table 5. Component comparison of the social networks

| Network | VW Social Network | | Online Social Networks [14] | | |
	All_Edges (99%)	All_Except_1_Edge (99%)	Fighter's Club	Got Love	Hugged
Number of Components	593	6042	29	13461	4018
Largest Component %	99.87%	98.73%	91%	92.10%	86.70%

Joint Degree Distribution.

Figure 3(a) shows the joint degree distribution of the network of all edges included (*All_Edges*). A clear diagonal is very distinctive and shows a clear connection between nodes of similar degree. High degree nodes connect to high degree nodes, and low degree nodes connect strongly to other low degree nodes. Another fact about the network that can be discerned from Figure 3(a) is the somewhat substantial cutoff of the diagonal and the relatively sparse bottom right corner of the image. This is due to the fact that for presentation we have removed degrees of one thousand or higher from the image but the maximum group size used for the 99th percentile is 457.

In comparison the network including edges of all weights, Figure 3(b) shows the joint degree distribution gray map of the same network with edges of weight one removed (*All_Except_1_Edge*). Significant differences include the absence of the substantial diagonal through the higher node degrees. This suggests that many of the nodes degrees consist of edges with a weight of one. It makes sense that as only a small portion of edges in the full edge network exists in this representation, but really shows that the high degree avatars are connected to many other avatars simply because they belong to one or a few relatively large groups.

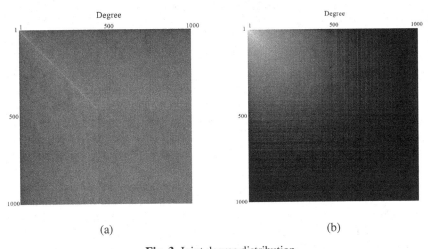

(a)   (b)

Fig. 3. Joint degree distribution

k-Nearest Neighbor Distributions.

In Figure 4(a) and 4(b) we have the kNN scatter plot of the virtual world social network with all edges included (*All_Edges*) and the virtual world social network with all edges with a weight greater than one (*All_Except_1_Edge*). The kNN value of an undirected network is the mapping of the degree of a node and the average degree of all nodes with the initial degree's neighbors. The green line in the figure above represents the average kNN for all degrees less than the current value. The behavior of this line is a good indicator of the behavior of higher degree nodes. If the average increases then so does the propensity of higher degree nodes in the social network to connect to other higher degree nodes. The axis scales vary quite a bit between the two networks because the removal of edges of weight one decreases the degree of all nodes significantly due to over 96% of the edges being removed.

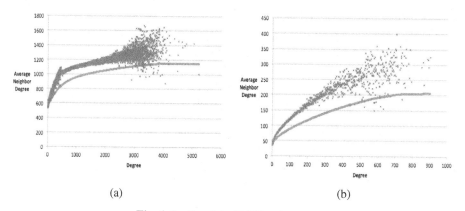

(a)   (b)

Fig. 4. Scatter plot of kNN and average

4.4   Discussion

In social network analysis, it is important to understand how the entities of the network are related to one another via links or shared groups. From the Table 3, we can see the number of avatars and links when we consider all edges between avatars (*All_Edges*) are very large. These numbers decrease significantly at all main percentile values if we take into account only edges that have weight greater than one (*All_Except_1_Edge*). Hence, there are a large number of avatars sharing only one common group with other avatars.

The more complex statistics show that the social networks of *All_Edges* and *All_Except_1_Edge* are reasonably close to those measured from other online social networks. The component analysis shows that while several isolated clusters of avatars exist, the main component contains the vast majority of avatars. The same fact is found in other online social networks. Numerous clusters of nodes exist in the social network but most of the nodes in the network are contained within the largest component.

The k-Nearest Neighbor scatterplot of the two social networks (*All_Edge* and *All_Except_1_Edge*) presented in the Figure 4 provides additional insight to the connectivity of nodes with similar degree. As both averages increase it shows that avatars tend to be related more with avatars that have a higher degree, or higher amount of avatars, in which they share groups in common with. The sparseness of the lower right corner of the Figure 4(b) suggests that the degrees of nodes are significantly lowered when edges with a weight of one are removed. The diagonal of similar degree nodes connecting to each other still persists through this network as in *All_Edges* network as seen in Figure 4(a). This hints that the removal of all edges with a weight of one does not necessarily affect the behavior of nodes tendency to connect to nodes of similar degree. This finding agrees with that found in the kNN distributions.

The social network found underlying Orkut is most similar to our network including all edges. As Orkut is the only undirected online social network of the four it suggests that perhaps undirected online social networks are naturally more compact and closer together than directed networks.

5   Conclusions

Understanding how avatars in virtual worlds behave when they connect to these worlds is important in terms of characterizing avatars behaviors and interests. The goal of this research study is to build and evaluate a social network of avatars that were crawled from Second Life virtual worlds. We have made a comparative study of important graph measures that can give a view about how social networks are differently formed in virtual worlds and in online networking.

The analysis of the virtual world social network compared to social networks found in online social networking sites has elucidated a few facts. Although the friendship data between avatars was unavailable to be crawled, the link definition of groups in common between pairs of avatars is a sound approach as the networks generate were not extremely different in their basic structural properties. Additionally, the fact that a user is using an avatar to interact with a 3D environment does not necessarily affect their social behaviors.

While this research outlines some of the basic structural properties and metrics to evaluate a virtual world social network derived from Second Life, there are several metrics we would like to explore (i.e., the clustering coefficient, associativity, and scale free metric) and use to further compare our virtual world social network with others. We also hope to analyze how the link definition affects the connectivity of the avatars in the social network.

Acknowledgments. This research is supported by the NSF grant number 1050801 III: EAGER: *Mapping Three Dimensional Virtual Worlds.*

References

1. Adamic, L.A., Adar, E.: Friends and neighbors on the web. Social Networks 25(3), 211–230 (2003)
2. Boyd, D.M., Ellison, N.B.: Social Network Sites: Definition, History, and Scholarship. Journal of Computer-Mediated Communication 13, 210–230 (2007)
3. Clauset, A., Shalizi, C.R., Newman, M.E.J.: Power-law distributions in empirical data. SIAM Rev. 51(4), 661–703 (2009)
4. Doodson, J.: The Relationship and Differences Between Physical and Virtual World Personality. University of Bath (2009)
5. Eno, J., Gauch, S., Thompson, C.: Intelligent Crawling in Virtual Worlds. In: Int. Joint Conf. on Web Intelligence and Intelligent Agent Technologies, WI-IAT 2009, vol. 3, pp. 555–558 (2009)
6. Eno, J., Stafford, G., Gauch, S., Thompson, C.: Hybrid User Preference Models for a Virtual World. In: User Modeling, Adaptation, and Personalization, Girona, Spain, pp. 87–98 (2011)
7. Gjoka, M., Kurant, M., Butts, C.T., Markopoulou, A.: Walking in Facebook: A Case Study of Unbiased Sampling of OSNs. In: INFOCOM 2010, pp. 1–9 (2010)
8. Hage, P., Harary, F.: Eccentricity and centrality in networks. Social Networks 17(1), 57–63 (1995)
9. Leskovec, J., Faloutsos, C.: Sampling from large graphs. In: Proc. of the 12th ACM SIGKDD Int. Conf. on Knowledge Discovery and Data Mining (KDD 2006), pp. 631–636. ACM, New York (2006)
10. Lewis, K., Kaufman, J., Gonzalez, M., Wimmer, A., Christakis, N.: Tastes, ties, and time: A new social network dataset using Facebook.com. Social Networks 30(4), 330–342 (2008)
11. Matsuo, Y., Hamasaki, M., Takeda, H., Mori, J., Bollegara, D., Nakamura, Y., Nishimura, T., Hasida, K., Ishizuka, M.: Spinning Multiple Social Networks for Semantic Web. In: Proc. of the Twenty-First National Conf. on Artificial Intelligence (AAAI 2006), pp. 1381–1386 (2006)
12. Mennecke, B., McNeill, D., Ganis, M., Roche, E.M., Bray, D.A., Konsynski, B., Townsend, A.M., Lester, J.: Second Life and Other Virtual Worlds: A Roadmap for Research. Communications of the Association for IS 18(28) (2008)
13. Mislove, A., Marcon, M., Gummadi, K.P., Druschel, P., Bhattacharjee, B.: Measurement and analysis of online social networks. In: Proc. of the 7th ACM SIGCOMM Conf. on Internet Measurement (IMC 2007), pp. 29–42. ACM, New York (2007)
14. Nazir, A., Raza, S., Chuah, C.-N.: Unveiling Facebook: a measurement study of social network based applications. In: Proc. of the 8th ACM SIGCOMM Conf. on Internet Measurement (IMC 2008), pp. 43–56. ACM, New York (2008)

Active Media Framework for Network Processing Components

Ichiro Satoh

National Institute of Informatics
2-1-2 Hitotsubashi, Chiyoda-ku, Tokyo 101-8430, Japan
ichiro@nii.ac.jp

Abstract. This paper presents a framework for building active media content and defining network processing for it. It makes two novel contributions. The first introduces network processing as first-class objects like active media content in the sense that components are introduced as the only constituent of our network processing for components as well as active content. It enablesan active media content to be composed of one or more active media or network processing components and to migrate between these components, which may be running on different computers. It also offered several basic operations for network processing, e.g., carrying, forwarding, duplication, and synchronization. The operations can be treated as active media components; they can be dynamically deployed at local or remote computers through GUI-based manipulations. It therefore allows an end-user to easily and rapidly configure network processing in the same way as if he/she had edited the documents. We constructed a prototype implementation of this infrastructure and its applications.

1 Introduction

Active media should be able to access a variety of content on the Internet as well as local content inside them, and to be exchanged between users through networks, including the Internet. Active media are often required to define their network processing. For example, a workflow management system distributes the documents among employees through the routes specified in documents. Some documents may require their preferred secure communications to enable them to be exchanged between users.

End-users often want to define the network processing of documents for them to be able to accomplish application-specific tasks. However, the customization and management of networking processing is too complex and difficult for end-users to achieve. Some confidential documents also need to pass through their preferred secure protocols. For example, electronic mails that exchange documents between users through networks and web servers enable us to share documents stored on remote computers. However, the existing network processing of documents, e.g., electronic mails and world-wide-web, just treat documents as data, i.e., first class objects. Therefore, the network processing of documents tends to be content-dependent and application-specific.

This paper proposes a component framework as a solution to these problems. It enables an enriched content to be composed of active media content, e.g., text and images. The framework introduces the notion of self-contained components in the sense that not

R. Huang et al. (Eds.): AMT 2012, LNCS 7669, pp. 423–432, 2012.

only the content of each component but also its codes to view and edit the content are embedded in the component to solve various problems, including content rights management, with existing content-distribution. It also enables network protocols for active media to be implemented by a set of visual content. The framework enables active media to define their own itineraries and migrate under their own control. Furthermore, active media can be executed and transmitted as first-class objects to their destinations. The framework introduces components for network processing as active components, so that it allows an end-user to easily and rapidly configure network processing in the same way as if he/she had edited the content.

2   Example Scenario

Ringi-sho is a document circulated to obtain managerial permission from stakeholders and has been widely used for round robin decisions in most Japanese companies.[1] A proposer creates a circular document, called ringi-sho, about his/her proposal and is circulatedly the document to each of the stakeholders, e.g., section managers, who are related with the proposal, so that the document receives approval stamps from all the stakeholders on the proposal. Such a document may have images in addition to text and its network processing depends on its stakeholders, so that it should support a variety of content and define its own delivery.

Existing ringi-sho documents have been delivered to stakeholders through existing network processing, e.g., e-mail, but we have had some problems. Since the stakeholders of documents depend on the documents, each of the stakeholders who receives them must select the next destination and send the documents to the selected destination explicitly. If one of the stakeholders cannot access his/her e-mail due to taking some time off, they are blocked in his/her mail spool. Ringi-sho documents may have active media content in future.

2.1   Related Work

Building systems from software components has already proven useful in the development of large-scale software [8]. Many frameworks for software components have been developed, e.g., COM [5] and JavaBeans [3]. These existing frameworks aim at defining the behaviors of distributed computing, i.e., server-side and client side processing, by combining software components. Therefore, these frameworks are suitable for professional developers.

The framework presented in this paper, therefore, has been designed independently of existing component frameworks for distributed computing or compound documents. This is because it permits document-centric components to migrate themselves over a network and process other components as first-class objects [1], e.g., migrating or saving them to other computers or on secondary disks. These features enable our components to have direct access to novel and powerful features that existing components do

---

[1] In a typical Japanese company, each manager reads and approves more than ten ringi-sho documents every business day.

not have. End-users can also easily customize their network processing of documents through user-friendly manipulations to edit visual components, and they can control their own network processing according to their content. Further, this is open to existing component frameworks. In fact, it can use typical Java-based components, e.g., Java Beans and Applets, as its components.

Several (non-component-based) attempts have been made to support active documents, e.g., Active Mail [2] and HyperNews [4], but these have aimed at particular application-specific documents, such as electronic mail and newspapers, so that they have not supported varied and complex content. The fuseONE system [10] is composed of GUI-based control panels to control appliances from active documents, i.e., GUI-based buttons and toggle switches. Like other compound document frameworks, these cannot transmit codes for viewing and editing documents. Several researchers have explored active networking technologies [9] to customize distributed computing, particularly network processing between computers. However, these existing technologies have focused on configurations for low-level network processing, e.g., routing and QoS protocols, so that they are not suitable for end-users.

We presented a compound document framework to provide software components designed for compound documents [6]. Although the previous framework was an early prototype of the framework presented here and it enabled components to carry and forward other components over a distributed system, the previous work was designed for distributed documents under the documents' control, whereas the framework presented in this paper supports various kinds of networking for documents. The framework presented in our previous paper [7]. was for context-aware multimedia content on digital signage in public/private spaces. It was designed to dynamically deploy multimedia content at computers according to the locations of their target users.

3   Basic Approach

Our framework provides a solution to the above problems. The idea behind the framework is to treat network processing for active media as first-class objects in active media, in the sense that the framework does not distinguish between software components for active media and components for network processing. The framework has a composition, self-contained components, and a media defined network.

- **Composition:** The framework must be composed of an active media component of nested components that can display visual parts, e.g., text, images, and windows, and that can enable us to edit components in-place without opening a separate window for each component.
- **Self-contained components:** Components should be distributed and operated without the need for any applications on their current computers. That is, when a computer receives a component, it must be able to view or edit it, including its inner components, even when the computer lacks applications.
- **Media-defined network:** Each component is an autonomous programmable entity and can determine which components or computers it will go to according to its program code and content, and then migrate to that destination.

The first allows a hierarchy of nested components to correspond to visual parts, e.g., text, images, and windows. When a component is contained by another component, the former is still an individual component so that it can be removed from the latter. The third is useful to deploy and execute components at remote computers, which have no program codes for the components. The third is supported by using mobile agent technology, where mobile agents are autonomous programs that can travel between computers under their own control. The framework supports four types of components as follows:

- The **active media component** stores active media content and programs to play the content within itself. It displays this content in the estate assigned by its container component by using its own programs. When an active component contains other components, it is responsible for managing the estates of its inner components within its own estate.
- The **forwarder component** can automatically migrate its visiting components to another component running on the same computer. It can also process its visiting components as first-class objects before it forwards them.
- The **transmitter component** is allocated at a component or computer and can automatically transmit its inner components to its remote destinations by using the protocols defined inside it.
- The **carrier component** can contain one or more components, e.g., active media ones, and carry them to its destination or along its itinerary. Since it can can treat its inner components, it can explicitly restrict or transform its inner components.

Each component can freely move into any other component except itself or its descendant components, as long as the destination component accepts it. The destination may be at a different computer. Each container component is responsible for automatically offering its own services and resources to its inner components and controlling its inner components.

4   Design and Implementation

This framework consists of two parts: runtime systems and components. The framework itself is independent of any programming language, but its prototype implementation is constructed on Java. It can exchange components between runtime systems, even when their underlying systems, e.g., operating systems and hardware, are different, because Java VM conceals differences between the underlying systems.

4.1   Runtime System

Each runtime system is a middleware system for managing and executing components. Each runtime system governs all the components inside it and provides them with APIs for components in addition to Java's classes. It assigns one or more threads to each component and interrupts them before the component migrates, terminates, or is saved. Each component can request its current runtime system to terminate, save, and migrate itself and its inner components to the destination that it wants to migrate to.

A hierarchy of components is maintained in the form of a tree structure, which has the component tree nodes of components (Fig. 1). Each node contains its target component, its attributes, and its containment relationship inside it and provides interfaces between its component and the runtime system. When a component is created in a runtime system, the system creates a node for the newly created component and runs the node inside it. When a component migrates to another location, the runtime system migrates its node with the component. When each runtime system saves or migrates a component over a network, it marshals the component, the component's inner components, and information about their containment relationships and visual layouts, called component nodes, into a bit-stream and then later unmarshals the components and information from the bit-stream. The runtime system then transmits the marshaled component to its destination through an extension of the HTTP protocol.

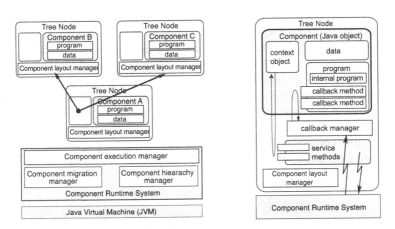

Fig. 1. Component hierarchy and structure of components

4.2   Component

Each component is defined as a subclass of AMTComponent or NetComponent written in Java, where the first supports active media components and the second supports carrier or transmitter components. Each component also defines the protocols that let components embedded in it to communicate with one another.

Active Media Component: Each active media component stores its active media content for playing or editing the content inside it. It displays this content in the estate assigned by its container component by using its own program. When an active media component contains other components, it is responsible for managing the estates of its inner component within its own estate. The AMTComponent class (or its subclasses) specifies attributes, e.g., its minimum size and preferable size, and the maximum size of the visible estate of its component in the estate is controlled by the node of the component that contains it. Drag-and-drop is one of the most common manipulations for assembling visible components.

Forwarder Component: When a forwarder component receives other components, it automatically redirects the contained components to its destinations on the same computer. Components are forwarded as a transformation of the subtree structure of the component hierarchy. The current implementation provides two extensions of the forwarder component, called *duplicator* and *synchronizer*. The first can receive another component and then create a copy of its visiting component including all instance variables. The cloned component has the same content as the original. The second can strand its inner components until it can determine whether specified conditions can be satisfied, e.g., the number of inner components, the arrivals of specified components, and time constraints. A typical synchronizer component defines a group of moving components, as a barrier synchronization mechanism for parallel processes. It strands the visiting components inside it, until it receives all the components within the group.

Transmitter Component: A transmitter component at the source computer and its coexisting transmitter component at the destination computer establish a point-to-point channel between them by using a favorite communication protocol defined inside them. After a transmitter component contains a component (with nested components), the transmitter suspends its containing component and requests the runtime system to serialize the state and code of the contained component. It next sends the serialized component to a coexisting transmitter component located at the destination. The transmitter component at the destination receives the data and then reconstructs the component. Existing transmitter components support plain TCP/IP, HTTP, ftp, and SSL to exchange their containing components with their coexisting ones.

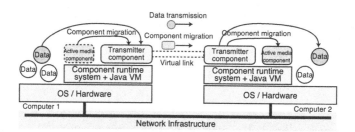

Fig. 2. Channel between transmitter components

Carrier Component: A carrier component can carry one or more components to its destination, which may be running on a different computer and is constructed as a hierarchical mobile agent that contains one or more mobile agents corresponding to components inside it (Fig. 3). Moreover, carrier components can encapsulate or restrict their inner components, because they can control them while carrying them, and they can provide a secret-key-based cryptographic procedure to protect these inner components against illegal access or modifications.

Remarks. The above components cover most basic functions to implement active media or network protocols. We can easily extend them by overwriting their Java classes.

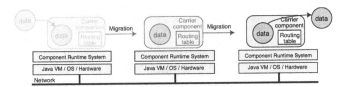

Fig. 3. Carrier component

5   Current Status

We implemented the framework using Java language (JDK 1.5 or later versions), and we developed various components for active media and network processing. Since the Java virtual machine and libraries abstract away differences between the underlying systems, e.g., operating systems and hardware, components can migrate between and be executed on runtime systems running on different computers, whose underlying systems may be different. We conducted a basic experiment on component migration with computers (Intel Core 5i 2 GHz with MacOS X 10.7 and Oracle's JDK 1.7) connected through 1-Gbps Ethernet. The time for component migration measured from one container to another in the same hierarchy was 5 ms, including the cost of drawing the visible content of the moving component and checking whether the component was permitted to enter the destination component. The cost of component migration measured between two connected computers was measured at 34 ms. The moving component was a simple text viewer and its size (sum of code and data) was about 9 KB (zip-compressed).

6   Applications

We developed various components based on this framework. This section introduces several of these and their uses.

6.1   Active Media Electronic Mail System

The first example is an electronic mail system based on the framework that consisted of two main components: an inbox component and letter components. The inbox component provides a window component that can contain two components. The first is a history of received mail and the second offers a visual space for displaying content selected from the history. The latter component corresponds to a letter. It can carry various active media components for accessing text, graphics, and animation. The component can deliver itself and its inner components to an inbox component at the receiver computer. Since the inbox document is the root of the letter component, when the document is stored and moved, all the components embedded in the document are stored and moved with the document. Figure 4 is a time-constrained component that contains one or more active media components corresponding to its content. The component makes its inner components invisible a specified time after it arrives and is displayed at the destination.[2]

---

[2] Assume the self-destructing terminals or tapes in the American *Mission: Impossible* television series and movies.

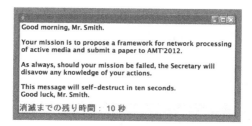

Fig. 4. Self-destructing letter

6.2 Application-Specific Distribution of Documents

The second application is a *Ringi* document described in the scenario described in Section 3. Figure 5 shows a container component that contains active media components. The component can carry its inner components to managers and gather managerial approvals from them. It has a table for specifying multiple destinations and sequentially migrates itself with its inner active media components to the destinations. When a manager receives the component, if he/she permits the proposal described in the inner active media components, he/she drags and drops his/her stamp component on his/her frame in the table. Next, the carrier components migrates themselves and their inner components to one of the managers who did not approve them. If they do not receive stamps from a manager for a specified time after they have arrived at their computers, they treat him/her as being absent and then migrate to another manager. When they can receives managerial approvals from all their specified managers or a manager disapproves them, they return to the computers of their proposers.

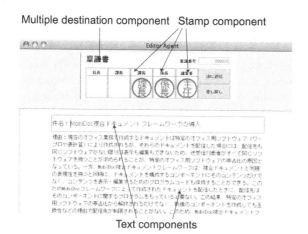

Multiple destination component   Stamp component

Text components

Fig. 5. Document circulated among multiple managers to obtain managerial permission

6.3 File Hosting Component

Commercial file hosting services, e.g., Dropbox, Sugarsync, and Evernote, have become currently popular. We developed a component that supports the APIs of Dropbox and Evernote. A user drags and drops components or files on the icons of a component on a document consisting of active document components, and the component automatically transmits the dropped components and files to Dropbox's or Evernote's file hosting services. When a user clicks on the icon, he/she can see the list of components and files stored at the services.

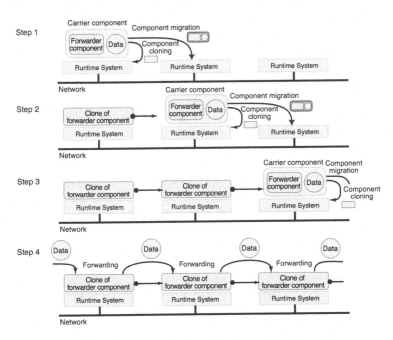

Fig. 6. Stream communication with forwarder components

6.4 Stream Communication

The fourth application is packet-based stream communication through multiple hops. Assume that data are decomposed into multiple pieces as multiple packets and are transmitted to its destination as a sequence. Unlike existing packet communications, the framework does not need to assign the destination to all packets that contain the pieces. Instead, it encapsulates the head piece of of the sequence in the carrier component that is designed to carry the piece and a forwarder component to the destination or immediate nodes, whichever is closest to the destination. Figure 6 shows the mechanism for stream communication by forwarder components deployed by carrier components. Whenever the carrier component migrates, it leaves a clone of its inner forwarder component at the current node and assigns the address for the next hop to the destination of the forwarder component remaining behind. The carrier component migrates to the next hop with its data and inner forwarder component until it arrives at the destination. As a result, forwarder components are left along the path from the source to the destination. When any of the forwarder components that are left receive the remaining pieces of data, they transmit these to their own next hops, i.e., their own neighboring forwarder components. That is, the remaining pieces can automatically follow the traverse of the head piece through the forwarder components at the immediate nodes until they arrive at the destination. Therefore, this approach raises the possibility of improving the performance of packet-based communication, because the header of each packet needs to be evaluated before the packet is forwarded to the next hop at each immediate node, whereas the approach can directly forward the remaining pieces of data to the next hop at each immediate node. Note that this system uses carrier components as a self-deployment method to dynamically distribute and install forwarder components at nodes for a stream session.

7   Conclusion

We presented a framework for building active media content and defining their network processing. It introduced the notion of a component hierarchy and mobile components. This notion enabled active media content to be composed of one or more various components and to migrate between these components, which may run on different computers, under its own control. The framework enabled active media content to pass other content from/to other components or computers. Components were introduced as the only constituent of our network processing for components. We also constructed a prototype implementation of this infrastructure and its applications.

References

1. Friedman, D.P., Wand, M., Haynes, C.T.: Essentials of Programming Languages. MIT Press (1992)
2. Goldberg, Y., Safran, M., Shapiro, E.: Active Mail - A Framework for Implementing Groupware. In: Proceedings of ACM CSCW 1992, pp. 75–83. ACM Press (1992)
3. Hamilton, G.: The JavaBeans Specification. Sun Microsystems (1997), http://java.sun.com/beans
4. Morin, J.: HyperNews, a Hypermedia Electronic-Newspaper Environment based on Agents. In: Proceedings of HICSS-31, pp. 58–67 (1998)
5. Rogerson, D.: Inside COM. Microsoft Press (1997)
6. Satoh, I.: Network Processing of Documents, for Documents, by Documents. In: Alonso, G. (ed.) Middleware 2005. LNCS, vol. 3790, pp. 421–430. Springer, Heidelberg (2005)
7. Satoh, I.: A Framework for Context-Aware Digital Signage. In: Zhong, N., Callaghan, V., Ghorbani, A.A., Hu, B. (eds.) AMT 2011. LNCS, vol. 6890, pp. 251–262. Springer, Heidelberg (2011)
8. Szyperski, C.: Component Software, 2nd edn. Addison-Wesley (2002)
9. Tennenhouse, D.L., et al.: A Survey of Active Network Research. IEEE Communication Magazine 35(1) (1997)
10. Werle, P., Kilander, F., Jonsson, M., Lönqvist, P., Jansson, C.G.: A Ubiquitous Service Environment with Active Documents for Teamwork Support. In: Abowd, G.D., Brumitt, B., Shafer, S. (eds.) UbiComp 2001. LNCS, vol. 2201, pp. 139–155. Springer, Heidelberg (2001)

Semantic Network Monitoring and Control over Heterogeneous Network Models and Protocols[*][**]

Christopher J. Matheus[1], Aidan Boran[1], Dominic Carr[2], Rem Collier[2], Barnard Kroon[2], Olga Murdoch[2], Gregory M.P. O'Hare[2], and Michael O'Grady[2]

[1] Bell Labs, Blanchardstown, Ireland
chris.matheus@gmail.com
[2] CLARITY: Centre for Sensor Web Technologies,
School of Computer Science & Informatics,
University College Dublin, Dublin, Ireland
Gregory.OHare@ucd.ie

Abstract. To accommodate the proliferation of heterogeneous network models and protocols we propose the use of semantic technologies to enable an abstract treatment of networks. Network adapters are employed to lift network specific data into a semantic representation that is grounded in an upper level "NetCore" ontology. Semantic reasoning integrates the disparate network models and protocols into a common RDF-based data model that network applications can be written against without requiring intimate knowledge of the various low level-network details. The system permits the automatic discovery of new devices, the monitoring of device state and the invocation of device actions in a generic fashion that works across network types, including non-telecommunication networks such as social networks. A prototype system called SNoMAC is described that employs the proposed approach operating over UPnP, TR-069 and SIXTH network models and protocols. A major benefit of this approach is that the addition of new models/protocols requires relatively little effort and merely involves the development of a new network adapter based on an ontology grounded in NetCore.

Keywords: semantic computing, sensing web, network monitoring, home area networks.

1 Introduction

The Internet of Things (IoT) is expected to encompass over 15 billion devices by 2015. In addition to shear volume, there is the added complexity of dealing with the unchecked proliferation of new network data models and protocols. To deal

---

[*] This work was partly funded by the Industrial Development Authority of Ireland.
[**] This work was partly funded by Science Foundation Ireland (SFI) under grant 07/CE/I1147.

R. Huang et al. (Eds.): AMT 2012, LNCS 7669, pp. 433–444, 2012.

with these issues network and active media applications will not only require high performance and scalability but will also need the means for quickly and dynamically evolving to accommodate the changing universe of devices. Doing this effectively necessitates a new approach for integrating network models and protocols that facilitates the management of devices and does so at various levels of abstraction so they can be handled generically as collections of devices while maintaining the specifics necessary to monitor and control them. This paper advocates an approach to solving this problem that leverages the benefits afforded by semantic web technologies.

Semantic web technologies enable the definition of formal data models called "ontologies" that provide a number of conceptual and computational benefits, including: data model alignment, heterogenous data integration, built-in data abstraction mechanisms, basic forms of automated inferencing, dynamic meta-modeling and automatic consistency checking. With recent advancements in formal reasoning and querying capabilities (e.g., OWL 2 [8], SPARQL 1.1 [6]), semantic technologies are ready to be taken seriously in production environments, including those involving heterogeneous network monitoring and control.

The Bell Labs' SNoMAC (Semantic Network Monitoring, Analysis and Control) research effort is implementing a semantic-based approach to the dynamic monitoring, analysis and control of heterogenous telecommunication networks. The SNoMAC solution formally describes networked devices based on a layered collection of extensible ontologies grounded in an upper-level ontology called NetCore. Together these ontologies make it possible to hide or indeed expose device details to the extent necessary for the problem at hand. In this way device- and protocol-specific information can be semantically encapsulated at the lowest levels, with higher layers focusing on more generic networked device characteristics. As a result, applications can be written against SNoMAC's formal semantic API, ignoring lower-level details that may change as standards evolve and new ones are added. To demonstrate this concept in practice, Bell Labs is working with CLARITY: Centre for Sensor Web Technologies[1] at University College Dublin (UCD) to integrate SNoMAC with their SIXTH sensor middleware [10]. This paper provides an overview of the SNoMAC approach and describes a functioning prototype that detects, monitors and controls devices in a home area network involving UPnP [5], TR-069 [4] and SIXTH-based devices.

2   SNoMAC Overview

SNoMAC is designed specifically for 1) the auto-discovery of new network devices, 2) the remote setting of state variables to configure device functionality, 3) the execution of commands on the devices and 4) basic network analysis. An overview of the SNoMAC concept and prototype architecture is depicted in Figure 1. Network Adapters (one for each supported protocol) interface with the network both to "lift" device data into a semantic representation before sending it to the server and to invoke device-specific commands coming from the client via

---

[1] http://www.clarity-centre.org/

the server. Such adapters interface with the SNoMAC server through two simple REST-based Web service APIs supported by the server's Listener-Controller Interface; the API implemented by the network adapter is for processing requests issued by the server (e.g., get/set state variables, invoke commands) and the other is implemented by the server to receive pushed updates from the adapters as the devices in the network change. The information that is exchanged is in the form of RDF annotations based on a network-specific ontology that is aligned with the NetCore ontology (i.e., all network ontology classes and properties are sub-classes or sub-properties of entities defined in NetCore).

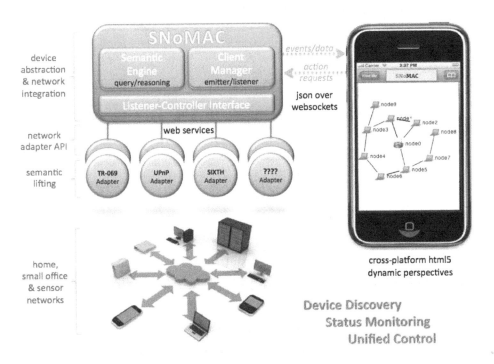

Fig. 1. Conceptual Overview of SNoMAC

The SNoMAC server initiates connections with available network adapters by requesting information for all known devices and issuing a request to be informed of subsequent device changes (i.e., node additions, deletions and state updates). Information about devices is stored in the Semantic Engine, which includes an RDF data store plus the means by which to automatically infer new information about the devices based upon meta data stored in their corresponding ontologies.

Web-based clients connect to the SNoMAC server through the Client Manager. Communication from the server to the client includes device information, action results and event notifications triggered by changes in the network. The clients communicate back to the server using "action requests" representing device specific commands that are propagated to the designated device where it is executed. Users interact with devices by clicking on their graphical representations to bring up a menu of available commands; these commands include getting and setting device variables (e.g., volume) and device-appropriate actions (e.g., "increase volume").

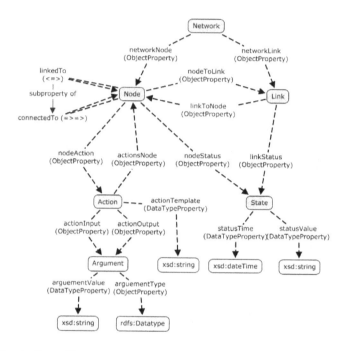

Fig. 2. Base Classes and Properties of the NetCore OWL Ontology

2.1 Ontologies

In Computer Science "an ontology is a specification of a conceptualization" [7]. We use the term more specifically here to refer to RDFS [2] and OWL [8] data models that formally define the classes, properties, individuals and their interrelationships relevant to a particular problem domain (e.g. networks). SNoMAC is built around the NetCore ontology shown in Figure 2. This formal OWL 2 ontology is designed to describe arbitrary networks at the highest level of abstraction, and as such consists of a relatively small number of primary classes: Network, Node, Link, State and Action.

The *Node* and *Link* classes make up the essential core of the ontology and are used to represent individuals that constitute network nodes and connections between them, respectively. Individual *Node* and *Link* instances are identified as members of one or more *Networks* through the *networkNode* and *networkLink* properties. While links can be represented as instances of the *Link* class, which allows them to be annotated with additional information (e.g., *bandwidth*, *isActive*), they can also (or alternatively) be represented using the symmetric property *linkedTo*, which allows them to be treated as OWL properties and thus can leverage optional property characteristics (e.g., symmetry, propertyChains). *linkedTo* is defined as a sub-property of *connectedTo*, which means that a node that is *linkedTo* another node is also *connectedTo* that node. *connectedTo*, unlike *linkedTo*, is a transitive property, meaning that if $n1$ is *connectedTo* $n2$ and $n2$ is *connectedTo* $n3$ then it can be inferred that $n1$ is *connectedTo* $n3$.

To go beyond basic network topology the ontology includes *State* and *Action* classes that represent state variables (for *Nodes* or *Links*) and executable actions (for *Nodes*), respectively. State variables may be read-only or read and writable (i.e., configuration variables). Actions represent parameterized functions that can be executed on a node, with the actionTemplate representing the functions call signature.

The NetCore ontology is intended to be extended and specialized to encompass various telecommunication networks. It is the inheritance afforded by the alignment of the upper-level NetCore ontology and the lower level telco ontologies that permits applications to deal with specific telecommunication devices in a generic fashion, ignoring low-level details until they are required to invoke some action. The details of of the low level models/protocols are handled by the specific network adapters that employ their own ontologies. As an example, the base classes and properties for the partial TR-069 ontology developed at Bell Labs for SNoMAC is shown in Figure 3. The UPnP ontology used in SNoMAC comes from [11], which has been align with the NetCore ontology. A similar ontology for SIXTH has been developed by UCD.

2.2   Semantic Benefits

Formal ontologies are used in SNoMAC to provide the following benefits:

- a formal API that can help ensure programers interpret the model consistently through automated consistency checking and a precise definition of what can be inferred from data;
- hierarchical abstractions that can more easily hide or reveal low level modeling and implementation details;
- automated inferencing of class and property membership that makes it trivial to inherit higher abstractions;
- ability to encode certain types of axioms that automatically detect specific conditions;

The following examples illustrate some of these benefits.

Abstraction and Specialization. Inheritance is the cornerstone of semantic reasoning and is heavily leveraged in SNoMAC. Simply put, it involves automatically inferring an individual's class or property membership based on an ontology's class and property hierarchies. For example, if an ontology states that *FemtoCell* is a subclass of *TR069:Device* and *TR069:Device* is a subclass of *NetCore:Node* and we state that *femtocell1* is a member of the class of *FemtoCell*

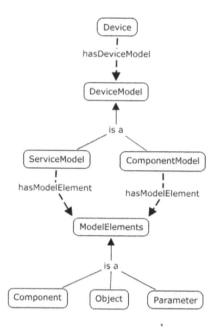

Fig. 3. Base Classes and Properties of the TR-069 Ontology

then it can automatically be inferred that $femtocell1$ is both a *TR069:Device* and a *NetCore:Node*, even if this information isn't explicitly stated in our data description of $femtocell1$. This kind of subsumption reasoning can also be used to infer property membership as in the example described above of how all *linkedTo* relationships imply a similar *connectedTo* relationship.

This form of inheritance reasoning provides the primary mechanism for supporting abstraction and data integration in SNoMAC. By aligning the NetCore class and property hierarchies with those of the network protocol specific ontologies, we can treat all devices as NetCore devices regardless of which network protocol they use. The fact that a low-level device is in fact an abstract NetCore device is automatically inferred by the system through the alignment of NetCore with the specified network ontologies.

Transitivity. As described above in the *connectedTo* example, a transitive property allows a reasoner to infer a property relationship between two entities (e.g., *connectedTo*) if they can be related through intermediary entities using the same property. The transitive characteristic of the *connectedTo* property allows for minimal specification of node connectivity in the data – specifically only the direct *linkedTo* connections need be provided – while permitting the automatic inferencing of total network connectivity if and when needed.

Property Chaining. Similar to transitive properties, property chains allow properties to be defined by defining a path of property connections chained from one entity to another. The classic example of this is the definition of the relationship *uncleOf* as the chaining of the *brotherOf* and *parentOf* properties: if *Bob* is the *brotherOf John* and *John* is the parent of *Sue* then *Bob* is the *uncleOf Sue*. In NetCore, a property chain is used to relate the property *linkedTo* to the chained properties *nodeToLink* and *linkToNode*[2]. Accordingly, two nodes *n1* and *n2* are automatically inferred to be *linkedTo* each other whenever there is a link instance *l* that connects the two by the relationships *n1 nodeToLink l* and *l linkToNode n2*. Because *linkedTo* is a sub-property of *conncetedTo*, this property chain also allows the automatic inferencing of full network connectivity (as defined by *connectedTo*) without even mentioning either *linkedTo* or *connectedTo* in the data representation of the network.

Complex Axioms. It is possible to define complex axioms in OWL from which we can automatically infer more interesting and useful relationships. In the following example a technique referred to as "man-man" [13] is used to automatically infer actual functional connectivity (as opposed to the logically stated connectivity of the network design); such a relationship could be instrumental in identifying and isolating network faults.

The axioms below specify that two nodes are *activelyConnected* (line 7) if all of the intervening nodes along the logical path that connects them (i.e., the *connectedTo* path) are actually active (line 3). This is accomplished by creating a property called *isActiveSelf* that takes on the value of an individual node in the event that the node is an *ActiveThing* (line2). An *ActiveThing* is any thing (such as a node) that has the property *isActive* equal to true (line 1). This *isActiveSelf* property is an *ObjectProperty* and as such can be used in a property chain with *connectedTo* to define what it means for two nodes to be *activelyConnected* (line 3). Then, whenever a node *n2* is know to be active (line 4) and the logical connections between *n1* & *n2* and *n2* & *n3* are declared (lines 5 and 6) it can be inferred that *n1* and *n3* are *activelyConnected*:

$$ActiveThing \equiv isActive\ some\ true \tag{1}$$

$$ActiveThing \subseteq isActiveSelf\ some\ self \tag{2}$$

$$connectedTo \circ isActiveSelf \circ connectedTo \subseteq activelyConnected \tag{3}$$

$$n2\ isActive\ true \tag{4}$$

$$n1\ connectedTo\ n2 \tag{5}$$

$$n2\ connectedTo\ n3 \tag{6}$$

$$\overline{n1\ activelyConnectedTo\ n3} \tag{7}$$

3   The SNoMAC Prototype

A SNoMAC prototype has been developed that targets the task of monitoring and controlling a Home Area Network that includes UPnP devices, TR-069 devices and and an array of sensors managed by SIXTH. Specific use cases that

---

[2] Although not graphically depicted in Figure 2 this property chain is defined in the NetCore OWL ontology.

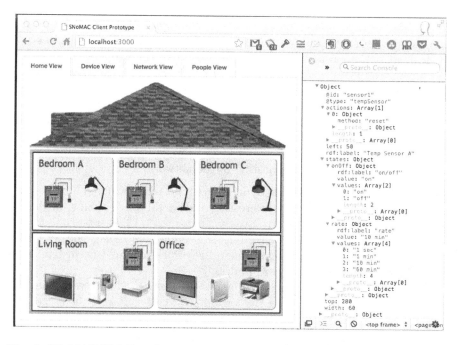

Fig. 4. SNoMAC Web Interface with JSON-LD RDF Objects shown in the Console

the system can handle include discovering when new devices are added or removed, remotely getting and setting device state variables, issuing commands to specific devices followed by processing their results and seamlessly handling the addition of new network types through the introduction of additional network adapters. The core of the prototype has been implemented as a node.js [3] application that connects to independent network adapters through a Web service API using JSON-LD[4] to exchange serialized RDF data about the networked devices. Clients connect to the server via websockets and pass event-driven messages using JSON-LD. An RDF datastore and SPARQL [6] engine (based on the node.js module rdfstore [5]) has been installed in the prototype and loaded with the NetCore ontology. OWL 2 inference rules have been implemented as SPARQL to enable the basic reasoning required to support semantics built into NetCore.

A desktop Web interface for the prototype has been implemented using HTML5/JavaScript. The interface has been designed to provide four views on the network (Home, Device, Network and People), although only the Home View is fully implemented. The Home View, which is shown in 4, depicts the devices distributed across the rooms of a house. The set of devices depicted by the interface currently includes temperature sensors, lamps, femtocells, WiFi router, a media server, computers and printers. Mousing over a device brings up a menu

---

[3] http://nodejs.org
[4] http://json-ld.org
[5] https://github.com/antoniogarrote/rdfstore-js

that permits the viewing and setting of state variables as well as the invocation of device specific actions. For example, when a user mouses over a Lamp and the pop up menu displays one of the state variables as "on/off: on", the user can select the submenu next to the variable and opt to turn the device off, resulting in a command being sent back to the server where it is forwarded on to the appropriate network adapter to turn the device off; when the state of the device changes the network adapter pushes the updated state back to the server and it is passed on to the client where the state of the lamp changes accordingly (i.e. it turns the state variable to "off" and the icon goes dark).

3.1 SIXTH Integration

SIXTH [10] is an extensible, scalable and intelligent Sensor Middleware based upon the Open Services Gateway initiative framework (OSGi) [6] which has been developed by the CLARITY team at UCD.

As illustrated in Figure 5, the primary SIXTH components include:

- a propertiesbased sensor model for control of, and access to, sensing devices;
- adaptors that provide sensor specific implementations to a standardised interface;
- higherlevel APIs for access to SIXTH's core functionality covering data access;
- retasking, notification, security and discovery;
- data processing service layers which build upon the APIs;
- integrated multiagent system to support insitu reasoning and WSN management;

SIXTH provisions extensibility by accommodating the dynamic addition of new functionality at runtime. Connection mechanisms to heterogeneous Sensor Networks are implemented as plug and play components termed *Adaptors*, and such functionality may be injected into an already running SIXTH instance in order to connect into the associated network. This scheme can also be used to incorporate dynamic service level behaviour such as a custom data retention policy or a sensor data aggregator function to supplant or complaint existing functionality.

Within SIXTH we encompass sensing of data from web resources (cyber-sensing) as a first-class citizen alongside direct sensing of physical world phenomena. SIXTH can dynamically configure data streams, referred to as CyberSensors, that can be both public and personalised (e.g., RSS, Twitter, Gowalla, Facebook and FourSquare). We thereby move toward a unified view of the Sensor Web. SIXTH is targeted at adaptive sensor networks and focused on runtime (re)tasking of sensor nodes. Cyber-sensing streams can be dynamically configured and the details and differences between these are hidden from the end user. SIXTH is primarily a gateway side middleware effort (software resides on the host machine) this decision was made so as to facilitate support for heterogeneous hardware, even in the case where we have no involvement in the programming

---

[6] http://http://www.osgi.org/

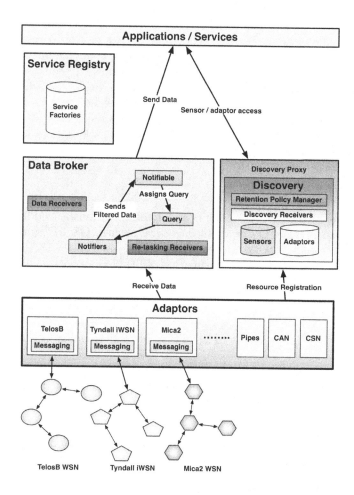

Fig. 5. The SIXTH Sensor Middleware System Architecture

of the network, within the SIXTH system an adaptor can be defined for any network. Despite this, efforts have been invested in developing an intelligent, agent driven, sensor network deployment running Agent Factory Micro Edition has been developed to run alongside the SIXTH gateway [14].

The SIXTH system architecture can be seen in Figure 5. Within this high level architecture several important concepts are evident. The Data Broker module is collectively responsible for distribution and management of sensor data. The discovery sub-system provides notification of resource status i.e. node timeout, adaptor creation etc) and management notice dissemination base upon credentialed security policies. The service registry in turn provides access to service modules e.g. sensor transformation services, aggregation services etc.

To create a communication bridge between SIXTH and SNoMAC, a software component was developed conforming to the SNoMAC Network Adapter API.

The Restlet[7] engine for Java 5 was used to enable RESTful communication. A SIXTH sensor device ontology was developed in alignment with NetCore. The initial development effort took less than three days and resulted in a complete working system capable of passing the entire suite of API tests except action commands. While this work represents just one case study, it does provide support for the claim that the addition of new network models/protocols to SNoMAC – and thereby enabling the semantic benefits that may accrue – is relatively simple and striaght forward.

4   Related Work

Bröring et al [3] has undertaken similar work from the perspective of the Sensor Web Enablement initiative, and present a framework that uses semantically enabled matchmaking to allow Plug & Play usage of Sensors within a network using SensorML standards. Similarly, Aloulou et al [1] present an OSGi based framework, also focussed on enabling plug & play sensor integration with a semantic reasoning engine, within an Ambient Assisted Living scenario.

Liu and Zhao [9] identify and describe how multiple sensors can be combined and used to fulfil various end user needs using semantic matchmaking, and further expanded this concept with Semantic Streams [15]. Similar work has been undertaken by Tran [12].

Although SNoMAC leverages a similar semantic approach, what sets it apart and marks it as novel is the aim to present the relevant semantic information to the end user through the interface as per Figure 4. This allows the end user to navigate the network at the appropriate level of abstraction, and allows them to make well informed changes at the best possible level.

5   Conclusion

The rapid evolution of communication networks presents challenges that require a new approach to network monitoring and control. Semantic technologies offer the means to encapsulate network details at various levels of abstraction making it easier to develop solutions that can adapt to changing data models and protocols. SNoMAC, and the NetCore ontology on which it is predicated, provide an example of how this can be effectively achieved today, at least in the domain of home area networks. We plan to continue developing the UPnP and TR-069 ontologies to increase the level of coverage over the functionality of the devices based on those models/protocols. We also intend to further explore the extent to which the SNoMAC approach can be seamlessly applied to other network models/protocols, e.g., ZigBe, Bluetooth, Bonjour and OMA-D. This future work will include experiments to help quantify the level of effort required to add new network adapters to SNoMAC. Network analysis using semantic techniques will also be explored more deeply.

---

[7] http://www.restlet.org/

References   ✎

1. Aloulou, H., Mokhtari, M., Tiberghien, T., Biswas, J., Kenneth, L.J.H.: A Semantic Plug&Play Based Framework for Ambient Assisted Living. In: Donnelly, M., Paggetti, C., Nugent, C., Mokhtari, M. (eds.) ICOST 2012. LNCS, vol. 7251, pp. 165–172. Springer, Heidelberg (2012)
2. Brickley, D., Guha, R.V.: Rdf vocabulary description language 1.0: Rdf schema (February 2004)
3. Bröring, A., Maué, P., Janowicz, K., Nüst, D., Malewski, C.: Semantically-enabled sensor plug and play for the sensor web. Sensors 11(8), 7568–7605 (2011)
4. Cruz, T., Simoes, P., Batista, P., Almeida, J., Monteiro, E., Bastos, F.: CWMP extensions for enhanced management of domestic network services. In: Proceedings of the 2010 IEEE 35th Conference on Local Computer Networks, LCN 2010, pp. 180–183. IEEE Computer Society, Washington, DC (2010)
5. UPnP Forum. UPnP Device Architecture 1.0. (April 2008), http://www.upnp.org/specs/arch/UPnP-arch-DeviceArchitecture-v1.1.pdf
6. Garlik, S.H., Seaborne, A., Prud'hommeaux, E.: SPARQL 1.1 Query Language, http://www.w3.org/TR/sparql11-query/
7. Gruber, T.R.: A translation approach to portable ontology specifications. Knowledge Acquisition 5, 199–220 (1993)
8. Hitzler, P., Krötzsch, M., Parsia, B., Patel-Schneider, P.F., Rudolph, S.: OWL 2 Web Ontology Language Primer. W3C Recommendation, World Wide Web Consortium (October 2009)
9. Liu, J., Zhao, F.: Towards semantic services for sensor-rich information systems. In: 2nd International Conference on Broadband Networks, BroadNets 2005, vol. 2, pp. 967–974 (October 2005)
10. O'Hare, G.M.P., Muldoon, C., O'Grady, M.J., Collier, R.E.M.W., Murdoch, O., Carr, D.: Sensor web interaction. International Journal on Artificial Intelligence Tools 21(2), 1240006 (2012)
11. Togias, K., Goumopoulos, C., Kameas, A.: Ontology-based representation of upnp devices and services for dynamic context-aware ubiquitous computing applications. In: International Conference on Communication Theory, Reliability, and Quality of Service, pp. 220–225 (2010)
12. Tran, K.-N., Compton, M., Wu, J., Gore, R.: Short paper: Semantic sensor composition. In: Proceedings of the 3rd International Workshop on Semantic Sensor Networks (2010)
13. Tsarkov, D., Sattler, U., Stevens, R.: A solution for the man-man problem in the family history knowledge base. In: Hoekstra, R., Patel-Schneider, P.F. (eds.) OWLED. CEUR Workshop Proceedings, vol. 529. CEUR-WS.org (2009)
14. Tynan, R., Muldoon, C., O'Hare, G.M.P., O'Grady, M.J.: Coordinated intelligent power management and the heterogeneous sensing coverage problem. Comput. J. 54(3), 490–502 (2011)
15. Whitehouse, K., Zhao, F., Liu, J.: Semantic streams: A framework for declarative queries and automatic data interpretation. Technical report (2005)

Opinion Dynamics on Triad Scale Free Network

Li Qianqian[1], Liu Yijun[1,*], Tian Ruya[1,2], and Ma Ning[1,2]

[1] Institute of Policy and Management, Chinese Academy of Sciences,
Beijing 100190, China
lqqcindy@gmail.com, yijunliu@casipm.ac.cn
[2] Graduate University of Chinese Academy of Sciences,
Beijing 100190, China
tianruya@126.com, maning2004437070@163.com

Abstract. In this paper, we investigate the opinion dynamics model of social impact theory on triad scale free network with power law degree distribution and tunable clustering coefficient. Based on this opinion dynamic model, we try to observe the clustering coefficient influence on opinion formation by adjusting the triad formation parameter. Simulation result shows that by adjusting triad scale free network parameters, a large clustering coefficient favors development of a consensus. In particular, when in the system with initial opinion proportion of +1, p_+=0.5, a consensus seems to be never reached for triad scale free network with any clustering coefficient.

Keywords: opinion dynamics, social impact theory, triad scale free network.

1 Introduction

Recently, there has been a growing interest in study of complex phenomena in social field, whereas statistical physics, mathematics, computer science are very popular research tools. In particular, one of very significant research area is opinion dynamics, which explain how the society reach consensus.

The dynamics of opinion formation is a non-linear phenomenon, both personal view interaction and collective behavior emergence play important role in the underlying mechanism. The individual's opinion may likely be affected by its nearest neighbors. In fact, some classic opinion dynamical models were based on personal interaction between their neighbors. For example, Sznajd model [1, 2], Deffuantmodel [3], KHmodel [4] and Galam's Majority Rule model [5].

Most of the time, the topological properties of network govern the dynamical behavior of complex system. Studies on topological structure have been an intriguing issue. A few example include food web [6], actor collaborating network [7], paper citation network [8], stock market network [9].

At the end of 20th century, there have been two milestone progress in network science: 1) Watts and Strogatzs proposed WS small-world network model [10], which

---

* Corresponding author.

R. Huang et al. (Eds.): AMT 2012, LNCS 7669, pp. 445–450, 2012.

explains those systems having highly clustered and small characteristic path length; 2) Barabási and Albert proposed BA scale free network model [8], which described the networks with power-law degree distribution.

Since then, some of opinion dynamical models, such as Sznajd model, Deffuant model, KH model et al. have been studied in the context of complex network.

Social network often have highly clustering coefficient. If person A knows B and C, then person B and C are more likely to know each other. Moreover, many empirical results discovered the fat-tail property in human behavior[11] The WS model shows a high clustering but without the power-law degree distribution, while the BA model with the scale free nature does not possess the high clustering. Therefore, Holme and Kim proposed a triad scale free network model [12], which has both the perfect power-law degree distribution and highly clustering. We regard this triad scale free network is more suitable for modeling the internet opinion formation net: 1) when there is a hot spot in public opinion, many netizens will post their view. The participating netizens will enlarge the network, which shows the growing property of the network; 2) views of opinion leaders will attract the public attention and influence the attitudes and behavior change of their followers, which can be modeled as the preferential attachment mechanism of BA scale free network, that is, a "rich-get-richer" phenomenon; 3) the communication between the followers of opinion leaders will increase the highly clustering of network. For example, you can always see the "reply 18 floor", etc..

In this paper, we investigate the opinion formation model of social impact theory based on triad scale free network. In this model, we consider both the neighborhood opinion and their influence strength.

2   Triad Scale Free Network

Holme and Kim introduced triad formation mechanism to increase clustering coefficient of network. Combining triad formation and BA scale free network could generate both highly clustering and power-law degree distribution.

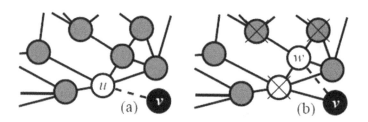

Fig. 1. preferential attachment and triad formation. (a) preferential attachment step: the new added vertex v attaches to vertex u with a probability to its degree; (b) triad formation step: the new added vertex v attaches to w in the neighborhood of one linked to in the previous preferential attachment step.

We below describe the model of triad scale free network [12] which established based on BA model [8] but has high clustering.

— Initial condition: starting with m_0 of vertices;
— Growth: we add a new vertex v with $m(\leq m_0)$ edges that link the new vertex to m different vertices already present in system;
— Preferential Attachment(PA): we assume that the probability Π that a new vertex v will be connected to vertex u depends on the connectivity k_u of that vertex, so that $\Pi_v = \frac{k_v}{\Sigma_j k_j}$;
— Triad Formation(TF): if we already link v to u in the previous PA step, then add one more edge from v to a randomly chosen neighbor of u. If all neighbors of u were already connected to v, do a PA step instead.

Briefly speaking, when add a new vertex to the existing network, we first perform one PA step, and then perform a TF step with probability p_t or a PA step with probability $(1-p_t)$. When $p_t = 0$, this model reduces to the original BA model.

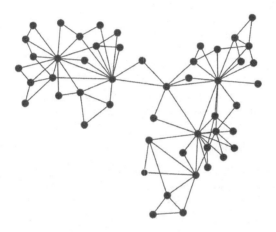

Fig. 2. Example of a scale free network. The number of Nodes is 50 with triad formation probability $p_t = 0.6$ and $m_0 = 5$, $m = 2$. So that the new added vertex is linked twice. In order to preserve the clarity of the network, the size of the network has been to kept small. This plot has been realized with the ORA software [13].

3   Opinion Dynamical Model

In complex network, we use node to represent individual(agent), edge linked one node to another represent information propagation relationship(we consider undirected network only). We set up a triad scale free network according given parameters. Once the network has been completely constructed. We establish social impact opinion formation model the network. This opinion formation model was based on psychology theory of social impact[14] which describes how individuals feel the presence of their peers and how they in turn influence other individuals. Hølyst, Kacperski and Schweitzer developed a opinion formation model based on social impact theory[15]. We modify the impact function I_i of the opinion dynamical model.

The initial condition is a population of N agents. Agent i is characterized by an opinion $\sigma_i = \pm 1$ and by clustering coefficient $c_i \in [0,1]$. $\sigma_i = 1$ represents individual i supports a viewpoint, whereas $\sigma_i = -1$ represents individual i opposes a viewpoint.

The total impact I_i that an individual experiences from its neighbors(social environment also) is

$$I_i = \sigma_i c_i \sum_{j=1}^{N_i} c_j \sigma_j \tag{1}$$

Where N_i :the number of i-th agent, c_i: the clustering coefficient of i-th agent.

$I_i > 0$ represents individual i gets supportiveness from its neighbors, whereas $I_i < 0$ represents the neighbors of individual i have opposite opinion with i.

Opinions of individuals may change asynchronously(asynchronous dynamics) in discrete time steps according to the rule

$$\sigma_i(t+1) = \begin{cases} \sigma_i(t), \ I_i \geq 0, & \text{with probability } \frac{e^{-I_i/T}}{e^{-I_i/T}+e^{I_i/T}} \\ -\sigma_i(t), \ I_i < 0, & \text{with probability } \frac{e^{I_i/T}}{e^{-I_i/T}+e^{I_i/T}} \end{cases} \tag{2}$$

Where the parameter T is interpreted as a "social temperature" describing a degree of randomness (disturbing factors) in the behavior of individuals.

4   Numerical Simulations

At first, we establish a triad scale free network and update opinion of individuals asynchronously, i.e., we choose an agent randomly, calculate its social environment I_i and update its opinion according to the above rule(see Equation(2)).

In order to describe the evolution process of the model, we employ a parameter, average opinion:

$$\bar{\sigma} = \frac{\sum_{i=1}^{N} \sigma_i}{N} \tag{3}$$

When $N_+ = N_- = N/2$, $\bar{\sigma}=0$. In Fig.3, we present the average opinion dynamic process for different triad formation probability. From the Fig.(3)(a) we can find when the initial opinion proportion of +1, $p_+ > 0.5$, the average opinion dynamic result on a high clustering coefficient scale free network is larger than the result on a low clustering coefficient scale free network. From the Fig.(3)(b) we can find when the initial opinion proportion of +1, $p_+ < 0.5$, the average opinion dynamic result on a high clustering coefficient scale free network is smaller than the result on a low clustering coefficient scale free network. Especially, from the Fig.(3)(c), we find that when the initial opinion proportion of +1, $p_+ = 0.5$, the opinion dynamic result is approximately to 0 on a triad scale free network with any clustering coefficient.

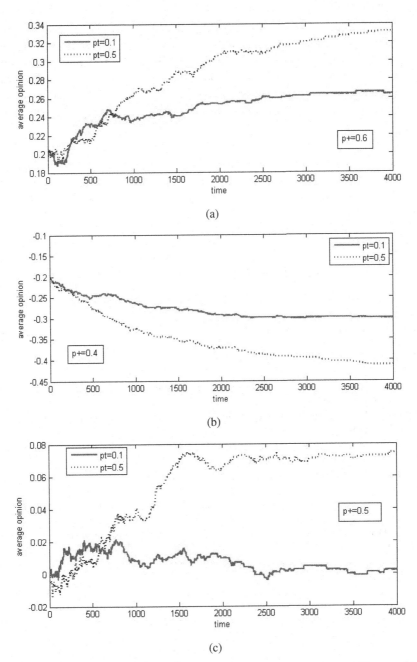

(a)

(b)

(c)

Fig. 3. relationship between opinion dynamics and triad formation probability p_t. (a) $T=2$, $N=500$, $m=3$, $p_+=0.6$ of opinion +1; (b) $T=2$, $N=500$, $m=3$, $p_+=0.4$ of opinion +1; (c) $T=2$, $N=500$, $m=3$, $p_+=0.5$, of opinion +1. The results are obtained by averaging over 100 independent realizations.

5   Conclusion

In the present work we have constructed a triad scale free network in order to approach the real network topology of online information propagation. The decision updating is governed by a social environment. By simulating this opinion formation model on a triad scale free network, we observe that in scale free network, a large clustering coefficient scale free network favors the development of dominant opinion when the initial opinion proportion +1, $p_+ \neq 0.5$. However, when $p_+ = 0.5$, non-consensus can be observed.

Acknowledgments. The authors gratefully acknowledge the support of National Natural Science Foundation of China (Grant No. 91024010) and Projects of Young Scientist Funds of Institute of Policy and Management, Chinese Academy of Sciences (Y200571Q01).

References

1. Sznajd-Weron, K., Sznajd, J.: Opinion evolution in closed community. Int. J. Mod. Phys. C 11, 1157–1165 (2000)
2. Sznajd-Weron, K.: Sznajd model and its applications. Arxiv preprint physics/0503239 (2005)
3. Deffuant, G., Neau, D., Amblard, F., Weisbuch, G.: Mixing beliefs among interacting agents. Adv. Complex Syst. 3, 87–98 (2000)
4. Hegselmann, R., Krause, U.: Opinion dynamics and bounded confidence: models, analysis and simulation. Journal of Artificial Societies and Social Simulation 5 (2002)
5. Galam, S., Zucker, J.D.: From individual choice to group decision-making. Physica A 287, 644–659 (2000)
6. Camacho, J., Guimerà, R., Nunes Amaral, L.A.: Robust patterns in food web structure. Phys. Rev. Lett. 88, 228102 (2002)
7. Newman, M.E.J., Strogatz, S.H., Watts, D.J.: Random graphs with arbitrary degree distributions and their applications. Phys. Rev. E 64, 026118 (2001)
8. Barabási, A.L., Albert, R.: Emergence of scaling in random networks. Science 286, 509–512 (1999)
9. Bonanno, G., Caldarelli, G., Lillo, F., Mantegna, R.N.: Topology of correlation-based minimal spanning trees in real and model markets. Phys. Rev. E 68 (2003)
10. Watts, D.J., Strogatz, S.H.: Collective dynamics of 'small-world' networks. Nature 393, 440–442 (1998)
11. Han, X.P., Wang, B.H., Zhou, T.: Researches of human dynamics. Complex Systems and Complexity Science 7 (2010)
12. Holme, P., Kim, B.J.: Growing scale-free networks with tunable clustering. Phys. Rev. E 65 (2002)
13. Carley, K.M., Reminga, J.: Ora: Organization risk analyzer. DTIC Document (2004)
14. Latane, B.: The psychology of social impact. American Psychologist 36, 343 (1981)
15. Lyst, J.A.H.O., Kacperski, K., Schweitzer, F.: Social impact models of opinion dynamics. Annual Reviews of Computational Physics 9, 253–273 (2002)

Distribution of Node Characteristics in Complex Networks of Tree Class*

Ying Tan[1,2], Hong Luo[3], and Shou-Li Peng[1]

[1] Center for Nonlinear Complex Systems, Department of Physics, Yunnan University,
Kunming, 650091, China
sl_peng@126.com
[2] Statistics and Mathematics College, Yunnan University of Finance and Economics,
Kunming, 650228, China
[3] School of Adult Education, Yunnan University, Kunming, 650091, China

Abstract. Based on the work of Park-Barabási (PB) we research in detail the (D,H)-phase diagram which describes the correlation and interplay among nodes of complex systems. To do this, we provide a frame of mathematical description, it includes: carrying out symbolization to the assortment of nodes, obtaining symbolic assertive matrix and enumeration formula. Applying the frame to two kinds of tree graphs we find that there exists vivid self-similar motif in the core domain of (D,H)-phase diagram. In order to draw the phase boundary we use a mixed curve of both Cassini oval and ellipse. The stationary of (D,H)-phase diagram is confirmed, but we also have seen a trend that the phase boundary has a phenomenon of little compression when the size of system increases. Finally, we suggest a new classification method to decide dyadic configuration of (D,H)-phase diagram and put it to use in the tree systems.

Keywords: (D,H)-phase diagram, dyadic effect, assortative matrix, Binary-tree, Dur-tree.

1 Introduction

Recent advances in the complex system (network) have developed and transferred from concerning the topological maps of systems to a description of characteristics of the topological maps, because we need answer an important question: how the system's components connect to each other [1]. In real complex systems the node play a twofold role. On one hand, it constructs topology, and simultaneously it also carries fundamental information about itself in the system. The case are quotidian such as in a RNA–DNA interaction network, each node gene has its biological functions [2,3]; collaborators in the science network can be classified based on their interesting subjects [4–6]. In a trade network, each company can be assigned goods parameters that represent its products or services. The various node properties would have some match, but not to be distributed at random in the network, they would possess certain correlation with

---

* Supported in part by NSFC(60962009).

R. Huang et al. (Eds.): AMT 2012, LNCS 7669, pp. 451–462, 2012.

the underlying structure. Two interactions may be driven such correlations: the new links prefer to choose the node characteristics, or the node characteristics could attract other nodes to be linked . We call the phenomenon as "assortative mixing" [7]. For a fixed system, the number of links between nodes sharing a common property is larger than the expectation of random distribution of the node characteristics on the network [8–12], a phenomenon called the dyadic effect [1]. Thus the original work of Park-Barabási (PB) open a new way to approach the interplay between node characteristics and network structure [5–7,13–17] in the network system. PB have answered several fundamental questions, for example, $\frac{x(x+1)}{2}$ parameters are necessary to mathematically describe the statistical distribution of node characteristics in a network where x is the number of the properties the nodes possess. They confirm the existence of dyadic effect by using the (D, H)-phase diagram which indicates the existent domain of two components: homogeneous and heterogeneous dyads. This effect is beyond dyadic; i.e., two different configurations of node characteristics are equivalent because there indeed exists many distinct configurations at the same (D, H)-phase parameter a. In this paper we attempt to deepen the recognition of (D, H)-phase diagram which is a basic tool to explore node characteristics distribution.

PB have concisely and perfectly described the idea of (D, H)-phase diagram. We present a mathematical frame i.e. symbolic dyad matrix of a graph system (nevertheless deterministic and/or nondeterministic), whose function is to formally describe dyads and count them simply (see section 2).

Conversing to that PB focus particularly on the aspect of practical significance of these questions, we would rather bias the non-practical aspect so that we choose a simplest graph system i.e. tree as our learn object.

The models of real complex systems (or network) has experienced a development from the stochastic model (ER) to deterministic model (the small world (WS), scale-free (BA) and their various combinations and variants) in the recent decade [18–21]. Correspondingly, the degree distribution for characterizing nodes has also experienced a great change from probabilistic Poisson (ER) to the various power-law (BA, pseudo-fractal, transfractal, recursive fractal), even simplest trivial constant (perfect fractal). Here a rational development of system model implies not only the transition from non-deterministic to deterministic but also displays a trend for simplifying the degree distribution. Following the trend we choose the tree as our work object. The tree has diversity in self-similarity combinations, for instance, the binary tree (Hata, or T-tree) which can be generated by an automatic sequence; Cayley tree, the symmetric tree which can simulate plants by the grammar of Lindenmayer system. However, these simple fractals with self-similarity can not be a good model of real complex system, because they have constantly degree distribution and a constant Hausdorff dimension. In order to improve this poor situation researchers explore a new pseudo-fractal field [22], transfractal [23]. Our researchers contribute in this aspect by many important works (see Rev. [24,25]). The models in the pseudo-fractals field can approach to real system by their power-law and infinite Hausdorff dimension. So we choose two representatives: binary tree and DUR-tree [26,27] with multiple branch from fractal and pseudo-fractal models. Although the deterministic self-similar models may only have theoretical meaning, it can approach to real system when its branches (edges) are deleted in the random or non-random way, for example, search tree, decision tree, recognition tree in the social or management systems. Applying the idea we will assort the node characteristics to the fractal binary tree and pseudo-fractal DUR-tree in the non-random and random way . In the following we start our work.

2   Foundamental Mathematical Description

2.1   Assortive Matrix of Graph

In order to determine to what extent the node characteristics correlate with the network structure, we give each node the property characterized by two values, 1 or 0 for simplicity. Let us call $n_1(n_0)$ the number of nodes with property 1 (0) so that the total number of nodes $Ng = n_1 + n_0$. There are three kinds of dyads defined as a link with its two end nodes in the network: (1-1),(1-0) and (0-0) if we regard (0-1) to be the same as (1-0). We need count the number m_{11}, m_{10} and m_{00} of three types of dyad. This is our goal. By employing the correlative conceptions of graph theory, the adjacency matrix is a crucial tool. We further suggest a symbolic adjacency matrix which is a mathematical expression of these dyads. The symbolization of properties $0, 1$ is defined as $0 \rightarrow w$ (white), $1 \rightarrow b$ (black), then all the types of dyad become ww, wb, bw, bb, see Fig.1 (noting that black or white node is represented by red or green circle respectively in the figure, and for undirected link, wb is the same as bw). Thus for each node we assort a

Fig. 1. 4 types of dyad

symbol to it. We have two kinds of way to arrange this match for the network with fixed nodes number Ng . Firstly we assort a b, w-symbolic sequence with the length Ng to the nodes according to a non-random way where the attractable or repulsive property among nodes can be considered. Secondly we assign a set of $\{b, w\}$-symbolic sequence with the same length to the nodes in a random way where their statistic properties of the networks can be considered. For the two ways we denote the symbolic sequence as rs and make its Descartes direct product as

$$Rs = rs \otimes rs$$

where the symbolic assortive matrix Rs with size $Ng \times Ng$ contains all information of dyads.

2.2   Symbolic Adjacent Matrix

The symbolic assortive matrix Rs tells the whole match of dyads. Furthermore we make Hadamard product "$\circ$" for Rs and the adjacency matrix A of our graph system as

$$D = A \circ Rs \tag{1}$$

It is easily confirmed that the enumeration formula of dyads reads as

$$\frac{1}{2} \sum_{1 \leq i,j \leq Ng} d_{ij} = m_{11}bb + m_{10}bw + m_{00}ww \tag{2}$$

where $D = [d_{ij}]$ is a symbolic adjacent matrix with dyad entry. For second assortment way we have a set of symbolic sequences, it may be realized by a set of $0, 1$ random integer sequences whose cardinal is T, thus there is a set of matrix $D(T)$. If T is larger enough to approximately ergodic, one can obtain the expectation of dyads with probability theory [8, 28]

$$\begin{cases} <m_{11}> = \dfrac{n_1(n_1 - 1)p}{2} \\ <m_{10}> = n_1(Ng - n_1)p \\ p = \dfrac{2M}{Ng(Ng - 1)} \\ M = m_{11} + m_{10} + m_{00} \end{cases} \tag{3}$$

where M is the edge number of the graph of the system. When T is larger enough we have a great deal of symbolic matrixes which count dyads, it will lead to a statistic ensemble of symbolic Laplacian matrix. This is a useful tool to study the random behavior of complex system. So far the content of our results is universal to any graph system.

2.3   Classification of Dyad Effect

Classification in Park-Barabási's Paper. In [1], Park-Barabási introduce two quantities (D, H), they will divide the dyads effect into four classes:

(a) Dyadic: this class expresses dyadic effect and reflects the clusterity with the same property;

(b) Anti-dyadic: this anti-dyadic effect reflects the clusterity with the another property;

(c) Heterophobic: this class indicates that the clusters of two class (a), (b) have all large quantity, the cluster with same property is closed;

(d) Heterophilic: this class indicates that the clusters of two class (a), (b) are sparse, there is many alone dyads which can not form dyadic effect. All these configurations will be reflected in the (D, H)-phase graph.

The Effect of Parameter a. Although the (D, H)-phase diagram can reflect four configurations, when the number of nodes rapidly increase with the generation of fractal (or pseudo-fractal) in the big network, the parameter $a = n_1/Ng$

can not overcome the difficulty for exhausting the details of the node characteristic distribution of webs. So the Park-Barabási's method which extract the boundary of phase graph by using the Mattheyses and Fiduccia (MF) program is successful [1]. They find that the phase boundary is stationary, or system-size independent, so the study of networks can work on small size. Additionally, the algorithm for extracting the phase boundary is NP hard. We want to search a simple method to obtain the phase boundary for our simple fractal models. We find also that the phase boundary of all binary (m)tree fractals and DUR-(m)Tree pseudo-fractals are stationary and can be depicted by a Cassini ellipse-oval curve (CEO curve, for abbreviation) with one parameter a we made. For the models with small size we analysis the fine construction of the inner domain of the phase diagram in detail and uncover some information of regularity.

3 Distribution of Node Characteristics

3.1 Binary Tree and Dur-Tree

The binary tree with m-branches can easily be generated by iteration of automatic sequence, see Fig.2,3. The initiator $G(0)$ is one link with two nodes, when obtained the $(n-1)$-th generation graph $G(n-1)$, then $G(n)$ is made of m copies of $G(n-1)$ on $G(1)$. Fig.2 and Fig.3 display the first three generation of Binary-tree with $m=1$ and the first two generation of Binary-tree with $m=6$ respectively.

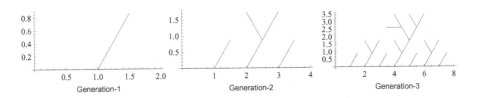

Fig. 2. Binary-tree with $m=1$(for $n=1,2,3$)

The nodes number of Binary-tree with m is $Ng=(2m+1)^n+1$, where m is the number of "branches" arising from the base of a single "stem", and the generation n is a control factor.

Another type of tree we discuss is Deterministic Uniform Recursive Tree (Dur-tree), Fig.4 shows the generating process: the initiator (denotes by $G(0)$) is also one link with two nodes, when obtained the $(n-1)$-th generation graph $G(n-1)$, for each node of $G(n-1)$, introduce m new nodes which link with each old node respectively, then $G(n)$ is generated. Their degree distribution is $p_k=(m+1)^{-k}$ with power-law. Fig.4 displays the Dur-tree with $m=6$ from Generation-1 to Generation-3:

For different m and generation n, we will denote the responding tree graph as $G(n)$-mB-Binary-tree (Dur-tree) in the flowing ("B" means branch). We first discuss (D,H)-phase diagram of 1B-and 6B-Binary-tree.

Fig. 3. Binary-tree with $m=6$ (for $n = 1, 2$)

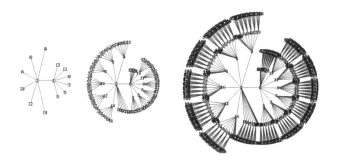

Fig. 4. Binary-tree with $m=6$ (for $n = 1, 2, 3$)

3.2   Region of the Whole (D,H)-Phase Diagram

Considering a fixed system with Ng nodes. Applying the enumeration formula of dyads in Section2, we compute the expectation of bb and bw dyads ($< m_{11} >$ and $< m_{10} >$) then the value of (D, H) for one time experiment about the assortment of b, w can be obtained. By repeating assortments for T times, we can construct whole (D, H)-phase diagram by the information of $(D(t), H(t))$ for $0 \leq t \leq T$. Fig.5(1-3) display the (D, H)-phase diagram of Binary-tree for: (1) 6-th Generation of 1B; (2) 4-th Generation of 2B and (3) 2-nd Generation

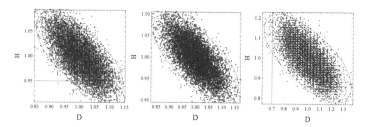

Fig. 5. (D, H)-Phase Diagram of Binary-tree. All stochastic of assortment is $T = 10000$ times. (1) $G(6)$-1B-Binary-tree, $Ng = 730$, CEO parameter: $C = 0.08$, $A = 0.138564065$, $\tan\theta = -0.459000676$; (2) $G(4)$-2B-Binary-tree, $Ng = 626$, CEO parameter: $C = 0.09$, $A = 0.155884573$, $\tan\theta = -0.466678941$; (3) $G(2)$-6B-Binary-tree, $Ng = 170$, CEO parameter: $C = 0.2$, $A = 0.346410162$, $\tan\theta = -0.477215110$.

of 6B, and Fig.6(1-3) shows the (D, H)-phase diagram of Dur-tree for: (1) 6-th Generation of 1B; (2) 4-th Generation of 2B and (3) 3-rd Generation of 6B, all stochastic of assortment is $T = 10000$ times.

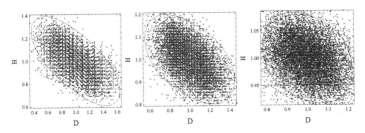

Fig. 6. (D, H)-Phase Diagram of Dur-tree under 10,000 times assortments. (1) $G(6)$-1B-Binary-tree, $Ng = 128$, CEO parameter: $C = 0.25$, $A = 0.433012702$, $\tan\theta = -0.345045039$; (2)$G(4)$-2B-Dur-tree, $Ng = 162$, CEO parameter: $C = 0.23$, $A = 0.415692194$, $\tan\theta = -0.2506613954$; (3)$G(3)$-6B-Dur-tree, $Ng = 686$, CEO parameter: $C = 0.11$, $A = 0.190525589$, $\tan\theta = -0.250661394$

It can be seen from above figures that for the two kind fractal-tree and pseudo-fractal Dur-tree, whatever the generation n or the branch with m copies, their core part of(D, H)-phase diagram have the almost same construction. In order to draw a smooth phase boundary, We make up a parameter equation as the mixed curve of both Cassini oval and ellipse (CEO). The expression of ellipse is

$$\begin{cases} x = r\cos t \\ y = \frac{r}{2}\sin t \end{cases}, \quad t \in [0, 2\pi] \tag{4}$$

and the expression of Cassini-oval is

$$r = \sqrt{C^2 \cos(2t) + \sqrt{A^4 - C^4 \sin^2(2t)}}, \quad t \in [0, 2\pi]$$

For simplicity, let $A = \sqrt{3}C$, then one parameter determine the boundary.

When draw the CEO curve, we follow a standard criterion: the points outside the phase boundary must be less than 1/1000 to the number of the inner points. The perihelion axis of every CEO curve have a inclination with respect to the horizontal axis, obviously the central point located at the point (1,1) of the (D, H)-phase diagram. The tangents of inclination of various (D, H) phase diagrams have only small variation. The investigation of the variation and its rule is a future task.

3.3   The Influence of Parameter a

An important observation is that many similar core motifs exist in the inner domain of various (D, H)-phase diagrams. This is the result of influence of parameter a. In order to make clear the influence of values of parameter a to

(D, H)-phase diagram, we take $G2$-6B-Binary-tree and $G6$-1B-Dur-tree for example. For $T10000$, we give all the different values of $< m_{11}(t) >, < m_{10}(t) >$ for all $1 \leq t \leq T$, which have several values corresponding to the different colored points in Fig.7:

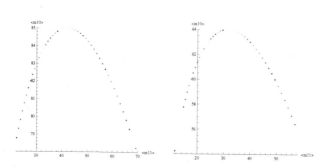

Fig. 7. Colored phase Diagram to $(< m_{11} >, < m_{10} >)$: (1) $G2$-6B-Binary-tree and (2) $G6$-1B-Dur-tree, $T = 10000$

For these points $(< m_{11} >, < m_{10} >)$ of expectation of dyads we fit these data as parabola equations which are

$$Y = 62.44688685972182 + 1.0250837053692212X - 0.011658847869553839X^2$$

for $(< m_{11} >, < m_{10} >)$ of $G2$-6B-Binary-tree and

$$Y = 45.90739717769711 + 1.0686151855788824X - 0.015763678476019977X^2$$

for $G2$-1B-Dur-tree. These equations would help to obtain further information of the structure of (D, H)-phase diagram such as graphic compression, variation of inclination.

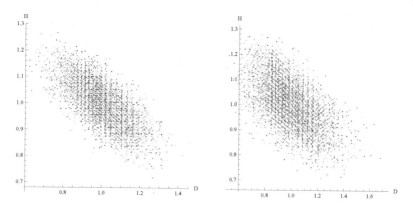

Fig. 8. Colored (D, H)-Phase Diagram: (1) $G2$-6B-Binary-tree and (2) $G2$-1B-Dur-tree, $T = 10000$

Now we analyze the whole (D, H)-phase diagram. If we endow the same color to all phase points with the same a parameter value, then Fig.8 lay out a colorful phase diagram with vivid layer construction i.e. motif.

These phase diagrams all have self-similarity, which obviously come from self-similarity of fractal themselves. We find by many experiments that the phase points and the expectation of dyads will rapidly increase as n and T increase. At this time CEO curve is compressed in the direction of short axis and the motif with self-similarity also encounter the same compression so that the CEO curve change as lathy shape.

4   A New Classification

In this section we give a new method of classification with the mean star coefficient of graph. For the tree the star is a representation of node correlation. Fig.9 illustrates various homogenous and heterogenous dyads. We come to dis-

Fig. 9. For the tree graph the star have two cases: 1 homogenous dyad; 3 heterogenous dyads and conjugacy class

cuss an assertive graph with black and white colors. After assortment of t times, the black (white) subgraph has $n_1(t)$ $(n_0(t))$ nodes and $n_1(t) + n_0(t) = Ng$. Calculating their mean star coefficient

$$cSubG(b) = \frac{1}{n_1(t)} \sum_{i=1}^{n_1(t)} d_i(b), \quad cSubG(w) = \frac{1}{n_0(t)} \sum_{i=1}^{n_0(t)} d_i(w) \tag{5}$$

where $d_i(b)$ and $d_i(w)$ are degrees of each nodes in subgraph $G(b)$ (black) and subgraph $G(w)$ (white), they reflect the mean of magnitude of homogenous points after one time assortment. We illustrate this by an example in case of G4-1B-Binary-tree (Fig.10(1-3)). When asserting color T times, Fig.11 displays the symmetric distribution of $SubG_4(b)$ and $SubG_4(w)$ of $G(4)$ -1B-Binary-tree at $T = 10000$. By deleting T, we get the relationship graph of $cSubG(b)$-$cSubG(w)$ with clear motif. According to Fig.12, the area with high density distribution: ρ_2, ρ_4 indicate the dyadic and anti-dyadic effect exist substantively; and the area with lowest density: ρ_3 shows the probability of common existence of both high dyadic and high anti-dyadic effect is sparse. ρ_1 indicates the quadrant where the dyadic effect is unsuccessful. The density distribution approximates to Gaussian shape because assertive operation is Bernoulli. It is shown from mentioned above that the new classification method can be able to work well.

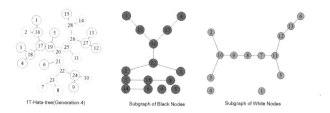

Fig. 10. (red denotes black point, green denotes white point) The subgraph $SubG_4(b)$ and $SubG_4(w)$ have assertive color respectively. The mean star coefficients of black and white subgraph are $\left(\frac{4}{3}, \frac{20}{13}\right)$ respectively.

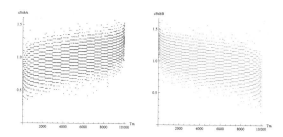

Fig. 11. Star coefficient of $SubG_4(b)$ and $SubG_4(w)$ at $T = 10,000$

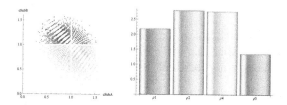

Fig. 12. (1) $cSubG(b)$-$cSubG(w)$ relationship; (2) the density distribution, where ρ_1, ρ_2, ρ_3 and ρ_4 locate on 3rd, 4th, 1st and 2nd quadrant respectively

5   Conclusion and Discussion

We have researched in detail the (D, H)-phase diagram. The frame of mathematical description works well, it provides symbolic assertive matrixes and enumeration formula. Applying the frame to two kinds of tree graphs we find that there exists vivid self-similar motif in their core domain of (D, H)-phase diagram. In order to draw the phase boundary we use a mixed curve of both Cassini oval and ellipse. The stationary of (D, H)-phase diagram is confirmed, but we have also seen a trend that the phase boundary has phenomenon of little compression when the size of system increases. Finally, we suggest a new classification method to decide dyadic configuration of (D, H)-phase diagram. It displays that the mean density distribution of domain of the dyadic effect may approximate the shape of Gauss. It is worthy to note that our mathematic frame is tree-independent, it can apply to all the complex systems whose model is graph. If T approaches thermodynamic limit or ergodic, the statistic ensemble of symbolic assertive matrix may shed a little light-spot about stochastic or deterministic symbolic dynamics [24, 29, 30] of complex systems.

References

1. Park, J., Barabási, A.-L.: Distribution of node characteristics in complex networks. PNAS 104(46), 17916–17920 (2007)
2. Barabási, A.-L., Oltvai, Z.N.: Understanding the cell's functional organization. Nat. Rev. Genet. 5, 101–113 (2004)
3. Albert, R.: Scale-free networks in cell biology. J. Cell Sci. 118, 4947–4957 (2005)
4. Adamic, L.A.: The Small World Web. In: Abiteboul, S., Vercoustre, A.-M. (eds.) ECDL 1999. LNCS, vol. 1696, pp. 443–452. Springer, Heidelberg (1999)
5. Dumais, S., Chen, H.: Hierarchical classification of Web content. In: Proc. of the 23rd Annual Int. ACM SIGIR Conf. Res. Dev. in Information Retr., pp. 256–263. ACM Press, New York (2000)
6. Menczer, F.: Growing and Navigating the Small World Web by Local Content. Proc. Natl. Acad. Sci. USA 99(22), 14014–14019 (2002)
7. Newman, M.E.J.: Assortative Mixing in Networks. Phys. Rev. Lett. 89, 208701 (2002)
8. Alba, R.D.: A graph-theoretic definition of a sociometric clique. J. Math. Soc. 3, 113–126 (1973)
9. Seidman, S.B.: Network structure and minimum degree. Social Networks 5(3), 269–287 (1983)
10. Sailer, L.D., Gaulin, S.J.C.: Proximity, sociality, and observation: the definition of social groups. Amer. Anthropologist 86, 91–98 (1984)
11. Borgatti, S.P., Everett, M.G., Shirey, P.R.: LS sets, lambda sets and other cohesive subsets. Social Networks 12(4), 337–357 (1990)
12. White, D.R., Harary, F.: The Cohesiveness of Blocks In Social Networks: Node Connectivity and Conditional Density. Sociological Methodology 31(1), 305–359 (2001)
13. Shrum, W., Cheek Jr., N.H., Hunter, S.M.: Friendship in school: Gender and racial homophily. Sociology of Education 61(4), 227–239 (1988)
14. Moody, J.: Race, School Integration, and Friendship Segregation in America. Amer. J. Sociology 107(3), 679–716 (2001)
15. Adamic, L.A., Glance, N.: The political blogosphere and the 2004 U.S. election: divided they blog. In: Proc. 3rd Int. Workshop on Link Discovery, pp. 36–43 (2005)
16. Yook, S.-H., Oltvai, Z.N., Barabási, A.-L.: Functional and topological characterization of protein interaction networks. Proteomics 4(4), 928–942 (2004)
17. Christakis, N.A., Fowler, J.H.: The Spread of Obesity in a Large Social Network over 32 Years. New England J. Medicine 357, 370–379 (2007)
18. Albert, R., Barabási, A.-L.: Rev. Mod. Phys. 74, 47–97 (2002)
19. Goltsev, A.V., Dorogovtsev, S.N., Mendes, J.F.F.: Critical phenomena in networks. Phys. Rev. E 67, 026123 (2003)
20. Barabási, A.L., Albert, R.: Emergence of Scaling in Random Networks. Science 15(286), 509–512 (1999)

21. Watts, D.J., Strogatz, S.H.: Collective dynamics of 'small-world' networks. Nature 393, 440–442 (1998)
22. Dorogovtsev, S.N., Goltsev, A.V., Mendes, J.F.F.: Pseudofractal scale-free web. Phys. Rev. E 65, 066122 (2002)
23. Rozenfeld, H.D., Havliv, S., Ben-Avraham, D.: Fractal and transfractal recursive scale-free nets. New J. Phys. 9, 175 (2007)
24. Zhang, Z.Z., Zhou, S.G., Fang, J.Q.: Recent progress of deterministic models for complex networks. Complex Systems and Complexity Science 5(4), 29–46 (2008)
25. Wang, B.H., Zhou, T., Wang, W.X., Yang, H.J., et al.: Several Directions in Complex System Research. Complex Systems and Complexity Science 21(5) (2008)
26. Zhang, Z.Z., Zhou, S.G., Qi, Y., et al.: Topologies and Laplacian spectra of a deterministic uniform recursive tree. Eur. Phys. J. B 63(4), 507–513 (2008)
27. Fang, J.: On complex network pyramids and their generality and complexity. In: Proc. of 4th Nat. Forum on Net. Sci. and Grad. Stu. Summer School, vol. 191, pp. 204–221. China Center of Adv. Sci. and Tech. (2008)
28. Fienberg, S.E., Meyer, M.M., Wasserman, S.: J. Amer. Stat. Asso. 80(389), 51–67 (1985)
29. Peng, S.-L., Zhang, X.-S.: The Generalized Milnor-Thurston Conjecture and Equal Topological Entropy Class in Symbolic Dynamics of Order Topological Space of Three Letters. Commun. Math. Phys. 213(2), 381–411 (2000)
30. Liu, H.-Z., Zhou, Z., Peng, S.-L.: Star transformations and their genealogical varieties in symbolic dynamics of four letters. J. Phys. A: Math. Gen. 31, 8431 (1998)

Dynamic Mergers Drive Industrial Competition Evolution: A Network Analysis Perspective

Rui Hou[1,*], Jianmei Yang[2], and Canzhong Yao[3]

[1] School of Management, Guangdong University of Technology, Guangzhou 510520, China
hour@gdut.edu.cn
[2] School of Business Administration, South China University of Technology, Guangzhou 510641, China
[3] School of Economics and Commerce, South China University of Technology, Guangzhou 510641, China

Abstract. This paper presents a novel method to explore the relationship between dynamics mergers and the evolution of industrial competition by introducing complex network tool. By taking China's beer industry as an example, we established Markets-Firms bipartite time series networks, weighted Markets-Firms time series networks and industrial competition time series networks by using the data from 1992 to 2009. Through analyzing the changes of topology index of these networks, we find that dynamic mergers play a key important role in the evolution of industrial competition. The results show that dynamic mergers promote the fragmented local markets to be consolidated into a global market for China's beer industry, and it also shows the process that competitive relationship among the rivals turns from some single segmented markets to a global cross-market. This paper gives a new view to observe the changing of industrial competition driven by dynamic mergers.

Keywords: dynamic merger, industry competition, complex networks, China's beer industry.

1   Introduction

Among all the competitive actions, merger activity, which will cause the disappearance of some firms and the increase of industry concentration ratio, would bring a more thorough change on the structure of industrial competition relationship. Any firm can exert an influence to the current industrial competition relationship by adopting the means of merger action to attack against rivals for consolidating their own market or invading rival's market. Changes of the industrial competitiveness structure driven by merger, even small ones, will be perceived by the rival and stimulate the rival's counterattack [1]. To make the initiator suffer the same loss, a subsequent merger action will be an ideal behavior of counter-attack. Consequently it might trigger dynamic mergers or sequential mergers and an industry merger wave may be formed [2-4]. Dynamic mergers are the most effective and direct means to change the inter-firm competitiveness in an industry quickly. Therefore, it is very valuable to

---

* Corresponding author.

R. Huang et al. (Eds.): AMT 2012, LNCS 7669, pp. 463–473, 2012.

investigate the relationship between dynamic mergers and the evolution of industrial competitiveness for both the firms which are involved in competition in one industry and the policy-makers in governments who are responsible for formulating the industrial competition policies.

To elaborate this process, we run a case of beer industry in China and construct Market-Firm bipartite time series networks, bipartite weighted time series networks and industrial competition relationship time series networks by introducing the research tool of complex network [5, 6]. We analysis the evolution process of industrial competition structure driven by dynamic mergers and observe changes of topology index of the networks. Since this evolution is caused by dynamic cross-market mergers, more attention is to be paid toward analyzing on how dynamic cross-market mergers do affect the evolution process of the entire industrial competition structure.

2   Dynamic Mergers in China's Beer Industry

The beer industry in China has the obvious characteristic of multi-market competition after experiencing a period of mergers and acquisitions and industry restructuring. Some beer firms seek to enter new potential markets for expanding their market share, usually by taking the means of mergers and acquisitions to achieve this goal. Along with more and more cross-market merger activities token place in China's beer industry, a multi-market competition context is formed gradually.

With more and more industry restructuring activities, the industry competition is much fierce and a few of beer firms begin to consolidate the fragmental markets by taking over target firms in other regional markets. Consequently, a group of well-known national brands formed represented by Tsingdao beer, Snow beer, Yanjing beer and Budweiser beer, etc. At the same time, some local beer brands represented by Pear river beer, King star beer, Chongqing beer, Jingwei and Harbin beer, etc., and are formed. According to the Competitive Dynamics theory [1], a competitive attack, by using the means of mergers, may be motivate a counter-attack response coming from the rival, also by merger. This may cause a chain reaction. As a result, an industry merger wave may occur [7-9]. Moreover, empirical studies have showed that bank mergers as scale-free coagulation [10]. Actually, more than 330 merger activities have taken place in China's beer industry during the last two decades. So, we can say that, the process of China's beer industry restructuring is the one that the bigger beer firms with national brands competed with each other and rushed to take over the smaller beer firms for much more market share.

3   Dynamic Mergers and Evolution of Industrial Competition

3.1   Industrial Competition Time Series Networks

Economic theory suggests that manufacturers who provide the same or similar products are treated as competitors, regardless of the size of their scales, which means that there will be a large number of competitors in one industry. By taking China's ceramic industry as a case, Yang et al [11, 12] developed a new method for industrial competition analysis and gave a novel discussion on the relationship between industry competition structure and firm's competitive actions from the complex network perspective. Here, nodes in the networks are on behalf of rivals, and edge represents the competitive relationship between any two nodes. However, the existing literatures on industrial competition network analysis usually assume that the networks are static, in other words, nodes and edges of the competition networks are fixed and unchanged [11, 12].

Dynamic mergers will bring a continuous changing for the industrial competition network all the time. Hence, the industrial competition network driven by dynamic mergers belongs to a time-dependent network or dynamic complex network [13, 14]. If a network's topological structure index stay constant all the time, then we could think it as a time-independent network, otherwise we could regard it as a time-dependent network.

Let industrial competition network within a certain period be a snapshot, then a set of time series networks are observable if we string together these snapshots. Thereby, we could explore the evolution process of industrial competition networks driven by dynamic mergers intuitively by observing the changing characteristics of these time series networks. To better describe the evolution process of industrial competition networks mentioned above, two kinds of time series networks, Market-Firm bipartite time series networks and industrial competition relationship time series networks are established in this paper.

3.2   Data

To establish the networks mentioned above, two kinds of data about China's beer industry must be collected yearly. Since the earliest data about beer firm mergers in China was recorded in 1993, we can construct the two kinds of time series networks mentioned above by introducing the data from 1992 to 2009.The resources of data are following. Firstly, official publications like *the yearbook of China's wine industry*, *the yearbook of China's food industry* and *the investment report of China's beer industry* issued by consulting company. Secondly, the Internet including the website of China's beer industry, homepages of beer firms, online news and medium reports, etc. To test the reliability of the data, a mutual check was taken between the data got from the publications and the Internet.

4   Market-Firm Bipartite Time Series Networks

4.1   Network-Building Rules

Bipartite network is the network that can reflect the link relationship between two kinds of nodes. In this paper, we build a series of Market-Firm bipartite networks by using the data of China's beer industry in different years and we want to find out the evolutional laws of the relationship between beer firms and their markets, which was driven by dynamic mergers.

Market node: Market node represents the independent regional and local geographic market. In China, due to the historical and administrative reasons, the local governments formulated some trade protection policies which separated the beer industry market into several local segmented markets. According to the realistic situations, we can take administrative provinces, municipalities directly under the central government as local segmentation markets.

Firm node: Firm node represents the general beer firms.

Edges: Edge represents a subordinate relationship. Here, more directly, it denotes that a firm exists in a specific market. If firm N exist in the market M, then we can draw an edge between firm N and market M. Besides, if firm N_i enters the market M_m by means of taking over another firm N_j, which exists in market M_m, we should also draw an edge between firm N_i and market M_m, as shown in Fig.1.

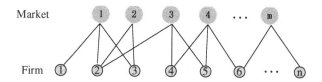

Fig. 1. Market-Firm bipartite network

Based on this rules, one Market-Firm bipartite network can be established by using the data of the specific year. So, we can get a series of Market-Firm bipartite networks a range from 1992 to 2009. To better show the influence of dynamic mergers upon the evolution process of Market-Firm bipartite networks, we extracted three bipartite network pictorial diagrams G_{1992}, G_{2000} and G_{2009} (Fig.2).

Fig. 2. Evolution process of Market-Firms bipartite networks

In Fig.2, the green nodes represent the local segmented markets in China's beer industry, while the red nodes denote the beer firms. The thickness of line/edge linked green nodes and red nodes are on behalf of firm's production capacities in a specific beer market. As can be seen from Fig.2, the relationship between markets and firms of China's beer industry has experienced a dramatically change from 1992 to 2009. A scattered network in 1992 has been evolved into a full connected network in 2009. We can also find that the vast majority firms all most existed in the local segmentation beer markets in the earlier stage, but some bigger firms began to enter into other segmented markets by the means of cross-market mergers slowly.

4.2   Changes of Bipartite Networks Topological Index

There are two kinds of nodes in bipartite networks, top node (market node) and bottom node (firm node). The degree of top node is defined as the number of edges linked to a specific top node, in other words, it is equal to the number of beer firms existing in that market. By the same taken, the degree of bottom node is defined as the number of edges linked to a specific bottom node, namely, it is equal to the number of segmented markets in which a specific firm entered.

In these time series networks of G_{1992}-G_{2009}, the degree of top node started from the peak of 42 in 1992, decreased to the 32 in 2009 gradually. At the same time, the average degree of nodes also decreased from 12.5 to 11.3. These changes indicate that the number of firms in each segmented beer market kept decreasing with the increase of the cross-market mergers. As to the bottom node, its degree started from 2 in 1992 and then rose up to 20 in 2009, but the average degree of nodes only grew from 1.0 in 1992 to 1.4 in 2009, which indicates that only a few firms have implemented a large number of cross-market merger activities while most of other firms only did a few merger activities.

Both the top node degree and the bottom node degree of all the bipartite networks from 1992 to 2009 follow stretched exponential distribution. The distribution index of top node degree ranges from 1.161-1.422, while that of the bottom node degree ranges from 0.290-0.348. Fig.3 (a) describes the changes of stretched exponential index of the top node degree distribution for those networks from 1992 to 2009, so does Fig.3 (b) for the bottom node.

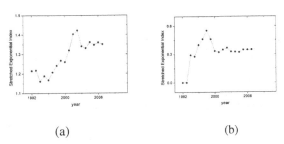

(a)   (b)

Fig. 3. changing curve of Top (a) and Bottom (b) node degree distribution index

4.3   Changes of Weighted Bipartite Networks Topological Index

With increase of cross-market merger activities, production capacities of the beer firms are also on the rise. Thus, the total production capacity of all firms in one single segmented market is increasing too. To describe the changes of the beer firm's production capacity and the total production capacity of a specific segmented beer market induced by dynamic merger activities, we need to build a series weighted Market-Firm bipartite networks. Here, the weight is equal to the production capacity. Calculating results show that the minimum weight of top node increased from 1 to 10 and the maximum also increased steadily from 299 in 1992 to 663 in 2009. The average weight of the top node had an increase from 76 to 220, which indicates that the total production capacity of each segmented market grew all the time. The minimum weight of the bottom node stayed at 1 and the maximum increased dramatically from 80 in 1992 to 1105 in 2009. The average weight of the bottom node grew from 6.2 to 28.8, which shows that the average production capacity of all the firms grew rapidly in this period.

The weights of both the top node and the bottom node follow power law distributions. Moreover, power law index of the top node weights (production capacity of market) ranged from 1.091 to 1.291, and power law index of the bottom node weights (production capacity of firm) ranged from -1.549 to -0.818. As shown in Fig.4 (a) and (b), the weighted Market-Firm bipartite networks have been changing all the time.

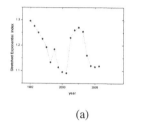
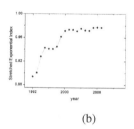

(a)   (b)

Fig. 4. changing curve of top (a) and bottom (b) node weight distribution index

5   Industrial Competition Time Series Networks

5.1   Network-Building Rules

The Market-Firm bipartite networks describe the connection information between the market node and the firm node, but it is not able to display competitive relationship among the firms in China beer industry intuitively. So, it is necessary to convert the bipartite network into a single particle network, namely, industrial competition network, by projecting the firm nodes to the market nodes.

In fact, the projecting rule is to redefine the competitive relationship between two beer firms. Here, the definition of competitiveness between two firms is that they share the same beer market. If a firm exists or enters a local segmented beer market by the means of merger, then we can draw an edge between the firm and any firm in that segmented market. According to the definition of market commonality and resource similarity originated by Chen (1996), if two firms pursue the same customers in the same market, there will be a competitive relationship between the two ones. In this paper, we assume that all the firms existing in one administrative area and sharing the same one market are rivals with each other.

Based on the ideas mentioned above, we can get a set of time series industrial competition networks from 1992 to 2009 by transforming the bipartite networks to single particle networks. We hope to explore the influence of dynamic mergers upon the evolution of industrial competition relationship by analyzing the changes of network topology index.

5.2   Changes of Networks Topological Index

5.2.1   Changes of the Number of Node and Edge

The industrial competition networks in 1992, 2000 and 2009 are selected to observe the evolution process of inter-firm competitive relationships in China's beer industry driven by dynamic mergers intuitively (Fig.5). From the evolutionary process, we can find some changing information on some specific beer firm's competitive composition, which is the main factor triggering merger action.

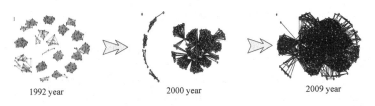

1992 year   2000 year   2009 year

Fig. 5. topology evolution process of industrial competition networks

In this kind of network, the node stands for beer firm and the edge for competitive relationship between the two rivals. The number of nodes in these networks had an annual decrease because some beer firms were taken over by other firms and disappeared, so does the change of the total number of edges in these whole networks. Fig.6 shows the changes of number of firms and edges in these industrial competition networks from 1992 to 2009.

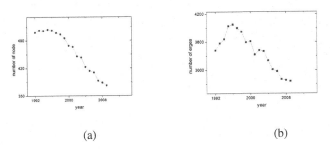

(a)   (b)

Fig. 6. changing curve of node's number (a) and edge's number (b)

5.2.2   Changes of Node's Degree

In a local segmented beer market, if ignoring the new entrants, the number of average competitors, namely the node degree in network, will decrease with the ongoing of merger activities. However, in the multi-market competition setting, on the contrary, the number of average rivals will increase. The reason is that a few of beer firms enter some potential markets by implementing cross-market merger activities and have to encounter more and more new rivals in those entered markets, while in the local segmented market the firms only need to confront the local rivals.

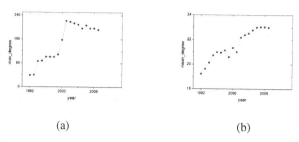

(a)   (b)

Fig. 7. changing curve of maximum node degree (a) and average node degree (b)

Fig.7 shows the changes of node degree and average degree of these industrial competition networks from 1992 to 2009. In 1992, the maximum node degree was 12 and the average degree was 2.1, while these two variables in 2009 are 128 and 7.6 respectively. From Fig.7 we can see that firms compete with more rivals after dynamic mergers. On average, the number of competitors is increase from 2.1 in 1992 to 7.6 in 2009. These changes show that dynamic mergers have speed up the evolution of the industrial competition networks and accelerate the intensity of competition among rivals in China's beer industry.

5.2.3   Changes of Degree Distribution Entropy

The variable, degree distribution entropy, is employed to describe the homogeneity of node degree distribution in networks. According to its definition, the higher is the homogeneity of node degree distribution, the larger is the degree distribution entropy, vice versa. By introducing the calculating method of degree distribution entropy, we got those entropy values for each network from 1992 to 2009 and their variations showed on Fig.8.

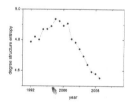

Fig. 8. changing curve of node degree entropy

From Fig.8 we can find that the entropy of node degree distribution for industrial competition network kept increase until 2000 and then began to decrease continuously. Before 2000, the number of competitors which different beer firms faced were little different, but in the next years, the differences of rival amount for each beer firm are obvious.

5.2.4   Changes of Clustering Coefficient

Clustering coefficient here is the topology index which is used to describe the transitivity of networks. Here, in this paper, this variable means that the probability of a rival's rival is also a rival. The larger the clustering coefficient is, the better the connectivity of network shows, vice versa.

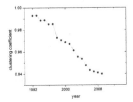   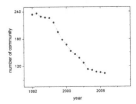

Fig. 9. changing curve of clustering coefficient   **Fig. 10.** changing curve of the number of communities

From Fig. 9 we can observe that dynamic mergers lead to decrease of clustering coefficient in the single particle networks while in Market-Firm bipartite networks whose clustering coefficient value almost reaches to the peak 1. Generally, there is a reducing trend in regard with clustering coefficient, whose value is 0.993 in 1992 and reduced to 0.940 in 2009.

5.2.5   Changes of Communities

Community structure is one of the key properties of complex networks and it plays an important role in their topology and function. Community has an obvious characteristic that the connections between nodes within community are relatively close, but the connections between different communities are relatively sparse. Based on the definition and the detection algorithm of community structure in complex networks, we got the communities in China's beer industrial competition networks for each year. And we also found that the number of communities in industrial competition networks has reduced from 234 to 104 quickly, from 1992 to 2009, which is shown in Fig. 10.

In the earlier stage in China's beer industry, the administrative interventions from local governments brought about so many local segmented beer markets, small size of beer firms and fierce inter-firms competition in those segmentation markets. However, an integrated industrial competition network gradually emerged with more and more cross-market dynamic mergers. Along with the frequent cross-market merger activities, the isolated communities in the industrial competition network, which formed in the earlier stage, began to consolidated and vanished, while the number of communities decreased continuously.

6   Conclusions

Dynamic mergers have played an extremely important role in the evolution of China's beer industrial competition relationship. On the one hand, dynamic mergers have speed up the transformation of China's beer industry market from some local and fragmental markets into an integrated and unified one. On the other hand, they also bring the rapid evolution of competitive landscape for China's beer firms from a local single segmentation market competition to a global multi-market competition. This paper presents a novel and intuitive method on describing this evolution process of industrial competition driven by dynamic mergers. By taking China's beer industry as an example, we find some laws and features emerging during the evolution of China's beer industrial competition.

Acknowledgment. This work was supported in part by the National Science Foundation of China under Grant 71103044, Research Foundation for the Major Project of the Human and Social Science Research Base of Guangdong Province Higher Education under Grant 10JDXM63005.

References

1. Chen, M.-J.: Competitor analysis and interfirm rivalry toward a theoretical integration. Academy of Management Review 21(1), 100–134 (1996)
2. Motta, M., Vasconcelos, H.: Efficiency gains and myopic antitrust authority in a dynamic merger game. International Journal of Industrial Organization 23(9-10), 777–801 (2005)
3. Nilsson, T., Sørgard, L.: Sequential horizontal mergers. European Economic Review 42, 1683–1702 (1998)
4. Hou, R., Yang, J., Yao, C.: A new conceptional model for merger waves: industry competition diffusion based on complex networks. In: The 2nd International Conference on Networks and Digital Society (ICNDS), Wenzhou, May 30-31 (2010)
5. Watts, D.J., Strogatz, S.H.: Collective dynamics of small-world networks. Nature 393, 440–442 (1998)
6. Barabasi, A.L., Albert, R.: Emergence of scaling in random networks. Science 286, 509–512 (1999)
7. Town, R.J.: Merger waves and the structure of merger and acquisition and time-series. Journal of Applied Economics 7, S83–S100 (1992)
8. Toxvaerd, F.: Strategic merger waves: a theory of musical chairs. Journal of Economic Theory 140, 1–26 (2008)
9. Gärtner, D., Halbheer, D.: Are There Waves in Merger Activity After All? International Journal of Industrial Organization 27, 708–718 (2009)
10. Pushkin, D.O., Aref, H.: Bank mergers as scale-free coagulation. Physica A: Statistical Mechanics and its Applications 336(3-4), 571–584 (2004)
11. Yang, J., Lu, L.P., Xie, W., et al.: On Competitive Relationship Networks. A New Method for Industrial Competition Analysis. Physica A 382(2), 704–714 (2007)

12. Yang, J.-M., Wang, W.-J., Chen, G.-G.: A two-level Complex Network Model and its Application. Physica A 388(12), 2435–2449 (2009)
13. Marro, J., Torres, J.J., Cortes, J.M.: Complex behavior in a network with time-dependent connections and silent nodes. Journal of Statistical Mechanics 2, 1–15 (2008)
14. Braha, D., Bar-Yam, Y.: Time-Dependent Complex Networks. Dynamic Centrality, Dynamic Motifs, and Cycles of Social Interaction. In: Understanding Complex System, pp. 39–50 (2009)

A Study of Collective Action Threshold Model Based on Utility and Psychological Theories

Zhenpeng Li[1,2] and Xijin Tang[1]

[1] Institute of Systems Science, Academy of Mathematics and Systems Science,
Chinese Academy of Sciences, Beijing 100190, P.R. China
[2] College of Mathematics and Computer Science, Dali University, Dali,
Yunnan 671003, P.R. China
{lizhenpeng,xjtang}@amss.ac.cn

Abstract. In this paper, we extend Granovetter's classic threshold model by adding both utility and psychological threshold. We conduct simulations with the presented model while also considering the spatial factor and friendship influence strength. We observe that the equilibrium of collective dynamics is not closely related to the friendship impact. With no utility and psychological threshold, the equilibrium state of the model is sensitive to the fluctuation of the collective threshold distribution and displays critical phenomena. By comparison, the equilibrium state with considering utility and psychological threshold looks positively robust. Furthermore, we observe that both cases demonstrate group bi-polarization pattern with the increase of standard deviation of the threshold.

Keywords: psychological threshold, utility, collective action, threshold model.

1   Introduction

Models of collective actions are developed for a variety of situations such as innovation[1], migration, and rumor diffusion, riot behavior, strikes and voting [2,4]. López-Pintado and Watts (2008) classified the existing models into two main categories, heuristic and utility models, respectively [7].   The heuristic models, which mainly include Bass's model of diffusion (Bass, 1969) [3] and threshold model of adoption (Granovetter, 1978) [5] correspond to plausible descriptions of how an individual may adopt a new product or practice as a function of the adoptions of others. The collective action can be viewed as the "domino effect" in threshold models, a quite usual phenomenon from our empirical observation for many radical events. For example, a riot ignited from a small group of radical actors might activate "collective consciousness" of a bulk of people. The main advantage of threshold model is concise and feasible; that is, once the activation rule is specified, the equilibrium, or even non-equilibrium of the collective action is relatively straightforward to compute.

Yet lack of general assumptions of the micro mechanisms results in qualitatively different properties for collective action, then studying the principles of collective dynamics from economic or psychological roots for micro details of the process

R. Huang et al. (Eds.): AMT 2012, LNCS 7669, pp. 474–482, 2012.

seems to be meaningful exploration, such as the interdisciplinary work on group decision making [4].

In economics, utility is a representation of preferences over some set of goods and
services, typically expressed by a utility function about the individual preferences.
Utility models address the psychological or economic considerations along the decision making process. Later Amos Tversky and Daniel Kahneman developed prospect
theory, assigning behavioral implications to value (utility) function to explain
irrational human economic choices [6].

The parameters in utility models are interpretable and policy implications are, at
least in principle, clear. However, the complicated model design and absence of unified forms inhibit its further development.

In this paper, we present a new model of collective action by taking advantages of
both threshold and utility models. Moreover, another two factors, the spatial factor
and friendship influence are also considered. The rest of the paper is organized as
follows: in Section 2 we brief the classic Granovetter's threshold model mechanism;
in Section 3 we provide the new model with utility and psychological threshold.   In
Section 4 we compare our new model with the classic Granovetter's model through
numeric simulation. Section 5 is our conclusion remarks.

2   Granovetter's Threshold Model

Granovetter's threshold model is one of the classic models about collective actions,
such as riots and strikes etc. The model assumes that the possibility that one actor
would join the collective action depends on the proportion of the actors who have
participated in the action. In one social group, each member has one's specific activation threshold, and the threshold of the whole group is subject to certain probability
distribution. The threshold for an instigator is zero; for the radical is lower than that
for the conservative. The strict mathematic form of threshold model is as shown in
Equation (1)

$$F(x) = \int_0^x f(x)dx \tag{1}$$

where $f(x)$ is the probability distribution of the group threshold x,   and $F(x)$ is the
corresponding cumulative distribution function, i.e., $F(x)$ stands for the proportion of
those actors whose thresholds are equal or less than x.   We assume that at the certain
time step t, the ratio of the actors who have joined the collective action is $r(t)$, then at
step $t + 1$ the proportion of actors who join the action is $r(t + 1) = F(r(t))$. When
$r(t + 1) = r(t)$, it is said that collective action reaches the equilibrium state [4].

Next we analyze the mechanism of collective action from economic and psychological aspects and investigate the collective action equilibrium while combing spatial
and friendship factors during numeric simulation.

3   The Threshold Model with Both Utility and Psychological Threshold

In the collective action, the decision that each individual joins the collective action depends on the tradeoff between his/her benefit and cost. For example, the reason that a radical instigator has lower activation threshold is that his/her active action could bring more economic or political benefits than others, i.e. participating the action could bring more benefit than cost. For this reason a jobless person might join the strike with a higher possibility than actors with stable living.   So the thresholds of all actors are heterogeneous, since the intention, background, benefit and cost of each actor are different.   Here we abstract those differences among actors from economic utility point of view, formally for certain collective action. Actor i has benefit b_i and cost c_i, the corresponding utility u_i for i is defined as

$$u_i = b_i - c_i \tag{2}$$

Except the tradeoff between benefit and cost, another factor is local social network information or local social signal. For example, friends have high impact than ordinaries. Provided that actor i has a neighborhood with size N_i, $w_{ij} = 2,1$ stands for the influence strength between friends and ordinary actors respectively. Let $N = \{1, \dots n\}$ be a finite but large set of individuals and $a_i \in \{0, 1\}$ be the common set of actions. That is, every individual makes a binary decision (e.g., whether or not to adopt a certain behavior), then we have local social signal or local social pressure defined in [7] as following

$$k_i = \frac{\sum_{j \in N} w_{ij} a_j}{N - 1} \tag{3}$$

Equ.(3) denotes the local information that actor i may access to. Value function of local social signal is named the network effect or network externalities. The externalities arise when the utility assigned to an action explicitly depends on the absolute or relative number of individuals choosing the action. Our analogy is that actor's utility will increase if he/she observes more and more people participate into the action, while the possible punishment or cost will decrease.

In this paper, we specify the value function form as $v(x) = 1 - e^{-\gamma x}, x \geq 0$, where γ denotes the parameter of risk aversion. According to the above analysis, it is straightforward to show that the utility function can be expressed as

$$u_i = b_i - c_i + v(k_i) \tag{4}$$

Based on classic Granovetter's threshold model, we adopt the psychological threshold which is applied to measure the critical point of some psychological stimulation. As one experimental verified concept, the psychological threshold illustrates that human psychological feeling keeps relatively stable until some stimuli reach the critical level. We use psychological threshold to describe actor would not participate the collective action if his/her utility value is less than his/her psychological acceptable critical level. Through this principle we bring economic and psychological impliations into the classic Granovetter's model. Our primary assumption is that collective action is rooted from psychological acceptable basis with utility connotation.   It is difficult to quantify the complicated and varying human decision-making process. However, the threshold mechanism illustrates comparatively a simple and effective way, whatever the stimuli are from group pressure, learning, utility motivation or other physiological, psychological or social factors.

Assume that each actor has different psychological tolerable threshold p_i. To measure the individual difference between practical utility and acceptable threshold p_i, we adopt a satisfying level e_i --- a behavioral tendency proposed by March and Simon while satisfying is suboptimal when judged by forward-looking game-theoretic criteria [8]. It may be more effective in leading agents out of social traps than other more sophisticated decision rules [9]. Here we define the satisfying level e_i as the following concise form

$$e_i = u_i - p_i \tag{5}$$

Obviously $e_i \geq 0$ means that the utility value of actor i who joins the action is larger than or equal to the actor i's corresponding psychological acceptable threshold. If we assume p_i is subject to uniform distribution $m(x)$ with interval $[a, b]$, the expected satisfying level of actor i is defined as

$$E_i = \int_a^b e_i m(x) dx \tag{6}$$

Provided that group utility threshold T_i subject to normal distribution $f(x)$, when the individual expected satisfying level $E_i \geq T_i$, actor i would like to participate the action, conversely i would not join the action.

Next we conduct simulations toward the above model with considering spatial and friendship influence. We suppose group utility is subject to normal distribution (this distribution is more interesting, because it gives a good description of population averages, for example central limit theorem).

4   Simulation and Results Analysis

For simulation we choose $N=100$ and put actors on periodic lattice with local eight neighbors while friend influence strength equals to 2 and ordinary impact is 1. Three variables, b_i (benefit), c_i (cost) and p_i (threshold) ~ $U(0,1)$ and $\gamma = 1.5$. Fig.1 shows the equilibrium state of collective action by setting friendship density be 0.1 and without considering utility and psychological threshold. In Fig. 1, the x-axis represents group average activation threshold, the y-axis for the standard deviation of group activation threshold, the pixels represent the number of actors who enter into the action when collective action reaches the equilibrium state. This diagram shows that there exists a critical standard deviation for every mean between 2 and 50. The corresponding critical curve is clearly distinguishable and seems to grow more or less linearly with the mean. When the mean is close to 50, the collective action is in unstable state, any small perturbation may lead to some unexpected equilibrium.

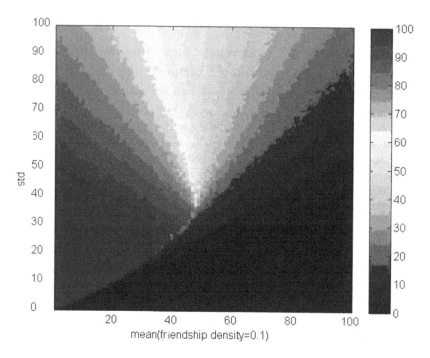

Fig. 1. Collective action equilibrium state vs. activation mean and standard deviation, without considering utility and psychological threshold, and with friendship density= 0.1

From Fig.2, we observe other interesting critical points around 12 and 33 of activation standard deviation. This equilibrium state remains unchanged until the standard deviation is close to 33. We also find that the friendship density or friend influence strength seem have no evident impact on the final equilibrium of collective action (in Fig.2 the density of friendship varies from 0 to 1; different color represents different friend density).

For comparison, we conduct simulation by considering utility and psychological threshold, with results as shown in Fig.3. We fix friend density 0.1 since it does not affect final equilibrium results.

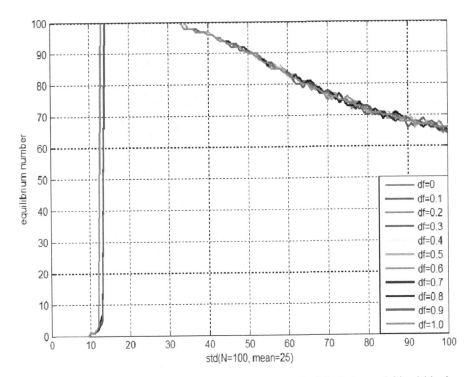

Fig. 2. Collective action equilibrium state vs. activation standard deviation and friendship density (activation mean is fixed at 25)

Fig.3 shows that within wide range of activation mean (from 1 to 50), the fluctuation of activation standard deviation results in unstable collective equilibrium. We also observe a vague critical line between standard deviation and mean, which is not as sharp as shown in Fig.1. The critical phenomena do not appear at activation mean equal to 50 as illustrated in Fig.1, either.

Furthermore, in order to investigate the equilibrium dynamics of collective action with and without utility and psychological threshold implication, we undertake simulations for each case[1]. The simulation on the threshold model of collective action without adopting utility and psychological implication demonstrates that group unstable critical phenomenon appears when the average activation threshold is less than 0.5 (in proportion) and standard deviation is close to 0.25 (in proportion) as shown in Fig.4. When the activation mean is 0.5 and the standard deviation is 0.25 the equilibrium number of collective action suddenly jumps from 12 to 100, i.e. all actors join the action.

---

[1] Since friends density does not significantly affect the collective action equilibrium, for the simulations in Fig.4 and Fig.5 we fix density at 0.1.

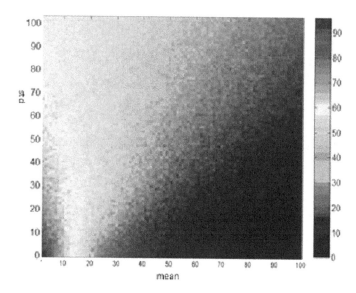

Fig. 3. Collective action equilibrium states under different activation mean and standard deviation when considering utility and psychological thresholds (friendship density 0.1)

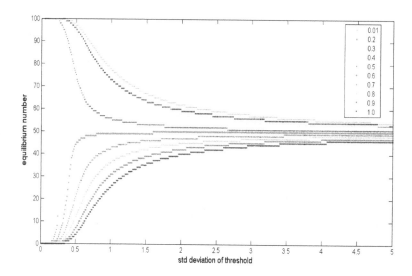

Fig. 4. Given average activation threshold, the impact on equilibrium state as a result of the variation of threshold standard deviation without considering utility and psychological thresholds

In distinct contrast to the above results, the simulation of the model with the involvement of utility implication and psychological threshold indicates that collective action equilibrium reveals stable transitional state, i.e. no critical phenomenon is found as shown in Fig.5.

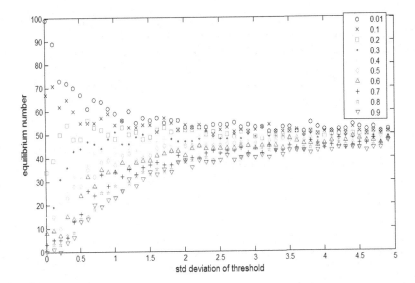

Fig. 5. Given average activation threshold, the impact on equilibrium state as a result of the variation of threshold standard deviation by considering utility and psychological thresholds

However the unexpected result is that both cases show common trend, the collective action displays bi-polarization pattern (the ratio between number of join and not join is approximate to 50% to 50%) with the increase of standard deviation at a variety of the activation mean. As respect to the collective stability of the equilibrium sate and the possibility of phase transition, Galam (2005) presented a general sequential probabilistic frame which shows how various different micro rules lead to the same either ordered or disordered phase [10].

5 Conclusion

In this paper, we present a new model by adopting both utility and psychological thresholds on the basis of Granovetter's threshold model. We investigate the collective action equilibrium through numeric simulations with consideration of spatial and friendship influence strength. We find that the equilibrium is not closely related to friendship impact. With no considering utility and psychological factors, the equilibrium of the model is sensitive to the variation of group threshold distribution and displays critical phenomena. On the contrary, when considering utility and psychological factors, the collective equilibrium is more robust to the threshold distribution.

Furthermore, we observe that both cases demonstrate group bi-polarization pattern with the increase of the standard deviation. Our preliminary conclusion is that the classic threshold model is more appropriate for describing riot, strike and similar unexpected outbreak of social events and the model of utility and psychological threshold is suitable to depict human economic behaviors such as technological spreading.

Granovetter's classic threshold model and other literatures about collective models may lack clear economic, management or psychological implications. We suppose that the collective action models desperately need more interdisciplinary contribution to analyze or explain human collective behaviors such as that from economics, sociology, social psychology, anthropology etc. More experimental and practical evidences are also needed to verify the effectiveness of the proposed model.

Acknowledgments. This research was supported by National Basic Research Program of China under Grant No. 2010CB731405 and National Natural Science Foundation of China under Grant No.71171187.

References

1. Roger, E.M.: Diffusion of Innovations, 3rd edn. Free Press, New York (1983)
2. Granovetter, M., Soong, R.: Threshold Models of Diffusion and Collective Behavior. Journal of Mathematical Sociology 9, 165–179 (1983)
3. Bass, F.M.: A new product growth model for consumer durables. Management Science 15, 215–227 (1969)
4. Galam, S., Moscovici, S.: Towards a theory of collective phenomena: Consensus and attitude changes in groups. European Journal of Social Psychology 21, 49–74 (1991)
5. Granovetter, M.S.: Threshold models of collective behavior. American Journal of Sociology 83, 1420–1443 (1978)
6. Kahneman, D., Amos, T.: Prospect Theory: An Analysis of Decision under Risk. Econometrica XLVII, 263–291 (1979)
7. Lopez-Pintado, D., Watts, D.J.: Social influence, binary decisions and collective dynamics. Rationality and Society 20(399), 797–817 (2008)
8. March, J.G., Simon, H.A.: Organizations. Wiley, New York (1958)
9. Macy, M.W., Flache, A.: Learning dynamics in social dilemmas. Proc. Nat. Acad. Sci. 99(3), 7229–7236 (2002)
10. Galam, S.: Local dynamics vs. social mechanisms: A unifying frame. Europhysics Letters 70, 705–711 (2005)

Recognition of Online Opinion Leaders Based on Social Network Analysis

Ma Ning[1,2], Liu Yijun[1,*], Tian Ruya[1,2], and Li Qianqian[1]

[1] Institute of Policy and Management Chinese Academy of Sciences, Beijing 100190, China
maning2004437070@163.com, yijunliu@casipm.ac.cn
[2] Graduate University of Chinese Academy of Sciences, Beijing 100049, China

Abstract. Opinion leaders on the internet play an important role in promoting the formation of online public opinion, which can influence the direction of public opinion. In this paper, we use social network analysis and content analysis to recognize the opinion leaders of online communities. First of all, we propose "Eight Degree" attribute indexes based on the characteristics of opinion leaders. Then, we construct an attribute matrix of the topic participants, and use the comprehensive evaluation to recognize opinion leaders. Finally, we divide the opinion leaders into the crucial active figure and the important position figure; we can also explore the potential opinion leader deeply by deleting the opinion leaders. The theoretical significance and practical value of this method have been demonstrated by a case study.

Keywords: Social Network Analysis, opinion leader, online community.

1   Introduction

With the development of the Internet, network public opinion becomes the mapping of public opinion in real social life, which is a direct reflection of public opinion. Internet has the characteristics of quickness, interaction, openness, fictitiousness and abruptness, so it becomes a free place for internet users to express their opinions and to communicate with each other. On the one hand, internet can bring together many of the views and suggestions which are beneficial to the development of society; but on the other hand, the Internet also constitutes a threat to the social security that cannot be ignored, as it contains all kinds of information, such as reactionary, violence, rumors, attacking and defaming, as well as much false information [1]. Relevant government departments should guide and interfere in the public opinion online, which is significant to defuse the public opinion crisis and maintain the social safety.

"Opinion leaders" was originally proposed by Lazarsfeld, Berelson and Gaudet in the 1940s. Opinion leaders give their influential comments and opinions, put forward guiding ideas, agitate and guide the public to understand social problems [2]. When

---

* The contact author.

R. Huang et al. (Eds.): AMT 2012, LNCS 7669, pp. 483–492, 2012.

the significant events happen at home and abroad, internet users arouse strong repercussions and high attention immediately. Opinion leaders are the most important person among these internet users, they play an important role in promoting the formation of online public opinion, they can also influence the direction of public opinion and guide the partial opinion to become the public opinion [3]. To identify the online opinion leaders is a prerequisite for guiding and interfering in the public opinion on the internet, so the identification of opinion leaders is very important and meaningful.

Social network analysis (SNA) was first proposed by Simmel in the 1960s [4]. Social networks reflect a kind of social structure, consist of actors and ties (one or more kind of mutual relations between actors) [5]. Social network analysis is widely used in quantification research on the different fields of sociology, such as economics, management, psychology, and so on. Social network analysis method can also better grasp the structure of the social networks and the relationship of the members in public opinion online. There are some studies based on social network analysis to recognize the opinion leaders, but some of the studies select more subjective attribute indexes, which need the expert estimation method to assign the value[6-8]; some of the studies use fewer network indexes, which could not fully recognize the different kinds of opinion leaders [9, 10].

Based on the above, we use social network analysis and content analysis to recognize the opinion leaders of online communities. First of all, we propose "Eight Degree" attribute indexes based on the characteristics of opinion leaders. Then, we construct an attribute matrix of topic participants, and use the comprehensive evaluation to recognize different kinds of opinion leaders, such as the crucial active figure and the important position figure. Finally, we verify the reliability of this method by an example.

2   Recognition Opinion Leaders Based on SNA

2.1   Construction of Social Network

Identifying the main content of the events is the first work, we could determine the keywords of one event by its news topic, title, content on the internet. Then, we use the public opinion information monitoring system to search for the relevant forum posts, in order to ensure the completeness and accuracy of search results, we could adjust the keywords according to the search results repeatedly; Finally, we construct the social network of this event according to the reply relationships between different posts, nodes in the social network stand for the internet users who participate in the discussion of this event, edges stand for the reply relationships between internet users.

2.2 "Eight Degree" Attribute Indexes

Based on the existing studies on online opinion leaders, we propose "Eight Degree" attribute indexes, which could recognize the online opinion leaders more objectively and comprehensively.

(1) The Connected Degree, Co_i

$Co_i=In\text{-}d_i+Out\text{-}d_i$, $In\text{-}d_i$ (in-degree) stands for the total number of posts who received; $Out\text{-}d_i$ (out-degree) stands for the total number of posts who replied to others. This index can measure the heat level and the enthusiasm of the users to participate in the discussion.

(2) The Attention Degree, At_i

$At_i=\sum C_i$, $\sum C_i$ stands for the number of clicks, this number of each post can be got by the public opinion monitoring platform directly.

(3) The Activity Degree, Ac_i

$Ac_i=\sum F_i$, $\sum F_i$ stands for the frequency of discussion in one event. The more contents the participant posted during the event, the more active the participant was.

(4) The Influence Degree, In_i

$In_i=\sum Ag_i/\sum Re_i$, $\sum Ag_i$ (the number of agreements) stands for the number of posts which have the same attitude with the main-creator; $\sum Re_i$ (the number of responses) stands for the number of posts who had received. The types of attitude are positive, neutral, negative respectively.

(5) The Diffusion Degree, Di_i

$Di_i=\sum A_i/\sum P_i$, $\sum A_i$ stands for the number of people who replies to the post i, $\sum P_i$ stands for the number of posts which reply to the post i, this number can measure the extensive of replies. The greater the number, the greater the diffusion of this post.

(6) The Centrality Degree, Ce_i

The centrality degree is a quantitative indicator which can measure the location of each node in the network topology, indicating the depth of the node in the network. This index can be used to find the core members of the network.

(7) The Posts length (Degree), T_i

$T_i=\sum W_i$, W_i is the total number of words in the post i. Generally, the longer the post, the more influential the post.

(8) The Change of Average Path Length (Degree), ΔL_i

$\Delta L_i=L-L_i$, L stands for the average path length, L_i stands for the new average path length after the node i has been removed. The node who has lager ΔL_i play a very important role in the connection of the whole network, that is the key node.

2.3 The Attribute Matrix

The attribute matrix of the participants about some online public opinion can be constructed based on the "Eight Degree" attribute indexes mentioned above. The example of such attribute matrix is shown in Table 1.

Table 1. The attribute matrix

	D_1	D_2	D_3	D_4	D_5	D_6	D_7	D_8
a_1	d_{11}	...	...	d_{14}	...	...	...	d_{18}
...	...	...	...	...	...	...	...	...
a_i	d_{i1}	...	...	d_{i4}	...	...	...	d_{i8}
...	...	...	...	...	...	...	...	...
a_n	d_{n1}	...	...	d_{n4}	...	...	...	d_{n8}

In this attribute matrix, the set $A=\{a_1, a_2, ..., a_n\}$ denotes all the *Agent* who participate in the discussion of some online public opinion; the set $D=\{d_{i1}, d_{i2}, ..., d_{im}\}$ denotes the attribute indexes of *Agent* a_i, and d_{ij} is the *j*-th attribute for the *i*-th Agent. In this article, we propose eight attribute indexes, so $m=8$. It should be noted that with the development of the research, we can define more attribute indexes of *Agent* a_i.

2.4   The Comprehensive Evaluation Method

Based on the above research, we make two assumptions as follows: (1) the marginal value of each attribute of the topic participants is linear, the importance of the participants is relative to the value of the attribute, and the value of every two attributes are independent; (2) the different attributes can be compensated completely, that is to say, regardless of the different effectiveness of each attribute, they still can be compensated [7].   Under these assumptions, the initial data can be normalized to 0-1, then we can evaluate the importance of each topic participant by the average value of all the attribute values, combined with the structure of the network and other auxiliary methods, we can recognize the opinion leaders among all the topic participants. Let average value is E_i, z_{ij} stands for d_{ij} after normalization, the greater the value of E_i of the topic participant is more likely to be the opinion leader.

$$E_i = \frac{1}{m}\sum_{j=1}^{m} z_{ij}, \quad i = 1, 2, \cdots, n$$

2.5   Categories of Opinion Leaders

The comprehensive evaluation and sorting methods are used to measure the "Eight Degree" attribute indexes of all the topic participants, the higher the sorting of the topic participants are the opinion leaders. According to the different meanings and functions of different attribute indexes, we sort different categories of attribute indexes, and recognize the important position figure and the crucial active figure among opinion leaders, such as the connected degree, the centrality degree, the diffusion degree and the change of average path length can be used to measure the importance of the position in the social network of topic participants, so we can use these indexes

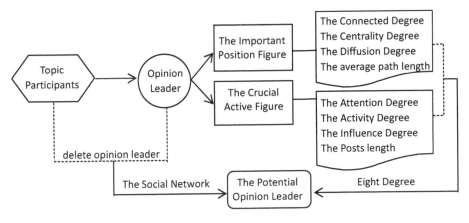

Fig. 1. The recognition mechanism of different opinion leaders

to recognize the important position figure; The activity degree, the attention degree, the influence degree and the posts length can be used to measure the activity of the participants involved in the topic, so we can use these indexes to recognize the crucial active figure. Finally, we can construct a new attribute matrix of the other participants by deleting the opinion leaders, combined with the structure of the new network, we could explore the potential opinion leader deeply (Fig. 1).

3   The Case Application

3.1   Brief Introduction of the Case

The former chairmen QianYunhui of LeQing in Zhejiang province died in a traffic accident in the morning of December 25, 2010. The death was adjudged as an ordinary traffic accident by the local police, but many internet users believed that there were still a few questionable points in the case, so the case aroused widespread concern and discussion online. Subsequently, internet users had a heated discussion of the questions that whether the witness had got a large sum of "shut-up fees" and whether the survey team leader had accept a bribe. The heated discussion was kept high until the driver was arrested formally. Based on the public opinion monitoring platform, we collected 1645 posts about this event from the major forums (Fig.2), and the number of topic participants is 318. Based on these initial data, we construct the social network of "QianYunhui event", in which the nodes stand for all participants, while edges stand for the reply relations, and arrows indicate the reply direction (Fig. 3).

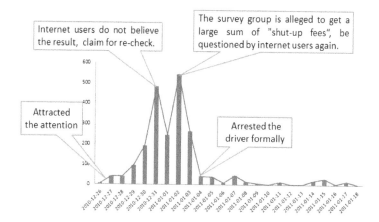

Fig. 2. The posts trendline of "QianYunhui Event"

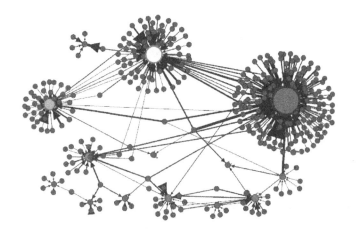

Fig. 3. The social network of "QianYunhui Event"

3.2   Calculation of "Eight Degree"

There are 318 topics participants in this case, and 27 of them are main-creators. Compare and contrast the main-creator and the rep-creator, we can find that main-creators play a decisive role in the development and evolution of the case, so we use Social Network Analysis method to recognize the opinion leaders among main-creators. Part of initial data about "Eight Degree" attribute indexes are shown in Table 2.

After calculated the average value of "Eight Degree" attribute indexes of every main-creator, we can acquire the comprehensive evaluation of opinion leaders; then recognize the important position figure and the crucial active figure among opinion leaders by sorting the different categories of attribute indexes (Table 3).

Table 2. The initial data of attribute indexes of main-creators (Part)

The main-creator	D_1	D_2	D_3	D_4	D_5	D_6	D_7	D_8
cb1949	25	800	1	0.600	0.200	13	728	0.001
deif0646	4	487	1	1.000	0.250	4	56	0.000
lhl050412	24	2516	1	0.625	0.500	12	592	0.054
ly776a	21	1111	1	0.476	0.286	9	753	0.001
......	......	......	......	......	......	......	......	......
BuGeiLi	799	65284	1	0.473	0.208	26	2956	0.009
DaWeiWang	2	212	1	1.000	0.500	2	29	0.000
TanZhiBiHui	48	1851	2	0.511	0.362	8	92	0.017
HongCan2010	3	36763	1	1.000	0.667	2	955	0.000
......	......	......	......	......	......	......	......	......
LiuYiMing	171	26861	58	0.339	0.202	112	2762	0.003
WangXueMei	136	40902	1	0.639	0.338	76	170	0.006
ZhongHuaQinNian	53	9639	1	0.283	0.302	12	870	0.002
ZhouLubao	347	35485	1	0.501	0.246	18	731	0.013

D_1: The Connected Degree; D_2: The Attention Degree; D_3: The Activity Degree; D_4: The Influence Degree; D_5: The Diffusion Degree; D_6: The Centrality Degree; D_7: The Posts length; D_8: The change of average path length.

Table 3. All kinds of important people of top ten

Rank-ings	Opinion Leader		The Important Position Figure		The Crucial Active Figure	
	Figure	Value	Figure	Value	Figure	Value
1	LiuYiMing	0.469	lhl050412	0.425	LiuYiMing	0.606
2	BuGeiLi	0.466	BuGeiLi	0.367	BuGeiLi	0.566
3	WangXueMei	0.386	LiuYiMing	0.332	HongCan2010	0.470
4	HongCan2010	0.344	WangXueMei	0.313	WangXueMei	0.460
5	lhl050412	0.336	testant	0.258	JinRiGuanZhu	0.435
6	JinRiGuanZhu	0.289	ZhouLubao	0.241	LinXiao	0.348
7	ZhouLubao	0.270	WangHaiTingFeng	0.230	ZhouLubao	0.298
8	testant	0.258	HongCan2010	0.218	stander	0.296
9	stander	0.245	stander	0.194	HuoShaoTiger	0.287

3.3   Analysis and Disscussion

As shown in Table 3, the main opinion leaders in "QianYunhui event" are "Liu YiMing", "Bu GeiLi" and "Wang XueMei" (the circles in Fig. 4), the contents of their posts are shown in Table 4, and the attitude of "Liu YiMing" and "Bu GeiLi" is negative, while "Wang XueMei" is neutral. The contents of their posts are just the focuses of dispute in this case, and the posting time of "Bu GeiLi" and "Liu YiMing" just the day before the two peaks in the posts trendline (Fig. 2), they play a great important role in promoting the development of the event. Despite the opinion leaders, we also identify the important position figure and the crucial active figure. "lhl050412" ranked first among the important position figures (the triangle in Fig. 4), he not only connected the main opinion leaders directly, but also had some reply supporters and his approval rating was 100%. "HongCan 2010" ranked first among the crucial active figures (the square in Fig.4), although his connectivity degree was low, his post length was longer, and his post acquired large number of clicks, the attention-degree was 36763, so he had "hidden" effect on the development of the event.

The attitude of the overwhelming majority of the posts about this event is negative, this brought negative impact on the local government and police, so it is necessary to guide and interfere in the public opinion online. The author thinks that effective interventions need to monitor forum, blog, micro-blogging all-day, this is very important to find new questions online in time, for example, there are 77% posts about this case published in FengHuang forum. Then, we can delete or sink the negative posts and negative opinion leaders, for example, in this case, if we remove the negative views of negative opinion leaders, the topic participates in the social network will be reduced by 69.5% and the reply relations will be reduced by 70.9%, the potential opinion leaders maybe emerge during this process (Fig. 5). At the same time, we should pay attention to the distribution of the topic participants, beware of the transformation from "the Internet World" to "the Real World".

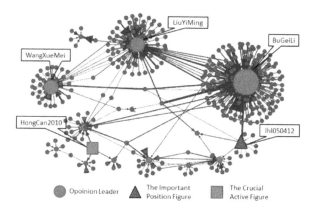

Fig. 4. The main opinion leaders of "QianYunhui"

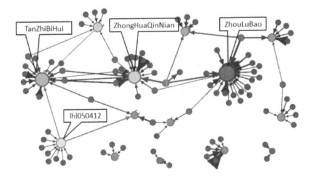

Fig. 5. The potential opinion leaders of "QianYunhui"

Table 4. The post contents of main opinion leaders

Oplinion Leader	Post Date	The Main Post Content	Attitude
LiuYiMing	2011-01-01	The witness and the survey team leaders had got a large sum of "shut-up fees"	Nagative
BuGeiLi	2010-12-30	Brought the video of police action to light: analyse the accident objectively	Neutral
WangXueMei	2011-01-13	The latest advances of QianYuehui event: we do not believe the result, please check again.	Negative

4   Results

In this study, we construct a social network of the online communities, based on the characteristics of the opinion leaders and the combination of social network analysis and content analysis method, we propose "Eight Degree" attribute indexes to identify the online opinion leaders, including the connected degree, the attention degree, the activity degree, the influence degree, the diffusion degree, the centrality degree, the posts length and the change of average path length. Then, we use the comprehensive evaluation to recognize different kinds of opinion leaders, such as the crucial active figure, the important position figure and the potential opinion leaders.

Further work is to improve and refine the attribute indexes which can be used to recognize opinion leaders and other important people, and analyze how to find new attribute indexes from the social network, these will make the recognition of opinion leaders more accurate and complete. In addition, Gatekeepers' deleting or sinking posts of negative opinion leaders is not enough to interfere in and manage the online public opinion. Positive opinion leaders should be consciously fostered to guide online public opinion to the right direction, which requires more deeply simulation studies on the propagation mechanisms of online public opinion and action principles of opinion leaders.

Acknowledgments. The authors gratefully acknowledge the support of National Natural Science Foundation of China (Grant No. 91024010).

References

1. Luo, Y.: A study on the opinion mining of Internet public opinion. University of Electronic Science and Technology of China, Chengdu (2010)
2. Lazarsfeld, P.F., Berelson, B., Gaudet, H.: The People's Choice: How the Voter Makes up His Mind in a Presidential Campaign. Columbia University Press, New York (1948)
3. Liu, J.M.: Public Opinion Propagation. Tsinghua University Press, Beijing (2001)
4. Luo, J.D.: Social network analysis. Social Sciences Academic Press, Beijing (2005)
5. Linton, C.F.: The Development of Social Network Analysis: A Study in the Sociology of Science. Empirical Press (2004)
6. Hu, Y., Zhang, C.B., Wang, Z.X., et al.: Leader formation model during public opinion formation in internet. Journal of Sichuan University (Natural Science Edition) 45(2), 347–351 (2008)
7. Ding, X.F., Hu, Y., Zhao, W., et al.: A Study on the Characters of the Public Opinion Leader in Web BBS. Journal of Sichuan University (Engineering Science Edition) 42(2), 145–149 (2010)
8. Jiang, C.Q., Zhu, Y.S., Ding, Y.: On Discovery of Opinion Leaders Based on UGC. Journal of Intelligence 30(10), 67–71 (2011)
9. Li, Z.Z., Ding, Z.H.: Exploring Online Opinion Leadership Based on Social Network Analysis——Public Opinion of College Student Employment Taken for Example. Journal of Intelligence 30(11), 67–71 (2011)
10. Yu, C.Y., Li, S., Zhao, W., et al.: Identification of Internet Opinion Leaders Based on Social Network Analysis. Science & Technology Information 12, 39, 36–37 (2011)

Critical Infrastructure Management for Telecommunication Networks

Haibo Wang[1], Bahram Alidaee[2], and Wei Wang[1]

[1] Sanchez School of Business, Texas A&M International University,
Laredo, Texas, USA
[2] School of Business Administration, The University of Mississippi, University,
Mississippi, USA

Abstract. Telecommunication network infrastructures such as cables, satellites, and cellular towers, play an important role in maintaining the stability of society worldwide. The protection of these critical infrastructures and their supporting structure become highly challenged to both public and private organizations. The understanding of interdependency of these infrastructures is the essential step to protect these infrastructures. This paper presents a critical infrastructure detection model to discover the interdependency based on the model from social network and new telecommunication pathways, while this study focuses on social theory into computational constructs. The policy and procedure of protecting critical infrastructures are discussed, and computational results from the proposed model are presented.

Keywords: Critical Infrastructure Management, Social Network, Critical Infrastructure Detection.

1 Introduction

New technology in telecommunication allows networks to be spread sparsely over large territories, provide exceptional cover and have tremendous impact on people's life. The use of telecommunication and the value of the network have grown exponentially in the past few decades, and our dependence on telecommunication networks become the most important factor in how people survive during a regional or global crisis. Therefore, telecommunication network infrastructures such as cables, satellites, and cellular towers, play an important role in maintaining the stability of society worldwide. The protection of these critical infrastructures and their supporting structures become highly challenged to both public and private organizations [1]. Recent catastrophic events such as the 9/11 terrorist attacks, Indian Ocean tsunami, Hurricane Katrina, Wenchuan earthquake in 2008, Haiti earthquake in 2010 and Japan earthquake and tsunami in 2011 have highlighted the importance of protecting critical infrastructure systems during anthropogenic or natural disasters. The failure of a telecommunication network will delay the relief effort due to the inability to send the information to responders. In general, there are three types of telecommunication

R. Huang et al. (Eds.): AMT 2012, LNCS 7669, pp. 493–501, 2012.

network failures: The destruction or temporary block out of critical network components such as underground fiber optical cables, data centers, cellular towers/antennas, satellites, and server farm. In addition, the destruction of supporting components such as power station, transformers and cooling system as well as network traffic congestion are the other two failures. The causes of these failures are various from case to case. In some cases, there is no physical damage to the critical infrastructures directly such as temporary block out of satellites and cellular towers by solar storm [2]. It is an expensive task to maintain the highly reliable telecommunication network infrastructures because of its complexity.

In general, critical infrastructures are defined as the systems and assets whether physical or virtual having a debilitating impact on security, national economic security, national public health or safety, or any combination of those matters [3]. The modeling and analysis of the dependencies and interdependencies among the critical components within the same critical infrastructure system are highly complicated and their behaviors are highly nonlinear. Much effort today in critical infrastructure management has been focusing on developing models to simulate the behavior of critical infrastructure systems and to discover the interdependencies and vulnerabilities among the critical infrastructure system. Different modeling approaches have been proposed including agent-based model [4], game theory based models [5], risk management model [6] and mathematical models. The survey on infrastructure interdependency research can be found in [7].

In this study, the authors present a critical infrastructure detection model for telecommunication network, which is based on the notation of most connected nodes in the graph or network. The strength of critical nodes in a network is discussed by Granovetter [8] and have been studied in the fields of vaccination strategy [9, 10], public utility system [11], human pathway [12] and drug design [13].

The rest of this paper is organized as follows: in the next section, we discuss the complex network disruption problem and introduce the critical infrastructure detection model which can be solved by the heuristics based on Tabu Search. Then we present the computation results. Finally, we draw some conclusions regarding the critical telecommunication infrastructure management.

2   Complex Network and Critical Infrastructure Detection

The complex networks represent the dynamic and evolutionary systems surround us today. While the study of complex networks from different fields often turn out unexpected results. These new concepts can be applied to many real world applications including telecommunication networks, transportation networks, social collaboration networks, biological network such as protein network, neural networks, metabolic network and many others. The survey of complex networks can be found in [14].

When we consider complex networks, researchers often raise questions on two important issues: disruption problem and influence problem. For example, in the telecommunication network, the disruption problem will be how to stop the spreading of a new virus over a telecommunication network. One can identify the critical components of the network and take them offline to break the path and connectivity, which is the best strategy to stop the new virus in a practical situation. In the influence problem, if the objective is to prevent communication on a telecommunication network during the cyber warfare, then the efficient way of doing so would be to jam the critical components. However, two problems are often interacted and co-existed in many situations. For example, to prevent the telecommunication network failure during the disaster, the task is to identify critical components with high availability and minimum downtime to reduce the loss of human life and property. The creation of useful redundancies improves self-healing structure in the complex network.

The development of telecommunication network infrastructure has mirrored the human social network, which is considered as a complex network [15]. The vulnerability of the complex network has been studied extensively [16]. The knowledge from the growing body of research in human social network helps researchers to understand the nature and behavior of telecommunication network infrastructure. However, the physical interdependency of the telecommunication network infrastructure gives us the opportunity to model and simulate the behavior of this type of complex network. The objective of this study is to detect the critical infrastructure in the telecommunication networks and measure the fragmentation or disruption of the network caused by the failure.

The following notations are adapted through this paper:

V is a number of facilities in telecommunication network
E is the set of all connectivity in the network
K is subset of facilities who failure results in maximum network disruption
x_{ij} is the pair-wise connectivity between facility i and j

This paper considers the critical infrastructure detection problem as follows: Given a telecommunication network with V facilities as nodes and an integer K, the objective is to find a set of K facilities in this network whose failure results in the maximum network disruption. This problem can be considered as a special case of Minimum-K-overlap problem [17]. Thus a natural, nonlinear formulation of this problem is:

$$\text{Min } \sum_{i,j \in V} \sum_{k=1}^{K} x_{ik} x_{jk} \qquad (1)$$

$$\text{s.t. } \sum_{k=1}^{K+1} x_{ik} = 1, \quad \forall i \in V, \qquad (2)$$

$$x_{ik} + X_{jl} \leq 1, \forall(i, j) \in E, \text{and } k \neq l = 1, \dots, K, \tag{3}$$

$$\sum_{i \in V} x_{i,K+1} \leq K \tag{4}$$

$$x_{jk} \in \{0,1\}, \forall i \in V \text{ and } k = 1, \dots, K + 1.$$

Where x_{ij} is equal to 1 if the facility i and j are in the same sub-network and 0 otherwise. It is easy to see that the linear constraints (2) and (3) given above can be replaced by adding penalty functions to the objective (1) as being a valid infeasibility penalty. By adding such penalty functions from the objective function of a minimization problem the model given above can be conveniently simplified to produce the following streamlined formulation.

$$\text{Min } \sum_{i,j \in V} \sum_{k=1}^{K} x_{ik} x_{jk} + M\left(\sum_{k=1}^{K+1} x_{ik} - 1\right)^2 + M \sum_{i,j \in E} \sum_{k \neq l = 1}^{K} x_{ik} x_{jl} \tag{5}$$

$$\text{s. t.} \sum_{i \in V} x_{i,K+1} \leq K \tag{6}$$

$$x_{jk} \in \{0,1\}, \forall i \in V \text{ and } k = 1, \dots, K + 1.$$

Equation (5) can be re-cast into an equivalent instance of xQx:

$$\text{Min } xQx \tag{7}$$

$$\text{s. t.} \sum_{i \in V} x_{i,K+1} \leq K \tag{8}$$

$$x_{jk} \in \{0,1\}, \forall i \in V \text{ and } k = 1, \dots, K + 1.$$

where the Q matrix results directly from the construction described above. This model is of the form of constrained binary quadratic program (CBQP). This slightly reformulated model, CBQP, can be readily solved by a basic Tabu Search (TS) methodology designed for the generic cardinality constrained binary quadratic program. This TS procedure is implemented by the use of strategic oscillation, which constitutes one of the primary strategies of TS. We employ various strategic oscillation using critical events and span parameters described in [18]. In addition, reader is referred to references [19-31] for a description of some of the more successful methods on solving xQx formulated problems.

3   Experimental Results

To evaluate the model proposed in this paper, we use two data sets. The first data set is the well-known terrorist network by Krebs [32], which is a well-known example in other studies [33-37] and included in this study for demonstrational proposes.

This data set is the map of 911 hijackers and co-conspirators with their direct and indirect association (See Figure 1). In [32], the values of network centrality metrics such as Degrees, Closeness and Betweenness are provided and discussed. However, in our study, we use the binary variable to describe the relationship such as simple connectivity without considering the other factors such as trust ties, resources and distance. In the real world problem, these factors play important roles for network disruption and will be considered in our future study as the weight to each node. We apply the xQx formulation and the TS heuristic to this network with 62 nodes and 153 edges. With a chosen value of k=20, we obtain the optimal solution with objective function value equal to 20. To measure the maximum network disruption, we use the measure of fragmentation proposed by [35]:

$$F = 1 - \frac{\sum_k s_k(s_k - 1)}{n(n-1)} \text{ , where } s_k$$

is the size of subnetwork. The F value for the terrorist network after deleting 20 nodes is 0.9657, where value of 1 means all remaining nodes isolated.

The second data set is GIS-based infrastructure data from FEMA HAZUS MR2. According to Pederson et al [7], the components in telecommunication infrastructure include High frequency radios, satellites, telephone poles, transformers (power station), wireless towers and other supporting facilities. We select GIS data from HAZUS on 8 cities with populations from 50,000 to 200,000 in the US and construct a network based on the GIS data. The number of facilities and the number of nodes deletion from optimal solution are given in Table 1. In Table 1, the value of F and the computation time for each problem are also reported.

Table 1. Critical Telecommunication Infrastructure for 8 cities

Problem ID	Total Number of Facilities	Optimal Number of Critical Nodes (k-value)	F Value of k Deletions	Computational Time (CPU in Seconds)
DE	32	9	0.951	0.03
VI	49	15	0.936	0.17
MI	57	19	0.881	0.31
TY	72	21	0.968	0.55
OD	98	23	0.953	0.92
WA	123	34	0.971	1.04
MC	147	34	0.979	1.06
BR	171	44	0.983	1.19

Fig. 1. The terrorist network structures by Krebs [32]

We see from Table 1 that in 7 out of 8 cities with less than 30% of nodes deletion (failure of facilities) can cause maximum network disruption with more than 90% fragmentation in the telecommunication network. In a real world setting, even a small number of facility outages can cause disruption in the telecommunication network depending on the fragmentation threshold value. Decision makers can set the value from case to case based on the situations. However, to prevent the disruption in the telecommunication network, we need to invest resource to add redundancy to the critical components. With the restriction of resource available to strengthen the network, we can use different k values to compute the fragmentation values and find the minimum value of k which can cause disruption with fragmentation values higher than the threshold. In Table2, we report the fragmentation value F with different k values for DE problem with 32 facilities, where $k = 9$ is the optimal solution from the proposed model.

Table 2. The measure of fragmentation under different k values

Number of Deleted Critical Nodes (k value)	F Value	Computational Time in Seconds
2	0.273	0.05
3	0.364	0.05
4	0.509	0.05
5	0.722	0.04
6	0.865	0.03
7	0.910	0.03
8	0.928	0.03
9	0.951	0.03

4   Conclusion

In this paper, we proposed a method of critical infrastructure detection in telecommunication networks and reported the results of network disruption in terms of fragmentation values caused by the facility outage through the deletion of critical nodes. The heuristic method used for solving xQx can produce good quality solutions with very short time, which is important to decision makers to evaluate the telecommunication network status in real time during a disaster. The model is also very robust with different k values considering the restriction on resource for critical infrastructure management and gives decision makers flexibility in creating useful redundancies to improve self-healing structure in the telecommunication network. Furthermore, it would be interesting to study the problem by including other important factors in the measurement of interdependency to see how these factors contribute to network disruptions in telecommunication infrastructure.

References

1. Government Accountability Office, Defense Critical Infrastructure: GAO-08-373R, in GAO Reports: U.S. Government Accountability Office, p. 1 (2008)
2. Eagleman, D.: Four ways the Internet could go down (July 10, 2012), http://www.cnn.com/2012/07/10/tech/web/internet-down-eagleman
3. Golicic, S.L., McCarthy, T.M., Mentzer, J.T.: Conducting a Market Opportunity Analysis for Air Cargo Operations. Transportation Journal 42, 5–15 (2003)
4. Schoenwald, D.A., Barton, D.C., Ehlen, M.A.: An agent-based simulation laboratory for economics and infrastructure interdependency. In: Proceedings of the 2004 American Control Conference (2004)
5. Svendsen, N.K., Wolthusen, S.D.: Graph Models of Critical Infrastructure Interdependencies. In: Bandara, A.K., Burgess, M. (eds.) AIMS 2007. LNCS, vol. 4543, pp. 208–211. Springer, Heidelberg (2007)
6. G. A. Office, Homeland Security: Key Elements of a Risk Management Approach (2001)

7. Pederson, P., Dudenhoeffer, D., Hartley, S., Permann, M.: Critical Infrastructure Interdependency Modeling: A Survey for U.S. and International Research. I.N. Laboratory (2006)
8. Granovetter, M.: The strength of weak ties. American Journal of Sociology 78, 1360–1380 (1973)
9. Cohen, R., Havlin, S., Ben-Avraham, D.: Efficient Immunization Strategies for Computer Networks and Populations. Physical Review Letters 91, 247901 (2003)
10. Nicolaides, C., Cueto-Felgueroso, L., González, M.C., Juanes, R.: A Metric of Influential Spreading during Contagion Dynamics through the Air Transportation Network. PLoS One 7, e40961 (2012)
11. Chassin, D.P., Malard, J.M., Posse, C., Gangopadhyaya, A., Lu, N.: Modeling power systems as complex adaptive systems. Pacific Northwest National Laboratory, Richland, Wash. (2004)
12. Naamati, G., Friedman, Y., Balaga, O., Linial, M.: Susceptibility of the Human Pathways Graphs to Fragmentation by Small Sets of microRNAs. Bioinformatics (February 10, 2012)
13. Singhal, M., Resat, H.: A domain-based approach to predict protein-protein interactions. BMC Bioinformatics 8, 1–19 (2007)
14. Boccaletti, S., Latora, V., Moreno, Y., Chavez, M., Hwang, D.U.: Complex networks: Structure and dynamics. Physics Reports 424, 175–308 (2006)
15. Freeman, L.: Centrality in social networks: Conceptual clarification. Social Networks 1, 215–239 (1979)
16. Holme, P., Kim, B.J., Yoon, C.N., Han, S.K.: Attack vulnerability of complex networks. Physical Review E 65, 056109 (2002)
17. Narayanan, H., Roy, S., Patkar, S.: Approximation Algorithms for Min-k-Overlap Problems Using the Principal Lattice of Partitions Approach. Journal of Algorithms 21, 306–330 (1996)
18. Wang, H., Alidaee, B., Glover, F., Kochenberger, G.: Solving Group Technology Problems via Clique Partitioning. International Journal of Flexible Manufacturing Systems 18, 77–97 (2006)
19. Billionet, A., Sutter, A.: Minimization of a Quadratic Pseudo-Boolean Function. European Journal of Operational Research 78, 106–115 (1994)
20. Alkhamis, T.M., Hasan, M., Ahmed, M.A.: Simulated annealing for the unconstrained quadratic pseudo-Boolean function. European Journal of Operational Research 108, 641–652 (1998)
21. Amini, M., Alidaee, B., Kochenberger, G.: A Scatter Search Approach to Unconstrained Quadratic Binary Programs. In: Corne, D., Glover (eds.) New Methods in Optimization, pp. 317–330. McGraw-Hill (1999)
22. Beasley, E.: Heuristic Algorithms for the Unconstrained Binary Quadratic Programming Problem. Working Paper (1998)
23. Boros, E., Hammer, P.: Pseudo-Boolean Optimization. Discrete Applied Mathematics 123(1-3), 155–225 (2002)
24. Boros, E., Hammer, P., Sun, X.: The DDT Method for Quadratic 0-1 Minimization. Rutcor Research Center RRR, 39–89 (1989)
25. Glover, F., Alidaee, B., Rego, C., Kochenberger, G.: One-pass heuristics for large-scale unconstrained binary quadratic problems. European Journal of Operational Research 137, 272–287 (2002)

26. Glover, F., Kochenberger, G., Alidaee, B., Amini, M.: Tabu Search with Critical Event Memory: An Enhanced Application for Binary Quadratic Programs. In: Voss, S.M.S., Osman, I., Roucairol, C. (eds.) Meta-Heuristics, Advances and Trends in Local Search Paradigms for Optimization, pp. 93–109. Kluwer (1999)

27. Glover, F., Kochenberger, G.A., Alidaee, B.: Adaptive memory tabu search for binary quadratic programs. Management Science 44, 336 (1998)

28. Katayama, K., Tani, M., Narihisa, H.: Solving Large Binary Quadratic Programming Problems by an Effective Genetic Local Search Algorithm. In: Proceedings of the Genetic and Evolutionary Computation Conference, GECCO 2000 (2000)

29. Lodi, A., Allemand, K., Liebling, T.: An evolutionary heuristic for quadratic 0-1 programming. European Journal of Operational Research 119, 662–670 (1999)

30. Merz, P., Freisleben, B.: Genetic Algorithms for Binary Quadratic Programming. In: Proceedings of the 1999 International Genetic and Evolutionary Computation Conference (GECCO 1999), pp. 417–424 (1999)

31. Palubeckis, G.: A Heuristic-Branch and Bound Algorithm for the Unconstrained Quadratic Zero-One Programming Problem. Computing, 284–301 (1995)

32. Krebs, V.: Uncloaking Terrorist Networks. First Monday 7 (2002), http://firstmonday.org/htbin/cgiwrap/bin/ojs/index.php/fm/article/viewArticle/941/863

33. Saxena, S., Santhanam, K., Basu, A.: Application of Social Network Analysis (SNA) to terrorist networks in Jammu & Kashmir. Strategic Analysis 28, 84–101 (2004)

34. Tyler, J.R., Wilkinson, D.M., Huberman, B.A.: E-Mail as Spectroscopy: Automated Discovery of Community Structure within Organizations. The Information Society 21, 143–153 (2005)

35. Borgatti, S.: Identifying sets of key players in a social network. Computational & Mathematical Organization Theory 12, 21–34 (2006)

36. Huberman, B., Adamic, L., Ben-Naim, E., Frauenfelder, H., Toroczkai, Z.: Information Dynamics in the Networked World Complex Networks, vol. 650, pp. 371–398. Springer, Heidelberg (2004)

37. Baumes, J., Goldberg, M., Magdon-Ismail, M., Al Wallace, W.: Discovering Hidden Groups in Communication Networks. In: Chen, H., Moore, R., Zeng, D.D., Leavitt, J. (eds.) ISI 2004. LNCS, vol. 3073, pp. 378–389. Springer, Heidelberg (2004)

Developing Self-Organizing Systems by Policy-Based Self-Organizing Multi-Agent Systems

Yi Guo[1], Xinjun Mao[1], Cuiyun Hu[1], Junwen Yin[1], and Xinzhong Zhu[2]

[1] Department of Computer Science and Technology,
National University of Defense Technology, Changsha, China
[2] College of Mathematics, Physics and Information Engineering,
Zhejiang Normal University, Jin Hua, China
{Berniegy,Xinjun_mao}@gmail.com

Abstract. Designing suitable interaction behaviors among agent to satisfy the system level requirements is a key issue when designers develop self-organizing systems. Such system is typically designed in an iterative way. However, this method is insufficient once the system requirements become unpredicted at the design-time. In this paper, we propose a policy-based approach to deal with this issue. In our approach, policy is the high-level abstraction to specify self-organizing mechanism as well as the mediator to adjust the local interaction behavior of agents, which enables developer to encapsulate self-organizing implementation mechanism and adapt to the changes of self-organizing requirements. We also study the corresponding implementation approach based on agent technology, including policy management and implementation, the corresponding agent architecture. Based on these implementation technologies, a running environment is provided to support the development of such system. Finally, a case is studied by the running environment to illustrate the effectiveness of our approach.

Keywords: Self-Organizing, Multi-Agent Systems, Software Engineering.

1 Introduction

Currently, self-organizing systems become an intriguing option at the design of information system application, and get more and more attention from academic researchers and industry practitioners. Today in many domains, the question is no longer whether to use self-organization, the challenge before us is how to construct, understand, and manage the systems [1]. Therefore, the software engineering technology for such systems has been extensively discussed in recent years and researchers have carried out many studies, for example development methods, design patterns, system models and architectures [2].

Due to the absence of centralized control node in the self-organizing system, an entity of the system only interacts with its local environment, which leaves a significant gap between the local interaction and global system characteristic, and brings obstacles which are put in evidence when the designers develop such systems.

R. Huang et al. (Eds.): AMT 2012, LNCS 7669, pp. 502–512, 2012.
© Springer-Verlag Berlin Heidelberg 2012

Therefore iterative design is used in many development methodologies [2], for example SodekoVS [6], CUP [7], and the General Methodology [8]. However, as pointed out by [9], a major trend of self-organizing system is to be deployed in the uncertain environment, and unvaried self-organizing process no longer satisfies the requirements of such systems. On the one hand, self-organizing systems require new approaches for run-time verification that can continuously monitor and adjust the system's performance as it reorganizes itself in response to unanticipated requirements. On the other hand, many self-organizing systems need to run uninterruptedly. But the way of iterative designing and deploying is hard to be used on such systems once the system was deployed.

To overcome these challenges, this study proposes a policy-based approach for the development of self-organizing systems. The approach enables developer to abstract the self-organizing mechanisms as policies of the system and change the policies at runtime when the self-organizing result can not satisfy the requirements. The rest of this paper is organized as follows: Section 2 introduces our policy approach and presents a meta-model for the PSOMAS (Policy based Self-Organizing Multi-agent Systems). Section 3 proposes the corresponding implementation technology as well as a running environment for such systems. A case study is illustrated in section 4 to show the effectiveness of our approach. In section 5, related works are introduced. Conclusions and future works are discussed in section 6.

2   Policy-Based Self-Organizing Multi-Agent Systems

One of the important reasons of difficult to adjust the behaviors of entities at runtime is that the self-organizing mechanisms are abstracted as the behaviors of the entities and usually hard coded in the software entities. We therefore should make changes of such abstracting method and implementation of self-organizing mechanism, aiming to support the adjustment of it at runtime. In human society, policies are used to restrict and guide the behaviors of people, and human society often presents self-organizing process in terms of the change of policies. For example the economic policies often result in the changes of macro economic index, which is owing to the people's economic behaviors (like stock transaction, saving money) affected by such policies. There is a compact relationship between the policy and the self-organization: Policies drive the entities to perform certain behaviors as well as these behaviors make system emerges the self-organizing process. Based on this relationship, we can make a further abstraction for the self-organizing mechanisms, which consider that the self-organizing behaviors of entities are driven by the policies of the system at runtime, rather than the hard-coded behavior of the entities.

As self-organizing systems are a sort of complex systems, they are usually engineered with agent metaphor by many researchers, which views the whole system as MAS (Multi-Agent Systems) and the software agents as basic components to construct the systems [1] [10]. In our approach, we also use the agent technologies to support our study.

Generally, a PSOMAS (Policy-based Self-Organizing MAS) consists of agents and system policies. An agent has a set of properties, acquired policies, and a set of actions. The properties represent the internal state and the perceived environment information of the agent, and the actions are selected to be performed by agent when the agent performs behaviors. A behavior of an agent can be seen as a process of actions performing, agent determines what action should be performed when certain conditions are satisfied. Moreover the behaviors are also affected by system policies. Each agent in PSOMAS is situated in a certain environment and only interacts with its local environment. There is no central control agent in PSOMAS. Additionally, there may be many software entities existing in the PSOMAS besides agent, like objects, as [12] discussed, it is easy to see that objects can be considered as special cases of agents in degenerate form when the subset of influential agents contains all agents in the environment. From this perspective, the environment of an agent consists of a group of agents. Figure1 shows the meta-model of the PSOMAS.

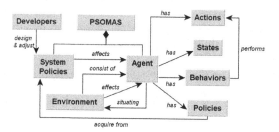

Fig. 1. Meta-Model of PSOMAS

Policy should provide the enough capability to exactly describe the behaviors of the agents. We distinguish two types of policies: *Obligation* and *Prohibition*. The *Obligation* policy represents the action that agents need to perform or the state the agents need to get. The *Prohibition* policy represents the action that agents must not to perform or the state not to get. Formally, policies can be described by EBNF as follows:

Policy:= *Obligation* (*IF* Self-condition *WHEN* Env-event *DO* (Action | State))
    | *Prohibition*(*IF* Self-condition *WHEN* Env-event *DO* (Action | State)

Self-condition represents the satisfied condition of the agent itself. *Env-event* represents the event from the environment of the agent. *"Obligation (IF Self-condition WHEN Env-event DO (Action / State))"* means that when the *Self-condition* of the agent is satisfied and *Env-event* is happened in the environment, the agent need to perform the *Action* or need to keep the *State*. *Prohibition* means the opposite semantics. *Self-condition* and *Env-event* can be viewed as *policy-conditions*.

3   Implementation Technology of Policy-Based Self-Organizing MAS

In this section we discuss the relevant implementation technology and propose a running environment for the PSOMAS.

3.1   Policy Management System

In the PSOMAS, the developers need to deploy the policies at runtime; the agents also need to acquire the policies from the system. PMS (Policy Management System) of the PSOMAS is used to manage the policies of system include loading policies from developers and help agents to acquire policies. Each PSOMAS has one PMS, which shown by Figure 2. The *Interpreter* is used to interpret the system policies to the runtime objects, which can be perceived and understood by the software agents at runtime. *Policy Pool* (PP) can be seen as a container which contains all of the policies of the system. *Policy Adapter* (PA) loads the interpreted policies from *Interpreter* and deposits them into PP. PA also modifies the policies in PP according to the changes of system policies. All of the agents in PSOMAS must register on the PMS and acquire policies from PMS at initialization. The RA (Register Adapter) is responsible for the registering of agents and informing policies to them.

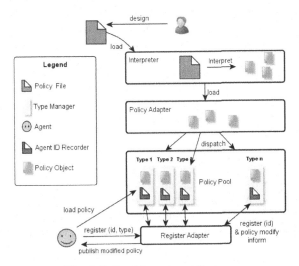

Fig. 2. Policy Management System

In self-organizing systems, there are often some different types of agents in the system, for example a P2P system consists of file downloader and file provider. They have different actions, properties, and behaviors. Agents in different types should be restricted and guided by different policies. The developer therefore should distinguish these agent types when they designing policies. Accordingly, in the PMS, policies should be classified and managed according to different agent types. To realize it, the PP consists of a group of *TypeManager* (TM), each of which is responsible for storing the policies for one agent type and recodes the agent ID which registered on it. When developers deploy policies to the system, PA dispatches the interpreted policies to different TMs as well as the agent must acquire the policies from the TM of its own type.

There are two scenarios that the PMS need to loads policies from the *Interpreter*. One is the system initialization and the other is the adjustment of policies. At the system initialization, PMS loads the interpreted policy objects from *Interpreter* and dispatches them into different TMs. Each TM has an agent ID recorder, which records the ID of registered agent who registers on it. When an agent registers on the PMS, RA receives the registering application, and registers the agent's ID on the recorder of corresponding TM according to the type of the registering agent. Then RA informs the registering agent that which TM the agent is registered on. The agent will load policies from this TM. For example considering agent *a* belonging to the type *A*, when *a* registers to PMS, its ID and type (i.e. *a, A*) are included in the register application information. RA then registers agent *a* on the TM *A* and informs *a* that it has been registered on TM *A*. Then *a* loads polices from TM *A*.

When developers modify the policies of the system, PA compares the changed policies with the old policies in each TM and modifies them respectively. Three kinds of changes can be distinguished in this process: deleting a policy, adding a policy, and substituting a policy. The last can be seen as a combination of the first two operations. When PA finishes this process, it informs the RA that which TMs are modified as well as what policies were changed. After receiving this information, RA acquires all IDs of agent from the recorder of each modified TMs, and publishes the modified policies to these agents. For example developer adds two policies for type *A*, PA loads the new policies from *Interpreter* and adds them to the TM *A*, then informs RA about this modification. RA then acquires all recorded agent IDs from the recorder of TM *A* and publishes the two added polices to these agents.

3.2   Model of BDIP Agent

The BDI architecture of software agents had been accepted for a long time both in academe and industry. We consider that the BDI architecture is useful to analyze the autonomous behaviors of rational agents and easier to accept the policy as a new decision component than other architectures (e.g. Reactive architecture). Therefore, by

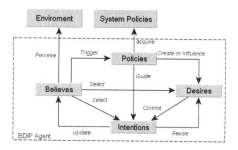

Fig. 3. BDIP Agent Model

the extension of the BDI architecture [11], we propose the BDIP (*Belief, Desire, Intention* and *Policy*) agent architecture (See Figure 3) to support the design of agent in PSOMAS. The perceived policies are deposited in the *Policies* component, and triggered by the *policy-conditions* which specified in the *Belief*. Policies can affect both the *Desire* component and the *Intention* component. For the *Desire* component, a new goal can be created or some of goals are prohibited by policy. For the *Intention* component, the execution of committed plan will be guided by the policy, for example some actions are forbidden to be executed, or the priorities of some actions are risen. The relationship among *Belief, Desire* and *Intention* is same as [11].

3.3   Running Environment for PSOMAS

We implemented a running environment for PSOMAS, named PSOMASE (PSOMAS running Environment), which is based on the Jadex 0.96 platform [13]. The architecture of PSOMASE is shown in Figure 4. The implementation of BDIP agent is of the extension of the Jadex agent. It consists of two components: an agent definition file (ADF) for the specification of internal beliefs, goals and plans as well as their initial values and procedural plan code. For defining ADFs, an XML language is used that follows the Jadex BDI meta-model specified in XML Schema. The procedural part of plans (the plan bodies) is realized in Java and has access to the BDI facilities of an agent through an application program interface. On the other hand, we extended the Jadex agent and implemented three functions for the BDIP agent: registering to PMS, loading policy from PMS, and policy deliberation. But these functions were realized as the background process, developer only needs to design the ADF and the corresponding Java class of plans when they design the software BDIP agent. Policy in PSOMASE is also designed as XML file, which provides a simple and clear way for the developers to represent the policies as we discussed in section 2. The XML policy file is interpreted to the Java classes at runtime by the *Interpreter* which implemented by Binding technology [14].

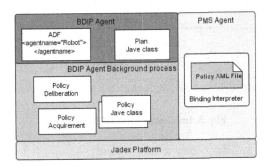

Fig. 4. PSOMASE Archtecture

The PMS is designed as a special agent. The implementation model of PMS agent is shown in Figure 5, where the dashed arrows mean "trigger". Five events are specified in the PMS agent including *InitEvent, InformAgentEvent, ModifyEvent, ModifyPublishEvent,* and *RegisterEvent*. The *Initevent* and *ModifyEvent* are triggered by *Interpreter*, which is used to inform PMS to loads policies. The former is triggered at the initialization, and the later is triggered if the developer changes the policies. *ModifyPublishEvent* is triggered by a plan which is used to modify the *PolicyPool*.

This event is broadcasted to all the corresponding agents who belong to the type of the changed policy and the changed policies are included in this event. *RegisterEvent* is triggered by the registering agent. Agent ID and the type of the agent are included in this event. After the registering process, PMS triggers *InformAgentEvent* and informs to the registering agent. The agent then loads the policies from PMS. Goals are driving forces for the actions of the agent. There are two goals belonging to PMS agent: *RecordAgentGoal* and *ModifyPoolGoal*. These two goals are both implemented as *AchieveGoal* [13] of the Jadex, which means the goal will be pursued as a target by plans as long as the specified event for this goal is triggered. *RecordAgentGoal* represents that the PMS agent registers the agent on itself if it received a registering application from an agent. The goal is triggered by the *RegisterEvent*. *RecordAgentGoal* has two sub-goals which are *InformAgentGoal* and *EnrllAgentGoal*. The former is pursued by the *InformAgentPlan* as well as the later is pursued by *EnrollAgentPlan*. *ModifyPoolGoal* means that if developers change the policies of the system, the PMS agent modifies the *PolicyPool* according to these changes, and triggers the *ModifyPublishEvent*, which will be broadcasted to the agents. Plan represents the agent's means to pursue its goals or responses for the events. It consists of a plan head and a corresponding plan body. The plan head is declared in the XML whereas the body is realized in a concrete Java class. According to the design of events and goals in PMS agent, four plans are implemented by Java class (See Figure 5).

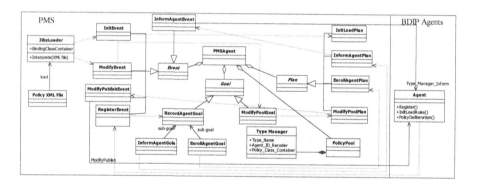

Fig. 5. Implementation UML of PMS

4   Case Study

In order to illustrate the effectiveness of our approach, we implement a case in a GUI approach by PSOMASE. In this case, a group of robots are sent out to the surface of Mars to find the ore resource and carry the ore back to base as soon as possible. In the first scenario, we dispose only one ore resource in the environment, and in the second scenario a lot of ore resources are disposed. We assume that the users of the robots do not make sure how many ore resources in the environment, so they firstly consider the resources are infrequent in the environment and need all of the robots to carry the ore as longs one of them find it.

The basic behaviors of robots are implemented as two goals and corresponding plans. One goal is *ExploringGoal*, which drives the robot randomly walk in the environment to find the ore resources. The other is *CarryingGoal*, which means carrying ore to the base as long as acquired a resource position. Based on the above design, one policy is deployed as follows:

Obligation(*IF* Find_Ore_at_Some_Position *WHEN* $ *DO* Broadcast (Position))

This policy means the robot should broadcasts its position to other robots as soon as it finds an ore resource. The symbol $ means no matter what event happened in the environment. Driven by the *CarryingGoal* and the above policy, the robots go to the ore resource and carry ore from there so long as one of them finds the resource.

From the running result of the first scenario we can see that the above behaviors design makes the robots working effectively. A snapshot of the first scenario is shown in figure 6. The gray transparent pane is the base. The number below the base is the amount of ore in the base which has been carried by robots. The yellow transparent cycle of a robot presents the scope that the robot can explored. The gray points mean the ore resource which has been found by robots and the number below the gray points mean the remained reserves of the ore resource.

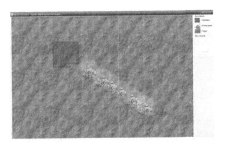

Fig. 6. Snapshot of One Ore Resource

However, if the robots were deployed in the second scenario, the transportation becomes more and more inefficient. The reason is that when the robots are carrying ore from one ore resource, if a new ore resource is found on their way, all of the robots will terminate their transportation and go to the new resource. With the increasing of the found resources, all of the robots will come and go among the found resource, and the transportation will be delayed. In order to resolve this problem, developer should make the robots firstly collect some ore resource positions whatever the position is found by itself or received from others. Then select the nearest position to carry ore. Therefore users should add the follow three new policies:

Obligation (*IF* Searching *WHEN* Others_Send_Message or Find_Ore_Resource *DO* storage_message)
Obligation (*IF* Message_number=MAX *WHEN* $ *DO* TakeOre_Nearest)
Prohibition (*IF* Message_number <MAX]*WHEN* $ *DO* RespondMessage)

By the changes of the policies; the robots carry the nearest ore resource as long as they have collected enough resource positions. The snapshot of the second scenario are shown in figure 7, the red points mean the ore resources which have not been found by robots. Figure 7 (a) represents that the robots have not collected enough positions of ore resource and they still keep searching in the environment. In figure 7 (b) the robots have collected enough positions of the ore resources. They carry the ore from the nearest resource which is chosen from the collected positions. From this case study we can see that the agents in the PSOMAS are affected by both its behaviors design and the policy of the system.   Once the self-organizing requirements are changed, developer could adjust the behaviors of agent by changing the policies to satisfy the new requirements runtime. This method avoids the developer to stop the running of the system and to redesign agents.

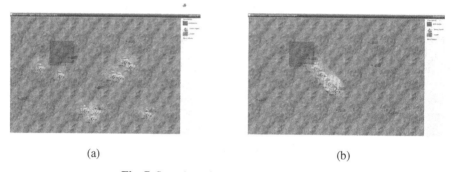

(a)   (b)

Fig. 7. Snapshot of The Mulit-Ore Resource

5   Related Work

In order to abstract and represent the self-organizing mechanisms, the notion of design patterns provides a useful way so that they can subsequently be recombined in novel ways [3]. The approach has been applied to a number of mechanisms, including market based control [4], gradient fields [5] etc. However, these works only abstract self-organizing mechanism for development of the system, but they can not support the adjustment of self-organizing mechanism on-line. Systematic development of self-organizing applications demands approaches to validate that systems exhibit the intended macro characteristics. Currently, two different approaches are exploited: formal verification of self-organizing dynamics [15] and systematic simulation studies [16]. However, self-organizing system will only be acceptable in an industrial application if one can give guarantees about their macroscopic behavior at runtime. If the environment of the system is uncertain, the formal verification and simulation is insufficient. Notwithstanding few works engaged on the runtime adjustment, but some researchers have studied the measurements of self-organizing system at either local or global scales at runtime. These researches [17][8] have discussed configuration or parameter tuning to accomplish this task. In both approaches it is necessary to have a manual external observation entity that evaluates the results and does parameter tuning. [18] goes further and has self-configuration using an intelligent verification tool. Our work can be seen as a complementarity for these works, which is used to adjust the behaviors of agent as long as the measured values are unexpected.

6   Conclusion and Future Work

This paper proposes a policy-based approach to developing self-organizing systems, which affect the emergences of the systems by restricting or directing the behaviors of agents in term of policies. We introduce the policy as the abstraction of the self-organizing mechanism and propose a system meta-model for the PSOMAS. In order to facilitate the development of such systems, we proposed a technical framework which includes two key technologies: PMS and BDIP agent. PMS is used to manage the policies in the system and help agent to acquire the policies at initialization or when the policies are changed by developers. A BDIP agent could acquire policies from the system and adjusts its behaviors according to the acquired policies. We have implemented a running environment for developing of PSOMAS, Finally a case of the self-organizing robots is studied, with which we can find that once the self-organizing requirements are changed at runtime, developer could adjust the behaviors of agent by changing of the policies to satisfy the new requirements at the runtime. This method avoids the developer to stop the running of the system and to redesign agents.

The work presented here can be extended in the following aspects: i) PSOMASE will be enhanced by providing more development tools such as code editor, debug tools, and policy analyzing component to analyze the conflict among the deployed policies. ii) Our policy approach can be optimized by the specific applications. iii) The development methodology of PSOMAS is in our consideration. We will study the systematic development methodology solution to help the developer to analyze; design and implementation the policy based self-organizing systems.

Acknowledgement. The authors gratefully acknowledge the financial support from NSFC under granted number 61070034, 61133001and 90818028, program for new century excellent talents in university, and top key discipline of computer software and theory in Zhejiang Normal University.

References

1. Serugendo, G.D.M., Gleizes, M.-P., Karageorgos, A.: Self-Organsation and Emergence in MAS: An Overview. Informatic 30, 45–54 (2006)
2. Guo, Y., Mao, X., Hu, C.: A Survey of Engineering for Self-Organization Systems. In: Proc. of the 23rd International Conference on Software Engineering & Knowledge Engineering, pp. 527–531 (2011)
3. Simonin, Charpillet, F., Buffet, O., Glad, A.: Engineering Self-Orgnizing Systems (2011)
4. De Wolf, T., Holvoet, T.: Design Patterns for Decentralised Coordination in Self-organising Emergent Systems. In: Proc. of the 4th International Workshop on Engineering Self-Organising Appications, pp. 28–49 (2006)
5. Sasinger, H., Bauer, B., Denzinger, J.: Design Pattern for Self-Organizing Emergent Systems Based on Digital Infochemicals. In: Proc. IEEE Conf. and Workshops on Engineering of Autonomic and Autonomous System, pp. 15–55 (2009)
6. Sudeikat, J., Braubach, L., Pokahr, A., Renz, W., Lamersdorf, W.: Systematically engieering Self-Organizing systems: The SodekoVS Approach. Electronic Communications of the EASST 17 (2009)

7. De Wolf, T.: Analysing and Engnieering Self-Organising Emergent Applications. Leuven University, Belgium (2007)
8. Gershenson, C.: Design and control of Self-Organizing Systems. Faculty of Science and Center Leo Apostel for Interdisciplinary Studies. Vrije Univ. Brussels, Belgium (2007)
9. Parunak, H.V.D., Brueckner, S.A.: Software Engineering for Self-Organizing Systems. In: 12th International Workshop on Agent-Oriented Software Engineering (2011)
10. Zhu, H.: SLABS: A Formal Specification Language for Agent-Based Systems. International Journal of Software Engineering and Knowledge Engineering 11(5), 529–559 (2001)
11. Rao, S., Georgeff, M.P.: Modeling Rational Agents within a BDI-Architecture. In: 2nd International Conference on Principles of Knowledge Representation and Reasoning, pp. 473–484 (1991)
12. Pokahr, Braubach, L., Lamersdorf, W.: A Flexible BDI Architecture Supporting Extensibility. In: 2005 IEEE/WIC/ACM International Conference on Intelligent Agent Technology, pp. 379–385 (2005)
13. Braubach, L., Pokahr, A., Lamersdorf, W.: Jadex: A BDI Agent System Combining Middleware and Reasoning. In: Software Agent-Based Applications, Platforms and Development Kits. Springer, Berlin (2005)
14. http://www.sourceforge.net/projects/jibx/
15. Randles, M., Zhu, H., Bendiab, A.T.: A Formal Approach to the Engineering of Emergence and its Recurrence. In: Proc. of 2nd International Workshop on Engineering Emergence in Decentralised Autonomic Systems, pp. 1–10 (2007)
16. Edmonds, B.: Using the Experimental Method to Produce Reliable Self-Organised Systems. In: Brueckner, S.A., Di Marzo Serugendo, G., Karageorgos, A., Nagpal, R. (eds.) ESOA 2005. LNCS (LNAI), vol. 3464, pp. 84–99. Springer, Heidelberg (2005)
17. Gardelli, L., Viroli, M., Omicini, A.: On the Role of Simulations in Engineering Self-organising MAS: The Case of an Intrusion Detection System in TuCSoN. In: Brueckner, S.A., Di Marzo Serugendo, G., Hales, D., Zambonelli, F. (eds.) ESOA 2005. LNCS (LNAI), vol. 3910, pp. 153–166. Springer, Heidelberg (2006)
18. Soares, A., Gatti, A.C., Carlos, J.P.: Towards Verifying and Optimizing Self-Organizing Systems through an Autonomic Convergence Method. Electronic Publication, http://www.lbd.dcc.ufmg.br/bdbcomp/servlet/Trabalho?id=7572

On Prioritized 2-tuple Ordered Weighted Averaging Operators

Cuiping Wei[1], Xijin Tang[2], and Xiaojie Wang[1]

[1] Management College, Qufu Normal University,
Shandong Rizhao 276826, P.R. China
[2] Academy of Mathematics and Systems Science,
Chinese Academy of Sciences, Beijing 100190, P.R. China

Abstract. This paper deals with linguistic aggregation problems where there exists a prioritization relationship over attributes. We propose a prioritized 2-tuple ordered weighted averaging (PTOWA) operator and study its properties. We then use this operator and a TOWA operator to aggregate satisfactions of attributes for alternatives.

Keywords : Multi-attribute decision making, linguistic terms, PTOWA operator.

1 Introduction

In multi-attribute decision making (MADM), due to the complexity and uncertainty of the objective things, as well as the fuzziness of the human mind, some attributes are suitable to be evaluated in the form of language[1]-[7]. For example, when evaluating the comprehensive qualities of the students or the performance of cars, the decision makers prefer to use 'excellent', 'good' and 'poor' to give an evaluation. For linguistic information aggregation, various linguistic aggregation operators have been proposed, including linguistic OWA operator [1], induced-linguistic OWA operator [2], linguistic WOWA operator [3], etc. In the aggregation process of these operators, the results do not exactly match any of the initial linguistic terms. Therefore, an approximation process has to be developed to express the result in the initial expression domain, but leads to the loss of information and lack of precision. Herrera and Martínez [4] presented an analysis method based upon 2-tuple for linguistic aggregation. Then they proposed 2-tuple weighted average (TWA) operator and 2-tuple ordered weighted averaging (TOWA) operator [4], and successfully applied the TOWA operator to multigranular hierarchical linguistic contexts in a multi-expert decision making problem [5]. Many achievements have been taken in MADM by using these linguistic aggregation operators.

It is important to see that the above linguistic aggregation operators have the ability to trade off between attributes. While in some situations where there exists a prioritization relationship over the attributes, we do not want to allow this kind of compensation. Yager studied this kind of problem where decision information is described by real numbers. He pointed out that the importance

R. Huang et al. (Eds.): AMT 2012, LNCS 7669, pp. 513–519, 2012.

weights of lower priority attributes were based on the satisfaction of alternative to the higher priority attributes [8]. Based on this idea, Yager proposed the prioritized average (PA) operator [9] and the prioritized ordered weighted averaging (POWA) operator [10]. Wei and Tang [12] introduced two averaging operators, a generalized PA operator and a generalized POWA operator. In the case with one attribute in each priority category, the two operators reduce to the PA operator and the POWA operator proposed by Yager.

Motivated by the above-mentioned studies, we consider linguistic aggregation problems where there exists a prioritization relationship over the attributes. This paper is structured as follows. In section 2, we shall make a brief review of 2-tuple and its related operators. In section 3, we propose a prioritized 2-tuple ordered weighted averaging (PTOWA) operator and discuss its properties. We then use this operator and a TOWA operator to aggregate satisfactions of attributes by alternatives. The paper is concluded in section 4.

2   2-tuple Linguistic Representation Model and Related Operators

For MADM problems with some qualitative attributes, we need to use a linguistic term set to describe the decision information. Herrera and Martínez [4] introduced a finite and totally ordered discrete linguistic term set: $S = \{s_\alpha | \alpha = 0, 1, \cdots, \tau\}$, whose cardinality value is odd. For example, a set of seven linguistic terms S could be

$$S = \{s_0 = \text{extremely poor}, \ s_1 = \text{very poor}, \ s_2 = \text{poor}, \ s_3 = \text{fair},$$
$$s_4 = \text{good}, \ s_5 = \text{very good}, \ s_6 = \text{extremely good}\}.$$

Furthermore, Herrera defined 2-tuple to aggregate linguistic information.

Definition 1. [4] Let $S = \{s_0, s_1, \cdots, s_\tau\}$ be a linguistic term set , then the 2-tuple can be obtained by the translation function θ:

$$\theta: \ S \to S \times [-0.5, 0.5), \ \theta(s_i) = (s_i, 0), \text{ for any } s_i \in S. \tag{1}$$

Definition 2. [4] Let $S = \{s_0, s_1, \cdots, s_\tau\}$ be a linguistic term set, $s_i \in S$ and $\beta \in [0, \tau]$, a value representing the result of a symbolic aggregation operation, then the 2-tuple can be obtained with the following function:

$$\Delta: \ [0, \tau] \to S \times [-0.5, 0.5), \ \Delta(\beta) = (s_i, \alpha) = \begin{cases} s_i, & i = \text{round}(\beta) \\ \alpha = \beta - i, & i \in [-0.5, 0.5), \end{cases} \tag{2}$$

where round $(\cdot)$ is the usual round operation.

Definition 3. [4] Let $S = \{s_0, s_1, \cdots, s_\tau\}$ be a linguistic term set, $s_i \in S$ and (s_i, α) be a 2-tuple. There is always a Δ^{-1} function such that from a 2-tuple it returns its equivalent numerical value $\beta \in [0, \tau]$:

$$\Delta^{-1}: S \times [-0.5, 0.5) \to [0, \tau], \Delta^{-1}(s_i, \alpha) = i + \alpha = \beta. \tag{3}$$

Let (s_i, α_1) and (s_j, α_2) be two 2-tuples. Then they should have the properties as follows:

(1) There exists an order: if $i > j$ then (s_i, α_1) is bigger than (s_j, α_2); if $i = j$ then
    a) if $\alpha_1 = \alpha_2$, then (s_i, α_1) and (s_j, α_2) represent the same information;
    b) if $\alpha_1 > \alpha_2$, then (s_i, α_1) is bigger than (s_j, α_2);
    c) if $\alpha_1 < \alpha_2$, then (s_i, α_1) is smaller than (s_j, α_2).
(2) There exists a negative operator: $\mathrm{Neg}(s_i, \alpha) = \Delta(\tau - (\Delta^{-1}(s_i, \alpha)))$, where (s_i, α) is an arbitrary 2-tuple, $\tau + 1$ is the cardinality of S, $S = \{s_0, s_1, \cdots, s_\tau\}$.
(3) There exists a minimization and a maximization operator:
$\max\{(s_i, \alpha_1), (s_j, \alpha_2)\} = (s_i, \alpha_1)$, $\min\{(s_i, \alpha_1), (s_j, \alpha_2)\} = (s_j, \alpha_2)$, if $(s_i, \alpha_1) \geq (s_j, \alpha_2)$.

Definition 4. [4] Let $\{(b_1, \alpha_1), (b_2, \alpha_2) \cdots, (b_n, \alpha_n)\}$ be a set of 2-tuples, the 2-tuple ordered weighted averaging (TOWA) operator is defined as

$$\mathrm{TOWA}\{(b_1, \alpha_1), (b_2, \alpha_2), \cdots, (b_n, \alpha_n)\} = \Delta\left(\sum_{j=1}^{n} w_j \beta_j^*\right), \tag{4}$$

where $w = (w_1, w_2, \cdots, w_n)^T$ is the related weighting vector of TOWA operator, such that $w_j \geq 0$, $\sum_{j=1}^{n} w_j = 1$. β_j^* is the jth largest of the values β_i and $\beta_i = \Delta^{-1}(b_i, \alpha_i)$, $i = 1, 2, \cdots n$.

3   PTOWA Operator

For a linguistic MADM problem, we assume that we have a collection of attributes $C = \{C_1, C_2, \cdots, C_n\}$ and there is a prioritization between the attributes expressed by the linear ordering $C_1 > C_2 > \cdots > C_n$. For any alternative x and attribute C_j, we assume that $C_j(x) \in S(x \in X)$ indicates the satisfaction of attribute C_j by alternative x, where $S = \{s_0, s_1, \cdots, s_\tau\}$ is a linguistic term set and τ is an even.

For each attribute, we transform $C_j(x)$ into a 2-tuple, denoted by a_j. According to the prioritization relationship between attributes and the satisfaction a_j, we first obtain the importance weighting vector $u = (u_1, u_2, \cdots, u_n)^T$ of the attributes. For each attribute we assume T_j is its 2-tuple weight. T_j is defined as

$$\text{(i) } T_1 = (s_\tau, 0); \quad \text{(ii) } T_j = \min\{T_{j-1}, a_{j-1}\}, \quad j = 2, 3, \cdots, n. \tag{5}$$

Transform T_j into its equivalent value, and then we get the normalized importance weights

$$u_j = \frac{\Delta^{-1}(T_j)}{\sum_{j=1}^{n} \Delta^{-1}(T_j)}, \quad j = 1, 2 \cdots, n. \tag{6}$$

Now, we obtain the importance weighting vector $u = (u_1, u_2, \cdots, u_n)^T$ of the attributes which reflects the prioritization relationship. For a given alternative x, when using TOWA operator to aggregation its satisfaction to each attribute we must be able to consider the importance weight associated with each attribute. Yager [11] suggested an approach to performing this type of aggregation by using OWA operator. We now used this approach for the case using TOWA operator.

We consider the situation when we start with a weighting vector $w = (w_1, w_2, \cdots, w_n)^T$, such that $w_j \geq 0$ and $\sum_{j=1}^{n} w_j = 1$, of the TOWA operator. We modify these weights $w_j, j = 1, 2 \cdots, n$, to include the weighting vector $u = (u_1, u_2, \cdots, u_n)^T$ of the prioritized attributes. Yager [11] and Torra [3] suggested an approach to obtain these modified weights. They suggested modeling a BUM function, a mapping $f : [0, 1] \rightarrow [0, 1]$ satisfying $f(0) = 0, f(1) = 1$, and $f(x) \geq f(y)$ if $x > y$, as a piecewise linear function. It is suggested that the function f interpolates the points $(\frac{i}{n}, \sum_{j<i} w_j)$. With this, we can obtain

$$f(x) = \sum_{k=1}^{j-1} w_k + w_j(nx - (j-1)), \quad \frac{j-1}{n} \leq x \leq \frac{j}{n}. \tag{7}$$

Using this function we can obtain the modified weights $v_j (j = 1, 2, \cdots n)$. We assume $ind(j)$ is the index of the jth largest of a_j. Thus $a_{ind(j)}$ is the jth largest of a_j and $u_{ind(j)}$ is its associated importance weight. Let $R_0 = 0$, $R_j = \sum_{k=1}^{j} u_{ind(k)}$. Then we can calculate the modified weights v_j by

$$v_j = f(R_j) - f(R_{j-1}), \quad j = 1, 2 \cdots, n. \tag{8}$$

The modified weights $v_j (j = 1, 2, \cdots n)$ take into account both the w_j and individual importance weights u_j of the attributes. We now use the modified weights to aggregate the satisfactions of attributes by an alternative. We define a function as follows.

Definition 5. Let $a_j = (b_j, \alpha_j)$ $(j = 1, 2 \cdots, n)$ be the satisfactions of attributes C_j by an alternative, and there is a prioritization between the attributes expressed by the ordering $C_1 > C_2 > \cdots > C_n$. The prioritized 2-tuple ordered weighted averaging (PTOWA) operator is defined as

$$\text{PTOWA}\{(b_1, \alpha_1), (b_2, \alpha_2), \cdots, (b_n, \alpha_n)\} = \Delta \left(\sum_{j=1}^{n} v_j \Delta^{-1} \left(a_{ind(j)} \right) \right), \tag{9}$$

where $a_{ind(j)}$ represents the jth largest of a_j , $v = (v_1, v_2, \cdots, v_n)^T$ is the related weighting vector of PTOWA operator satisfying $v_j \geq 0 (j = 1, 2, \cdots, n)$ and $\sum_{j=1}^{n} v_j = 1$, and v can be obtained by Eq. (8).

For convenience of notation, we denote $\text{PTOWA}\{(b_1, \alpha_1), (b_2, \alpha_2), \cdots, (b_n, \alpha_n)\}$ $= (\tilde{b}, \tilde{\alpha})$. We can easily prove that the PTOWA operator satisfies the following properties.

Proposition 1. (Boundedness) Let $\{(b_1, \alpha_1), (b_2, \alpha_2), \cdots, (b_n, \alpha_n)\}$ be a set of 2-tuples, then we have $\min\limits_{j} \{(b_j, \alpha_j)\} \leq (\tilde{b}, \tilde{\alpha}) \leq \max\limits_{j} \{(b_j, \alpha_j)\}$.

Proposition 2. (Idempotency) Let $\{(b_1, \alpha_1), (b_2, \alpha_2), \cdots, (b_n, \alpha_n)\}$ be a set of 2-tuples, if $(b_j, \alpha_j) = (b, \alpha)$, $j = 1, 2, \cdots, n$. Then we obtain $(\tilde{b}, \tilde{\alpha}) = (b, \alpha)$.

In the preceding we consider the situation that there is a prioritization between the attributes expressed by the linear ordering $C_1 > C_2 > \cdots > C_n$. Here we assume that the collection $C = \{C_1, C_2, \cdots, C_n\}$ of attributes is partitioned into q distinct categories, $H_1, H_2, \cdots, H_q$ such that $H_i = \{C_{i1}, C_{i2}, \cdots, C_{in_i}\}$. Here C_{ij} are the attributes in category H_i, $C = \bigcup\limits_{i=1}^{q} H_i$ and $n = \sum\limits_{i=1}^{q} n_i$. We assume a prioritization between these categories $H_1 > H_2 > \cdots > H_n$. The attributes in the class H_i have a higher priority than those in H_k if $i < k$. We assume that for any alternative $x \in X$, we have for each attribute C_{ij} a linguistic term $C_{ij}(x) \in S$ indicating its satisfaction to attribute C_{ij}.

We now give a method to aggregate the satisfactions of attributes by alternative x based on the PTOWA operator and the TOWA operator:

1) Aggregate the satisfactions of each category H_i based on the TOWA operator. For each attribute, we transform $C_{ij}(x)$ into a 2-tuple, denoted by a_{ij}. We associate each priority class H_i a TOWA weighting vector $W_i = (w_{i1}, w_{i2}, \cdots, w_{in_i})^T$, such that $w_{ij} \geq 0$ and $\sum\limits_{j=1}^{n_i} w_{ij} = 1$. Using this we calculate the aggregation value a_i of each category H_i:

$$a_i = \text{TOWA}(a_{i1}, a_{i2}, \cdots, a_{in_i}) = \Delta \left(\sum_{j=1}^{n_i} w_{ij} \beta_{ij}^* \right),$$

where β_{ij}^* is the jth largest of the values β_{ik} and $\beta_{ik} = \Delta^{-1}(a_{ik})$, $k = 1, 2 \cdots, n_i$.

2) Calculate the importance weight of each category H_i. We assume T_j is its 2-tuple weight. T_j is defined as $T_1 = (s_\tau, 0)$; $T_j = \min\{T_{j-1}, a_{j-1}\}$, $j = 2, 3 \cdots q$.

Transform T_j into its equivalent value, and then we get the normalized importance weights $u_j = \frac{\Delta^{-1}(T_j)}{\sum_{j=1}^{q} \Delta^{-1}(T_j)}$, $j = 1, 2 \cdots, q$.

3) Calculate the PTOWA aggregation value for alternative x:

$$\text{PTOWA}(a_1, a_2, \cdots, a_q) = \Delta \left(\sum_{j=1}^{q} v_j \Delta^{-1} \left(a_{ind(j)} \right) \right),$$

where $v = (v_1, v_2, \cdots, v_q)^T$ is the related weighting vector of PTOWA operator. Then we can use the PTOWA aggregation value to rank the alternatives.

Example 1. Consider the following prioritized collection of attributes : $H_1 = \{C_{11}, C_{12}\}$, $H_2 = \{C_{21}\}$, $H_3 = \{C_{31}, C_{32}, C_{33}\}$. We assume there is a prioritization ordering $H_1 > H_2 > H_3$ between these categories and the linguistic term set S is defined as

$$S = \{s_0 = \text{extremely poor}, \ s_1 = \text{very poor}, \ s_2 = \text{poor}, \ s_3 = \text{fair},$$
$$s_4 = \text{good}, \ s_5 = \text{very good}, \ s_6 = \text{extremely good}\}.$$

Assume for alternative x we have
$C_{11}(x) = s_3$, $C_{21}(x) = s_4$, $C_{22}(x) = s_6$, $C_{31}(x) = s_3$, $C_{32}(x) = s_4$, $C_{33}(x) = s_1$.

We now using the above method to aggregate the satisfactions of attributes for alternative x. We associate with each priority class H_i an OWA weighting vector W_i as follow: $W_1 = (1)$, $W_2 = (0.5, 0.5)$, $W_3 = (\frac{1}{6}, \frac{2}{3}, \frac{1}{6})$. For priority class H_i, by sept 1), we can get the TOWA aggregation values $a_1 = (s_3, 0)$, $a_2 = hboxTOWA\{(s_4, 0), (s_6, 0)\} = (s_5, 0)$ and $a_3 = \text{TOWA}\{(s_3, 0), (s_4, 0), (s_1, 0)\} = (s_2, 0.83)$.

By step 2), we get $T_1 = (s_6, 0)$, $T_2 = \min\{T_1, a_1\} = (s_3, 0)$, $T_3 = \min\{T_2, a_2\} = (s_3, 0)$. With this we have $u_1 = 0.5$, $u_2 = u_3 = 0.25$.

Compare the 2-tuples a_1, a_2 and a_3, we have $a_2 > a_1 > a_3$. Thus,

$$ind(1) = 2, \qquad ind(2) = 1, \qquad ind(3) = 3;$$
$$a_{ind(1)} = (s_5, 0), \quad a_{ind(2)} = (s_3, 0), \quad a_{ind(3)} = (s_2, 0.83);$$
$$u_{ind(1)} = 0.25, \qquad u_{ind(2)} = 0.5, \qquad u_{ind(3)} = 0.25.$$

Now we assume the scope of the aggregation is expressed by a weighting vector $w = (0.2, 0.3, 0.5)^T$ of TOWA. By Eq. (7), we get the function such that

$$f(x) = \begin{cases} 0.6x, & 0 \le x \le \frac{1}{3}; \\ 0.2 + 0.3(3x - 1), & \frac{1}{3} \le x \le \frac{2}{3}; \\ 0.5 + 0.5(3x - 2), & \frac{2}{3} \le x \le 1. \end{cases}$$

Since $R_0 = 0$, $R_1 = 0.25$, $R_2 = 0.75$, $R_3 = 1$, we have $f(R_0) = 0$, $f(R_1) = 0.15$, $f(R_2) = 0.625$, $f(R_3) = 1$. Then by Eq. (8), we get the modified weights $v_1 = 0.15$, $v_2 = 0.475$, $v_3 = 0.375$.

Using the PTOWA operator, we get the overall satisfaction $(\tilde{b}, \tilde{a})$ of alternative x:
$(\tilde{b}, \tilde{a}) = \text{PTOWA}(a_1, a_2, a_3) = \Delta\left(\sum_{j=1}^{3} v_j \times \Delta^{-1}\left(a_{ind(j)}\right)\right) = \Delta(3.236) = (s_3, 0.236)$.

4   Concluding

For linguistic aggregation problems where there exists a prioritization relationship between the attributes, we propose a prioritized 2-tuple ordered weighted averaging (PTOWA) operator. Based on the PTOWA operator and the TOWA operator, we give a method to aggregate the satisfactions of attributes for an alternative. A numerical example is given to illustrate the feasibility and effectiveness of the proposed method.

Acknowledgement. The authors would like to thank the anonymous referees for their valuable suggestions to revise the original paper. The work was partly supported by the National Natural Science Foundation of China (71171187, 11071142), the National Basic Research Program of China (2010CB731405), Ministry of Education Foundation of Humanities and Social Sciences (10YJC630269).

References

1. Herrera, F., Verdegay, J.L.: Linguistic assessments in group decision. In: Proceeding of the 1st European Congress on Fuzzy and Intelligent Technologies, Aachen, pp. 941–948 (1993)
2. Herrera, F., Herrera-Viedma, E.: Aggregation operators for linguistic weighted information. IEEE Transactions on Systems, Man and Cybernetics-Part A 27(5), 646–656 (1997)
3. Torra, V.: The weighted OWA operator. International Journal of Intelligent Systems 12(2), 153–166 (1997)
4. Herrera, F., Martínez, L.: A 2-tuple fuzzy linguistic representation model for computing with words. IEEE Transaction on Fuzzy Systems 8(6), 746–762 (2000)
5. Herrera, F., Martínez, L.: A model based on linguistic 2-tuples for dealing with multi-granularity hierarchical linguistic contexts in multi-expert decision making. IEEE Transactions on Systems, Man and Cybernetics-Part B 31(2), 227–234 (2001)
6. Xu, Z.S.: Linguistic aggregation operators: an overview. In: Bustince, H., Herrera, F., Montero, J. (eds.) Fuzzy Sets and Their Extensions: Representation, Aggregation and Models, pp. 163–181. Springer, Berlin (2007)
7. Wei, C.P., Feng, X.Q., Zhang, Y.Z.: Method for measuring the satisfactory consistency of a linguistic judement matrix. Systems Engineering Theory and Practice 29(1), 104–110 (2009)
8. Yager, R.R.: Modeling prioritized multi-attribute decision making. IEEE Transactions on Systems, Man, and Cybernetics Part B, Cybernetics 34, 2396–2404 (2004)
9. Yager, R.R.: Prioritized aggregation operators. International Journal of Approximate Reasoning 48(1), 263–274 (2008)
10. Yager, R.R.: Prioritized OWA aggregation. Fuzzy Optim Decis Making 8(3), 245–262 (2009)
11. Yager, R.R.: On the inclusion of importances in OWA aggregation. In: Yager, R.R., Kacprzyk, J. (eds.) The Ordered Weighted Averaging Operators: Theory and Applications, pp. 41–59. Kluwer, Norwell (1997)
12. Wei, C.P., Tang, X.J.: Generalized prioritized aggregation operators. International Journal of Intelligent Systems 27(6), 578–589 (2012)

A Novel Collaboration Partner Model
Based on the Personal Relationships of SNS

Chengjiu Yin[1], Jane Yin-Kim Yau[2], and Yoshiyuki Tabata[1]

[1] Research Institute for Information Technology, Kyushu University,
Hakozaki 6-10-1, Higashi-ku, Fukuoka, 812-8581 Japan
{yin,tabata}@cc.kyushu-u.ac.jp
[2] Dept of Computer Science, Malmö University, Sweden
jane.yau@mah.se

Abstract. In this paper, we describe a novel model for locating appropriate 'helpers' for users based on the Chain of Friends (CoF) personal relationship in a SNS system, in order to locate appropriate 'helpers' for different users. This model is called SESNMM (Search Engine for Social Networked Mobile Model) and allows individual users, located in remote locations, to participate in a collaborative online community, via our SESNMM-based system. Such typical helpers are willing to help other users solve their tasks/problems and it is intended that both the users and helpers gain knowledge from these interactive online sessions. We have applied this model for inviting PC members of an international conference – namely LTLE 2012. The results showed that our model is very effective for discovering collaboration partners, locating useful helpers, finding users with similar interests in order to create communities for providing future and longer-term helping and teaching exchange.

Keywords: Social Networking, Personal Relationship, collaboration, Search Engine.

1 Introduction

A social network has been defined by Boyd and Ellison (2008) as a web-based service which allows individuals to a) construct a profile within a bounded system, b) generate a list of other users with whom they share a connection, and c) view and connect to others via this list (e.g. a friend of a friend in Facebook)[1]. Different members within such a social networked site may have different mutual trust and closeness amongst one another. In particular, one of the fastest growing and most popular of social networking sites is the Social Networks Service (SNS). Such social interactions and access to social information resources on SNSs can be facilitated by mobile devices anytime and anywhere[2].

In this paper, we propose a novel model for locating appropriate 'helpers' for users based on the Chain of Friends (CoF) personal relationship in a SNS system, in order to locate appropriate 'helpers' for users. The model is called SESNMM (Search Engine for Social Networked Mobile Model). It builds on our previous work in foreign language learning exchange via SNS [3] where we developed a system for

R. Huang et al. (Eds.): AMT 2012, LNCS 7669, pp. 520–527, 2012.

facilitating local Japanese students who wish to learn English, to find recommended native speakers across the globe that are willing to perform language exchange with them.

Every person/user in a social network has a personal relationship connection (e.g., "Friend", or "Friend of a Friend" in Facebook). In our SESNMM model, we depict these as shown in Figure 1 – direct Friend of User, or Acquaintances of A, Friend of B, or Acquaintances of C. These connections form the so-called "Chain of Friends" (CoF), from User to D.

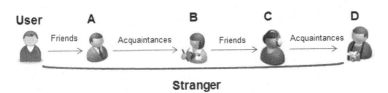

Fig. 1. Chain of Friends

Utilizing these personal relationships, our SESNMM model was designed to locate appropriate 'helpers' for individual users to help them with their tasks/ problems. For example, if the system can recommend a 'best' request CoF for users and locate a helper to the user, even if they are *Strangers*. This is a novel approach using personal relationships to get help from others even if they are *Strangers*.

This paper is organized as follows. In the next section, related work is presented. Then the recommendation algorithms we use in our system is described which is for finding appropriate 'helpers' and locating the 'best' request CoF to users. The process of the development of our SESNMM model is depicted. Finally, we present the conclusions and future work.

2   Related Works

An e-learning support system based on a location-based social network was developed by Kang and Choi (2011) [4]. These authors argued that the social interactions facilitated by such a system decrease the rate of drop-out in an e-learning course where there are normally minimal interactions between participants.

Relating to such systems is the so-called Help Network which was created before the born of the social networking technologies. A number of studies have demonstrated that one of the most effective channels for gathering information and expertise within an organization or institution is its informal network of collaborators, colleagues and friends (i.e. Help Network) [5]. Such a Help Network coupled with social networking technologies has huge potential in generating and sharing a rich wealth of information and learning resources as well as in providing a mechanism for instant real-time communication with others around the globe. Additionally, research suggests that social networks can potentially be useful for learners to solve problems because learners have access a) to many 'helpers' who can and may be willing to help them solve and complete the tasks, or b) lots of relevant and useful information/learning materials [6].

3   Locating the Chain of Friends (CoF) in SESNMM Model

Our SESNMM model provides a connection between friends and acquaintances; this is also known as a Help Network system [7]. Our model can help users locate an appropriate collaborator/helper, who can in turn help them solve their tasks/problems, if any. If an appropriate helper is a Stranger without any connections, then locating a helper for this user may normally be a problem; this forms one of our research questions in this paper. In this section, we describe the algorithms we use in our model for locating appropriate 'helpers' and the 'best' request CoF to users.

3.1   Algorithm for Finding an Appropriate Collaborator/Helper

Our model uses the users' self-administered profiles (including personal information, study interests, schedules, and past actions) in order to locate an appropriate helper (who also has such a profile) who has the ability to solve the problem. Additionally, for each problem, the user enters some related keywords so that a match from the user and a helper can be found suited to the problem.

We designed a formula for calculating the appropriate degree. Consider that n is the number of the keywords that the user inputs, and compare with the other person's profile, schedules, interests and actions, the number of the matched keywords is n_m. It is assumed that the Level of Matched Keywords (LMK) is calculated as follows:

$$\left(LMK = \frac{n - n_m}{n} \right), \quad \text{where} \quad 0 \le LMK \le 1$$

In case of LMK value is equal or close to zero, then the person will be recommended as an appropriate helper who is close to the user's request.

3.2   Getting Help from a Stranger

We have previously carried out a survey that when we seek help/learning support from others and concluded that: 1) The more intimate the personal relationships are, the easier it is to get help/support; 2) The more simple things are, the easier it is to get help/support [8]. Through this conclusion, we found a way for utilizing the CoF to get help from a stranger. Our model can recommend an appropriate request CoF to the user. In the case that there are many request CoFs, the model recommends a 'best' request CoF according to the strength and personal relationship to the user.

For example, A is a Japanese person and studying Chinese. A wrote a blog in Chinese and wants someone to correct it for him, so he inputs the keyword "Language Learning, Chinese" and searches it on the system, and then the system becomes aware of the persons who have relations with the keyword and recommends an appropriate request CoF for A.

As shown in the Figure 2, C and B can speak Chinese, C is a friend of M, B is a internet friend of M, and M is a friend of A. There are 3 Request CoFs:

- For the case 1, A asks for the help to B or C directly. But as they are not friends or acquaintances, it is difficult that he would get help from B or C.
- For the case 2, C is a friend of M and Chinese is her mother tongue.
- For the case 3, B is an internet friend of M and he is a Chinese learner.

Comparing the case 2 with the case 3, the system recommends the case 3 for A. Then A asks M to introduce his friend C to him.

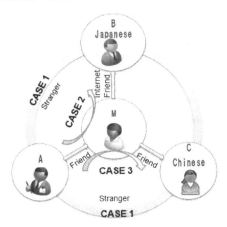

Fig. 2. Request CoFs

3.3   Algorithm for Recommending a 'Best' Request CoF

A 'best' CoF should not be only with a close personal relationship between these persons, but also the small number of the persons in the chain. It is the conditions to determine whether the CoF is appropriate or not.

Table 1. Categories of Personal Relationships

Relationship	Level	Definition and Explanation
a) Family	6	They are family members such as father, mother, brother or sister.
b) Relatives	5	They are very close to the learner such as boy/girlfriend, relatives or close family friends.
c) Friends	4	They are persons whom the leaner has met and talked with frequently such as friends, classmate or teachers.
d) Acquaintances	3	They are persons whom the leaner has met and talked with for a few times.
e) Internet friends	2	They are persons whom the learner has never met before, but has talked for many times online.
f) Strangers	1	They are persons whom the learner has never met before, either online or offline.

The following table describes the categories of personal relationships, which we utilize in locating CoFs in our SESNMM mode. The table contains six categories of personal relationships, from an intimate relationship to an unfamiliar relationship, as shown in Table 1.

According to the "six degrees of separation" theory (Milgram, 1967), we can know a social network typically comprises a person's set of direct and indirect personal relationships, and the length of the CoF is no more than six persons. We designed a formula for calculating CoF Adequacy (CoFA). Consider that k is the number of the intermediaries in the CoF, m_k is level of the personal relationship which was self-administered by the learner for the person k in their profile, and we get a formula for calculating the CoF Adequacy (CoFA) in the following:

$$\left(CoFA = \sum_{k=1}^{n} \left(\frac{k}{6} * \frac{6-m_k}{6} \right) \right), \text{ where } 1 \le m \le 6, \ n = \{1,2,3...\}, \text{ and } k = \{1,2,3...n\}$$

In case of CoFA value is more close to zero then the CoF is more appropriate, n is a natural number, and k is a natural number from 1 to n.

4   A Social Networked Collaboration Partner Model- SESNMM

The main aim of our model is to facilitate collaboration and set up an online collaborative community amongst remote individual users via social networked mobile technologies.

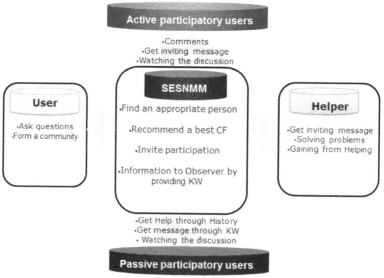

SNS-based Collaboration model for problem solving

Fig. 3. SESNMM- A Model for facilitating a collaborative online helping community

Figure 3 shows our model depicting how individual users get help in an online community.

1. *Overview*: Figure 3 above illustrates that via the environment, how mutual help among users can be facilitated, and especially how users may find an appropriate partner(s) or 'helper(s)' in the online community for collaboration (or in order to solve some problem tasks collaboratively with others). An appropriate request CoF can be recommended upon their request by utilizing their personal relationships to determine which helpers can support the user to get help more easily. The system also sends messages to other persons, then a helping group is formed automatically to solve the problem. The methods of locating the CoF were explained above. Alternatively, the user can also create a community to discuss the problem. After the problem is solved, the helping group will be dissolved. The user will be alone again. The collaborative model is hence transformed back into the individual model.

On the left side of the diagram, one can see that in the original setting, a user is at home or in school individually. If he/she encounters a problem that he/she cannot resolve, then he/she can access a social networked environment to seek help from others online. The user's activity status is changed from individual to collaborative/group helping. Whereas users have the possibilities to gain help from peers in the offline setting, we argue that using such an online social networked model, users will have a larger and diverse set of 'peers/helpers' who may be willing to offer help, anytime, anywhere and on any topic.

2. *Roles*: There are four roles to the model – *user, helper, participation,* and *observer*. Each role has its own objectives which are fulfilled when the collaborative helping processes take place.

Users are these who have problems. Through the SESNMM based online system, they can solve the problems with others who may also correct their mistakes. They will acquire knowledge in the process. The users' mistakes/misconceptions can always be potentially corrected, which ensure that users acquire the correct knowledge.

Helpers are professors who are recommended by the system. Their roles are to help solve others' problems. This is gaining by helping which can enhance their understanding deeper by helping and teaching others.

Active participatory users are users who participate in the online community who want to enhance their understanding deeper by actively observing the discussion between the users and helpers. They can participate in the discussion if desired.

Passive participatory users are users who want to enhance their understanding deeper by actively observing the discussion between users and helpers. Users' mistakes can be regarded as good learning materials/lessons to learn from. These mistakes can help other users' in-depth understanding.

3. *Problem-solving methods in SESNMM*: There are two methods in which users can get help for solving their problems in our SESNMM model – one is via an indirect connection (i.e. it can be considered as a stranger) and the other one is via the online community.

1) The problem-solving method via an indirect connection works as follows. After a user has posted a problem in the SESNMM based system, the system automatically a) forms a learning group to solve this problem, b) extrapolates an appropriate helper

who has the ability to solve the problem, and c) recommends a best Chain of Friends (CoF) to help the user solve this problem. Additionally, the system invites some appropriate users to attend the group as participatory users, who also may have ability to solve the problem. These users may give comments relating to the problem from different points of views during the collaborative helping process. There may be others attending this helping session such as the Observers. Knowledge awareness functions are provided by the system so that users receive messages when there are helping sessions of topics related to their interests.

2) The problem-solving method via the online community works as follows. After a user has created his own helping community and posted his/her problem, the system automatically invites *Participants* as in the previous method. This community is not concerned with whether the problem can be solved completely but rather its focus is on facilitating users to actively participate, and gather more knowledge. Participants and Observers can give their opinions and comments according to their own experiences. All users can put forward their own point of views from different angles of the problem. Finally, the user collates these points of views to retrieve a solution to his problem.

5   Conclusions and Future Work

We proposed a social networked collaboration model-SESNMM, which can facilitate collaboration and set up an online collaborative community amongst remote individual users via social networked mobile technologies. This model can locate appropriate 'helpers' for users based on the Chain of Friends (CoF) personal relationship in a SNS system, in order to locate appropriate 'helpers' for users. In this way, personal direct and indirect relationships can be utilized for pedagogical purposes, a network of friendships can be potentially enhanced, and knowledge sharing and creation can be supported and expanded.

Additionally, our model forms a SNS among users who have similar interests so that their own communities can be created for further and future learning and teaching exchange.

We have applied this model for inviting PC members of the 2012 LTLE international conference. The results showed that our model is very effective for discovering collaboration partner.

References

[1] Boyd, D.M., Ellison, N.B.: Social network sites: Definition, history, and scholarship. Journal of Computer-Mediated Communication 13(1), 210–230 (2008)
[2] Tan, Q., Kinshuk, Jeng, Y.-L., Huang, Y.-M.: A collaborative mobile virtual system based on location based dynamic grouping. In: IEEE International Conference on Advanced Learning Technologies, pp. 16–18 (2010)

[3] Yin, C., Dong, Y., Tabata, Y., Ogata, H.: Recommendation of helpers based on personal connections in mobile learning. In: International Conference on Wireless, Mobile and Ubiquitous Technologies in Education, Kagawa, Japan, March 27-30, pp. 137–141 (2012)

[4] Kang, J.-M., Choi, S.-Y.: An e-learning support system based on location-based social network. International Journal of Advancements in Computing Technology 3(7), 55–63 (2011)

[5] Eveland, D.J., Brown, W., Mattocks, J.: The Role of "Help Networks" in Facilitating Use of CSCW Tools. In: Proceedings of Computer Supported Cooperative Work, pp. 265–274. ACM Press (1994)

[6] El-Bishouty, M.M., Ogata, H., Rahman, S., Yano, Y.: Social Knowledge Awareness Map for Computer Supported Ubiquitous Learning Environment. Educational Technology & Society 13(4), 27–37 (2010)

[7] Milgram, S.: The Small World Problem. Psychology Today, 60–67 (1967)

An Innovative Way for Mining Clinical and Administrative Healthcare Data

Siu Hung Keith Lo and Maiga Chang

School of Computing and Information Systems, Athabasca University, Canada
keithshlo@yahoo.com, maiga.chang@gmail.com

Abstract. A novel method of "predicting" sitter case attribute value is presented in this paper. The method allows users to choose two attributes, seed and target attribute, and to predict the target attribute value of the forthcoming sitter case. The method first retrieves string sequences of the seed attribute according to filters the users set. Then, it finds the words in the sequences and calculates the term frequencies of the words. With the term frequencies, the proposed method uses vector space model to measure the similarity between the testing sequences and the benchmark sequence. At the end, the testing sequence which has highest Cosine similarity value is chosen and the filtering value the method uses to generate the testing sequence is the predicted result. These predicted results allow hospitals to adjust their strategies on resource assignments to better handle patient needs.

Keywords: Healthcare, Regular Expression, Data Mining, Sitter, Hospital Networks.

1 Introduction

Sitter is an external on-call resource hired to "watch" patients who are at risk and need constant supervision. In case something happens to the patient, the sitter informs nurses for intervention. Due to the shortage of in-hospital medical staffs, sitters are hired to free up staffs' time, letting them focus on the jobs which require more skills. It is more cost effective to use sitters to watch patients because

- medical knowledge is often not required, sitters have lower wage than health professionals
- sitters are on-call basis; hospitals call the sitters in and only pay for the particular shifts.

With immense number of patients in public hospitals, reviewing long histories of patient charts and doing analysis can be tedious work. Although statistical reports are being generated to report usages, not much is being done to turn data into real knowledge (i.e., discover patterns and relationships between different clinical and non-clinical elements). Currently, the only feeback on sitter usage are the sitter numbers and dollars spent, some valuable information may still be buried. Such information may be important indications to correlate different clinical and

R. Huang et al. (Eds.): AMT 2012, LNCS 7669, pp. 528–533, 2012.

non-clinical factors, which may be critical to provide both management and practitioners to improve the quality of healthcare.

The purpose of this research is to to analyze sitter usage data in relation to non-nominative patient information and to predict the consequent results directly from previous case sequence without the understanding of the meaning of attributes. Regular expression is applied to the design of the proposed method. The predicted results can be used as a reference to provide healthcare administrators to fine tune staff proportion to better respond patient needs or/and adjust certain procedures to carry out treatments more effectively.

2   Literature Review

In healthcare setting, a lot of information about patient episodes are recorded in various systems. Reports are being generated but they mostly contain only counts, sums and groupings of collected data. Although some manipulations are being done to those reports to facilitate data representation, they are mostly visual appeals or pivot tables, which do not necessarily provide more knowledge or discovery of new information.

According to Fayyad, Piatetsky-Shapiro and Smyth [1], the tremendous amount of data collected and stored in large and numerous data repositories has far exceeded human's ability for comprehension without machine aided analysis. It is almost unavoidable to have data entry errors or inconsistencies in huge data sets. With data mining techniques, outliers can be spotted out and further analysis can be done to determine if those are erratic entries.

Many meaningful patterns can be analyzed and extracted from regular expressions [2]. Regular expression is a metalanguage that describes finite-state automata used for string pattern recognition [3]. It is also a way of describing complex patterns in texts. It has been used to extract information in biomedical field and provided an alternative approach to do complex semantic parsing [9][10][11]. Its advantage is to use shorter and simpler representation for presenting long sequences which contains repeated patterns.

Interesting patterns can be discovered by the recommender system and may assist healthcare institutions to alert health practitioners about some higher risk patients and to find out the reasons of why some patients require higher cost of care and length of stay. For example, some patients with aggressive behaviors can be related to side effects of certain treatments and medications.

3   Clinical and Administrative Healthcare Data

Clinical administrative data such as sitter data is not being used sufficiently. Without discorvering knowledge with data mining techniques, a lot of information may still be hidden. With only numbers showing on statistical reports, more complex questions about patient care cannot be answered. For example, there may be relationships between gender, culture, length of stay, diagnosis, and locations that can affect patient's need for sitters.

According to the hospital's guidelines, no sitter orders can last for more than one entire shift. For patients who need sitter supervision for more than one shift, additional orders must be placed. In every order, the hospital must choose a primary reason from a pre-defined list to explain why the patient requires sitter's further supervision. The reasons can be one of the followings: Agitation, avoid use of 4-point restraints, avoid use of other type of restraint, avoid use of posey vest/jacket, away without leave, behavior problem, constant observation in 4-point restraints, Delirium, Dementia, Disorientation, eating disorder, Psychosis, risk of falls, Suicidal, Trauma, violent, youth protection, and other. The data analyzed by this research includes sitter order's date, department's mission (a.k.a. sub-division), sitter's shift (Day, Evening, Night), primary reason (patient's problem), units that placed the order, health professional who placed the order, health professional (supervisor level) who authorized the order, patient's medical record number, patient's family and first name, patient's gender, patient's primary spoken language, and patient's bed number and location.

The sitter system depends on the hospital's admission, discharge and transfer (ADT) system to get more detailed patient information. Other than the medical record card number and basic information about the patients, the sitter system does not store any other patient specific information but the sitter cases. With data consolidation between the sitter system and hospital's ADT system, it is able to provide the research team the following anonymous data like patient's date of birth (and it allows us to calculate the patient's age), gender, marital status, preferred language, municipality, diagnosis, admission type, admission and discharge date (and it allows us to calculate the length that the patient stays at hospital), and discharge location.

4   Regular Expression Based Data Mining

The research team uses regular expression to summarize and present sitter cases' attribute sequence for a time period. The proposed method can then predict the attribute value that the forthcoming case may have by expanding the regular expression presented the particular cluster, i.e., the regular expression can be seen as a sort of deduction rules. Furthermore, the method applies vector space model to calculate the similarity of two regular expressions. The closer two regular expressions are similar to each other may imply that two sequences may have hidden relationships. Therefore, the method is capable of using an attribute's regular expression to predict the follow-up value of the other attribute.

The objective of the proposed method is to predict the attribute value of forthcoming sitter case based on the attribute values of past sitter cases. Due to all sitter cases have its date and shift stamp, they can be seen as sequential records. Sitter cases after data pre-processing consist of multiple attributes, as Table 1 lists.

Table 1. Sitter cases

Mission	Site	Shift	Reason	Age	Gender	Marital Status	Lang	Adm Type	Length of stay	Discharge Location
Surgery	RVH	Night	Away without Leave	70-79	M	SINGLE_ADULT	French	ER	20-29	Home
ER	RVH	Day	Disorientation	60-69	F	SINGLE_ADULT	French	Stretcher	0-9	Hospital
ER	RVH	Evening	Agitation	70-79	F	SINGLE_ADULT	French	Stretcher	0-9	Hospital
ER	RVH	Night	Disorientation	50-59	F	SINGLE_ADULT	French	Stretcher	0-9	Hospital
Medicine	MGH	Night	Suicidal	80-89	M	MARRIED_ADULT	English	ER	0-9	Home

The proposed method uses regular expression to summarize the values of particular attribute in the case sequence. Before the regular expression can be applied, we need to find a string sequence to represent the cases. The particular attribute is called "seed" which is the attribute the user wants the method to use the seed to predict another attribute's forthcoming value. To facilitate the representation of the sequence elements, a single alphabet index is being used to represent each attribute value. For example, take Reason attribute in Table 1 as the seed, the sequence, EJAJO, is being produced if the following codes are assigned to represent different reasons: E for Away without Leave, J for Disorientation, A for Agitation, and O for Suicidal (O).

From searching a particular attribute and value in the dataset, a string sequence of the chosen seed can be found. A sequence may be similar to some other sequences found by searching for different attributes and values in the dataset. The similarity of two sequences may imply that the follow-up attribute values in two sequences (i.e., the future attribute values predicted from the expansion of the sequence) may have hidden relationship. The proposed method uses the sequence similarity to discover relationships between the values of seed and target attribute. In other words, we assume that the symbolic sequence of an attribute's values may contain hints to reveal another attribute's values. For example, a series of sitter reasons (i.e., the seed attribute) can be used as a predictor to predict length of stays (i.e., the target attribute).

The method does the following steps to predict target attribute's value:

1. deciding seed attribute–seed attribute can be chosen by users;
2. deciding target attribute–target attribute can be chosen by users;
3. filtering the dataset with particular conditions (i.e., specific attributes and values)–the conditions can be chosen by users;
4. generating string sequence (i.e., the benchmark sequence) of the seed attribute values from the filtered dataset;
5. generating string sequences (i.e., the testing sequences) of the seed attribute values from the filtered dataset by using all possible values that the target attribute has as additional filtering criteria;
6. finding words of different lengths for all sequences include the benchmark and the testing sequences;
7. calculating the similarity values for each testing sequences from the benchmark sequence;
8. and, choosing the testing sequence which has highest similarity and use its value as the predicted value for the target attribute.

The proposed method uses word matching technique to determine whether two sequences are similar. A word is a repeated character sequence. A word finding engine within regular expression approach has been developed to find out possible words in different lengths. Once the words are found, they are stored in a dictionary.

By calculating the term frequencies of each sequence and convert them into vector space, with normalized vectors of term frequencies, it can be applied to see how close a string testing sequence is to the benchmark string sequence. Cosine similarity measure [4] has been widely used in clinical analysis to compare sequences generated by data collection tools with timestamps [5][6][7]. It has also been proven to be a robust metric for scoring the similarity between two strings, and it is increasingly being used in text mining related queries [8].

5   Evaluation and Discussion

The research team evaluates the proposed regular expression based data mining method with the data which consists of all the sitter usages within the hospital network, consisting of five hospitals (4 adult sites and 1 child & adolescent site), for the entire years of 2008, 2009 and 2010. To evaluate the accuracy of the prediction results suggested by the proposed method, we compare the predicted results with the known records existed in the dataset. In general, results are quite promising with fair accuracies.

The proposed innovative method help users predict the target attribute value of a forthcoming sitter case based on their chosen seed attributes and criteria. The method doesn't need to know the meanings of attributes and to do complicate calculations like information value and entropy. It simply generates string sequences, finds the words in the sequences, and measures the Cosine similarity a testing sequence has with the benchmark sequence. At the end, the method gives the user its prediction value for the target attribute with the filtering value used to generate the most similar testing sequence against the benchmark sequence.

Such prediction method is important to hospitals. The administrative personnel can prepare the hospital ready for the potential sitter requests, moreover, they can allocate necessary resources like beds and medical professionals with particular skills for the forthcoming patients. The proposed method can also be used to do prediction for the dataset from other disciplines and areas, as long as the dataset is sequential and the attributes used for seed and target attributes are categorical or can be transformed to categorical attributes.

Acknowledgement. The authors wish to thank the support of Athabasca University and the Academic Research Funding.

References

1. Fayyad, U., Piatetsky-Shapiro, G., Smyth, P.: The KDD Process for Extracting Useful Knowledge from Volumes of Data. Journal of ACM Communications 39(11), 27–34 (1996)

2. Lin, C.-H., Hsiao, H.-S.: Hierarchical State Machine Architecture for Regular Expression Pattern Matching. In: 19th ACM Great Lakes Symposium on VLSI, Boston, MA, USA, pp. 133–136 (2009)
3. Jurafsky, D., Martin, J.-H.: Speech and Language Processing: An Introduction to Natural Language Processing. In: Computational Linguistics and Speech Recognition. Prentice Hall, Upper Saddle River (2000)
4. Zhao, Y., Karypis, G.: Evaluation of hierarchical clustering algorithms for document datasets. In: 11th International Conference on Information and Knowledge Management, McLean, VA, USA, pp. 515–524 (2002)
5. Augustyniak, P.: Optimal Coding of Vectorcardiographic Sequences Using Spatial Prediction. Journal of IEEE Transactions of Information Technology in Biomedicine 11(3), 305–311 (2007)
6. Bratsas, C., Hatzizisis, I., Bamidis, P., Quaresma, P., Maglaveras, N.: Similarity Estimation among OWL Descriptions of Computational Cardiology Problems in a Knowledge Base. Journal of IEEE Computers in Cardiology 32(5), 243–246 (2005)
7. Chen, C.-M., Hong, C.-M., Huang, C.-M., Lee, T.-H.: Web-based Remote Human Pulse Monitoring System with Intelligent Data Analysis for Home Healthcare. Cybernetics and Intelligent Systems, 636–641 (2008)
8. Subhashini, R., Kumar, V.J.S.: Evaluating the Performance of Similarity Measures Used in Document Clustering and Information Retrieval. In: 1st International Conference on Integrated Intelligent Computing, Bangalore, India, pp. 27–31 (2010)
9. Grishman, R.: Information Extraction: Techniques and Challenges. International Summer School on Information Extraction: A Multidisciplinary Approach to an Emerging Information Technology, Rome, Italy, pp. 10–27 (1997)
10. Mutalik, P.-G., Deshpande, A., Nadkarni, P.-M.: Use of general-purpose negation detection to augment concept indexing of medical documents. Journal of the American Medical Informatics Association 8(6), 598–609 (2001)
11. Chapman, W.-W., Bridewell, W., Hanbury, P., Cooper, G.-F., Buchanan, B.-G.: A simple algorithm for identifying negated findings and diseases in discharge summaries. Journal of Biomedical Informatics 34(5), 301–310 (2001)

An Adaptive Recommendation System for Museum Navigation

Jason C. Hung[1,*], Chun-Hong Huang[2], and Victoria Hsu[3]

[1] Department of Information Management
Overseas Chinese University, Taichung, Taiwan
jhungc.hung@gmail.com
[2] Department of Computer Information and Network Engineering
Lunghwa University of Science and Technology, Taoyuan County, Taiwan
ch.huang@mail.lhu.edu.tw
[3] Department of Computer Science and Information Engineering
Tamkang University, Taipei, Taiwan
saintvoice.1981@gmail.com

Abstract. People like to attend exhibition activities, but hard to enter into the information effectively. We build new system with wireless internet and mobile device to guide visitor into the core information initiatively and effectively. The mobile guide system could classify visitor base on exhibition information and personal information that provide more suitable for users. Our system combined with semantic web technology to connect items data which users' markup the type or property information in our system to created human portfolio. Our system is in compliance with human portfolio and metadata method to provide user information automatically and appropriately.

Keywords: Mobile guide, Mobile device, Wireless internet, Semantic web, Human portfolio.

1 Introduction

Many people visit the exhibitions or museums for their leisure time. Most of the museums and the exhibitions will provide the corresponding information to people for their visiting. Now, the technologies of mobile devices and wireless network could provide visitors their own style visiting via mobile devices which devices may be provided from the organizers or their own.

This research aims to propose a scheme of the guide system which first proposes a data storage format such that the exhibition organizers can store all the exhibition data simply. Second, the visitors can simply describe some personal information, which will be evaluated by the best appropriate recommendation method (BAR) we proposed in this paper, and this guide system will provide the contents or information that is fit for the visitor by wireless network technique. And this information will be shown on the mobile device to the visitor.

---

* Corresponding author.

R. Huang et al. (Eds.): AMT 2012, LNCS 7669, pp. 534–542, 2012.

Mobile devices are small computational equipment [1]. Users can get information from internet or telecommunication networks and execute some program by these devices, e.g. cell phone, PDA, notebook, iPad and so on. Metadata of the information is very important to mobile devices. Metadata is first defined in the conference Metadata workshop [2] and applied to data storage, data retrieving and so on. Dublin Core is a simple [3], efficient and popular metadata standard. It can fast organize the network resources, improve the precise of data search and retrieving, provide a metadata format to describe the network resources by many experts from different areas, and the network resources will be divided to 15 categories.

Categories for the description of works of art, CDWA, are a popular metadata definition to art exhibitions and museums categories [4]. It is proposed by Art Information Task Force, AITF, of J. Paul Getty Trust. CDWA provides a scheme to describe the content of works of art such that we can establish a database of the works of art by these describes. There are 27 main categories and 233 subcategories in CDWA.

After establishing the metadata, the ontology and the semantic web will be the critical techniques to develop our BAR method. Ontology was used to some specified and existed type or the well-described statements in philosophy [5][6]. In computer science, ontology represents knowledge as a set of concepts within a domain, and the relationships between those concepts. Common components of ontologies include individuals, classes, attributes, relations, function terms, restrictions, rules, axioms and events. Semantic web is the concept proposed by Tim Berners-Lee in W3C [7]. The main idea of semantic web is to let computers can "understand" the text files on the internet, that is, to know the semantics of the text files. By using the techniques of semantic web, the search engine can use a unique and precisely vocabulary and mark to the text files they searched without confusing.

The rest of this paper is organized as follows. In Section 2 we describe the details of the appropriate recommendation system and the BAR method we developed. In Section 3 we show the interfaces of the developed system. The remaking conclusion will be provided in Section 4.

2   Appropriate Recommendation System

2.1   System Procedure

For an exhibition or a museum, we first establish the database of the works of art by Dublin Core and CDWA. A visitor has to describe some of his/her personal data to the recommendation system before using this recommendation system. The recommendation system evaluates these personal data by the BAR method and finds some works of art will be recommended to the visitor. Then, the visitor will get the information about the recommended works of art via the mobile device he/she takes.

With this recommendation system in the mobile device, the visitor can mark the works of art he/she likes during the visiting. The recommendation system will record the works of art to the database of the visitor's profile. The more marks the visitor made, the more precisely our recommendation system will be, by the BAR evaluation. Figure 1 is the procedure of the recommendation system.

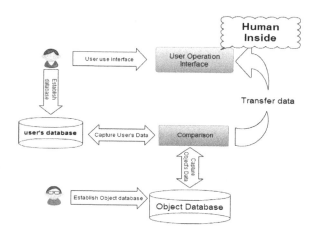

Fig. 1. Procedure of the Recommendation System

2.2   Database Establishment

The database of the recommendation system includes the following tables: art_detail, art_relation, type, human_relation, human.

By CDWA, the table art_detail stores the information about the works of art including: title, author, date, format, material, and description. The table human stores the visitors' personal information including: name, sex, birthday, telephone number, e-mail address, address, and education degree. The table type stores the information of all kinds of types in this system. The table art_relation stores the information about the types of the works of art in the exhibition. The table human_relation stores the relationship of the visitor and the work of art. The visitor likes a work of art and mark it in the system that will be store in this table. Figure 2 represents the database structure of the recommendation system.

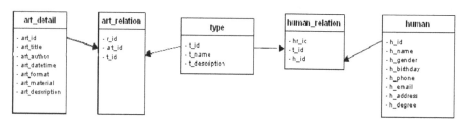

Fig. 2. Database Structure of the Recommendation System

2.3   Best Appropriate Recommendation Method

The BAR method includes two phases: 1) when the visitor establishes his/her personal data initially, BAR evaluates to determine the initial weights for recommendation; 2) when the visitor mark a work of art, BAR evaluates the attributes of the work of art and the personal data to update the weights to make   more precisely commendation.

In phase 1, we first evaluate the initial weights by the following variables of personal data:

```
UA   user's age
UD   user's education degree
UI   user's interest
WUA  weight of user's age
WUD  weight of user's education degree
WUI  weight of user's interest
```

$P = \{H_i \mid 1 \le i \le n\}$ a set of human_portfilio database where Hi means the ith human data from human_portfilio database.

The user's age is divided into 3 intervals: less than 19, from 20 to 40, greater than 40, and we assign weight 1, 2, 3, to each interval respectively. The user's education degree is classified by primary school, junior high school, senior high school, collage, educated school, and we assign weight from 1 to 5 for each degree respectively. The user's interest options include art, music, sport etc. and are divided into art related and not art related, and assign weight 2 and 1 respectively.

```
Input  user's personal data values in database
Output user's weight values WP
Set  new user, WP = 0, WUA = 0, WUD = 0,
WUI = 0
// evaluate weight of user's age
If  0<UA•19
  WUA = WUA + 1
else If  20•UA•39
  WUA = WUA + 2
else
  WUA = WUA + 3
// evaluate weight of user's education degree
If  UD = "primary school"
  WUD = WUD + 1
else If  UD = "junior high school"
  WUD = WUD + 2
else If  UD = "senior high school"
WUD = WUD + 3
else If  UD = "collage"
  WUD = WUD + 4
else
WUD = WUD + 5
// evaluate weight of user's interest
If  UI = art related options
WUI = WUI + 2
else
WUI = WUI + 1
```

```
// add all weights
WP = WUA + WUD +WUI
Store  WP to user's profile
```

The initial weight is evaluated in phase 1 and the recommendation system recommends the works of art to the visitor depending on it before the visitor marks some works of art he/she likes. Then, when the visitor starts to visit the exhibition and marks some works of art he/she likes, the recommendation system receives the marks and BAR begins evaluating to update the weight of the visitor. That is the phase 2 of BAR method. Some variable definitions used in phase 2 BAR is shown as follows.

$A = \{A_i \mid 1 \le i \le n\}$ set of art database where Ai means the ith author data from author database.
$S = \{S_i \mid 1 \le i \le n\}$ a set of art database where Si means the ith style data from style database.
$MAI = \{VA_i \mid VA_i \in A, 1 \le i \le k\}$ author's identification of the work of art which user marked
$MSI = \{VS_i \mid VS_i \in S, 1 \le i \le k\}$ style identification of the work of art which user marked
$RAI = \{RA_i \mid 1 \le i \le n\}$ Record the frequency of author's identification of the work of art which user marked
$RSI = \{RS_i \mid 1 \le i \le n\}$ Record the frequency of style identification of the work of art which user marked

In phase 2 of BAR method, when the visitor marks a work of art, the recommendation system retrieves the attributes author and style from database. After the visitor finishing this visiting, the recommendation system evaluates that which author and which style the visitor marked most, then store this author and this style information in the visitor database. Therefore, when the visitor visits another exhibition next, the recommendation system can make good recommendation by these data.

2.4   Linking Semantic Web

The mobile device which is taken by the visitor during the visiting can receive the marks he/she made and provides the information and recommends of the works of art to the visitor. Moreover, by using the techniques of semantic web and wireless network, the visitor can get more information by the mobile device by connecting to other websites.

```
Input  user marked author MAI,
user marked style MSI
Output the highest author and style value
For i = 1 to i = k  do
Set  RAi = 0, RSi = 0
```

```
// count the frequency of authors and styles that the
visitor marked
For i = 1 to i = k  do
For j = 1 to j = n  do
IF  VAi = Aj  do  RAi += 1
IF  VSi = Sj  do  RSi += 1

// find which author and which style that the visitor
likes most
```

$$RAI_{\max} = \arg\max\left\{RA_i \mid 1 \le i \le n\right\}$$

$$RSI_{\max} = \arg\max\left\{RS_i \mid 1 \le i \le n\right\}$$

```
Store  RAImax, RSImax
```

3   Interfaces of the Appropriate Recommendation System

3.1   System Interface

The visitor has to login the appropriate recommendation system first to use this system such that the system will generate the visitor's database and store the visitor's information. The login interface of the recommendation system is shown in Figure 3. At the first time the visitor uses the appropriate recommend system, the visitor has to enter some personal information to complete the registration procedure. The registration interface is shown as Figure 4.

Fig. 3. Login Interface

When the visitor completes the registration procedure, the appropriate recommendation system will evaluate the visitor's personal information by BAR method to generate the visitor's weight. Then the visitors will be classified into 3 groups: high-end visitor, mid-end visitor and low-end visitor. A visitor is classified to be a high-end visitor if his/her personal weight is between 7 and 10. A visitor is classified to be a mid-end visitor if his/her personal weight is between 4 and 6. A visitor is classified to be a low-end visitor if his/her personal weight is between 1 and 3.

The appropriate recommendation system shows different information of a work of art to the visitor depending on which group that the visitor belongs. For a high-end visitor, the recommendation system shows all the information of the work of art on the visitor's mobile device, including image, title, author, date, format, description

新使用者請加入會員

姓名：

密碼：

確認密碼：

性別：　　男　　女

生日：

電話：

E-mail：

住址：

學歷：　請選擇學歷　▼

職業：

興趣：

☐ 藝術	☐ 文學	☐ 休閒旅遊	☐ 娛樂
☐ 生活	☐ 時尚	☐ 運動	☐ 教育
☐ 科技	☐ 科學	☐ 電腦	☐ 醫享
☐ 商業	☐ 金融	☐ 政治	☐ 社會科學

[送出] [重新填寫]

Fig. 4. Registration Interface

and preserving location. For a mid-end visitor, the recommendation system shows parts of the information of the work of art on the visitor's mobile device, including image, title, author, date, format, description. For a low-end visitor, the recommendation system shows the basic information of the work of art on the visitor's mobile device, including image, title, author, date.

The interface also provides bottoms to the visitor to mark that he/she likes the work of art or not. As described before, this information will be retrieved by the system and evaluated by phase 2 of BAR method to generate and update the personal information. Therefore the appropriate recommendation system will provide the suitable information to the visitors.

3.2   Comparison to other Navigation Systems

Recently there are many ways to mobile navigation including navigation by staff, navigation by paper notes, navigation by voice and navigation by multimedia. The mobile devices used in all kinds of navigation uses RFID technique mostly. These navigation systems using RFID are belongs to passive navigation systems. That is, the visitor gets the exhibition information by sensing the mobile devices. The appropriate recommendation system we designed is an active system. The visitor can get the information and mark his/her opinion about the work of the art by this system. The system also recommends the works of art to visitor by using the personal information and opinion of the activity.

Different from most recent navigation systems using RFID devices, the appropriate recommendation system we designed can be used in most popular mobile devices, e.g. smart phone, PDA, notebook, iPad and so on. Visitors can install this recommendation system in their own mobile devices and use it to any exhibitions if the information of the exhibitions are in the database of the system.

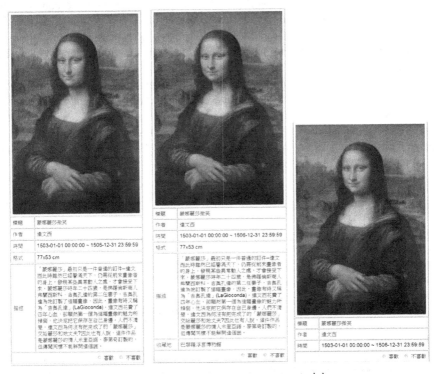

Fig. 5. Information Shown to different level visitor

The appropriate recommendation system provides information depending on the visitor groups. That is different from most recent navigation systems which provide all information to all visitors. The appropriate recommendation system provides suitable information to the visitors. Table 1 lists the comparison of the appropriate recommendation system and the other navigation system.

Table 1. Comparison of appropriate recommendation system and other navigation system

	Recent navigation system	*Appropriate recommendation system*
Mobile device	Specified RFID devices	Popular personal mobile devices
Content provide	passive	Active
Content	All information to all visitors	Information depending on visitor group

4   Concluding Remarks

Using navigation system by mobile devices is very popular for exhibitions and museums recently. The wireless network technique, the semantic network technique and personal mobile devices are also well-developed. Visitors can get more information by different mobile devices than before. The appropriate recommendation system we designed can provide suitable information to visitors fast and convenient by their own mobile devices.

The appropriate recommendation system can be improved by considering the visitors' own experiences in the BAR method evaluation such that the system can recommend works of art to visitors more precisely. The appropriate recommendation system can be extended to be a community system. The visitors can share and exchange their experiences and make more commands to the exhibitions or museums on the system. These are all the future researches.

References

1. Roschell, J.: Unlocking the learning value of wireless mobile devices. Journal of Computer Assisted Learning 19, 260–272 (2003)
2. http://zh.wikipedia.org/wiki/Metadata
3. Hillman, D.: Using Dublin Core (2005),
   http://dublincore.org/documents/usageguide/#whatis
4. Agbabian, M.S., Masri, S.F., Nigbor, R.L., Ginell, W.S.: Seismic damage mitigation concepts for art objects in museums. In: Proceeding of Ninth World Conference on Earthquake Engineering (1988)
5. Gruber, T.R.: A Translation Approach to Portable Ontology Specifications. Knowledge Systems Laboratory, 199–220 (1993)
6. Arvidsson, F., Flycht-Eriksson, A.: Ontologies I (2008)
7. Berners-Lee, T., Connolly, D., Kagal, L., Scharf, Y., Hendler, J.: N3logic: A logical framework for the World Wide Web. Theory and Practice of Logic Programming 8, 249–269 (2008)

Adaptive SVM-Based Classification Systems Based on the Improved Endocrine-Based PSO Algorithm

Kuan-Cheng Lin[1], Sheng-Hwa Hsu[1], and Jason C. Hung[2]

[1] Department of Management Information Systems
National Chung Hsing University, Taichung, Taiwan
{kclin,g099029005}@nchu.edu.tw
[2] Department of Information Management
Overseas Chinese University, Taichung, Taiwan
jhung@ocu.edu.tw

Abstract. In this study, we proposed an wrapped feature selection and SVM's kernel parameters optimization scheme using Improved Artificial Endocrine System to get an optimal support vector machines classification system. By taking the advantage of the mechanisms of hormone action in Artificial Endocrine System, we can avoid to obtain local optimums and oscillations. We used the UCI database to evaluate the performance of the proposed scheme with the previous methods. The experiment results indicated that the proposed scheme can avoid local optimum and also reduce feature numbers significantly with a good-enough accuracy in high-complexity datasets. Moreover, by decreasing the number of unnecessary features, we can even improve the accuracy of classification.

Keywords: feature selection, classification, Artificial Endocrine System, support vector machine.

1 Introduction

Because of the amount of data in database is rapidly increasing, traditional data processing methods are not suitable for data mining. To find out valuable information from large amount of data accurately and efficiently, researchers developed many data mining methods based on machine learning. Feature selection optimization has been an important issue in data mining, especially for classification problems, because it could reduce unnecessary feature numbers to increase the efficiency of classifiers and even improve the classification accuracy. Feature selection optimization has been applied to many different areas, such as pattern recognition[1], bioinformatics[2] and text categorization[3].

With the evolution of computer science, solutions of classification problems have evolved from traditional manual processing to machine learning algorithms, such as Artificial Neural Network(ANN)[4], k-Nearest Neighbor algorithm(k-NN)[5] and Naive Bayes algorithm[6]. Artificial Neural Network imitates the communication mechanism of biological neural networks. ANN use neurons to compose a communication network and change the weights of the interconnections by training process to

R. Huang et al. (Eds.): AMT 2012, LNCS 7669, pp. 543–552, 2012.
© Springer-Verlag Berlin Heidelberg 2012

solve classification problems. The k-NN algorithm method is based on the closest training examples in the feature space. The k-NN algorithm classifies objects determine to the closest training examples in the feature space. The object is simply assigned to the class of its nearest neighbor in k-NN classifiers. The naïve Bayes classifier is mainly based on Bayes' theorem, using probabilistic method to perform classification tasks. Traditional classification methods in machine learning also have the same disadvantage: because their theories basically come from the law of large numbers, the number of samples can affect the results, that is, sample number must reach a certain amount to learn effectively, too many sample will lead to slow learning or an over-fitting situation. For the above reasons, Vapnik proposed a machine learning method, Support Vector Machine(SVM)[7,9-12], based on Vapnik-Chervonenkis(VC) theory[7] and Structure Risk Minimization(SRM)[8].

SVM is a supervised learning method, it corrects the weight values continually during the process to make the result meet our expectations. SRM can assure that empirical risk minimized and lower the VC dimension during learning process to control the expected risk. SVM results are related to several factors, such as: data integrity, feature selection and parameters of the training classifier. In the above factors, last two can influence each other. Using feature selection optimization to reduce irrelevant, redundant and noisy type of data can lower the dimension of the data and improve computing efficiency, even generate a better result. Feature selection optimization with classifiers can be divided into two types: filter approach and wrapper approach. Wrapper approach is more often used than the filter approach, because of its higher accuracy. There are many kinds of wrapper algorithms, such as: Genetic Algorithm(GA)[13,17,18], Particle Swarm Optimization(PSO)[14,19] and Cat Swarm Optimization(CSO)[15]. The GA algorithm mimicked four main procedures of natural chromosomes evolution to solve optimization problems, they were inheritance, mutation, selection, and crossover. In the PSO algorithm, every particle moved based on the balance of individual optimal value and social optimal value to approach the best value in solution space. In the CSO algorithm, the candidate solutions were transformed into two groups of cats, update their locations and velocity gradually to approximate the optimal solution. In all the methods mentioned above, PSO is the most widely used method because of its faster convergence speed, less parameter settings and greater environment adaptive capability. Without proper parameter settings of PSO, the algorithm will lead to local convergence. Besides feature selection, parameter optimization could also improve classification results [16].

Within the advanced research of machine learning, Nature-inspired Computation (NC) has drawn people's attention[20]. The definition of Nature-inspired Computation is to mimic the characteristics of the physiological systems and animal behaves in nature, and transform their mechanisms into algorithm models. The main characteristics of NC are strong self-adaptability and learning ability. Compared to traditional statistical algorithms, NC is more suitable to solve complex problems, such as optimization problems, computer networks and wireless communication. In recent years, with the development of endocrinology, Artificial Endocrine System (AES) has become a popular research issue[21]. Using   decentralized control mechanism characteristic of AES, we can improve self-adaptability and stability in feature selection process. Experiment results showed that PSO combined with AES was better than traditional PSO [21]. Therefore, we proposed a novel feature selection scheme using AES with SVM to solve classification problems. We adjusted the updating mechanism of the PSO algorithm by adding a mechanism of hormone regulation. Because of the mechanism, particles were more sensitive to the environment while moving, and avoided to local convergence. Through the proposed scheme, the classification process will become more efficient and accurate. The experiments will used UCI datasets to verify the performance of the proposed scheme.

2   Related Works

2.1   Particle Swarm Optimization

PSO is an algorithm that mimics simple society model with swarm intelligence concept, it was proposed by Kennedy and Eberhart in 1995[26]. Through observations of animal social behavior, they found that the way animals exchange information in the group provided their advantages on survival and evolution. In the early stage of the PSO, they simulate the situation of bird swarm finding their food. At first, each bird began to fly to their own best direction by their instinct or experience. If one of them found a better spot, it will used their unique message transmission mechanism to gather the group, eventually led the group to fly to better place for foraging.

PSO use a group of possible solutions moving in the solution space to find out the best solutions. In the optimization process, every particle is responsible for searching the best solution in the part of the region as a moving target; meanwhile, the best solution would compare to the best solutions found by other particles. This method can help us find the best solution in whole area and lead the group heading to the direction of the best solution. The main procedure of PSO is stated as follows:

1. Initialization: Randomize the initial position and speed of each particle.
2. Evaluation: Use fitness function to evaluate the fitness value of each particle.
3. Finding the Pbest: Compare current fitness value of each particle to individual historical best fitness. If the current fitness is better than historical best fitness, set the current particle to be Pbest.
4. Finding the Gbest: Compare the individual best fitness value to global best fitness value. If the individual best fitness is better than global best fitness, set the individual best particle to be the Gbest.
5. Updating the statuses of the particles according to equation (1).

$$X_i(k+1) =$$
$$X_i(k) + \omega V_i(k) + c_1 rand(\cdot) \cdot \left(X_{pbest_i}(k) - X_i(k)\right) + c_2 rand(\cdot)\left(X_{gbest} - X_i(k)\right)$$
$$\tag{1}$$

$X_i(k)$ and $V_i(k)$ are the position and speed of the i particle in k iteration. c1 and c2 are learning parameters, ω is a weighting parameter.

2.2   Endocrine PSO Algorithm

EPSO generally refers to a system inspired by biological endocrine system or the principle of endocrine regulation mechanism. Biological endocrine system is composed by the endocrine gland and endocrine cells. Through the mutual cooperation of the endocrine system, nervous system and immune system, makes the physiological function of organisms can adapt to changes in vivo and in vitro environment. T.Ogata[24] proposed a endocrine system model for the affective identification function for robots in 1996, he used hormones to maintain the operation of the robot and had better adaptability. Neal and Timmis proposed an AES based internal control mechanism[25], through the stimulation of the internal environment or the external environment, AES secretes enough hormones for the whole system, eventually the system would maintain in both best balance and best adaptability.

Traditional PSO has two main disadvantages: local convergence and slow convergence speed, so Endocrine-based Particle Swarm Optimization (EPSO) was proposed in [27]. EPSO used the hormone regulation mechanism of endocrine system to correct the way PSO update, giving small amount of hormone to better particles to maintain stability; in contrast, giving large amount of hormone to worse particles to let them grow faster. In order to avoid update rate is too large, hormone updates are limited to a pre-set range. And taking into account the effect between hormones, ring type of particle swarm is used.

Equation(2) and (3) are hormone's update formulas. $EM(S)$ represents the status of hormone, f_{max} is the best fitness of the group, f_i represents the fitness of the i particle, f_{i-1} and f_{i+1} are the fitness of the i-1 particle and the i+1 particle, f_{avg} is the average fitness of the group. To limit the update range we used the atan function, $fun_1(x) = atan(x)$, $fun_2(x) = atan(-x)$. $EM_i(k+1)$ represents the amount of hormone of i particle in generation k+1, $E_i(k)$ is the amount of hormone of the i particle in generation k, $rand(\cdot)$ is a random number between 0 to 1, c_3 is a coefficient of update range, c_4 is the hormone-scale variable, c_3 and c_4 are set to 0.05 in our research.

$$EM(S) = fun_1\left(\frac{f_{max}-f_i}{f_{max}-f_{avg}}\right) \cdot \left[\frac{\pi}{2} + fun_2(f_i - \frac{f_{i-1}+f_{i+1}}{2})\right] \quad (2)$$

$$EM_i(k+1) = c_4 E_i(k) + c_3 rand(\cdot)EM(S). \quad (3)$$

Equation(4) is the adjusted PSO formula with hormone regulation mechanism, $X_i(k+1)$ and $X_i(k)$ are the position of the i particle in generation k+1 and k, ω is a weighting parameter, $V_i(k)$ is the velocity of the i particle in generation k, $rand(\cdot)$ is a random number between 0 to 1, $X_{pbest_i}(k)$ is the best position of the i particle in generation k, X_{gbest} is the global best position, $EM_i(k+1)$ represents the amount of hormone of i particle in generation k+1. α is set to prevent local convergence, if the performance do not improve, $\alpha=0.1$; else $\alpha=0$.

$$X_i(k+1) =$$
$$X_i(k) + \omega V_i(k) + c_1 rand(\cdot) \cdot \left(X_{pbest_i}(k) - X_i(k)\right) + c_2 rand(\cdot)\left(X_{gbest} - X_i(k)\right) + EM_i(k+1) + \alpha \cdot RAND(\cdot) \quad (4)$$

The main steps of EPSO are described as follows :

1. Randomly generate the initial position, velocity and hormone amount of each particle.
2. Calculate the fitness of every particle and compare it to the best fitness in their historical position, set the particle to pbest if the fitness is the best value than ever. If it is the best value in whole area, we also set the particle to gbest.
3. Update the position, velocity and hormone amount by the method which combined with hormone regulation mechanism.
4. Determine if the algorithm meet the termination criteria; if not, go back to step 2.
5. Output the best solution and terminate the algorithm.

3   Feature Selection and SVM Parameter Optimization by Using EPSO

To improve the update scheme of PSO, the hormone regulation mechanism of AES can avoid the PSO algorithm falling into local optimum. We can improve the performance of a SVM-based classifier with a optimized feature subset. Finally, we can get a best classifier model and it can be used to classify unknown datasets. Our proposed scheme use EPSO combined with SVM to increase our classification accuracy. The steps of using EPSO on feature selection are as follows:

1. Initialization: Randomly generate the parameters of each particle: position, speed and hormone amount. Each particle represents a feature subset. The number of dimension in solution space equals to original feature number of dataset. The moving range of each particle are limited in 0~1, the position of particles in each dimension determine whether the feature is selected, if the position value is greater than 0.5, the feature will be selected, if not, the feature will not be selected.
2. Evaluation: Convert the particles into feature subsets, use classifier to evaluate the performance of each set by classification accuracy. Then update the individual best particle and global best particle.
3. Termination criterion determination: If the termination criterion is satisfied, proceed to Step 5., if it's not, proceed to Step 4..
4. Updating the status of each particle: According to the EPSO updating equations, update the position, speed and hormone amount of each particle. Back to Step 2.
5. Algorithm termination: Stop the algorithm cycle and output the optimized feature subset.

If we can minimize the feature number with a good-enough level of accuracy, the process of machine learning will be more efficient. Therefore, to make sure the algorithm can be applied to practical use, we added a feature number comparison step in global best updating process. When the individual best accuracy equals to global best accuracy, we compare their feature numbers. If the feature number of the individual best is smaller than global best, we set the individual best solution to global best solution. This step can enhance the effect of EPSO on feature number minimizing.

Figure 1 illustrates the mechanism of the proposed scheme. First, we divide the original dataset into two subsets: training dataset and testing dataset. Then we input the training dataset, feature subset and kernel parameters selected by EPSO feature selection operation into SVM to build a classifier model. Testing dataset is then used to test the performance of trained-SVM classifier model. If the termination criterion is satisfied, output the result and stop the algorithm; if it's not, back to the EPSO feature selection operation.

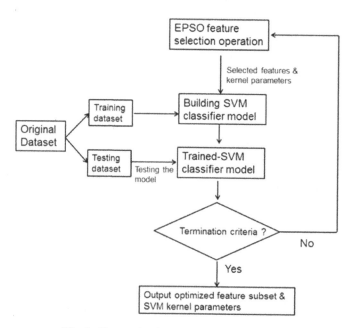

Fig. 1. The mechanism of the proposed scheme

4   Experiments

The experiments will compare the performance of PSO+SVM and EPSO+SVM classification systems. The datasets of UCI database[23] are chose to verify the proposed scheme. Table 1 displays the dataset used in the experiments. The operating system of experiments is Windows 7, and the hardware is including AMD Phenom II X4 CPU, 3.25 GB of RAM. The program is coded with Dev C++ and LIBSVM[22]. To compare the results of EPSO and PSO, we set all the parameters of each scheme to the same except the update process. C1, C2 = 2, particle numbers = 40, C3 and C4 of EPSO were set to 0.05. The 5-fold validation was used to measure the performance.

Table 1. Experiment Datasets from the UCI repository

No.	Dataset	No. of classes	No. of features	No. of instances
1	German	2	24	1000
2	Australian	2	14	690
3	Pima-Indian diabetes	2	8	768
4	Heart disease	2	13	270
5	Breast cancer	2	10	699
6	Ionosphere	2	34	351
7	Iris	3	4	105
8	Sonar	2	60	208
9	Vehicle	4	18	846
10	Vowel	11	10	528
11	Satellite	7	36	6435

4.1 Experiment 1: Feature Reduction

To verify the proposed scheme can reduce the number of feature, we compare the average of 5-fold cross validation. If the accuracy didn't improve in one hour, the algorithm would stop and output the result. The result is shown in Table 2. According to the result, the proposed scheme could reduce the number of selected features and maintain the accuracy in a good-enough level.

Table 2.

Dataset	PSO+SVM		IEPSO+SVM	
	Average Accuracy rate (%)	Average No. of selected features	Average Accuracy rate (%)	Average No. of selected features
German	0.8±0.016	21±2	0.801±0.022	17.4±2.5
Australian	0.8884±0.031	11±2.82	0.9014±0.033	8.4±2.07
Heart disease	0.8962±0.028	11±1.87	0.9296±0.02	9.2±1.09
Breast cancer	0.9842±0.012	6.4±1.94	0.9857±0.011	5.8±0.44
Ionosphere	0.9771±0.021	24.8±5.06	0.9885±0.015	19.8±2.48
Sonar	0.856±0.046	43.8±7.88	0.9186±0.059	34±4.84
Vehicle	0.8888±0.024	16.2±1.3	0.8924±0.023	13.6±1.51
Vowel	0.702±0.066	9.2±2.16	0.7494±0.051	9.4±1.67
Satellite	0.9827±0.015	21.4±8.47	0.9885±0.018	21.2±3.42

4.2 Experiment 2: Consideration of Time Cost

We use in experiment 2, Since time cost is important in practical consideration, the computing time is set to the termination criteria. If the time of computing reached 2 hours, the algorithm would stop and output the result. The result is shown in Table 3.

Through the results, it is obvious that the EPSO+SVM is better than traditional PSO+SVM scheme with higher accuracy and less feature number in most of the dataset. And the results show that EPSO+SVM can avoid getting into a local optimum situation.

Table 3. Consideration of Time Cost

Dataset	PSO+SVM		IEPSO+SVM	
	Average Accuracy rate (%)	Average No. of selected features	Average Accuracy rate (%)	Average No. of selected features
German	0.798±0.015	21.4±1.51	0.807±0.014	12.8±2.16
Australian	0.8971±0.03	9.8±4.02	0.8811±0.036	7.6±1.94
Heart disease	0.9±0.03	10.4±1.67	0.9185±0.016	8±1.87
Breast cancer	0.9842±0.012	7.2±2.38	0.98±0.013	4±1.58
Ionosphere	0.9743±0.018	23.4±6.14	0.9857±0.014	14.4±2.5
Sonar	0.8558±0.053	44.6±9.39	0.9283±0.067	35.6±3.78
Vehicle	0.8924±0.026	16±1.87	0.8983±0.021	12±1.73
Vowel	0.7212±0.029	10.4+2.3	0.7737±0.037	6.6±3.2
Satellite	0.9827±0.015	22.8±5.93	0.9827±0.015	14.4±2.07

5   Conclusion

In this study, we used EPSO+SVM to perform the classification work, by using this scheme, we can get a higher accuracy than traditional PSO+SVM scheme, and we also reduced the number of selected features. By adding the hormone regulation mechanism, EPSO can avoid local shocks, and find out the most optimal solution to build the classification model. In the future, if we can apply this scheme in other machine learning area, we can achieve both better efficiency and better effectiveness.

References

1. Belacel, N., Boulassel, M.R.: Multicriteria Fuzzy Classification Procedure Profit: Methodology and Medical Application. Fuzzy Sets and Systems 141, 203–217 (2004)
2. Stevens, R., Goble, C., Baker, P., Brass, A.: A Classification of Tasks in Bioinformatics. Bioinformatics 17(2), 180–188 (2001)
3. Mitra, V., Wang, C.J., Banerjee, S.: Text Classification: A Least Square Support Vector Machine Approach. Applied Soft Computing 7(3), 908–914 (2007)
4. Zhang, G.P.: Neural Networks for Classification: A Survey. IEEE Transactions on Systems, Man, and Cybernetics—Part C: Applications and Reviews 30(4), 451–462 (2000)
5. Hwang, W.-J., Wen, K.-W.: Fast kNN classification algorithm based on partial distance search. Electronics Letters 34(21), 2062–2063 (1998)

6. Lewis, D.D.: Naive (Bayes) at Forty: The Independence Assumption in Information Retrieval. In: Nédellec, C., Rouveirol, C. (eds.) ECML 1998. LNCS, vol. 1398, pp. 4–15. Springer, Heidelberg (1998)

7. Vapnik, V.N.: The Nature of Statistical Learning Theory. Springer, NY

8. Vapnik, V., Cortes, C.: Support-Vector Networks. Machine Learning 20(3), 273–297 (1995)

9. Lee, Y.-J., Mangasarian, O.L.: RSVM: Reduced Support Vector Machines. Data Mining Institute, Computer Science Department, University of Wisconsin (2001)

10. Chapelle, O., Haffner, P., Vapnik, V.N.: Support Vector Machines for Histogram-Based Image Classification. IEEE Transactions on Neural Networks 10(5), 1055–1064 (1999)

11. Sebald, D.J., Bucklew, J.A.: Support Vector Machine Techniques for Nonlinear Equalization. IEEE Transactions on Signal Processing 48(11), 3217–3226 (2000)

12. Suykens, J.A.K., Vandewalle, J.: Least Squares Support Vector Machine Classifiers. Neural Processing Letters 9, 293–300 (1999)

13. Hussein, F., Kharma, N., Ward, R.: Genetic Algorithm for Feature Selection and Weighting, a Review and Study. Pattern Recognition Letters 10(5), 335–347 (1989)

14. Tu, C.-J., Chuang, L.-Y., Chang, J.-Y., Yang, C.-H.: Feature Selection using PSO-SVM. IAENG International Journal of Computer Science 33, 111–116 (2007)

15. Lin, K.-C., Chien, H.-Y.: CSO-based Feature Selection and Parameter Optimization for Support Vector Machine. In: JCPC Joint Conferences on Pervasive Computing, pp. 783–788 (2009)

16. Staelin, C.: Parameter Selection for Support Vector Machines. HP Labortory (2003)

17. Liu, S., Jia, C.Y., Ma, H.: A New Weighted Support Vector Machine with GA-based Parameter Selection. In: Proceedings of the 4th International Conference on Machine Learning and Cybernetics, vol. 7, pp. 4351–4355 (2005)

18. Huang, C.L., Wang, C.J.: A GA-based Feature Selection and Parameters Optimization for Support Vector Machines. Expert Systems with Application 31(2), 231–240 (2006)

19. Lin, S.W., Ying, K.C., Chen, S.C., Lee, Z.J.: Particle Swarm Optimization for Parameter Determination and Feature Selection of Support Vector Machines. Expert Systems with Application 35(4), 1817–1824 (2008)

20. Wang, L., Kang, Q., Wu, Q.-D.: Nature-inspired Computation - Effective Realization of Artificial Intelligence. Systems Engineering-Theory and Practice 5(27) (2007)

21. Chen, D.-B., Zhao, C.-X.: Particle Swarm Optimization based on Endocrine Regulation Mechanism. Control Theory and Applications 24(6), 126–134 (2007)

22. Hsu, C.W., Chang, C.C., Lin, C.J.: A Practical Guide to Support Vector Classification (2003), http://www.csie.ntu.edu.tw/~cjlin/papers/guide/guide.pdf

23. Hettich, S., Blake, C.L., Merz, C.J.: UCI Repository of Machine Learning Databases. Department of Information and Computer Science, University of California, Irvine, CA (1998), http://www.ics.uci.edu/~mlearn/MLRepository.html

24. Sugano, S., Ogata, T.: Emergence of Mind in Robots for Human Interface-Research Methodology and Robot Model. In: The Proceedings of IEEE International Conference on Robotics and Automation, pp. 1191–1198 (1966)

25. Neal, M., Timmis, J.: Timidity: A Useful Emotional Mechanism for robot control. Informatica 27, 197–204 (2003)
26. Kennedy, J., Eberhart, R.: Particle Swarm Optimization. In: The Proceedings of IEEE International Conference on Neural Networks IV, pp. 1942–1948 (1955)
27. Chen, D.-B., Zou, F.: A Multi-Objective Endocrine PSO Algorithm. In: The Proceedings of 2009 1st International Conference on Information Science and Engineering, December 26-28 (2009)

A Probability Model for Recognition of Dynamic Gesture Based on a Finger-Worn Device

Yinghui Zhou[1], Zixue Cheng[2], Lei Jing[2], and Junbo Wang[2]

[1] Graduate School of Computer Science and Engineering
[2] School of Computer Science and Engineering
University of Aizu, Aizuwakamatsu, Fukushima, Japan
{d8131104,z-cheng,leijing,j-wang}@u-aizu.ac.jp

Abstract. Gesture recognition based on body-worn devices can be used for healthy improvement and life support. Many methods have been proposed for gesture recognition. However, most of them are concerned about recognition accuracy only. In some practical applications, real-time performance of a recognition method on a wearable device is also a key problem. In the paper, we propose a probability model for accurate and real-time recognition of dynamic gestures. The model learns from HMM but difference in the sense that our model builds probability matrices of feature distribution for each observation points instead of each gesture, which reduces the number of probability matrix to improve processing efficiency. A gesture can be recognized by the way of look-up table to search maximum similarity to pre-stored gestures in the matrices. To verify the model, eight kinds of one-stroke finger gestures are taken as the target of recognition. Result shows reasonable recognition accuracy and computational complexity.

Keywords: Wearable Computing, Gesture Recognition, One-stroke Gestures, Probability Model.

1   Introduction

Recognition of dynamic gesture based on wearable devices has become a hot topic in many application fields such as health care, daily life support, industrial application, entertainment, etc. [1]. The goal of the applications is to sense and recognize human gestures with a natural way for providing corresponding feedbacks. For example, behavior monitoring in medicine is executed through daily gesture recognition, which could give a warning when an abnormal activity happened [2]. In these applications, gesture recognition is a core problem that is a key pre-condition of providing accurate services.

There are various methods to handle gesture recognition in the research field of machine learning. The common way is to pre-define gesture classes, and then an unknown gesture is recognized by determining it belonging to which gesture classes through signal analysis. Therefore, the recognition problem can be regarded as a classification problem. Preece et al. [3] divided the classification methods based on

R. Huang et al. (Eds.): AMT 2012, LNCS 7669, pp. 553–560, 2012.

body-worn sensors into three categories: threshold-based, supervised learning, and unsupervised learning algorithms. Threshold-based classification is to compare a signal feature with a predetermined threshold to detect whether a particular activity happened. This approach can be used to distinguish static posture and dynamic gestures, which has been successfully applied to fall detection [4]. Supervised approaches need numerous labeled samples to train the classifier for pre-defined gesture classes. The trained classifier can be used to label an unknown activity by feature analysis. Unsupervised learning refers to find hidden structure from unlabeled sensor data, which is used to identify clusters of related patterns in the feature space. Therefore, unsupervised approaches commonly combined with a supervised scheme to get better classification result [5].

Gesture recognition in most of researches adopted supervised approaches such as Decision Tree, K-Nearest Neighbor, Naïve Bayes, Hidden Markov Models (HMM), etc.. Among the approaches, HMM is a state model-based approach. One of the advantages of HMM is that it can easily be extended to deal with the task of temporal classification. Therefore, HMM has been successfully applied to speech recognition, and is becoming a hot topic for gesture recognition [6][7]. A HMM can be described by three probability measures:

$\lambda = (\pi, A, B)$;

π: the initial state distribution;
A : the state transition matrix;
B : the observation symbol matrix;

Given the observation sequence of an unknown gesture and a HMM model, we can label the gesture by searching the most likely sequence of state transitions.

HMM is an effective method for gesture recognition since it can handle temporal variations of gestures. However, HMM needs make a lot of assumptions about the data and also needs to set a number of parameters. Moreover, we need to build a HMM model for each gesture so that the number of matrix would increase with gesture class, which make the classification processing become complexity. Therefore it is hard to be used for real-time applications on body-worn devices with limited computing resources and storage space.

In this paper, we present a probability model for gesture recognition based on a body-worn device. The model takes HMM as a reference to build probability matrices for state analysis of observation points. But differing from HMM method that needs to build a HMM model for each gesture class, our method only builds one model for all classes. Moreover, each HMM model corresponding to a gesture class includes three probability matrices so that the number of matrix would increase with the gesture classes. But our model builds a matrix for each observation point, which indicates that the number of probability matrix would be constant as long as keeping the number of observation point unchanged. Therefore, the advantage of the model is that we can just concentrate on limited probability matrices to determine an unknown gesture by a process of look-up-table.

The paper is arranged as follows. The following section gives an introduction on the model including model description and model application. Section III describes a process of model generation including data collection and parameter definition. Section IV introduces an experiment of using a trained model to classify finger gestures and gives the evaluation. The last section gives the conclusion and future works.

2   Probability Based Recognition Model

The model is built from the viewpoint of gesture analysis. Since a gesture could be expressed as a temporal signal, the features of a signal at some time points should reflect the characteristics of the signal. We call the time points as observation points. For different gesture, the features on these points should be different. Therefore, we can explore the differences to classify the gestures.

The model consists of probability matrices at observation points to show the state of each kind of gesture at each observation point. An unknown gesture could be classified through searching the most likely state sequence corresponding to these observation points. In this section, we will introduce the model from three aspects: 1) what elements are there in the model; 2) how to generate and train a model; 3) how to use the model for gesture recognition.

2.1   Elements of the Model

The core of the model is a set of probability matrices that include following components.

The model consists of multiple probability matrices. m denotes the number of probability matrices, which is decided by the number of observation points. The positions of observation points is of importance, since the features at these points can reflect the basic characteristics of a gesture signal.

Fig. 1 shows a gesture signal where m observation points are selected for generating probability matrices. We denote the observation point set as $S = \{s_1, s_2, \ldots s_m\}$. At each point s_k, we build a probability matrix M_k to reflect the state of signal feature at the point. Therefore, to all points, a matrix set can be represented as $M = \{M_1, M_2, \ldots M_m\}$.

Regarding each matrix M_k, it can be characterized by the number of rows, the number of columns, and elements of the matrix.

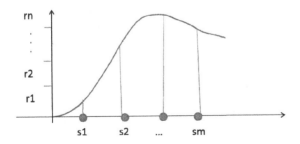

Fig. 1. An example of a gesture signal

The number of row is denoted with n. The model divides the whole range of feature values of gesture signals such as accelerometer-based acceleration values, mean, or energy into several small intervals. The number of intervals is n. The purpose of such a processing is to analyze which intervals the features values of a kind of gestures would mainly distribute in. The different distributions could be used for gestures classification. We denote the interval set as $R = \{r_1, r_2, ..., r_n\}$ as shown in Fig. 1, which divides the whole range of the acceleration values into n intervals. Therefore, each row in the matrix reflects the states of signal in corresponding interval.

The number of column is denoted with c. The parameter c is determined by the number of gesture classes. That is to say, each column reflects the states of a kind of gesture in different feature interval of current observation point. The gesture set is expressed as $A = \{a_1, a_2, ..., a_c\}$.

The last component also the most important one is $p_{k\text{-}ij}$, i.e. the element of matrix. $p_{k\text{-}ij}$ reflects the state of the j-th gesture in i-th interval from the viewpoint of the k-th observation point. In the model, $p_{k\text{-}ij}$ denotes occurring probability that the feature value of j-th gesture in k-th observation point falls into i-th interval, which can be obtained from training samples. The element set is expressed as

$$P_k = \{p_{k-ij} \mid 0 \le p_{k-ij} \le 1\} .$$

According to above discussion, we descript the model as:

$$\lambda = (M, R, A, P)$$

2.2   Model Generation and Training

Model generation is to determine the four components of model m, n, c, and $p_{k\text{-}ij}$, which depends on training samples.

Before training the model, we suppose the parameter c is known. That means the training process is a supervised learning method that we have known the number of gesture classes and labeled each training sample to corresponding class. The parameters m and n could be obtained through the way of samples training, namely to find the optimal observation points and intervals of feature values. But in this paper, we do not focus on the problem of how to determine the optimal parameters m and n. We determine them by empirical analysis to search the reasonable observation points and intervals that can reflect the basic characteristics of a gesture signal. Therefore, to the four components m, n, c, and $p_{k\text{-}ij}$, model training is to determine elements of each matrix, $p_{k\text{-}ij}$.

Given NS_j training samples that belong to the same kind of gesture j, each sample is a gesture signal. We observe the feature values of a sample in each observation point $\{s_i \mid s_i \in S\}$, and judge which interval the feature values fall into. The times t_{ij} of falling into an interval would be recorded to get the probability $p_{k\text{-}ij}$, namely

$$p_{k-ij} = \frac{t_{ij}}{NS_j} ,$$

and to any j in k,

$$\sum_i p_{k-ij} = 1$$

Therefore, trained matrices can be expressed as Fig. 2.

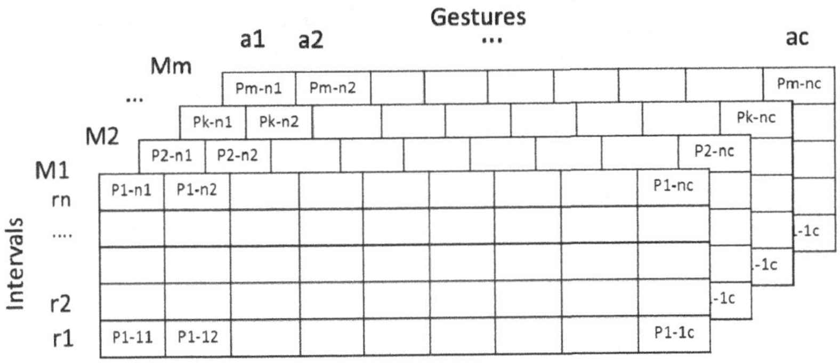

Fig. 2. Trained Probability Matrices of Model

2.3   Gesture Recognition

Given an unknown gesture signal and the model λ, we first need to calculate the feature values of the signal in observation points $S = \{s_1, s_2, ..., s_m\}$ to get an observation sequence, $O = \{O_1 O_2 ... O_m\}$.

To any element O_k in the observation sequence, we judge which interval it belongs to, such as i-th interval. Then we can look-up the probability, p_{k-ij}, corresponding to the interval in matrix M_k. Thus, to whole observation sequence O, an occurring probability corresponding to a gesture j, $P(O)_j$, could be obtained through each probability value at i-th interval of k-th matrix, which reflects a degree of the observation sequence matching a reference gesture. We define the $P(O)_j$ as:

$$P(O)_j = \sum_{k=1}^{m} p_{k-ij}$$

To a multiple axis-based sensing device like a 3-axis accelerometer, the probability can be calculated by

$$P(O)_j = \sum_{asix} \sum_{k=1}^{m} (p_{k-ij})_{asix}$$

Therefore, to all reference gesture classes, the unknown gesture would belong to the one with maximum similarity, i.e.

$$O \in \arg\max_j (P(O)_j)$$

3   Model Training

3.1   Data Collection

In the research, our recognition objects are one-stroke finger gestures that are a kind of dynamic gestures. The characteristic of these gestures is no more than one degree of freedom for each finger joint in one direction [8]. One-stroke gestures are atom gestures that could not further divided into small one. The reason of studying these gestures is that they are naturally from our daily lives and could express some basic intentions, such as "push" a switch to turn on a light. Moreover, if the atom gestures could be recognized, some complex gestures could be detected by combining them. Figure 3 shows 6 kinds of finger gestures, which are the studying objects in the research. The 6 kinds of gestures represent "Crook", "Unbend", "Up", "Left-Rotate", "Right-Rotate", and "Shift", respectively.

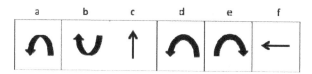

Fig. 3. six kinds of one-stroke finger gestures

We collect the gesture signals by a finger-worn device named Magic Ring [9]. Its core unit is a 3-axis accelerometer with sensitivity scale of ±1.5g. The sampling rate is 50HZ. The start and end of each gesture is recorded manually. Four subjects jointed our experiment of data collection. They are the students of local university without preliminary knowledge on the research. Each subject is asked to wear the device to do all gestures for 5 times as a natural and relax posture. All gesture data would be transported into a PC for data preprocessing and gesture recognition.

3.2   Parameter Definition

Parameter m denotes the number of probability matrix, i.e. the number of observation points. The more the points are set, the better the recognition accuracy might be. But a lot of observation points would cause the increase of probability matrices, so that the computational cost would increase. In order to reach a trade-off between the accuracy and the complexity of recognition, we select some representative points as less as possible, which can generally reflect the characteristics of a signal. In the research, the observation points refer to the sampling points. Because gesture signals vary with persons in duration even for the same person, we adopt an interpolation algorithm to unify the durations within a same range, one second, i.e. 50 samples. And then, through the observation and analysis of gesture signals, the number of observation points, m, is set as 4 and their positions are 5th, 15th, 30th, and 45th time point respectively. Therefore, the set S can be expressed as:

$$S = \{s_1, s_2, s_3, s_4\} = \{5^{th}, 15^{th}, 30^{th}, 45^{th}\}.$$

Parameter n denotes the number of row, i.e. the number of intervals of feature values. In the research, we adopt the processed acceleration value as the feature. Here, the processed data means data normalization. The purpose of such a processing is same as that of data interpolation, which can make the acceleration values lie within a same range to improve recognition accuracy. The range of processed acceleration values would be further divided into several intervals. In the research, we normalize the range of acceleration value from -500 to 500 and divide the range into 6 intervals. The interval set is set as:

$$R = \{r1, r2, ..., r6\}$$
$$= \{[-500, -300), [-300,-100), [-100, 0), [0, 100), [100, 300), [300, 500)\}$$

Parameter c denotes the number of column, i.e. the number of gesture classes. In the research, we defined 6 kinds of finger gesture. Therefore, c is 6.

$p_{k\text{-}ij}$ denotes the element of matrix, i.e. the occurring probability of the acceleration value of j-th gesture in k-th observation point occurring in the i-th interval. The probability values are obtained through training gesture samples.

4   Recognition and Evaluation

The 120 samples from 4 subjects are used for model training and testing. We adopted 2-fold LOOCV (Leave One Out Cross Validation) method for statistical analysis of performing results. The method means that the whole data set is divided into 2 subsets. The model is trained by using the whole data set except one subset. The processing is repeated 2 times, and excludes a different subset each time.

The recognition accuracy is shown as a confusion matrix in Table 1. Both the row and the column represent the classes of gestures. Each element in the matrix refers to the number of a class of corresponding to the row being classified as a class of corresponding to the column. For example, M(3,3)=11 indicates that within 20 samples of gesture "c", 11 samples are recognized accurately. But M(3,1)=1 indicates that within 20 samples of gesture "c", 1 samples are recognized as gesture "a".

It can be seen from the table, the average of accuracy is 85%. The computational complexity is $O(mnc)$, where m is the number of observation points, n is the number

Table 1. Confusion Matrix of Gesture Recognition

gestures	a	b	c	d	e	f
a	20	0	0	0	0	0
b	0	20	0	0	0	0
c	1	1	11	6	1	0
d	0	0	0	16	2	2
e	0	0	0	0	20	0
f	2	0	0	2	1	15

of interval, and c is the number of gesture class. The computational complexity is lower than that of HMM algorithm of $O(m^2 nc)$.

5   Conclusion

Aiming at the problem of online recognition of dynamic gesture using a finger-worn device, this paper proposed a probability model based on the feature distribution of gesture signal. The model learns from HMM, which classifies a gesture through analyzing its feature distributions in several observation points to find the most likely observation sequence. Comparing with the HMM model, an important advantage of our model is that can reduce the computational complexity of training and testing on the premise of reasonable accuracy, since we just need to concentrate the limited probability matrices to classify an object.

In future, we would consider the problem of defining the initial matrix and determining the optimal parameters. Regarding the experiment evaluation, we would add more gestures to test the performance of the model and use multi-fold cross validation to test the feasibility. In further, we would give quantitative analysis on the computing delay of the model.

Reference

1. Ward, J.A., Lukowicz, P., Troster, G., Starner, T.E.: Activity Recognition of Assembly Tasks Using Body-worn Microphones and Accelerometers. IEEE Trans. on Pattern Analysis and Machine Intelligence 28(10), 1553–1567 (2006)
2. Zhou, Y., Cheng, Z., Hasegawa, T., Jing, L., et al.: Detection of Daily Activities and Analysis of Abnormal Behaviors Based on a Finger-worn Device. In: Proc. of 5th IET International Conference on Ubi-Media Computing (U-Media 2012), Xining, China (August 2012)
3. Preece, S.J., Goulermas, J.Y., Kenney, L.P.J., et al.: Activity identification using body-mounted sensors–a review of classification techniques. Physiological Measurement 30(4), R1–R33 (2009)
4. Bourke, A.K., Lyons, G.M.: A Threshold-based Fall-Detection Algorithm Using A Bi-axial Gyroscope Sensor. Medical Engineering & Physics 30(1), 84–90 (2008)
5. Dy, J.G., Brodley, C.E.: Feature Subset Selection and Order Identification for Unsupervised Learning. Journal of Machine Learning Research 5, 845–889 (2004)
6. Rabiner, L.R.: A Tutorial on Hidden Markov Models and Selected Applications in Speech Recognition. Proceedings of the IEEE 77(2) (1989)
7. Lee, H., Kim, J.H.: An HMM-Based Threshold Model Approach for Gesture Recognition. IEEE Trans. on Pattern Analysis and Machine Intelligence 21(10), 961–973 (1999)
8. Jing, L., Zhou, Y., Cheng, Z., Wang, J.: A Recognition Method for One-stroke Finger Gestures Using a MEMS 3D Accelerometer. IEICE Transaction on Information E94-D(5), 1062–1072 (2011)
9. Jing, L., Zhou, Y., Cheng, Z., Huang, T.: Magic Ring: A Finger-worn Device for Multiple Appliances Control using Static Finger Gestures. Sensors 12(5), 5775–5790 (2012)

Design of a Situation-Aware System for Abnormal Activity Detection of Elderly People

Junbo Wang[1], Zixue Cheng[1], Mengqiao Zhang[2], Yinghui Zhou[2], and Lei Jing[1]

[1] School of Computer Science and Engineering, The University of Aizu, Japan
{j-wang,z-cheng,leijing}@u-aizu.ac.jp
[2] Graduate School of Computer Science and Engineering,
The University of Aizu, Japan
{m5142105,d8131104}@u-aizu.ac.jp

Abstract. Internet of Things (IoT) is becoming one of hottest research topics. Elderly care is one of important applications in IoT, to grasp the situations around the elder people and then corresponding information can be sent to the care-givers to support the elder people. Abnormal activity detection is a particularly important task in the field, since the services should be immediately provided in such cases. Otherwise the elder people may be in danger. The existing approaches to this problem use some basic living patterns of the elder people, e.g. mobility per day, to detect abnormal activities. However, the detail abnormal activities in various specific situations cannot be detected, e.g., whether there is some abnormal activity when the elder people go to toilet, sleeps or eats something. To solve the above problem, in the paper, we propose a situation-aware abnormality detection system based on SVDD for the elder people. An experiment has been performed focusing on feasibility of the method and accuracy of the system to detect situations and abnormities from real sensors.

Keywords: IoT, Elderly care, Situation-Aware, Abnormal activity detection, SVDD.

1 Introduction

Internet of Things (IoT) is becoming one of the hottest research topics recently. The basic idea of IoT is that, variety of smart objects augmented with various abilities, e.g. sensing, wireless communication, processing etc, are able to interact with each other and cooperate with their neighbors to reach common goals.

There are many applications in IoT, e.g. smart home/office, smart grid, smart city etc. With increase of the number of the elder people, well and effectively taking care of them is becoming a big social problem in the advanced country, e.g. Japan, and also is becoming a very important research topic in IoT. Currently, Japan has highest proportion of elderly citizen over the world, 21% over the age of 65 [1]. In additional, care for the elder people living alone is a particularly important research topic, since they are isolated with society, and lack of support from family. According to Tokyo medical examiner's office, people over 65 who died alone in their residence stood at 2,211 in 2008 [2].

R. Huang et al. (Eds.): AMT 2012, LNCS 7669, pp. 561–571, 2012.

To take care of the elder people, abnormal activity detection is particular important, since the services should be immediately provided to them to avoid dangers. And some disease often occurs with the abnormal activities. For example, some disease can be predicted if the elder people go to the restroom too many times at night, e.g., hyperplasia of prostate, chronic nephritis, bladder inflammation, diabetes, and so on.

To detect the abnormal activities of the elder people, various researches have been performed [7-10]. However, the detail abnormal activities in various specific situations cannot be detected in the exiting researches. For example, the abnormal activities when the elder people go to toilet, sleep, or eat something, cannot be detected. Grasping and detecting the above detail abnormal activities can greatly help the care-givers take care of the elder people. For example, based on detection of the abnormal activity, some possible disease can be predicted by doctors and services, e.g. warning message, can be immediately provided to the elder people and care-givers.

To solve the above problem, in the paper, we propose a situation-aware abnormality detection system based on SVDD for the elder people.

Situation-aware is a very important research topic in the research paradigm of ubiquitous computing, which detects the situations around the user and provides the corresponding services or automatically changes environment to adapt the user. In the paper, not only the general situations can be detected, but the abnormal activities in the situations also can be recognized to support the users.

To detect abnormal activities in various situations, first the features, based on which the abnormal activities can be predicted, are analyzed and designed. Then a zone-based situation detection system is proposed to detect the situations around the user. After that, the method to detect abnormal activities in the situations based on designed features and SVDD (Support Vector Data Description) is presented. Finally, we perform the experiments to evaluate the feasibility of the method and accuracy of the system. Through the experiment result we can see that the accuracy of the system is acceptable for detecting situations and abnormality.

The rest of the paper is organized as follows. Section 2 introduces the related works. Section 3 presents the basic ideas and feature design. Section 4 presents situation detection and abnormal activity recognition. The details of experiment results have been presented in Section 5. And finally we conclude the paper in Section 6.

2   Related Works

Activity recognition has been focused on in various researches based on variety of classification methods [3-6], e.g. hidden Markov models [5], support vector machine [6] and Naive Bayesian classifier etc. In [5], a HMM-based human activity recognition method was implemented using single triaxial accelerometer based on a proposed HMM structure, which allows both backward and skip transitions. Classification of the human activities is performed with support vector machine in [6], based on autoregressive model. The result for four activities, i.e. running, still, jumping and walking is shown better than traditional time domain and frequency features. However, most of these works focused on the normal activities of human, e.g. walking, running, jumping, etc., but cannot be directly used in detecting abnormal activities due to the different features and training dataset.

Meanwhile, various researches have been performed on abnormal activity detection. Park et el. proposed a similarity based method to detect human abnormal behavior in [7]. They modeled episodes as a series of event, and computed similarity between the happened episodes with the ones in history. Finally the abnormal behavior can be recognized if the similarity is larger than a threshold. However, the time for computing similarity may not meet the requirement in real time system when there are so many episodes in history. Especially when the number of evens in the current episodes and the number in the history episodes are different, since the similarity between the current episode and all subsequence in history episode will be computed. A threshold based method has been proposed in [8] to detect human abnormal movement in wireless sensor network. However, the optimal threshold is hard to be found for customized services of each user.

And some abnormal activity detection methods have been proposed based on outlier detection method in machine learning, e.g. support vector machine (SVM) and support vector data description (SVDD). In [9] an abnormal human activity detection method was proposed based on one-class SVM. First, they adopted a set of HMMs to model the normal traces and after transforming the n training traces into a set of feature, they trained a one-class SVM to detect abnormal activity. The method was further enhanced in [10] by employing HDP-HMM and Fisher kernel. However, the methods are lack for detail analysis of features of the elder people's abnormal activities. Therefore, they cannot be directly used in the elderly care without detail design of features for the abnormal activities of the elder people. In [11], a method to detect abnormal living patterns for the elder people was proposed based on SVDD. The support vector data description (SVDD) method is proposed by Tax and Duin [12][13], inspired by SVM. In [11], they first analyzed the features for daily behavior pattern classification in details, based on which some kinds of disease can be predicted. And then the abnormal activity is recognized based on trained SVDD using training dataset. However, the method cannot detect the abnormal activities in some specific situation. For example, the abnormal activities when the elder people go to toilet, sleep, or eat something, cannot be detected we discussed above. Without detection of such detail abnormal activities, care-givers or doctors are hard to predict the related disease and give them suitable care immediately.

3   Basic Ideas and Feature Design

3.1   Basic Ideas

Fig. 1 shows basic ideas of the system. First, the situation of the elder people is detected in a smart home. To get precise location information of the elder people, we have developed a u-tiles sensor network [14][15], which is built behind a floor. Through u-tiles, location and position relation between the elder people and surrounding objects can be grasped by it. Furthermore, we classify the area in the smart home into various meaningful zones. The situation of the elder people will be detected based on zones and other detail information, e.g. status of the home appliances, actions of the user, etc.

After that, abnormal activities in the detected situations will be recognized by SVDD with various features analyzed in Subsection 3.2. Finally, the system sends the corresponding information to care-givers, e.g. remote family and staff in care center.

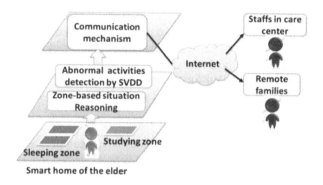

Fig. 1. Basic ideas and structure of the system

3.2   Feature Design

To detect abnormal activities of the elder people, first we should analyze the features, based on which some disease may be predicted. Here we classify the features into two types, i.e. basic feature (BF) and special feature (SF). BF represents some common features, which can be used to detect abnormality in most of situations. SF is used to represent some special features just for some specific situations.

In the paper, we mainly consider the following situations for the elder people as shown in Table 1, which often happen around the elder people.

First, we consider the situation of taking medicine. Taking correct and prescribed medications great improve the health of the elder people [17]. Many elder people use drugs to manage chronic conditions such as heart disease, lung disease, arthritis, pain and depression. And many elder people use multiple medications at the same time to maintain the effectiveness of the drugs [18]. The abnormal behaviors, e.g. forgetting to take medicine or take medicines in a wrong order, may bring very serious consequences. Thus we designed the situation features as shown in Table 1 to detect the abnormal activities when the elder people taking medicine. Based on BF designed in Table 1, we can detect the abnormal activities, when the elder people do not take medicine on time or forgot to take medicine. In SF, we want to detect whether the elder people take medicines in a wrong order, or whether the interval between having a meal and taking medicine is suitable, which are important for taking medicine [18].

The second situation in Table 1 is for going to the restroom, based on which some problem can be detected, e.g. renal function. For example, if the elder people go to the restroom too many times at night, the elder people may have some diseases, e.g. chronic nephritis, bladder inflammation, diabetes, and so on.

Third, insomnia is another kind of unhealthy activity for the elder people, which may be related to the disease in neurasthenia. Long-term insomnia may cause nerve cell aging and disorders of metabolic. And taking too much sleep is also not good for the elder people since it will affect metabolic. Therefore, in the third situation, we want to detect whether there is abnormality when the elder people is in the sleeping situation. To achieve the goal, we selected two BFs which are start time and end time/duration to detect whether the elder people sleep on time and whether they have suitable sleeping hours. Furthermore, we design four SFs, which are getting up times, duration of getting up, naps in the daytime and duration of naps to evaluate the sleeping quality. If the elder people get up too many times or the duration of getting up is too long, they may get insomnia with high possibility. And if the elder people have too many naps or the duration of naps is too long, there is a high possibility that they have a bad quality sleeping last night. Therefore, we use the above four SFs coupling with two BFs to detect abnormal activities in sleeping situation.

Fourth, eating is also a very important aspect to represent health condition of the elder people. With growth of age, the elder people may be not sensitive in taste and smell, and thus may not have a good appetite. Then the observation of eating for the elder people can let the remote family and staffs in care center quickly get the health condition of the elder people, e.g. whether having enough and balanced foods.

Besides the above situations which may be related to diseases and health condition of the elder people, whether the elder people take an appropriate body exercise should also be considered in the system. Therefore, we have designed the features to detect the abnormality for the situations of taking a walk and having some works in yards as shown in Table 1.

Table 1. Feature design for various situations to detect abnormal activities of the situation

Situation	Situation Features			
	Basic Features			Special Features
	Start time	Duration/end time	Times per day	
Taking medicine	O	×	O	Order, Interval with a meal
Going to the restroom	×	O	O	Times per hour and at night
Sleeping	O	O	×	Getting up times and duration, Naps in daytime and Duration of Naps
Eating	O	O	O	Types of foods
Taking a walk	O	O	O	Times per week
Having some works in yards	×	O	×	Times per week

4   Situation and Abnormal Activity Detection

In the section, we mainly present how to detect the situations around the elder people and the abnormal activities in the situations.

4.1   Hardware for Detection

The system mainly consists of two parts. The first part is u-tiles sensor network to detect the position and trace of the elder people. And the second part is smart plug developed by the NTT Company Japan [16].

First, let's introduce the u-tiles sensor network [14, 15]. The basic idea of u-tiles sensor network is to detect precise location and position relation between the elder people and surrounding objects by embedding various sensors, including pressure sensors and RFID antennas under the floor. Fig. 2 shows u-tiles sensor network, where a piece of u-tiles is almost 40×40cm. Under each piece of u-tiles, there are pressure sensors and RFID antennas which are connected with a reader through a PIC based switch. The IDs of users and objects can be read by each piece of u-tiles, and the position relation between users and objects can be grasped. The detail of u-tiles hardware environment can be found in [14][15].

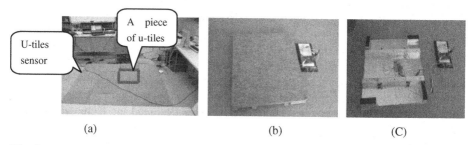

(a)   (b)   (C)

Fig. 2. U-tiles hardware environment. (a) Total system of U-tiles System, (b) and (c) one piece of tile connect with ZigBee wireless communication module.

Meanwhile, to let u-tiles sensor network be easily and freely implemented in a smart home, we enhanced it by connecting some of pieces of u-tiles with wireless communication modules as shown in Fig. 2 (b) and (c). Therefore, pieces of u-tiles can be implemented freely in a room as shown in Fig. 3 to detect position relation.

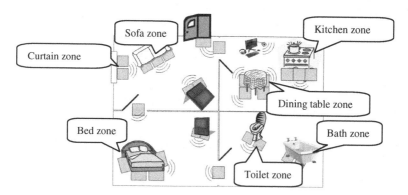

Fig. 3. System image to build a smart home based on U-tiles and zones

Fig. 3 shows an image of a smart home implementing u-tiles sensor network. U-tiles can be implemented in the zones where users often have activities on, e.g. around the bed, in front of sofa as shown in Fig. 3. The pieces of u-tiles can send the detected information to the PC through wireless communication modules.

Fig. 4. Implementation of smart plug to control home appliances, e.g., a lamp

Besides u-tiles sensor network, we have implemented smart plug developed by NTT Company in our system to detect and control the status of home appliances, i.e. turning on or off of a lamp as shown in Fig. 4.

4.2   Zone Based Situation Reasoning

With the hardware introduced above, in this subsection we will present how to detect the situation around the elder people. As shown in Fig. 4, we classify the smart home into various zones. Zone is a meaningful area related with some specific location, furniture, and home appliances, in which some specific activities may happen with high possibility. For example, when the elder person is in the sofa zone and the TV is on, he may be watching TV. And then by further considering other factors, e.g. environment factors, action etc., the detail situation around the elder people can be grasped. In the paper, we mainly consider the following four factors to represent a situation, i.e. zone, status of related home appliances, actions of the elder people and environment. We developed a wearable device called magic ring, by which some basic gestures/actions can be detected [19]. Then a situation s can be represent by the following four tuple,

$$s = <z, D, A, E>, \text{ where}$$

z is used to represent the current zone of the elder people;
D is a set including the status of related devices;
A is a set including current actions of the elder people;
E is a set including environment factors.

Table 2 shows detection rules based on the zone-based situation definition.

For example, if the system detects that the elder person is in the toilet zone and the light of the bathroom is on, the situation of the elder person can be recognized as "going to the restroom".

Table 2. Situation detection rules in the system

Situation	Detection rules			
	Z	D	A	E
Taking medicine	Pill box zone	Pill box is being opened	N/A	N/A
Going to the restroom	Toilet zone	Light in bathroom is on	N/A	N/A
Sleeping	Bed zone	Light in bedroom is off	Almost no action	No light
Eating	Dining table zone	N/A	Eating	N/A
Taking a walk	Out of the yards	N/A	Walking	N/A
Having some works in yards	Yards	N/A	Some actions	N/A

4.3   Abnormal Activity Recognition Based on SVDD

After detecting situations around the user, we should recognize whether there is abnormal activity in these situations. In the paper, we use SVDD to recognize abnormal activity.

SVDD is inspired by Support Vector Machine (SVM) which is a method of Machine Learning. SVDD tries to construct a boundary around the target data by enclosing the target data within a minimum sphere. The boundary of a dataset can be used to detect target data or outlier. The detail can be found in [12, 13].

Differing from other abnormal activity detection researches, we create several hyper spheres based on the different situations with various features as shown in 1. For each kind of situation, a hyper sphere is created to recognized abnormal activity.

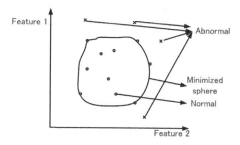

Fig. 5. An image of SVDD to create sphere based on a training data set

Fig. 5 shows an image of SVDD to detect abnormality. First, support we have some training sets of normal activities in such situation as little circles shown in Fig. 6, with two features, i.e. feature 1 and feature 2. Then we can draw a hyper sphere to include all of normal activities in training set with center a and radius R, which can be presented as follows,

$$F(R,a) = R^2 + C\sum_i \xi_i \qquad (1)$$

with the constraints that all normal activities are within the sphere,

$$\|x_i - a\|^2 = R^2 + \xi_i, \quad \xi_i \geq 0 \quad \forall i \tag{2}$$

Then we minimize the volume of the sphere, i.e. equation (1), by minimizing R^2, and with the constraints in equation (2) with Lagrange multipliers. The detail can be found in [12, 13].

Support the minimized sphere can be computed as shown in Fig. 6. Then the outliers as crosses in Fig. 6, are recognized as abnormal activities. If there are N features for a situation, the sphere should be N dimensions.

5   Accuracy of the System While Detecting Situations and Abnormalities from Real Sensors

Here we perform an experiment to evaluate the accuracy of the system while detecting situations and abnormal activities from real sensors. To achieve the goal, we recruited five subjects. For each subject, we performed an experiment using almost 40 minutes. First, we give five minutes introduction to each subject. And then we take 10 minutes to let each subject design their activities to simulate some basic activities at home for one day when they are sleepless at night. The insomnia is possibly due to take too many naps in the daytime. Finally, we let each subject take almost 24 minutes to simulate one day's activities. In the experiment, we take a video to record the situations of the subject and then compare them with the situations detected by the system. Fig. 6 shows the situations of a subject, where SinV is used to represent the situation recorded in video, and DS is used to represent the detected situation. X axis shows time and Y axis represents situation numbers shown in Table 3.

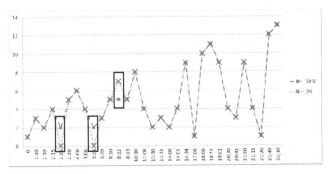

Fig. 6. Situations detected in the smart home through the experiment of the subject

From Fig. 7 we can see that, the two curves are almost overlapped except 3 points, which means the system has three error detections for this subject. The error detections happen at 3:26, 5:38 and 8:12. Two of them happen when the subject goes through the kitchen, which shows our system may have error when the user moves too quickly. In the future we are planning to solve the problem by employing more sensitive sensors. Through all experiment of 5 subjects, the system have totally detected

146 activities and correctly detected 136 by comparing with the video. The accuracy of the system in the experiment is 93.2%. And in the experiment of 5 subjects, all abnormal activities can be detected in the system and the system sends a warning service to remote family.

Table 3. Situations designed in the experiment to be detected

Situations	No.
In kitchen	1
go through kitchen	2
In bookroom	3
In Toilet	4
Have a Nap	5
Getting up from nap and go to kitchen	6
Getting up from nap and go to Bookroom	7
Getting up from nap and go to toilet	8
Watching TV on Sofa	9
Sleep	10
Switch on the light	11
Back to sleep	12
Getup from sleep	13
null	0

6   Conclusion

Elderly care is a particular important problem in our society and a very hot research topic in the field. In the paper, we have proposed a situation-aware system to detect abnormal activities of the elder people using SVDD. From the experiment we can see that the method work well to detect abnormal activities. In the future, we are going to implement our system in a home to detect the situations and evaluate our method. Meanwhile, we are going to evaluate the system to detect other situations and abnormal activities.

References

1. Asia: Japan: Most Elderly Nation. The New York Times. (July 01, 2006) (retrieved January 07, 2008)
2. http://search.japantimes.co.jp/cgi-bin/nn20100721f1.html
3. Mannini, A., Sabatini, A.M.: Machine learning methods for classifying human physical activity from on-body accelerometers. Journal of Sensors 10, 1154–1175 (2010)
4. Preece, S.J., Goulermas, J.Y., Kenney, L.P.J., Howard, D.: A Comparison of Feature Extraction Methods for the Classification of Dynamic Activities From Accelerometer Data. IEEE Trans. on Biomedical Engineering 56(3) (March 2009)

5. Han, C.W., Kang, S.J., Kim, N.S.: Implementation of HMM-Based Human Activity Recognition Using Single Trivial Accelerometer. IEICE Trans. on Fundamentals E93-A(7) (July 2010)
6. He, Z., Jin, L.: Activity Recognition From Acceleration Data Using AR Model Representation And SVM. In: Proc. of the Seventh International Conference on Machine Learning and Cybernetics, Kunming, July 12-15 (2008)
7. Park, K., Lin, Y., Metsis, V., Le, Z., Makedon, F.: Abnormal Human Behavior Detection in Assisted Living Environments. In: Proc. of PETRA 2010, Samos, Greece, June 23-25 (2010)
8. Burchfield, T.R., Venkatesan, S.: Accelerometer-Based Human Abnormal Movement Detection in Wireless Sensor Networks. In: Proc. of HealthNet 2007, San Juan, Puerto Rico, USA, June 11 (2007)
9. Yin, J., Yang, Q., Pan, J.: Sensor-Based Abnormal Human-Activity Detection. IEEE Trans. on Knowledge and Data Engineering 20(8) (August 2008)
10. Hu, D.H., Zhang, X., Yin, J., Zheng, V.W., Yang, Q.: Abnormal Activity Recognition Based on HDP-HMM Models. In: Proc. of IJCAI 2009, USA (2009)
11. Shin, J.H., Lee, B., Park, K.S.: Detection of Abnormal Living Patterns for Elderly Living Alone Using Support Vector Data Description. IEEE Trans. on Information Technology in Biomedicine 15(3) (May 2011)
12. Tax, D.M.J., Duin, R.P.W.: Support Vector Data Description. Mach. Learn. 54(1), 45–66 (2004)
13. Tax, D.M.J., Duin, R.P.W.: Support Vector Domain Description. Pattern Recognition, Letter 20(11-13), 1191–1199 (1999)
14. Ichizawa, T., Tosa, M., Kansen, M., Yishiyama, A., Cheng, Z.: The Attribute and Position based Ubiquitous Development Environment Using Antennas with an Automatically Switch, IPSJ SIG Technical Reports, No. 14, pp. 109–114 (2006) ISSN 0919-6072
15. Wang, J., Cheng, Z., Jing, L., Ota, K., Kansen, M.: A Two-stage Composition Method for Danger-aware Services based on Context Similarity. IEICE Transactions on Information and Systems E93-D(6), 1521–1539 (2010)
16. http://www.intellilink.co.jp/solutions/green/products/xechno-tap.html
17. Nanada, C., Fanale, J.E., Kronhom, P.: The Role of Medication Noncompliance and Adverse Drug Reactions in Hospitalizations of the elderly peoplely. Archives of Internal Medicine 150(4), 841–846 (1990)
18. Medications and the Older Adult, http://www.chryslerretirees.com/benefits2/hrdocs.chrysler.com/dashboard/dash/dash_images/medication.pdf
19. Jing, L., Zhou, Y., Cheng, Z., Wang, J.: A Recognition Method for One-Stroke Finger Gestures Using a MEMS 3D Accelerometer. IEICE Transactions on Information and Systems E94.D(5), 1062–1072 (2011)

Revealing Cultural Influences in Human Computer Interaction by Analyzing Big Data in Interactions

Rüdiger Heimgärtner[1] and Harald Kindermann[2]

[1] Intercultural User Interface Consulting (IUIC), Undorf, Germany
ruediger.heimgaertner@iuic.de
[2] Upper Austria University of Applied Sciences, Steyr, Austria
harald.kindermann@fh-steyr.at

Abstract. Understanding human information needs worldwide requires the analysis of much data and adequate statistical analysis methods. Big interaction data from empirical studies regarding cultural human computer interaction (HCI) was analyzed using statistical methods to develop a model of culturally influenced HCI. There are significant differences in HCI style depending on the cultural imprint of the user. Having knowledge about the relationship between culture and HCI using this model, the local human information needs can be predicted for a worldwide scope.

Keywords: Culture, Relation, Data Analysis, Global, Local, Intercultural Interaction Analysis, Tools, Big Data, Modeling, Data Mining, Statistics, ANOVA.

1   Introduction and Related Work

Rapidly progressing globalization requires consequent adaptation of the methods and processes applied in Human Computer Interaction (HCI) design to the relevant cultural needs. HCI must be attuned to the intended cultural domain; user group and context, which have been common in computer science for about a quarter of a century (cf. [1]). To do so, the HCI designer must be able to immerse himself in these cultural domains, user groups and contexts in order to extract the relevant requirements for HCI design. If a situation is known, prejudices do not crop up but rather proper evaluations develop, which speeds up the design process and make it specific. Intercultural User Interface Design (IUID) is designated to match these requirements by integrating several disciplines. For example it integrates information technology with cultural studies within this prominent and complex field. However, the relations between HCI and culture are not yet well elaborated, even if there are some initial approaches (cf. e.g., [2]; [3]; [4]; [5]; [6]). A proper taxonomy of all approaches, methods and processes in IUID is also still absent, even if there are some clues (cf. e.g., [7]). [8] reviewed culture in information systems research to clear the path toward a theory of information technology culture conflict. [9] suggested a three-perspective model of culture and information systems as well as their development

R. Huang et al. (Eds.): AMT 2012, LNCS 7669, pp. 572–583, 2012.
© Springer-Verlag Berlin Heidelberg 2012

and use. In 1997, [10] presented an explanatory theory for management information systems in the Chinese business culture. In 2000, [11] provided an empirical study for the development of a framework for the elicitation of cultural influence in product usage to shed light on culture and context. [12] demonstrated a path towards culture-centered design. In 2002, a prominent method for cultural design and its integration in technical product design was elaborated on by [13]. [14] described the integration of culture in the design of ICTs. However, the cultural influences in HCI design on the interaction level have not been investigated in detail so far.

Many kinds of culturally influenced interaction patterns are only recognizable over time which requires the collection of big data. Hence, enhanced algorithms and tools must be used for data analysis (cf. [15]). Therefore, designing tools for non-experts to do their own analysis with big data in interactions will be a prominent task for interaction designers in the future (cf. [16]). Even if there are several tools for user interaction logging (e.g. ObSys for recording and visualization of windows messages, cf. [17]), none of the existing tools provide explicitly cultural usability metrics to measure culturally imprinted interaction behavior. This is because until now the connections between HCI and culture have not been systematically collected and prepared for such tools.

This paper describes the intercultural interaction analysis tool (IIA tool) and some statistical methods to elucidate the connection between HCI and culture by analyzing big data (cf. e.g., [18]). First, a study collecting big interaction data is presented. Then, the application of appropriate statistical methods to analyze the connections between culture and HCI based on the big interaction data is explained and the results and challenges are addressed.

However, before it is possible to analyze the interaction of the user with the system, it is necessary to measure dynamic aspects in HCI using automated analysis tools by recording the user behavior chronologically without gaps. For this purpose, the IIA tool can be used (cf. [19]). Then the culturally caused differences in HCI can be recognized by analyzing the collected quantitative data using statistical methods (e.g. ANOVA, Kruskal-Wallis test, post hoc tests, etc.). From this point usability metrics (cf. [20]) of high empirical value for measuring quantitative variables in culturally influenced HCI can be derived.

2   Assumed Relationships between Cultural Dimensions and HCI Dimensions

[21] investigated Hofstede's cultural consequences on information systems research and found out that this is an uneasy and incomplete partnership, which has been confirmed by [6]. However, [22] provided concrete indices describing the cultural characteristics of nations, which can be used as first quantitative reference points which can be compared to the quantitative cultural interactions indicators constituting the HCI dimensions introduced by [23]. In addition, several basic assumptions were derived from the work of [24] regarding the connection between cultural dimensions and HCI dimensions (cf. [23]). The empirical hypotheses primarily concern quite basic user behavior, which can be described by the following cultural dimensions: Time orientation, density of information networks, communication speed and action

chains (sequential actions). In this regard, it is reasonable to assume that HCI dimensions like information speed (distribution speed and appearance frequency of information units), information density (number and distance of information units) or information structure (order of information units) stand in relation to the culturally different basic behavior patterns of the users. If this is the case, the differences, which [24] discovered, also imply differences in information speed, information density and information frequency. Thereby, also interaction frequency and interaction style should be affected. For instance, based on the action chain dimension of [24], it can be assumed that replying to questions by German users is linear (i.e. answered consecutively) in comparison to Chinese users. Furthermore, due to the high task orientation, the number of dialog steps until the completion of the task could be lower for German users. In addition, it can be assumed for German users that the number of interactions is higher during completion of a task, such as the number of uses of optional functions and using help functions or adjusting colors etc. because of their desire to work very exactly. However, the number of mouse movements or mouse clicks by German users should turn out to be lower than for Chinese users according to higher uncertainty avoidance and strong task orientation of German users (cf. [25] and [26]). For these reasons, one interaction step during completion of a test task (and thereby complete test duration) might last longer for German than for Chinese users. Finally, the speed of moving the mouse should also be lower for German users in accordance to higher uncertainty avoidance, lower communication speed, and low relationship orientation.

3   Big Interaction Data from Two Online Studies

Two online studies separated in date by one year (the first in 2006 and the second in 2007) served to identify the preferences of users according to their cultural background (especially regarding their interaction behavior) (cf. [23]). Randomly selected employees from SiemensVDO (today Continental Automotive GmbH) all over the world were invited by email to do the test session using a special tool for measuring the interaction behavior developed by the author (cf. [19]). About 10GB of data were collected. Out of the 14500 test persons invited to participate in the larger second main study, 2803 downloaded and started the test. This return rate of 19.3% is sufficient for reasonable statistical analysis. 66.8% of the tests were aborted. The remaining 33.1% of the tests were completed by the test persons and analyzed by the author (916 valid data sets). Only these complete and valid data sets were analyzed using the IIA data analysis module and the statistic program SPSS.

Table 1 gives an impression of the characterization of the test participants in the two online studies regarding the distribution of samples, nationality, test language, age, gender and experience with PC.

Table 1. Characteristics of the Test Participants in the Two Online Studies

Study	Sample Size	Test Language (C/G)	Nationality (Chinese/German/US/Canadian)	Age (<20,20-30,30-40,>40)	Gender (m/f)	PC experience (no/beginner/adv./exp.)
1	102	34 : 44	35 : 39 : 9 : 8	1 : 30 : 37 : 6	83 : 19	2 : 18 : 52 : 30
2	916	105 : 375	100 : 345 : 133 : 64	6 : 140 : 284 : 138	746 : 170	13 : 159 : 487 : 257

The randomly selected participants in the online studies at SV were employees locally situated in China, the USA, Canada, Germany, France, Romania, Mexico, the Czech Republic, India and Brazil (most in Germany, the USA and in China). Participants were matched as closely to the requirements of having some PC experience and a basic education level that matches the job requirements of SiemensVDO. About 18% were female, 82% male. Their age was between 18 and 64. They all shared the company culture of SiemensVDO. The differences in HCI in these studies were repeatedly analyzed in relation to two groups of test persons according to the selected test languages (Chinese (C) and German (G)) in order to reduce data analysis costs.

In the following the presentation concentrates on the more representative second main study, because about nine times more valid test data sets were available (n2 : n1 = 916: 102 = 8.98). Furthermore, the second study mostly mirrored the results of the first study.

3.1   Data Collection with the IIA Tool: Setup, Test Setting and Usage

A special intercultural interaction analysis tool (IIA tool) for measuring the interaction behavior of the user has been developed by [19]. The interaction of culturally different users doing the same test task can be observed (using the same test conditions i.e. the same hard- and software, environment conditions, language and test tasks) as well as requiring the same experience with the use of the system. Logging data of dialogs, debugging and HCI event triggering while using the system are highly valuable ([27]). This data can be logged during usability tests according to certain user tasks. The IIA tool provides data collection, data analysis and data evaluation. It serves to record and analyze the user's interaction with the system in order to identify culturally influenced variables such as color, positioning, information density and interaction speed as well as their values. The collection and preparation of the data is done automatically for the most part by the IIA data collection module. The data is stored in databases in a format that is immediately usable by the IIA data analysis module, which does subsequent data conversion or preparation. Common statistic programs like SPSS can be deployed to apply statistical methods (cf. [28]). The IIA data evaluation module enables classification using neural networks to cross-validate the results from data analysis (cf. [29]).

A test participant (SV employees) downloaded the IIA data collection module via the corporate intranet of SV locally on his computer, started the tool and completed all the test tasks. Before closing the tool, the collected data was transferred automatically onto a non-public and secure network drive on a server at SiemensVDO by the IIA tool.

The tests were carried out on the computer belonging to test persons, in which the test system had been automatically installed after confirmation of the participation of the test persons in the test. Therefore, the user could use his own PC and was not under time pressure to do the test because there was no need to use a foreign test device within predefined time slots. In the case of difficulties in understanding, the test participants could use the online help. The individual test meetings lasted for approximately 45-90 minutes and always took place using the same scheme.

Every meeting recorded all system events regarding usage of the keyboard and mouse as well as taking a screenshot after finishing every test task. Furthermore, the collected data was prepared for subsequent statistical analysis in SPSS using CSV format. The IIA analysis module instructs the data loader module to extract, load and prepare the data from the general database for analysis.

The method of asking many users online by letting them do test tasks (use cases) and then collecting the qualitative data (user preferences) which emerged quantitatively through this process has mainly been used for Chinese (C) and German (G) speaking employees of SiemensVDO worldwide. It is assumed that the language is strongly associated with nationality. A test session with the IIA data collection module comprised five main parts: collecting the demographic data, executing the test tasks, surveying the cultural values of the user, cross-checking the test results of the user and debriefing. After the start of the IIA data collection module, the user first had to select his preferred test language, i.e. the language in which the test takes place.

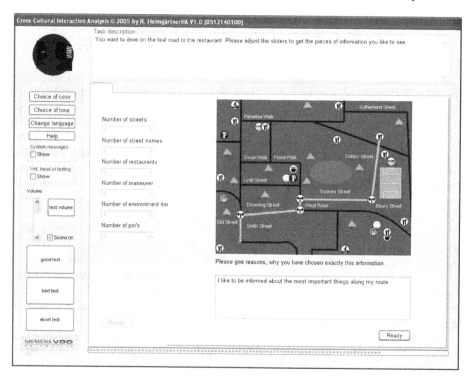

Fig. 1. Screenshot of the "map display test task" during the test session with the IIA data collection module

During the whole test session, the IIA tool records the interaction between users and the system (cf. [30]), e.g., mouse moves, clicks, interaction breaks or the values and changes of the slide bars set up by the users in order to analyze the interactional patterns of users from different cultures. Thereby, all levels of the interaction model (physical, lexical, syntactical, semantic, pragmatic and intentional, cf. [31]) necessary for dialog and interaction design can be analyzed using the IIA tool (cf. [19]). The IIA tool allows the measurement of numeric values for information speed, information density and interaction speed in relation to the user. These are hypothetically correlated to cultural variables concerning the surface level such as the number or position of pictures in the layout or affecting the interaction level such as the frequency of voice guidance as used e.g., in the maneuver guidance test task or in the map display test task as indicated in Figure 1.

3.2 Data Analysis Concept – IIA Tool as Method Framework for Data Analysis

Processing of the gathered data was carried out largely automatically via the IIA data collection module itself due to the well elaborated data structure concept. The data was checked for plausibility and entered in a reasonably predefined format into a result database. A part of the IIA data analysis module (integrated in the IIA data collection module) computes the most important coefficients regarding information speed, information density, information order, information serialization, interaction speed and interaction exactness as well as relationship-orientation, uncertainty avoidance and power distance. The outputs of the IIA analysis module are network diagrams presenting the average parameters of the data collection, comparing different nationalities, countries or cultures. Several statistical methods had been applied to analyze the collected data. One-way ANOVA as a statistical method for comparing the means of two or more independent samples was used to identify significant cultural differences in variables, which are distributed normally. The interactional differences between the user groups separated by the primary cultural imprint (e.g., using test language or nationality) can be identified using the Tukey-HSD-Post-Hoc-Test after one-way ANOVA. The analysis of the empirically collected data compares the average values using the IIA data analysis module. For more details on the IIA tool and the test setting, please refer to [19] and [6].

4   Revealing Cultural Influences in HCI

Cultures are orientation systems for group members (cf. [32]). The characteristics of cultures can be described using cultural dimensions (cf. [25]). The interaction behavior of the user with the system can be described with HCI dimensions (cf. [6]). The aim is to determine the connection between cultural dimensions and HCI dimensions and their values using appropriate methods and tools to yield a model for culturally influenced HCI that serves to predict the HCI style of members of any culture. With this knowledge, relevant design recommendation can be derived to develop user interfaces with high usability. Quantitative values for the indices of the cultural dimensions are available from empirical studies by [25]. Further quantitative values for the indicators of the HCI dimensions are also available (cf. [6]). The 916 valid data sets together with the values of the indices of Hofstede were used to determine the connections between the values of the cultural and HCI dimensions. The following methods and tools are used to yield cultural differences in HCI and to determine the relationship between culture and HCI.

In both online studies, some values of the implemented variables in the IIA tool showed significant differences, which represent differences in user interaction according to the different cultural backgrounds of the users. Therefore, these variables can be called "cultural interaction indicators" (CIIs).

The variables in the valid test data sets are not distributed comparably between the first and the second online study. Therefore, in part, the same variables have been analyzed either by ANOVA or by Kruskal-Wallis-test (indicated with F or χ^2 in Table 2). Table 2 presents the most interesting cultural interaction indicators for intercultural HCI design because they represent "directly visible" and "directly hidden" cultural variables that can be derived from the quantitative results of the two online studies. The level of significance is referenced with asterisks (* p<0.05; ** p<0.01).

Table 2. Cultural Interaction Indicators (CIIs) found in both studies most relevant for intercultural HCI design

Cultural Interaction Indicator (CII)	Type	First study	Second study	Absolute mean values (Second study)
MG.Car Speed	NVIV	F(2,102)=8.857**	χ^2 (2,916)=29.090**	(C): 43.67, (G): 27.41
MG.Message Distance	NVIV	F(2,102)=7.645**	F(2,916)=16.241**	(C): 112.69, (G): 141.05
MD.Number Of POI	VIV	F(2,102)=3.143*	χ^2 (2,916)=32.170**	(C): 35.26, (G): 19.91
Maximal Open Tasks	VIV	χ^2 (2,102)=12.543**	F(2,916)=15.140**	(C): 4.64, (G): 2.87
MG.Info Presentation Duration	NVIV	χ^2 (2,102)=17.354**	χ^2 (2,916)=82.944**	(C): 88.11, (G): 93.03
Number Of Characters	VIV	χ^2 (2,102)=16.452**	χ^2 (2,916)=67.637**	(C): 81.08, (G): 197.62

The significant cultural interaction indicators are the following: MG.Car Speed (χ^2 (2, 916) = 29.090**) means the driving speed of the simulated car in the "maneuver guidance test task" ((C) less than (G)). MG.Message Distance (F (2, 916) = 16.241**) denotes the temporal distance until showing the maneuver advice messages in the maneuver guidance test task. (C) expected about 30% more pre-advice ("in x m turn right") than (G) before turning right. This can be an indication of the higher information speed and higher information density in China compared to Germany, for example. MD.Number of POI (χ^2 (2, 916) = 32.17**) counts the number of points of interest (POI) set by the user in the map display test task. Information density increases with the number of POI and is two times higher for (C) than for (G). Max Open Tasks (F (2, 916) = 15.140**) represents the maximum number of open tasks in the working environment (i.e. running applications and icons in the Windows® task bar) during the test session with the IIA data collection module. (C) tend to work on more tasks simultaneously than (G) (ratio (C, G) = 1.7 : 1) which can possibly be explained by the way of work planning (polychronic vs. monochronic timing, cf. [33]) or the kind of thinking (mono-causal (sequential) vs. multi-causal (parallel) logic, cf. [2]). Number of characters (χ^2 (2) = 67.637**) contains the number of characters entered by the user during the maneuver guidance and map display test tasks in answering open questions ((C) < (G)). This is explained by the fact that the Chinese language needs considerably fewer characters to represent words than German.

In addition, the analysis of the log files of the data collection from the second online study using 1632 valid data sets revealed the following four best quantitatively significant cultural interaction indicators (cf. table 3).

Table 3. Cultural interaction indicators derived from log files of the second online data collection

Cultural interaction indicators (CII) derived from log files of the second study						
df = 2	*Oneway ANOVA*			*Kruskal-Wallis*		*Interpretation*
Name of CII	*F*	*p*	*h*	*χ^2*	*p*	*CII is significant*
Test Duration	11.53	0.000	0.404	54.508	0.000	yes, quantitatively
MouseMoves_norm	26.20	0.000	0.225	57.900	0.000	yes, quantitatively
KeyDowns_norm	27.31	0.000	0.318	59.451	0.000	yes, quantitatively
LeftButtonDowns_norm	28.84	0.000	0.266	59.471	0.000	yes, quantitatively

MouseLeftDown_norm ($F_{(2, 916)} = 28.84**$) count the number of mouse clicks with the left mouse key divided by the duration of the total test session. *MouseMove_norm* ($F_{(2, 916)} = 26.20**$) is the number of mouse move event recorded during the complete test session divided by test duration, representing the amount of mouse movements by the user. (C) did about 40% more mouse clicks and mouse moves than (G). At the same time, (G) took about 15% longer than (C) to do the test (*Test duration* ($F_{(2, 916)} = 11.53**$)). *Key Downs_norm* ($F_{(2, 916)} = 27.31*$) is the number of key press events during the complete test session representing the number of pressed keys on the keyboard by the user. (C) produced about 40% more key presses than (G).

All derived CIIs from the log files of the online studies using the IIA tool are quantitatively significant. The significant difference in the group means between (C) and (G) can probably be explained by the fact that the interaction data from the user is logged permanently into a history file and therefore every small difference is recorded. The maximal possible range of values of the HCIDs could thereby be used to reveal any differences in interaction behavior of the user.

The studies with the IIA tool comparing Chinese and German users revealed different interaction patterns according to the cultural background of the users that can be called "cultural interaction patterns" (CIPs). I.e., the cultural interaction indicators can be used to recognize the cultural interaction behavior of the user and to relate these CIPs to the characteristics of the user's culture. Thereby, they represent the individual interaction behavior of the user with the system over time formed by the agglomeration of the values of the CIIs.

5   Discussion

There are many variables involved in conducting cultural studies in HCI. Therefore, it must be ensured that the differences found really depend on cultural aspects and not on demographic causes, the experience of the user with the computer, machine, environment or the methods and means used for example. E.g., it is not trivial to recognize differences in the interaction behavior that are not culture dependent but have demographic causes (e.g. different information reception or other interaction style due to age differences).

Special techniques can be applied to avoid such problems, such as choosing reasonable samples, adjusting the collected data and using control variables. If the disturbing variables are known, they can be consciously observed and controlled as "controlled disturbing variables" in data processing. For example, age, gender and computer experience are all variables that can affect the results negatively because these variables are already influenced by culture and hence, contain implicitly cultural aspects that decrease the variability of the culturally explainable part of the variance in the values of the CIIs. Hence, ANOVA or the Kruskal-Wallis-Test was done to determine the significance level of those variables. Due to the distribution in the data set of these variables, the results from ANOVA or the Kruskal-Wallis-Test must be considered. The results indicate that the variable "age" is a significant variable that contributes to the classification of users according to their test languages. The influence of the disturbing variable of "age" on the validity and on the values of the cultural interaction indicators depend on the fact that the age of the test persons of the different countries was not distributed equally in the samples. For example, there were no Chinese test persons above the age of 39 in the first online study (n=102).

In addition, not all possible disturbing variables can be taken into account because of cultural complexity.

Moreover, the number of analyses and the availability of relevant sets of data are still relatively low for intercultural studies in HCI. In addition, it is problematic to completely bring cultural models into accordance with HCI design. A possible explanatory model containing CIIs necessarily differs from reality (because no model by definition covers all aspects of reality completely). Moreover, the correctness of such an explanatory model varies with the number of CIIs used for one HCI dimension (HCID). In this sense, the explanation strength can only still be weak because until now, each HCID has been represented only by a few CIIs and only some of those CIIs reveal very high separation power.

Furthermore, an enormous interpretation effort is necessary (even in doing quantitative studies) to get plausible and reliable results from which valid conclusions can be drawn. Hence, much work still remains for an adequate explanation model for cultural HCI.

6   Conclusion

The combination of different statistical methods to determine cultural differences and influences in HCI together with big interaction data represents an initial idea and the first step to ascertain the proper relationships between culture and HCI. The data analyzed for the qualitative and quantitative studies revealed a trend for the investigated cultures that permits a shift towards a model of culturally influenced HCI. With the right combination of interaction indicators representing the HCI dimensions, it is possible to capture interaction differences that are culturally imprinted (according to cultural aspects such as nationality, mother tongue, country of birth, etc. or to cultural dimensions).

There are coherences between the interaction of the users with the system and their cultural background. The cultural differences in HCI concern layout (complex vs. simple), information density (from high to low), personalization (from greater to lesser), language (symbols vs. characters), interaction speed (from high to low) and interaction frequency (from high to low).

In addition, the cultural differences found in HCI are quantitatively measurable by a computer system using a special combination of cultural interaction indicators representing cultural interaction patterns depending on the culturally imprinted interaction behavior of the user.

However, the results primarily concern the cultural differences in HCI related to use cases in driver navigation systems. Therefore, recommendations mainly emerge from the results for the design of intercultural user interfaces in driver navigation systems for the global market. Nevertheless, at least the following first clues about the design of user interfaces can be generated for Chinese and German users, which should be taken into account very carefully for the new software architecture of driver navigation systems or the expansion of available systems. For German users, the number of information units (e.g. the number of POI) presented simultaneously (e.g. at map display) should be lower than for Chinese users. Number, duration and frequency of information units (e.g. (system-) news or maneuver announcements) presented sequentially by the system should be appear less often for German users than for Chinese users. At the design stage of information systems, the frequency, speed and way of using the interaction equipment should be considered. For instance, Chinese users click the mouse buttons more frequently and less exactly in comparison to German users.

In conclusion, it is not trivial to derive general guidelines for intercultural HCI design from the results of the studies. On the contrary, it is a challenge to find out the reasons for the differences found in HCI. This awkward situation shall be improved upon in the future by deriving scientifically sound and practically relevant design guidelines from an explanation model of culturally influenced HCI. Revealing cultural influences in HCI by analyzing big data in interactions to finally create and use a model or even a theory for culturally influenced HCI should help to better understand culturally imprinted human information and interaction needs worldwide.

References

1. Winograd, T., Flores, F., Coy, W.: Erkenntnis, Maschinen, Verstehen: Zur Neugestaltung von Computersystemen. Rotbuch Verl., Berlin (1989)
2. Röse, K., Liu, L., Zühlke, D.: Design Issues in Mainland China: Demands for a Localized Human-Machine-Interaction Design. In: Johannsen, G. (ed.) 8th IFAC/IFIPS/IFORS/IEA Symposium on Analysis, Design, and Evaluation of Human-Machine Systems, pp. 17–22. Preprints, Kassel (2001)
3. Marcus, A.: Cross-Cultural User-Experience Design. Diagrammatic Representation and Inference, 16–24 (2006)

4. Vatrapu, R., Suthers, D.: Culture and Computers: A Review of the Concept of Culture and Implications for Intercultural Collaborative Online Learning. In: Ishida, T., Fussell, S.R., Vossen, P.T.J.M. (eds.) IWIC 2007. LNCS, vol. 4568, pp. 260–275. Springer, Heidelberg (2007)

5. Clemmensen, T.: Towards a Theory of Cultural Usability: A Comparison of ADA and CM-U Theory. In: Kurosu, M. (ed.) HCD 2009. LNCS, vol. 5619, pp. 416–425. Springer, Heidelberg (2009)

6. Heimgärtner, R.: Cultural Differences in Human-Computer Interaction. Oldenbourg Verlag (2012)

7. Clemmensen, T., Roese, K.: An Overview of a Decade of Journal Publications about Culture and Human-Computer Interaction (HCI). In: Katre, D., Orngreen, R., Yammiyavar, P., Clemmensen, T. (eds.) HWID 2009. IFIP AICT, vol. 316, pp. 98–112. Springer, Heidelberg (2010)

8. Leidner, D.E., Kayworth, T.: Review: a review of culture in information systems research: toward a theory of information technology culture conflict. MIS Q. 30, 357–399 (2006)

9. Kappos, A., Rivard, S.: A three-perspective model of culture, information systems, and their development and use. MIS Q. 32, 601–634 (2008)

10. Martinsons, M.G., Westwood, R.I.: Management information systems in the Chinese business culture: an explanatory theory. Inf. Manage. 32, 215–228 (1997)

11. Honold, P.: Culture and Context: An Empirical Study for the Development of a Framework for the Elicitation of Cultural Influence in Product Usage. International Journal of Human-Computer Interaction 12, 327–345 (2000)

12. Shen, S.-T., Woolley, M., Prior, S.: Towards culture-centred design. Interact. Comput. 18, 820–852 (2006)

13. Röse, K.: Methodik zur Gestaltung interkultureller Mensch-Maschine-Systeme in der Produktionstechnik. Univ., Kaiserslautern (2002)

14. Young, P.A.: Integrating Culture in the Design of ICTs. British Journal of Educational Technology 39, 6–17 (2008)

15. Holzinger, A.: On Knowledge Discovery and Interactive Intelligent Visualization of Biomedical Data - Challenges in Human-Computer Interaction & Biomedical Informatics. In: 9th International Joint Conference on e-Business and Telecommunications (ICETE 2012), pp. IS9–IS20 (2012)

16. Fisher, D., DeLine, R., Czerwinski, M., Drucker, S.: Interactions with big data analytics. Interactions 19(3), 50–59 (2012)

17. Gellner, M., Forbrig, P.: ObSys–a Tool for Visualizing Usability Evaluation Patterns with Mousemaps. Human Computer Interaction: Theory and Practice (2003)

18. Auinger, A., Aistleithner, A., Kindermann, H., Holzinger, A.: Conformity with User Expectations on the Web: Are There Cultural Differences for Design Principles? In: Marcus, A. (ed.) HCII 2011 and DUXU 2011, Part I. LNCS, vol. 6769, pp. 3–12. Springer, Heidelberg (2011)

19. Heimgärtner, R.: A Tool for Getting Cultural Differences in HCI. In: Asai, K. (ed.) Human Computer Interaction: New Developments, pp. 343–368. InTech, Rijeka (2008)

20. Nielsen, J.: Usability metrics. Alertbox (January 2001)

21. Ford, D., Connelly, C., Meister, D.: Information systems research and Hofstede's culture's consequences: An uneasy and incomplete partnership. IEEE Transactions on Engineering Management 50, 8–25 (2003)

22. Hofstede, G.: VSM94: Values Survey Module 1994 Manual. IRIC, Tilberg (1994)

23. Heimgärtner, R.: Cultural Differences in Human Computer Interaction: Results from Two Online Surveys. In: Oßwald, A. (ed.) Open Innovation, vol. 46, pp. 145–158. UVK, Konstanz (2007)
24. Hall, E.T.: The Silent Language. Doubleday, New York (1959)
25. Hofstede, G.H., Hofstede, G.J., Minkov, M.: Cultures and organizations: software of the mind. McGraw-Hill, Maidenhead (2010)
26. Halpin, A.W., Winer, B.J.: A factorial study of the leader behavior descriptions. In: Stogdill, R.M., Coons, A.E. (eds.) Leader Behavior: Its Description and Measurement. Bureau of Business Research. Ohio State University, Columbus (1957)
27. Kralisch, A.: The Impact of Culture and Language on the Use of the Internet Empirical Analyses of Behaviour and Attitudes, Berlin (2006)
28. Ho, R.: Handbook of univariate and multivariate data analysis and interpretation with SPSS. Chapman & Hall/CRC, Boca Raton (2006)
29. Haykin, S.: Neural networks: a comprehensive foundation. Prentice Hall (2008)
30. Gerken, J., Bak, P., Jetter, H.-C., Klinkhammer, D., Reiterer, H.: How to use interaction logs effectively for usability evaluation. In: BELIV 2008: Beyond Time and Errors - Novel Evaluation Methods for Information Visualization (a CHI 2008 Workshop) (2008)
31. Herczeg, M.: Interaktionsdesign: Gestaltung interaktiver und multimedialer Systeme. Oldenbourg, München (2006)
32. Thomas, A., Kinast, E.-U., Schroll-Machl, S.: Handbook of intercultural communication and cooperation. Basics and areas of application. Vandenhoeck & Ruprecht, Göttingen (2010)
33. Hall, E.T.: Beyond Culture. Anchor Books, New York (1976)

Predicting Student Exam's Scores by Analyzing Social Network Data

Michael Fire, Gilad Katz, Yuval Elovici, Bracha Shapira, and Lior Rokach

Telekom Innovation Laboratories and Information Systems Engineering Department,
Ben-Gurion University of the Negev, Beer-Sheva, Israel
{mickyfi,katzgila,elovici,bshapira,liorrk}@bgu.ac.il

Abstract. In this paper, we propose a novel method for the prediction of a person's success in an academic course. By extracting log data from the course's website and applying network analysis methods, we were able to model and visualize the social interactions among the students in a course. For our analysis, we extracted a variety of features by using both graph theory and social networks analysis. Finally, we successfully used several regression and machine learning techniques to predict the success of student in a course. An interesting fact uncovered by this research is that the proposed model has a shown a high correlation between the grade of a student and that of his "best" friend.

Keywords: Social Network Analysis, Data Mining, Score Prediction, Machine Learning, Web Log Analysis, Multi Graph.

1 Introduction

The ability to predict individual or group success in exams and courses has been researched in the past four decades [1, 2]. Accurately predicting students' exam or course grades has the potential to help students in various ways; by using accurate predictions we can detect early on students who have difficulties with the course materials and help them to improve. Moreover, using this kind of prediction technique can help in several other education-related areas [3]:

1. Discriminating among enrolment applicants.
2. Advising students on their majors.
3. Identifying productive programmers.
4. Identifying employees who might profit best from additional training.
5. Improving computer classes for non-CIS majors.
6. Determining the importance of oft-cited predictors of computer competency, such as gender or math ability.
7. Exploring the relationship between programming abilities and other cognitive reasoning processes.

One common approach for solving this type of prediction problem is to extract as many attributes as possible, sometimes as many as hundreds. By evaluating

R. Huang et al. (Eds.): AMT 2012, LNCS 7669, pp. 584–595, 2012.

the value of each attribute, researchers can attempt to predict exam grades or other variables using linear regression or multiple regression methods. Usually, when using regression, one tries to predict the dependent variables' values using independent attributes of different types. The number of independent variables is very large and includes age and gender [4,5], GPA grades [6], educational level of parents [7], emotional and social factors [8], and even the complexity measure of teachers' lecture notes [9]. Dependent variables include course grade [5,6,10, 11], exam grade [8,12] and even aptitude in computer programming [13]. Other methods used in tackling the grade prediction problem are the factor analysis or other classification schemes with statistical analysis [5,10,14,15]. For example, Rountree et al. [15] used Decision Trees analysis in order to identify combinations of factors that may assist in predicting success or failure in a CS1 class. In this paper, we present a method for building a social network(SN) of students in a course for the purpose of predicting their final exam grades. The method was tested on the course "Computer and Network Security", which was a mandatory course taught by two of the authors of this paper at Ben-Gurion University. The SN contains data collected from 185 participating students from two different departments. The course's SN was created by analyzing the implicit and explicit cooperation among the students while doing their homework assignments. Using this SN, we extracted various social attributes, such as the grade of the student's "best friend". We demonstrate - using linear regression - that the best friend's grade has a direct influence on a student's exam grade. We also demonstrate - by using multiple regression and machine learning (ML) - that other social parameters are also influential in determining a student's grade. Moreover, they can help predict which students are likely to fail the test.

2   Related Work

In the past four decades, a considerable amount of research has gone into detecting factors that affect students' exam and course grades. Most of these studies used regression in order to detect the relevant factors that influence students' grades. Petersen and Howe [11] tried to predict grades in introductory CS courses. They found that previous success in high school mathematics and science had a positive correlation with the CS course grade. Konvalina et al. [12] used students' high school performance and background in mathematics in order to predict final exam grades in the course "Introduction to Computer Science". Butcher and Muth [6] found a linear correlation between different ACT grades and the final CS course grade. Evans and Simkin [3] utilized a 100 question survey in order to predict students' grades. They discovered that the Myers-Briggs cognitive style [17] can assist in predicting students' grades. Chamillard [18] used students' performance in prior academic courses in order to predict their performance in a particular course. Bennedsen and Caspersen [8] studied the influence of emotional and social factors on students' learning outcomes in an introductory computer science courses. Surprisingly, they did not find a linear correlation between the exam grade and the social well-being and emotional health of the students. Recently, Joseph and Suzanne [2] examined peer tutoring's influence on students' performance in CS1 and CS2 courses. They found indications that peer tutoring had a positive impact on a students' course performances, but did not have a significant impact on the course final exam grades. In this paper we too use several regression and ML techniques in order to predict students' final exam grade in the course "Computer and Network Security". In order to carry out these predictions, we mainly used attributes extracted from the course's SN, created by us. Similar techniques that involve social-network analysis and regression were used by Christakis [19] in researching the spread of obesity, and by Altshuler et al. in predicting the individual parameters and social links of smart-phones users [20].

3   Methods and Experiments

To cope with the challenge of predicting students' grades, we extracted features from the course's SN and used regression and ML to analyze them.. We constructed the SN of the course by using different types of homework assignments. Some assignments were individual (and done online), while others required a group effort. After constructing the course's SNs, the next step was to extract the relevant features from these networks. We developed a Python code using the Networkx graph package[1] for this purpose. We then combined the SN graph features with other student data collected throughout the course, such as their grades in the assignments and in the final exam. Finally, we used the collected features in order to test our hypothesis with that a student's SN can influence that particular student's grades. The rest of this section describes in detail the steps taken in order to run our experiments and prove our hypothesis.

3.1   Constructing the Course Social Networks

Different Homework Assignments. We constructed the course SN using the course homework assignments which were a part of the students' final course grade. There were three types of assignments:

- **Online Assignments:** The students received five different online assignments. These assignments were individual and every student was expected to solve them without help. Every student got a different solution, according to his private email address. Despite the difference in the solution itself, all online assignments could be solved by using the same techniques. These assignments were stored in a dedicated website, with every access and every submission to the website recorded in the site's log. By analyzing these logs we were able to identify the students who had cooperated with each other.
- **Coding Assignment:** Students were given one large coding assignment to be completed in pairs. The assignment consisted of developing a web proxy.
- **Theoretical Writing Assignments:** The students received two different theoretical writing assignments. The first assignment was to be done in pairs, while the second could be done in a group of up to four students.

---

[1] http://networkx.lanl.gov/

3.2 Creating the Class Cooperation Social Network Graphs

While most papers about social networks analysis deal solely with information gathered "online", this study draws some of the information used for the generation of the network from the "real world" - social interactions which were conducted "off the grid". By using homework assignments and analyzing the website logs (which contained 10,759 entries), we were able to construct SNs of explicit and implicit cooperation among the students. The implicit connections are used to model all the social interactions that happened "offline" among the students - emails with questions, conversations in the lab while preparing the assignments and even forums dedicated for the course. These connections were very important, as we sought to model the social interactions within the student body. Another motivation for this type of input was the fact that the site offered no means of communications among the students - a thing we consider changing in future work. We defined the course's SN as a weighted multi-graph $G = <V, E>$, where V is the set of vertices in the multi-graph, and each vertex $v \in V$ contains unique information about one of the students that participated in the course. We defined E as the multi-set of links in the graph, with each link $e \in E$ defined to be a tuple $e = (u, v, t, w) \in E$, where $u, v \in V$, w is the weight of the link, and t can be one of the following values: $t \in \{$partner's links, same computer link, same time link$\}$.

We constructed the cooperation multi-graph links in the following manner. First, we defined the partner's links to be the explicit connection among the students who submitted theoretical or coding assignments together as partners. Secondly, we defined the same computer links to be the implicit connection among students who used the same computer while solving the online assignments. In order to discover the same computer links, we used the online assignments logs to extract the following data for each user: user name, IP address, and the browser user-agent string. We used the combination of user IP address and browser user-agent strings to fingerprint computers in almost unique manner [21]. By collecting computer fingerprints, we were able to conclude with a high degree of confidence which users had worked together on the same computer. We then were able to define a link between every pair of users that used the computer. Finally, we defined the same time links to be the second implicit connection among users who probably solved the assignments together but submitted their solution from different computers.

To uncover the same time links, we used the online assignments to extract the timestamp of each submission. We defined two users as connected to each other by a time link if the two had accessed our website in almost the same time on several occasions. For each link, we assigned a weight according to the number of times such a correlation between the users was found. For example, if two students submitted, as partners, three different assignments, then the weight of the partner's link between them would be three. In addition, we also added several information layers to the multi-graph based on the student's information. For each vertex, we added the following information: the student's department, assignments grades, and the final exam score. At the end of the construction

Fig. 1. The course social network multi-graph. Each link color represents connection of a different type; *red* is "partner's link", *yellow* is "same computer link" and *blue* is "same time link". The color of each vertex defines the student's department and each vertex size is in accordance with the student's final test score.

phase, a multi-graph with 184 vertices and 360 links was generated. We used the Cytoscape software[2] to visualize our multi-graph (see Figure 1). In the visualized graph, different link types were displayed by using different colors; red for "partner's link", yellow for "same computer link", and blue for "same time link". The strength of each link in the multi-graph was calculated according to the weight of the link between the vertices where stronger connections were visualized by bolder lines. Additionally, the following visualization elements were also added to the vertices of the multi-graph: a) Each vertex is represented by a different color according to the student's department; blue vertices for students from the Information Systems Engineering (ISE) department and red for students from the Computer Science (CS) department. b) Different vertex sizes were assigned according to the student's final test score; the higher the student's test grade, the bigger the size of his vertex. Using the course weighted multi-graph, we also created a graph describing the SN of the students who participated in the course. This network is a weighted graph where each link's strength is the accumulated weight of the links between a pair of students. Namely, we define $G' = < V, E' >$, where $E' = \{(u, v, w')|u, v \in V$ and $w' = \sum_{((u,v,t,w) \in E)} w\}$. Using the graphs mentioned above, we attempted to estimate the accuracy of our construction process and whether or not the topology and attributes of the graphs could be used in order to predict a student's final test grade.

Calculating the Graph Characteristics. We calculated several graph characteristics (see Table 1): a) general graph statistics (to be explained later on in this paper), b) the number of links of each type, and c) the number of graph components. To evaluate the integrity of our representation of the implicit and explicit links, we calculated several statistics. For each link type, we calculated the ratio of links among students from the same department and links among

---

[2] http://www.cytoscape.org/

Table 1. Multi-Graph Visualization Characteristics

Characteristics	Color	Amount
Total Links	-	360
Total Vertices	-	184
Total Vertices With Final Grade	-	169
ISE Students Vertices	Blue	109
CS Students Vertices	Red	75
Partner's Links	Red	240
Same Computer Links	Yellow	23
Same Time Link	Blue	96

students from different departments (inter and intra-departmental links). We also calculated the percentage of connections among vertices that were both explicit and implicit.

3.3 Predicting Students' Final Test Scores

Based on the SN graphs, we developed a Python code using the Networkx graph package. This code extracted a feature vector for each student based on the SN graphs described in the previous section. Using the extracted features together with the students' personal information, such as a student's department, and assignments score, we attempted to predict the student's final test grade (*final-grade*). This was done by applying several regression and ML techniques.

Features Extraction. Using the students' personal information combined with the SN graphs, we extracted the following features for each student $u \in V$:

Personal Information Features

- *Student assignments scores* the score the student received on each one of his assignments.
- *Students department* - the department the student belongs to.
- *Student Final Test Grade* - The student's final test grade between 0 and 100.

Topological Features

- *Student's degree (number of friends)* - used to define the student's centrality in the network and defined as $d(u) := |\{v|(u,v) \in E'\}|$
- *The number of friends of u by type* - using the multi-graph G, we can extract the number friends u has with respect to a specific connection type:

$$\text{partners-number}(u) := |\{(u,v,t,w) \in G | t = \text{'partners link'}\}|.$$

$$\text{same-time-number}(u) := |\{(u,v,t,w) \in G | t = \text{'same time link'}\}|.$$

$$\text{same-comp-number}(u) := |\{(u,v,t,w) \in G | t = \text{'same computer link'}\}|.$$

Friends Grades Features. We defined several functions and sets designed to make the friends features definitions clearer and easier to understand:

- *The best friend's score* - we attempted to determine the influence one's best friend has on one's scores. One's best friends are defined as the friends with the maximum connection weight to the student in $G^{'3}$

$$\text{best-friend}(u, G^{'}) := \text{bf}(u, G^{'}) :=$$
$$\text{first-element}(\{v | (u, v, w) \in G^{'} \text{ and } w \geq w^{'}, \forall (u, v^{'}, w^{'}) \in G^{'}\}).$$

- *The second best friend's score* the same as the previous feature. If a student is u's best friend after we remove the first best friend from G', then the second best friend of u is defined to be:

$$\text{second-best-friend}(u, G^{'}) := \text{sf}(u, G^{'}) := best - friend(u, G^{'}_{bf(u)})$$

where $G_u := < V - u, E >$.
- We define $w_((u, v), u, v \in V$ to be the weight of edge (u, v) in the weighted social graph, namely: $w_{(u,v)} := \{w | (u, v, w) \in E^{'}\}$.

Using the above definition we can define the following features:

- *The best friend test grade* - defined as:

$$\text{best-friend-grade}(u) := \text{final-grade}(\text{bf}(u))$$

- *The average and weighted average score of the two best friends* - defined as:

$$\text{two-best-friends-avg-grade}(u) := \frac{\text{final-grade}(\text{bf}(u)) + \text{final-grade}(\text{sf}(u))}{2}$$

, and

$$\text{two-best-friends-weighted-avg-grade}(u) :=$$
$$\frac{w_{u,bf(u)}\text{final-grade}(\text{bf}(u)) + w_{u,sf(u)}\text{final-grade}(\text{sf}(u))}{2}$$

- *Student's friends maximum score* - defined as

$$\text{max-fscore}(u) := max(\{\text{final-grade}(v) | (u, v) \in E^{'}\})$$

- *Student's friends minimum score* - defined as

$$\text{min-fscore}(u) := min(\{\text{final-grade}(v) | (u, v) \in E^{'}\})$$

- *Student's average and weighted average friends scores* - defined as

$$\text{avg-friends-score}(u) := \frac{\sum_{(u,v) \in E} \text{final-grade}(v)}{d(u)}$$

---

[3] If there is more than one "best" friend, we arbitrarily chose one vertex from the set.

Predicting Grades. Using the R-project statistical software[4], WEKA [22] (a popular suite of ML) and the features defined in the above subsections, we ran several regression and ML algorithms. Our goal was to find a correlation between the different features and the students' final test grade and attempt to predict who among the students will receive a score of below 60, thus failing the final test. Using regression, the following two experiments were run: a) a simple linear-regression in order to find a correlation between students' grades and their best friends' grades, and b) a multiple linear regression and step-wise regression in order to build a module for predicting the students' grades. Each of the modules was evaluated by calculating the regression *P-value* and *R-square* values. We also evaluated different supervised learning algorithms in an attempt to predict which students are most likely to fail in the final test. We used WEKAs C4.5 (J48) decision tree, IBk, NaiveBayes, SMO, Bagging, AdaBoostM1, Rotation-Forest and RandomForest implementations of the corresponding algorithms. The Bagging, AdaBoostM1, and RotationForest algorithms were evaluated using the J48 as the base classifier.

4 Results

In the following section, we present the results obtained using the methods described in the previous section. The results consist of three parts. First, we present the results of the statistical test used to validate the integrity of the constructed SN graph. Secondly, we present the results of the regression analysis techniques mentioned above. Finally, we present the results of the ML algorithms mentioned in Section 3.3.

4.1 Graph Construction Integrity

To assess the integrity of the constructed SNs graph, we determined how many of the links in the graph are among students of the same department. The results of the analysis showed that 99.58% of the explicit "partners links", 100% of the implicit "same computer links", and 68.75% of the implicit "same time links" were among students from the same department (see Figure 2). In addition, we calculated how many of our implicit links (i.e., "same computer link" and "same time link") were also explicit links (i.e., "partners links"). The results showed that 52% (12 out of 23) of the implicit "same computer links" and 32% (31 out of 97) of the implicit "same time links" were also explicit partners links.

These results are particularly significant considering the a-priori probabilities. With 109 students in the Information Systems Engineering department and 75 in the Computer Science department, the a-priori probability of intra-departmental connection ranges between 0.4 and 0.59 (as one can see, there are hardly any). Moreover, the a-priori probability of an implicit link to be explicit as well is 0.007, while more than half of our implicit links share this trait.

---

[4] http://www.r-project.org/

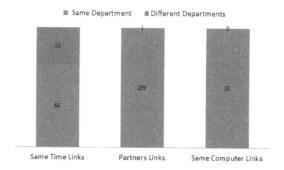

Fig. 2. The distribution of links between students by departments. Each bar illustrates the link type distribution between inter and intra-departmental links.

4.2 Regression Analysis Results

Using the R-project software, we ran several regression algorithms based on the full features vectors that we extracted from 163 students, out of which 41 failed the final exam (with a grade lower than 60). Using the regression algorithms, we generated and evaluated several prediction models in order to predict the students' final grades. These models were mainly based on their SN attributes.

Simple Linear Regression. Using a subset of the SN features described in Section 3.3, we attempted to use the simple linear regression model: $y = \beta_0 + \beta_1 x_1$. Four different features were tested in order to create this simple regression model: a) Using only the Best-Friend-Grade feature in our regression model produced a regression model with a positive slope of 0.2525 ($\beta_1 = 0.2525$) with R^2 (R-square) of 0.0553, *mean absolute error* (MAE) of 11.08 and a *p-value* of 0.002 (see Figure 3), b) Using the *Two-Best-Friends-Avg-Grade* feature produced a regression model with a positive slope of 0.3219, with R^2 of 0.0506, *MAE* of 11.045, and a *p-value* of 0.0038, c) Using the Final exercise grade (Final-Ex-Grade) feature produced a model with R^2 of 0.05458, MAE of 10.929, and *p-value* of 0.0027, and d) Using the number of friends with "same-computer" link type (same-comp-number) feature produced a model with a slope of -4.708 with R^2 of 0.019, and *p-value* of 0.079.

Multiple Regression. In this section, we present the results obtained using the features presented in Section 3.3 and the general multiple linear regression model: $y = \beta_0 + \beta_1 x_1 + \beta_2 x_2 + \ldots + \beta_n x_n$. We constructed a model to predict the final test grades of the students using different feature combinations. When using multiple regression algorithms with all the extracted features, the result was a model that predicted the students' final test grades with *Multiple R-squared* of 0.174, *MAE* of 10.377 and p-value of 0.009. When a regression model that uses only the four exercises grades was created, the result was a *p-value* of 0.02 with *MAE* 10.98. Next, we generated a regression prediction model using

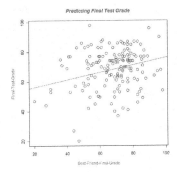

Fig. 3. Linear Regression. Predicting the student's final test grade using the "Best-Friend-Grade" feature.

backwards stepwise regression. This resulted in a worse *Multiple R-squared* value (with a value of 0.1445), but a slightly better p-value of 0.0001 and with *MAE* of 10.644. The model only used five features: the best friend final grade, first and third exercises grades, friend's maximum grade, Final-Ex-Grade, and same computer links number.

Machine Learning Results. Using WEKA, we ran ML algorithms in order to predict which of the 163 students will receive a grade of below 60 and fail the final test. Each of our classifiers was evaluated using 10-fold cross validation approach. We used the area-under-curve (AUC) measure in order to evaluate our results. As expected, the ensemble methods fared best, especially the Rotation Forest and Adaboost classifiers. When all the features were used, the Rotation Forest classifier performed best with an AUC of 0.672. Afterwards, we used T-tests with significance of 0.05 to compare between the different classifiers. According to T-test results, the RotationForest classifiers returned better AUC results than the naive ZeroR and the simple OneR classifiers.

5 Conclusions

To paraphrase an old saying, we believe that, "If you tell me what your friend's grades are, I will tell you what yours will be". We attempted to prove this hypothesis using the SNs of a course taught during the spring semester of 2011. We constructed the course's SN graphs using information we collected from the log of the course website and by analyzing the course homework assignments. We tested the integrity of our graphs by comparing the link distribution of the implicit links to that of the explicit links and looked for correlation. In addition, we also presented a visualization of the SN graphs of the course. Using this visualization, we were able to detect an interesting phenomenon; when looking closely into the SN graph (see Figure 1), one can see that students' final test grades are closely related to those of his friends' grades. We attempted to find an explanation to this phenomenon by using different regression and ML techniques.

Table 2. Regressions Results

Features	p-value	R^2	MAE
Best-Friend-Grade	0.002	0.055	11.08
Two-Best-Friends-Avg-Grade	0.004	0.051	11.05
Final-Ex-Grade	0.003	0.055	10.93
Four Exercises Grades	0.02	0.07	10.98
All-features	0.009	0.174	10.38
Backwards Stepwise Selected Features	0.0001	0.145	10.64

Using multiple linear regressions, we were able to prove that a correlation exists between a students' final grade and that of their friends. This correlation has a small p-value, which strongly supports this hypothesis. Furthermore, the regression models presented in this paper have R-squared values ranging from 0.145 to 0.174, and mean absolute error ranging from 10.38 to 10.64 (See Table 2). These values are not uncommon for human behavioural studies (this is especially true in the field of grade prediction, according to Evans and Smikin [3]). We also discovered interesting phenomenon where there is negative correlation between the final grade and the *same-comp-number* feature (with a slope of -4.708 and *p-value* of 0.079). This correlation may indicate that students whom cheat on their homework assignments and solve their assignments with their friends on the same computer tend to receive lower final test grades. Using supervised learning algorithms we created different classifiers that predicated which students are most likely fail the test. Our Rotation Forest classifier received an AUC of 0.672.

We believe this work has two main future research directions. The first is to use different classification methods combined with ML algorithms to identify students who are likely to fail their final exam in other courses. The second future research direction is to use the SN of the course combined with text analysis techniques (used on the homework assignments) to follow the diffusion of information across the network. We believe it could be very interesting to examine the correlation between homework plagiarism, knowledge diffusion patterns, and success in the course's final exam.

6   Availability

Anonymous version of the students' multi-graph SN topology is available for other researchers to use on our website http://proj.ise.bgu.ac.il/sns/.

References

1. Alspaugh, C.: Identification of some components of computer programming aptitude. Journal for Research in Mathematics Education, 89–98 (1972)
2. Cottam, J., Menzel, S., Greenblatt, J.: Tutoring for retention. In: Proceedings of the 42nd ACM Technical Symposium on Computer Science Education, pp. 213–218. ACM (2011)

3. Evans, G., Simkin, M.: What best predicts computer proficiency? Communications of the ACM 32(11), 1322–1327 (1989)
4. Deckro, R., Woundenberg, H.: Mba admission criteria and academic success. Decision Sciences 8(4), 765–769 (1977)
5. Cronan, T., Embry, P., White, S.: Identifying factors that influence performance of non-computing majors in the business computer information systems course. Journal of Research on Computing in Education 21(4), 431–446 (1989)
6. Butcher, D., Muth, W.: Predicting performance in an introductory computer science course. Communications of the ACM 28(3), 263–268 (1985)
7. Ting, S., Robinson, T.: First-year academic success: A prediction combining cognitive and psychosocial variables for caucasian and african american students. Journal of College Student Development (1998)
8. Bennedsen, J., Caspersen, M.: Optimists have more fun, but do they learn better? on the influence of emotional and social factors on learning introductory computer science. Computer Science Education 18(1), 1–16 (2008)
9. Keen, K., Etzkorn, L.: Predicting students' grades in computer science courses based on complexity measures of teacher's lecture notes. Journal of Computing Sciences in Colleges 24(5), 44–48 (2009)
10. Fowler, G., Glorfeld, L.: Predicting aptitude in introductory computing: A classification model. AEDS Journal 14(2), 96–109 (1981)
11. Petersen, C., Howe, T.: Predicting academic success in introduction to computers. AEDS Journal 12(4), 182–191 (1979)
12. Konvalina, J., et al.: Identifying factors influencing computer science aptitude and achievement. AEDS Journal 16(2), 106–112 (1983)
13. Hostetler, T.: Predicting student success in an introductory programming course. ACM SIGCSE Bulletin 15(3), 40–43 (1983)
14. Campbell, P., McCabe, G.: Predicting the success of freshmen in a computer science major. Communications of the ACM 27(11), 1108–1113 (1984)
15. Rountree, N., Rountree, J., Robins, A., Hannah, R.: Interacting factors that predict success and failure in a cs1 course. ACM SIGCSE Bulletin 36(4), 101–104 (2004)
16. Mazlack, L.: Identifying potential to acquire programming skill. Communications of the ACM 23(1), 14–17 (1980)
17. Allinson, C., Hayes, J.: The cognitive style index: A measure of intuition-analysis for organizational research. Journal of Management Studies 33(1), 119–135 (1996)
18. Chamillard, A.: Using student performance predictions in a computer science curriculum. ACM SIGCSE Bulletin 38(3), 260–264 (2006)
19. Christakis, N., Fowler, J.: The spread of obesity in a large social network over 32 years. New England Journal of Medicine 357(4), 370–379 (2007)
20. Altshuler, Y., Aharony, N., Fire, M., Elovici, Y., Pentland, A.: Incremental learning with accuracy prediction of social and individual properties from mobile-phone data. In: First International Workshop on Wide Spectrum Social Signal Processing (WS³P), Netherlands, Amsterdam (2012)
21. Eckersley, P.: How Unique Is Your Web Browser? In: Atallah, M.J., Hopper, N.J. (eds.) PETS 2010. LNCS, vol. 6205, pp. 1–18. Springer, Heidelberg (2010)
22. Hall, M., Frank, E., Holmes, G., Pfahringer, B., Reutemann, P., Witten, I.H.: The weka data mining software: an update. SIGKDD Explor. Newsl. 11, 10–18 (2009)

SPTrack: Visual Analysis of Information Flows within SELinux Policies and Attack Logs

Patrice Clemente[1], Bangaly Kaba[1], Jonathan Rouzaud-Cornabas[2],
Marc Alexandre[1], and Guillaume Aujay[1]

[1] ENSI de Bourges – LIFO
88 Bd Lahitolle, 18020 Bourges, France
Patrice.Clemente@ensi-bourges.fr
[2] LIP – INRIA – ENS Lyon
9 rue du Vercors, 69007 Lyon, France

Abstract. Analyzing and administrating system security policies is difficult as policies become larger and more complex every day. The paper present work toward analyzing security policies and sessions in terms of security properties. Our intuition was that combining both visualization tools that could benefit from the expert's eyes, and software analysis abilities, should lead to a new interesting way to study and manage security policies as well as users' sessions. Rather than trying to mine large and complex policies to find possible flaws within, work may concentrate on which potential flaws are really exploited by attackers.

Actually, the paper presents some methods and tools to visualize and manipulate large SELinux policies, with algorithms allowing to search for paths, such as information flows within policies.

The paper also introduces a complementary original approach to analyze and visualize real attack logs as session graphs or information flow graphs, or even aggregated multiple-sessions graphs.

Our wishes is that in the future, when those tools will be mature enough, security administrator can then confront the statical security view given by the security policy analysis and the dynamical and real-world view given by the parts of attacks that most often occurred.

1   Introduction

In the field of computer security, system administrators become more and more confronted with large and complex security policies without tools to analyze those policies. Even on security hardened systems enforcing Mandatory Access Control mechanisms like GrSecurity or SELinux, very precise and accurate policies always allow numerous security breaches. In [1], the authors show that on a finely defined SELinux security policy, there remain over than 1 million of potential security breaches, using indirect sequences of interactions on the system. For example, a user can connect on a system and can gain privileges exploiting a flaw in a legitimate program; he can then access to confidential root data. Thus, to analyze such potential flaws, one should need to see and analyze policies, in

R. Huang et al. (Eds.): AMT 2012, LNCS 7669, pp. 596–605, 2012.

the most natural way. As all flows rely on sequences of multiple interactions, the best way is to visualize them as visual graphs.

But visually analyzing policy graphs with thousands of nodes and edges is not that useful and easy. Users should be able to focus their attention of specific parts, properties or sub-paths in the policies. More than that, one can need to visualize policies in terms of security properties violations.

Thus, the paper focuses on an abstract view of confidentiality and integrity security properties: information flows. Our approach provides a method and a tool to track information flows within security policies.

Although the paper presents preliminary work, the approach should fulfill multiple objectives in the future: allowing to detect security breaches in the security policy and modify it in consequence.

Sometimes, analyzing policies is inefficient or lacks experimental or practical feedback. That is why in a second part, the paper faces the problem of system session logs visualization. It provide methods and tools to efficiently aggregate numerous system sessions into one graph, showing several interesting things. First, those graphs show events occurring often within particular sessions or within all sessions. The color of edges is related to that frequency. Applying that method to attack logs gathered from honeypots allow us to detect what potential flaws (of the policy) where actually exploited by real attackers. We concentrate on information flows during attacks. Those are valuable results, as we whish to be able in the future to react on those policies and concentrate on the more dangerous paths on the graph among the millions of flaws in the policy.

The paper is organized as follows. Section 2 surveys related work. Section 3 presents our approach for SELinux system policies visualization using interaction graphs, information flow graphs and tracking information flows within policies. Section 4 presents our approach for the visualization of SElinux system logs and, in particular, system logs of attacks. It introduces session graphs, multiple session graphs and multiple session information flow graphs in order to provide statistical views of attacks and information flows during attacks. Section 5 presents the tool and particular algorithms before concluding with perspectives in Section 6.

2 Related Work

2.1 System Policies Visualization

There have been some work done in the field of analysis and visualization of security policies. In [2], the author surveys more than 20 papers about security visualization, and the most parts of them are focused on policies for network tools. Only a part is related to operating systems: it deals with Rule based Resource Access Control (RBAC) security policies visualization [3], but do not provide any ways to track RBAC violation attempts for example. However, some other work on the visualization of system security policies exist. In [4], the authors deal with the hierarchical visualization of NTFS access control policies. Their approach, transforming Access Control Lists subtrees into (sub-)rectangles cannot be applied to whole operating systems policies. The authors of [5] provide a tool for the visualization and comparison of security policies, to determine their conflicts for example. But the paper do not deal with risks or flaws within a single policy alone. In [6], the authors provide a tool for analyzing and visualize security properties but only deals with small policies and do not confront visualization of static policies with real system execution, or attack sessions. The interesting work of [7] go a step further. The authors provide a tool to query about policies violation of simple security properties based on information flows, or others (e.g. separation of duties). But they do not provide a tool to render, explore and filter large policies. They neither consider the problem of confront policies with real system usage or attacks against it.

2.2 System Logs Visualization

Generally speaking, security monitoring or visualization often rely on network security. Visualization of system session logs neither attack logs do not make an exception to that. In [8], the author largely explain how to deal with information security visualization, but the attention is focused on network information. Even when talking about system events, it is focused on exploiting them to gain IPs of the attackers. In [9,10], the authors provides approach and a research prototype in order to deal with large amount of network data but not provide any consideration of system logs and data. Indeed, many work deal with very specific but deeply studied ways of visualizing networks logs and data, such as [11,12]. Many work, such as [13,14] provides interesting ideas for the visualization of attacks, such as displaying the attacks as a force directed graph or visualizing multiple attacks into one single representation. In [15] the approach is to interactively build the graph, which is quite impossible with large ones. In [16], the authors go a step further with providing statistical view of network attacks.

However, all work are strictly focused on network attacks, with logs collected from NIDS (Networks Intrusion Detection Systems). To our knowledge, there exists no work on the visualization of operating system logs of attacks based on HIDS (Host Based Intrusion Detection System) logs. We think this is a gap that should be filled regarding the objectives previously given: better comprehension of system attacks; further comparing system sessions with system policies; and comparing potential flaws in system policies with real exploits in system attacks. Those are our major motivations for this paper.

3   Visualizing Security Policies

In this section we present our method and related tool designed in order to allow the manipulation of real SELinux security policies, and Security Properties violations among them.

Figure 1 describes the functional architecture of our approach. Starting from a SELinux MAC (Mandatory Access Control) policy, we build a policy graph and possibly an information flow graph.

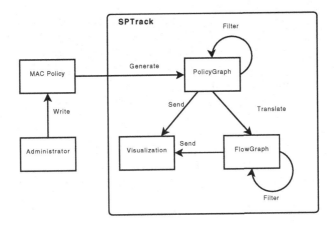

Fig. 1. Functional architecture

3.1   Security Policy Graph

On operating systems, there are entities performing interactions (i.e. operations) on other entities of the system. Under SELinux, such entities are called security contexts. The active ones, e.g. users, processes are called *subjects* and passive ones (files resources, sockets) are called *objects*. A SELinux security policy is defined by rules giving "security permissions" (e.g. read/write, exec) between subjects and objects, regarding the "security class" of the object (e.g. file). Security contexts are represented as a 3-uple of user, role an type related to a given entity. For example, many contexts are related to root. They can be for example like the following: user = *system_u* (i.e. acting as, or belong to), role = *object_r* (i.e. a passive entity), and type = *home_t* (i.e. the root's home directory). Such a context will represented as: *system_u : object_r : home_t*.

We represent SELinux security policies as graphs of contexts where nodes are security contexts subjects or objects and where edges are interactions permissions. More precisely, each interaction is a couple of (*security_class*, *security_permissions*).

A complete SELinux policy is very complex to define and to administrate : thousands of contexts and interactions. Obviously, it is difficult to visualize in a useful way. This is why we introduce information flow graphs.

3.2   Information Flow Graph

Moreover, as we focus here of information flows related security properties, we introduce the notion of information flow graph (IFG). An information flow graph related to a given policy is a sub-graph representing only interactions (edges) able to transfer information between contexts.

Criticality level of interactions. We use a mapping table to build this sub-graph regarding for each couple (*security_class*, *security_permission*) which

type of flow it is and how much its ***criticality*** level is. The criticality refers
to the potential erasability provided by the interaction (syscall) for the subject
to alter the integrity or the confidentiality of the object. For example, the interaction couple (*signal, send*) from a context a to another context b is seen as an
information flow from a to b with a low level criticality. On the other hand, the
context *user_u* (*user_u* : *user_r* : *user_t*) performing a (*file, write*) operation
to a context *user_home_t* (*user_u* : *user_r* : *user_home_t*) as a direct flow from
user_u to *user_home_t*, called a *write_like* operation, is highly critical because
it alter the integrity of the object. Typically, *read* and *write* syscall are the most
critical, whereas *signals* for example are the less ones. Typically, when considering only highly critical flows, the resulting sub-graph of information flows is
generally very smaller than the full one. The criticality level is used to color the
graph from the potential danger of the flow. An example is given in the next
subsection.

Such a vision is quite more interesting to focus on than any previous work.
We defined 4 levels of criticality regarding the values of our mapping table. In
terms of colors, the lowest criticality can be *green* (level 0 to 7), *blue* (8 to 15),
yellow (16 to 25), *red* (26 to 40): the most dangerous flows.

For example, in order to maintain an acceptable confidentiality property between any *root* context (*system_u* : *) and the *user* (*user_u* : *user_r* : *user_t*),
one would have to verify for any flow from *system_u* : * to any *user*'s exists above
the *yellow* level. For a completely restrictive property, no flow at all should be
possible.

Our visualization tool provides many classical search algorithms that takes
into account the criticality level of the edges.

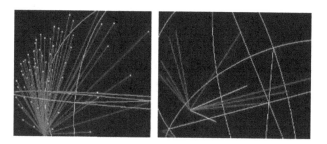

Fig. 2. Security policy graph vs information flow graph

The left part of Fig. 2 shows an example of the visualization of a small security
policy with only 71 nodes and 71 edges. The nodes' labels have been removed
in order to only show the reduction scale of using the IFG. The right part of the
figure presents the corresponding information flow subgraph, recolored according
to criticality values. On Fig. 2, only mid-critical (yellow) and highly critical (red)
flows were kept from the left side.

With large IFG, it can be more visually helpful to filter graphs to only paths between nodes (or superset of nodes), especially for information flow paths. For example, in Fig. 3, the graph is filtered in order to display only information flows between two security contexts (*root_u* : *object_r* : *nscd_var_run_t* and *sysadm_u* : *sysadm_r* : *bootloader_t*), and the labeling of nodes is made only on the highly critical level (in red). The red path here is composed of 4 nodes (i.e. includes 2 intermediary nodes).

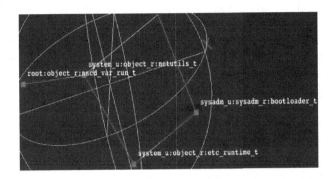

Fig. 3. A potential flow in a filtered policy

Thanks to that approach, the administrator can list the remaining security contexts involved in the information flows displayed. He can then select some of them to precisely analyze and track to/from what contexts can information flow occur with them.

For the real policy we used in our experiments, i.e., a SELinux Gentoo hardened policy, having 1703 security contexts and 175190 aggregated[1] edges between contexts, filtering should be guided by more concrete and actual information. That is why we worked on the analysis of real system sessions, and in particular, real attacks sessions, gathered on our high interaction honeypots. This is the purpose of the next section: visualizing system logs and in particular attacks system logs.

4   Visualizing Attack Session Logs

The purpose of this section is to present how we deal with attack session logs in order to create graph representing system sessions of attacks. Figure 4 describes the global system logs visualization architecture: the construction of sessions graphs (ITG), multiple session graphs and multiple session information flow graphs (see below).

Firstofall, we propose to factorize multiple sessions into one single graph. These session logs were gathered during two years from our high interaction

---

[1] Each edge contains all possible permissions between the 2 linked context, thus leading to an average compression ratio of the number of edges by 75%.

honeypots[2] and thus represents attack sessions, generated with SELinux under auditing mode. Each session logs has been formatted and stored on a large DB2 database with a unique session number. The database stores 130Gb of data divided into 200 millions lines, representing more than 31,000 attack sessions.

4.1 Session Graph

For each attack log, we are able to construct the corresponding so-called session graph. (SG). That graph contains all interactions made since the beginning of the attacks session (e.g. *login* via *ssh*) till its end (e.g. *logout*). The SG shares the same structure as the security policy graph detailed in the previous section. The main difference is that its edges represent interactions that were *actually* performed during the session.

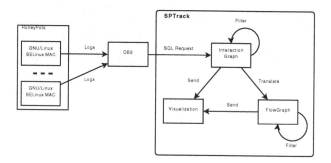

Fig. 4. Session visualization architecture

4.2 Multiple Session Graph

Sessions graphs are very useful to analyze session under our tool. Moreover, as we collected multiple attack sessions, we wanted to make some statistical visualization of multiple sessions, by using one single graph, called multiple session graph (MSG), synthesizing all sessions, with the hypothesis that this will let emerge repetitive behaviors of attackers, or very odd ones.

Our approach consider two criteria for the construction of the MSG, and in particular, for each edge of this graph: (a) the absolute number of occurrences of a given interaction among all the sessions where it occurs; (b) the number of sessions where this interaction occurs.

To ponder the first criterion, we took its logarithm as we thought that the number of sessions if more relevant for the consideration of all attackers behaviors. The following equation 1 is used to assign a score to each edge during the construction of the graph, i.e. during the parsing of the attack logs

---

[2] High interactions honeypots are real machines, not emulated ones, in order to capture attackers' activities.

stored on our database. *edge_number* represent the criterion (a) above, whereas *session_number* is second criterion (b).

$$score = \big(1 + \ln(edge\_number)\big) \times session\_number \qquad (1)$$

As a result for visual exploitation, we defined a color set related to four score ranges following a logarithmic scale, that can be intuitively explained as the following;

- blue: the interaction occurred only one time,
- green: the interaction occurred few times,
- yellow: the interaction occurred multiple times,
- red: the interaction occurred in all stored sessions.

In our experiments, 50 attack sessions that consist of 2 millions of lines of sessions logs were reduced to a graph with 150 nodes and 130 edges. Actually, the MSG aggregates as many sessions logs as we can have. In practice, it is easy to build the total graph of the 31,000 attacks sessions we gathered of the honeypots.

4.3 Multiple Sessions Information Flow Graph

In order to compare potential flaws in policies and real exploit by attackers, we introduce multiple session information flow graphs MSIFG. They are build following the same pattern as the policy IFG. We use the criticality value for each elementary operation. Higher the criticality of elementary operation is, higher the risk of a real information flow has happened on the system. The criticality level can be also used to reduce the graph size by only displaying the real sensitive operations (e.g. red edges) within all sessions. Thus, we are able to find critical paths within the aggregated sessions based on the computed score. Within figure 5, a path can be seen, between the SSH daemon (*system_u* :

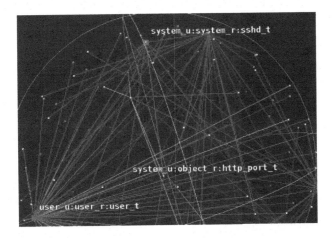

Fig. 5. Flow within a MSIFG

system_r : sshd_t) and user (*user_u : user_r : user_t*) through the the HTTP port (*system_u : object_r : http_port_t*) that goes. It seems to be a path of an attacker getting an interactive shell (*ssh*) via an *http* port. Such a path is quite surprising and would motivate the administrator to closer look at the entire session of that attack. He would then also be able to react adequatively. Moreover, he should track on the policy all possible paths between user and the SSH daemon for example.

5   Conclusion

In this paper, we have presented a global platform called SPTrack[3] for the visualization of both SELinux security policies and SELinux system logs. We have provided method and algorithms to apply the visualization tool to the detection and the visualization of possible information flows on the system regarding its security policy. To do that we transform interaction graphs into information flow graphs, where paths are information flows and edges' colors are criticality level of edges. In parallel, we have provided implemented algorithms to the visualization of SELinux system logs applied to attack sessions gathered from our high interaction honeypots. We have exhibited methods to aggregate thousands or even millions of attack sessions into one single MSG. By transforming MSGs into MSIFGs, we have finally provided ways to analyze the most/least usual information flows among numerous attacks, and furthermore we abled to track those flows given a criticality level.

Currently, our solution can only deal with security properties designed in terms of information flows. However, information flows can be sufficient to model many security properties (integrity and confidentiality of data, subjects, users, domains, binaries, groups). We are working to express more precisely what security property one can analyze. Moreover, we may extend the security properties available: the user should be able to specify which security property he wants to study in the policy/logs. Moreover, the user should be able to select a criticality threshold to those studies, in order to focus 'more dangerous' paths.

Those properties should also be non-information flow based properties, such as separation of duties (no modification of an object followed by its execution from the same user), or trusted path execution for examples (all execution are made within specific binaries repositories).

References

1. Briffaut, J., Lalande, J., Toinard, C.: Formalization of security properties: enforcement for mac operating systems and verification of dynamic mac policies. International Journal on Advances in Security 2(4), 325–343 (2010)
2. Tamassia, R., Palazzi, B., Papamanthou, C.: Graph Drawing for Security Visualization. In: Tollis, I.G., Patrignani, M. (eds.) GD 2008. LNCS, vol. 5417, pp. 2–13. Springer, Heidelberg (2009)

---

[3] SPTrack is based on Walrus open source code [17].

3. Montemayor, J., Freeman, A., Gersh, J., Llanso, T., Patrone, D.: Information visualization for Rule-Based Resource Access Control. In: Proc. of Int. Symposium on Usable Privacy and Security (SOUPS), Citeseer (2006)

4. Heitzmann, A., Palazzi, B., Papamanthou, C., Tamassia, R.: Effective visualization of file system access-control. Visualization for Computer Security, 18–25 (2008)

5. Rao, P., Ghinita, G., Bertino, E., Lobo, J.: Visualization for Access Control Policy Analysis Results Using Multi-level Grids. In: 2009 IEEE International Symposium on Policies for Distributed Systems and Networks, pp. 25–28. IEEE (July 2009)

6. Wahsheh, L.A., Leon, D.C.D., Alves-Foss, J.: Formal Verification and Visualization of Security Policies. Journal of Computers 3(6), 22–31 (2008)

7. Xu, W., Shehab, M., Ahn, G.J.: Visualization based policy analysis: case study in SELinux. In: SACMAT 2008: Proceedings of the 13th ACM Symposium on Access Control Models and Technologies, pp. 165–174. ACM, New York (2008)

8. Marty, R.: Applied Security Visualization. Addison-Wesley Professional (2008)

9. Kolano, P.Z.: A Scalable Aural-Visual Environment for Security Event Monitoring, Analysis, and Response. In: Bebis, G., Boyle, R., Parvin, B., Koracin, D., Paragios, N., Tanveer, S.-M., Ju, T., Liu, Z., Coquillart, S., Cruz-Neira, C., Müller, T., Malzbender, T. (eds.) ISVC 2007, Part I. LNCS, vol. 4841, pp. 564–575. Springer, Heidelberg (2007)

10. McPherson, J., Ma, K.L., Krystosk, P., Bartoletti, T., Christensen, M.: Portvis: a tool for port-based detection of security events. In: VizSEC/DMSEC 2004: Proceedings of the 2004 ACM Workshop on Visualization and Data Mining for Computer Security, pp. 73–81. ACM, New York (2004)

11. Ma, K.: Cyber security through visualization. In: Proceedings of the 2006 Asia-Pacific Symposium on Information Visualisation, vol. 60, p. 7. Australian Computer Society, Inc. (2006)

12. Ball, R., Fink, G., North, C.: Home-centric visualization of network traffic for security administration. In: Proceedings of the 2004 ACM Workshop on Visualization and Data Mining for Computer Security, pp. 55–64. ACM (2004)

13. Mansmann, F., Fischer, F., Keim, D.A., North, S.C.: Visual support for analyzing network traffic and intrusion detection events using TreeMap and graph representations. In: Proceedings of the Symposium on Computer Human Interaction for the Management of Information Technology, CHiMiT 2009, pp. 19–28. ACM Press, New York (2009)

14. Hideshima, Y., Koike, H.: Starmine: a visualization system for cyber attacks. In: APVis 2006: Proceedings of the 2006 Asia-Pacific Symposium on Information Visualisation, pp. 131–138. Australian Computer Society, Inc., Darlinghurst (2006)

15. Noel, S., Jajodia, S.: Managing attack graph complexity through visual hierarchical aggregation. In: Proceedings of the 2004 ACM Workshop on Visualization and Data Mining for Computer Security, VizSEC/DMSEC 2004, p. 109 (2004)

16. Luse, A., Scheibe, K., Townsend, A.: A Component-Based Framework for Visualization of Intrusion Detection Events. Information Security Journal: A Global Perspective 17(2), 95–107 (2008)

17. CAIDA: Walrus - Graph Visualization Tool (2009), http://www.caida.org/tools/visualization/walrus/

Using Mixed Node Publication Network Graphs for Analyzing Success in Interdisciplinary Teams

André Calero Valdez[1], Anne Kathrin Schaar[1], Martina Ziefle[1], Andreas Holzinger[2], Sabina Jeschke[3], and Christian Brecher[4]

[1] Human-Computer Interaction Center, RWTH Aachen University, Aachen, Germany
{calero-valdez,schaar,ziefle}@comm.rwth-aachen.de
[2] Institute for Medical Informatics, Statistics and Documentation, Medical University Graz
andreas.holzinger@medunigraz.at
[3] Institute of Information Management in Mechanical Engineering (IMA),
Center for Learning and Knowledge Management (ZLW), Assoc. Institute for
Management Cybernetics e.V. (IfU), RWTH Aachen University, Aachen, Germany
sabina.jeschke@ima-zlw-ifu.rwth-aachen.de
[4] Laboratory for Machine Tools and Production Engineering, RWTH Aachen University,
Aachen, Germany
c.brecher@wzl.rwth-aachen.de

Abstract. Large-scale research problems (e.g. health and aging, eonomics and production in high-wage countries) are typically complex, needing competencies and research input of different disciplines [1]. Hence, cooperative working in mixed teams is a common research procedure to meet multi-faceted research problems. Though, interdisciplinarity is – socially and scientifically – a challenge, not only in steering cooperation quality, but also in evaluating the interdisciplinary performance. In this paper we demonstrate how using mixed-node publication network graphs can be used in order to get insights into social structures of research groups. Explicating the published element of cooperation in a network graph reveals more than simple co-authorship graphs. The validity of the approach was tested on the 3-year publication outcome of an interdisciplinary research group. The approach was highly useful not only in demonstrating network properties like propinquity and homophily, but also in proposing a performance metric of interdisciplinarity. Furthermore we suggest applying the approach to a large research cluster as a method of self-management and enriching the graph with sociometric data to improve intelligibility of the graph.

Keywords: Publication Network Analysis, Sociometry, Interdisciplinarity, Research Cluster Assessment, Bibliometry, Visualization.

1 Introduction

Interdisciplinarity is a hyped term when it comes to directions of scientific research [2]. Inter- or transdisciplinary approaches promise breakthrough developments [3] by leveraging method competences from different fields in unison. Scientific teams have been shown to outperform solo authors in knowledge generation [4].

R. Huang et al. (Eds.): AMT 2012, LNCS 7669, pp. 606–617, 2012.
© Springer-Verlag Berlin Heidelberg 2012

In order to acquire funding for research scientists often need to look into interdisciplinary approaches to solve real world problems [5]. But interdisciplinarity cannot be achieved by simply combining researchers from different fields into a research group. In contrast, interdisciplinarity – though widely acknowledged as a reasonable research procedure from a technical point of view – suffers from diverse cognitive research models across team members, stemming from different knowledge domains, research languages, methods, models, and procedures. Aggravating, as team members are mostly not aware of threes different professional upbringings, team's cooperation is often not perceived as successful or effective by team members [6].

Efforts have been made to understand how interdisciplinarity must be learnt from a socio-cultural, social, cognitive perspective to gain insights on the learning processes of interdisciplinarity as a faculty [7]. It has been found that successful interdisciplinarity requires a conscious effort, time and resources to establish the required interpersonal relationships for effective communication [8]. Successful teams have also been shown to perform better at interdisciplinarity than newly formed teams [9].

But before one can select measures to improve communication effectiveness or interpersonal relationships it is necessary to determine what factors contribute to interdisciplinary success and furthermore what constitutes interdisciplinary success.

In traditional disciplinary research established and widely accepted methods of measuring success exist. But how can one translate measurements like the judgment of an established community for peer review if no established group of peers exists. Quality of outlets by measuring impact factors might also be inappropriate, because young interdisciplinary fields of research have no established outlets, and acclaimed disciplinary focused outlets might reject interdisciplinary publications due to misunderstanding or out of scope problems [10].

Assuming that publications are a measure of disciplinary success, publication cultures differ between disciplines leaving interdisciplinary research without a unified calibrated measure for success [11].

1.1   Using Publication Network Analysis to Manage Success

Understanding how families of scientific disciplines differ has already been analyzed by Publication Network Analyses [12]. Also flows of citations have been used to analyze development of a research field [13] in highly inter- and transdisciplinary field. Web-Based Data mining of publication data can be used to understand how scientific fields progress [14]. Using graph representation for publication analysis suggests itself because of the innate graph-like structure of publications. Inbetweenness Centrality of Journal Graphs has been used as a measurement for interdisciplinarity in outlets [15].

Even if publications are a valuable measuring tool for whole fields of research, how can one identify latent structures that lead to high quality scientific output in specific interdisciplinary teams. Understanding how groups of people are linked and how they can be affected has been studied in the early 50ies with sociometry [16].

Mapping qualitative data (e.g. who talks with whom) to graphs reveals important nodes and possible change agents to influence the whole social network.

But what are the implications for interdisciplinary teams? Can one do measuring and steering interdisciplinary research efforts by looking as sociometric data and publication networks [17]? Do similarities exist?

The idea of the quantified self [18] defines a new perspective that uses specific (mobile) applications for measuring parameters (vital or habitual) in order to allow self-management. Whenever something is measured intentionally, the outcome is altered during the measurement (by the awareness for the measurement). This effect is often applied in cognitive behavioral therapy by increasing awareness of the measured dimension. This improved awareness increases self-efficacy and thus improvement in behavior [19]. Can this approach be used to allow steering of research groups?

2   Visualizing Publication Networks

The idea for using mixed node graphs for publication network analysis came to us when trying to demonstrate the research efforts of a highly active interdisciplinary research group at RWTH Aachen University (http://www.humtec.rwth-aachen.de/ehealth). The group and its research program started in 2009 (funded by the excellence initiative of German federal and state governments). In order to make research efforts and its success transparent to the German Wissenschaftsrat (the highest scientific board in Germany), we tried to understand how we have worked, why we were successful and what had lead to this development. For this purpose we generated a visualization of our publication behavior. But in order to see the interdisciplinary efforts, we needed something different than simple co-author networks, because the output of the cooperation (namely the publication) should be a part of the representation as well, to match the users mental model [20]. The typical binary co-authorship relationship actually represents an n-ary relationship between n-1 authors and a publication. This is why we tried to use mixed node publication network graphs.

Graph theoretical analyses of bibliometric data usually use single node type network graphs (i.e. all nodes are authors or all nodes are publications). These mostly contain single typed edges (e.g. co-author relationship or citations). The use of mixed node publication network graphs allows a graph to contain more information (than a co-authorship graph) and can easily be reduced to one by using an injective mapping function. Making these entities part of the graph makes visual interpretation easier.

2.1   How the Mixed Node Publication Network Graph Is Constructed

The network graph G is constructed with mixed node types. A node either represents an author (A-Node) a publication (P-Node) or a discipline (D-Node). From a graph theory point of view nodes (i.e. vertices) are not regarded as differently. We define three sets representing authors, publications and disciplines:

$$A = \{ \; a \mid a \text{ is author in ehealth research group} \} \tag{1}$$

$$P = \{ p \mid p \text{ is a publication written by any } a \in A \} \tag{2}$$

$$D = \{ d \mid d \text{ is a discipline studied by any } a \in A \} \tag{3}$$

Then we can define three vertex-mappings f_a, f_p and f_d and three sets of vertices V_1, V_2 and V_3 as follows:

$$f_a : A \rightarrow V_1, f_a(a) = v; \; a \in A \wedge v \in V_1 \tag{4}$$

$$f_p : P \rightarrow V_2, f_p(p) = v; \; p \in P \wedge v \in V_2 \tag{5}$$

$$f_d : D \rightarrow V_3, f_d(d) = v; \; d \in D \wedge v \in V_3 \tag{6}$$

$$\text{with } V_1 \cap V_2 \cap V_3 = \varnothing \tag{7}$$

We define the sets E_1 and E_2 and a weight mapping ω as follows, using that f^{-1} is the inverse of f:

$$E_1 = \{ \; e \mid e=(v_1, v_2), \; v_1 \in V_1 \wedge v_2 \in V_2 \wedge if f_a^{-1}(v_1) \text{ is author of } f_p^{-1}(v_2) \} \tag{8}$$

$$E_2 = \{ \; e \mid e =(v_1, v_3), \; v_1 \in V_1 \wedge v_3 \in V_3 \wedge if f_a^{-1}(v_1) \text{ studied discipline } f_d^{-1}(v_3) \} \tag{9}$$

$$\omega : E \rightarrow \mathbb{R}, \; \omega(e) = 1, \text{ if } e \in E_1 \text{ and } \omega(e) = 0.5, \text{ if } e \in E_2 \tag{10}$$

Then we define two graphs as follow: G_r we call the reduced mixed node publication network graph and G_f we call a full mixed node publication network graph.

$$G_r = (V, E) \text{ with } V = V_1 \cup V_2 \text{ and } E = E_1 \tag{11}$$

$$G_f = (V, E), \text{ with } V = V_1 \cup V_2 \cup V_3 \text{ and } E = E_1 \cup E_2 \tag{12}$$

The reduced and full mixed node publication network graphs are representations of publication networks that can be visualized using standard graph visualization tools. G_r is a bipartite and G_f a tripartite graph.

2.2   Spatial Mapping of the Publication Graph

In order to allow visual analysis by a human person graphs need be lain out graphically. For this purpose we use the open-source software Gephi [21]. Gephi allows graph input by various means (e.g. HTTP-JSON interface) and different layout algorithms.

In this case 2D-spatial mapping is performed by Gephi using it's *Force-Atlas 2* algorithm. Graphs in Gephi allow additional information for graph elements. In particular color, size and labeling can be defined for edges and vertices (i.e. nodes).

For our visualization we set the size of P-Nodes to 10, A-nodes to 50 and D-Nodes to 100 (see fig. 1).

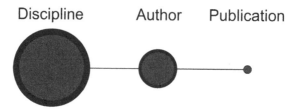

Fig. 1. Three node sizes represent disciplines Authors and publications. Edges represent relationships between nodes.

Using the *Force-Atlas 2* algorithm creates a visual representation according to the following rules:

- All nodes are attracted to the center.
- All nodes repel each other.
- All nodes that are connected by an edge attract each other, according to the weight of the edge (i.e. heavier edges equals stronger attraction).
- Optionally node sizes can be added to the repulsion to prevent visual overlapping.

This allows the following visual conclusions:

- Two A-Nodes are spatially closer if they publish together.
- Two A-nodes are spatially closer if they come from the same discipline.

2.3   Temporal Mapping of Graph

In order to understand development of publication networks temporal mappings or snapshots were required. The idea was that according to publication date nodes and edges were sequentially entered into the graph using the JSON interface of Gephi. This would allow programmatically based animated publication graphs.

Timing of insertion is structured (pauses between new years) to give the impression of stretching time, which allows the layout algorithm to further sort nodes spatially. This sorting and inserting is recorded into a video file. The resulting video of the sorting algorithm is then sped up until it fits into a 90-second clip.

In this clip nodes move according to the attractive forces of the continuously running layout-algorithm giving the impression of a birds-eye-view of moving people that group together. The human brain (even in its early infancy) tends to apply agenticity [22] (infer agents behind patterns) if objects move in atypical non-physical motions [23]. This further enhances the impression of persons moving to find their "peer group" in the publication network graph (see https://vimeo.com/48446978).

2.4   Benefits of Visually Mapping Mixed Node Publication Networks

Gephi allows for several different graph analyses of network graphs. Traditionally these are used with social network graphs (i.e. co-authorship graphs). Interpretation of graph statistics must be reevaluated for mixed node graphs. Graph statistics that are of interest in regard to publication networks are:

- *Number of Weakly Connected Components* [24] refers to the amount of components that are only weakly connected (i.e. only by directed edges in one direction). In an undirected graph they reflect the number of unconnected communities (i.e. subgraphs).
- *Graph density* reflects to the degree of how connected a network graph is. If the density is 1 all nodes are connected with each other. Higher density means that the network is better connected. For bipartite graphs (like G_r) maximal density is limited by:

$$\frac{1}{1+\left(\frac{n-1}{2m}+\frac{m-1}{2n}\right)}, \text{ when } n,m \text{ are the cardinalities of the two parts.}$$

- *Graph Diameter* refers to the maximal distance between any two nodes in a network. The smallest possible diameter is 2 (for G_r) and 3 (for G_f). When more than one discipline exists in a graph the smallest possible diameter can become 5 (if two authors of two different discipline publish tighter). Larger diameters mean that some authors in the network are not publishing together.
- *Average Path Length* refers to the average length from any node to all other nodes. Larger numbers can mean less cooperation or the existence of highly central nodes (that lie on many paths). It cannot be lower than 1 (for G_r) and 2 (for G_f).
- *Average Degree* refers to the average of outgoing edges in the graph, represents the average of publications per author mixed with the average of authors per publication. When using G_f one must be aware of the two confounding influences. The average number of authors per discipline and the average number of disciplines per author. This makes immediate interpretation of this value harder.
- *Betweenness Centrality, Closeness Centrality, Eigenvector Centrality and Eccentricity* [25] are measures for nodes indicating how important they are for finding short paths in the network. The Closeness Centrality reflects the average importance of a node when randomly spreading information to the whole network (which might be used to model communication flow), while Betweenness Centrality reflects the average importance of a node to find a shortest path between two specific nodes. Eigenvector Centrality measures the importance of a node for the total network. Central persons (i.e. Professors) should show high values in Betweenness Centrality, Closeness Centrality and Eigenvector Centrality. Eccentricity refers to the maximum possible distance to any other node for a specific node. It can only be smaller than the diameter and should be high in weakly connected nodes.
- *Modularity and Community Detection* [26] [27] can be used to identify groups in connected graphs that share more edges than randomness would predict. Modularity then measures the amount of how much higher the connections within a community are against connections between communities. Lower values mean that communities interact more with another.

The human mind is capable of analyses that are not computationally easy. Tasks that are relatively easy for the human brain but hard for computers are called "Human Intelligence Tasks" (HIT). Seeing structural changes in a network graph from a meta-perspective is one of those tasks (e.g. seeing whether two subgraphs are connected).

Especially interpreting measures like density and centrality is rather hard for mixed node graphs. Visualization makes the interpretation of these measures fundamentally easier. Enriching visualization with qualitative sociometrical data allows for high quality educated guessing in understanding a mixed node publication network graph[28].

3   Analyzing the Publication Network Visualization of the eHealth Research Group at RWTH Aachen University

Two types of analyses are possible: Graph Statistics from Gephi and informed pattern recognition from humans. Both are performed here as an example. As graph data publication data from the ehealth group is used as a full mixed node publication network graph. The term informed is used in this case because social anatomy of the group is well known by the author. The mixed node graph is shown here (see fig. 2). Furthermore nodes are colored according to the discipline to that they belong.

3.1   Graph Statistics

Applying the Gephi graph analysis reveals the following statistics. The graph contains 14 authors, 6 Disciplines and 198 publications. The average Degree is 3.009 and the diameter of the graph is 6. The average path length is 3.055 Graph density is .014. The graph only contains one weakly connected component, which has 8 communities and a modularity of .512. These results demonstrate a highly interconnected network with short paths between disciplines, authors and publications. In regard to centrality measures (closeness, inverse eccentricity, betweeness and eigenvector), two nodes are prominent P1 (1st place in all measures) and CS1 (2nd place in all measures). Nonetheless P1 and CS1 are dramatically different, as presented in the next section.

3.2   HIT-Analysis

When looking at the animated network graph certain additional factors become obvious, that are hard to see from the statistics point of view. Certain structures become visible which remain hidden from centrality measurements.

In this graph it is obvious that the node P1 plays a structurally important role, which is also predicted by the centrality measures. The node CS1 in contrast is predicted to play an important role, but visually remains on the outskirts of the graph. Looking at the social anatomy of the group reveals why CS1 is not located at the

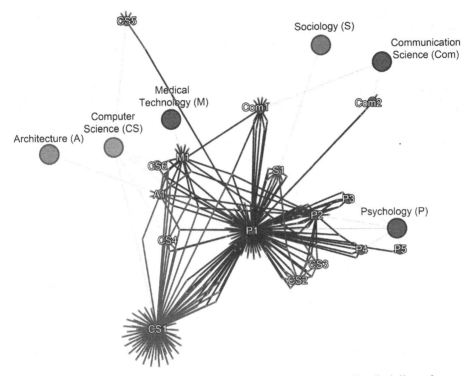

Fig. 2. Gf for the ehealth programme publication data over 3 years. Six disciplines, fourteen authors and 198 publications, (http://www.humtec.rwth-aachen.de/ehealth).

center. The person behind CS1 has had only bi-weekly attendance at the institute, and sits in a single-person office.

Typical social structures reveal themselves in a graph like propinquity and homophily assuming an underlying implicit multiplexity of the edges. Nodes that cluster together come from the same projects (e.g. Com1, A1, CS6, M1, P1), share offices (e.g. CS2, CS3, P2), come from similar discipline (e.g. P1, P2, P3, P4, P5), are friends in their free time (e.g. CS2, CS3) or apply similar methods in their research (e.g. CS2, CS3, P2). Interdisciplinary publication success becomes also visible by looking at the color distribution of the graph. Particularly the group of CS6, M1 and A1 have published very interdisciplinary.

Nodes that are also the outskirts of the graph (COM2 and CS5) are members of the team that have joined our team quite recently.

3.3   Additional Insights

Regarding the user's barriers and fears of new technology (e.g. [29] or [30]) is important. One could have expected that information visualization would have evoked negative and competitive feelings within the group. However, the contrary was the case. When demonstrating the visualization within the group reactions where positive throughout. Not a single member of the team focused on ranking member into a publishing-top-list or anything similar. In contrast members of the group were astonished to see how their publication behavior was so revealing about themselves. Thus, the visualization did evoke additional interest for the group and a hedonic gaming attitude on how to increase interdisciplinary publication behavior as a mean for further team cooperation. For example, some members firstly realized that there are members of the team that shared research interests with them, but have not published together yet. Looking at publications from a revealing of existing and unpublished insights point of view, proved itself to be very helpful. Members reported the visualization be a motivating factor for themselves. This shows that (1) information visualization in form of picturing publication networks can facilitate social behavior and increase team identity and (2) performance measurement does not provoke hostile team behavior, if (1) the reason for the performance visualization is made transparent and (2) the tool can be used as a self-control instrument of the group (rather than by heads only).

4   Cybernetic Application of Publication Network Visualization for Interdisciplinary Innovation Management

This lead to the question whether one can apply this approach in a cybernetic way to allow self-measurement to steer a scientific cluster? In order to test this idea, we first created a reduced mixed node publication network graph for the publications of the cluster of excellence (Integrative Production Technology for High-Wage Countries: http://www.production-research.de). The reduced graph was chosen, because no author information on disciplinarity was publically available (see fig. 3).

Applying the Gephi graph analysis reveals the following statistics. The graph shows an average degree of 3.766 and a Diameter of 23. The average path length is 8.08. Graph density is .005 (maximal theoretical possible density of this graph ~.18). Community analysis reveals 28 Communities and a modularity of .844.

In regard to node statistics two professors' nodes (located in the center of the graph) dominate the centrality measures with one exception. In regard to eigenvector centrality a node from the node-cluster in the top center ranks third. This node is a bridge node that has many strong ties within his group but also weak ties (which are important for allowing information between node-clusters) to another group.

One must wonder whether social analysis of this graph is possible? From various sources we have heard that the just reported bridge-node is also a person that is seen as interested in various topics, communicative and extroverted. This hint might lead to the conclusion that social structures are hidden in a graph but need to be studied on their won. This graph can only be analyzed and interpreted correctly if underlying social parameters are assessed. This could allow analyzing success factors of central nodes on the fly and allow steering by identifying networking agents or designing cluster specific seminars to enhance interconnectivity within a research cluster.

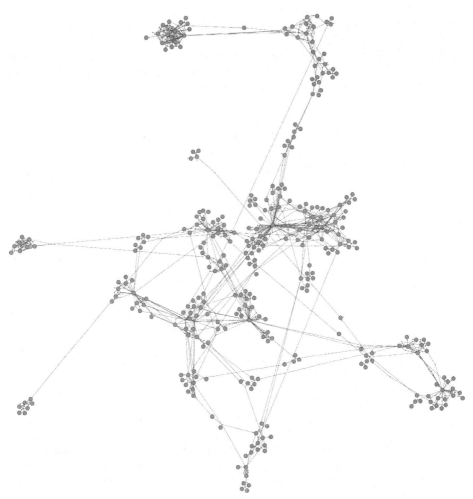

Fig. 3. Gr of publication data of a research cluster over five years of over 2500 publications and 274 Authors. Structures emerge immediately to the human eye.

5   Conclusion

Success factors for interdisciplinary research efforts can be measured by looking at publication network graphs. By using mixed node graphs important real world proper- ties are added to the graph, which simplify human interpretation, by making implicit relationships (i.e. co-authorship) explicit (by showing co-authored work).

But in order to give publication network graphs more meaning further data is re- quired. Parameters like impact factors, citation indices should be incorporated from a bibliometrical point of view. From a sociometric point of view properties like perso- nality traits, motivation types, method competences and many more need to mapped to gain further insights. Furthermore recognition of user's requirements for any social application is important [31].

From a graph theory point of view using mixed node graphs might break the interpretability of some of the used graph statistics but the enhancement of visibility outweighs this problem for the time being.

Acknowledgements. We would like to thank the members of the "ehealth research group" for allowing us to analyze their publication data, especially Philipp Brauner for his helpful advice. Furthermore we thank Dr. Ulrich Prahl and Eva Nick for generously supplying us with validated publication data of the CoE. This research was funded by the Excellence Initiative of the German federal and state governments.

References

1. Ziefle, M., Röcker, C., Wilkowska, W., Kasugai, K., Klack, L., Möllering, C., Beul, S.: A Multi-Disciplinary Approach to Ambient Assisted Living. In: Röcker, C., Ziefle, M. (eds.) E-Health, Assistive Technologies and Applications for Assisted Living, pp. 76–93. IGI Global, Hershey (2011)
2. Jacobs, J.A., Frickel, S.: Interdisciplinarity: A Critical Assessment. Annual Review of Sociology, pp. 43–65. Annual Reviews, Palo Alto (2009)
3. Bordons, M., Zulueta, M.A., Romero, F., Barrigón, S.: Measuring interdisciplinary collaboration within a university: The effects of the multidisciplinary research programme. Scientometrics 46, 383–398 (1999)
4. Wuchty, S., Jones, B.F., Uzzi, B.: The Increasing Dominance of Teams in Production of Knowledge. Science 316, 1036–1039 (2007)
5. Holzinger, A.: Successful Management of Research & Development. BoD–Books on Demand (2011)
6. Ziefle, M., Jakobs, E.M.: New challenges in Human Computer Interaction: Strategic Directions and Interdisciplinary Trends. In: Proceedings of the 4th International Conference on Competitive Manufacturing Technologies, pp. 389–398. University of Stellenbosch, South Africa (2010)
7. Lattuca, L.R.: Learning interdisciplinarity - Sociocultural perspectives on academic work. J. High. Educ. 73, 711–+ (2002)
8. Marzano, M., Carss, D.N., Bell, S.: Working to make interdisciplinarity work: Investing in communication and interpersonal relationships. J. Agric. Econ. 57, 185–197 (2006)
9. Cummings, J.N., Kiesler, S.: Who collaborates successfully?: prior experience reduces collaboration barriers in distributed interdisciplinary research. In: Proceedings of the 2008 ACM Conference on Computer Supported Cooperative Work, pp. 437–446. ACM, New York (2008)
10. Wilson, E.V., Lankton, N.K.: Interdisciplinary research and publication opportunities in information systems and healthcare. Communications of the Association for Information Systems 14, 332–343 (2004)
11. Nedbal, D., Auinger, A., Hochmeier, A., Holzinger, A.: A Systematic Success Factor Analysis in the Context of Enterprise 2.0: Results of an Exploratory Analysis Comprising Digital Immigrants and Digital Natives. E-Commerce and Web Technologies, 163–175 (2012)

12. Grahl, J., Sand, B., Schneider, M., Schwind, M.: Publication Network Analysis of an Academic Family in Information Systems. In: Heinzl, A., Buxmann, P., Wendt, O., Weitzel, T. (eds.) Theory-Guided Modeling and Empiricism in Information Systems Research, pp. 1–13. Physica-Verlag HD (2011)

13. Bassecoulard, E., Lelu, A., Zitt, M.: Mapping nanosciences by citation flows: A preliminary analysis. Scientometrics 70, 859–880 (2007)

14. Cronin, B.: Bibliometrics and beyond: some thoughts on web-based citation analysis. Journal of Information Science 27, 1–7 (2001)

15. Leydesdorff, L.: Betweenness centrality as an indicator of the interdisciplinarity of scientific journals. J. Am. Soc. Inf. Sci. Technol. 58, 1303–1319 (2007)

16. Moreno, J.L.: Foundations of sociometry: An introduction. Sociometry, 15–35 (1941)

17. Agasisti, T., Catalano, G., Landoni, P., Verganti, R.: Evaluating the performance of academic departments: an analysis of research-related output efficiency. Research Evaluation 21, 2–14 (2012)

18. Wolf, G., Carmichael, A., Kelly, K.: The quantified self. TED (2010), http://www.ted.com/talks/gary_wolf_the_quantified_self.html

19. Strecher, V.J., DeVellis, B.M., Becker, M.H., Rosenstock, I.M.: The Role of Self-Efficacy in Achieving Health Behavior Change. Health Educ. Behav. 13, 73–92 (1986)

20. Calero Valdez, A., Ziefle, M., Alagöz, F., Holzinger, A.: Mental Models of Menu Structures in Diabetes Assistants. In: Miesenberger, K., Klaus, J., Zagler, W., Karshmer, A. (eds.) ICCHP 2010, Part II. LNCS, vol. 6180, pp. 584–591. Springer, Heidelberg (2010)

21. Bastian, M., Heymann, S., Jacomy, M.: Gephi: An open source software for exploring and manipulating networks (2009)

22. Shermer, M.: Agenticity. Scientific American 300(6), 36 (2009)

23. Premack, D.: The infant's theory of self-propelled objects. Cognition 36, 1–16 (1990)

24. Tarjan, R.: Depth-first search and linear graph algorithms. In: 12th Annual Symposium on Switching and Automata Theory, pp. 114–121 (1971)

25. Brandes, U.: A faster algorithm for betweenness centrality*. Journal of Mathematical Sociology 25, 163–177 (2001)

26. Blondel, V.D., Guillaume, J.L., Lambiotte, R., Lefebvre, E.: Fast unfolding of communities in large networks. Journal of Statistical Mechanics: Theory and Experiment, P10008 (2008)

27. Lambiotte, R., Delvenne, J.C., Barahona, M.: Laplacian dynamics and multiscale modular structure in networks. Arxiv preprint arXiv:0812.1770 (2008)

28. Holzinger, A.: On Knowledge Discovery and Interactive Intelligent Visualization of Biomedical Data - Challenges in Human–Computer Interaction & Biomedical Informatics. In: Proceedings of the 9th International Joint Conference on e-Business and Telecommunications, Rome, Italy, pp. IS9–IS20 (2012)

29. Schaar, A.K., Ziefle, M.: Smart clothing: Perceived benefits vs. perceived fears. In: 2011 5th International Conference on Pervasive Computing Technologies for Healthcare (PervasiveHealth), pp. 601–608 (2011)

30. Schaar, A.K., Ziefle, M.: Potential of e-Travel Assistants to Increase Older Adults' Mobility. In: Leitner, G., Hitz, M., Holzinger, A. (eds.) USAB 2010. LNCS, vol. 6389, pp. 138–155. Springer, Heidelberg (2010)

31. Calero Valdez, A., Schaar, A.K., Ziefle, M.: State of the (net) work address Developing criteria for applying social networking to the work environment. Work: A Journal of Prevention, Assessment and Rehabilitation 41, 3459–3467 (2012)

On Text Preprocessing for Opinion Mining Outside of Laboratory Environments

Gerald Petz[1], Michał Karpowicz[1], Harald Fürschuß[1], Andreas Auinger[1], Stephan M. Winkler[2], Susanne Schaller[2], and Andreas Holzinger[3]

[1] University of Applied Sciences Upper Austria, Campus Steyr, Austria
{gerald.petz,harald.fuerschuss,michal.karpowicz,
andreas.auinger}@fh-steyr.at
[2] University of Applied Sciences Upper Austria, Campus Hagenberg, Austria
{stephan.winkler,susanne.schaller}@fh-hagenberg.at
[3] Medical University Graz, Medical Informatics, Statistics and Documentation, Austria
andreas.holzinger@medunigraz.at

Abstract. Opinion mining deals with scientific methods in order to find, extract and systematically analyze subjective information. When performing opinion mining to analyze content on the Web, challenges arise that usually do not occur in laboratory environments where prepared and preprocessed texts are used. This paper discusses preprocessing approaches that help coping with the emerging problems of sentiment analysis in real world situations. After outlining the identified shortcomings and presenting a general process model for opinion mining, promising solutions for language identification, content extraction and dealing with Internet slang are discussed.

Keywords: Opinion mining, sentiment analysis, text mining, content extraction, language detection, Internet slang, Web analytics.

1 Introduction and Motivation for Research

The rise of Web 2.0 led to many changes of Internet use; the shift in communication processes allow users to participate and contribute [1], [2]. Neither private users nor companies can avoid this new form of communication, so it is important to pay attention to these communication processes: Content created by users of Web 2.0 applications contains a multitude of market research information and opinions concerning products, brands and corporations [3], through which economic opportunities and risks can be recognized at a very early stage. New strategic, tactical and operational plans and measures can be embraced and new brand messages can be created [4]. One of the biggest challenges for qualitative market research in the Web 2.0 is to handle the variety of information [5] and the handling of the diversity of data, which are mostly weakly structured and non-standardized [6]. Since manual processing of this big data is not manageable, the need for the support of human intelligence by machine intelligence is necessary [7]. Such semi-automatic solutions can be found in the field

R. Huang et al. (Eds.): AMT 2012, LNCS 7669, pp. 618–629, 2012.
© Springer-Verlag Berlin Heidelberg 2012

of opinion mining, which deals with the identification and extraction of subjective opinions about certain topics, e.g. products or services [3].

While performing well in a laboratory environment, where specifically prepared texts are used [8], several problems arise, when using the prototype in real-world situations. These require additional processing to be performed in order to improve the quality of the opinion mining process. Three crucial shortcomings were recognized:

- Contrary to laboratory environments, the content acquired in real-world situations often features a multitude of languages [9, 10]. In the field of opinion mining, where language-specific tools and models are frequently utilized, this is a major problem, since the application of improper methods leads to incorrect or worse sentiment analysis results.
- While social networks, such as *Facebook* or *Twitter*, typically provide programming interfaces, which can be used to acquire clean data, the main content on webpages is almost always surrounded by advertisements, navigational components and other context-irrelevant elements. This so-called *boilerplate* should not be considered, since it provides no valuable information and can decrease the classification accuracy (as described in the context of web mining in [11]).
- User-generated content often includes so called Internet slang, such as emoticons or abbreviations, which is rarely considered in the context of opinion mining. By neglecting this growing phenomenon, a lot of hidden information about the author's sentiment gets lost [12].

The objective of this paper is to discuss techniques that can help coping with the challenges that arise when performing sentiment analysis outside of laboratory environments. In order to achieve this, the utilized prototype and its background are described in the following section. The subsequent section focuses on outlining the identified problems and providing promising solutions to overcome these shortcomings when performing opinion mining in real-world situations.

2   Background and Related Work

2.1   Opinion Mining

In general, opinion mining deals with scientific methods in order to find, extract and systematically analyze views on certain topics, e.g. products, companies or brands. Opinion mining has been studied by many researchers in recent years; one can identify several main research directions [13]: (1) *Sentiment classification*: The goal of sentiment classification is the classification of content according to its sentiment (positive, neutral or negative) about objects (e.g. [14, 15]). (2) *Feature-based opinion mining*: In feature-based opinion mining the sentiment about certain properties of objects (e.g. technical features of a digital camera) is analyzed at sentence level (e.g. [16]). (3) *Comparison-based opinion mining*: Another focus is the comparison-based opinion mining, dealing with the finding of sentences, in which comparisons of similar objects are made (e.g. [17]).

Different technical approaches for opinion mining can be identified: Sentiment classification based on document level and based on sentence level or on feature level. Most classification methods – both at document level and sentence level – are based on the identification of opinion words or phrases. Two approaches are usually used: corpus-based approaches (e.g. [18], [19], [20]) and dictionary-based / lexicon-based approaches (e.g. [16], [14], [21], [22]).

A multitude of algorithms can be used to build models that are able to classify the sentiment of sentences / documents. This includes common approaches such as support vector machines (SVM) [23], linear regression (LR) [24, 25], decision trees (DT) [26], artificial neural networks (ANN) [2], Latent Dirichlet Allocation (LDA) [27], Pointwise Mutual Information (PMI) [27], and genetic programming (GP) [29, 30]. A variety of other approaches and algorithms have been studied: e.g. Markov blanket classification [31], or joint sentiment / topic models (JST) [32].

2.2   Prototype

In the research projects *OPMIN 2.0* and *SENOMWEB*, a software prototype for an opinion mining using a feature-based approach was implemented. This implementation is based on a process model, shown in figure 1, which illustrates the process steps taken in order to find, extract and analyze web data with regard to their sentiment orientation ([33], [34]).

In the first two steps the relevant data sources, such as social networks, forums, websites or RSS-feeds, and appropriate methods for performing the analysis are chosen. The focus of the preprocessing step is the preparation and pre-structuring of the extracted content in order to be able to transform it into a processible structure using methods from the fields of natural language processing and information retrieval in the next phase. In the fifth step the sentiment analysis is performed to determine if a sentence is positive, negative or neutral (cf. section 4). Lastly an evaluation is done to examine the quality of the methods used. The prototype utilizes different kinds of text preprocessing algorithms as well as classification algorithms (e.g. binary classifiers as well as multi-class models [35]). Multi-class classification models assign one specific class (from a set of available classes) to each sample whereas binary classifiers are trained to label samples as belonging to a given class or not.

When using the prototype based on the process model outside of the laboratory environment, several shortcomings could be identified. Since the later steps, especially the analysis, rely on clean and correct content and make use of language-specific models, the recognized problems need to be solved at an early stage. Therefore the techniques and approaches that aim to overcome these issues need to be integrated in the preprocessing phase. Before the processing of Internet slang in the general context of preprocessing, the language must be identified and the relevant content has to be extracted. Thus techniques from the well-established research fields of language detection and content extraction are presented in the sections 3.1 and 3.2, while the subsequent preprocessing of the text is described in section 3.3.

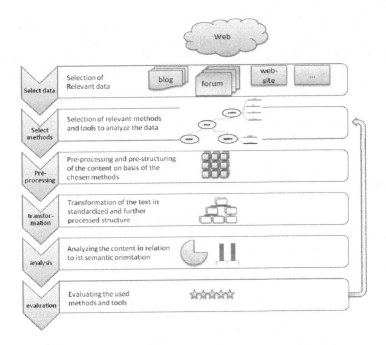

Fig. 1. Process model for opinion mining [34]

3   Approaches for Improving Sentiment Analysis

3.1   Language Identification

As stated above, opinion mining in Web 2.0 has to deal with different languages [10].
In order to perform multilingual opinion mining using language-specific models, that
can be obtained by language-specific training or by using methods that automatically
generate corpora for new languages using available resources (e.g. [36]), techniques
for automatic language identification are required. Being a well-established field of
research many possible solutions for determining the language of written text have
emerged, ranging from approaches based on data-compression techniques [37] to
methods that use language-specific short common words, such as determiners, con-
junctions or prepositions [38]. A set of techniques, which are widely utilized in the
field of language identification [39], are based on character *n-grams*, i.e. sequences of
characters. A representative approach using n-grams was proposed in [40]: First training data is used to create language-specific models, which describe the frequency of
the occurring character sequences. In order to classify a text, the frequencies of the
contained n-grams are calculated and compared to the ones in the available language
models. This way the likelihood of a text to be written in a certain language can be
determined, e.g. a document containing multiple occurrences of the character sequence CZE is more likely to be Polish than English.

This approach has numerous advantages: Not only is it easy to implement, it also is computationally cheap and has very small language models, which don't require a lot of training samples to construct [40]. While providing high accuracy on long texts [41], the technique doesn't perform well on short fragments (e.g. sentences) [38], especially when trying to detect similar languages. This deficiency is problematic in the context of opinion mining: Since user-generated content often contains multiple languages, e.g. sentences using English Internet slang between text in a user's native language, the identification needs to be performed on short texts.

A promising approach that performs better on short text fragments, is presented in [39]: Using dictionary language models obtained by training, each word gets a score based on how specific it is for a language. This value, referred to as relevance, is calculated by comparing the language-specific frequencies of a word with its frequency in a general, language neutral background corpus, which can be approximated by merging all language-specific corpora.

While requiring larger dictionary models and more training compared to approaches based on n-grams, the described dictionary method provides a number of benefits, which can be useful in the opinion mining process. In contrast to n-gram-based techniques, that only return the most likely language, the dictionary approach can return a set of possible languages as well as no language at all in case the identification was not successful. In the context of an opinion mining tool this would enable the involvement of a human to perform the classification, instead of using the most likely but wrong language. Additionally the authors of the dictionary approach describe a fast method for language segmentation, which can be used to identify single-language blocks in a text. This segmentation can improve the precision of language-specific algorithms, increasing the overall quality of multilingual opinion mining.

3.2   Content Extraction

The loss of precision caused by non-relevant data surrounding the main content is a major problem when performing opinion mining on webpages. While human readers can easily distinguish between the relevant parts of a page and the surrounding *boilerplate* (such as navigation components, advertisements or references to related articles), the automatic identification of the main content is a nontrivial task keeping researchers busy since the middle of the 1990s [3]. Because of the possible benefits that content extraction algorithms can provide to many domains – e.g. by improving the precision of search engines or enhancing the readability of web articles on mobile devices – numerous approaches have been developed. We evaluated the following approaches by using a German-language data set.

Techniques. The earliest approaches to solve the problem of extracting the main content of webpages were based on *wrappers*: programs that gather relevant data from a page according to a site-specific set of rules, which are either handwritten or automatically derived from a number of training documents [42]. Since the aforementioned rules need to be established for every website (resp. website template) specifically, wrapper-based solutions not only require a set of example pages or a developer manually defining extraction criteria, but also constant maintenance (in case of design changes etc.). Over the years much progress was made in the field of automatic content extraction: Many approaches have been developed, that resolve the shortcomings of wrapper-based techniques. Because of their multitude, an extensive evaluation of all methods would go beyond the scope of this section. Instead criteria are defined, that content extraction techniques have to meet to be eligible for the use in the opinion mining process:

- Techniques used in the opinion mining process should require little to no manual effort.
- In order to be able to process large numbers of documents, content extraction methods should not be costly regarding performance and resources.
- Eligible techniques must not be specific to certain websites. An approach should not need site-specific training, since the availability of example pages can't be guaranteed in the opinion mining process.
- The approaches should not be based on assumptions regarding structural information. Since web development is an ever-changing technology, no clues regarding the main content should be taken from certain HTML-elements.

According to these criteria three promising approaches have been chosen for further analysis and evaluation.

Content Extraction via Tag Ratios (CETR). [43] describes an approach for automatic content extraction, that is based on a simple observation: While typical boilerplate, such as navigation components or advertisements, contains much code and only few text, content sections mostly consist of text. According to this observation, document lines with a high text to HTML-tag ratio are likely to be part of the relevant content and vice-versa. In order to identify the main text, a histogram with the text-to-tag ratios of a document is constructed and smoothed, which is necessary to prevent the loss of content lines at the edges. [44] propose three methods to obtain the content from a document using the smoothed histogram: applying a threshold using standard deviation to determine if lines are content or boilerplate (*CETR-TM*), utilizing k-means clustering (*CETR-KM*) or constructing a 2D model using the histogram and the absolute smoothed derivatives of the tag ratios to perform 2D clustering (*CETR*).

Maximum Subsequence Segmentation (MSS). [45] proposes the use of maximum subsequence optimization to identify a document's main content. For this approach, a page is first tokenized (broken up into tags, words and symbols). A token-level classifier is then used to find a score for each token, which is based on trigrams (the token and its two successors) and its parent tag. Since the scores indicate the tokens' likelihood to be boilerplate, the relevant content can be identified by finding the maximum subsequence. In order to enable the determination of token scores, the classifier requires training on labeled example documents.

Boilerpipe. [46] analyzed language- and domain-independent boilerplate detection using a number of shallow text features, such as average word or sentence length.

They found out, that, after segmenting a document into atomic text blocks, high accuracy in non-content identification can be achieved through the use of solely two text features: *number of words* and *link density* per block. Based on this research *boilerpipe*, a library for automatic content extraction, has been developed by one of the authors of the original paper.

Evaluation. In this part the evaluation setup, including the data set, implementation and metrics used, as well as the results are presented.

Data set and implementation. For training and evaluation purposes a data set consisting of 700 German-language documents from 205 different sources was prepared. The example pages were arbitrary collected from German and Austrian news sites, whereby no distinction was made regarding the popularity and size of the sites or the topics covered. All documents in the data set were hand-labeled by a human annotator using special markers to flag the relevant content.

In order to perform automatic content extraction on the documents in the data set the respective authors' implementations[1] were obtained. While the *boilerpipe* library didn't need any configuration and was used with the extractor types *ArticleExtractor* (specialized for the use on web articles) and *DefaultExtractor*, the parameters needed for *CETR* were determined empirically: The best results could be achieved when using two clusters (*CETR* and *CETR-KM*) and setting the threshold-coefficient to 2.0 (*CETR-TM*). The *MSS* classifier was trained on two-thirds of the data set, leaving one third for the evaluation. In order to increase the accuracy of the evaluation, ten training-evaluation cycles were performed using randomly selected subsets of the whole data set, the results of which were averaged.

Metrics. To evaluate the techniques, various evaluation metrics are used, each of which illustrates a different aspect of the presented approaches and their performance.

Precision and recall - common measures used in the context of information retrieval [47] - can be used to evaluate content extraction techniques in order to depict the relevance of extracted content. In this paper these measures are calculated on token-level using a bag-of-words approach, where precision, recall and their harmonic mean (F_1-*score*) can be defined as

$$P = \frac{|W_h \cap W_a|}{|W_a|}, R = \frac{|W_h \cap W_a|}{|W_h|}, F_1 = 2 \times \frac{P \times R}{P+R} \tag{1}$$

with W_h being the words found by a human annotator and W_a being the words extracted automatically.

Other metrics that can be used when evaluating content extraction techniques are the Levenshtein edit distance, which represents the number of edit operations (insertions, deletions and substitutions) required to transform one sequence into another, as

---

[1]   *CETR:*http://www.cs.illinois.edu/~weninge1/cetr/,
*MSS:*http://jeffreypasternack.com/demos.aspx,
boilerpipe: http://code.google.com/p/boilerpipe/.

well as the alignment length, which additionally counts the required align operations. For the evaluation in this paper modified versions - working on token-level with no substitution operations allowed - of these metrics are utilized, analogous to the scoring system employed in the *CleanEval* shared task for cleaning webpages [48], that considers a text-only score [45] defined by

$$score(a,b) = 1 - \frac{editDistance(a,b)}{alignmentLength(a,b)} \qquad (2)$$

which, while being expensive to compute, provides a good measure for the quality of the extracted content. Since the described techniques should come into use in the opinion mining process, time is an important factor. To allow a comparison of the approaches, their computational time is measured using a randomly assembled set of 100 documents on which content extraction is performed using a *Dell Latitude E6520* laptop powered by an *Intel Core i5-2520M* processor with 4GB RAM.

Table 1. Results of the evaluation

Technique	Precision	Recall	F_1-Score	Text-only score	Time
Boilerpipe Art.	87.94	92.86	90.33	78.79	1.62 s
Boilerpipe Def.	72.49	90.65	80.56	61.95	0.7 s
CETR	69.26	91.77	78.94	60.05	1.97 s
CETR-KM	70.22	90.35	79.02	60.24	1.19 s
CETR-TM	70.44	90.72	79.30	61.05	0.76 s
MSS	91.29	93.56	92.41	81.72	4.1 s

Results. The results, presented in table 1, show that the extraction using *MSS* with a trained model performs best on the German data set. While having slightly worse scores, the *boilerpipe* article extractor not only runs more than twice as fast, it also requires no language-specific training or modifications, which makes this particular technique very suitable for multilingual opinion mining.

Although performing worst in this evaluation, the approaches based on tag-ratios can be a good and fast option when high recall is top priority. By adjusting the parameters, which were chosen empirically in order to obtain high *CleanEval* text-only scores in this evaluation, the recall rate can be enhanced at the cost of precision. This can be achieved by increasing the number of clusters (*CETR* and *CETR-KM*) or decreasing the threshold coefficient (*CETR-TM*).

3.3   Text Preprocessing

The frequent lack of terminal punctuation and grammar in user-generated content poses a problem for sentence-based sentiment analysis. Therefore text preprocessing plays a major role in opinion mining and further analysis approaches.

In order to obtain a satisfactory sentiment analysis result, the following text preprocessing steps are applied:

1. Splitting into sentences and words
2. Replacing acronyms, symbols and emoticons
3. Stemming

In the first step, text blocks are splitted into their sentences and furthermore each sentence into its words for better handling. Next, to make symbols and emoticons usable for sentiment analysis, a corpus with over 500 symbols and emoticons is utilized in order to make symbols and emoticons usable for sentiment analysis. The meaning of these symbols and emoticons was determined manually. All symbols and emoticons that are detected in the sentences are replaced by significant words that enable a proper sentiment analysis result (e.g. the emoticon : -) is replaced by the word *funny*). Additionally, in order to process acronyms in a useful way, a dictionary which includes the most commonly used abbreviations and its synonyms was built and is applied on each sentence. Finally a stemming tool called *TreeTagger* [49] is used to tag every single word and put it in its principal form. Therefore, the sentences are not only preprocessed and pre-structured, but also contain a unified structure that is used for building a matrix, which conduces as input for machine learning algorithms that are utilized for the analysis.

4   Conclusion and Further Research

This paper discusses problems and promising solutions that could be identified when using a prototypical opinion mining tool outside of laboratory environments. The presented approaches and techniques were put into the context of opinion mining and integrated into a general process model as preprocessing tasks. Further research is needed in order to fully assess the benefits added by utilizing these approaches. Therefore an extensive evaluation of an enhanced opinion mining tool being put to use in real-world situations is planned.

Acknowledgements. This work emerged from the research projects *OPMIN 2.0* and *SENOMWEB*. The project *SENOMWEB* is funded by the European Regional Development fund (*EFRE, Regio 13*). *OPMIN 2.0* is funded under the program *COIN – Cooperation & Innovation*, a joint initiative launched by the Austrian Ministry for Transport, Innovation and Technology (*BMVIT*) & the Ministry of Economy, Family and Youth (*BMWFJ*).

References

1. Alby, T.: Web 2.0. Konzepte, Anwendungen, Technologien, 3rd edn. Hanser, München (2008)
2. Nelles, O.: Nonlinear system identification: from classical approaches to neural networks and fuzzy models. Springer (2001)
3. Liu, B.: Web data mining. Exploring hyperlinks, contents, and usage data, 2nd edn. Data-centric systems and applications. Springer, Berlin (2008)

4. Steinecke, U., Straub, W.: Unstrukturierte Daten im Business Intelligence. Vorgehen, Ergebnisse und Erfahrungen in der praktischen Umsetzung. HMD - Praxis der Wirtschaftsinformatik 47(271), 91–101 (2010)

5. Guozheng, Z., Faming, Z., Fang, W., Jian, L.: Knowledge Creation in Marketing Based on Data Mining. In: International Conference on Intelligent Computation Technology and Automation (ICICTA), vol. 1, pp. 782–786 (2008)

6. Holzinger, A.: Weakly Structured Data in Health-Informatics. In: Proceedings of INTERACT 2011 International Conference on Human-Computer Interaction, Workshop: Promoting and Supporting Healthy Living by Design, pp. 5–7 (2011)

7. Holzinger, A.: On Knowledge Discovery and Interactive Intelligent Visualization of Biomedical Data. In: Proceedings of the 9th International Joint Conference on e-Business and Telecommunications (ICETE 2012), pp. IS9–IS20 (2012)

8. Holzinger, A., Geierhofer, R., Modritscher, F., Tatzl, R.: Semantic Information in Medical Information Systems: Utilization of Text Mining Techniques to Analyze Medical Diagnoses. Journal of Universal Computer Science 14(22), 3781–3795 (2008)

9. Maynard, D., Bontcheva, K., Rout, D.: Challenges in developing opinion mining tools for social media. In: Proceedings of @NLP can u tag #user_generated_content?! Workshop at LREC 2012, Istanbul, Turkey (May 2012)

10. Abbasi, A., Chen, H., Salem, A.: Sentiment analysis in multiple languages: Feature selection for opinion classification in Web forums. ACM Trans. Inf. Syst. 26(3), 12:1–12:34 (2008)

11. Yi, L., Liu, B.: Web page cleaning for web mining through feature weighting. In: Proceedings of the 18th International Joint Conference on Artificial Intelligence, pp. 43–48. Morgan Kaufmann Publishers Inc., San Francisco (2003)

12. Gamon, M.: Sentiment classification on customer feedback data: noisy data, large feature vectors, and the role of linguistic analysis. In: Proceedings of the 20th International Conference on Computational Linguistics (2004)

13. Kaiser, C.: Opinion Mining im Web 2.0 – Konzept und Fallbeispiel. HMD - Praxis der Wirtschaftsinformatik 46(268), 90–99 (2009)

14. Kim, S.-M., Hovy, E.: Determining the Sentiment of Opinions. In: Proceedings of 20th International Conference on Computational Linguistics, Geneva, Switzerland, pp. 1367–1373 (2004)

15. Nadali, S., Masrah, A.A.M., Rabiah, A.K.: Sentiment Classification of Customer Reviews Based on Fuzzy logic. In: Mahmood, A.K. (ed.) International Symposium in Information Technology (ITSim), pp. 1037–1044. IEEE (2010)

16. Hu, M., Liu, B.: Mining and summarizing customer reviews. In: Proceedings of the Tenth ACM SIGKDD International Conference on Knowledge Discovery and Data Mining, pp. 168–177 (2004)

17. Jindal, N., Liu, B.: Mining comparative sentences and relations. In: Proceedings of the 21st National Conference on Artificial Intelligence, vol. 2, pp. 1331–1336. AAAI Press (2006)

18. Hatzivassiloglou, V., Wiebe, J.: Effects of Adjective Orientation and Gradability on Sentence Subjectivity. In: Proceedings of the 18th Conference on Computational Linguistics, pp. 299–305 (2000)

19. Turney, P.D.: Thumbs Up or Thumbs Down? Semantic Orientation Applied to Unsupervised Classification of Reviews. In: Proceedings of the 40th Annual Meeting of the Association for Computational Linguistics, Philadelphia, PA, USA, pp. 417–424 (2002)

20. Wiebe, J., Mihalcea, R.: Word Sense and Subjectivity. In: Proceedings of the 21st International Conference on Computational Linguistics and the 44th Annual Meeting of the Association for Computational Linguistics, pp. 1065–1072 (2006)

21. Ding, X., Liu, B., Yu, P.S.: A Holistic Lexicon-Based Approach to Opinion Mining. In: International Conference on Web Search & Data Mining, Palo Alto, California, February 11-12. ACM, New York (2008)
22. Popescu, A.-M., Etzioni, O.: Extracting Product Features and Opinions from Reviews. In: Proceedings of Human Language Technology Conference, pp. 339–346 (2005)
23. Tong, S., Koller, D.: Support Vector Machine Active Learning with Applications to Text Classification. Journal of Machine Learning Research 2, 45–66 (2001)
24. Weisberg, S.: Applied linear regression, vol. 528. Wiley (2005)
25. Vapnik, V.: The nature of statistical learning theory. Springer (2000)
26. Witten, I., Frank, E.: Data Mining: Practical machine learning tools and techniques. Morgan Kaufmann (2005)
27. Kreuzthaler, M., Bloice, M.D., Faulstich, L., Simonic, K.M., Holzinger, A.: A Comparison of Different Retrieval Strategies Working on Medical Free Texts. Journal of Universal Computer Science 17(7), 1109–1133 (2011)
28. Holzinger, A., Simonic, K.M., Yildirim, P.: Disease-disease relationships for rheumatic diseases Web-based biomedical textmining and knowledge discovery to assist medical decision making. In: IEEE COMPSAC, pp. 573–580 (2012)
29. Koza, J.: Genetic programming II: automatic discovery of reusable programs (1994)
30. Affenzeller, M., Wagner, S., Winkler, S.: Genetic Algorithms and Genetic Programming: Modern Concepts and Practical Applications. Numerical Insights. Taylor & Francis (2009)
31. Bai, X.: Predicting consumer sentiments from online text. Decision Support Systems 50(4), 732–742 (2011)
32. Lin, C., He, Y.: Joint sentiment/topic model for sentiment analysis. In: Proceedings of the 18th ACM Conference on Information and Knowledge Management, pp. 375–384. ACM, New York (2009)
33. Faschang, P., Petz, G., Dorfer, V., Kern, T., Winkler, S.M.: An Approach to Mining Consumer's Opinion on the Web. In: 13th International Conference on Computer Aided Systems Theory, Eurocast 2011, pp. 37–39 (2011)
34. Faschang, P., Petz, G., Wimmer, M., Dorfer, V., Winkler, S.M.: Evaluation of Tools for Opinion Mining. In: EEE (ed.) Proceedings of the 2011 International Conference on E-Learning, E-Business, Enterprise Information Systems & E-Government, Las Vegas, pp. 3–9 (2011)
35. Schaller, S., Winkler, S.M., Dorfer, V., Petz, G., Fürschuß, H.: A Machine Learning Suite for Opinion Mining in Web. In: Proceedings of the 14th International Asia Pacific Conference on Computer Aided System Theory, IEEE APCast (2012)
36. Mihalcea, R., Banea, C., Wiebe, J.: Learning Multilingual Subjective Language via Cross-Lingual Projections. In: Proceedings of the 45th Annual Meeting of the Association of Computational Linguistics, pp. 976–983 (2007)
37. Benedetto, D., Caglioti, E., Loreto, V.: Language Trees and Zipping. Phys. Rev. Lett. 88(4), 48702 (2002), doi:10.1103/PhysRevLett.88.048702
38. Grefenstette, G.: Comparing two language identification schemes. In: Proceedings of the 3rd International Conference on Statistical Analysis of Textual Data (JADT 1995), pp. 263–268 (1995)
39. Řehůřek, R., Kolkus, M.: Language Identification on the Web: Extending the Dictionary Method. In: Gelbukh, A. (ed.) CICLing 2009. LNCS, vol. 5449, pp. 357–368. Springer, Heidelberg (2009)
40. Cavnar, W.B., Trenkle, J.M.: Trenkle: N-Gram-Based Text Categorization. In: Proceedings of SDAIR 1994, 3rd Annual Symposium on Document Analysis and Information Retrieval, pp. 161–175 (1994)

41. Dunning, T.: Statistical Identification of Language (1994)
42. Laender, A.H.F., Ribeiro-Neto, B.A., da Silva, A.S., Teixeira, J.S.: A brief survey of web data extraction tools. SIGMOD Rec. 31(2), 84–93 (2002), doi:10.1145/565117.565137
43. Weninger, T., Hsu, W.H.: Text Extraction from the Web via Text-to-Tag Ratio. In: Proceedings of the 2008 19th International Conference on Database and Expert Systems Application, pp. 23–28. IEEE Computer Society, Washington, DC (2008), doi:10.1109/DEXA.2008.12
44. Weninger, T., Hsu, W.H., Han, J.: CETR: content extraction via tag ratios. In: Proceedings of the 19th International Conference on World Wide Web, pp. 971–980. ACM, New York (2010), doi:10.1145/1772690.1772789
45. Pasternack, J., Roth, D.: Extracting article text from the web with maximum subsequence segmentation. In: Proceedings of the 18th International Conference on World Wide Web, pp. 971–980 (2009)
46. Kohlschütter, C., Fankhauser, P., Nejdl, W.: Boilerplate detection using shallow text features. In: Proceedings of the Third ACM International Conference on Web Search and Data Mining, pp. 441–450. ACM, New York (2010)
47. van Rijsbergen, C.J.: Information Retrieval, 2nd edn. Butterworth-Heinemann, Newton, MA, USA (1979)
48. Baroni, M., Chantree, F., Kilgarriff, A., Sharoff, S.: Cleaneval: a Competition for Cleaning Web Pages
49. Schmid, H.: TreeTagger - a language independent part-of-speech tagger, http://www.ims.uni-stuttgart.de/projekte/corplex/TreeTagger/ (accessed March 10, 2011)

Human Involvement in Designing an Information Quality Assessment Technique
– Demonstrated in a Healthcare Setting –

Shuyan Xie, Markus Helfert, and Lukasz Ostrowski

School of Computing, Dublin City University, Glasnevin,
Dublin 9, Ireland
{shuyan.xie,markus.helfert,lukasz.ostrowski}@computing.dcu.ie

Abstract. Information quality (IQ) has gained increasing importance in the last decade, yet results of assessing and improving IQ in practice are still rare. In this paper we employ a human-centered design approach and illustrate how human involvement in designing an IQ assessment technique can improve the resulting IQ assessment technique. We demonstrate the engagement with practitioners, users, and researchers during the design stages of the assessment technique. Using an emergency medical care (EMS) case our human-centered design approach is scoped to a healthcare setting. The design approach resulted in an improved assessment technique that assists increasing the quality of information exchanges. Our results showed the importance and impacts of human involvement during a design process.

Keywords: Information Quality, Design Science, Human-Centered Design, Information Exchange.

1 Introduction

Information quality (IQ) has become a critical concern of organizations and an active area of Management Information Systems research. The growth of technologies and the direct access of information have increased the need for, and awareness of, high-quality information in organizations [1-5]. Many have indicated that there is a relation between the quality of information and success of organizations. Poor IQ costs billions in society and economic impact [2, 3]. The challenges facing the IQ community within healthcare domain are immense as the tools and methods which collect, process, and use the healthcare-related information are in a constant state of flux. An effective quality based information system for healthcare considered far from sufficient, for example, it is reported between 44,000 and 98,000 [6].

Organizations have increasingly invested in tools and methods to improve IQ, even so, they often still find themselves stymied in their efforts [7]. In the search for IQ improvement approaches, limited attention has turned to the expertise of practitioners. At the same time, users and practitioners have long been acknowledged as important contributions to the success of innovative products and services [8, 9]. The ability of

R. Huang et al. (Eds.): AMT 2012, LNCS 7669, pp. 630–645, 2012.

practitioners to be such effective innovators has been ascribed to a combination of adequate technological expertise and superior knowledge of the user domain so-called use experience [10]. In addition, a general consensus has been reached in relation to the definition for IQ, sometimes used synonymously with data quality (DQ), as being information/data that is "fit for use" [3]. The definition strongly implied that user and practitioner involvement should be considered in IQ assessment.

Having known the importance of human involvement in the design of IQ approaches, most Information Quality (IQ) assessment approaches yet lack human engagement in the early stages of development. The purpose of this paper is, therefore, to propose a human-centered IQ assessment technique design, demonstrated in healthcare emergency medical service (EMS) setting.

Healthcare is known as a service involving various disciplines and its information management has long been a complicated issue. Indeed, as the EMS example shows, it is particularly a time and information critical service; however IQ is often not adequately addressed. Our previous research showed that IQ studies mostly focus low data level and lack business process view on an enterprise level [11]. By engaging practitioners in EMS field we have noticed similar limitations in IQ exercises in practice as well, especially in the quality aspect of information exchange. This observation corresponds to the findings in IQ literatures. The need of the IQ assurance approach arose because they experienced the symptoms of poor information control over routine tasks involving a mix of manual and automated processing concerning the delivery of patient care and patient handoffs (i.e. poor document control, poor data tracking, loss of unacceptable number of information etc.). To design an effective IQ assessment approach, our research focused to engage with practitioners in order to capture their expertise. Results reveal positive feedback towards the proposed conceptual level IQ assessment technique for information exchange.

The rest of the paper is organized as follows: Section 2 provides information of related work and the design approach. Section 3 presents the IQ challenges in EMS practice. Human-centered design approach for an IQ assessment technique development is demonstrated in section 4. Finally, this paper is concluded in section 5.

2   Information Quality Challenges in Emergency Medical Service

IQ is a well-established concept, and it has gained increasing attention during the last years in different fields with different foci. However, there is still a critical need for a methodology that assess how well organizations exercise and ensure IQ in today's dynamic environment. It is a challenging task particularly in the healthcare sector, where they deal with large quantities of vital life saving information. Practitioners in healthcare are facing increasing complexities, especially in the EMS setting that information is timely critical and handled across multiple organizations yet IQ is limitedly emphasized. Investment and improvement projects have been introduced in EMS setting but resulting in unsatisfying outcome [12]. The reason often is that the impractical design and implementation causes user resistance. Therefore, a human-centered design for IQ assessment technique is necessary.

2.1   Information Quality

In today's information era, the quality of accessed and stored information plays a huge role for an organization's operational efficiency. Information quality is commonly defined in the literature as "fit for use" [13], which implies that IQ is relative, as information considered appropriate for one use may not possess sufficient attributes for another use. The information users determine whether the quality of information is sufficient and satisfying. In this sense, involvement with users and practitioners for an effective IQ assessment technique development is essential.

IQ is described as a multi-dimensional concept including accuracy, completeness, timeliness etc. [3, 14] with varying attributed characteristics depending on an individual's philosophical and system interaction point of view. Shanks and Corbitt [15] contend that IQ should be assessed within the context of its generation, while [16] add that it needs to be assessed according to its intended use. Literature has proposed IQ assessment methodologies to address the importance of contexts [17-19]. Healthcare context has been addressed as one of the most challenging in IQ studies [20, 21]. On a daily basis, the media reports on the impact of poor IQ in the healthcare sector [22]. Information is generated, exchanged, and stored with involvement of various processes, actors, and locations etc. that are essential to understanding IQ. The traditional approach to ensure quality predominantly focuses in the technical aspects of quality paying little attentions the human side. This leads to unsatisfying results when the proposed programs being implemented in practice. To bridge the gap, we deploy a human-centered design process to develop an IQ assessment technique, demonstrated in healthcare domain.

2.2   Case Scenario: IQ Practice in EMS Setting

EMS is regarded one of the most essential and critical services in healthcare setting. It is a continuum of care that can be measured and mainly assessed on three units: Dispatch center, ambulance center, and emergency department (ED) [23]. In an emergency response, a large amount of information is created, transferred, and stored across organizations. Information exchange across organizations is an essential routine and the quality has direct impacts on the patient care outcome. Schooley and Horan [24] outline that EMS are time-critical information services – the medical necessity to deliver emergency services as rapidly as possible coupled with the dependence of these services upon timely information from multiple organizations. The information-critical element refers to the fact that this service is highly dependent upon information – from the nature and location of the incident, to the medical needs of the patient that should be attended to at the awaiting hospital [25, 26]. Our previous work stressed that quality information such as accurate information, complete information etc. should be addressed in EMS setting [27].

Healthcare governance is entitled to manage and ensure the quality of the information and information systems that is satisfying and efficient in use. EMS is complicated service that information exchange and sharing is challenging to manage. Information managers are faced with their emerging role in establishing quality management standards for information collection and application in the day-to-day delivery of health care. IQ assessment tools and methods have been introduced to healthcare organizations, for example, the Record Matching method based tool ChiceMaker was successfully implemented in New York City Department of Health and Mental Hygiene [28], and UK nation-wise developing a framework for healthcare information management that integrate patients data [29]. However, most practices are not applicable to the dynamic EMS setting, especially for the information exchange across organizations.

It is found that IQ of information exchange assurance in EMS setting is important yet studies are lacking. We, therefore, in this paper involve practitioners to design a suitable solution to address such limitations. The following phases are undertaken: Phase One – generate ideas by observing the limitations under current IQ assessment in EMS setting. Phase Two – explore a suitable approach to address the limitations. Phase Three – design an appropriate assessment technique by gathering the information from the previous two phases.

3   Human-Centered Design Approach

This section gives an overview of our theoretical development, which is grounded in IQ studies and human-centered design literature. It reveals the need for a new approach regarding the quality of information exchange assessment and improvement.

3.1   Motivation for Human Centered Design

The idea of human involvement is a widely accepted principle in the development of usable devices or systems [30]. A user or practitioner becomes involved when they consider an output for him/her to be of important or significant. This involvement leads to a more consistent behavior and higher satisfaction towards the output [31]. The benefits include the improvement brought to the quality, as it is built with more accurate information in practice. This results in greater acceptance as well as increased user participation [32].

Alam [33] investigated human involvement in new service development, and their findings support the importance of four key elements mentioned in literature:

- *Objective/purpose of involvement* – to develop successful approaches that improve the existing ones to solve the upfront problems.
- *Stage of involvement* - most important stages of the development process involving practitioners/users are idea generation, service/system/methodology/device design, and testing/pilot run.
- *Intensity of involvement* - the most preferred levels of involvement were extensive consultation and information and feedback.
- *Modes of involvement* – face-to-face interviews, user visits and meetings, observation and feedback, and focus group discussion are most common modes.

In this paper, we adopt those four elements to guide our development process with human involvement.

User involvement is perceived to be higher in the earlier stages of development, for example, analysis and design rather than the later implementation stages [30]. Fears that lack of human involvement in the design process, could lead to an illogical user interaction from the practice point of view exist [34]. Various IQ assurance methodologies and approaches have been proposed in IQ literature to ensure the quality of information for organizations [2, 3, 35-39]. However, human engagement at early design stages is not adequately addressed.

3.2   Human-Centered Design Science Approach

It is important to consider which methods are suitable and most beneficial for any particular point in the artifact development. We choose Design Science Research Methodology (DSRM). Design research involves the analysis of the use and performance of designed artifacts to understand, explain and very frequently to improve on the behavior of aspects of information systems. The design science paradigm seeks to extend the boundaries of human and organizational capabilities by creating new and innovative artifacts, including constructs, models, methods, and instantiations [40, 41]. In this paper, the artifact is presented as a method – an IQ assessment technique.

Ostrowski and Helfert [42] state that the act of designing new innovative artifacts does not occur in isolation, it is a process of constant engagement with practitioners. Construction of artifacts is a living process engaging practitioners from the field. The bilateral construction of an artifact falls within the scope of engaged scholarship [43]. In response to the modes of practitioner involvement mentioned in section 2.1, we select relevant methods: observation and feedback, interviews, focus group, user visits and meetings.

Observation and Feedback – Indirect approaches of observation, for example, ethnography or contextual inquiry is beneficial in a busy healthcare environment. Ethnography observes the user in their natural environment, and then interviews separately the involved personnel to get feedback and further insight to their observations. Contextual Inquiry is a more intrusive technique however, and in a busy healthcare environment it can be inappropriate to ask questions as tasks are being performed.

Interviews – can also take a variety of forms, and are adopted according to the type and purpose of analysis they contribute towards. Semi structured interviews are undertaken with separate experts. They allow for the rich exchange of insight with the participants [44].

Focus groups – a research technique, which collects data through the interaction of a group, under the researcher's defined topic [44]. Here, people share similar views and discuss these views to reach a consensus on the issue. We selected related information manager and director from healthcare organizations.

User visits and meetings – the users are invited to several meetings of the development team. The give input on different aspects of the design and development process.

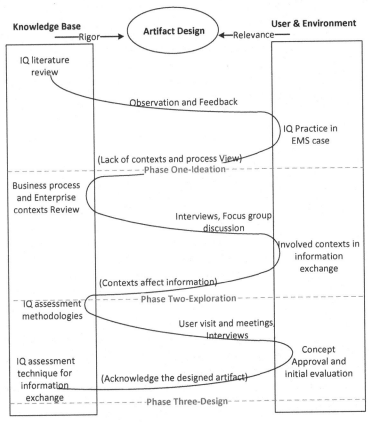

Fig. 1. An IQ assessment technique design by engaging practitioners

Under DSRM, we consistently and continuously engage practitioners and users, while the combination of literature facilitates the understanding of the construction. Fig. 1 presents the process of engaging practitioners in the artifact design with the modes of involvement at different phases.

- Phase One – Ideation: Limitations of IQ assessment approaches found in literature and EMS case. Through observation and feedback we can generate the ideas to solve the problems.
- Phase Two – Exploration: Explore ways to bridge the gap. Interviews and focus group discussion modes are carried out in seek of deeper and precise insights from users and practitioners.
- Phase Three – Design: Propose a technique that addresses the limitations through Interview and User visits and meetings to engage practitioners for the design and feedback.

4   Human Engagement in Developing an IQ Assessment Technique

4.1   Phase One – Ideation

As mentioned in section 3, traditional IQ research focus on static low-data level, based on statistic calculation. Higher business level is not adequately addressed, particularly for dynamic information exchanges. Such limitations are found in practice as well. We conduct field examinations with the EMS unit in Ireland and we view EMS as one enterprise that consists of several organizations. The need of the investigation arose because they experienced problems and challenges in information control over routine tasks.

4.1.1   Field Investigation

Dublin County's EMS is one of the largest emergency response units in Ireland. Currently they have encountered many challenges in assuring IQ in practice. Health Information and Quality Authority (HIQA) is an independent authority that is responsible for quality and performance across EMS organizations including 9-9-9 Control Room, fire services, ambulance services, and health care facilities. Within this role the HIQA strives to find ways to use information to integrate service performance. The EMS system in Dublin has distinguished itself as an early system integrator, pushing data about emergency incidents from computer-aided dispatch (CAD) systems to emergency responders, and a priority dispatch system has been integrated into the CAD system, which connects the Control Room and dispatch centres. However, even with all the IT investment, they still find themselves stymied in their efforts to effectively use the information in the dynamic service environment. One of the central challenges that HIQA faces is to regulate and assure the quality of the information exchange and information flow across EMS organizations. Although Prehospital Emergency Care Council (PHECC) is established for prehospital information system management, challenges to manage information exchange with hospital personnel are still up front. In essence, the authorities mandate certain levels of quality information exchange, compliance with designated emergency response times, health care provision protocols. Considering the information critical characteristic, the authority explores ways to assess and evaluate the quality of information to ensure the quality of the services delivered. However, the current IQ exercises are focused on static digital data that is entered and stored in the computers within each individual organization. For example, within the control room and ambulance stations that information is exchanged through CAD, and IQ assessment is based on the stored information in CAD. Information exchanged between the ambulance crew and hospital is not integrated or assessed at all, especially for the written or verbal information.

4.1.2   Idea Generation: IQ Assessment on an Enterprise Business Level

We observed that the emergency response involving multiple organizations collecting and sharing information processes related to the incident, the patient care, and service performance. For the purpose of this study, we focus on information exchanges across organizations during an emergency incident. Information flow, which is closely connected to the business (EMS delivery) processes. Information accumulates and changes in a dynamic manner as the service progresses across a series of system components and actors. Information itself is presented in different forms such as electronic, paper, and voice, facilitated with two-way radios, application interfaces, and written patient care records etc. From EMS case observation, it clearly presents that information exchange is a process-focused concept where information is connects to location, actor, and purpose etc. enterprise contexts on a business level. Although the exchanged information is linked to processes, the IQ assessment in practice does not consider business processes. In addition, quality of the exchanged information is affected by the contexts that facilitated information exchange across organizations, yet the IQ assessment stays in low data level that business level is not address.

The low-data level does not capture the requisite semantics to accurately communicate information across business processes. As a result, most of the IQ issues exist at the process and organizational boundaries. The top (or business) level is the focal point with the highest probability for discrepancy [11, 45].

Therefore, the idea to develop a novel approach for information exchange is generated: IQ assessment on a business level – connecting information to enterprise contexts by deploying business process concepts.

4.2   Phase Two – Exploration

From phase one, the direction for the IQ assessment technique design is generated. To connect information to other enterprise contexts, it is necessary to explore its feasibility and appropriateness. This section provides theoretical and practical foundations to support our assessment design.

4.2.1   Theoretical Development: Business Process and Enterprise Contexts

A business process is "a set of logically related tasks performed to achieve a defined business outcome" [46]. Business processes play the function of integrating the enterprise, where an aggregation of enterprise contexts that are composed of people, information, and technologies, performing functions agreed purposes, and responding to events [47]. Business process models (BPM) are described as graphic-oriented representations of sequences of activities, typically showing event, actions, and links in those sequences from end to end [48]. Business process model is particularly well suited to cross-functional perspective, classifying activities and identifies important elements in understanding the information exchange [49]. Based on these characteristics of BPM and the observed connections of information exchange to business processes, the academic panel agreed to select BMP as an approach to bring IQ assessment to a business level.

Based on semantics, pragmatics, and the activity theory, and some "contextual" approaches, Leppänen [50] distinguished the context domains of purpose, actor, action, object, facility, location, and time. Schooley and Horan [51] identified the most important three dimensions to analyze interorganizational services. It is founded that the operational dimension of contextual factors for information sharing are

business processes, technological resources, information across organizational boundaries, and **organizational goals and participants.** It is positively corresponded to our findings on 45 articles in information system and management literature of business process and enterprise contexts. As shown in Fig.2 that **business process, technology, information**, and **organization** are the mostly mentioned factors affect an enterprise performance

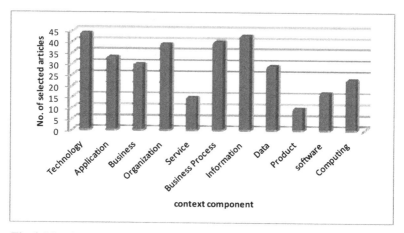

Fig. 2. Most important context factors for information exchange in an enterprise

Under these findings, the proposed IQ assessment technique starts with BPM, through which the context factors of **technology, process, and organization** can be captured and analyzed in relation to the quality of exchanged **information**.

4.2.2   IQ and Enterprise Contexts in Practice

Involvement with practitioners through interviews and focus group discussions allow us to get insight from their experience on how these contexts affect IQ in EMS practice. In addition, users and practitioners can provide either positive or negative feedback about the context approach for IQ by deploying BPM. We summarized the interview and discussion results as below.

Technology View: The Dublin EMS Agency has made significant efforts to collect and utilize incident data to manage service. The efforts have always been made on the separate and disparate information systems that support EMS. It was noticed that very little information is aggregated and shared across all organizations. What happens on the pre-hospital side is often not well known to those who operate within the hospital side.   Pre-hospital patient information was recorded in one system/location, and hospital patient information was recorded in another. It has been a problem of inconsistent, inaccurate, or incomplete patient records that occurred when the information was entered from the pre-hospital Patient Care Records (PCR) to the hospital patient records (HPR), which cause the patient medical history error and inefficient health care delivery.

Processes View: Standardized EMS delivery process is lacking, and attention on design and redesign emergency response processes from end-to-end perspective does not exist. We noticed that the efforts to ensure the data collection and transmission either automatically or manually. There are two processes that appeared most unnatural and problematic: one is that the non-standardized and unstructured communication process about the patient and the incident on way to hospital and in the emergency department (ED). That is where the information found incomplete, time consuming, and incorrect. Another process is re-entering the patient and treatment information based on the written PCR from the ambulance crew. High chances of inconsistent and inaccurate records occur during this manual process.

Organization View: The organizational participants discussed challenges to improving the flow of information. One issue was how a large proportion of incident information is transmitted via voice or hand-written communications and not captured in the system. The authority states that the EMS Agency attempts to address this matter with information technology such as mobile PCR system integrated to the hospitals. However, it has been noticed that hospital staff rarely use the system, and they continue to rely on the traditional methods of receiving incoming patient reports-voice "snapshot" from the paramedics in combination with paper reports. The original purpose behind the system was to advance patient information to the physician prior to patient arrival, but neither the paramedics nor the hospital staff adapts the system due to various reasons.

The practitioners and users agreed that those three enterprise context factors affect the quality of exchanged information, and also showed positive feedback towards BPM approach for IQ assessment.

From the ideation and exploration phases we are able to generate the most practical solutions to improve their current IQ exercise. This means that the following design phase is well supported in practice, as well as rich knowledge foundation. Table 1 summarizes the human involvement contributions towards the design process.

Table 1. Summary of first two phases in developing the IQ assessment technique

Phase	User Involvement/ Contribution
Ideation	Identify IQ Problems and challenges in practice.
Exploration	Seek for realistic and user satisfying solution approaches.

4.3   Phase Three – Design

As analyzed above, the purpose is to design an IQ assessment technique that connects to the contexts on an enterprise business level, based on this EMS case. The concept of business process approach is established in theory and practice during the first two phases. The challenge left is how to link static IQ assessment to the captured contexts of technology, process, and organization in dynamic BPM.

The most important enterprise contexts are identified, and for this research purpose, we consider following characteristics: 1) Organization which includes the organizational participates and rules to perform the processes, 2) Information including the information types and information format, 3) Technology that facilitates the tasks, and 4) Process includes the steps and procedures executed. The context information captured in BPM can be classified of who (organization), what (information type/format), when (process), how (technology). This allows IQ measurement based on the specified enterprise contexts.

Academic and practitioner researchers have produced several generic IQ frameworks [2, 52-54]. Typically, these use a small number of components or dimensions of IQ to group a larger number of IQ criteria or characteristics. IQ dimensions are "a set of IQ attributes that represent a single aspect or construct of IQ" [2]. By identifying different aspects or constructs of IQ it is then possible to either objectively or subjectively measure the quality of information against those aspects or constructs identified. Numerous dimensions including, for example, timeliness, accuracy, coherence, have been identified in the literature. Three functional forms for developing objective data quality metrics [3]. These are 1) simple ratio, 2) minimum or maximum operation and 3) weighted average. Each functional form is appropriate to a specific quality dimension. Pipino etc. [3] also suggest the use of a questionnaire to measure stakeholder perceptions of IQ dimensions lending further substance to this research's initial posit of the importance of an empirical approach.

In this study, we connect the measurable dimensions for information to the contexts that are extracted from BPM, shown in Fig. 3.

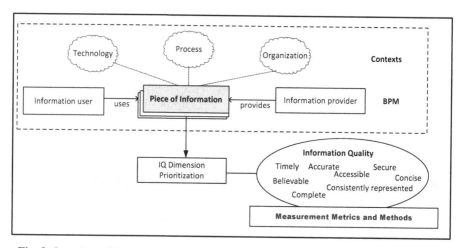

Fig. 3. Overview of the proposed quality assessment technique for exchanged information

In depth discussion formed the IQ assessment technique for information exchange, presented in Fig. 4. The designed technique adopted **Business Process Models** where rich enterprise context can be captured; designed an **Information Profile** where the information content is organized accordingly; and constructed an **IQ Analysis Framework** where IQ measurement and improvement is developed based on previous information. Business process model defines the exchanged information within an enterprise context because it overarches organization and application systems that interact with each other [47]. The Business Process Model then allows to assess statically the "right piece of information from the right source and in the right format is at the right place at the right time" [13], which will be structured in the information profile in form of what, when, who, and how under the dynamic information exchange processes. And finally the IQ Analyzing provides concrete IQ measurement and assessment methods and metrics for the exchanged information.

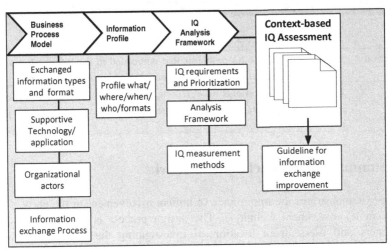

Fig. 4. Overview of the proposed quality assessment technique for exchanged information

4.3.1   Feedback on the Benefits of a Human-Centered Design Approach

The background analysis of information quality, business process and enterprise context, and the EMS case led us to identify several features that are important for understanding the quality assessment for information exchange. In order to gain the insights from academic researchers and practitioner, we visited related personnel – from the Technical University of Dortmund in Germany, Instituto Superior Técnico in Portugal, Dublin City University in Ireland, HIQA, Irish Prehospital Care Council and two hospitals – to discuss the designed IQ assessment technique.

95% of the participants agreed and provided positive views of this novel approach: a context based IQ assessment technique within an enterprise business level is valuable, and the need for such approach is largely due to numerous contextual challenges and barrier in information management. In addition, all participants reached to agreement that this design completed the human-centered design Principles [55]:1) *Early focus*. In design of this IQ assessment, early focus on users and tasks is addressed. As designers, we understood the users, their cognitive behavior, attitudes and the characteristics of their tasks; 2) *Active user participation*. We carefully selected users and practitioners emphasizing the most relevant and skills of typical practitioners and users for IQ exercise. This includes work domain experts and actual end-users; 3) *Multidisciplinary design teams*. In design the proposed technique, we closely involved relevant academic researchers and usability designers in the process. 4) *Integrated design*. The work practices, system examination, in person interaction, desk research etc are conducted in parallel.

The benefits of the human-centered design approach through the EMS case demonstration presented in many ways, shown in Table 2:

Table 2. Benefits of a human-centered design approach for IQ assessment technique

Phase	Benefits
Ideation	Users and practitioners provided useful and practical information and feedback on the problems and challenging.
Exploration	Users identified the IQ problems that embedded in the information exchange processes, which provided design approaches and offered a practical view.
Design	Involvement from fields to research, interaction for comments, feedback, suggestions, acceptance, leaning by designers. These lead to a more practical and satisfying outcome.

5   Summary and Concluding Remarks

This paper demonstrates the importance of human involvement in the early stages to develop an IQ assessment technique. The design process is closely engaged with practitioners and users, from ideation of investigating the existing problems in practice to design a novel artifact as a solution. Positive feedback has been received although more evaluation on the design process and results are needed. As for the implementation and testing of the designed artifact, human involvement is also essential.

In addition to examine human involvement in a design process, this study contributes to the design of an IQ assessment within an enterprise business level by involving practitioners, users, and academic researchers in the early stages. Complementary to our research focus of IQ studies, we had close and in depth interactions in field with EMS involved organizations from the very beginning of problem revealing to the problem solving. Overall, a human-centered design approach benefits the utter output:

- Users provided useful information and ideas;
- Users helped define the scope of the approach;
- Users and practitioners were satisfied and accepted the design, and therefore from practical view it will be an improved approach.

Additional contribution is that such design approach not only brought positive impacts in practice, but also in academic research and theories. The field of IQ research is enriched by providing a technique that considers a high level quality of information assurance for dynamic information exchange environment.

Validation will be carried out by continuous human-centered approach to fully apply and evaluate this technique in EMS case, and generalizing to other domains. A prototypical software tool can be developed in the future based on the technique features.

References

1. Falorsi, P.D., Filiberti, S., Pallara, S.: Improving the quality of toponymic data in the italian public administration. In: Proceedings of the ICDT Workshop on Data Quality in Cooperative Information Systems, DQCIS (2003)
2. Lee, Y.W., et al.: AIMQ: a methodology for information quality assessment. Information & Management 40(2), 133–146 (2002)
3. Pipino, L., Lee, Y., Wang, R.Y.: Data quality assessment. Commun. ACM 45(4), 211–218 (2002)
4. Richard, Y.W.: A product perspective on total data quality management. Commun. ACM 41(2), 58–65 (1998)
5. Shankaranarayan, G., Ziad, M., Wang, R.Y.: Managing Data Quality in Dynamic Decision Environments: An Information Product Approach. IGI Global (2003)
6. Leape Ll, B.D.M.: Five years after to err is human: What have we learned? JAMA: The Journal of the American Medical Association 293(19), 2384–2390 (2005)
7. Madnick, E.S., et al.: Overview and Framework for Data and Information Quality Research. J. Data and Information Quality 1(1), 1–22 (2009)
8. von Hippel, E.: Lead Users: A Source Of Novel Product Concepts. Management Science 32(7), 791–805 (1986)
9. Morrison, P.D., Roberts, J.H., Von Hippel, E.: Determinants of User Innovation and Innovation Sharing in a Local Market. Management Science 46(12), 1513 (2000)
10. Magnusson, P.R.: Exploring the Contributions of Involving Ordinary Users in Ideation of Technology-Based Services. Journal of Product Innovation Management 26(5), 578–593 (2009)
11. Xie, S., Helfert, M.: An Architectural Approach to Analyze Information Quality for Inter-organizational Service. In: International Conference on Enterprise Information Systems (ICEIS), Beijing, China, pp. 438–443 (2011)
12. Messelken, M., et al.: The Quality of Emergency Medical Care in Baden-Württemberg (Germany): Four Years in Focus. Dtsch Arztebl International 107(30), 523–530 (2010)
13. Wang, R.Y., Strong, D.M.: Beyond accuracy: what data quality means to data consumers. Journal of Management Information System 12(4), 5–33 (1996)
14. Ballou, D., et al.: Modeling Information Manufacturing Systems to Determine Information Product Quality. Management Science 44(4), 462–484 (1998)
15. Shanks, G., Corbitt, B.: Understanding Data Quality: Social and Cultural Aspects. In: Proc. 10th Australasian Conference on Information Systems (1999)
16. Katerattanakul, P., Siau, K.: Measuring information quality of web sites: development of an instrument. In: Proceedings of the 20th International Conference on Information Systems 1999, pp. 279–285. Association for Information Systems, Charlotte (1999)
17. Fehrenbacher, D.D., Helfert, M.: Contextual Factors Influencing Perceived Importance and Trade-offs of Information Quality. Communications of the Association for Information Systems 30(8) (2012)
18. Ge, M., Helfert, M.: A Review of Information Quality Research. In: A Review of Information Quality Research. MIT, Cambridge (2007)
19. Pham Thi, T.T., Helfert, M.: Modelling Information Manufacturing Systems. International Journal of Information Quality 1(1), 5–21 (2007)
20. Levis, M., Brady, M., Helfert, M.: Identifying Imformation Quality Problems in a Healthcare Scenario. In: Nunes, M., Isaías, P., Powell, P. (eds.) (2008)

21. Simonic, K.M., et al.: Optimizing Long-Term Treatment of Rheumatoid Arthritis with Systematic Documentation. In: Proceedings of the 5th International Conference on Pervasive Computing Technologies for Healthcare, Pervasive Health. IEEE, New York (2011)
22. Omachonu, V.K., Ross, J.E.: Principles of Total Quality. CRC Press (2004)
23. Dave, G., Parmar, K.: Emergency Medical Services and Disaster Management. Jaypee Brothers Publishers (2002)
24. Schooley, B., Horan, T.: End-to-End Enterprise Performance Management in the Public Sector through Inter-organizational Information Integration. Government Information Quarterly 24, 755–784 (2007)
25. Dawes, S.S., Prefontaine, L.: Understanding New Models of Collaboration for Delivery Government Services. Communications of the ACM 46(1), 40–42 (2003)
26. Horan, T.A., Marich, M., Schooley, B.: Time-critical information services: analysis and workshop findings on technology, organizational, and policy dimensions to emergency response and related e-governmental services. In: Proceedings of the 2006 International Conference on Digital Government Research, San Diego, California, pp. 115–123 (2006)
27. Xie, S., Helfert, M.: Assessing Information Quality Deficiencies in Emergency Medical Service. In: 15th Information Conference on Information Quality Conference, Arkansas, United States (2010)
28. Papadouka, V., et al.: Integrating the New York Citywide Immunization Registry and the Childhood Blood Lead Registry. Journal of Public Health Management and Practice 10, S72–S80 (2004)
29. Alshawi, S., Missi, F., Eldabi, T.: Healthcare information management: the integration of patients' data. Logistics Information Management 16(3/4), 286–295 (2003)
30. Kujala, S.: User involvement: A review of the benefits and challenges. Behaviour & Information Technology 22(1), 1–16 (2003)
31. Barki, H., Jon, H.: Rethinking the Concept of User Involvement. MIS Quarterly 13(1), 53–63 (1989)
32. Foster Jr., S.T., Franz, C.R.: User involvement during information systems development: a comparison of analyst and user perceptions of system acceptance. Journal of Engineering and Technology Management 16(3-4), 329–348 (1999)
33. Alam, I.: An exploratory investigation of user involvement in new service development. Journal of the Academy of Marketing Science 30(3), 250–261 (2002)
34. Allen, C.D., et al.: User involvement in the design process: why, when & how? In: Proceedings of the INTERACT 1993 and CHI 1993 Conference on Human Factors in Computing Systems, pp. 251–254. ACM, Amsterdam (1993)
35. Batini, C., et al.: A comprehensive data quality methodology for web and structured data. Int. J. Innov. Comput. Appl. 1(3), 205–218 (2008)
36. Batini, C., et al.: Methodologies for data quality assessment and improvement. ACM Comput. Surv. 41(3), 1–52 (2009)
37. De Amicis, F., Batini, C.: A methodology for data quality assessment on financial data. Studies in Communication Science 4(2), 115–136 (2004)
38. Richard, Y.W., Diane, M.S.: Beyond accuracy: what data quality means to data consumers. J. Manage. Inf. Syst. 12(4), 5–33 (1996)
39. Wang, R.Y., Storey, V.C., Firth, C.P.: A framework for analysis of data quality research. IEEE Transactions on Knowledge and Data Engineering 7(4), 623–640 (1995)
40. March, S.T., Smith, G.F.: Design and natural science research on information technology. Decis. Support Syst. 15(4), 251–266 (1995)

41. Hevner, A.R., et al.: Design Science in Informaitona System Research. MIS Quarterly 28(1), 75–105 (2004)
42. Ostrowski, L., Helfert, M.: Reference Model in Design Science Research to Gather and Model Information. In: AMCIS 2012 Proceedings. Seattle AISeL (2012)
43. Van de Ven, A.: Engaged Scholarship: A Guide for Organizational and Social Research. Oxford University Press, New York (2007)
44. Te'eni, D., et al.: The process of organizational communication: a model and field study. IEEE Transactions on Professional Communication 44(1), 6–20 (2001)
45. Eden, A.H., Kazman, R.: Architecture, design, implementation. In: Proceedings of the 25th International Conference on Software Engineering (2003)
46. Reijers, H.A. (ed.): Design and Control of Workflow Processes. LNCS, vol. 2617. Springer, Heidelberg (2003)
47. Aguilar-Saven, R.S.: Business process modelling: Review and framework. International Journal of Production Economics 90(2), 129 (2004)
48. Lu, R., Sadiq, W.: A Survey of Comparative Business Process Modeling Approaches. In: Abramowicz, W. (ed.) BIS 2007. LNCS, vol. 4439, pp. 82–94. Springer, Heidelberg (2007)
49. Christie, A.M., Earl, A.N., Kellner, M.I., Riddle, W.E.: A Reference Model for Process Technology. In: Montangero, C. (ed.) EWSPT 1996. LNCS, vol. 1149, pp. 1–17. Springer, Heidelberg (1996)
50. Leppänen, M.: A Context-Based Enterprise Ontology. In: Abramowicz, W. (ed.) BIS 2007. LNCS, vol. 4439, pp. 273–286. Springer, Heidelberg (2007)
51. Schooley, B.L., Horan, T.A.: Towards end-to-end government performance management: Case study of inter-organizational information integration in emergency medical services (EMS). Government Information Quarterly 24(4), 755–784 (2007)
52. Cappiello, C., et al.: Context Management for Adaptive Information Systems. Electronic Notes in Theoretical Computer Science 146(1), 69–84 (2006)
53. Ge, M., Helfert, M.: Data and Information Quality Assessment in Information Manufacturing Systems. In: Abramowicz, W., Fensel, D. (eds.) BIS 2008. LNBIP, vol. 7, pp. 380–389. Springer, Heidelberg (2008)
54. Jeusfeld, M.A., Quix, C., Jarke, M.: Design and Analysis of Quality Information for Data Warehouses. In: Ling, T.-W., Ram, S., Lee, M.L. (eds.) ER 1998. LNCS, vol. 1507, pp. 349–362. Springer, Heidelberg (1998)
55. Gulliksen, J., Göransson, B.: Reenginering the System Development Process for User-Centred Design. In: Proceeding of Interact 2001. IOS Press, Amsterdam (2001)

On Applying Approximate Entropy to ECG Signals for Knowledge Discovery on the Example of Big Sensor Data

Andreas Holzinger[1], Christof Stocker[1], Manuel Bruschi[1], Andreas Auinger[2], Hugo Silva[3], Hugo Gamboa[4], and Ana Fred[3]

[1] Medical University Graz, A-8036 Graz, Austria
Institute for Medical Informatics, Statistics & Documentation,
Research Unit Human-Computer Interaction
{a.holzinger,c.stocker,m.bruschi}@hci4all.at
[2] Upper Austria University of Applied Sciences, A-4400 Steyr, Austria
andreas.auinger@fh-steyr.at
[3] Technical University Lisbon, 1049-001, Lisbon, Portugal
IT - Instituto de Telecomunicações
{hsilva,afred}@lx.it.pt
[4] New University of Lisbon, 2829-516 Caparica, Portugal
Physics Department, Faculty of Sciences and Technology
hgamboa@plux.info

Abstract. Information entropy as a universal and fascinating statistical concept is helpful for numerous problems in the computational sciences. Approximate entropy (ApEn), introduced by Pincus (1991), can classify complex data in diverse settings. The capability to measure complexity from a relatively small amount of data holds promise for applications of ApEn in a variety of contexts. In this work we apply ApEn to ECG data. The data was acquired through an experiment to evaluate human concentration from 26 individuals. The challenge is to gain knowledge with only small ApEn windows while avoiding modeling artifacts. Our central hypothesis is that for intra subject information (e.g. tendencies, fluctuations) the ApEn window size can be significantly smaller than for inter subject classification. For that purpose we propose the term *truthfulness* to complement the statistical validity of a distribution, and show how truthfulness is able to establish trust in their local properties.

Keywords: Information entropy, ApEn, big data, knowledge discovery, ECG complexity.

1 Introduction and Motivation for Research

Research on heart rate variability (HRV) has attracted considerable attention in the fields of psychology and behavioral medicine. Still, quantification and interpretation of HRV remains a complex issue [1], since the heartbeat period variability is the result of the activity of vasomotor and respiratory centers,

R. Huang et al. (Eds.): AMT 2012, LNCS 7669, pp. 646–657, 2012.

of baroreflex and chemoreflex closed loop regulation, of cardiovascular reflexes mediated by vagal and sympathetic afferences and of vascular autoregulation [2]. This collection of mechanisms act over various frequencies (similar but not coincident), and contribute to the complexity of the signal [2]. To summarize, Batchinsky et al. [3] state that HRV refers to a collection of methods describing regular periodic oscillations in the heart rate, attributed to the vagal and/or sympathetic branches of the autonomic nervous system.

These findings underline how important it is to recognize the regulatory complexities and organ system interconnections, when drawing conclusions based on heart rate related signals such as the electrocardiogram (ECG).

A promising approach to extract knowledge out of heart rate signals is the heart rate complexity (HRC). It utilizes a statistical approach, with methods derived from nonlinear dynamics. The *structural complexity* of the sampled ECG signal is used to describe the *regulatory complexity* and organ system interconnections [4]. Note that complexity and variability are not necessarily the same [5]. A periodical signal, such as a sinus wave, is variable but not complex. This is a desirable property, since it allows complexity measures to ignore the complicated periodic oscillations to some extend.

One complexity measurement in particular is of high interest to us, which is the Approximate Entropy (ApEn) introduced by Steven Pincus [6] in 1991. ApEn is a statistic quantifying regularity and complexity in a wide variety of relatively short (greater than 100 points) and noisy time series data [7]. The development of ApEn was initially motivated by data length constraints commonly encountered in heart rate, electroencephalography (EEG) and endocrine hormone secretion data sets [8].

In the context of ECG data, ApEn was previously used by Batchinsky et al. [3] to investigate the changes in heart rate complexity in patients undergoing post-burn resuscitation. They used ApEn with relatively large windows (> 800 heart beats) of the RRI[1] time series, in order to perform classification of pathological conditions based on the absolute ApEn values. Thus they were aiming at inter-subject classification.

Our ultimate goal is to justify much smaller ApEn windows (< 100 heart beats), to gain intra-subject knowledge (e.g., tendencies and fluctuations) that are consistent for many subjects, without the need for classification based on the absolute ApEn values. To be more specific, we analyzed a number of ApEn distributions out of the *sampled ECG signal* from 26 subjects using ECG data collected while the subjects were submitted to a concentration test [9].

In this paper we investigate the lower bounds for a valid ApEn window size, as well as possible causes for sporadic deviations in small window distributions. For this purpose, we propose the term *truthfulness* to complement the statistical validity of a distribution, in order to establish trust in the window placements, as well as to give an indication for stability of the ApEn distribution.

---

[1] RRI stands for R-to-R interval and describes the latency between two consecutive R peaks in the ECG. It's inversely proportional to and can be used to derive the heart rate.

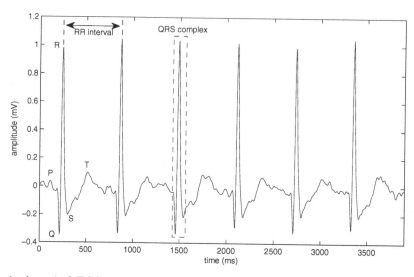

Fig. 1. A typical ECG signal; the first wave is annotated with the typical P-QRS-T complexes. Note that the *RR interval* is not constant (not even for resting subjects), since it oscillates periodically, shortening with inspiration and lengthening with expiration [10].

2   Background

As mentioned before, we make use of the electrocardiogram (ECG). Electrocardiography is an interpretation of the electrical activity of the heart over a period of time. It is a non-invasive procedure, using electrodes attached to the surface of the skin. The recording of the electrical activity is called electrocardiogram, and it is commonly used to measure the heart rate, the regularity of the beats, and characterize properties or injuries in the heart chamber. A clinical assessment of the ECG mostly relies on relatively simple measurements of the intrabeat timings and amplitudes [10].

Figure 1 shows some of the most common attributes analyzed in the ECG signal. The width of the QRS complex is representative of the time the ventricles need to depolarize and is typically lasting between 80 to 120 ms. It is interesting to note, that the lower the heart rate, the wider is the S-T complex [10].

One signal analysis approach commonly applied to ECG time series is frequency domain analysis, and uses the fast Fourier transform (FFT). From the FFT, two key frequency bands of the periodic oscillations in the ECG are typically identified as relevant in the analysis of RRI data, namely high-frequency (HF) power ($0.15 - 0.40$ Hz), and low-frequency (LF) power ($0.04 - 0.15$ Hz) [3]; a frequency range < 0.03 Hz, called very low frequency (VLF) power is also often found in literature.

The RRI indicates the beat-to-beat interval. Each consecutive RRI is of different length, unless the patient is paced. The reason for this is that the RRI oscillates periodically, shortening with inspiration and lengthening with expiration.

This phenomena is called respiratory sinus arrhythmia (RSA) [10]. The term arrhythmia might be misleading to some, since in this case it is not due to a medical condition, but it is a sign of a healthy organism. It occurs at the same frequency as the respiratory rate and has been shown to be a major component of HF [3]. The LF on the other hand is less specific, but has been related to both sympathetic and parasympathetic autonomic activity [11].

Artifacts in the ECG can lead to the spurious quantification of RRIs, which might result in substantial misinterpretation of the data. Results of Berntson and Stowell [12] revealed that even a single artifact, occurring within a 128 s interbeat interval series, can impart substantial spurious variance into all commonly analyzed frequency bands, including that associated with respiratory sinus arrhythmia. They emphasize the importance of artifact awareness for studies of heart period variability [12].

3   Related Work

Approximate Entropy found its way into many fields of application within the medical domain.

Acharya et al. [13] proposed a methodology for the automatic detection of normal, pre-ictal, and ictal conditions from recorded EEG signals. Beside Approximate Entropy, they extracted three additional entropy variations from the EEG signals, namely Sample Entropy (SampEn), Phase Entropy 1 and Phase Entropy 2. They fed those features to seven different classifiers, and were able to show that the Fuzzy classifier was able to differentiate the three classes with an accuracy of 98.1 %.

Hornero et al. [14] performed a complexity analysis of intracranial pressure dynamics during periods of severe intracranial hypertension. For that purpose they analyzed eleven episodes of intracranial hypertension from seven patients. They measured the changes in the intracranial pressure complexity by applying ApEn, as patients progressed from a state of normal intracranial pressure to intracranial hypertension, and found that a decreased complexity of intracranial pressure coincides with periods of intracranial hypertension in brain injury. Their approach is of particular interest to us, because they proposed classification based on ApEn tendencies instead of absolute values.

In the field of ECG analysis, Batchinsky et al. [3] recently performed a comprehensive analysis of the ECG and Artificial Neural Networks (ANN) to improve care in the Battlefield Critical Care Environment, by developing new decision support systems that take better advantage of the large data stream available from casualties. For that purpose they analyzed the HRC of 800-beat sections of the RRI time series from 262 patients by several groups of methods, including ApEn and SampEn. They concluded that based on ECG-derived noninvasive vital signs alone, it is possible to identify trauma patients who undergo Life-saving interventions using ANN with a high level of accuracy. Entropy was used to investigate the changes in heart rate complexity in patients undergoing post-burn resuscitation.

4   Methods and Materials

We are concerned with knowledge discovery [15], [16], [17] within ECG signals. For this purpose we model the continuous input signal given by the sensors as $U(t)$. After the digitalization process, the ECG signal $U(t)$ is considered as a discrete time series sampled at different points $t \in T$ over time, and will thus be referred to as ECG time series. Furthermore, to be consistent with Pincus [6], we denote data points of the ECG time series as $u(t)$. Note that the ECG time series we work with are considered *regularly* sampled, which basically means that the used sampling frequency was a constant.

4.1   Approximate Entropy (ApEn)

Approximate Entropy measures the logarithmic likelihood that runs of patterns that are close remain close on next incremental comparisons [6]. We state Pincus' definition [6], [8], for the family of statistics $\text{ApEn}(m, r, N)$, using our previous definition of the ECG time series:

Definition 1. *Fix m, a positive integer and r, a positive real number. Given a regularly sampled time series $U(t)$, a sequence of vectors $\boldsymbol{x}(1), \boldsymbol{x}(2), ..., \boldsymbol{x}(N - m + 1)$ in $\mathbb{R}^m$ is formed, defined by*

$$\boldsymbol{x}(i) = [u(t_i), u(t_{i+1}), ..., u(t_{i+m-1})] . \tag{1}$$

Define for each i, $1 \leq i \leq N - m + 1$,

$$C_i^m(r) = \frac{number\ of\ j\ such\ that\ d[\boldsymbol{x}(i), \boldsymbol{x}(j)] \leq r}{N - m + 1} , \tag{2}$$

with $d[\boldsymbol{x}(i), \boldsymbol{x}(j)]$ following Takens' [18] definition of a distance metric

$$d[\boldsymbol{x}(i), \boldsymbol{x}(j)] = \max_{k=1,2,...,m} \left(|u(t_{i+k-1}) - u(t_{j+k-1})| \right) . \tag{3}$$

Furthermore, define

$$\Phi^m(r) = (N - m + 1)^{-1} \sum_{i=1}^{N-m+1} \log C_i^m(r) , \tag{4}$$

*then the **Approximate Entropy** [6] is defined as*

$$\text{ApEn}(m, r, N) = \Phi^m(r) - \Phi^{m+1}(r) . \tag{5}$$

4.2   The Idea of Truthfulness

Our goal is to investigate the potential of using smaller ApEn windows (< 100 heart beats) on ECG time series data. With that in mind, we investigated potential pitfalls to ultimately validate a lower bound for heart beat windows. In this preliminary work we tried this without actually needing to interpret the data, but by taking a closer look at the *truthfulness* of the computed ApEn distributions.

We define the terms *ApEn window* and *ApEn distribution* as follows:

Definition 2. *Given a window size N and a time series U(t) containing K data points, we split the time series U(t) into $\lfloor K/N \rfloor$ equally spaced and adjacent parts of size N. Each part, in combination with the computed ApEn value for the part, is referred to as an ApEn window.*

Definition 3. *The sequential collection of all ApEn windows of a given time series U(t) is called the ApEn distribution of the time series U(t).*

With the basic terms defined we state the following definition for an ApEn windows truthfulness:

Definition 4. *The value of an ApEn window within an ApEn distribution is considered truthful for a defined distributional characteristic (e.g. linear or cubic), when sliding the window for an arbitrary amount of data points only influences the value according to the distribution and a specified tolerance.*

Note three things about this definition. First, it is not explicitly specified what rules the distribution follows (e.g. if it's linear or cubic). So we have to define what "according to the ApEn distribution" means. Secondly, this means that a window size of $N = 4$, for example, would cause nearly all windows to be considered truthful, since there is almost no interpolation, however it would not be useful, nor statistically valid. So it is important to note, that truthfulness does not imply usefulness, but the potential to be useful. And last, note that adjacent windows influence the truthfulness of each other. This is not considered to be a problem, since the source of the deviation could as well be between two adjacent windows.

Based on the definition of the truthfulness of a window, we define truthfulness of a ApEn distribution as follows:

Definition 5. *If all but a negligible amount of ApEn windows within an ApEn distribution are considered truthful, then the ApEn distribution itself is considered to be truthful.*

This is a loose definition, since it is not defined what is considered to be a negligible amount of ApEn windows. However, it enables one to have a tolerance, since in statistic measures there will almost always be outliers. It also enables one to deal with ECG artifacts to some extend, without having to unnecessarily increase the window size, just so that rare artifacts don't cause the distribution to be considered not truthful.

We propose the term truthfulness not as a substitute for statistical validity, e.g. as the term was used by Pincus [8], but as an extension. One does not imply the other. While the statistical validity is concerned with a single ApEn value for a given time series, truthfulness is concerned with multiple ApEn values for a given time series forming an ApEn distribution. In other words, the deviations in the window values do not describe uncertainty of the values, but instability concerning the time dimension.

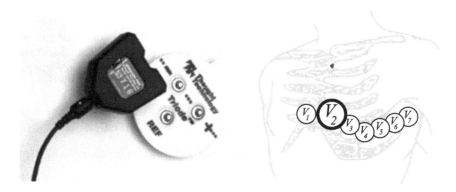

Fig. 2. The left image shows the ECG acquisition sensor. On the right you can see the lead V_2 sensor placement used in the experimental setup.

Our goal is to establish trust in the window placements, as well as to give an indication for stability of the ApEn distribution, and ultimately enabling one to extract knowledge from the properties of an ApEn distribution (e.g. tendencies, fluctuations).

Example 1. If moving some ApEn windows of an ApEn distribution for a negligible amount of data points causes the distribution to decrease over time, when it was increasing before, then the distribution is not considered to be truthful.

4.3   Data Acquisition

For these experiments we used the data of an investigation [9] performed by the co-authors in Lisbon, where 26 volunteering subjects[2] willingly participated in individual sessions (one per subject), during the course of which their ECG signal was recorded. Unlike the conventional acquisitions in a resting position, in each session the subject was asked to complete, on a computer, a task consisting of a concentration test, designed for an average completion time of 10 minutes.

Prior to initiating the task, subjects were equipped with the sensor and placed in front of the computer in a sitting position. No limitations on posture or motion during the activity were imposed, although the task was designed in such way that the subject only used the mouse to interact with the computer. To motivate the subjects commitment to the test, a performance score was computed and assigned to each subject.

For sensor placement the V_2 precordial derivation was chosen, located on the fourth intercostal space in the mid clavicular line, at the right of the sternum (Fig. 2 on the right). Ten-20 conductive paste was used to improve conductivity; for the same purpose, prior to the sensor placement, the selected area was prepared with abrasive gel.

---

[2] 18 males and 8 females between the ages of 18 and 31 years.

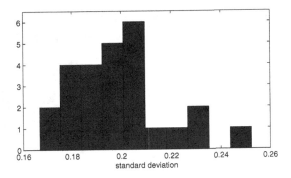

Fig. 3. The histogram of the standard deviation (SD) distribution of all ECG time series, in order to choose a common and valid r as ApEn parameter. The SDs range between 0.16 and 0.25.

Unlike the standard 12-lead ECG recording setup, which involves six sensors placed on the chest area and six other placed on the limbs, a one-lead surface mount setup was used. Acquisition was performed at a 256 Hz sampling rate using a ProComp2 encoder, and a gain 50 local differential triode (Fig. 2 on the left), with 2 cm inter-electrode spacing, and channel bandwidth of 0.05 – 100 Hz, both from Though Technology Ltd. The acquired signals were filtered using a zero phase forward and reverse 4th order Butterworth filter in the 2 – 30 Hz passing band, which proved to provide adequate results for our application.

5   Experimental Results and Lessons Learned

When choosing values for the ApEn parameters listed in (5), we tried to follow Pincus recommendation. Pincus [8] concluded that for $m = 1$ or $m = 2$, values of r between 0.1 SD (standard deviation) and 0.25 SD of the data produce good statistical validity of ApEn(m, r, N). He also states that ApEn can meaningfully be applied to $N \geq 1000$ data points [8]. However, the term data points is arbitrary, since it's amount depends on the sampling frequency of the signal.

We computed the standard deviation for all 26 ECG time series, as can be seen in Fig. 3, and we decided to use a common r of $r = 0.04$ with $m = 2$ for all ECG time series.

We started with a very small window size of $N = 1590$, which covers about 10 heart beats, in order to amplify possible side effects introduced by small windows. We moved the windows four times (two times to the left and two times to the right) for 30 data points each, which is less than 2 % of the window size and calculated the mean, as well as the SD for each group of five window positions. In order to establish the truthfulness of the ApEn values of the windows, we considered a linear characteristic and a tolerance of 5 % of the distance between consecutive window values.

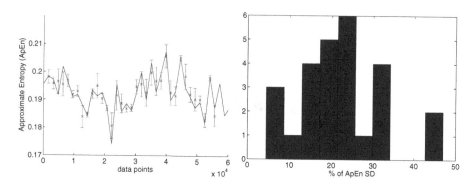

Fig. 4. The impact of the window placement to a window's ApEn value. The left shows part of the ApEn distribution of an ECG time series with the ApEn parameters $m = 2$, $r = 0.04$, $N = 1590$. The red error bars denote the standard deviation around the mean when sliding the windows four times (two times to the left and two times to the right) for 30 data points each, which is less than 2 % of the window size. The right shows that for all 26 ECG time series, the mean SD of all window groups amounts up to 45 % of the SD of its ApEn distribution.

Figure 4 shows, that the SD of the window groups is not necessarily correlated to the distance between consecutive window-values, but seem arbitrary. In fact, the mean SD of all window groups covers between 5 % and 45 % of the SD of its ApEn distribution. This is because the mean SD of all window groups ranges only between $2.4 \cdot 10^3$ and $3.8 \cdot 10^3$. So the impact is stronger for those ApEn distributions that have only a small SD. As expected the window size of $N = 1590$ (10 heart beats) is not considered to be truthful.

Note that the error bars in Fig. 4 do not describe the uncertainty of the concrete window values, but the deviation when slightly sliding the window. This means that for each slide, the underlying data is different, but still closely related in time.

Interestingly, by investigating further with different window sizes ranging between $N = 2000$ and $N = 10000$ data points, we found no notable correlation between the mean SD of all window groups and the SD of the its ApEn distribution.

Those findings strongly suggest that something is introducing deviations, that could for simplicity be described as noise, to the ApEn distributions. We expected it might be the influence of an additional QRS complex in every other window, since the amount of QRS complexes per window oscillates for ± 1 QRS complexes. We fabricated scenarios, where moving the window would result from covering exactly 10 - to covering exactly 11 QRS complexes. We found that the QRS complex has an influence, but even for a small window size only covering 10 QRS complexes it's only around 8 % of the mean SD of all window groups of its ApEn distribution. That means, that the majority of the source lies somewhere else.

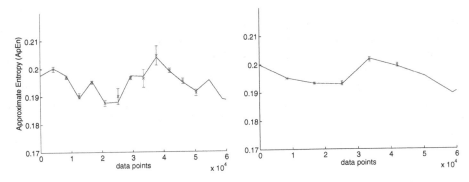

Fig. 5. The impact of the window placement to a window's ApEn value for larger N. The left shows part of the ApEn distribution of the same time series used in Fig. 4, with an ApEn window size of $N = 4200$ and the right with an ApEn window size of $N = 8400$. The red error bars denote the standard deviation around the mean when sliding the windows four times (two times to the left and two times to the right) for 2 % of its corresponding window size.

We now suspect, that the source of this "noise" might be the influence of the lower frequency bands when using very small windows, since a window would only contain parts of an oscillation period. The oscillation period ranges between 2.5 s and 35 s, which for us would be 640- to 9000 data points, or 4- to 50 heart beats. That, in turn, could cause ApEn to treat this influence as arbitrary complexity, because it has too few information available to recognize those as patterns.

We computed the ApEn distributions for larger windows, such as $N = 4200$ (~ 20 heart beats) and $N = 8400$ (~ 40 heart beats) to see how the deviation changes. Figure 5 shows parts of the resulting ApEn distributions of the same ECG time series. Increasing the window size results in an increased stability of the distribution and that in turn causes more windows to be considered truthful.

It is interesting to note, that if there is a window size that establishes the truthfulness for the ApEn distribution of one ECG time series, it does not imply that the window size is able to establish the truthfulness for the ApEn distribution of another ECG time series. This means that the lower bounds for truthfulness can be different, depending on the ECG time series.

Our results indicate that for the 26 ECG time series we have available, a window size of $N = 18000$, which is about 100 heart beats or approximately 70 seconds, is enough to establish the truthfulness for all corresponding ApEn distributions. Note that we considered a linear characteristic and chose a tolerance of 5 % of the distance between consecutive window values.

6   Conclusion and Future Research

We pointed out, that knowledge discovery within ApEn distributions of ECG time series has to overcome a lot of obstacles. We showed that the complex nature of an ECG signal renders it difficult to establish trust in the fluctuations of their ApEn distributions. For that purpose we introduced the term *truthfulness* as a complement to the statistical validity of the ApEn windows and to give an indication for stability of the ApEn distribution.

We believe that it is possible to validate intra subject information (e.g. tendencies, and fluctuations) with only small ApEn windows, while avoiding modeling artifacts, in order to gain knowledge of the underlying ECG time series. Our hypothesis is that this can be done with a higher resolution than usually applied for inter subject classification based on absolute values, if the truthfulness of the ApEn distribution can be established.

For future work, it would be interesting to investigate alternatives to the distribution assembling, e.g it would be interesting to use distances between windows that are smaller than the window size itself. This would result in the values of the windows being closely related to its neighbors, since they partially describe the same data, but we think this might enable one to have higher resolution of the distributions to some extend, while still being able to establish their truthfulness.

Acknowledgments. We thank the anonymous reviewers for their helpful comments. This work was partially funded by the Fundação para a Ciência e Tecnologia (FCT) under the grant SFRH/BD/65248/2009 and by the Institute for Systems and Technologies of Information, Control, and Communication (INSTICC), whose support the authors gratefully acknowledge.

References

1. Berntson, G.G., Bigger, J.T., Eckberg, D.L., Grossman, P., Kaufmann, P.G., Malik, M., Nagaraja, H.N., Porges, S.W., Saul, J.P., Stone, P.H., van der Molen, M.W.: Heart rate variability: origins, methods, and interpretive caveats. Psychophysiology 34(6), 623–648 (1997)
2. Porta, A., Guzzetti, S., Montano, N., Furlan, R., Pagani, M., Malliani, A., Cerutti, S.: Entropy, entropy rate, and pattern classification as tools to typify complexity in short heart period variability series. IEEE Transactions on Biomedical Engineering 48(11), 1282–1291 (2001)
3. Batchinsky, A.I., Salinas, J., Cancio, L.C.: Assessment of the need to perform life-saving interventions using comprehensive analysis of the electrocardiogram and artificial neural networks. In: RTO-MP-HFM-182: Use of Advanced Technologies and New Procedures in Medical Field Operations, vol. 39, pp. 1–16 (2010)
4. Buchman, T.G.: Nonlinear dynamics, complex systems, and the pathobiology of critical illness. Curr. Opin. Crit. Care 10(5), 378–382 (2004)
5. Holzinger, A., Popova, E., Peischl, B., Ziefle, M.: On Complexity Reduction of User Interfaces for Safety-Critical Systems. In: Quirchmayr, G., Basl, J., You, I., Xu, L., Weippl, E. (eds.) CD-ARES 2012. LNCS, vol. 7465, pp. 108–122. Springer, Heidelberg (2012)
6. Pincus, S.M.: Approximate entropy as a measure of system complexity. Proceedings of the National Academy of Sciences of the United States of America 88(6), 2297–2301 (1991)

7. Simonic, K., Holzinger, A., Bloice, M., Hermann, J.: Optimizing long-term treatment of rheumatoid arthritis with systematic documentation. In: 2011 5th International Conference on Pervasive Computing Technologies for Healthcare (PervasiveHealth), pp. 550–554 (May 2011)
8. Pincus, S.M.: Approximate entropy (ApEn) as a complexity measure. Chaos 5, 110–117 (1995)
9. Gamboa, H.: Multi-Modal Behavioral Biometrics Based on HCI and Electrophysiology. PhD thesis, Universidade Tecnica de Lisboa, Instituto Superior Tecnico (2008)
10. Clifford, G., Azuaje, F., McSharry, P.: Artech House engineering in medicine & biology series. In: ECG Statistics, Noise, Artifacts, and Missing Data, pp. 55–99. Artech House (2006)
11. Akselrod, S., Gordon, D., Ubel, F.A., Shannon, D.C., Berger, A.C., Cohen, R.J.: Power spectrum analysis of heart rate fluctuation: a quantitative probe of beat-to-beat cardiovascular control. Science 213(4504), 220–222 (1981)
12. Berntson, G.G., Stowell, J.R.: Ecg artifacts and heart period variability: don't miss a beat! Psychophysiology 35(1), 127–132 (1998)
13. Acharya, U.R., Molinari, F., Sree, S.V., Chattopadhyay, S., Ng, K.H., Suri, J.S.: Automated diagnosis of epileptic eeg using entropies. Biomedical Signal Processing and Control 7(4), 401–408 (2012)
14. Hornero, R., Aboy, M., Abasolo, D., McNames, J., Wakeland, W., Goldstein, B.: Complex analysis of intracranial hypertension using approximate entropy. Crit. Care Med. 34(1), 87–95 (2006)
15. Holzinger, A.: On knowledge discovery and interactive intelligent visualization of biomedical data - challenges in human computer interaction & biomedical informatics. In: Conference on e-Business and Telecommunications (ICETE 2012), Rome, Italy, pp. IS9–IS20 (2012)
16. Holzinger, A., Scherer, R., Seeber, M., Wagner, J., Müller-Putz, G.: Computational Sensemaking on Examples of Knowledge Discovery from Neuroscience Data: Towards Enhancing Stroke Rehabilitation. In: Böhm, C., Khuri, S., Lhotská, L., Renda, M.E. (eds.) ITBAM 2012. LNCS, vol. 7451, pp. 166–168. Springer, Heidelberg (2012)
17. Holzinger, A., Simonic, K., Yildirim, P.: Disease-disease relationships for rheumatic diseases: Web-based biomedical textmining and knowledge discovery to assist medical decision making. In: 36th International Conference on Computer Software and Applications, COMPSAC, Izmir, Turkey, pp. 573–580. IEEE (2012)
18. Takens, F.: Invariants related to dimension and entropy. In: Atas do 13, Col. brasiliero de Matematicas, Rio de Janeiro (1983)

Towards a Framework
Based on Single Trial Connectivity
for Enhancing Knowledge Discovery in BCI

Martin Billinger[1], Clemens Brunner[1], Reinhold Scherer[1],
Andreas Holzinger[2], and Gernot R. Müller-Putz[1]

[1] Institute for Knowledge Discovery, BCI Lab, Graz University of Technology, Austria
[2] Research Unit Human-Computer Interaction, Institute for Medical Informatics,
Statistics and Documentation, Medical University Graz, Austria
reinhold.scherer@tugraz.at

Abstract. We developed a framework for systematic evaluation of brain-computer interface (BCI) systems. This framework is intended to compare features extracted from a variety of spectral measures related to functional connectivity, effective connectivity, or instantaneous power. Different measures are treated in a consistent manner, allowing fair comparison within a repeated measures design. We applied the framework to BCI data from 14 subjects recorded on two days each, and demonstrated the framework's feasibility by confirming results from the literature. Furthermore, we could show that electrode selection becomes more focal in the second BCI session, but classification accuracy stays unchanged.

Keywords: brain-computer interface, band power, coherence, directed transfer function, connectivity, knowledge discovery.

1 Introduction

Diseases such as amyotrophic lateral sclerosis (ALS) or cerebral palsy (CP) disturb nervous system functions, limiting affected individuals in their abilities to interact with their environment. Voluntary movement or communication can be difficult or even impossible due to loss or impairment of motor functions. A brain-computer interface (BCI) could potentially help to improve quality of life for such individuals by allowing them to communicate or interact with their environment without relying on motor functions [21]. BCIs measure a user's brain signals and translate them into control commands for applications [10,14].

A commonly used brain signal for BCI control is the electroencephalogram (EEG) [16], which records cortical electrical activity from the scalp. The EEG contains several typical rhythmic activities such as the sensorimotor rhythm (SMR). Individuals can modulate SMR by motor imagery (MI), that is, imagining movement of their extremities, which causes power changes in specific frequency bands. These changes occur in different cortical areas, depending on which movement is imagined [15].

The state-of-the-art method for detecting different MI patterns is classification of band power (BP) features [3]. BP is the instantaneous power in pre-defined

R. Huang et al. (Eds.): AMT 2012, LNCS 7669, pp. 658–667, 2012.

frequency bands of single EEG channels. Although BP features allow a BCI to reliably detect different MI tasks [16], no information about the interaction of different brain areas can be obtained. However, such information could provide more insight in the neurological processes involved, and might improve classification of MI tasks.

The interaction of brain areas is expressed in functional or effective connectivity [6]. While functional connectivity measures correlated activation of brain areas, effective connectivity measures causal information flow. Numerous connectivity measures have been developed and proposed in BCI-related studies [4, 9, 12, 19]. Since these studies vary greatly in procedure, comparing their results is not easily possible. Hence, there is a clear need for a unified procedure to systematically compare feature extraction methods based on different measures of connectivity.

In this article, we present a framework for systematically evaluating BCI systems. This framework supports many different spectral connectivity measures, as well as non-connectivity measures such as BP. This allows fair comparison within a broad range of methods.

Although connectivity measures provide only three dimensional data (source, destination, frequency), they lead to a high dimensional feature space for classifying MI tasks. The number of features usually greatly outnumbers the number of available training samples. Thus, our framework also performs channel and frequency band selection based on statistical significance. This reduces the number of features and helps the researcher to interpret brain interactions relevant for BCI use [7, 8].

2   Methods

2.1   Connectivity Measures

Different measures related to connectivity can be derived from a vector autoregressive model (VAR) representation of the observed signals. The number of free parameters in a VAR model is M^2p, which limits the minimum estimation window length. The number of channels M, model order p, and window length N determine spatial, spectral and temporal resolution of derived measures.

Time resolution of VAR models can be increased by obtaining multiple realizations of the observed time series. This is a common procedure in neurophysiological studies, where repeated trials of a task are performed [13]. Unfortunately, MI BCIs are required to respond to single trials. Thus, a careful trade-off between temporal, spectral and spatial resolutions must be found. While we limit spatial resolution with automatic channel selection, model order and estimation window length are set manually.

2.2   BCI Framework

Fig. 1 shows an overview of the BCI framework. The outer cross-validation loop serves to test the framework as a whole. It resembles the flow of typical BCI operation, where parameters selected from available data are applied to novel data. The optimal time segment for classifier training during MI is determined in the inner cross-validation loop. A detailed description of each functional unit follows below.

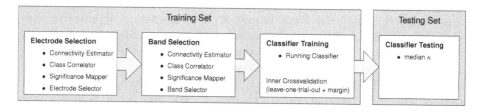

Fig. 1. Framework for the BCI based on functional or effective connectivity. The framework resembles the work-flow of a typical BCI, where the system is optimized on available data and applied to novel data, possibly in an online setting. This work-flow is embedded in a cross-validation procedure, where parts of the pre-recorded data serve as the novel testing set.

Connectivity Estimator. The Connectivity Estimator fits a sliding window VAR model to multiple EEG channels and subsequently extracts any required connectivity measures. For details about retrieving connectivity measures from VAR models see [18].

This results in a four-dimensional data set $c_k(i, j, f, t)$ for each trial k, which describes a measure of connectivity from channel j to i at frequency f and time t relative to the cue. BP can be estimated from the EEG's autospectra obtained by discrete Fourier transform (DFT). In this case $c_k(i, j, f, t) = 0$ for $i \neq j$.

Class Correlator. The correlation coefficient $r(i, j, f, t)$ is calculated between c_k and corresponding class labels y_k. $p(i, j, f, t)$-values are obtained from the asymptotic normal distribution, which are used for testing against the null-hypothesis of no correlation.

Significance Mapper. A modified version of the direct estimation of false discovery rate (FDR) [20] is used to determine which correlations are significant at a specified FDR. Similar to [1], instead of accepting all hypotheses with p-values below a certain threshold, only those p-values are accepted that form clusters of a minimum size in the t/f-plane.

Electrode Selector. Electrode selection is performed in two steps, with a ranking criteria e_l based on the squared correlation coefficient:

$$e_l = \sum_f \sum_t \left(\sum_i r^2(i, l, f, t) + \sum_j r^2(l, j, f, t) \right) \tag{1}$$

Electrodes l with the highest e_l are selected. First, only significant r are considered. If the procedure cannot find a pre-defined number of electrodes, more electrodes are added to the selection based on all r.

Band Selector. Band selection identifies a list of all channel pair frequency bands, in which the measure c_k correlates significantly with the class labels. Bands are identified for each channel pair (i, j) as the frequency ranges in which there is significantly correlated activity of at least two seconds. For functional connectivity measures, where $c_k(i, j, f, t) = c_k(j, i, f, t)$ or $c_k(i, j, f, t) = -c_k(j, i, f, t)$, only one of the channel pairs (i, j) and (j, i) is used.

If no frequency bands are found, the system chooses a set of default bands, which were defined as the frequency range of 5 to 25 Hz for each channel pair.

Classification. First, the optimal time segment for training the linear discriminant analysis (LDA) classifier is determined. This is accomplished by estimating the classification accuracy for each time segment with inner cross–validation. In a leave-one-out procedure, each trial is used as the validation set once and a margin of 10 trials before and after the validation trial is excluded from the training set [11].

Subsequently, the classifier is trained on the full training set in the time segment with the highest accuracy. Testing is performed by applying the classifier to the testing set. This is the only place in the framework where the testing set is used. Thus, the classifier is tested on completely unseen data.

2.3 Data Acquisition

Three electrooculogram (EOG) and 45 EEG channels were recorded. Electrodes were placed according to the international 10–20 System. Fig. 2 (left) shows the exact electrode positions. Signals were recorded at a sampling rate of 300 Hz using three synchronized g.USBamp amplifiers (g.tec, Guger Technologies OEG, Graz, Austria) with passive Ag/AgCl ring electrodes and filtered between 0.5 and 100 Hz. The notch filter was set to 50 Hz to suppress line noise.

14 healthy volunteers without prior experience in BCI control participated in a BCI experiment. On each of two separate days (sessions) six training runs and three feedback runs were performed. During feedback, the participants were instructed to control a virtual plane along a path with two different MI tasks (for details, see [2]). The training paradigm was based on the synchronous Graz BCI training paradigm [16], modified to visually resemble the continuous feedback paradigm. At the beginning of each session, an artifact run was performed to estimate the influence of artifacts such as eye movement and eye blinks on the EEG. Subsequently, four training runs were followed by one feedback run, two further training runs, and two final feedback runs. In this article, we only use the data recorded during training.

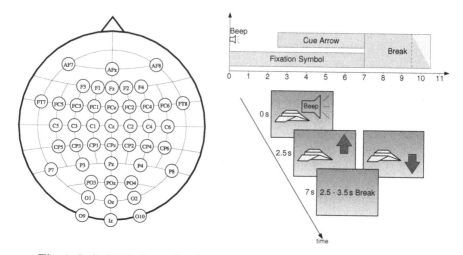

Fig. 2. Left: EEG electrode placement. Right: experimental paradigm.

From each session a total of 90 trials of right hand MI and 90 trials of feet MI are available. Fig. 2 (right) shows the timing of a trial. Trial duration is 7 s, followed by a break of 3 ± 0.5 s. Trial start was indicated by an acoustic beep and appearance of the fixation object. At $t = 2.5$ s an arrow appeared, pointing up or down, to cue the participant to perform hand or feet MI, respectively.

2.4   Data Analysis

Prior to application of the framework to the data, some pre-processing steps were performed. The EOG was removed from the EEG using a regression based approach [17]. Subsequently, the EEG data was down-sampled from 300 to 100 Hz.

The VAR model order was chosen as $p = 9$. This allows reasonable frequency resolution with reasonable window lengths. Window length was set to 3 s prior to electrode selection, and to 1.5 s after electrode selection. The number of electrodes selected was 10.

The framework was applied to the comparison of BP, coherence (COH), and directed transfer function (DTF) features. BP is known to work well with this kind of task and serves as the reference method. COH is a measure for functional connectivity, and DTF is a measure for effective connectivity. Each measure was independently applied to the electrode selection and the classification step.

The electrode selection step is evaluated using the concept of entropy. Entropy is a measure of uncertainty in a probability distribution, defined in (2), where p_l is the probability that each electrode l is selected from the set of all electrodes L.

$$E = -\sum_{l \in L} p_l \log_2(p_l) \tag{2}$$

The probabilities p_l are estimated from the electrode selection step in the outer cross-validation. An entropy of $E = 0$ corresponds to perfectly consistent selection, where always the same set of electrodes is selected. Increasing values of E indicate increasing uncertainty.

Classification performance is measured with Cohen's Kappa κ. To obtain a robust measure, κ is estimated for each time segment between cue and end of trial, and the median value of κ is reported as classification performance.

The results of electrode selection were analyzed using a repeated measures analysis of variance (ANOVA), with the dependent variable *entropy*, and the factors *selection* (measure used for selection) and *session*. Similarly, classification performance was analyzed with the dependent variable κ, and the factors *selection*, *method* (measure used for classification) and *session*. Sphericity corrections were applied when required. Factors found to be significant by the ANOVAs were subject to paired t-tests with Holm-Bonferroni correction.

3   Results

Table 1 lists the ANOVA results for electrode selection. Factors *selection* and *session* are both highly significant ($p < 0.01$), and there are no significant interaction effects between factors. Fig. 3 shows the results of post tests on both significant factors. BP has the lowest selection entropy of all methods. Also, entropy is significantly lower in the second session. Fig. 4 shows an example of electrode selection from both sessions for one subject. In the first session, the distribution of selection probabilities is more spread out than in the second session, which is indicated by a higher entropy value.

Table 1. ANOVA results for electrode selection. d_1 and d_2 are the between-group and within-group degrees of freedom of the F-statistic F.

Effect	d_1	d_2	F	p	sig
selection	2	26	13.956	0.00008	**
session	1	13	9.292	0.00933	**
selection:session	2	26	2.169	0.13458	

Table 2 lists the ANOVA results for classification. Factor selection is significant ($p < 0.05$), and factor method is highly significant ($p < 0.01$). All other main and interaction effects are not significant. Fig. 5 shows the results of post tests on both significant factors. Electrode selection with DTF leads to the highest classification performance, while classification with DTF is significantly worst. Differences between BP and COH are not significant.

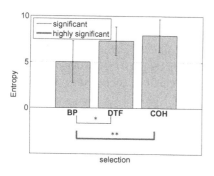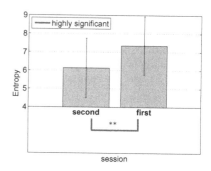

Fig. 3. Entropy of electrode selection. Left: comparison of connectivity measures (averaged over both sessions). Right: comparison of first and second session (averaged over all selection methods). The brackets below the bar charts indicate if differences are significant ($p < 0.05$, *) or highly significant ($p < 0.01$, **).

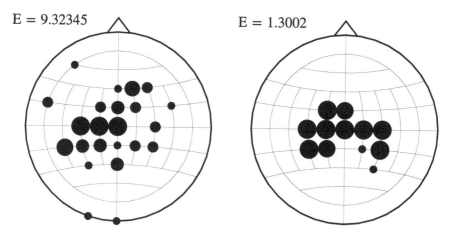

Fig. 4. Exemplary electrode selection probability for subject BV3, using COH. Left: first session. Right: second session. Bigger circles indicate higher probability of an electrode to be selected.

Table 2. ANOVA results for classification performance. The column $p[GG]$ contains Greenhouse-Geisser corrected p-values for factors that violate sphericity assumptions.

Effect	DFn	DFd	F	p	$p[GG]$	sig
selection	2	26	5.386	0.01104		*
method	2	26	17.980	0.00001	0.00015	**
session	1	13	0.758	0.39964		
selection:method	4	52	0.534	0.71136		
selection:session	2	26	0.458	0.63760		
method:session	2	26	1.130	0.33849		
selection:method:session	4	52	0.549	0.70077		

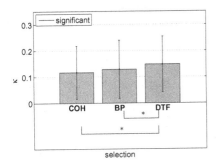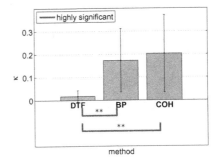

Fig. 5. Classification performance (Cohen's Kappa). Left: comparison of connectivity measures used for electrode selection. Right comparison of connectivity measures used for classification. The brackets below are explained in Fig. 3.

4   Discussion

We proposed a framework for simulating offline BCIs based on spectral features. Features are selected from connectivity or band power measures. To avoid overfitting, the whole work flow is embedded in nested cross-validation procedures. The framework is designed to be modular. Thus, components can easily be replaced or added. For example, the simple LDA could be replaced by more sophisticated classifiers. Additional connectivity measures or pre-processing steps can be added with little effort.

Our simulation results show that BP and COH work equally well for classification. This finding is in line with [9], who argue that this is caused by a bias towards zero-phase between electrodes. Although DTF provides the worst features for classification, it gives the best electrode selection in terms of classification performance. This is an interesting finding that will require further investigation.

The issue of zero-phase bias is inherent to autoregressive models, since instantaneous terms are not directly modeled. This is addressed in [5], who propose an extension to VAR models to include an instantaneous term. An alternative could be to use independent component analysis (ICA) for pre-processing, which maximizes independence between components, and effectively removes zero-phase components.

Electrode selection entropy is lower in the second session. This could indicate a training effect induced by the feedback. Activation becomes more focused as the subjects get accustomed to the task. However, this effect has no influence on classification accuracy.

The framework helps researchers to manage their knowledge at various levels. On the methodological level, any number of measures, each providing a high dimensional feature space, are reduced to their respective classification performance. Statistical comparison allows to easily comprehend their suitability for BCI use (Figs. 3, 5). On the level of individual measures, relevant electrodes are identified from the multitude of possible connectivities between all channel pairs in all frequency bins (Fig. 4). Furthermore, the subset of actual connectivities can be further analysed, deepening the understanding of brain connectivity in respect to specific tasks.

5 Conclusions

We could demonstrate that the proposed framework for BCI evaluation works correctly by confirming some of the results of [9]. Thus, the framework could be used to determine the best set of methods and electrodes to be used in individuals or patient groups.

The DTF, a measure of effective connectivity, appears to be useful for electrode selection. This indicates that connectivity measures can provide useful information for BCIs. However, to realize a BCI based on connectivity features, more work is required to address the issue of zero-phase bias between EEG channels.

Acknowledgments. This work was supported by the FP7 Framework EU Research Project ABC (No. 287774) and the FWF Project Coupling Measures for BCIs (P20848-N15). This paper only reflects the authors' views and funding agencies are not liable for any use that may be made of the information contained herein.

References

1. Billinger, M., Kaiser, V., Neuper, C., Brunner, C.: Automatic frequency band selection for BCIs with ERDS difference maps. In: Proceedings of the 5th International Brain–Computer Interface Conference (2011)
2. Billinger, M., Neuper, C., Müller-Putz, G.R., Brunner, C.: User-centric performance estimation in a continuous online BCI. In: Proceedings of the 3rd TOBI Workshop (2012)
3. Brunner, C., Billinger, M., Vidaurre, C., Neuper, C.: A comparison of univariate, vector, bilinear autoregressive, and band power features for brain computer interfaces. Medical and Biological Engineering and Computing 49, 1337–1346 (2011), http://dx.doi.org/10.1007/s11517-011-0828-x, doi:10.1007/s11517-011-0828-x
4. Brunner, C., Scherer, R., Graimann, B., Supp, G., Pfurtscheller, G.: Online control of a brain-computer interface using phase synchronization. IEEE Transactions on Biomedical Engineering 53, 2501–2506 (2006)
5. Erla, S., Faes, L., Tranquillini, E., Orrico, D., Nollo, G.: Multivariate autoregressive model with instantaneous effects to improve brain connectivity estimation. International Journal of Bioelectromagnetism 11(2), 74–79 (2009)
6. Friston, K.J.: Functional and effective connectivity in neuroimaging: A synthesis. Hum. Brain Mapping 2, 56–78 (1994)

7. Holzinger, A.: On knowledge discovery and interactive intelligent visualization of biomedical data - challenges in human-computer interaction & biomedical informatics. In: Proceedings of the 9th International Joint Conference on e-Business and Telecommunications (ICETE 2012), Rome, Italy, pp. IS9–IS20 (2012)

8. Holzinger, A., Scherer, R., Seeber, M., Wagner, J., Müller-Putz, G.: Computational Sensemaking on Examples of Knowledge Discovery from Neuroscience Data: Towards Enhancing Stroke Rehabilitation. In: Böhm, C., Khuri, S., Lhotská, L., Renda, M.E. (eds.) ITBAM 2012. LNCS, vol. 7451, pp. 166–168. Springer, Heidelberg (2012)

9. Krusienski, D.J., McFarland, D.J., Wolpaw, J.R.: Value of amplitude, phase, and coherence features for a sensorimotor rhythm-based brain computer interface. Brain Research Bulletin 87(1), 130–134 (2012)

10. Kübler, A., Furdea, A., Halder, S., Hammer, E.M., Nijboer, F., Kotchoubey, B.: A brain-computer interface controlled auditory event-related potential (P300) spelling system for locked-in patients. Annals of the New York Acadamy of Sciences 1157, 90–100 (2009)

11. Lemm, S., Blankertz, B., Dickhaus, T., Müller, K.R.: Introduction to machine learning for brain imaging. NeuroImage 56(2), 387–399 (2011),
http://www.sciencedirect.com/science/article/pii/S1053811910014163

12. Lim, J.H., Hwang, H.J., Jung, Y.J., Im, C.H.: Feature extraction for brain–computer interface (BCI) based on the functional causality analysis of brain signals. In: Proceedings of the 5th International Brain–Computer Interface Conference (2011)

13. Möller, E., Schack, B., Vath, N., Witte, H.: Fitting of one ARMA model to multiple trials increases the time resolution of instantaneous coherence. Biological Cybernetics 89, 303–312 (2003)

14. Müller-Putz, G.R., Scherer, R., Pfurtscheller, G., Rupp, R.: Brain-computer interfaces for control of neuroprostheses: from synchronous to asynchronous mode of operation. Biomedizinische Technik 51, 57–63 (2006)

15. Neuper, C., Pfurtscheller, G.: Event-related dynamics of cortical rhythms: frequency-specific features and functional correlates. International Journal of Psychophysiology 43, 41–58 (2001)

16. Pfurtscheller, G., Neuper, C.: Motor imagery and direct brain-computer communication. Proceedings of the IEEE 89, 1123–1134 (2001)

17. Schlögl, A., Keinrath, C., Zimmermann, D., Scherer, R., Leeb, R., Pfurtscheller, G.: A fully automated correction method of EOG artifacts in EEG recordings. Clinical Neurophysiology 118, 98–104 (2007)

18. Schlögl, A., Supp, G.: Analyzing event-related EEG data with multivariate autoregressive parameters. In: Neuper, C., Klimesch, W. (eds.) Event-related Dynamics of Brain Oscillations, pp. 135–147. Elsevier (2006)

19. Shoker, L., Sanei, S., Sumich, A.: Distinguishing between left and right finger movement from eeg using svm. In: 27th Annual International Conference on Engineering in Medicine and Biology Society, IEEE-EMBS 2005, pp. 5420–5423 (January 2005)

20. Storey, J.D.: A direct approach to false discovery rates. Journal of the Royal Statistical Society: Series B (Statistical Methodology) 64(3), 479–498 (2002),
http://dx.doi.org/10.1111/1467-9868.00346

21. Wolpaw, J.R., Birbaumer, N., McFarland, D.J., Pfurtscheller, G., Vaughan, T.M.: Brain-computer interfaces for communication and control. Clinical Neurophysiology 113, 767–791 (2002)

Author Index